ASTRONOMY:
THE COSMIC JOURNEY

Astronomy: The Cosmic Journey

1991 EDITION

WILLIAM K. HARTMANN

Wadsworth Publishing Company
Belmont, California
A Division of Wadsworth, Inc.

Astronomy Editor: Anne Scanlan-Rohrer
Development Editor: Mary Arbogast
Editorial Assistant: Leslie With
Production Editor: Gary Mcdonald
Designer: MaryEllen Podgorski
Print Buyer: Karen Hunt
Art Editor: Marta Kongsle
Art Assistants: Bobbie Broyer, Amy Yates
Copy Editors: Don Yoder, Charles Hibbard
Technical Illustrators: Catherine Brandel, Victor Royer, Darwen and Vally Hennings, Lisa Sliter, Salinda Tyson
Color Separators: Rainbow Graphic Arts Co., Ltd.; GTS
Compositor: Graphic Typesetting Service, Los Angeles

Frontispiece: The nebula NGC 1977 in the constellation Orion. See Figure 20-7 on page 441.
Cover: The brightest part of the Milky Way, toward the galactic center (see pages 514–515 for other views), rises over a telescope dome at Mauna Kea Observatory, Hawaii. Four "foreground" objects in our solar system also are prominent: the Moon (right of dome), Venus (above dome), and Saturn and Mars (right of Venus). Special filter creates "cross-hair" diffraction pattern on bright light sources. Camera tracked on rising motion of stars during the exposure, causing slight blurring of dome and horizon. (24-mm wide-angle lens at f2.8, 6-min exposure on Ektachrome ISO 1600 film; photo by author.)
Back Cover: A great advance in planetary studies came in 1990 when the Magellan spacecraft began detailed mapping of the surface of Earth's sister planet, Venus. Some of the most striking features are large craters, believed to be caused by meteorite impacts. Three large craters (diameters 37–50 km, or 23–30 mi) are shown in this view. Between the craters is a volcanic plain with unusual straight fractures. See page 195. (NASA.)

Printed in the United States of America   48

1 2 3 4 5 6 7 8 9 10—95 94 93 92 91

**Library of Congress
Cataloging-in-Publication Data**

Hartmann, William K.
    Astronomy : the cosmic journey / William K. Hartmann. — 1991 ed.
       p.   cm.
    Includes bibliographical references (p.
    Includes index.
    ISBN 0-534-14946-4
    1. Astronomy.   I. Title.
QB45.H32   1991
520—dc20   90-12963

For Teachers and Students

With human footprints on the Moon, radio telescopes listening for messages from alien creatures (who may or may not exist), technicians looking for celestial and planetary sources of energy to support our civilization, orbiting telescopes' data hinting at planetary systems around other stars, and political groups trying to figure out how to save humanity from nuclear warfare that would damage life and climate on a planet-wide scale, an astronomy book published today enters a world different from the one that greeted books a generation ago. Astronomy has broadened to involve our basic circumstances and our enigmatic future in the universe. With eclipses and space missions broadcast live, with the USSR occupying a permanent space station, and with American and Soviet leaders discussing cooperative missions to Mars, astronomy offers adventure for all people—an outward exploratory thrust that may one day be seen as an alternative to mindless consumerism, ideological bickering, and wars to control dwindling resources on a closed, finite Earth.

Today's astronomy students not only seek an up-to-date summary of astronomical facts; they ask, as people have asked for ages, about our basic relationships to the rest of the universe. They may study astronomy partly to seek points of contact between science and other human endeavors: philosophy, history, politics, environmental action, even the arts and religion.

Science fiction writers and the special effects artists on recent films help today's students realize that the unseen worlds of space are real places—not abstract concepts. Today's students are citizens of a more real, more vast cosmos than conceptualized by students of a decade ago.

In designing this book, the Wadsworth editors and I have tried to respond to these developments. Rather than jumping at the start into the murky waters of cosmology, I have begun with the viewpoint of ancient people on Earth and worked outward across the universe. This method of organization automatically (if loosely) reflects the order of humanity's discoveries about astronomy and provides a unifying theme of increasing distance and scale.

This arrangement aims to give an unfolding, ever-expanding panorama of our cosmic environment. We hope it unfolds like a story in which each chapter provides not only a new facet but also a growing understanding of the relationships among the elements of the whole.

The subtitle refers to three separate cosmic journeys that we undertake simultaneously. First, we travel through historical time, where we see how humans slowly and sometimes painfully evolved our present picture of the universe. Second, we journey through space, where we see how our expanding frontiers have revealed the geography of the universe. Beginning with an Earth-centered view, we study the Earth–Moon system, the surrounding system of planets, the more distant surrounding stars, our own vast galaxy, and the encompassing universe of other galaxies. Finally, we travel back through cosmic time. Familiar features of the Earth are typically only a few hundred million years old. The solar system is about 4.6 billion years old. Our galaxy is roughly 13 or 14 billion years old. The universe itself began (or began to reach its present form) an estimated 14 billion years ago.

Because astronomy touches many areas of life and philosophy, I have allowed the text to encompass a wide range of relevant topics, including space exploration, financing of science, cosmic sources of energy, the checkout-counter's barrage of astrology and other pseudoscience, and the possibility of life on other worlds, as well as the conventional "hard science" of astronomy. This variety of topics shows how basic scientific research touches all areas of life—I hope in a way that lets readers ponder the relation between science and priorities in our society.

Using This Book

The arrangement of text material into nine parts and twenty-eight chapters, plus a section of optional enrichment essays, should give instructors some flexibility in tailoring a course according to their interest. For example, those who are not much interested in historical development could use Part A only as assigned outside reading.

Each part gives some historical background, describes recent discoveries and theories, and then discusses advances that might occur if society continues to support research. This more or less chronological approach has several purposes. Since there is often a certain logic to the order in which discoveries were made, historical emphasis may help readers remember the facts. Second, historical discussion allows us to introduce basic concepts in a more interesting way than by reciting definitions. It makes life richer to realize that some seemingly modern concepts descend from knowledge of ancient millennia and have thousand-year-old names. Third, there is a widespread fallacy that the only progress worth mentioning is that of the last few decades. Astronomy, of all subjects, shows clearly that, to paraphrase Newton, we see as far as we do because we stand on the shoulders of past generations. Exploration of the universe is a continuing human enterprise. As we try to maintain and improve our civilization, that is an important lesson for a science course to teach.

Another principle we have followed is to treat astronomical objects in an *evolutionary* way, to show the sequence of development of matter in the universe. Stars, pulsars, black holes, and other celestial bodies are linked in evolutionary discussion, rather than listed as different types of objects detected by different observational techniques. I have also not hesitated to mention nonscientific approaches to cosmology and evolution, such as "creationist" concepts recently encouraged in two states' school systems by their state legislatures but later thrown out after courtroom battles.

The New Edition

This 1991 edition is a substantial update of the 1989 fourth edition. As a result of Wadsworth's extensive surveys of teachers the fourth edition contains numerous changes in presentation since the third edition. The discovery of the solar system through the Copernican revolution is now described in Chapter 3. This arrangement provides an unbroken chronology of the events leading to our current view of the solar system. Following the teaching style of many instructors, we concentrate more of the basic physical principles in two early chapters, on gravity (4) and light (5). Scientific exploration of the Earth–Moon system (6 and 7) now leads directly into our examination of the solar system, and the outer planets beyond Jupiter get their own chapter (12) to accommodate the recent explosion of data about them. The discussion of stellar evolution (18 and 19) has been further clarified, including current thought on evolution in multiple star systems (21). The galaxies section has been refocused, discussing galactic evolution in Chapter 24 and active galactic nuclei, colliding galaxies, and quasars in Chapter 25. The book has been completely updated throughout.

In the main part of the text, mathematics is almost nonexistent. The book can thus be used for a descriptive course. Nine basic equations are distributed through the book in optional boxes. The general content of each box is included qualitatively in the text, but the boxes introduce a higher level of physics and math, allowing a more quantitative course to be taught. Sample calculations using these equations now appear in every box, and the *Advanced Problems* at the ends of chapters use the optional equations. The nine basic equations are described in more detail below.

Because color imagery plays an increasingly important role in astronomy, we use color figures throughout the book. Emphasis is on true-color imagery to give students the best conception of astronomical objects' appearance, though false-color image enhancement techniques are also described.

More specifically, teaching aids incorporated in the book include the following.

Illustrations and Text

1. In addition to the classic large telescope photos and recent NASA photos, I have included three other categories of illustrations:

 a. Photos from recently published research papers, kindly provided by various authors and institutions.

 b. Photos by amateurs with small and intermediate instruments, often used to show sky locations of well-known objects in the large-telescope photos. These can help readers to visualize and locate these objects in the sky, a difficult task if based on classic large-telescope photos alone. Photographic data provided with many of these pictures may be used in setting up student projects in sky photography.

 c. Scientifically realistic paintings show how various objects might look firsthand to observers in space. Discussion of features shown in the paintings illustrates a synthesis of scientific data from various sources.

2. Key concepts are shown in **boldface** type. These are repeated in *Concepts* lists at the ends of chapters and defined in a *Glossary* at the end of the book.

3. The nine optional basic equations are introduced in the text as needed. Teachers offering a more quantitative course can integrate them into the course work; teachers offering more descriptive courses may skip over them. They are set off in boxes for optional use. The nine boxes discuss:

 I. *The Small-Angle Equation,* useful for calculating apparent sizes of objects at known distances.

 II. *Newton's Universal Law of Gravitation,* illustrating the simplicity of gravitational attraction between bodies throughout the universe.

 III. *Calculating Circular and Escape Velocities,* useful for deriving speeds or masses in coorbiting systems (planetary, binary star, galactic).

 IV. *Measuring Temperatures of Astronomical Bodies: Wien's Law,* which shows how radiation measurements can reveal the temperatures of distant objects.

 V. *The Definition of Mean Density,* a simple concept for gaining information about the nature of material inside planets and stars.

 VI. *Typical Velocities of Atoms and Molecules in a Gas,* by which we characterize temperature, as well as collision energies when the atoms or molecules smash into each other. These energies, in turn, control the types of chemical or nuclear reactions that can occur.

 VII. *The Doppler Effect: Approach and Recession Velocities,* which shows how spectral measures can reveal radial velocities of distant objects.

 VIII. *The Stefan–Boltzmann Law: Rate of Energy Radiation,* which shows how temperature and luminosity measurements can reveal sizes of radiating sources.

 IX. *The Relativistic Doppler Shift,* a modification of equation VII, to explain phenomena that occur at high speeds approaching the speed of light.

4. Limited numbers of references to technical and nontechnical sources appear in the text. They are there partly to help students and teachers find more material for projects and partly to help instructors emphasize that statements should be verifiable. These sources are included in an expanded *References* section.

End-of-Chapter Materials

1. *Chapter Summaries* review basic ideas of the chapter and sometimes synthesize material from several preceding chapters.

2. *Concepts* lists include the important concepts appearing in **boldface** in the text. Reviewing the *Concepts* is a good way for the student to review the content of each chapter.

3. *Problems* are aimed at students with nonmathematical backgrounds.

4. *Advanced Problems* usually involve simple arithmetic or algebra and are often applications of the nine basic equations. These can be omitted in nonmathematical courses.

5. *Projects* are intended for class use where modest observatory or planetarium facilities are available. The intent is to get students to do astronomical observing or experimenting.

Supplementary Material

1. *Enrichment Essays* can be used or not as instructors wish. These include essays on *Pseudoscience and Nonscience* and *Astronomical Coordinates and Timekeeping Systems.*

2. *Appendixes* are included on *Powers of Ten, Units of Measurement,* and *Supplemental Aids in Studying Astronomy.*

3. The *Glossary* defines all terms included in the *Concepts* lists, as well as other key terms.

4. The *References* section includes all sources mentioned in the text. Nontechnical references useful for student papers are starred (*); widely available journals and magazines are emphasized in this group, including most astronomy articles appearing in *Scientific American* in recent years.

5. The *Index* includes names and terms.

6. *Star Maps* for the seasons are found after the index. Since more detailed, larger maps are usually available in classrooms or laboratories, these star maps have been simplified, emphasizing the plane of the solar system and the plane of the galaxy and indicating major constellations mentioned or illustrated in the text.

Acknowledgments

My thanks go to many people who helped produce this book. I have tried to incorporate as many of their suggested corrections and improvements as possible; final responsibility for weaknesses remains mine.

The names of those reviewers who criticized parts of various editions for research comprehensiveness and teaching potential are listed on the following page. I also thank a wide circle of colleagues, especially in Tucson and at the University of Hawaii, for helpful discussions.

In locating and providing photographs, Dale Cruikshank, Walter Feibelman, Steven Larson, Alfred McEwen, Alan Stockton, Don Strittmatter, Alar Toomre, and Dave Webb were helpful. Don Bane and Jurrie Van der Woude (Jet Propulsion Laboratory), David Moore (National Optical Astronomy Observatory), Margaret Weems (National Radio Astronomy Observatory), and David Malin (Anglo-Australian Observatory) gave kind assistance with their institutional files. I especially thank Chesley Bonestell, Dennis Davidson, Don Dixon, David Egge, James Hervat, Paul Hudson, Pamela Lee, Ron Miller, Jim Nichols, Mark Paternostro, Kim Poor, Adolf Schaller, Andrei Sokolov, Richard Sternbach, and Brian Sullivan for stimulating discussion about paintings of astronomical subjects. Floyd Herbert and Mike Morrow helped me in making several wide-angle sky photos with clock-driven mounts.

The staff at Wadsworth Publishing Company has worked hard to create a useful and beautiful product, and made the job a pleasure at the same time. I thank Anne Scanlan-Rohrer, Mary Arbogast, Gary Mcdonald, Carolyn Deacy, and Marta Kongsle for their friendship and professional help in shepherding the 1991 edition through various editorial stages. Thanks to all of them as well for exceptional work on production and design, cheerful dispositions, hospitality, and care for the subject, not to mention the harried author.

William K. Hartmann
Tucson

Reviewers

I appreciate the help and advice of the following reviewers and advisors who have made comments or suggestions on various portions of this and previous editions, or who have given helpful suggestions about teaching astronomy with this book.

Helmut A. Abt

Bill T. Adams, Jr.

Richard A. Bartels

Henry E. Bass

Lee Bonneau

David R. Brown

Edward Budd

Warren Campbell

Michael Castelaz

Robert J. Chambers

Clark R. Chapman

Kwan-Yu Chen

Clark G. Christensen

Dale P. Cruikshank

Donald R. Davis

Robert J. Doyle

Robert J. Dukes, Jr.

John Fix

Benjamin C. Friedrich

F. Trevor Gamble

Owen Gingerich

Edward Ginsberg

Jeff Goldstein

Richard Greenberg

Hubert E. Harber

Alan Harris

Thomas G. Harrison

Gayle Hartmann

Floyd Herbert

George H. Herbig

Ulrich O. Hermann

J. R. Houck

William W. Hunt

R. I. Kollgard

Nathan Krumm

James LoPresto

Emil J. Michael

Leonard Muldawer

Gerald H. Newsom

Charles J. Peterson

Terry P. Roark

John L. Safko

John M. Samaras

John Schopp

Richard Schwartz

Robert F. Sears, Jr.

Myron A. Smith

Richard C. Smith

Ronald Smith

Michael L. Stewart

Kjersti Swanson

Donald J. Taylor

James E. Thomas

Scott D. Tremaine

Peter D. Usher

Stuart J. Weidenschilling

Walter G. Wesley

Ray J. Weymann

Richard M. Williamon

R. E. Williams

Warren Young

Contents
in Brief

Detailed Contents

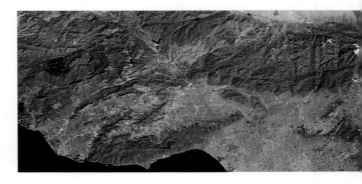

PART C

EXPLORING THE EARTH–MOON SYSTEM 119

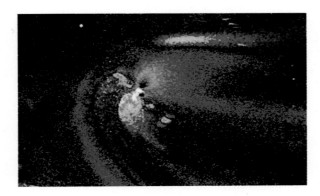

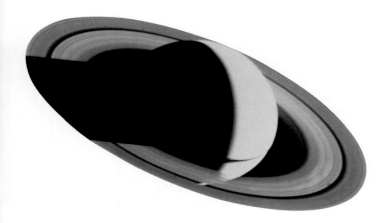

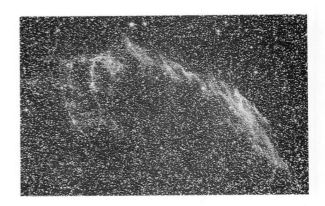

ASTRONOMY:
THE COSMIC JOURNEY

About the author: William K. Hartmann is a senior scientist and founding member of the Planetary Science Institute of Science Applications International, in Tucson. He is known for research on the formation and evolution of the solar system, coauthoring the most widely accepted theory of lunar origin, and co-discovering "asteroid" Chiron's cometary outburst in 1988. He was a co-Investigator on NASA's Mariner 9 mission to Mars, and has served on advisory committees for NASA and the National Academy of Science. His paintings of planetary landscapes and other astronomical subjects have been published and collected internationally. He was selected for the NASA Fine Arts Program team to cover the Galileo launch in 1989, and has collaborated on space art projects in the USSR. In addition to *Astronomy: The Cosmic Journey,* he is the author of *The Cosmic Voyage* (Wadsworth) and *Moons and Planets* (Wadsworth). Hartmann has collaborated on four popular, illustrated books on astronomy: *The Grand Tour, Out of the Cradle, Cycles of Fire,* and a Soviet-American astronomical art collection, *In the Stream of Stars.* He has also published a book of his text and photos on the Sonoran Desert, *Desert Heart.* Asteroid 3341 was named after him in 1987.

Invitation to the Cosmic Journey

If you awoke one day to discover that you had been put on a strange island, your first project after getting food and water would be to try to find out where you were. This is just what has happened to all of us. We have all been born on an island in space: the Earth. We are all passengers on a cosmic journey. Astronomy is the process of finding out where we are—our place in space and our position in the unfolding history of the universe.

This definition of astronomy would not always have satisfied people. At one time we thought we knew where we were: at the center of the universe, with the Sun, stars, planets, and satellites all revolving around us. Modern astronomy contradicts that old theory. The definition is not fully satisfactory now, either, because it doesn't really explain what astronomers do.

OUR DEFINITION OF ASTRONOMY

Oddly enough, as science progresses, astronomy becomes harder to define. When astronomy was restricted to observations by earthbound viewers, it was easily defined as "the science of objects in the sky." But now that we have actually begun to explore space, other disciplines have become involved. Is the astronaut who chips a rock sample off the Moon practicing astronomy? What about the researcher who studies the sample once it reaches the Earth? What of the nuclear physicist who measures properties of nuclear reactions going on at the center of stars? What about the biologist interested in life on Mars? Their work should be included in our definition of astronomy.

There is a popular misconception that as various scientists pursue new research, knowledge proliferates into an endless complex of specialized disciplines. Indeed, students are usually taught this way, with advanced courses probing into narrower and narrower specialties. According to this idea, research is like climbing a

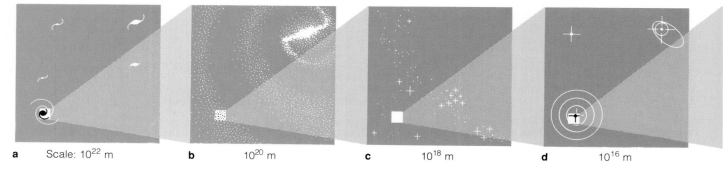

a   Scale: 10^{22} m   b 10^{20} m   c 10^{18} m   d 10^{16} m

Figure I-1 A journey through the universe from large-scale structure of galaxies to the human scale. The change in scale from one picture to another ranges from a factor of 100 to a factor of 100 000. In cartoon form, **a** shows galaxies; **b** and **c,** parts of our galaxy; **d,** our solar system and the nearest star; **e,** part of our solar system; **f** the Earth–Moon system; and **g,** human observers. Scale in meters is given in terms of powers of ten.

tree, starting near the trunk and proceeding to ever more specialized branches of knowledge.

But science can also be compared to working our way *down* the tree of knowledge. At first we see a bewildering variety of seemingly unrelated phenomena—twigs of knowledge. After analyzing these twigs, we see that they meet in a branch. As workers map details on the branch, someone with vision may discover that it is joined to another branch. As we work our way toward the trunk, more and more branches join. By grasping the relations between the major branches close to the trunk, we can better understand the higher branches and twigs. This is why different specialties are connecting with astronomical research; astronomy is becoming more general.

A historical example illustrates this process. Around 1600, Galileo and others conducted many experiments to understand seemingly unrelated types of motion, such as the motion of falling objects and the motion of the Moon. The wide variety of motions and accelerations seemed impossible to explain in a simple way. However, by about 1680, Isaac Newton realized that they were connected by simple relationships between mass, acceleration, and gravitational attraction. As a result of Newton's insight, ordinary college students can learn to understand motions better than the Galileos of the past!

In the same way, students can grasp the relationship between the Earth and the cosmos better than the explorers who mapped astronomical frontiers in the past. Better than any generation in human history, we can begin to sense where we are, at least in relation to other physical objects in the universe.

This book treats astronomy not as a narrow set of

academic observational results or abstruse physical laws, but as a voyage of exploration with practical effects on humanity. A broad view of the major discoveries of the last years, the last decades, and even the last centuries is at least as important as the latest factual detail, because the broad view helps us understand how scientific research will continue to affect us and perhaps offer us new options for living. In an era of energy crises, food shortages, and environmental threats, when badly applied technology has created such horrors as biological weapons and hydrogen bombs, educated people are called on to make judgments about their own actions—the kinds of jobs they have, the materials they consume, and the actions of organizations they work for. In short, we are called on to judge what kind of civilization should continue. Thus we all need to know how our world is affected by its cosmic surroundings and to understand the scientific and social procedures available for obtaining and extending that knowledge.

Therefore, this book is organized according to the following definition of **astronomy:**

Astronomy is the study of all matter and energy in the universe, emphasizing the concentration of this matter and energy in evolving bodies such as planets, stars, and galaxies, and fully recognizing that we observers— humanity—are part of the universe, and that our home, the Earth, is only one of the many places in the universe, but also the special point from which our voyage of exploration has started.

The reason for emphasizing humanity is that astronomy influences how we think about ourselves and our

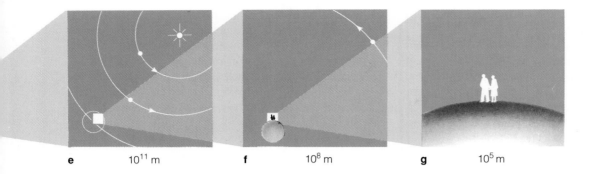

e 10^{11} m   f 10^8 m   g 10^5 m

role in the universe. The growing perception of environmental problems in the 1970s and 1980s has shown that we *must* look at the Earth as a single astronomical body in order to survive. As the poet Archibald MacLeish said after one of the Apollo flights, we have now seen the Earth as "small and blue and beautiful" and ourselves as "riders on the Earth together." Thus this book regards astronomy as including space flights and studies of rocks from our neighbor planets, just as much as telescopic observations of remote galaxies.

A SURVEY OF THE UNIVERSE

The ensuing chapters start with subjects on Earth and then move outward through space. Before beginning, however, we give a brief preview of the universe, starting at the largest scale.

The **universe** is everything that exists. To the best of our knowledge, the universe consists of untold thousands of clusters of galaxies (Figure I-1). **Galaxies** are swarms of billions of stars. Galaxies differ in form. Some are football shaped, some irregular. Our galaxy, like many others, is a disk with about a hundred billion stars arrayed in curving spiral arms. Most galaxies, including ours, are surrounded by a halo of **globular star clusters,** each a spheroidal mass of hundreds of thousands of stars. If we made a model of our galaxy big enough to cover North America, Earth would not be as big as a basketball, or even a BB. It would be about the size of a large molecule.

Scattered inside our galaxy, mostly in the spiral arms, are groupings of stars called **open star clusters** and

clouds of dust and gas called **nebulae.** Individual stars are scattered at random. Each **star** is an enormous ball of gas, mostly hydrogen. In the center of a typical star, atoms are jammed together so closely that their nuclei interact to create the heat and light of the star.

Three hundred million billion kilometers (200 000 000 000 000 000 mi) from the center of our galaxy and 40 thousand billion kilometers from the nearest star is the **Sun.** A hydrogen sphere more than a million kilometers in diameter, the Sun is the center of the **solar system**—the small system made up of the Sun and its family of orbiting rocky and icy **planets.** Circling around most planets are smaller bodies called **satellites.** Between planets are various other small bodies called **asteroids** and **comets.**

About 150 million kilometers (km) from the Sun is the **Earth.** The **Moon,** Earth's satellite, is only 384 000 km away. The Earth itself is nearly 13 000 km across, a tiny speck among myriads of larger and smaller specks in the universe.

A WORD ABOUT MATHEMATICS

Our preceding survey of the universe shows how hard it is to express astronomical quantities in ordinary units. For this reason the system of expressing numbers as **powers of ten** is useful. For example, instead of being written as the unwieldy number 300 000 000 000 000 000 km, the distance from the Sun to the galactic center becomes simply 3×10^{17} km; the distance from the Earth to the Sun is 1.5×10^8 km. A reader uncertain about this system should consult Appendix 1 at the end of the book. We will use this system occasionally, as convenience dictates.[1]

When using any equation to calculate physical quantities, you must remember to use a consistent set of units. For example, the English system of inches, ounces, and so on must not mix in the same equation with the metric system. The metric system, long used by scientists, is becoming universal and we will use it here, often giving English equivalents. The metric system is *much* easier to use than the English system because its units come in multiples of 10. One centimeter = 10 millimeters, 1 meter = 100 centimeters, and so on. Gone are complicated conversions like 1 mile = 5280 feet. Specifically, in this book we will give preference to the **"SI" metric system**—the units adopted in the Systeme International (International System), such as meters, kilograms, and so on. In some cases, astronomical units such as "parsecs" will be explained and adopted. These units of measurement are summarized in Appendix II.

A NOTE ABOUT NAMES OF PEOPLE

One thing that distinguishes science from the pseudo-science now so popular is that scientific assertions are backed by evidence. Because much of this evidence has already been published elsewhere, it is not always discussed in detail, but is merely referenced. This is usually done by listing the name of the author and the date of publication (for example, "Smith, 1988"); the reference is then listed in the back of the book so that a reader who wants more information can locate the source. With so many wildly distorted claims being published today as if they were scientific, this referencing of evidence is important in helping you recognize whether you have reliable material in your hands. Whether in science, business, or politics, readers and listeners should learn to demand documentation of assertions and, if in doubt, look up the documents to judge their quality. Aside from promoting this intellectual tradition, we believe that the extensive bibliography will be useful to teachers and to students preparing term papers.

Students are sometimes distressed to find these names, thinking they are supposed to learn names and dates. The names are only to document assertions, not to be memorized. The emphasis in this book is on the exciting nature of the universe that astronomy is revealing, not on the astronomers themselves. Many of the Greek and Renaissance astronomers mentioned in Chapters 2 and 3 are major figures in our Western intellectual heritage, however, and students should learn about them.

A HINT ON USING THIS BOOK

When reading a text, many students underline everything that seems important. My publishers and I have taken care of part of that job by putting terms that represent important concepts in boldface and then listing them in order at the end of the chapter. A useful study procedure is to read a chapter through and then go over the concept list. Try to define each term and use it in a sentence. If you have trouble with a term, go back and check it in the chapter or look it up in the glossary. If you are comfortable with all the terms in each concept list, you will have gone a long way toward mastering the material and will be able to enjoy many articles about astronomy in newspapers and magazines.

FACE TO FACE WITH THE UNIVERSE

The universe is immense; we are perhaps audacious even to attempt studying it. Yet something noble in the human spirit urges us to seek knowledge of the universe and our place in it.

Throughout history, scientific discoveries in the heavens have sometimes illuminated and sometimes conflicted with religious, metaphysical, and philosophical conceptions. Persecution, legal action, even wars have resulted from the challenge to old beliefs by new knowledge. In this book we present scientific theories and the *evidence* for them with the understanding that our knowledge is limited and slowly improving. New evidence is what usually distinguishes new knowledge from old superstition.

Astronomy gives us a new cosmic perspective from which many conflicts of the past, often founded on amazingly few facts, now seem inconsequential. For example, we have given up the medieval arguments over how many angels can dance on the head of a pin, and stopped persecuting those who advocate the theory

---

[1]We also use the term *billion* here to mean 1 000 000 000, or 10^9, which is the usual American meaning. The British use *billion* to mean 1 000 000 000 000, or 10^{12}.

that the Earth moves around the Sun. Unfortunately, world news demonstrates that we still seem unable to give up warring over ideologies, whether Christian vs. Moslem vs. Jew or capitalist vs. communist. Astronomy, however, unifies humanity by showing us that we are all in this together, facing the unknown—a word that aptly describes much of the universe, its past, and its future.

The universe is no abstract concept; it consists of real, physical places. This book offers descriptions of those places, often with photos or scientifically realistic paintings to show how we think they look.

Suppose that we were able to explore these places and that in all these vast reaches, among all the stars and galaxies, we found no one else—no living creatures anywhere except ourselves. Or suppose that we discovered nonhuman intelligent life—alien civilizations, perhaps advanced and incomprehensible or perhaps very simple. Our contacts with them could alter our history far more profoundly than the first contacts between Native Americans and Europeans in 1492.

These are two wildly different possibilities. One is likely to be true. Either boggles the mind. We are just barely entering the century in which we may be able, if research continues, to decide which scenario is more realistic.

CONCEPTS

astronomy	planet
universe	satellite
galaxy	asteroid
globular star cluster	comet
open star cluster	Earth
nebula	Moon
star	powers of 10
Sun	SI metric system
solar system	

PROBLEM

1. Scientific discoveries are sometimes described as conflicting with religious or philosophical beliefs. Before continuing this book, examine your present beliefs or expectations and comment on which of the following views best match your own:

a. There are concepts in astronomy that conflict with my own religious or philosophical views.

b. Astronomical concepts are consistent with my religious or philosophical views; there is no problem.

c. If there is a conflict, scientific hypotheses will evolve and eventually become consistent with my religious or philosophical views.

d. Science measures the reality of the physical world only; there is no need for this to be consistent with an inner psychological or spiritual reality.

e. Science tries to reflect how all nature behaves; thus science and a valid religious philosophy of daily life should be consistent with each other if they are to contribute to a well-integrated personality.

The Early Discoveries

Prehistoric
"standing stones"
erected around 2000 B.C.
in what is now the English village
of Avebury symbolize early humanity's
attempts to interact with the cosmos.
Similar constructions nearby at Stonehenge are aligned on the Sun's position
at sunrise on the date of the summer
solstice. (Photo by author.)

CHAPTER 1

Prehistoric Astronomy: Origins of Science and Superstition

In 1906 the American astronomer Percival Lowell wrote:

Smoke from multiplying factories . . . has joined with electric lighting to help put out the stars. These concomitants of an advancing civilization have succeeded above the dreams of the most earth-centered in shutting off sight of the beyond, so that today few city-bred children have any conception of the glories of the heavens which made of the Chaldean shepherds astronomers in spite of themselves.

This observation seems even more apt today. It is difficult to realize how important the sky once was in daily affairs. In addition, we have forgotten how much we depend on ideas that may seem obvious today but were developed only through centuries of struggle to understand nature.

There are several reasons to review the ancient discoveries. First, they were often very basic. We forget how hard they were to recognize. For example, we think it obvious that the Earth moves round the Sun, but early scholars were severely criticized for saying so.

Second, discoveries were often made just when society reached a point where they *could* be made, because the needed technical devices or theoretical concepts had just been invented. Therefore, a historical approach helps place important concepts in a logical order.

Third, many common ideas today come from astronomical traditions of the past. For example, although we favor a decimal numerical system, we have 24 hours in a day instead of 10, and 360 degrees in a circle instead of 100. As we will see in this chapter, the reason comes from ancient astronomy and mathematics.

THE EARLIEST ASTRONOMY: MOTIVES AND ARTIFACTS (c. 30 000 B.C.)

Archaeology and anthropological studies of present-day primitive tribes[1] shed light on the earliest astronomical practices. Imagine yourself an intelligent hunter of 30 000 ÿ (years) ago. Agriculture is not yet practiced. You have to depend on hunting and gathering for food. Celestial phenomena are not your most immediate concern, but you know that certain celestial cycles are important. You want to know when berries on the mountain will ripen and when migratory birds will arrive at a nearby lake. Such knowledge requires cumulative day counts and knowledge of the seasons, best derived from sky phenomena. When you travel far, you want to be able to find your way home. Celestial objects are your only reliable guides in unfamiliar landscapes.

If you are a woman, you also want to know in what season the child you may carry will be born and when your next menstrual period will occur. Living mostly outdoors, women undoubtedly used the $29\frac{1}{2}$-d (days) cycle of the Moon's phases (for example, from one full moon to the next) to keep track of their 27- to 30-d menstrual cycles. Some experiments indicate that menstrual cycles of women sleeping in the light of an artificial moon become synchronized to its phases, and some scientists have even speculated that the human menstrual cycle evolved to match the period of the Moon's phases (Luce, 1975).

For such reasons, early people began to keep records of events in the sky. There may be actual physical evidence of the earliest such records. Curious notations, found on thousands of artifacts dating as far back as 30 000 B.C., scattered over Europe, Africa, and the Soviet Union, may be the beginnings of calendar systems—perhaps counts of the number of days between phases of the Moon, such as new moon (the date when the Moon is between the Earth and the Sun or, more loosely, the date of the thinnest crescent, Figure 1-1) and full moon (Marshack, 1972). The makers of these tools were not the brutish cave dwellers of some stereotypes; their craftsmanship is demonstrated by the magnificent paintings of their prey—mammoths, boars, reindeer—that they left on European cave walls.

---

[1]Time grows short for these studies. The last few pockets of primitive culture are being discovered and integrated into surrounding cultures by scientific expeditions and even guided tours.

Figure 1-1 Paleolithic evidence suggests that ancient people tabulated the phases of the Moon. The configuration shown here, called "the old moon in the new moon's arms," is seen in the evening sky a few days after a new moon, when the portion not lit by the Sun is dimly lit by light reflected off the Earth. (Photo by author.)

This interpretation is supported by contemporary calendar sticks made by modern aborigines. For example, a lunar calendar stick carved by Indian Ocean islanders shows a series of grooves almost identical to the Paleolithic examples of 30 000 y earlier. And North American Indian calendar sticks similarly record historical events and other natural events by sequences of carvings.

The calendar stick in Figure 1-2, for example, records an earthquake and other events. Since the Indians who made it had no written language, it can be "read" only by the carver. It is dramatic evidence of how records of astronomical events may have first accumulated, thus allowing the discovery of astronomical cycles.

CALENDAR REFINEMENTS (10 000–3000 B.C.)

Around 10 000 B.C., the domestication of animals and the cultivation of crops began. This agricultural revolution brought two new incentives for understanding the sky. First, to know when to plant, people had to be able

Figure 1-2 Portion of a calendar stick carved by Tohono O'odham Indians of Arizona. The stick, begun in 1841 and passed on to the carver's son, carries a record of social and natural events, including an earthquake, for the years 1841 to 1939. (Calendar stick from collection of Arizona State Museum, University of Arizona. Photo by author.)

to predict seasonal changes days in advance. Second, in the stable villages made possible by agriculture, people kept records; the mere existence of records gave thinkers an incentive and opportunity to discover seasonal and annual cycles of celestial events.

This stage of development is shown in historical records of Indians of the American Southwest. Arizona Indians, for example, named the 12 lunar cycles (months) that made up the year—among them a "Green Moon" (March, plants leafing) and a "Hungry Moon" (May, no wild foods). This was obviously the beginning of a calendar system. But since each lunar cycle is about $29\frac{1}{2}$ d, 12 such cycles would be only 354 d, not a full year. Thus the lunar-based calendar must be readjusted each year.

People began to get a better sense of the yearly cycle by counting days or Moon cycles between certain dramatic annual events. In Egypt, for instance, the Nile flooded at a certain time each year. At high northern latitudes people looked eagerly for the Sun to return to the northern sky after the cold, short days of winter. In some early societies, day counts of this cycle gave estimates of about 360 d, later refined to 365 and then $365\frac{1}{4}$.

Accumulated observational experience revealed the correlation between this annual cycle of seasons and the apparent movements of the Sun. The Sun does not follow the same path across the sky every day. In the summer at northern latitudes (for example, Europe and the United States), the Sun is high at midday; in the winter, it is low in the southern sky at midday. Thus

careful observations of the Sun's motion relative to other features of the sky would have helped determine the exact number of days in the year and the exact dates of such observations.

Organized observations of this kind were made throughout the world thousands of years ago. Mesopotamian calendars were being designed by 4000 to 3000 B.C. In recent decades, anthropologists have found modern pretechnological people using similar means to create sophisticated systems of time measurement.

OTHER EARLY DISCOVERIES

Before describing ways in which these principles were utilized and extended, we pause to describe other aspects of the sky discovered or defined by early people. These aspects are important aids in describing many astronomical phenomena.

North Celestial Pole Camp out for a night and you will discover that the stars slowly wheel around a single point in the sky. In the Northern Hemisphere that single point is called the north celestial pole; in the Southern Hemisphere, the south celestial pole. The wheeling of the stars is due to Earth's rotation on its axis, and therefore one complete circuit takes about 24 h (hours). The two **celestial poles** can be thought of as the projections of the Earth's axis onto the sky, or the spots targeted by vertical searchlights at the North and South Poles. You can visualize the sky as a giant dome overhead, as seen in Figure 1-3. Recognizing the celestial pole helped in navigation, since it indicates the north or south direction and can be used to measure an observer's latitude (a point to which we will return shortly).

North Star A bright star, **Polaris** (shown in Figure 1-4) happens to lie near the north celestial pole. Thus it stands almost still all night above the northern horizon as other stars wheel around it. It serves as an excellent reference beacon.[2] For nearly 1000 y, Polaris has been

[2]Note that a magnetic compass points only approximately north, because Earth's magnetic field is not aligned with true north. Because the field varies, the compass needle also varies in direction by small amounts from year to year. In any case, the compass was not available to the ancients. It was apparently first invented by the Olmecs in Mexico around 1400–1000 B.C. Magnetic materials were later found by the Greeks around 600 B.C., and compasses were recorded in China around 300 B.C. to A.D. 100 (Carlson, 1975).

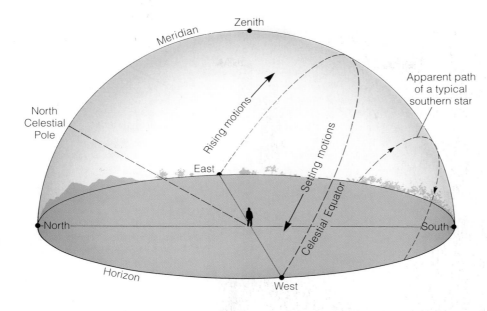

Figure 1-3 General properties of the sky as defined by the Earth's daily rotation. The sky is shown as seen by observers at midnorthern latitudes, such as the United States.

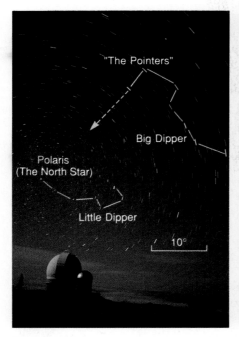

a   b

Figure 1-4 A view of the northern sky showing the Big Dipper and Little Dipper and the pointers at the end of the Big Dipper's bowl, which help viewers find the North Star. **a** Ten-minute time exposure with a stationary camera on a tripod shows star trails as Earth rotates through about $2\frac{1}{2}°$. **b** Four-minute exposure with a small motor driving the camera to track the stars represents approximately the naked-eye view. Notice that the observatory dome is slightly blurred in this exposure, as the camera itself has rotated through 1° to track the stars. (Both photos by author at Mauna Kea Observatory, Hawaii, with 24-mm wide-angle lens, f2.8, on Ektachrome ISO 1600 film.)

located within a few degrees of the north celestial pole, but in other times other stars have been nearer the pole. If a bright star is located near the north celestial pole, it is called the **North Star.**

Zenith and Meridian Wherever an observer stands, the point directly overhead is called the **zenith.** The

meridian is an imaginary north–south arc from horizon to horizon through the celestial pole and the zenith. The zenith and meridian are shown in Figure 1-3. This line marks the highest point that each star reaches above the horizon on any night; it is the dividing line between the rising stars and setting stars. For this reason it is especially useful in timekeeping. In the daytime, the

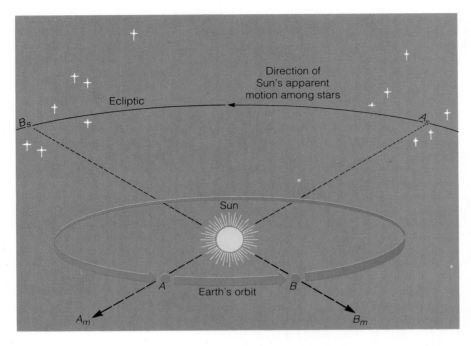

Figure 1-5 Explanation of observations shown in Figure 1-6. As Earth moves in a few weeks from point A to point B in its orbit, the Sun first appears among background stars A_s and then against stars B_s. Similarly, the midnight sky, opposite the Sun, first contains stars in directions A_m and later those in direction B_m.

moment when the Sun crosses the meridian is called noon; in very early times, therefore, people used this astronomical concept to divide the day.

Celestial Equator   As shown in Figure 1-3, the imaginary circle lying 90° from both the north and the south celestial poles is the **celestial equator**. It is the projection of the Earth's equator onto the sky and is parallel to the daily east-to-west paths of the stars as they wheel around the celestial poles. The celestial equator helps divide the sky into easily recognizable portions.

The 360° Circle   Our division of the circle into 360 units, or degrees, follows a Babylonian practice of about 3000 y ago. It probably comes from the ancient estimate of 360 d for one year, and from the fact that 360 is easily divisible into 2, 3, 4, 5, or 6 equal parts. The decimal (10-based) system of counting and writing numbers later replaced the sexagesimal (60-based) system in most uses *except* the expression of angles and time.

The Ecliptic   By observing the relative positions of Sun and stars at dawn or dusk, one can establish that the Sun appears to shift nearly a degree to the east

each day relative to the stars, as shown in Figures 1-5 and 1-6. (This can be confirmed by noting that the Sun "moves" 360° in 365 d.) The cause of this shift is Earth's motion around the Sun, also seen in Figure 1-5. Ancient observers found that the Sun traces the *same path* through the stars year after year. The path differs from the celestial equator (being tipped to it by $23\frac{1}{2}°$ and crossing it at two points). This path of the Sun among the stars is a circle extending all the way around the sky and is called the **ecliptic**. We now know that the ecliptic is the same as the plane of Earth's orbit around the Sun, as can be visualized in Figure 1-5.

The Planets   Ancient observers—priest-astronomers and wakeful shepherds alike—found that five of the brighter starlike objects were not fixed like the rest. They came to be known in the Western world as **planets**, from the Greek word for "wanderers."[3] Not until after

_____

[3]Many of our technical terms have Greek roots because modern Western traditions are directly linked to Greek civilization through the Roman and medieval worlds. However, many of the concepts discussed here were discovered in pre-Greek times and later renamed by the Greeks. Roman and medieval writers, often unaware of the earlier work, revered the Greek thinkers and kept their terms.

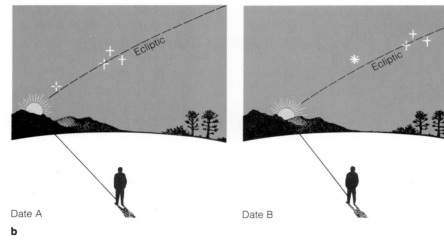

Date A   Date B

b

Figure 1-6 Observation of heliacal risings. **a** Shortly before sunrise some stars are already well up in the sky. Others, just above the horizon, are barely visible in the orange glow of dawn. The latter stars have just risen; in a few moments the sky will be too bright to see them. On the previous day at sunrise, they were lost in the glow. Therefore, this is the first day of the year on which these stars are visible; it is called the date of heliacal rising. (Photo by author at Mauna Kea Observatory, Hawaii; fixed 35-mm camera; wide-angle lens; 1-min exposure at f2.8 on commercially available 3M ASA 1000 film.) **b** The same principle in diagrammatic form. On date A, the star shown by dashed marks is too close to the Sun to be visible at sunrise. A day or so later, on date B, it is far enough from the Sun to be visible for a few moments at sunrise—the date of heliacal rising.

the telescope was invented, in about 1610, was there any proof that these planets were globes like our Earth or that three faint ones had gone undiscovered. For this reason the ancients spoke of five planets, whereas we speak of nine (counting our own).

The Zodiac Ancient observers tracking the paths of the planets from night to night among the stars discovered that the planets never stray out of a zone about 18° wide centered on the ecliptic. Many star patterns,

or constellations, along this zone were said to resemble animals, so the zone came to be called the **zodiac** (after Greek for "animals"—the same root as *zoo*). The constellations located within the 18° zone came to be known as the signs of the zodiac. They include such familiar examples as Scorpius (the scorpion) and Leo (the lion)— see the star maps on the last pages of the book.

Heliacal Risings and Settings Any given star rises and sets slightly earlier each night. The **heliacal** (he-

LIE-ah-cal) **rising** of a star *occurs on the first day each year when the star can be seen just before dawn.* (*Heliacal* means "near the sun," from the Greek *helios,* "sun.") One day earlier, the star rises a few minutes *after* the sky gets too bright for the star to be seen. On the day of heliacal rising, the star glimmers near the horizon just a minute or so before the sky gets too bright, as shown in Figure 1-6. On the day after, the star rises slightly earlier and can be seen for several minutes before the sky lightens. Similarly, **heliacal setting** *occurs on the last day each year when the star can be seen at dusk.* On the next day, by the time darkness falls, the star has already set.

Why would anyone living in, say, 4000 B.C. care about the heliacal rising or setting of a certain star? The answer is that each heliacal rising or setting occurs on the same date every year. Depending on the star, that date might be the first day of spring or the day when planting should start. Without clocks, calendars, and morning newspapers on the doorstep, heliacal risings and settings would tell the date with an accuracy of a day or two. It is scarcely surprising that the ancient Egyptian calendar began with the heliacal rising of Sirius, which marked the beginning of the Nile's annual flooding.

Similarly, agriculture in Java was once scheduled according to the heliacal rising of Orion's belt. Australian aborigines begin their spring when the Pleiades, a prominent star cluster, rise in the evening sky. An Arizona Indian tribe also divided their agricultural year according to the Pleiades (Castetter and Bell, 1942):

Pleiades rising in summer, start planting;
At the zenith at dawn, too late to plant more;
Past the zenith, time for corn harvest;
One quarter down from the zenith, time for deer
   hunting;
Setting, time for harvest feast.

ORIGIN OF THE CONSTELLATIONS

As memory aids for learning and locating stars, early observers named groups of stars for their resemblance to familiar animals, objects, and mythical characters— such as Scorpius (which really resembles a scorpion, see Figure 1-7), Cygnus the swan, or Orion (the hunter), and many others. These groups are **constellations.** People in different cultures saw different patterns, often derived from their mythologies; for example, the ancient Chinese constellations differ from Western constellations. The star group called Ursa Major was seen as a bear in Europe, Asia, North America, and even ancient Egypt, where there are no bears; for this reason, Gingerich (1984) suggests the bear identification may go all the way back to ice-age Euro-Asia, from where it spread.

Most of our own familiar constellations came out of the Near East. As early as 3300 B.C., Leo and Taurus often appear in struggle on Mesopotamian artifacts. Star symbols sometimes shown on Leo's shoulders indicate that the artisans had the constellation in mind rather than a real lion. The struggle motif may refer to Leo chasing Taurus out of the sky: As Leo rises, Taurus sets. These images originated when the risings and settings of Leo and Taurus coincided with equinox and solstice dates. Later, they gained a mythical importance of their own and the lion-bull struggle persisted in art until A.D. 1200 (Hartner, 1965).

Many constellations are known from *historic* records at least as early as 420 B.C., but they evidently date from much earlier (Gingerich, 1984). The English astronomer Michael Ovenden (1966) did some astronomical sleuthing to find out who originally formalized most of our present constellations, when, and why. The most ancient constellations fill only the northern sky, leaving an empty zone around the south celestial pole.[4] This suggests that the ancient constellation makers couldn't see this southern zone from their northern latitude. As shown in Figure 1-8, for an observer at any given northern latitude $L°$, the sky is divided into three zones: the *north* **circumpolar zone,** $L°$ from the north celestial pole, containing stars that never set; the **equatorial zone,** containing stars that rise and set; and the *south* circumpolar zone, $L°$ from the south celestial pole, containing stars that never rise. The size of the south circumpolar zone is the clue to the northern latitude $L°$ of the ancient observers. (To see this, study Figures 1-8 and 1-9 together. Figure 1-10 shows the north cir-

---

[4]Constellations now filling this southern zone, such as the Southern Cross and the Telescope, were not charted until European navigators sailed these waters in the 1600s and 1700s. In 1930 the International Astronomical Union revised the boundaries of all constellations for easier record keeping. Ancient names were kept, but the old, irregular boundaries were replaced by neater, geometric ones.

Figure 1-7 Color view of the summer constellation Scorpius, rising in the east. Field of view top to bottom is about 80° in this wide-angle photo. Scorpius is one of the few constellations that closely resemble their namesakes, in this case a scorpion with two curving claws at top and curved tail at bottom. A double star marks the stinger at the end of the tail. In the heart of the scorpion the brightest star, Antares, is distinctly redder than adjacent stars—a difference that can be seen with the naked eye. This time exposure also shows the softly glowing starfields and dark dust clouds of the Milky Way, which are visible to the naked eye on a very dark night far from city lights. (Photo by author, Mauna Kea Observatory, 25-min guided time exposure, 24-mm lens, f2.8, Ektachrome ISO 1600 film.)

cumpolar zone from a latitude of 20°.)[5] Using this and other clues, Ovenden obtained a latitude of $36 \pm \frac{1}{2}°$ N for the ancient constellation makers.

He estimated the date of their activity by using a more complicated fact about the sky. Because of lunar and solar forces acting on the Earth, the spinning Earth slowly wobbles in space like a spinning top, and the axis thus points in different directions during different millennia. This wobble is called **precession**. Due to precession, the positions of the north and south celestial poles and celestial equator slowly drift among the stars, as seen in Figure 1-11. During the 26 000 y required for one such circuit, different stars become the North Star.[6] The ancient constellations show a rough symmetry around the star Alpha Draconis, about 25° from Polaris. Since this star was the north star around 2600 B.C., Ovenden concluded that most constellations were designed about 2600 ± 800 B.C.

These findings clarify the purpose of the constellations. Many people have supposed the constellations were dreamed up by uneducated shepherds or sailors who amused themselves by finding images in random star groupings. Now it appears that the original constellations were more carefully designed to help people learn the positions of the celestial pole, equator, and coordinates of about 2600 ± 800 B.C. The purpose may have been to help teach navigation. This idea accounts for the seemingly forced representations of some constellations. For example, Hydra, a peculiar, elongated constellation of 25 faint stars, lies accurately along the celestial equator of about 3000 B.C. Certain additional evidence suggests the designers were seafarers.

What seafaring civilizations lay near 36° N in the period 2600 ± 800 B.C.? The Minoan sailors of Crete (35° N) are the most likely suspects. Vestiges of Minoan culture were transmitted to the Greeks and thence to us, consistent with this theory. In summary, many constellations may be Minoan creations handed down to us from around 2600 B.C., with still earlier elements incorporated into them. We should not assume that "it all started with the Greeks." We have received knowledge and traditions from even earlier societies!

As shown in the four star maps at the end of this book, different constellations are visible in the evening sky during different seasons. Today they no longer have mythical significance, but it is pleasant to greet them each year as old friends who signal the arrival of each new season.

---

[5]To make such a photo yourself, see Project 11 at the end of this chapter.

[6]This effect of *precession* is detailed in the next chapter, where Figure 2-10 shows Alpha Draconis as the North Star in 2600 B.C.

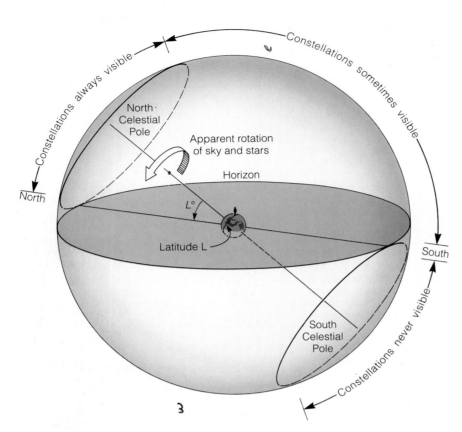

Figure 1-8 For an observer at latitude *L* in the Northern Hemisphere, no stars and constellations within *L*° of the north celestial pole ever set. Stars in a middle zone rise and set, and a southern zone within *L*° of the south celestial pole is never seen.

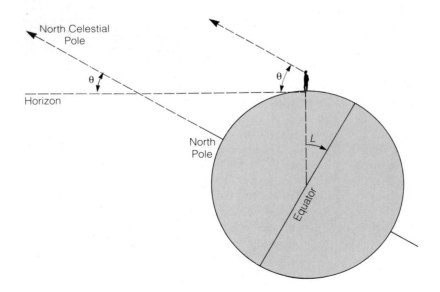

Figure 1-9 How to determine your latitude astronomically. Angle *L* is the latitude by definition. The elevation θ of the north celestial pole above the horizon equals the latitude *L*. To understand the principle better, try drawing the diagram for an observer at the equator (θ = *L* = 0°) and at the pole (θ = *L* ° = 90°). For convenience, keep the horizon line horizontal and the observer at the top of the globe.

a

b

Figure 1-10 Due to Earth's rotation, a stationary camera making a time exposure at night toward the northern or southern sky will reveal the circular tracks of the stars across the sky. **a** Northern sky. Polaris is the bright star near the center of the circular star traces. Because this photo is made at latitude 20°N, stars within 20° of the north celestial pole never set. **b** Southern sky. In this photo from the same location, stars within 20° of the south celestial pole never rise. The stars in the south make rainbowlike arcs across the sky as Earth turns during the night. (Both photos by author from Mauna Kea Observatory, illuminated by moonlight. Stationary 35-mm camera with 15-mm wide-angle fisheye lens; approx. 20-min exposure on commercially available 3M ASA 1000 film.)

THE SEASONS: SOLSTICES, EQUINOXES, AND THEIR APPLICATIONS

Most people think the Sun rises in the east and sets in the west. This is roughly true, but it is only an approximation. As the seasons progress, the Sun rises and sets at different points on the horizon. Consider an observer in the United States or elsewhere in the Northern Hemisphere. In summer, the Sun rises in the *north*east, passes high overhead at noon, and sets in the *north*west. In winter, the Sun rises in the *south*east,

passes low in the southern sky at noon, and sets in the *south*west. Only on two dates, the beginning of spring and the beginning of autumn, does the Sun rise exactly due east and set exactly due west (see Figure 1-3, where such a path is shown).

The reasons for this have to do with the **cause of the seasons:** the tip of the Earth's equator by $23\frac{1}{2}°$ to the plane of the ecliptic, as seen in Figure 1-12. As Earth moves around the Sun during a given year, the axis always stays pointed in the same direction—always toward the North Star (or, to be more precise, toward the north celestial pole). Thus, as seen in the figure,

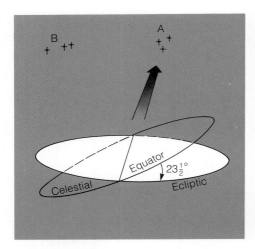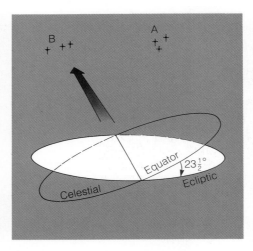

Figure 1-11 The effect of precession, or rotation of the plane of the celestial equator. At different dates, A and B, during the 26 000-y precessional cycle, the intersection of the celestial equator and ecliptic is seen in different positions, A and B, among the stars.

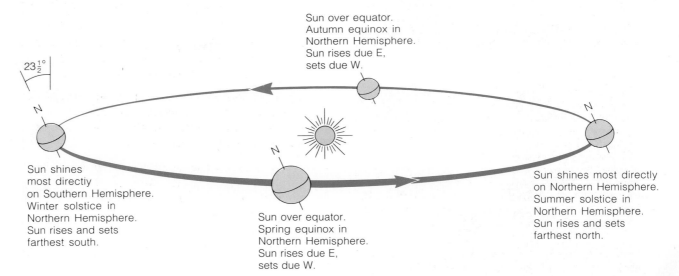

Sun over equator.
Autumn equinox in
Northern Hemisphere.
Sun rises due E,
sets due W.

Sun shines
most directly
on Southern Hemisphere.
Winter solstice in
Northern Hemisphere.
Sun rises and sets
farthest south.

Sun over equator.
Spring equinox in
Northern Hemisphere.
Sun rises due E,
sets due W.

Sun shines most directly
on Northern Hemisphere.
Summer solstice in
Northern Hemisphere.
Sun rises and sets
farthest north.

Figure 1-12 The cause of the seasons. Earth's axis is tipped at $23\frac{1}{2}°$ to its orbit plane in a constant direction as it moves around the Sun. Therefore the Sun shines more on the Northern Hemisphere during one season (northern summer) and more on the Southern Hemisphere 6 mo later (northern winter). See text for further discussion.

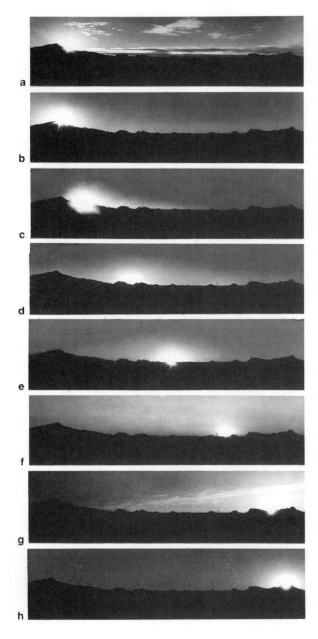

Figure 1-13 Movement of the sunset position along the horizon during an 8-mo period. Moving south in **a** the Sun reaches winter solstice in **b.** Here the setting position reverses direction and proceeds north again in **c,** reaching the vernal (spring) equinox in **e.** Summer solstice is reached in **h.** The author's former home, from which these pictures were made, happened to occupy the focus of a "natural Stonehenge," where the two solstice positions were marked by prominent mountain peaks. Such situations may have triggered the prehistoric idea of building solstice markers to reset calendars twice a year. (Photos by author.)

the Northern Hemisphere is tipped toward the Sun for part of the year and away from the Sun during the other part of the year. When the Northern Hemisphere is tipped toward the Sun, the Sun rises and sets further around toward the northern part of the horizon; the Sun also rises higher in the sky and shines more directly down on us; the weather grows warmer and we call the seasons spring and summer.

Thus there are four special days each year, called **equinoxes** and **solstices:**

Spring Equinox (First Day of Spring, About March 21)   Sun crosses the celestial equator moving north.[7] Rises due east, sets due west. An important day to ancients in the Northern Hemisphere because it marked the return of the Sun to "their" sky, bringing warmth. Also called the vernal equinox.

Summer Solstice (First Day of Summer, About June 22)   Sun reaches point farthest north of celestial equator. Rises and sets farthest north. Most hours of daylight in Northern Hemisphere, ensuring warm weather. Most important day of year in many ancient northern calendars.

Autumn Equinox (First Day of Autumn, About September 23)   Sun again crosses celestial equator, moving south. Colder weather on the way.

Winter Solstice (First Day of Winter, About December 22)   Sun farthest south of celestial equator. Rises and sets farthest south. Fewest hours of daylight in Northern Hemisphere.

Part of this sequence can be seen in Figure 1-13, which is a series of photos looking west at sunset and taken one month apart. In the first photos, around winter solstice, the Sun sets in the southwest (left), but as we move toward summer, the sunset position slides to the north (right). We might call this cyclical shift in the sunset and sunrise positions the **solstice principle.**

Ancient Applications of the Solstice Principle

These concepts were very important to primitive people because they had no dated newspapers arriving on their doorsteps and no calendars to hang on their walls. Observations of solstices and equinoxes offered a way

---

[7]As a memory aid, recall that the Sun crosses the *equa*tor on the *equi*nox.

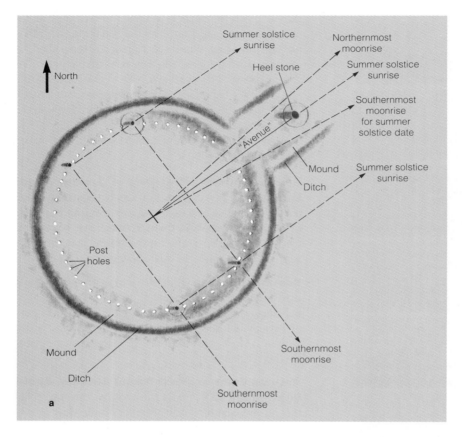

North

Summer solstice
sunrise

Heel stone

Northernmost
moonrise

Summer solstice
sunrise

Southernmost
moonrise
for summer
solstice date

"Avenue"

Mound

Ditch

Summer solstice
sunrise

Post
holes

Mound

Ditch

Southernmost
moonrise

Southernmost
moonrise

a

Figure 1-14 a Aerial plan of the original Stonehenge construction around 2500 B.C. An outer ditch, mound, and ring of posts or stones were cut by an avenue leading toward summer solstice sunrise. As seen from the center, the summer solstice Sun rose over the heel stone, partway down the avenue. Selected moonrises could also be observed and timed. **b** Aerial photo shows modern Stonehenge. Central ring of giant stones was added around 2000 B.C. White path is a twentieth-century addition. The avenue leads out of the right of the picture, and the heel stone is near the road. (Georg Gerster.)

b

Figure 1-15 Stonehenge (rear) and its heel stone (foreground), which marks the position of sunrise on the date of summer solstice. (Photo by author.)

to reset the calendar each year. For example, if you lived in an agricultural community, and tradition dictated that the best time for planting was mid-April, how would you know when mid-April had arrived? Priest-astronomers might have begun counting days at the time of spring equinox and might know that the optimum planting time was 24 d after equinox.

Several lines of evidence indicate that such ideas were practiced. For example, Spanish explorers recorded that Peruvian Inca Indians determined the date of solstice in this way and conducted ceremonies at that time, sending out runners to spread the word through the Inca empire of the beginning of a new year. Similarly, Pueblo Indians in the Southwest observed sunrise and sunset positions to determine the solstice or equinox dates.

But one of the most extraordinary lines of evidence reaches back much earlier in time. Around 2800–2200 B.C. at **Stonehenge,** in England, prehistoric builders constructed a large circular embankment with a broad avenue leading outward about $\frac{1}{2}$ km ($\frac{1}{4}$ mi) directly toward the horizon position of summer solstice sunrise, as shown in Figure 1-14. Subsequently, as late as 2100 to 1500 B.C., additional huge stones were brought in to construct a monument in the center of the ring. These included 30 to 50-ton stone blocks brought as far as 30 km (20 mi) and some 5-ton stones hauled about 380 km (240 mi). These stones can be seen in Figure 1-15. A large stone was placed in the avenue in such a way that

an observer at the center would see the Sun rise over the stone *on the day of summer solstice* as shown in Figure 1-16. Archaeologists have found additional post holes near this stone, suggesting that a grid of wooden posts might have allowed "fine-tuning" of the observations to determine on which day the Sun rose furthest north. As shown in the map in Figure 1-14, other astronomical alignments have also been found in Stonehenge, suggesting that the builders may have been much more sophisticated, keeping track of motions of the Moon as well as the Sun and possibly predicting eclipses, though this idea remains controversial.

Of course, we don't know whether Stonehenge was used as a practical observatory to *measure* dates of astronomical events like summer solstice or whether it was more of a ceremonial temple where, perhaps, solstice-related rituals were held. Many other prehistoric standing-stone monuments exist in Great Britain, but astronomical orientations are clear only for some of them (cf. Figure 1-17). There is some evidence that the earliest builders at Stonehenge had precise observing in mind and that builders in later centuries were more interested in elaborate ritual, because the huge stones that were among the latest additions seem to show less clear astronomical alignment. Perhaps they were added just to enhance a sense of mystic magnificence to what had become a quasi-religious temple. In any case, more than 4000 y ago the builders of Stonehenge had a strong awareness of the solstices.

The fact that Stonehenge is designed around a solstice alignment was pointed out as long ago as 1740 by English scholar Dr. W. Stukely. Yet in the 1880s and 1890s, when American astronomer Samuel Langley and English astronomer J. N. Lockyer spoke of Stonehenge as an astronomical observatory, most scholars thought ancient people could not have been sophisticated enough to orient large structures this way. Today we have much more evidence that early civilizations did make sophisticated use of the solstice principle and other astronomical knowledge.

For example, one of the largest temples ever built, the Temple of Amon-Re in Karnak, Egypt, built around 1400 B.C., covered twice the area of St. Peter's in Rome. Its central hallway, about 370 m (1200 ft) long, was aligned within $\frac{1}{2}°$ of the summer solstice sunset position (Figure 1-18). Other Egyptian temples had similar orientations, whereas still others were oriented toward heliacal risings and settings of certain stars. Because of precession, the rising and setting points of stars shift

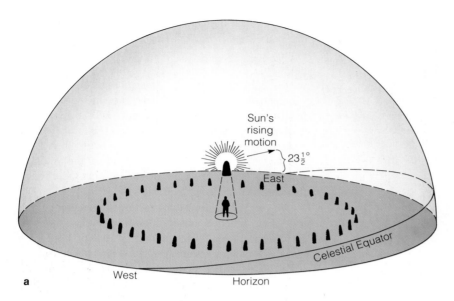

a

b

Figure 1-16 The Stonehenge principle. **a** The Sun's position at solstice is 23½° from the celestial equator. The rising (or setting) Sun on this date is thus as far as possible from due east or west, respectively. This position on the horizon is marked by a pillar or other prominent marker. Additional markers in a ring could allow observations of other selected risings or settings during the year. **b** Sunrise on the morning of summer solstice from the center of Stonehenge, showing the Sun's position over the top of the distant heel stone. (Georg Gerster, Photo Researchers, Inc.)

slowly over the centuries. Some of these temples show evidence of rebuilding as if to correct for such shifts, even though the rebuilding often ruined the symmetry of the temple by changing the angle of the central hall. The Egyptians are known to have made astronomical observations: Chicago's Oriental Institute has part of a device for charting star positions that was made by the young Pharaoh Tutankhamen (the famous "King Tut") around 1350 B.C.

Archaeologists have recently published a flood of additional analyses of temples throughout the world with suspected astronomical alignments. Among these are temples of Angkor Wat, Kampuchea (Cambodia) (c. A.D. 1150); Tihuanaco, Bolivia (A.D. 375–725); the Caracol (Figure 1-19) and other sites in Mexico (c. A.D. 1000); Casa Grande, Arizona (c. 1350); Cuzco, Peru (c. 1500); and Medicine Wheel, Wyoming (1500–1760). Evidence includes special alignments or window designs that mark solstices or other events and legends or historical records of astronomical observations or ceremonies being

Figure 1-17 Circles of standing stones are dotted throughout Great Britain and northern Europe. Astronomical alignments at most sites are less clear than at Stonehenge and are still being studied. While some may have been for calendrical functions, others may have had nonastronomical purposes. These giant stones were erected in prehistoric times as part of a vast double-ring complex some miles north of Stonehenge at Avebury. The Avebury complex is so large that the rings of stones enclose part of a village. The purpose of the complex is uncertain. (Photo by author.)

Figure 1-18 The ruins of the long hall in the Egyptian temple of Amon-Re, built around 1400 B.C. Nineteenth-century astronomer-archaeologist Norman Lockyer used this illustration when he pointed out in 1894 that this main hall is aligned toward the summer solstice sunset position.

carried out at the sites. Such evidence of this sort indicates that numerous astronomical traditions of calendar-keeping sprang up independently around the world.

New Year's Day and the Solstice Principle in Ancient Europe

Why does our New Year start on January 1? Does anything special happen on that date in terms of Earth's motion around the Sun? No. Ancient Rome and certain other cultures began the New Year in mid- to late March, the time of the spring equinox, when the Sun moved into the northern half of the sky and warmer weather arrived. But in many northern European cultures a big celebration of the New Year came on about December 22, winter solstice, when the days began to get longer and the long, cold, gloomy northern nights began to shorten. In 152 B.C. the Roman Senate reportedly moved New Year's Day from March 15 to January 1 to allow a military campaign scheduled for the new year to begin

early. Most authorities agree that the early Christian church, uncertain of the historical birthdate of Jesus, chose December 25 for the Christ-mass celebration to co-opt major pagan celebrations of the equinox and New Year season.

Interestingly, vestiges of the ancient March solstice New Year's Day persisted in historic times. Some older European communities used to start the New Year on March 25. The early Christian church had a feast day on this date to commemorate the conception of Jesus—possibly another co-opting of the pagan calendar. In *Tess of the D'Urbervilles,* novelist Thomas Hardy records that farm labor contracts in nineteenth-century England were still calculated from that date.

Even Christian cathedrals inherited some of the ancient solstice/equinox traditions. The old (330–1506) and new (1506–) cathedrals of St. Peter in Rome were reportedly oriented so accurately to the east that on the morning of the spring equinox "the great doors were thrown open and as the Sun rose, its rays passed through

Figure 1-19 The Mayan ruin "Caracol," built around A.D. 1000 in Chichen Itzá, Yucatan. Its windows include astronomical sight lines and were probably designed to facilitate the extensive observations of Venus and other bodies, recorded by Mayan astronomer-priests. (Anthony Aveni.)

the outer door, then through the inner door, and penetrating straight through the nave, illuminated the High Altar" (Lockyer, 1894). Certain other Christian cathedrals and missions were oriented toward sunrise on the day of the saint to whom they were dedicated. This sounds suspiciously like temple orientation traditions in ancient Egypt and elsewhere. Probably these ancient traditions were absorbed by the early Christians from their cultures. Ghostly fingers of ancient practices reach into the present to touch us!

Modern Applications of the Solstice Principle

A totally different application of the solstice principle is appearing in modern times. As fossil energy supplies dwindle, we are more concerned about utilizing solar energy in our buildings. The cheapest way is by clever building orientation and roof overhang to let sunlight in during winter and keep it out during summer.[8] In a sunny climate, most of the heat input through windows and walls comes in afternoon and morning when the Sun is low and strikes these surfaces full on. (Midday heat input through roofs must be minimized by thick insulation.) As shown in Figure 1-20, house design can utilize the solstice principle. Southeast- and southwest-facing windows, for example, can be unshaded to provide "free" warmth around winter solstice, and northwest-facing windows can be covered with external shields

[8]This is called passive solar utilization because it requires no further machinery or energy consumption. Active solar utilization implies further machines, such as fans to distribute warm air once it is heated by the Sun.

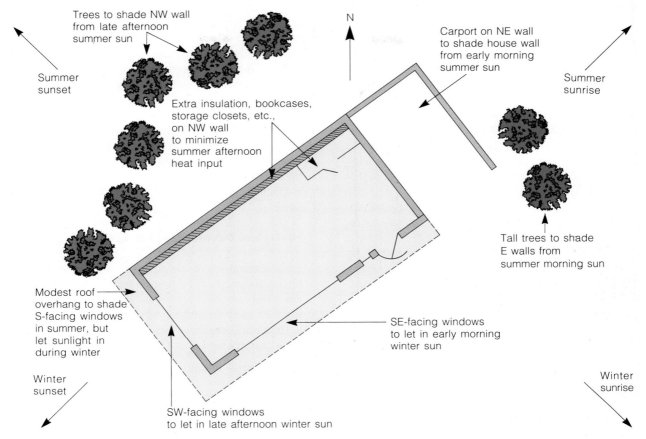

Figure 1-20 The solstice principle as applied to environmental architecture and landscaping. Northeast and northwest walls are shaded to reduce heat from the summer morning and afternoon sun. Southern windows allow sunshine to enter and heat the house in winter.

in summer to prevent late afternoon sunlight from pouring unwanted heat into the house.

ASTROLOGY: ANCIENT ORIGINS OF A SUPERSTITION

If you lived in a society where the calendar was determined by observations of the stars, where priest-astronomers counted days from the summer solstice and studied heliacal risings in order to announce when to plant crops, and where the death of a king was remembered as happening in "the year when Mars passed through the constellation of Gemini," you could see how easy it might be to make a serious philosophical error.

Instead of realizing that the stars offer a practical way to date and coordinate human affairs, you might come to believe that the stars *control* or at least influence human affairs. This superstition is called **astrology**. We call it a superstition for the same reason that we say it is superstitious to believe in invisible fairies in your garden: No experimental evidence supports the belief, and predictions based on the belief have no better than random accuracy.

An occupational hazard of being an astronomer is to be introduced at parties as an astrologer. This mistake is annoying because astrology is not part of modern astronomy but a pseudoscience associated with astronomy as it was practiced 3000 y ago.

The pseudoscientific nature of astrology can be

understood best by exploring astrology's roots. Astrology can be traced back at least 3000 y, when it flourished with other ancient magical beliefs. A common form of ancient magic was to associate patterns in nature with patterns of human events. Most people today would scoff at having their futures read from flight patterns of migrating birds or patterns in the bloody intestines of freshly sacrificed animals. Yet these were popular forms of divination in Babylon and Rome when astrology was flourishing, and they have the *same basic logic*. In astrology, the pattern in nature is that made by the planets as they move through the imaginary figures of the constellations.[9]

Archaeological research suggests that early people correctly used patterns in the sky in a practical way to foretell *natural* events, such as the change in seasons. Egyptian priest-astronomers, as we saw, designed their calendar to begin with the heliacal rising of Sirius, which was also the date of the Nile's flooding. Observation of the heliacal rising would thus help predict the flood, an important prediction for agricultural planning.

When priest-astronomers were observing signs in the sky to calibrate the calendar or determine when to plant crops, it must have been easy for the "man in the street" to conclude mistakenly that celestial signs forecast *human* events.

Human minds are quick to assume that if one event follows another, the second is caused by the first. Sometimes this is true and sometimes not. Centuries ago, logicians identified and named this error with the Latin phrase *"post hoc, ergo propter hoc"* ("following this, therefore *because* of this"). Yet this type of error is still common today. Historical evidence suggests that astrology grew out of this error. For example, if a king died a few days after an eclipse, some early observers reasoned that an eclipse causes or foretells the death of a king.

There is some evidence to suggest that early astronomical knowledge later degenerated into myth and pseudoscience. For example, the best astronomical alignments at Stonehenge were probably those constructed first, around 2500 B.C. Later construction, around 1800 B.C., muddled the design. In the nineteenth and twentieth centuries we see the monument used for mystical rites by costumed Druids and self-proclaimed witches. Here is an example of the loss of knowledge of original astronomical purpose.

In the same way, records of astronomical events, originally gathered by accurate observation, may have degenerated as later interpreters added mystical interpretations of astronomical events to accounts of historical incidents, describing them as if one controlled the other.

Historical Development of Astrology

Striking examples of this error can be found in ancient historical records. A Babylonian text of about 1600 B.C. describes motions of Venus and historical events together, assuming that the same events will occur again in the future whenever Venus presents the same patterns (paraphrased from Pannekoek, 1961):

If in the month of Abu, on the sixth day, Venus appears in the east, rains will be in the heavens and there will be devastation. Venus remains in the east until the tenth day of Nisannu and disappears on the eleventh. Three months she is not seen. On the eleventh of Duzu, Venus flares up in the west. Hostility will be in the land; the crops will prosper.

Similarly, a letter of an Assyrian astrologer to his king in 668 B.C. shows the same misconception still current a thousand years later (Pannekoek, 1961):

When Jupiter appears in the third month the land will be devastated and corn will be expensive. . . . When Jupiter enters Orion the gods will devour the land.

Mesopotamian astrology became a popular fad in the Greek world. Around 350 B.C. a Greek philosophical school, called *catasterism*, identified individual stars with certain of the humanlike Greek gods. Of course, implicit in the concept of a god is the idea that divine will prevails over human will. So it was a short step to the idea that the stars determine the fate of humans. The practice of forecasting personal fortunes spread from heads of state to sophisticated Greek households seeking diversion and finally to commoners.

Astrology also spread rapidly eastward. In a Chinese example as recent as 1882, an unsuccessful astrological

---

[9]An interesting example of this all-too-easy mental connection— between changing patterns in the sky and mystic patterns of human relationships—was given by French diarist Anaïs Nin during a 1935 visit to a planetarium. A practicing psychoanalyst, she had been thinking of the "human tangles" of her patients. Her diary records: "The planetarium . . . restored to me a sense of space and I could detach myself from the haunting patients. . . . I saw, instead of stars, relationships moving like constellations, moving away and towards each other . . . turning, moving, according to an invisible design, according to influences we have not yet been able to measure, analyze, contain."

attempt to stave off the dire consequence of a comet is blamed on government bungling (Stephenson and Clark, 1977):

When a [comet] was seen last year, an imperial decree was written to the palace and court officers, ordering them to perform their respective duties conscientiously. In . . . this month, the comet was seen again in the southeast. This must be due to the frequent mistakes committed by those employed in the administration.

In medieval times, astrology was banned by the Church as magic, yet it persisted, deeply ingrained in people's thinking. For example, historian Owen Gingerich (1967) documents how astrology was apparently used to pick dates for laying cornerstones of palaces and churches, and how a palace built in 1563 was oriented toward sunset on the date of the battle it commemorated. These traditions suggest a holdover from the ancient patterns of astronomical temple orientation at Stonehenge and in Egypt, but with the original purpose forgotten and replaced by ritual.

Since the Renaissance, astrology has continued to flourish alongside scientific thought. Today the ancient magic of 3000 y ago arrives on our doorsteps in the astrology columns of our daily papers!

Problems with Astrology

A test of astrology is whether it is consistent with observations. The basic practice is to forecast the influences for a particular day by casting a *horoscope*. The horoscope is usually presented as a circular chart, showing the positions of the planets, Sun, and Moon with respect to the constellation signs at the moment being studied. For example, since Venus was the goddess of love, prominence of Venus might indicate a loving influence. The prominence of the war god Mars might suggest an aggressive influence. This technique follows the rules laid down by astrologers centuries ago.

But application of the ancient rules creates an embarrassing problem for astrologers today. Astrologers nearly 4000 y ago designated 12 "signs," or portions of the zodiac, corresponding to the 12 zodiacal constellations. Like the planetary gods and goddesses, each constellation and its sign were said to have a certain personality trait. In 1867 B.C., the point where the Sun crossed the celestial equator on March 21 (spring equinox) and entered the Northern Hemisphere was in the constellation Aries and was called the "first point

of Aries." Thus a person born between March 21 and April 19 was said to be an "Aries," since the Sun lay in this constellation and thus in this sign. The original astrologers of that era thought this pattern was fixed. But around 130 B.C., Hipparchus discovered precession, which shifts the first point of Aries relative to the constellations. Thus the signs no longer correspond to their respective constellations!

Astrologers during Ptolemy's time tried to patch up the scheme. They defined the signs as 30° intervals measured from the March equinox point and claimed that the signs were more important than the actual constellations. Astrologers still claim that a person born between March 21 and April 19 is an Aries, but the Sun during most of this period is really in Pisces! Thus astrology has become removed from the realities of the sky.

Another embarrassment is that we now know that the stars move independently; thousands of years from now the constellation patterns themselves will have changed. Although one could patch up astrology further by inventing new constellations and associated psychological characteristics to cover this problem, the discrepancy shows that astrology as it is now practiced is not consistent with observed reality.

Another example will arise in the near future. When astronauts land on Mars, they will see no war god Mars in the sky. They will be standing on him! Instead, they will see a new planet, Earth, crossing the zodiac. Will astrologers invent a new personality influence associated with Earth's position in the Martian sky when the first baby is born on Mars? The old astrological rules will hardly apply!

Does Astrology Work?

A more scientific test of astrology is to examine the success of its predictions. It has failed in many spectacular cases. For example, a grouping of all known planets in Libra in 1186 led astrologers to predict disastrous storms, because Libra was associated astrologically with the wind. People dug storm cellars in Germany; the archbishop of Canterbury ordered fasting; the palace in Byzantium was walled up; people fled to caves in the Near East. But the conjunction of planets passed without incident. Similarly, in 1524, all known planets clustered in Aquarius, the Water Bearer. This time astrologers predicted a second Deluge, but the month of the conjunction passed without disaster and was reportedly drier than usual (Ashbrook, 1973).

Similarly, pseudoscientists and astrologers predicted cataclysmic earthquakes and other disasters due to the so-called Jupiter effect—increased gravitational stresses caused by a rough alignment of Jupiter, Earth, and other planets on the same side of the Sun in March 1982. The month and year came and went without the "predicted" cataclysms.

Some statistical studies of astrological predictions have been reported to see if these predictions are more successful than random predictions. Jerome (1975) reviewed several of these studies and concluded that "legitimate statistical studies of astrology have found absolutely no correlation between the positions and motions of the celestial bodies and the lives of men."

Still another test would be to search for forces exerted by planets or stars by which astrological influences could be manifested. Some modern apologists for astrology argue that these forces might be like gravity. So far, no such mysterious forces have been detected. Furthermore, we can calculate that the gravitational influences of nearby objects on a newborn baby, such as that of the mother during birth, are greater than the gravitational influences of the planets!

What Should We Do?

Because its logic is identical to that of ancient magic, because it is inconsistent with the twentieth-century sky, and because it fails to predict accurately, we conclude that astrology does not warrant our belief or our attention in planning our lives.

We might just smile at astrology as another amusing human foible if it weren't for the nagging suspicion that it threatens healthy civilization. For example, a 1975 Gallup poll indicated that about 12% of all Americans take astrology quite seriously. In an age of space exploration and serious nuclear threats, do we really want ancient magic to be an influence in human affairs? It is sad to realize that millions of people seeking advice about real problems in their lives spend their (sometimes meager) money on the products of astrologers and pious con artists. As astronomer Bart Bok (1975) points out: "The astrologer can refine his interpretations to any desired extent—the end product becoming increasingly more expensive as further items are added." Nowadays, computers are used to calculate planetary positions while casting horoscopes, giving a scientific gloss that misleads those who confuse scientific tools with scientific method. Thus naive members of the pub-

lic are fooled into thinking that astrology has some scientific basis.

Philosophically, the continuing practice of astrology helps remind us that the human mind has an amazing capacity to come up with ideas—many of which are wrong—and then to develop intricate systems of thought based on those ideas without ever establishing that the ideas are correct.[10]

Essay A, in the back of this book, describes further examples of pseudoscience that are merchandised as scientific truth in books and tabloids at supermarket checkout stands. Essay A also discusses in more detail how to distinguish science from pseudoscience—not through the hypotheses themselves, but rather through the way *evidence* is used to confirm or refute the hypothesis.

If we agree that astrology is little more than ancient superstition that wastes people's money and mental energy, should we try to suppress it? Newspapers routinely reject suggestions that they drop astrology columns, which they claim are harmless entertainment that promotes sales. Of course, most advice in astrology columns is innocuous. It might be said of astrology, like fairy stories, that if you believe it hard enough, it will work. For example, if you read morning advice to avoid frivolous expenses today, you may indeed be more prudent than usual. But one would hope we could find better sources of inspiration than ancient magic that is pseudoscience at best and fraudulent waste at worst. In a free country the answer is not suppression. In a free country with a free press, we hope that educated, pragmatic citizens will pick and enjoy the best from among competing published materials, and look for evidence of accuracy before committing themselves to a system of belief.

---

[10]I once met an industrious young man who had amassed a library of books on arcane details of various authors' astrological systems. His goal was to make more accurate astrological forecasts to guide his own life and to perfect a gambling system he could apply in Las Vegas. He had once had a string of successful bets in a football pool and was convinced that a fatal bet in which he lost all his winnings was merely the result of an erroneous reading of the horoscope.

As a student I rented a room from an ill, elderly woman of low income. I was aghast to discover that she had accumulated a substantial library of books and tracts purchased from a radio and TV evangelist. The books included drawings of various horned and tailed "demons," which, according to the books, were responsible for various illnesses. It was depressing to see the energy and money she needed for life being squandered in this way.

Figure 1-21 The March 1970 total solar eclipse as seen from the village of Atatlan, Mexico. The sky has darkened dramatically. (Photo by author.)

ECLIPSES: OCCASIONS FOR AWE

So far we have treated the Moon and the Sun as independent bodies that can be tracked for practical purposes. Cycles of the Moon divide the year into about 12 equal months; solstices and equinoxes, defined by sunrise positions, divide the year into four seasons. Ancient people who discovered these facts may have gained a sense of well-being: At least *some* things in their environment could be counted on from year to year. Calendars could be made and agriculture regulated. The predictable, friendly Sun could be worshiped as a deity, always providing light and warmth.

What, then, could be made of **eclipses,** the sporadic occasions when the Sun or more commonly the Moon disappears while above the horizon? In the few minutes of a total solar eclipse the sky turns dark, stars can be seen in daytime, and an uncanny chill and gloom settle over the land, as indicated in Figures 1-21 through 1-23. During the hours of a total lunar eclipse, the Moon turns blood-red. Small wonder that the Greek root of eclipse, *ekleipsis,* means abandonment or that the ancient Chinese pictured a solar eclipse as a dragon trying to devour the Sun. Eclipses must have seemed an unpredictable menace to the scheme of things.

We know eclipses awed early people. In Greece, the poet Archilochus, observing the solar eclipse of April

Figure 1-22 The "diamond ring effect" during a solar eclipse. The Sun's pearly glowing atmosphere makes a ring around the Moon. The Sun is almost totally eclipsed, but the last gleam of light from the Sun's disk reaches us past the edge of the Moon, adding a brilliant "diamond" to the ring. (Photographed in India in 1980; National Optical Astronomy Observatories.)

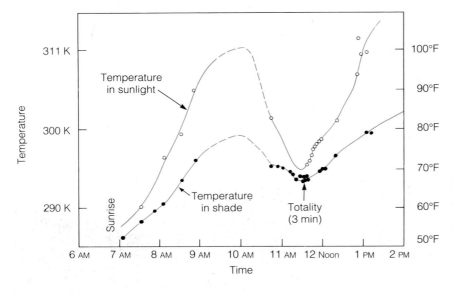

Figure 1-23 Temperatures recorded during the total eclipse of the Sun in Mexico, March 1970. The peculiar cooling would have been just one of many awe-inspiring effects of eclipses experienced by ancient witnesses.

6, 648 B.C., was moved to write: "Nothing can be sworn impossible . . . since Zeus, father of the Olympians, made night from midday, hiding the light of the shining Sun, and sore fear came upon men." Herodotus reports that a war between the Lydians and the Medes stopped when an eclipse surprised the competing armies during a battle in the 580s B.C. A hasty peace was cemented by a double marriage of couples from the opposing camps.

Knowledge of what causes eclipses allayed people's fears of them. And if people knew how to predict eclipses they could become very powerful. For instance, in February 1504, Christopher Columbus was having trouble convincing Jamaicans that they were obliged to feed him and his crew. Seeing that his almanac foretold an eclipse of the Moon on February 24, Columbus warned the Indians that the Christian god would punish them by turning the Moon to blood. The scoffing natives changed their attitude when the eclipse began on schedule.

This event may have inspired Mark Twain, whose Connecticut Yankee established his authority in King Arthur's court by predicting a total solar eclipse and then "commanding" the eclipse to end. This imaginative scene suggests the political prestige gained by some of the first priest-astronomers who discovered how to predict eclipses.[11]

---

[11]Even animals are affected by eclipses. I have seen roosters crowing and birds flocking to trees at midday because of the gloom during a solar eclipse, causing the creatures to believe evening has fallen. There is a story about an eclipse in Colorado during which an astronomer arrived too late to get a prime observing

In addition to providing us with interesting historical reading, the ancient astronomers who recorded eclipses centuries ago provided another valuable service: Their records are still used today to help analyze a slight increase in the length of the day, at a variable rate of about 0.002 s (second) per century (Stephenson, 1982).

Cause and Prediction of Eclipses

Whatever their emotions and motives, some early people learned how to predict eclipses—and this was an early step toward recognizing that we live on a world among worlds orbiting in space.

To understand how eclipses can be predicted, we should first understand the **causes of eclipses.**

Two important things to remember about eclipses are: (1) They happen when the shadow of one celestial body falls on another, and (2) what you see during an eclipse depends on your position with respect to the shadow. If you see an eclipse at all, you are either in the shadow (a celestial body has come between you and the source of light) or you are looking at the shadow from a distance as it falls on some surface.

Earthbound observers see two types of eclipses: solar and lunar. **Solar eclipses** occur when the Moon, on its $29\frac{1}{2}$-d journey around the Earth, happens to pass

site. He hurriedly set up his equipment in an empty chicken coop to protect his instruments from the wind, and then spent most of the eclipse trying to shoo away the chickens, who dutifully reported to the roost when darkness fell.

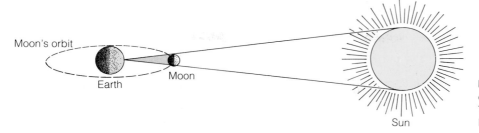

Figure 1-24 Geometry of an eclipse of the Sun (not to scale). The shadow of the Moon falls on Earth.

between Earth and the Sun as illustrated in Figure 1-24. A **total solar eclipse** occurs if the Moon completely covers the Sun as seen by an earthbound observer. A **partial solar eclipse,** shown in Figure 1-25, occurs if the Moon is "off center" and covers only part of the Sun. By coincidence, the Moon happens to have the same angular size as the Sun—about $\frac{1}{2}^{\circ}$, or a little less than the angular size of a little fingernail at arm's length.[12] Therefore, the Moon usually just covers the Sun during a solar eclipse. But if the Moon happens to be at the farthest point in its orbit, it has a smaller angular size than usual and doesn't quite cover the Sun. This causes an **annular solar eclipse,** in which the observer sees a ring (Latin *annulus*) of light, which is the rim of the Sun, surrounding the Moon.

Lunar eclipses occur when the Moon, on its journey around Earth, passes through a point exactly on the opposite side of Earth from the Sun. This point lies in the shadow cast by Earth (Figure 1-26). It takes the Moon a few hours to pass through Earth's shadow, during which time the Moon usually turns an astonishing copper-red because the only sunlight reaching the Moon passes through Earth's atmosphere, as seen in Figure 1-27. In effect the Moon is illuminated only by the colored light of a sunset.

Eclipses can happen in other parts of the solar system. For example, Figure 1-28 is a photo of a solar eclipse as seen from the Moon's surface, and Figure 1-29 shows an eclipse on Mars.

Umbral and Penumbral Shadows

In solar and lunar eclipses, the shadows of Earth and the Moon are not sharply defined. Because the Sun as seen from Earth has an angular size of $\frac{1}{2}^{\circ}$, sunlight is

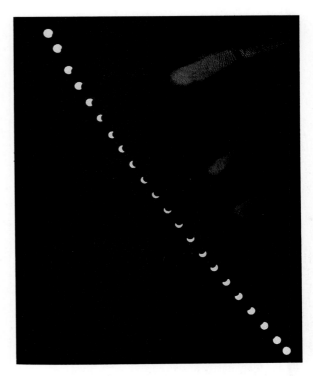

Figure 1-25 Multiple exposure showing the Sun's movement across the sky during a partial solar eclipse on July 10, 1972. At maximum, 72% of the Sun was covered by the Moon. Photos were made 6 min apart with a fixed 35-mm camera and a dark filter. (NASA photo by A. K. Stober.)

slightly diffuse; its rays come from slightly different directions.

Thus shadows—including eclipse shadows—cast by sunlit objects near Earth are not sharply defined. The inner, "core" area of a shadow—the part that receives no light at all—is named from the Latin word for shadow, **umbra.** The outer, fuzzy boundary is called the **penumbra.** An observer in the umbra sees the Sun entirely obscured; an observer in the penumbra sees

---

[12]If this concept is unclear, see the discussion of the small-angle equation in Optional Basic Equation I in Chapter 2.

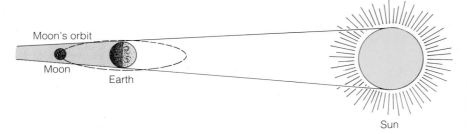

Moon's orbit

Moon

Earth

Sun

Figure 1-26 Geometry of an eclipse of the Moon (not to scale). The Moon passes through Earth's shadow.

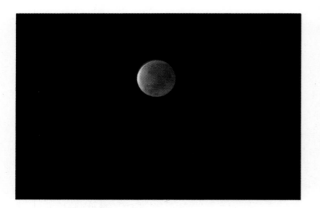

Figure 1-27 A nearly total lunar eclipse. Sunset-colored reddish light refracted through Earth's atmosphere colors the dim, umbral part of the shadow (right). The silvery crescent on the left is in the penumbral part of the shadow, partly lit by the Sun. This photo gives a good impression of the visual appearance of a lunar eclipse. (Photo by Stephen M. Larson.)

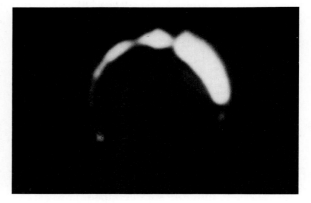

Figure 1-28 An eclipse of the Sun by Earth, photographed from the surface of the Moon. This television image, made by Surveyor III in 1967, shows a ring of light that is Earth's atmosphere, back-lit by the Sun, which lies behind the black disk of Earth. The brightest light is coming through the atmosphere over eastern Asia; local clouds affect the amount of light transmitted. (NASA.)

the Sun only partly obscured. Figure 1-30 shows some varieties of solar eclipses, illustrating these principles. If you hold your hand at eye level above smooth ground in sunlight, its shadow will have a central, dark umbra and a penumbra about a centimeter wide. If you hold your hand high and spread your fingers, their shadows will be indistinct. An ant in the umbra would see a total solar eclipse; an ant in the penumbra would see a partial solar eclipse.

The relative sizes of umbra and penumbra depend on the distance between the shadow-casting body and the surface on which the shadow appears. During a lunar eclipse, Earth's shadow on the Moon has an umbra several times the Moon's diameter and a considerably larger penumbra. Total lunar eclipses, with the Moon in the umbra, can last up to $1\frac{3}{4}$ h.

On the other hand, the Moon's umbral shadow on Earth does not exceed a 267-km (166-mi) diameter. Due to the motion of this shadow across Earth, total solar eclipses cannot last more than $7\frac{1}{2}$ min. In an annular eclipse, the Moon is too far away to produce any umbral shadow, and the eclipse has no true total phase, as seen in Figure 1-30 (case C).

Frequency of Eclipses

To be able to predict eclipses, we must understand the intervals between them. Discovery of these intervals several thousand years ago marks the beginning of our ability to predict seemingly mysterious celestial events. To understand the technique, imagine a viewpoint in space. If the Moon's orbit around Earth lay exactly on

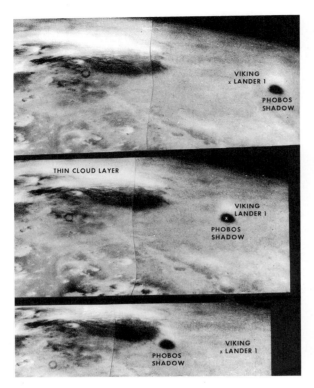

Figure 1-29 Oblique orbital views of an eclipse happening on the surface of Mars. The penumbral shadow of the small Martian moon Phobos moves from right to left. The shadow, about 90 km wide, passed over the site of the Viking I lander, helping scientists locate the lander's position. (NASA Viking I orbiter photo.)

the plane of Earth's orbit around the Sun, the Moon would pass exactly between Earth and the Sun every time around, as can be visualized from Figures 1-24 and 1-26. It would also enter the Earth's shadow on every pass, and a solar and a lunar eclipse would occur every month.

But the Moon's orbit is tipped by an angle of 5° out of Earth's orbital plane, as sketched in Figure 1-31. Therefore the Moon is likely to pass "above" or "below" the Earth–Sun line and Earth's shadow. Since the two orbital planes must, by simple geometry, intersect in a line, there is a line (NN′ in Figure 1-31) that contains the **nodes,** the only two points where the Moon passes through Earth's orbital plane, or ecliptic. Thus we now can understand the origin of the ancient word *ecliptic:* An eclipse can happen only when the Moon is passing

through the ecliptic. That the ancients gave this name to the Sun's annual path shows their interest in eclipses. The line between the two nodes, NN′, is called the **line of nodes.** To produce an eclipse, the Moon must be near one of the nodes, and even then *an eclipse will occur only if the line of nodes is pointing at the Sun.*

When is the line of nodes pointed at the Sun? If no external forces acted on the Moon's orbit, the orbital plane would stay fixed with respect to the stars and the line of nodes would line up with the Sun twice a year, as can be seen in Figure 1-32. Eclipses *could* occur at these two possible times each year, but *only* if the Moon moved through points N and N′ at these moments. Predicting eclipses would require alertness during only two intervals per year.

But there is one last complication. The Moon's orbit is not fixed with respect to the stars. It is disturbed by various forces. As a result of these forces, the Moon's orbit shows what is called a **regression of nodes:** The orbital plane swings slowly around, always keeping its 5° tilt to Earth's orbital plane. This regression of nodes is analogous to the precession of the Earth's axis. The line of nodes NN′ rotates slowly with respect to the stars, taking 18.61 y to complete one rotation. Therefore the line of nodes aligns itself with the Sun twice in a period somewhat less than a year. This period, called the *eclipse year,* turns out to be 346.6 d.

Thus several simultaneous cycles must be in phase to produce an eclipse: the 29.5-d cycle of lunar revolution with respect to the Sun; the 1-y cycle of Earth's revolution; and the 18.6-y lunar regressional cycle. These overlapping cycles cause subtle periodicities in the occurrence of eclipses. These periodicities helped ancient astronomers discover how to predict eclipses, as we will see in the next sections.

Discovery of the Saros Cycle (c. 1000 B.C.?)

From the information just given, we can understand one periodicity among eclipses. The Moon passes nearest the Sun's direction every 29.5 d (month), and the node N lines up with the Sun every 346.6 d (eclipse year). A lunar or solar eclipse occurs if the two cycles coincide. How often do they coincide? Using figures more exact than those just given, we find that 223 lunar months equal 6585.321 d, while 19 eclipse years equal 6585.781 d. Thus the two cycles come almost exactly into phase (to within only 0.46 d) every 6585 d, or 18 y, 11 d.

This interval is called the **saros cycle** (a Greek

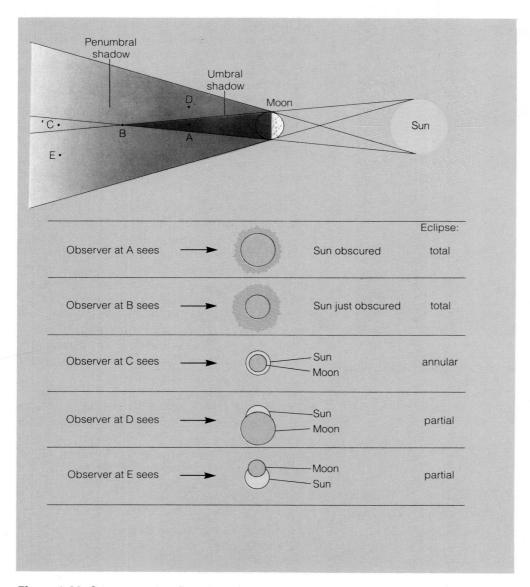

Figure 1-30 Geometry and configurations of solar eclipses by the Moon. Top diagram shows umbral and penumbral shadows of the Moon. Observers at different points see different kinds of eclipses, as shown here. By chance, the tip of the umbra (B) lies very close to Earth's surface.

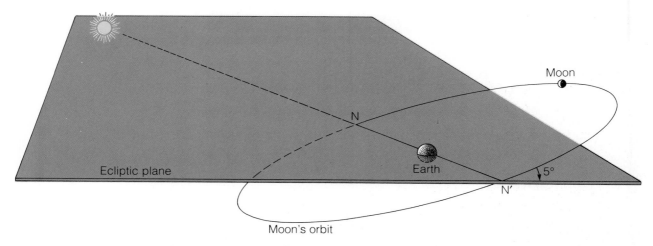

Figure 1-31 The Moon's orbit lies out of the ecliptic plane. Thus only at points N and N' can Earth, Moon, and Sun be aligned to produce an eclipse.

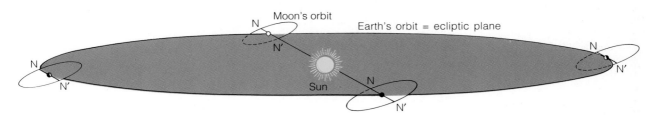

Figure 1-32 As Earth moves around the Sun, there are only two periods each year when the nodal line NN' of the Moon's orbit aligns with the Sun so that eclipses can occur.

name from an earlier Assyrian–Babylonian word). Early astronomers discovered that if, in a given year, a particular sequence of eclipses occurred, a similar sequence would probably occur after one saros. (Shorter periodicities, such as 41 and 47 months, also exist but are less accurate. Various periodicities were discovered and used by early astronomers to predict eclipses.)

Because the Moon's umbral shadow on Earth is so small, a fixed observer has only a small chance of seeing any given solar eclipse. But because Earth's umbra is large and the Moon can be seen from the whole night hemisphere of Earth, half the Earth will see every lunar eclipse (barring cloud cover). Total lunar eclipses are therefore relatively common for any observer, and they were thus easier than solar eclipses for ancient astronomers to predict using the saros and other cycles. Mod-

ern astronomers use computers and orbital theory, not cycles, to predict all eclipses accurately.

On the average a given location may witness a lunar eclipse nearly every year and a partial solar eclipse nearly every other year, but a total solar eclipse only about once every four centuries. Many of the dates are related by cycles.

Let us now return to the viewpoint of ancient earth-bound peoples. Once records were made and kept, astronomers could benefit from their community's past observations of eclipses. If they discovered periodicities, they could make rough predictions. We know this happened by 700 B.C. in the Mediterranean area and probably by about A.D. 500 in Central America, because records found in Assyrian libraries and Mayan cities discuss the prediction of eclipses.

TABLE 1·1

Total Solar Eclipses, A.D. 1991–2020

Date	Duration of Totality (min)	Selected Regions Where Totality Visible
1991 Jul. 11	7.1	Hawaii, Central America, Brazil
1992 Jun. 30	5.4	S. Atlantic
1994 Nov. 3	4.6	Chile, Brazil
1995 Oct. 24	2.4	Iran, India, Vietnam
1997 Mar. 9	2.8	NE Asia
1998 Feb. 26	4.4	Central America
1999 Aug. 11	2.6	Europe, India
2001 Jun. 21	4.9	S. Africa
2002 Dec. 4	2.1	S. Africa, Australia
2003 Nov. 23	2.0	Antarctica
2005 Apr. 8	0.7	S. Pacific
2006 Mar. 29	4.1	Africa, USSR
2008 Aug. 1	2.4	Siberia, China
2009 Jul. 22	6.6	India, China, S. Pacific
2010 Jul. 11	5.3	S. Pacific
2012 Nov. 13	4.0	Australia, S. Pacific
2013 Nov. 3	1.7	Africa
2015 Mar. 20	2.8	N. Atlantic, Arctic
2016 Mar. 9	4.2	S.E. Asia, Pacific
2017 Aug. 21	2.7	United States
2019 Jul. 2	4.5	Pacific, S. America
2020 Dec. 14	2.1	S. Pacific, S. America

Note: Brackets show pairs of eclipses separated by one saros cycle. Sequences of eclipses separated by one saros cycle are often visible, at least in partial form, from the same site. In addition, 22 annular solar eclipses occur in 1991–2020, including one in the central US on May 10, 1994, with 6.2 minutes duration.

Three Examples of Eclipses in History

Thales (c. 580 B.C.) Legend states that one of the first known astronomers, Thales of Miletus, predicted the eclipse that stopped the battle between warring Greek factions in the 580s B.C. Thales may have made this prediction by knowing of the saros cycle from Mesopotamian records, or he may have discovered a useful 3-saros periodicity of 669 lunar months, which would have been prominent in the eclipse records of his region for the preceding 125 y.

You can repeat the discovery of the saros cycle by studying Table 1-1, which lists total solar eclipses for the last quarter of this century. Table 1-2 lists some current *lunar* eclipses.

Mayans (c. A.D. 400) We know that Mayan astronomers of Mexico and Guatemala recorded and predicted eclipses long before European contact. One of three priceless Mayan manuscripts left after the Spanish conquest is a record of observed and predicted solar eclipses as well as other astronomical information, such as the motions of Venus (Figure 1-19). Tragically, because the Spaniards burned 27 other Mayan manuscripts in 1562, we may never know the extent of Mayan knowledge.

While the Mayan calendrical system had a 365-d yearly count and a so-called long count, or cumulative tally of days since a starting date about 3000 y previously, it also had a cyclic count useful for predicting eclipses. A combination of word names and number names for days (such as our usage *July 4*) produced a 260-d cycle called the Mayan Sacred Round. In other

TABLE 1·2

Selected Lunar Eclipses over the United States, 1991–2000

1991, December 21 (Partial)
1992, December 10 (Partial)
1993, November 29 (Total)
1994, May 25 (Partial)
1996, September 27 (Total)
1997, March 24 (Partial)
2000, January 21 (Total)

words, the day names repeated every 260 d. As we have seen, the nodes of the Moon's orbit are aligned with the Sun (in position for potential eclipses) once every 173.31 d. Eclipses could thus occur three times in 519.93 d, less than 2 h short of two Mayan Sacred Rounds. The Sacred Rounds thus warned Mayan priests of possible eclipse dates.

Why did the Mayans (and perhaps others) develop an eclipse-based calendar? One motivation may have been the fact that *five* solar eclipses were visible in the Mayan area during the 14 y from A.D. 331 to 344: a total eclipse in 331, a partial eclipse in 335, near total eclipses in 338 and 344, and an annular eclipse in 342 (Harber, 1969).

Such a cluster of eclipses may have impelled the Mayan priest-astronomers to keep records so that they could predict the Sun's next disappearance. In fact, the first astronomical observations in the Mayan manuscripts date from about this time. Interest in eclipses may have been further spurred in A.D. 495, when two partial solar eclipses were visible in Central America only 30 d apart, with a lunar eclipse between them (Owen, 1975). Mayan astronomy was supported by the state and highly organized. For example, a convention of prehistoric Central American astronomers was held at Copán, Honduras (probably on May 12, 485), to discuss the calendrical system.

Although the Mayan priests probably could not predict every eclipse, once they got their calendrical system working they might have had the best of both worlds. When they correctly predicted an eclipse, their power and knowledge were proved; when predicted eclipses did not occur, they could attribute this happy omen to the power of their rituals. The successes and failures of Mayan astronomy may thus have worked together to nourish the cultural traditions and help sustain the society for centuries.

We should also note that some scholars attribute the Moon alignments at Stonehenge (Figure 1-14a) to eclipse-predicting efforts dating as early as 2000 B.C., but this conclusion is quite controversial.

The Crucifixion An interesting example of the use of eclipses to date historical events is given by Humphreys and Waddington (1983) and Schaefer (1989), who attempt to find the date of the crucifixion of Jesus. References in the books of Matthew, Mark, Luke, and John suggest the date of Friday, Nisan 14 or 15 in the Hebrew calendar, on or just before Passover. Calculating back through calendar revisions to find years when this date fell on a Friday, the scholars restrict the number of possible dates to four Fridays between A.D. 27, and A.D. 34. Additional historical references suggest restricting the date further to April 7, 30, or April 3, 33. At this point the scholars invoke a speech of Peter, reported in Acts, which refers to a blood-red Moon, possibly referring to an event at about the time of the crucifixion. Calculating the dates of lunar eclipses, they find that a partially eclipsed Moon rose over Jerusalem on the night of Friday, April 3, A.D. 33, which they conclude was the date of the crucifixion.

SUMMARY

Unaided by telescopes, unknown geniuses of prehistoric times made many of the most basic astronomical discoveries. They (1) recognized and tracked five planets; (2) discovered the celestial poles and the four directional coordinates defined by daily star motions; (3) learned to use heliacal and solstitial risings and settings to formulate calendars; (4) designed constellations as memory aids for learning the sky; (5) recognized the ecliptic and the zodiac as the paths of the Sun and planets, respectively; (6) may have recognized some effects of precession, which causes stars to shift their positions relative to the celestial poles and the celestial equator; and (7) discovered eclipse-related cycles.

Stable societies encouraged astronomy, and vice versa. Astronomy probably contributed to civilization by creating calendars and encouraging record keeping. The discovery of eclipse cycles in particular required record keeping over many years.

Around 2600 B.C. there may have been a golden age in the Mediterranean world, when modern constellations were sighted and named on or near Crete, the Pyramid of

Khufu was carefully oriented in Egypt, and the Stonehenge solstitial observatory copied some of these ideas in England. By around 1400 B.C. building temples with astronomical alignments was a well-developed art in Egypt.

A similar development under way in America since the first millennium of the Christian era declined due to uncertain causes (climatic change?) shortly before the European invasion.

While the recognition of solstices, eclipse cycles, and so on did have applications in ancient societies, it did not lead to visualization of Earth's movement through space. As we will see in the next chapter, the first glimmers of a Sun-centered system came in Greek times, around 200 B.C. These concepts were not fully clarified until the most recent "moments" in the long story of human civilization. Of the 150 generations that have lived since astronomy emerged around 3000 B.C., only the last dozen generations (since about A.D. 1600) have well understood the concept of Earth as a ball moving around the Sun.

CONCEPTS

celestial poles	Stonehenge
Polaris	astrology
North Star	eclipse
zenith	causes of eclipses
meridian	solar eclipse
celestial equator	total solar eclipse
ecliptic	partial solar eclipse
planet	annular solar eclipse
zodiac	lunar eclipse
heliacal rising	umbra
heliacal setting	penumbra
constellation	node
circumpolar zone	line of nodes of Moon's orbit
equatorial zone	
precession	regression of nodes of Moon's orbit
cause of the seasons	
equinox	saros cycle
solstice	
solstice principle	

PROBLEMS

1. Suppose you are standing facing north at night. Describe, as a consequence of the Earth's rotation, the apparent direction of motion of each of the following:
   a. A star just above the north celestial pole
   b. A star just below the north celestial pole

c. A star to the left of the north celestial pole
d. A star on the northern horizon

2. Answer the parts of Problem 1 but for an observer in the Southern Hemisphere facing south and looking near the south celestial pole.

3. What is your latitude? What is the approximate elevation of Polaris above your horizon?

4. What is the radius of the north circumpolar zone of constellations as seen from your latitude?

5. How many degrees from the south celestial pole is the southernmost constellation that you can see from your latitude?

6. Solar eclipses are slightly more common than lunar eclipses, but many more people have observed lunar eclipses than solar eclipses. Why?

7. Does the Moon cast an umbral shadow on Earth during an annular solar eclipse? Why or why not?

8. How would a lunar eclipse look if Earth had no atmosphere? Compare the Earth's appearance from the Moon during such an imaginary eclipse with its actual appearance from the Moon.

9. Why would lunar and solar eclipses each occur once per month if the Moon's orbit lay exactly in the ecliptic plane?

10. Compare ancient European and American cultures of around A.D. 200 to 1000.
   a. How much did each know about eclipses?
   b. How many years apart did they achieve a comparable ability to predict eclipses?
   c. When, if at all, did they begin to understand the causes of eclipses?
   d. During this period, were the two cultures' advances in technology similar to their astronomical advances?
   e. Do you think that either of these two types of advances (or the two combined) offer a valid measure of cultural achievement?

ADVANCED PROBLEM

11. Using the following diagram, prove that the angle of elevation λ of the north celestial pole above the horizon

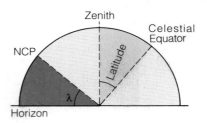

equals the latitude of the observer. Why does the angle from the celestial equator to the zenith equal the latitude?

PROJECTS

1. Starting on the date of the new moon (shown on many calendars), observe the sky at dusk and record whether the Moon is visible. Repeat each evening for several weeks and record the Moon's appearance. Repeat at the next new moon. How many days is the Moon visible between the new moon and the first quarter? Between the first quarter and the full moon? How might these counts relate to clusters of grooves, such as 3, 6, 4, 8, . . . , found on prehistoric tools? Can you prove or only speculate that the prehistoric records are lunar calendars?

2. From the preceding project, determine how much later the Moon sets or rises each night.

3. If a planetarium is available, arrange for a demonstration showing the following:
  a. The position of the north celestial pole
  b. The position of the celestial equator
  c. The daily motion of the stars
  d. The prominent constellations
  e. The daily motion of the Sun with respect to the stars
  f. The position of the ecliptic
  g. Planetary motions

4. Measure the angle of elevation λ of the North Star above the horizon. A protractor can be used as shown to make the measurement. Compare the result with your latitude.

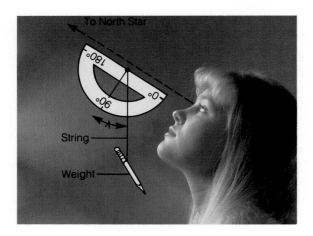

5. Identify a bright planet and draw its position among nearby stars each night for several weeks. (A sketch covering about 10° × 10° should suffice.) Does the planet move with respect to the stars? How many degrees per day? (The latter result will differ from one planet to another and from one week to another.)

6. Find a viewing area with a clear western horizon and determine the date of heliacal setting for some bright star or star group. If several students work independently on the same star, compare results. With how many days' uncertainty can the heliacal setting date be identified?

7. From a viewing area with a clear western horizon, chart sunset positions with respect to distant hills, trees, or buildings for several days and demonstrate the motion of the sunset point from day to day. Do the same for a few days around winter or summer solstice and demonstrate the reversed drift of the sunset position. How accurately can you measure the solstice date in this way? Does the Sun approach the horizon vertically or at an angle?

8. Using a light bulb across the room for the Sun, a small ball for the Moon, and yourself as a terrestrial observer, simulate total, partial, and annular solar eclipses.

9. Using the same props, demonstrate a total eclipse of the Moon. Show why a lunar eclipse occurs only during a full moon.

10. If an eclipse of the Moon occurs while you are taking this course (see Table 1-2), observe it. Compare the visibility of surface features on the Moon before the eclipse, in the penumbra, and in the umbra. Confirm that the curved shadow of Earth defines a disk bigger than the Moon. What color is the Moon and why? Would the Moon be lighter or darker during eclipse if there were an unusually large number of storm clouds around the "rim" of the Earth, as seen from the Moon?

11. Make a star trail photo like that of Figure 1-10 by setting your camera at night on some interesting foreground, opening the lens wide open, focusing on infinity, and exposing on a "fast" film, like Tri-X, for an hour or more.

CHAPTER 2

Historic Advances:
Worlds in the Sky

In their epic *Gilgamesh,* the Mesopotamians forecast that their own works would vanish as the wind. As they predicted, their names were lost and many of their discoveries degenerated into myth. But some of their knowledge of nature reached the Greeks.

The Greeks, with their blend of practicality and imagination, expanded this knowledge rapidly. Historical records give the names and sometimes the biographies of many Greek astronomers. Partly because of their work, we now accept without question the Earth's roundness, the nearness of the spherical Moon, and other commonplace but subtle ideas.

If you doubt the sophistication of these early scientists, try to measure the diameter of Earth and the relative distances between Earth, the Moon, and the Sun without any telescopes or electronic devices. *They* did.

Because discoveries often depend on an underlying conceptual or philosophical framework, in this chapter we look at the attitudes toward nature that the Greeks inherited and developed. We finish with some notes on astronomy in non-Western cultures.

EARLY COSMOLOGIES
AND ABSTRACT THINKING
(2500–100 B.C.)

Neither Greek scholars nor their systematic observations burst upon the Mediterranean scene from nowhere. The *idea* of thinking about abstract physical concepts can be traced back to early **cosmologies,** or theories about the origin and nature of the universe. For example, certain Egyptian cosmologies assigned different roles to different godlike personages who interacted with the real world. The Memphite theology (c. 2500 B.C.) spoke of an intelligence that organized the "divine order" of the universe (see Figure 2-1). According to Wilson

Figure 2-1 An ancient Egyptian conception of the universe. Stars are distributed over the body of the sky goddess, who arches over Earth. Some Egyptian drawings showed several goddesses, arching one over another—an idea that may have carried over to Greek and medieval times, when the planets and stars were visualized as distributed in concentric shells. (Giraudon/Art Resource.)

(1951), this theory's "insistence that there was a creative and controlling intelligence, which fashioned the phenomena of nature and which provided, from the beginning, rule and rationale, was a high peak of pre-Greek thinking." This kind of thinking represented a step toward astronomical science, because it assigned different gods, intelligences, or forces to different kinds of natural events and sought relations between them. Early naturalists expressed these relations in myths; later naturalists expressed them in "laws," or generalizations derived from repeated observations.

It is interesting to trace how cosmological thought evolved. Many ideas seem to have carried over directly from the world of 3000 y ago to us today. For instance, the dominant god among all Egyptian gods came to be Amon-Re, to whom the solstitial temple at Karnak was dedicated (see Chapter 1). Around 1350 B.C. the revolutionary Akhenaten, husband of the famous Queen Nefertiti, created a new religion based on the idea of a single god, Aton, the solar disk. Akhenaten's heresy was short-lived because he was overthrown by priests of the old religion. They installed a new young pharaoh, Tutankhamen, whose treasure-filled tomb astonished the world when it was discovered in 1922.

Out of this world, around 1300 B.C., escaped a tribe of nomads whose book of monotheistic religious thought is the core of Western religious principles today. Some scholars believe their Psalm 104 may be a direct transcription of a hymn to the Sun written by Pharaoh Akhenaten himself and still preserved (Pritchard, 1955). The pharaoh praised the Sun in almost biblical terms: "How manifold it is, what you have made! . . . You created the world according to your desire. . . ."

The Hebrews recorded a similar cosmology: "In the beginning God created the heavens and the earth. . . ." This cosmological theory asserted that Earth was created in six stages, or days: (1) light, day, night; (2) sky; (3) dry land, ocean, plants; (4) Sun, Moon, stars; (5) sea creatures, birds; and (6) humans. This cosmology dominated much Western art and science (Figure 2-2).

Cosmological theories of this type, which were especially common in the Middle East, stimulated new questions about relations between phenomena in the universe. Out of this thinking came an important new idea. Regardless of the question of gods, facts could be learned about nature by systematic observations and experiments from which repeatable results could be obtained. A naturalist in Alexandria could get the same results as a naturalist in Athens. **Science** (from Latin "to know") is simply the process of learning about nature by applying this technique: Questions are formulated that can be answered by observations or experiments, which are then carried out.

This system of scientific observation and recording of nature was developed to the greatest extent in Greece. In addition to richly provocative Egyptian and Hebrew philosophies, the Greek world received a legacy of astronomical concepts, such as the ecliptic, the zodiac,

Figure 2-2 English painter William Blake's version of the Western creation myth reflects the idea that the universe was created according to rational principles. This idea contributed to the birth of modern science, for it allowed philosophers to study the evolution of the universe without considering sudden changes in natural laws that might be caused by a pantheon of capricious gods. (Photograph courtesy of the Bettmann Archive, Inc.)

solstices, the saros cycle, as well as generations of astronomical observations of eclipses and planet motions from Mesopotamian (and perhaps European) sources. This inheritance, plus the Greeks' inclination to philosophize about natural phenomena, led to a Greek renaissance.

THE SYSTEM OF ANGULAR MEASUREMENT

One of the most important inheritances that the Greeks received from the earlier world was the sexigesimal (60-based) system of measuring angles and time, which was mentioned in Chapter 1. Just as there are 60 min of time

in one hour and 60 s in a minute, the system of angular measurement uses the following definitions and symbols:

$$1\,\textbf{degree} = 1° = \tfrac{1}{360}\text{th of a circle}$$
$$60\,\textbf{minutes of arc} = 60' = 1°$$
$$60\,\textbf{seconds of arc} = 60'' = 1'$$

To give a better idea of 1 second of arc: It is the angular size of a tennis ball seen at a distance of about 8 mi.

Applying this numerical system for measuring angles, the Greeks developed not only rules of geometry (such as Euclid's geometry and Pythagoras' famous theorem about the hypotenuse of a right triangle), but also ways of measuring phenomena in the sky. With sighting devices, Greeks and other early observers measured the positions of planets relative to the fixed pattern of background stars and the number of degrees of the Sun above the southern horizon as it crossed the meridian at noon in different seasons. These measures first revealed the *detailed* systematics of the movements of celestial bodies.

An important concept is the difference between linear measure and angular measure. **Linear measure** gives the actual length of something in linear units such as inches, meters, or miles. **Angular measure** gives the angle covered by an object (or, alternatively, the apparent separation between two objects) at a given distance from the observer, in angular units such as degrees. The verb **subtend** refers to the angle covered by such an object: For example, we might say a distant object subtends 1°. A useful rule of thumb is that your thumbnail at arm's length subtends about 1°. The disks of the Sun and Moon always subtend $\tfrac{1}{2}°$. The pointers in the Big Dipper (see Figure 1-4) subtend about 5°. As the Greeks knew, when you see a distant object (such as a ship at sea or the Moon), you can directly measure not its linear size or distance, but only its angular size. We unconsciously *infer* linear distance of many objects by recognizing the object (such as a ship) and knowing roughly how big it is; we similarly *infer* the linear size by estimating an object's distance and judging it must be as big as a house, a dog, and so on. The treachery of such inferences shows up, for example, when people report unfamiliar aerial objects such as a bright meteor, called a fireball. People commonly report that a fireball "looked as big as a dinner plate," but this statement is literally meaningless; they really perceive only angular size, not linear size. To specify angular size correctly, they would have to say, "It looked as big as a dinner plate at a distance of 50 feet" or "It looked twice the

Figure 2-3 These views toward the northeast from the center of Stonehenge (compare map, Figure 1-14) illustrate angular measurements applied to photography with different lenses. **a** This view using an ultra-wide-angle "fisheye" lens subtends a horizontal angle of 120°. **b** This view with a standard wide-angle lens subtends 65°. **c** This view with a normal lens subtends 40°; this is about the field of view of most snapshots, postcard views, paintings, and so on. **d** This telephoto view subtends only 15°. (Photos by author with a 35-mm camera. Lens focal lengths of a–d: 15 mm, 24 mm, 50 mm, and 135 mm, respectively. Curvature of pillars at edges of a is distortion common with ultra-wide-angle lenses.)

angular size of the Moon." Similarly, the common report that a fireball "must have landed just over the hill" is almost always wrong. The speaker misjudges the distance because he *assumes* a certain linear size after he *observes* only the angular size. Fireballs are typically in the upper atmosphere 60 mi from observers!

As we will see in a moment, Greek thinkers carefully separated angular measures from linear measures, and used angular measures together with clever logic to estimate linear sizes and distances of the Sun, Moon, and other objects.

The word **resolution** refers to the smallest angular sizes that can be discriminated with optical systems. The human eye, for example, can resolve an angle of about 2′; thus, we can see details covering about $\frac{1}{15}$ of the lunar disk. The largest planetary disks subtend only

about 1′ and are thus too small for the eye to resolve.

These principles help us analyze photographs and other images. For example, as shown in Figure 2-3, a typical snapshot subtends an angle of about 40°, about the portion of a scene that the eye concentrates on. A 35-mm camera is so named because it uses film 35 mm wide and makes a negative about 35 mm in width. With modern films, the typical 40°-wide snapshot can resolve 2′ details and thus presents a view comparable to what the eye sees. However, 40° views made with smaller cameras (such as instamatic and disk cameras) are limited by the inherent graininess of film and often do not resolve details as small as 2′. Thus they look grainy and do not resolve as much as the eye can see from the same spot. TV images suffer a similar limitation, resolving less detail than the eye can see in the typical 40°

view. Conversely, modern motion pictures photographed and projected with 70-mm film have a dramatically realistic presence because, like life, they can present the eye with a 40° view containing more detail than the eye can resolve.

The ancient system of angular measure, using degrees, minutes of arc, and seconds of arc, is still used today by surveyors, engineers, navigators, astronomers, and others. Modern large telescopes can often resolve details as small as 0.5″, but rarely can do better than this because of the shimmering quality of heat waves in the atmosphere. Satellite telescopes that have been launched into orbit have been able to resolve still smaller angles. Later in this book we will encounter more applications of angular measurement.

EARLY GREEK ASTRONOMY (c. 600 B.C. to A.D. 150)

Around 600 B.C. the Greeks began vigorously applying logic and observation to learn about the universe. They talked more of tangible physical "elements" and less of metaphysical relations. They used geometric principles to measure cosmic distances as well as farmyards.

One of the first known Greek thinkers was Thales of Miletus (a Greek-dominated town in present-day Turkey). Living about 636–546 B.C., Thales was a noted statesman, geometer, and astronomer. He is best known for predicting the peacemaking solar eclipse mentioned in Chapter 1. He probably knew some Mesopotamian astronomical concepts—perhaps the saros cycle, the lengths of seasons, and the daily changing position of the Sun among the constellations of the ecliptic. Thales also reportedly speculated that the Sun and stars were not gods, as was then usually thought, but balls of fire. Of course, Thales could not prove his idea, but he got other Greeks thinking in terms of tractable, physical ideas.

Thales' school produced several notable thinkers. Anaximander (611–547 B.C.) made astronomical and geographical maps; speculated on the relative distances of the Sun, Moon, and planets from our Earth; and argued that the matter from which things are made is an eternal substance. Heraclitus (535?–475? B.C.) made this remarkable comment:

This ordered cosmos, which is the same for all, was not created by any one of the gods or by mankind, but it was ever and shall be ever-living Fire, kindled in measure and quenched in measure. . . . The fairest universe is but a dust-heap piled up at random.

The Pythagoreans: A Spherical, Moving Earth (c. 500 B.C.)

Pythagoras (flourished 540–510 B.C.), famous for his theorem on right triangles, was also one of the first experimental scientists. Pythagoras proposed the unusual idea that Earth is spherical. He may have gotten this idea by studying the phases of the Moon. The line separating the lit side of the Moon (or any planetary body) from the unlit side (the **terminator**) changes its curvature as the Moon's phases progress, thus revealing that the Moon is spherical rather than flat, as shown in Figure 2-4. By analogy, then, the Earth and other bodies would also be spherical.

In southern Italy, Pythagoras founded a school that had wide influence around 450 B.C. It is unclear, however, which thinkers should be credited with which ideas in this school. Pythagoras himself put the Earth at the center of the universe, but later Pythagoreans proposed that it moves, like the Moon and the planets, around a distant center. The universe is spherical with a central "fire" containing a force that controls all motion. Around it, in order outward from the center, move Earth, the Moon, the Sun, the five planets, and the stars. This system predates by more than 2000 y Copernicus' revolutionary, correct model of the planets moving around the Sun (see Chapter 3). The idea of a spherical Earth persisted among some Greeks, though it was not universally accepted.

Anaxagoras (500?–428 B.C.) is credited with deducing the true cause of eclipses. Thereafter, the observed roundness of the Earth's shadow on the Moon (Figure 2-4) undoubtedly helped to establish the theory that Earth itself is a spherical body. After residing in Athens for 30 y, Anaxagoras was charged with impiety and banished for saying that the Sun was an incandescent "stone" even larger than Greece.

Plato (c. 400 B.C.)

Though known primarily as a philosopher, not a scientist, Plato (427?–347 B.C.) reasoned that astronomy contributed to the civilization of humanity. In *Timaeus*, he said that philosophy came from astronomy:

| a | Gibbous | Full | Gibbous | Quarter | Crescent | b |

Figure 2-4 Evidence of the Moon's true nature. **a** The phases correctly suggested to some Greeks that the Moon is not a disk but a sphere illuminated by the Sun. **b** The Earth's curved shadow on the Moon during every lunar eclipse suggested that Earth too is spherical. **c** Contrary to popular conception, the Moon is visible in the day during part of the month, as well as at night. Thus its phase can be studied in relation to the Sun, showing that the phases match those of a sphere illuminated by the Sun. (Photo by author, Sonora, Mexico.)

c

Had we never seen the stars, the sun, and the heavens, none of the words we have spoken about the universe would have been uttered. But now the sight of day and night, and the revolutions of the years, have created number and given us a concept of time as well as the power of enquiring about the nature of the universe: and from this source we have derived philosophy. No greater good ever was or will be given by the gods to mortal man.[1]

---

[1]In the 1800s Ralph Waldo Emerson extended this thought with the following lines: "If the stars should appear one night in a thousand years, how would men believe and adore, and preserve for many generations the remembrance of the city of God?" Science fiction writer Isaac Asimov ironically reversed this thought in a 1941 story called "Nightfall." He imagined a planet in a multistar system with six suns. Only once in 2050 y did the orbits of all the suns bring them all to one side of the planet, plunging the other side into darkness so that the stars came out. Instead of inspiring worship, this event so frightened the inhabitants that they burned everything around them to give them light. As a result, civilization on this planet was burned to the ground every 2050 y and had to start over again.

Aristotle: The Earth Again at the Center (c. 350 B.C.)

The most influential Greek scientist-philosopher was Aristotle (384–322 B.C.). His views were built on earlier knowledge but were biased in favor of absolute symmetry, simplicity, and an abstract idea of perfection. Aristotle's universe was spherical and finite, with the Earth at the center. Planets and other bodies moved in a multitude of spherical shells centered on the Earth. The shells were supposed to turn with varying rates, which explained the observed changeable motions of the planets.

Aristotle is credited with founding modern scientific investigation. His school at Lyceum (a grove near Athens) contained a library, a zoo, and lavish physical and biological research equipment paid for by his onetime pupil Alexander the Great, then ruler of Greece. In the Middle Ages, when research lapsed, Aristotle came to be regarded as the final authority, and so his rejection of the Pythagorean idea and his placement of Earth at the center of the solar system turned out to delay progress in astronomy. However, there was little reason for him to choose a Sun-centered over an Earth-centered system, since both views were consistent with observations known in his time.

Aristotle was right about several important astronomical ideas, however.

1. He thought the Moon is spherical.

2. He argued that the Sun is farther away than the Moon because:

　a. The Moon's crescent phase shows that it passes between the Earth and the Sun.

　b. The Sun appears to move more slowly in the sky than the Moon. (This second argument is not rigorous, but the first is.)

3. He thought the Earth is spherical because:

OPTIONAL　BASIC　EQUATION　I

The Small-Angle Equation

Angles and linear measures can be combined in an extremely useful and simple equation called the **small-angle equation,** which involves the angular size of an object, its linear size, and its distance. If any two of these quantities are known, the third can be calculated. Let us call the angular size α, expressed in seconds of arc. Let the diameter of the object be d and its distance D. Then the small-angle equation is

$$\frac{\alpha}{206\,265} = \frac{d}{D}$$

The number 206 265 is called a *constant of proportionality;* it stays the same in all applications of the equation.*

Consider an example. Suppose a friend who is 2.0 m tall is standing across a field, where he subtends an angle of $\frac{1}{2}°$, or $1800''$, as shown in Figure 2-5. How far away is he? We want to solve the equation for D. Rearranging the equation, we have

$$D = 206\,265\,d/\alpha$$

Using SI metric units (see Appendix 2), we would write $d = 2.0$ m. Thus, expressing the equation in SI metric units and powers of 10, we would have

$$D = \frac{206\,265\,d}{\alpha}$$

*The number 206 265 is actually the number of seconds of arc in 1 radian (rad). A radian is an angle of about $57°.3$, defined as the angle subtended at the center of a circle by one radius of the circle laid along the circumference. The radian has many applications in geometry.

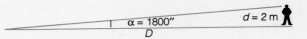

Figure 2-5 An application of the small-angle equation. If your friend is 2 m tall and subtends an angle of $\frac{1}{2}°$ (or $1800''$), his distance is D is 230 m.

$$D = \frac{2.06\,(10^5)\,2}{1.8\,(10^3)} = 2.3 \times 10^2\,\text{m} = 230\,\text{m}$$

Your friend is about one-sixth mi away.

As the Greeks realized, exactly the same geometry can be used to investigate astronomical distances. The Greeks could not get good measurements of the Moon's diameter, but only its angular size α, which is roughly $\frac{1}{2}°$, or $1800''$. If we use the modern knowledge that the Moon is about 3500 km in diameter, we can estimate its distance just as we did for the friend's distance above (Figure 2-6. In SI metric units, d would be 3.5×10^6 m. The equation would read:

$$D = \frac{206\,265\,d}{\alpha} \simeq \frac{2.06\,(10^5)\,3.5\,(10^6)}{1.8\,(10^3)}$$

$$\simeq 4 \times 10^8\,\text{m} \simeq 4 \times 10^5\,\text{km}$$

or about 400 000 km.

Several mathematical notes should be observed. First, the symbol $\simeq$ means "approximately equal to"; it is useful whenever approximate values (such as $\frac{1}{2}°$) are involved.

Second, this calculation shows the economy of

a. The curvature of the Moon's terminator rules out its being a disk, and the Earth is likely to be like the Moon in this respect.

b. As a traveler goes north, more of the northern sky is exposed while the southern stars sink below the horizon—a circumstance that would not arise on a flat Earth.

The apparent motions of the Sun, the Moon, and the stars around Earth could be explained, said Aris-

totle, either by their actually moving around us or by Earth moving. But Aristotle concluded that Earth is stationary and gave a very powerful argument. If Earth were moving, we ought to be able to see changes in the relative configurations of the various stars, just as, if you walk down a path, you see changes in the relative positions of nearby and distant trees. If you line up a tree in the middle distance with a very distant tree and then step to one side, the nearby tree will seem to shift

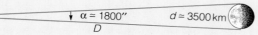

Figure 2-6 The same geometry as in Figure 2-6 can be applied to measure interplanetary distances, such as the distance to the Moon.

writing the numbers as powers of 10—for example, 3.5×10^6 m instead of 3 500 000 m.

Third, the answer is given only to an accuracy of one significant figure. **Significant figures** are the number of digits known for certain in a quantity. For example, π to one significant figure is 3; to three significant figures, 3.14. Generally, an answer to a calculation should have no more significant figures than does the least accurate number in the equation.† In this case that number was the angular size of the Moon, $\frac{1}{2}°$, or 0.5°, with only one significant figure.

A fourth note is that, following our rule given in the prologue, we convert data to SI metric units. (In our small-angle equation this is unnecessary, since the dimensions of d and D are the same and cancel out; but it is a useful habit.)

The small-angle equation has many applications in astronomy. It lets us calculate the size of distant objects, once we know their distance.

†This statement is especially important to users of electronic hand calculators, which blindly print answers with seven or eight significant figures, even when the accuracy of the input is only two significant figures.

Sample Problem 1. How big are the smallest craters we can see on the Moon with a backyard telescope?

Solution: Many backyard telescopes can resolve angular detail as small as one second of arc, or 1″. The Moon is 384 000 km away. So we are asking: How big is an object that subtends 1″ at that distance? The student should use the small-angle equation to solve for d and confirm that such a telescope can show craters as small as $1\frac{1}{4}$ km across (roughly a mile across) on the Moon.

Sample Problem 2. The Sun has a diameter of about 1.4 million km and is 150 million km away. Prove that it subtends an angle of roughly $\frac{1}{2}°$.

Solution: We need to solve the equation for the angle α. This gives

$$\alpha = \frac{d}{D} \ 206 \ 265$$

Substituting the values for diameter d and distance D, we get 1925″. Since there are 3600″ in one degree, this would be equivalent to 0.5348°. Because the d and D values were given to only two significant figures, the third and fourth digits in the answers cited above are meaningless and we must round off to two significant figures: 0.53°, or about $\frac{1}{2}°$.

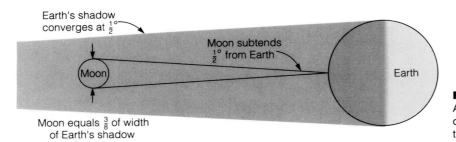

Earth's shadow
converges at $\frac{1}{2}^{\circ}$

Moon subtends
$\frac{1}{2}^{\circ}$ from Earth

Moon

Earth

Moon equals $\frac{3}{8}$ of width
of Earth's shadow

Figure 2-7 The geometry by which Aristarchus estimated the relative sizes of the Moon and Earth (not drawn to true scale).

to the side of the distant one. Such a shift in position due to motion is called **parallax**, or a *parallactic shift.* If Earth were moving in a straight line, we would see a continuous parallactic shift of the nearer stars with respect to more distant stars; and if Earth were moving around some distant center, we would see a periodic parallactic shift back and forth among the stars. But a visual survey of the stars and the constellations over time showed no evidence of such a shift. So, reasoned Aristotle, Earth must not move.

The reasoning was sound, but the stars are too far away to produce noticeable parallactic shifts for the unaided eye during a human lifetime. In the same way, distant mountains show little parallactic shift from a car speeding down an interstate highway, even though nearby trees whiz by in seconds. Stellar parallaxes were sought for years and not discovered until 1838.

Aristotle died shortly after being forced to leave Athens for allegedly teaching that prayer and sacrifices to the Greek gods were useless.

Aristarchus: Relative Distances and Sizes of the Moon and the Sun (250 B.C.)

Aristarchus (310?–230? B.C.) of Samos (an island off present-day Turkey) cleverly extended the Greek methods of seeking quantitative data. His only surviving work is "On the Sizes and Distances of the Sun and Moon," although his other astronomical works are quoted by other Greek authors.

He devised a way to measure the relative distances of the Sun and Moon from Earth, based on the geometry of the Moon's orbit and phases. From this he correctly inferred that the Sun is much farther away than the Moon.

Aristarchus also formulated a way to measure the relative sizes of Earth and Moon. We can reconstruct

his calculations, which used what he knew about eclipses. As shown in Figure 2-7, Aristarchus imagined a total eclipse of the Moon in progress. He drew a circle to represent the Earth. Because he knew the Sun's angular diameter is $\frac{1}{2}^{\circ}$, he could represent Earth's umbral shadow by lines leading away from Earth and converging at an angle of $\frac{1}{2}^{\circ}$. Aristarchus also knew that the Moon looks $\frac{3}{8}$ as big as Earth's shadow, so that wherever he placed the Moon in the Earth's shadow, its diameter would have to be $\frac{3}{8}$ the distance across the shadow at that point. But where to place the Moon's circle? Aristarchus knew that the Moon, like the Sun, has an angular diameter of $\frac{1}{2}^{\circ}$ as seen from Earth. These criteria specified the position and size of the Moon, as trial and error with a diagram like Figure 2-7 will show.

What about the Sun? Aristarchus estimated that the Sun is 18 to 20 times as far away as the Moon, based on observations of the moment when the Moon's disk was half illuminated. He then applied the observation that the Sun's angular diameter is $\frac{1}{2}^{\circ}$, like the Moon's, thus fixing its relative size. Using these methods, Aristarchus concluded that the Moon is one-third as big as Earth and that the Sun is about seven times as big as Earth. The correct figures are closer to one-fourth and 100, but Aristarchus was on the right track.

Aristarchus made still another contribution. Because he thought the Sun is much bigger than Earth, he guessed (without many supporting observations) that the Sun, not Earth, must be the central body in the system. For this an outraged critic declared he should be indicted for impiety.

Although Aristarchus made some quantitative errors, he was nonetheless far ahead of later scholars who thought Earth is flat. Aristarchus correctly visualized the Moon in orbit around a spherical Earth and Earth in orbit around the Sun, and he developed a method of measuring interplanetary distances. These ideas were not confirmed for another 2000 y!

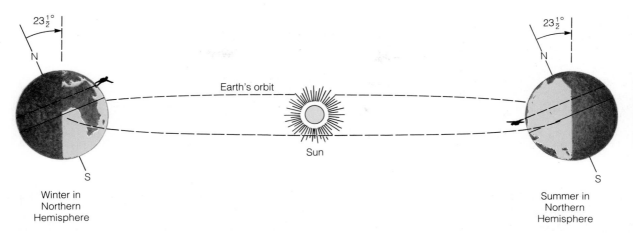

Figure 2-8 The seasons are caused by the $23\frac{1}{2}°$ angle of tilt between Earth's north pole and the north pole of the plane of Earth's orbit around the Sun. As shown by the human figure, an observer at a fixed northern latitude finds the noontime Sun more nearly overhead in summer than in winter. This difference, which would not arise if the tilt were zero, makes summer days hotter and longer. Eratosthenes measured the difference between the summer and winter noontime Sun's elevation and was able to use it to deduce the $23\frac{1}{2}°$ tilt angle. (The angular difference between the summer and winter noontime Sun elevations is twice the tilt angle—a statement the student may try to verify.)

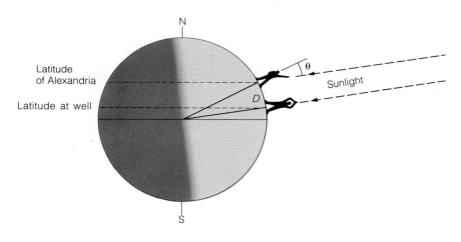

Figure 2-9 The geometry of Eratosthenes' measurement of the size of Earth. When the Sun was directly over a certain well, Eratosthenes measured the Sun's angle θ from the zenith at Alexandria, a known distance D away. He found D was $\frac{1}{50}$ of the way around Earth. He could thus find Earth's circumference and diameter.

Eratosthenes: Earth's Size (200 B.C.)

As Greece declined and Rome prospered, Greek scholars became resident intellectuals in many parts of the Mediterranean world. Eratosthenes (276?–192? B.C.) was a researcher and librarian at the great **Alexandrian library** in Egypt. He reportedly completed a catalog of the 675 brightest stars and measured the $23\frac{1}{2}°$ inclination of the Earth's polar axis to the ecliptic pole, as shown in Figure 2-8. As described in Figure 1-12, this is the tilt that causes our seasons.

Eratosthenes is most famous for using angular geometric relations to measure Earth's size. Told that at summer solstice the Sun shone directly down a well near Aswan, he noted that the Sun's direction was off vertical by $\frac{1}{50}$ of a circle on the same date at Alexandria (Figure 2-9). He realized this difference had to be due to the curvature of the Earth and concluded that Earth's circumference was 50 times the distance from Alexandria to the site of the well. Measuring that distance, he multiplied by 50 and got an estimate of the Earth's circumference. His estimate was probably within 20% of

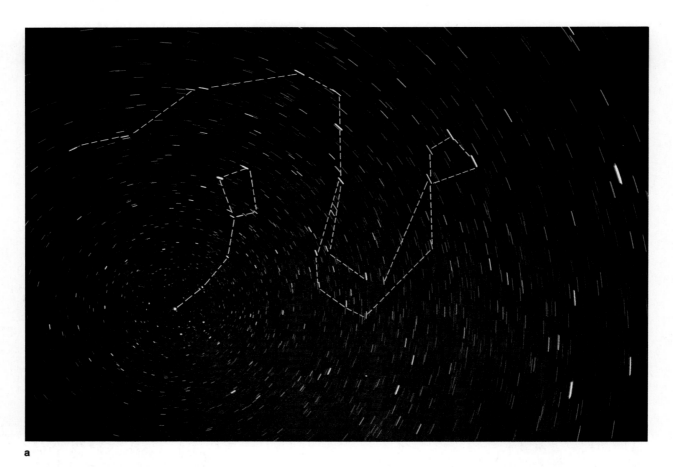

a

Figure 2-10 a This color photo shows stars wheeling around the North Star in a 10-min time exposure. Constellation patterns of the Little Dipper and Draco the Dragon are shown. (Wide-angle photo by author, 24-mm lens, f2.8, Ektachrome ISO 1600 film.)
b (*Opposite*) Map of the same region identifies constellations and shows changing position of the north celestial pole in other centuries. Because of precession, the north celestial pole around 2500 B.C. lay near Alpha Draconis. In A.D. 8000 it will lie near the star Alderamin, and in A.D. 15,000, near the very bright star Vega.

the right answer. This Greek master clearly understood the shape and approximate size of Earth 1700 y before Columbus!

Hipparchus: Star Maps and Precession (c. 130 B.C.)

From his observatory on the island of Rhodes, Hipparchus (160?–125? B.C.) observed the positions of astronomical bodies as accurately as possible and compiled a catalog of some 850 stars. His exhaustive observa-

tions—all done, of course, without a telescope—along with material he inherited from Babylon, enabled him to predict with reasonable accuracy the position of the Sun and Moon for any date. Hipparchus has been called antiquity's greatest astronomer.

The most important discovery attributed to Hipparchus is *precession* (though astronomers of earlier centuries may have been aware of its effects; see Chapter 1). Comparing his own measurements of star positions with materials handed down to him from centuries before, Hipparchus found that, with respect to the back-

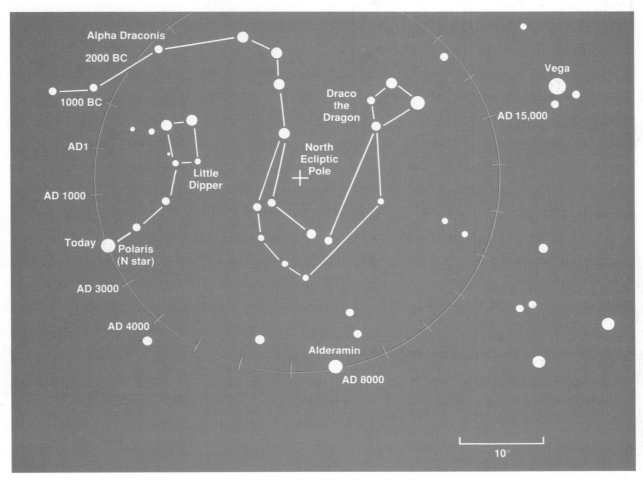

b

ground stars, there had been curious shifts in the positions of the north celestial pole, the vernal and autumnal equinoxes, and other coordinates. The whole celestial equator was oriented somewhat differently with respect to the stars! Could the old maps be wrong? Hipparchus concluded instead that the whole coordinate system of the celestial equator and the poles was drifting slowly with respect to the distant stars. This drift came to be known as **precession** or *precession of the equinoxes*.

In modern terms, *precession is the result of a wobble of the spinning Earth due to forces produced by the Sun and the Moon.* Just as a spinning top describes a conical wobble when it is pulled downward by the force of the Earth's gravity, the spinning Earth's polar axes describe a conical wobble with respect to the fixed stars; hence

each celestial pole describes a circle among the stars, shown in Figure 2-10b. The circle, $23\frac{1}{2}°$ in angular radius, is centered on the ecliptic poles. Thus, as mentioned in the last chapter, in different millennia different stars become the North Star; a complete cycle takes about 26 000 y. As seen in Figure 2-10b the star Alpha Draconis was the North Star around 2500 B.C., when Stonehenge was being built. All star coordinates, which are measured with respect to the celestial equator, therefore change slightly each year.

Hipparchus also contributed to the description of solar and planetary motions. Although a few of Hipparchus' contemporaries accepted Aristarchus' idea that the Earth moves around the Sun, Hipparchus and most other astronomers thought this model unnecessarily complex. His own observations showed that the Sun's apparent motion with respect to the stars is not uniform from day to day; he therefore concluded (incorrectly) that the Sun moves around the Earth in a circular orbit whose center is slightly offset from the Earth.

Like so many ancient works, most of Hipparchus' writings have disappeared except in others' reports—according to which Hipparchus also studied the relative distance from Earth to the Moon and the Sun. He calculated that the Moon is $29\frac{1}{2}$ Earth diameters away, close to the correct value of about 30. Hipparchus apparently realized that the Sun must be much farther away than Aristarchus' estimate of 18 to 20 times the lunar distance.

Ptolemy: Planetary Motions (A.D. 150)

Claudius Ptolemy (flourished c. A.D. 140) was another scholar associated with the Alexandrian library. His fame as an astronomer is based on a 13-volume work, *The Mathematical Collection.* Passed on to the Arabs after the destruction of the library, the work became known as *al-Megiste* (The Greatest). European translations were called the *Almagest,* and the book was famous for more than a thousand years.

Ptolemy extended Hipparchus' star catalog to 1022 entries, correcting older reported star positions to compensate for precession.[2] But his best-known contribution was a method for predicting the positions of the Sun, Moon, and planets, called the **epicycle theory,** or *Ptolemaic theory.* Following Hipparchus, Ptolemy incorrectly assumed that the Earth is near the center of the planetary system, as shown in Figure 2-11. In order outward from the Earth, he placed in *circular* orbits the Moon, Mercury, Venus, the Sun, Mars, Jupiter, and Saturn. To explain why planets apparently do not move at uniform rates, Ptolemy devised combinations of circular motions. Each planet, he said, moves in a circle called the *deferent,* whose center is offset from Earth. The planet does not move strictly on this circle but in a smaller circle called an *epicycle,* whose center moves along the deferent just as the Moon moves around Earth while the Earth moves around the Sun

---

[2]Correction for precession continues today. For precise setting of a large, modern telescope, a star's coordinates published for "epoch 1950" or any other year must be corrected for the current date, or the star could be missed. The correction is generally done by a computer operating as part of the telescope.

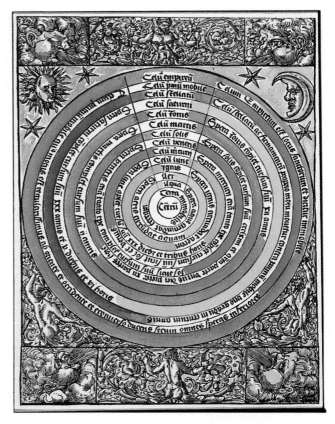

Figure 2-11 A medieval conception of the universe from 1537 shows the persistence of Aristotle and Ptolemy's ideas. The central sphere is labeled "Terra immobilis," or immovable Earth. Around it are shells of water, air, and fire and then shells carrying the moon, sun, planets, and stars. (The Granger Collection, New York.)

(Figure 2-12). The epicycle theory was simply a case of adding "wheels within wheels" until there were enough wheels to explain the observed irregular movements of the planets. Astronomers who applied this theory and its later changes during the Middle Ages could predict the actual positions of planets within a few degrees. At that time, such predictions were used more often for casting astrological horoscopes than for astronomical studies of the universe!

In spite of its usefulness in ancient times, Ptolemy's theory was incorrect. Today we know that planets move in elliptical orbits around the Sun, not in circles around the Earth. Ptolemy has been criticized for abandoning Aristarchus' Sun-centered system and for biasing his theory toward a supposed "perfection" of the circle,

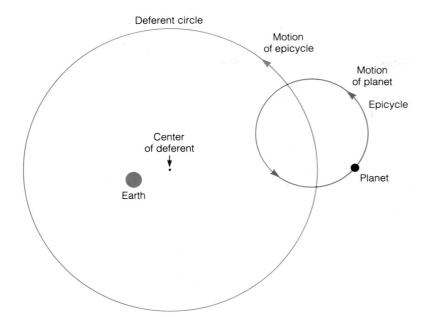

Figure 2-12 Ptolemy's system for explaining planetary motions. Ptolemy thought each planet moved in a circular epicycle, whose center moved in a circular orbit (deferent) around a point near the Earth.

thus delaying the introduction of the true elliptical orbits. But the system fitted the observations available in Ptolemy's time. Aristotle's old argument still seemed true: The Earth could not move around a distant center, because no one had observed parallax. Certainly no one had yet realized they were observing elliptical motions. And Ptolemy's system worked fairly well. The trouble with Ptolemy's choice of Aristotle over Aristarchus lies not in its being a mistake, but in its historical effects: The *Almagest* became the Bible of ancient astronomy, and the erroneous Earth-centered system held sway for 1400 y.

The Loss of Greek Thought (c. A.D. 500)

Alexandria, where Cleopatra first fascinated Julius Caesar and Mark Antony around 47 B.C., was the world's intellectual center by A.D. 250. With Rome's fall and the world in disorder in 410, maintaining the great library of ancient discoveries became more and more difficult.

Among the last guardians of the old knowledge in Alexandria was one of the first known woman astronomers, Hypatia (c. A.D. 375–415). Widely admired for her intelligence and beauty, she wrote a commentary on Ptolemy's work and invented astronomical navigation devices, but was murdered by a mob during one of the riots that plagued Alexandria during its decline. In A.D. 640, after a 14-mo siege by the Arabs, Alexandria fell.

The library buildings were burned and the best collection of Greek knowledge was lost. Because there was no printing, there were few other reference books. Table 2–1 summarizes the dramatic Greek advances.

ANCIENT ASTRONOMY BEYOND THE MEDITERRANEAN

With the fall of Alexandria in A.D. 640, in the West the rate of new discoveries declined and Europe slipped into the Dark Ages. But intellectual progress occurred in other cultures.

Islamic Astronomy

Much of the Alexandrian knowledge passed into the hands of the Arabs. About a century after Muhammad, Islamic leaders in the new capital of Baghdad began to sponsor translations of old Greek texts around A.D. 760. The next known measurement of the Earth's circumference was made near Baghdad in 820 and was only 4% too large. Similarly, Arab astronomer Muhammad al-Battani (c. 850–929; known later in Europe as Albategnius) made only a 4% error in his measurement of the eccentricity, or noncircularity, of the Earth's orbit. (He would have called it the Sun's orbit around the Earth.) By A.D. 1000 the Islamic empire had spread to Spain, and astronomical tables were published with the 0° ref-

TABLE 2·1

Astronomical Discoveries of the Greeks

Observation	Inference	Observer Commonly Quoted
Curved lunar terminator	Moon round	Pythagoreans
Round shadows during lunar eclipses	Earth round	Pythagoreans
Crescent phases of Moon	Moon between Earth and Sun	Aristotle
Different stars at zenith at different latitudes	Earth round	Aristotle
No evident stellar parallax observed by naked eye	Distance Earth moves is small compared with distances to stars	Aristotle
Relative sizes and angles of Moon and Earth's shadow	Moon smaller than Earth; Sun bigger than Earth	Aristarchus
Angle from first quarter to last quarter moon slightly less than 180°	Sun tens of times farther away from Earth than Moon	Aristarchus
Relation between angular shift of zenith distance of Sun and linear distance traveled on Earth	Calculable circumference of Earth	Eratosthenes
North celestial pole's shift with respect to constellations	Precession	Hipparchus

erence longitude in Córdoba (rather than in Greenwich, England, as in the modern longitude system, introduced when Britannia ruled the waves).

Although it is the Mesopotamian–Greek–Arab development of astronomy that most clearly contributed to modern Western conceptions, other astronomical centers, initially independent in their development, eventually influenced the West.

Astronomy in India: A Hidden Influence

Astronomical practices in India date back to about 1500 B.C. The first known astronomy text, describing planetary motions and eclipses and dividing the ecliptic into 27 or 28 sections, appeared around 600 B.C. By this time India had contact with the Mesopotamian and Greek worlds, and influences probably traveled both ways.

Texts dating from around A.D. 450 use Greek computational methods and refer to the longitudes of both Alexandria and Benares, a major Indian astronomical center. Arabs who later visited India wrote of Brahmagupta (588–660?) as one of the greatest Indian astronomers, who reportedly helped introduce Greco-Indian astronomy to the Arabs.

Unfortunately, most records of this fertile early period of Indian astronomy were destroyed during invasions in the 1100s. The great center at Benares was destroyed in 1194, and various university libraries of Buddhist and other ancient literature were burned in religious wars. A massive observatory—one of the world's five major observatories by the 1700s—was reestablished at Benares in later centuries (and damaged again by invading religious fanatics).

Astronomy in China: An Independent Worldview

In legend, Chinese astronomers were predicting eclipses before 2000 B.C.; scholars estimate the time as being closer to 1000 B.C. Thus Chinese astronomy flourished about the same time as Greek astronomy. Both were probably influenced by the same Middle Eastern cultures. Excellent Chinese observations include the world's best lists of the mysterious "guest stars" (exploding stars that we will encounter in later chapters) from 100 B.C. to the present, still consulted by modern astronomers.

Chinese conceptions of the universe were strikingly modern. As early as 120 B.C. a Chinese text stated (quoted by Needham, 1959): "All time that has passed from antiquity until now is called *chou;* all space in every direction . . . is called *yü.*" The Chinese term for the universe, *yü-chou,* was thus similar to the modern concept of space–time, a four-dimensional continuum encompassing all that has existed.

Another Chinese statement of this period is apt (quoted by Needham, 1959): "The Earth is constantly in motion, never stopping, but men do not know it; they are like people sitting in a huge boat with the windows closed; the boat moves but those inside feel nothing." Contrast this with Aristotle's view of the same era, which, for most Westerners, put a stationary Earth at the universe's center. Unfortunately, these advanced ideas had little influence on Western astronomy until after the Renaissance.

The Chinese astronomer Yü Hsi reportedly discovered precession independently. He also said in A.D. 336, "I think that the heavens are infinitely high." At that time Western astronomy was about to enter the Dark Ages. Ptolemaic groundwork was being laid for the medieval conception that planets moved in as many as 10 separate physical shells that surrounded the Earth and held up the stars, as shown in Figure 2-12.

Of course, some of these Chinese ideas were only speculations. Yet they did record observations, too. Ancient Chinese observations of Halley's comet, for example, were used by astronomers in 1986, during the comet's recent approach, to refine our knowledge of its orbit.

Astronomy in Polynesia: Stars as Guideposts

Though they had cosmological myths, the Polynesians were not so much interested in understanding the relations of worlds in space as in plotting the relations of islands in the trackless Pacific. Driven by this need, Polynesian navigators developed amazing skills. Lacking the magnetic compass, they steered toward chosen directions by learning where certain bright stars rise and set on the ocean horizon. Those directions remain constant for each star (except for the very slow changes of precession occurring over several generations). Modern Polynesians can sail such a constant setting that they can accurately measure subtle nighttime shifts in wind direction on the open ocean even though they themselves are moved by the wind (Lewis, 1973).

Polynesian navigators found their latitude by noting which stars passed directly over their masts. For example, in A.D. 1000, if Aldebaran passed overhead, the boat was about 5° south of Hawaii's latitude; if Arcturus, about 100 mi north. This method was accurate to about 0.2° to 0.5°, or about 12 to 30 mi. To reach, say, Hawaii from Samoa, one could simply sail north to the correct latitude and then east until the island was sighted.

American Indian Astronomy: Native Science Cut Short

Many people still underestimate the sophistication of American Indian civilization. Native Americans, as noted in Chapter 1, built astronomically aligned observatories, tracked and recorded planetary positions, and devised calendars based on eclipse cycles. Much of this "protoscience" developed between A.D. 100 and 1200, especially in Central America. The influence of Central American civilization reached as far north as the "frontiers" of the Mississippi Valley and the New Mexico–Arizona deserts. Cultural decline occurred around 1300 throughout Central America and the northern frontiers, perhaps due to climate changes. Many astronomical traditions were preserved, however, and European explorers found Native Americans in various regions conducting astronomical observations of solstice dates, planetary positions, and so on.

a   b

Figure 2-13 A prehistoric American Indian observatory, Casa Grande, Arizona, dating from c. A.D. 1350. Windows, originally on an upper floor in this four-story adobe structure, were cut in different shapes, apparently to facilitate astronomical observations. **a** A late afternoon view of a "solstice window." **b** A view through the same window at sunset on the summer solstice shows it was built so that the sunset position on the horizon was revealed by a diagonal view through the cylindrical shape. This orientation allowed determination of the day of the solstice each year. Similar methods were used by Pueblo Indians in historic times to calibrate their calendars. (Photos by author.)

Particularly interesting is the astronomy flourishing in the American Southwest when the Spanish arrived, because it has been preserved and practiced until modern times in the New Mexico pueblos and Arizona Hopi villages. Ethnographers of the 1800s and early 1900s recorded many of the ancient traditions passed down by Indian priests. With interest in archaeoastronomy surging since the 1970s, many of these records are being restudied in order to interpret possible astronomical functions of ancient Indian ruins. Zeilik (1985) gives a good survey of Pueblo astronomy. Priests were charged with studying sunrise and sunset positions in order to *predict* dates such as solstices for religious ceremonies and to set times for planting crops. Using the principles of Stonehenge, they determined dates either by observing the Sun's position on the horizon from a certain spot relative to distant mountains or by measuring the Sun's rising or setting position by observing the pattern of light and shadow cast through windows onto special markers placed in opposing walls. The latter method supports the theory that specially shaped windows found in the ruins of Casa Grande, built in central Arizona around A.D. 1350, were solstice observing sites, as shown dramatically in Figure 2-13.

Closer to the equator, the Mayans created a strange calendar that was still in use when the Europeans arrived.

They celebrated the beginning of the new year on July 26! What could have led them to this choice? Once again, it was astronomical observation, but in a non-European tradition. In the tropics ($23\frac{1}{2}°$ north latitude to $23\frac{1}{2}°$ south), but not in Europe, the Sun can pass directly overhead at noon. Perhaps because their horizons were obscured by dense jungle, the Mayans paid as much attention to *zenith* observations as to observations of the Sun's position on the horizon. Near the latitude of a major Mayan site, Edzna, in Yucatan, the Sun passes through the zenith, on its way to more southern latitudes, at noon on July 26. Recent archaeological studies show that Edzna was a major city of some 20 000 people in the first centuries A.D. In the courtyard in front of the main, five-story pyramid, a cleverly designed stone pedestal (Figure 2-14) allowed priests to measure the important "New Year's Day" when the Sun passed through the zenith (Thomsen, 1984; Malmstrom, 1987, private communication). Probably it was in or near this prehistoric city that the early Mayan astronomers first codified, and then ceremonialized, July 26 as their New Year's Day.

The Mayans' "alien" astronomy is perhaps the most fascinating example of incipient Native American science. Although it survived until historic times and produced written records of complex planetary observa-

a

b

Figure 2-14 The Mayan site of Edzna lies almost exactly at latitude where the Sun passes overhead on the date of the beginning of the Mayan new year, July 26. In the main temple courtyard **a**, the Mayans erected a stone device to measure this event. At the precise moment when the Sun shines straight down from the zenith, the shadow of the top of the stone covers the entire shaft. At other dates and times (as in photo **b**), part of the shaft is in sunlight. (Photos courtesy V. H. Malmstrom, Dartmouth College.)

tions, astronomical conferences, eclipse predictions, and calendars, all of this was stamped out by the żealous Europeans who were destroying non-Christian practices.

SUMMARY

We have seen how various cultures, preliterate and technological alike, moved toward certain concepts about Earth as a world among other worlds in space. Some of these concepts were purely practical; some, abstract. These movements came in fits and starts and were scattered throughout the world. Progress toward knowledge has not been continuous. Cultures have advanced and regressed, depending on their stability and vigor.

Many of the discoveries reviewed here dealt with the relation of Earth, the Moon, and the planets. Table 2-1 furnishes a good review of key observations by the Greeks. Their advances, among all those of antiquity, were most important in influencing Western scientific thought. Although conceptual models of the universe differed from culture to culture, all cultures moved toward discovering astronomical relationships.

CONCEPTS

cosmology

science

degree

minute of arc

second of arc

linear measure

angular measure

subtend

resolution

terminator

parallax

small-angle equation (from Optional Basic Equation)

significant figures (from Optional Basic Equation)

Alexandrian library

precession

epicycle theory

PROBLEMS

1. How critical can we be of early theorists who believed the Earth is at the center of the universe? Explain, considering the following questions:

a. Did they have any basis for not putting the Earth at the center?

b. Did either possibility fit the available observations?

c. Did any of the Greeks *prove* that any celestial bodies do or do not revolve around a central Earth?

2. As the Moon goes through its phases:

a. Why is its terminator usually curved?

b. At what lunar phase is the terminator straight?

c. At what lunar phase is the terminator not seen?

3. Contrast the types of astronomical observations available to the Greeks with those available today.

4. Do you agree with the quotation from Plato implying that astronomical events may have influenced the beginnings of philosophizing about the universe? Could astronomical phenomena have influenced religious concepts?

5. Do you think Aristotle's faith in symmetry and "perfection" helped or hindered his investigations of the universe?

6. Why is the Moon's umbral shadow on the Earth much smaller than the Moon itself? (*Hint:* See Figure 2-7.)

7. How does Hipparchus' discovery of precession prove the existence of earlier, careful astronomical records of stars' positions?

8. Do you think destructive events such as the pillaging of the Alexandrian library or the sacking of the observatory and library at Benares are significant or insignificant in world history? (The answer, of course, requires that you define *significant*.)

9. Does the Sun pass through the zenith every day on the equator? If not, on what dates does it do so?

ADVANCED PROBLEMS

10. At latitude 40° N will the Sun ever pass through the zenith? At latitude $23\frac{1}{2}$° N? If so, on what date(s)?

11. The angular diameter of the Sun is roughly $\frac{1}{2}$° and its distance from Earth is about 150 million km.

a. Use the small-angle equation to estimate the Sun's diameter.

b. Could Aristarchus determine the ratio of the Sun's diameter to its distance?

c. Why could the Greeks, such as Aristarchus, not determine the Sun's linear diameter or distance?

12. American TV pictures have about 435 resolution elements (sometimes called picture elements, or "pixels") along the width of the picture.

a. Assuming a TV picture is photographed with a lens giving the standard "snapshot" field of view of 40°, calculate the angular size of the smallest details resolved, and compare this with what the eye could see from the same viewpoint.

b. European television systems mostly have 20 to 50% more scanning lines across the screen and more resolution elements than American TVs (except in England, where the number is about 23% less). Comment on the resolution and sharpness of these TV images.

13. A backyard telescope with aperture greater than about 15 cm (6 in.) can reveal features with an angular diameter 1". The Moon is about 400 000 km away. What size lunar crater could you see with such a telescope?

14. If you landed on a small satellite and found that walking in a 1-km straight line caused the stars in front of you to rise 1° farther above the horizon (while stars overhead also shifted by 1°), what would be the satellite's circumference? What would be its diameter?

15. The photo in Figure 2-4c was taken in the Northern Hemisphere. What time of day was it? (*Hint:* Consider what time of day it would have been if the Moon were in a crescent phase with the horns of the crescent pointing down to the left.)

PROJECTS

1. Using a distant, strong light source such as the Sun or a light bulb, a small ball to represent the Moon, and your eye to represent a terrestrial observer, show that crescent phases of the Moon prove that it passes between the Earth and the Sun.

2. If you travel far enough during vacation to change your latitude significantly, compare measurements of the elevation of the North Star (or the Sun during daytime, taking care not to stare directly at it) made from various points in your trip. A sighting device like that described for the problems in Chapter 1 can be used. (Point the device at the Sun by watching its shadow; don't look directly at the Sun.) How accurately can latitude be determined in this way? Measure the number of kilometers or miles corresponding to your change of latitude, and estimate the circumference of the Earth. (This method is similar to Eratosthenes'.) The project can be done as a class effort, with different people's reports of elevation angles plotted against their latitude to give a curve showing how elevation angle changes with latitude. Coordinate with your instructor.

3. With a camera and fairly fast black-and-white film, such as Tri-X, make time exposures of the night sky. Try different exposures, such as 1 min, 5 min, and 1 h. Make one series including the North Star, one toward the eastern or western horizon, and one toward the southern horizon. Explain the patterns made by the trails.

4. During a camping trip or late-evening outing, pick an equatorial constellation in the sky and follow its motion. Make a series of sketches at different hours, showing its position with respect to the horizon. Do the same for a circumpolar constellation and contrast the results.

5. Using star maps, trace out the position of the celestial equator in the sky. Compare this with the position of the Sun's path, the ecliptic.

CHAPTER 3

Discovering the Layout of the Solar System

Astronomy waxed and waned in various parts of the world for a thousand years between A.D. 500 and 1500. But throughout that time, no one realized that the dots of light drifting among the stars from night to night were worlds like Earth, spherical globes pursuing their own orbits around the Sun. The planets were just . . . dots of light, gods in the sky. The revolutionary change in this view was destined to come from Europe. As a result, we now conceive of the **solar system** as the Sun and all the planets orbiting around it, together with innumerable, small interplanetary bodies, called asteroids and comets, also orbiting around the Sun.

The change came from an explosion in European knowledge during the Renaissance—the two or three centuries of intense inquiry and exploration culminating in the 1500s.[1] Already, in 1492, Columbus's findings had led to the realization that the world was not the bottom half of a Heaven-above-Earth-below universe, but rather a sphere. By 1522, Magellan's crew had sailed around it. How did Renaissance naturalists figure out what lay in the sky beyond?

CLUES TO THE SOLAR SYSTEM'S CONFIGURATION

Centuries of accumulating observations of planets' positions in the sky (gathered mostly to refine the calculations of astrologers) led to new ideas about the arrangement of the planets, Earth, and Sun.

We will examine, first, some clues that could be discovered even before the telescope was invented. Then we will record how the true nature of the solar system

---

[1]It is intriguing to speculate how history might have differed if a different culture had made these breakthroughs first—say, the Mayans or Aztecs!

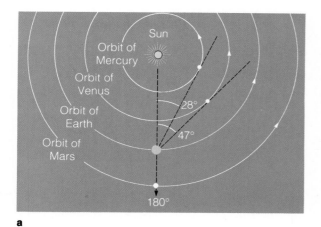

a

b

Figure 3-1 Mercury and Venus always appear close to the Sun, and therefore can be seen well only in the dawn or evening sky. **a** As seen from Earth, Mercury can never be more than 28° from the Sun, and Venus never more than 47°. Sometimes they can transit in front of the Sun, as seen from Earth. Exterior planets, however, can appear at opposition, 180° from the Sun. **b** Typical appearance of Venus in the dawn or dusk sky. (Photo by author, 24-mm wide-angle lens, f2.8, 25 s on 3M ISO 1000 film.)

was discovered. One important clue is that, to an earthly observer, the planet called Mercury never strays more than 28° from the Sun and Venus never strays more than 47°. This means that Mercury and Venus are always fairly close to the direction of the Sun. They are usually in the daytime sky, although, of course, the sky is too bright for them to be seen. (Bright Venus can sometimes be seen in the daytime sky if you know exactly where to look.) All other planets can appear at any angular distance from the Sun along the zodiac. This observation indicates that Mercury and Venus lie closer to the Sun than Earth does, whereas the other six planets are outside Earth's orbit (see Figure 3-1a). For this reason Mercury and Venus came to be called **inferior planets**, while more distant planets are called **superior planets**.

Final proof of this arrangement may have come when ancient astronomers observed actual **transits**, or passages of an inferior planet directly between the Earth and the Sun. Only a planet closer to the Sun than Earth could pass between Earth and the Sun. During a transit of Venus, a sharp observer looking through fog or a smoked glass can see Venus as a tiny black spot moving across the Sun, but Mercury is too small to see without a telescope. Transits of Mercury are more common. The next transit of Mercury is November 14, 1999, and the next transit of Venus June 8, 2004.

Retrograde Motion

Superior planets slowly shift position from night to night, usually west-to-east relative to the background stars. Of the superior planets, Mars most dramatically exhibits a motion that puzzled all early analysts of the solar system. Observers who plotted its position from night to night found that as Mars approaches a point opposite the Sun in the midnight sky, it slows down, reverses itself, and drifts westward in so-called **retrograde motion** for some days before resuming its normal eastward (prograde) motion.

We now know that Mars does not really reverse its motion in its orbit around the Sun. The appearance is an illusion caused by the motion of Mars *relative to earthly observers*, as shown in Figure 3-2. The Earth moves around the Sun faster than Mars does, following an "inside track," closer to the Sun. Mars seems to go backward whenever we overtake and pass it.

PROBLEMS WITH THE PTOLEMAIC MODEL

In Chapter 2, we showed how ancient scientists arrived at a solar system model with planets moving in deferent circles and epicycles around Earth, as was shown in

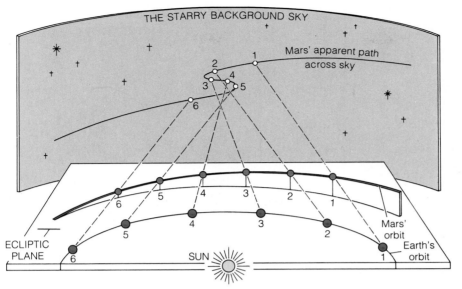

Figure 3-2 A schematic model showing why Mars appears to have a complicated path across the sky. Positions of Earth and Mars are shown for six dates. Note that Mars moves slower than Earth and at a slight angle to Earth's orbit. As Earth overtakes Mars, Mars' apparent path relative to the background stars changes speed and direction. This phenomenon revealed major problems with the Ptolemaic theory of the solar system and led to Kepler's discovery of elliptical planetary orbits around the Sun.

Figure 2-11. But the relative motions of inferior and superior planets, especially Mars, were destined to cause trouble. Inferior planet motions were explained by putting Mercury's and Venus' orbits between Earth and the Sun and adjusting their deferent and epicycle rates of motion so they would never get more than 47° and 28° respectively from the Sun. Mars' motions were explained by adjusting its epicycle motion so it would sometimes appear to be moving east to west.

The main goal of this model was to predict the positions of planets, often for astrological applications. The models attempted to break down the seemingly irregular motions of the planets into combinations of circular motions or uniform cycles. Eudoxus (c. 360 B.C.) reportedly represented planetary motions by a combination of motions of 27 rotating concentric spheres carrying the planets; Aristotle (c. 360 B.C.) reportedly used 55 spheres; Apollonius of Perga (c. 220 B.C.) reportedly increased the number of possible combinations of cycles by introducing epicycles; and Ptolemy perfected the **epicycle model** around A.D. 140. Ptolemy's version of the epicycle model is known as the **Ptolemaic model** of the solar system and is seen in Figure 3-3.

In the *Almagest*, Ptolemy showed how a properly chosen set of epicycles could account for all observations of planetary positions accumulated by his time, and he made relatively accurate predictions of planetary

positions. Arab astronomers introduced this scheme into Europe in the early Middle Ages. By the 1200s and 1300s, however, astronomers in Damascus realized that further adjustments were needed and discussed adding small epicycles to the main epicycles.

In 1252, King Alfonso X of Castile, acting as a medieval version of the National Science Foundation, supported a 10-y project conducted by Arab and Jewish astronomers to calculate the extensive Alfonsine Tables, an almanac of predicted planetary positions based on the Ptolemaic model. These tables became the basis of planetary predictions for the next three centuries. However, the solar system now seemed very complicated because of the many overlapping cycles. (On seeing the complexity of the Ptolemaic epicycles, Alfonso is said to have remarked that had he been present at the creation, he could have suggested a simpler arrangement!) New epicycles had to be added to make positions agree with observations, and the calculations became ever more complicated.

Around 1340 the English scholar William of Occam enunciated his famous principle, called **Occam's razor**, applicable to all branches of science and philosophy. It was called a "razor" because it helped scientists in any given field cut through a thicket of competing theories to find the best theory. Paraphrased in brief, it said:

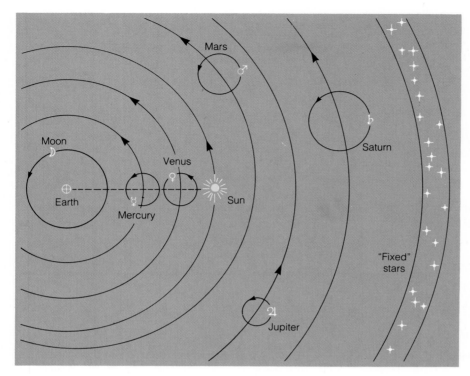

Figure 3-3 The solar system as it might have been conceived by a Ptolemaic astronomer between A.D. 100 and 1500. The diagram is actually simplified, since the Ptolemaic astronomers imagined even more complex hierarchies of multiple epicycles and off-center motions to explain planetary movements. The symbols on the chart are ancient astronomical (and astrological) signs for the planets.

> **Among competing theories, the best theory is the simplest theory—that is, the one with the fewest assumptions.**

In other words: *The best theory is the one requiring the fewest modifications in order to fit the available observations.* This idea may have contributed to suspicion of the tacked-on epicycles. Some European scholars, concerned about how the planets moved and remained suspended, regressed to Aristotle's more simplistic idea of spherical shells (see Figure 2-11). Purbach (c. 1460) thought the spheres were made of transparent crystal with special hollows for the epicycles; Fracastoro (c. 1550) proposed a scheme with 79 spheres!

According to astronomical historian Owen Gingerich (1973a, 1973b), the growing dissatisfaction with the Ptolemaic system was aesthetic (a result of its increasing unwieldiness) as much as intellectual (a dissatisfaction with its results). In addition, obvious errors of a few degrees had accumulated in the predicted positions. One observer remarked that a new model would be more "pleasing to the mind." This observer was Nicolaus Copernicus.

THE COPERNICAN REVOLUTION

The **Copernican revolution** was an intellectual revolution that abolished the old theory of an Earth-centered universe with the discovery that the Sun is at the center of the solar system, with Earth moving around it. The Copernican revolution took about a century and a half, from roughly 1540 to 1690. It involved five very famous scientists: Copernicus, Tycho, Kepler, Galileo, and Newton.[2]

Copernicus' Theory

Nicolaus Copernicus (Figure 3-4) was born February 14, 1473, the son of a Polish merchant. During his university education in Italy, he became excited by the surging scientific thought of that country. He associated with several astronomers and mathematicians and made his first astronomical observations at age 24. A few years

---

[2]For historical reasons, Tycho and Galileo are commonly referred to by their first names.

Figure 3-4 Nicolaus Copernicus (1473–1543). Copernicus is holding a model with a central Sun circled by Earth, and Earth by the Moon. (The Bettmann Archive.)

later, a cathedral post gave him the economic security to continue his observations. At age 31 he observed a rare conjunction that brought all five known planets as well as the Moon into the constellation of Cancer. He found that their positions departed by several degrees from predictions in his set of Alfonsine Tables.

Familiar with several classical alternatives to Ptolemy's system, Copernicus analyzed planetary motions by various methods, including use of small "epicyclets." He soon realized that the solar system would be simpler and the prediction of planetary positions easier if the Sun were placed at the center and Earth placed as one of the Sun's orbiting planets. In 1512 he circulated a short comment (*Commentariolus*) containing the essence of his new thesis: The Sun is the center of the solar system, the planets move around it, and the stars are immeasurably more distant. This comment was not widely distributed, however, and few of Copernicus' acquaintances realized that the work he was pursuing would scandalize and revolutionize the medieval world.

He continued his studies but, fearing controversy, delayed publication for many years. Finally, encouraged

by visiting colleagues, including some in the clergy, he allowed the *Commentariolus* to be more widely circulated. News of Copernicus' work spread rapidly. Late in life, Copernicus prepared a synthesis of all his work, *De Revolutionibus (On Revolutions*, 1543). In this book he laid out and explained the evidence about the solar system's arrangement:

Venus and Mercury revolve around the Sun and cannot go farther away from it than the circles of their orbits permit [for example, the 47° figure mentioned earlier for Venus]. . . .

According to this theory, then, Mercury's orbit should be included inside the orbit of Venus. . . .

If, acting upon this supposition, we connect Saturn, Jupiter, and Mars with the same center, keeping in mind the greater extent of their orbits . . . we cannot fail to see the explanation of the regular order of their motions.

This proves sufficiently that their center belongs to the Sun.

Having thus laid out the correct nature of the solar system, Copernicus tackled the solar system's relation to the stars:

The extent of the universe . . . is so great that, whereas the distance of the Earth from the Sun is considerable in comparison with the other planetary orbits, it disappears when compared to the sphere of the fixed stars. I hold this to be more easily comprehensible than when the mind is confused by an almost endless number of circles, which is necessarily the case with those who keep the Earth in the middle of the universe.

Although this may appear incomprehensible and contrary to the opinion of many, I shall, if God wills, make it clearer than the Sun, at least to those who are not ignorant of mathematics.

Now the stage was set for turmoil. Church officials and most intellectuals held that the Earth was at the center. The printer of *De Revolutionibus*, a Lutheran minister, had tried to defuse the situation by extending its title to *On the Revolutions of Celestial Orbs*, as if to imply that the Earth was not necessarily included. He had also inserted a preface stating that the new theory need not be accepted as physical reality but could be seen merely as a convenient model for calculating planetary positions. From a philosophical viewpoint, we might accept this, but it did not deter medieval critics. Already

Copernicus himself had come under fire from Protestant fundamentalists: In 1539 Martin Luther had called him "that fool [who would] reverse the entire art of astronomy. . . . Joshua bade *the Sun* and not the Earth to stand still."

In a world of strong dogmas, tampering with established ideas is dangerous. The Reformation era of ideological clashes was no exception. In the 1530s Michael Servetus had been criticized for certain writings on astrology and astronomy; in 1553 he was burned at the stake as a heretic for professing a mysterious theology that offended both Protestants and Catholics.[3]

Copernicus himself missed the height of the violent debate. He was ill in his last year and the first copies of his book were reportedly delivered to him on the day of his death, in 1543, at age 70. But the Copernican revolution was under way.

About 1584, a 36-year-old Italian theologian and naturalist, Giordano Bruno, became known for tracts that combined science and theology, vigorously defending the Copernican view against academics. His message included exhortations that warring Protestant and Catholic factions in Europe could be brought together by a better appreciation of the new scientific view of Earth's and humanity's place in the universe (Lerner and Gosselin, 1973, 1986).

Bruno expanded on the Copernican cosmology. The stars, he said, were all worlds like the Sun. Many planets might orbit around them, offering abodes for other races. Bruno traveled in Europe, lecturing on the theological implications of astronomy and science, often with remarkable imagery that was at once poetic and scientific. For example, he discussed optics using "light" both in the physical and in the theological sense as "divine light." When he returned to Italy in 1592, he was arrested by the Inquisition, a church court established to detect and punish heresy. In 1600, after eight years of investigation of his philosophical and political views, he was burned at the stake.

In 1575, a correspondent wrote to the astronomer Tycho Brahe: "No attack on Christianity is more dangerous than the infinite size and depth of the heavens."

---

[3]Both Protestants and Catholics were involved in outrageous suppression. It was John Calvin himself who masterminded Servetus' execution, although, in a fit of moderation, he recommended beheading instead of burning. Servetus, a man of wide learning and varied interests, had improved geographic data on the Holy Land and discovered blood circulation in the lungs.

Was the Earth to be taken as merely a minor province of the universe? Where was heaven? In 1616, the Catholic church banned reading of *De Revolutionibus* "until corrected." It was corrected in 1620 by removing nine sentences asserting that it contained actual fact, not just theory.

These sad incidents illustrate the continuing problem of reconciling differences. As Will Durant wrote:

The heliocentric astronomy compelled men to reconceive God in less provincial, less anthropomorphic terms; it gave theology the strongest challenge in the history of religion. Hence the Copernican revolution was far profounder than the Reformation; it made the differences between Catholic and Protestant dogmas seem trivial.

Tycho Brahe's Sky Castle

Tycho Brahe (Figure 3-5) was a flamboyant naturalist who wore a silver nose to cover a dueling mutilation. With funds from the king of Denmark, he built the first modern European observatory, named Uraniborg (Sky Castle), at his island home near Copenhagen. From his observations, all made with the naked eye (the telescope had not yet been invented), he made catalogs of star and planet positions. By demonstrating that stars and other bodies show no angular shift in position as our position shifts with rotation of Earth, Tycho proved that stars and planets were many times farther away than the Moon, for which he *could* detect a shift (called *parallax*).

At age 16 Tycho noticed errors in predicted planetary positions in the same Alfonsine Tables that Copernicus had used. At age 25, in 1572, Tycho saw a temporarily bright exploding star (further discussion in Chapter 19). By demonstrating that it had no parallactic shift, he disproved the popular belief that it was an object in the Earth's atmosphere or near the Earth–Moon system. In 1577 he observed a bright comet and showed that it too was a remote object, far beyond the Moon.

These discoveries were critical in overturning pre-Copernican theories. They meant that new objects could appear in the supposedly unchangeable heavens and that planets could not be attached to crystalline spheres, because such spheres would be smashed by the comets.

These observations inspired Tycho to catalog the precise positions of the stars and planets, which he did between 1576 and 1596. Unable to convince himself that the Earth could move, Tycho invented a compromise

Figure 3-5 Tycho Brahe (1546–1601), as shown in an old print. Silver plate on his nose covers a dueling scar.

Figure 3-6 Johannes Kepler (1571–1630). (The Bettmann Archive.)

solar system in which the Earth was central and stationary, but other planets were placed in the correct sequence from the Sun.

His pension withdrawn by the king of Denmark, Tycho moved to Prague in 1599, where he was joined in 1600 by a 30-year-old assistant named Johannes Kepler. When Tycho died in 1601, Kepler inherited the great compendium of Tycho's observations, with all its potential for fruitful analysis.

Kepler's Laws

Devoutly religious and a believer in astrology, **Johannes Kepler** (Figure 3-6) was sure that planetary motions must be governed by hidden regularities—"the harmony of the spheres." With Tycho's material, Kepler first went to work on the orbit of Mars, which presented the most notable case of retrograde motion. This movement had plagued astronomical theorists since Ptolemy. He found something astonishing: After all the centuries of debate over the arrangement of circular orbits, the orbit that fitted Mars' motion best was not a circle at

all, but an ellipse. **Ellipses** are roughly egg-shaped figures that can range from nearly circular to highly flattened, elongated loops. Each ellipse is symmetric around two inner points called **foci** (singular **focus**). Kepler found that Mars' orbit is an ellipse which is almost circular and that the Sun lay exactly at one focus. Eventually, this was found to be true of every planet's orbit. We will describe these elliptical, or so-called Keplerian, orbits in more detail in the next chapter.

Kepler went on to discover two other related principles, and these *three laws of planetary motion* were published in two books, *New Astronomy* (1609) and *The Harmony of the Worlds* (1619). **Kepler's laws** describe how the planets move (without attributing this motion to any more general physical laws), show that the Sun is the central body, and allow accurate prediction of planetary positions:

> **1. Each planet moves in an ellipse with the Sun at one focus.**
>
> **2. The line between the Sun and the planet sweeps over equal areas in equal time intervals.**

3. The ratio of the cube of the semimajor axis to the square of the period (of revolution) is the same for each planet. (This is sometimes called the harmonic law.)[4]

Although the planetary orbits are ellipses, they are only slightly elliptical—that is, they are nearly circular. This is why the Ptolemaic system of circles worked as well as it did.

In the case of the Earth, the semimajor axis, or average distance from the Sun, is 1 **astronomical unit** (AU) and the other planets' distances are measured in multiples of this unit. Working in astronomical units and years, we can easily confirm the third law numerically, using a as the semimajor axis and P as the period. For the Earth,

$$\frac{a^3}{P^2} = \frac{(1\,\text{AU})^3}{(1\,\text{y})^2} = \frac{1}{1} = 1.00$$

The same formation works for other planets.

Kepler's laws explained the apparent retrograde motion of Mars: One consequence of the laws, taken together, is that any planet moves faster than any other planet further away from the Sun. Thus Earth moves faster in its orbit than Mars (see Figure 3-2). Therefore, Earth catches up to Mars like the faster driver on the inside track at a race. As explained earlier in this chapter, this creates the illusion of retrograde motion.

Galileo's Observations

Kepler's laws might not have gone so far in establishing the Copernican model of the solar system had it not been for the contemporary invention of the telescope and for extensive observations by an Italian scientist, **Galileo Galilei** (Figure 3-7). Unlike Kepler, Galileo had a superbly practical turn of mind. For example, after reportedly watching the regular swing of a lamp in the Pisa cathedral, he applied the periodic motion of the pendulum to regulate clocks. As early as 1597, Galileo wrote to Kepler: "Like you, I accepted the Copernican position several years ago. . . . I have not dared until now to bring [my writings on this] into the open."

Galileo perfected the telescope and began astronomical observations with it in late 1609. By 1610 he had made some of the most important observations ever.

For example, he found four satellites revolving around Jupiter—proving at last that some bodies do not revolve around the Earth. Also, he found that the planet Venus undergoes a variety of phases, from crescent to nearly full. The full phase implies a situation with Venus on the far side of the Sun. Note in Figure 3-3 that in the Ptolemaic theory, Venus' epicycle was between the Sun and Earth, implying that only crescent phases could exist. *Here, then, was proof that the Ptolemaic model of Venus' orbit was wrong.* The Copernican model, however, fit the observation. As a final example, Galileo discovered mountains on the Moon and spots on the Sun, showing that these were not polished celestial orbs, as the ancients surmised. In particular, Galileo emphasized that the Moon was a *world*, with geological features, like Earth. These discoveries electrified European intellectuals. Other early telescope users, such as Thomas Harriot in England, duplicated Galileo's observations of Jupiter's moons, Venus' phases, lunar mountains, and sunspots by the end of 1610, but were not as widely known as Galileo.

Because Galileo wrote in Italian rather than Latin, he built a popular following outside the universities. Academics and churchmen saw Galileo as a threat, and soon began criticizing him. His invitations to reactionary academics and churchmen to look through his telescope and see for themselves led nowhere. Some looked and said they saw nothing; some refused to look; some said that if the telescope had been worth anything, the Greeks would have invented it.

From 1613 to 1633 Galileo was in frequent contact with church authorities, even in Rome. In 1616 a cardinal read Galileo an order that he must not "hold or defend" Copernican theory, though he could discuss it as a "mathematical supposition." In 1632, Galileo's great book *Dialogue of the Two Chief World Systems* appeared. It featured a fictionalized debate between Copernican and Ptolemaic advocates. In 1633, 69-year-old Galileo was ordered to Rome to stand trial before the Inquisition, where a curious episode occurred. The court produced a purported copy of the cardinal's order of 1616 telling Galileo not to "hold or defend" Copernicanism. The Inquisition's copy, still in the files today, also ordered him not to "teach" or "discuss" it in speech or in writing (contrary to the cardinal's 1616 order). The document in the Inquisition files is not the original and lacks the names of the alleged witnesses. Historians now suspect this document was a fraud created to frame Galileo.

The Inquisition jurors were inclined to be lenient only if Galileo repudiated his work. The elderly Galileo saw no point in getting himself killed; his book was already

---

[4]The major axis of an ellipse is its longest diameter; the semimajor axis is half that. Since most planets in the solar system have nearly circular orbits, the semimajor axes of their orbits are essentially the orbital radii.

Figure 3-7   Galileo Galilei (1564–1642). (The Bettmann Archive.)

published and he had faith that intelligent people could see plain truth through telescopes or in print. So he recited a prepared recantation and was sentenced to prison, a sentence commuted by the pope to house arrest on Galileo's own estate, where he died in 1642.

Newton's Synthesis

In spite of the Inquisition, the evidence for a Sun-centered solar system accumulated so rapidly that the Copernican revolution was almost complete. The main element still lacking was an overall theoretical scheme that would draw together Kepler's empirical laws of planetary motion into a concise physical explanation of the behavior of the solar system. To be intellectually satisfying, this theory needed to start with a few universal principles and show that the Keplerian orbits and the Galileian satellite motions *had* to exist as a consequence of these principles. The man who achieved this synthesis—the man usually deemed the greatest physicist who ever lived—was **Isaac Newton** (Figure 3-8).

Isaac Newton was a father of physics and astronomy. Between the ages of 23 and 25, while attending Cambridge, he almost single-handedly developed calculus, discovered the principle of gravitational attraction and certain properties of light, and invented the reflecting telescope (in which a curved mirror replaces a lens). Newton once said that he made his discoveries "by always

Figure 3-8   Isaac Newton (1642–1727). (The Granger Collection, New York.)

thinking about them," a trait that no doubt contributed to his reputation for absentmindedness.

At age 41, Newton began writing his famous *Principia*, a revolutionary compendium of physics, and published it three years later in 1687. He became president of the Royal Society at 60, died at 84 in 1727, and was buried in Westminster Abbey. Of Newton, Alexander Pope wrote:

Nature and Nature's laws lay hid in night:
God said, "Let Newton be!" and all was light.

In 1726, an acquaintance gave this account of Newton's early thoughts on gravity, based on conversations with Newton (Ball, 1972):

The first thoughts [were those] he had when he retired from Cambridge in 1666 on account of the plague. As he sat alone in a garden, he fell into a speculation on the power of gravity: that as this power is not found sensibly diminished at the remotest [height] to which we can rise, neither at the tops of the loftiest buildings, nor even on the summits of the highest mountains, it appeared to him reasonable to conclude that this power must extend much farther than was usually thought; why not as high as the moon, said he to himself?

The Moon must be attracted to Earth by some force, Newton thought, because it does not travel in a straight line, as it would if no force were pulling on it. Reasoning in this way, Newton was able to deduce one of the most important discoveries in the history of science. It is called **Newton's law of universal gravitation:**

> **Every particle in the universe attracts every other particle with a force proportional to the product of their masses and inversely proportional to the square of the distance between them.**

We will explore some of the ramifications of this law in the next chapter. Here we will simply note that the attraction of every particle for every other particle gave a quantitative explanation at last of why the planets follow orbits around the Sun instead of flying off into interstellar space in a straight line: The massive Sun, in the center of our solar system, *attracts* the planets. If the Sun suddenly vanished, the planets would indeed fly away!

Once Newton discovered that masses attract each other gravitationally, he concluded that gravity is the *only* force involved in keeping the planets moving around the Sun. But this solution led to another riddle for natural philosophers of the day: How could the Sun influence the planets in their Keplerian orbits if it never touched them and always stayed at such a great distance from them? And how could the planets stay in the sky without spheres to hold them up?

Newton answered all these questions of "action at a distance" with three simple laws of motion and his law of gravity, the basis of most modern physics except for the corrections made necessary by work on relativity during the present century. These laws were enunciated in Newton's book *Principia* in 1687. *They are quite unlike Kepler's three laws.* They are not merely empirical rules based on observation, but *fundamental postulates* from which Kepler's laws and many other phenomena can be predicted. **Newton's laws of motion** are:

> **1. A body at rest stays at rest and a body in motion moves at constant speed in a straight line unless a net force acts on it.**
>
> **2. For every force acting on a body, there is a corresponding acceleration proportional to and in the direction of the force and inversely proportional to the mass of the body. In other words, force = mass × acceleration.**
>
> **3. For every force (sometimes called action) on one body, there is an equal and opposite force (called reaction) acting on another body.**

The properties of Keplerian orbits follow from Newton's laws. An important exercise in advanced astronomy courses is to derive all three of Kepler's laws from Newton's laws. This exercise shows that if Newton's laws are true, the Copernican theory and Kepler's laws also have to be true.

Newton's laws thus tidied up the miscellaneous observations of preceding centuries and completed the Copernican revolution. If the force of gravitation had a *different* form than given by Newton's universal law, then orbits might be nonelliptical. The law of gravitation, when combined with the first and second laws of motion, shows why planets do not move in straight lines but are always deflected toward the Sun. The third law explains why rockets work and ultimately led to artificial satellites and spacecraft orbiting moons and planets in Keplerian orbits, as we will see in the next chapter.

By the time of Newton's death, at age 84 in 1727, the solar system was conceived essentially as we see it today, lacking only the discovery of the three outer planets, Uranus, Neptune, and Pluto. Subsequent astronomical observations have shown that Newton's laws also apply in all other parts of the universe that we can see. They correctly predict properties of certain pairs of stars that orbit around each other, and properties of stars orbiting around galaxies.

The Solar System at the End of the Copernican Revolution

A fantastic philosophical and scientific advance was wrought by Copernicus, Tycho, Kepler, Galileo, and Newton, as summarized in Table 3-1. To see its effect, compare Figure 3-9 to Figure 3-3. Figure 3-9 contrasts the solar system, as perceived after Kepler's orbital work and Galileo's observations, with the Ptolemaic view in Figure 3-3. Earth is no longer at the center but relegated to an orbit like any other planet's. Mercury and Venus are no longer trapped on deferents between Earth and Sun. None of the planets follow epicycles. Jupiter has moons of its own and Saturn has rings.

TABLE 3·1

Five Key Figures in the Copernican Revolution

Nicolaus Copernicus	1473–1543	Proposed circular motions of planets around Sun
Tycho Brahe	1546–1601	Recorded planets' positions
Johannes Kepler	1571–1630	Analyzed Tycho's records; deduced elliptical orbits and laws of planetary motion
Galileo Galilei	1564–1642	Made telescopic discoveries supporting Copernican model
Isaac Newton	1642–1727	Formulated laws of gravity and used them to explain elliptical planetary orbits

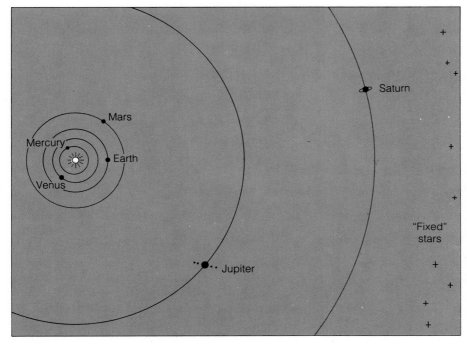

Figure 3-9 The solar system as it might have been conceived by an astronomer around 1700, at the end of the Copernican revolution. This diagram shows the true orbits of the then-known planets to true scale. The view is essentially the correct, modern view, except that the outermost planets (Uranus, Neptune, and Pluto) and the asteroids had not yet been discovered. Compare with Figure 3-3 to see the change from the Ptolemaic view.

BODE'S RULE

A curious relationship discovered by the German astronomer Titius and popularized by his colleague Johann Bode, in 1772, is helpful in memorizing the distances of the planets from the sun. **Bode's rule** is: Write down a row of 4s, one for each planet, and add the sequence, 0, 3, 6, 12, 24, and so on, doubling each time, as shown in Table 3-2. By dividing the sums by 10, you get the number of astronomical units between each planet and the Sun.

Because Bode's rule, unlike Kepler's laws, does not necessarily follow from Newton's laws, it is considered more descriptive than explanatory, and not a law of physics. It is merely a handy way to remember planetary positions. Nonetheless, it indicates that the planets formed in such a way that each planet was nearly twice as far from the Sun as the next inner planet. Apparently,

	Mercury	Venus	Earth	Mars	Asteroids	Jupiter	Saturn	Uranus	Neptune	Pluto
TABLE 3·2 Bode's Rule: Distances of Planets from the Sun										
	4	4	4	4	4	4	4	4	4	4
	0	3	6	12	24	48	96	192	—	384
Predicted Distance	0.4	0.7	1.0	1.6	2.8	5.2	10.0	19.6	—	38.8
Actual Distance	0.4	0.7	1.0	1.5	2.8	5.2	9.5	19.2	30.0	39.4

Note: All distances are expressed in astronomical units (1 AU = average distance of Earth from the Sun).

the planets had to be far enough apart that the gravitational forces of one did not disturb the formative process of its neighbor.

In 1781 Bode's rule was strengthened with the discovery of Uranus at its predicted position, about twice as far from the Sun as Saturn. Astronomers then noted that the rule predicted a planet between Mars and Jupiter. German observers, nicknamed "celestial police," set out to find the missing planet, but an Italian observer beat them to it. The Italian discovered the first and largest asteroid, Ceres, at just the right distance! Ceres might have become known as the smallest planet, except that within a few years the "celestial police" found three more asteroids at about the same distance. Today we know of thousands of asteroids between Mars and Jupiter. The asteroids more or less confirmed Bode's rule, though in this case the planet-forming process did not go to completion.

In 1846, the discovery of Neptune somewhat reduced the credibility of Bode's rule by putting a planet outside a predicted position, though the 1930 discovery of Pluto did put a small planet at roughly the next predicted position.

THE SOLAR SYSTEM AS WE KNOW IT TODAY

Figure 3-10 shows the solar system as we know it today. It is more complex and interesting than the simple system of Sun, Earth, and five other planets known to the ancients. First, we see the orbits of eight major planets (solid curves) spaced evenly and more or less obeying Bode's rule. Four small planets, including Earth, are close to the Sun; four much larger "giant" planets orbit further away from the Sun. We will study the properties of all these planets in later chapters.

Second, we note a number of interplanetary bodies. Many of these are asteroids—rocky bodies only a fraction the size of the smallest planets and located mostly in a belt between Mars and Jupiter. Others are comets. A comet's orbit, similar to that of Halley's comet, is shown at top left.

Third, we note that the so-called ninth planet, Pluto, has a considerably more elliptical orbit than other planets. Moreover, Pluto crosses inside Neptune's orbit, and is the only planet that crosses another's orbit. Pluto is also much smaller than the other planets—even smaller than our Moon and only about 60% as big as Mercury. Its orbit is also unusually inclined, running well "above" the plane of the solar system, where the other planets' orbits are concentrated. Pluto is thus unusual among the planets. Astronomers have long realized that many comets orbit beyond Neptune, and in 1977 astronomers discovered an object (now called Chiron) one-tenth as big as Pluto on an eccentric orbit between Uranus and Saturn. For these reasons, astronomers are beginning to suspect that they should classify Pluto not as a true planet but only the largest known example of an interplanetary body. More Pluto-sized bodies may be discovered in Pluto's region. Pluto is about three times as big as the largest known asteroid, a 1000-km-diameter object named Ceres, in the asteroid belt. Thus while Pluto has held its status as the ninth planet since its discovery in 1930, it may be "demoted" in the near future, depending on new discoveries. We can go on calling Pluto the ninth planet for now, but let us remember that we have not yet learned everything there is to know about the solar system!

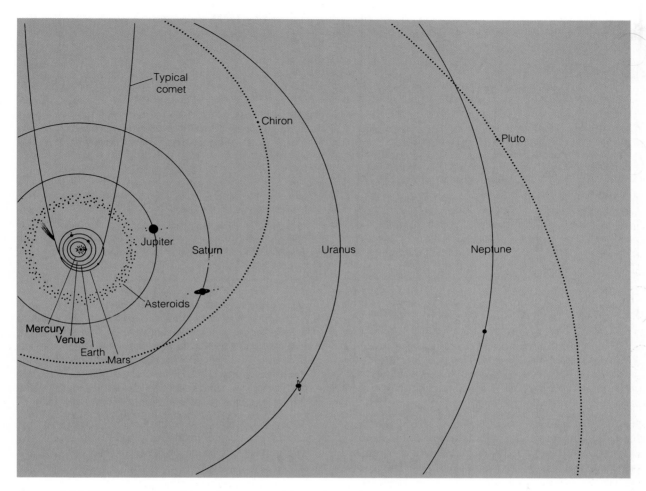

Figure 3-10 The arrangement of the solar system as it is now known, shown approximately to scale. Orbits of the nine main planets and a typical comet are shown, plus positions of typical asteroids. The orbits of Pluto and the unusual asteroidlike object Chiron are also shown as dotted. Note eccentricities of the orbits of Pluto, Mercury, and Mars.

SUMMARY

The planets are tiny specks circling the Sun. If you backed off far enough to see the system as a whole, the outer giants would hardly be noticeable and the inner planets would be lost in the glare of the Sun. This conception of the solar system was accepted only after one of the major intellectual upheavals in human history took place about four centuries ago. The key to this Copernican revolution was the work of five scientists listed in Table 3-1, which you should review. Of special importance were Kepler's three laws, which described how planets moved, and Newton's laws of motion and gravity, which revealed the underlying forces that explain Kepler's laws.

CONCEPTS

solar system

inferior planet

superior planet

transit

retrograde motion

epicycle model

Ptolemaic model

Occam's razor

Copernican revolution

Tycho Brahe

Johannes Kepler

ellipse

focus

Kepler's laws

astronomical unit

Galileo Galilei

Isaac Newton

Newton's law of universal gravitation

Newton's laws of motion

Bode's rule

PROBLEMS

1. To an observer north of the plane of the solar system, do the planets appear to revolve around the Sun clockwise or counterclockwise? Which way to an observer south of the plane? (*Hint*: All planets revolve in the same direction as the earth rotates, from west to east.)

2. Which planet moves the largest number of degrees per day in its orbit around the Sun? Which the least?

3. Can the full moon ever occult Venus (pass between Earth and Venus)? Draw a sketch to show why or why not.

4. One of Galileo's telescopic discoveries was that Venus, like the Moon, goes through a complete cycle of phases, from narrow crescent to full. How did this disprove the Ptolemaic model, which restricts Venus to positions between the Sun and the Earth?

5. Which planet can come closer to Jupiter: Earth or Uranus? (*Hint:* Use Bode's rule.)

6. Does Mars' actual orbital motion around the Sun change in any special way during the time it exhibits retrograde motion?

7. If Venus, Earth, Mars, and Jupiter are in a straight line on the same side of the Sun, what phenomenon does an observer on Earth see? What would an observer on Mars see? An observer on (or near) Jupiter?

8. Do you think Copernicus and other major figures in the Copernican revolution would have viewed themselves as revolutionaries? Why? Contrast the causes of scientific revolutions and political revolutions. What roles do factual discoveries, strong personalities, controversy, and publicity play in each? Which are more dangerous to human life? Which have the more lasting effects?

ADVANCED PROBLEMS

9. A small telescope will show the disks of Mars and Jupiter when they are closest to Earth. Assume that Mars approaches within 60 million kilometers and Jupiter within 630 million kilometers. Mars has a diameter of 6787 km, and Jupiter's diameter is 142 800 km. Use the small-angle equation to calculate the angular sizes of the two planets in seconds of arc. Which appears larger in angular size in the telescope?

10. Suppose we have a large telescope that, on a good night, resolves details as small as $\frac{1}{2}$ second of arc. Based on Problem 9, what are the smallest features it could reveal on Mars?

PROJECTS

1. On a piece of typewriter paper try to make a scale drawing of the orbits of the planets, based on orbital radii listed in Table 8-1. Which orbits are hard to show clearly? What size dots could represent Earth and Jupiter at this scale?

2. Make a large wall chart showing the orbits of the planets out to Saturn in scale. Mark the motion of each planet in one-day or one-week intervals, as appropriate. Using the *Astronomical Almanac,* an astronomy magazine, or a similar source, locate each of the planets' relative positions for the current date. Update the chart during the semester and watch for alignments that represent conjunctions, elongations, and so on. Confirm these in the night sky.

3. If a planetarium is nearby, arrange for a demonstration of planetary motions from night to night. Demonstrate the apparent retrograde motion of Mars.

4. With a telescope of at least 2-in aperture, examine Jupiter and confirm that it is attended by four prominent satellites, as discovered by Galileo. (On any given night, one or more satellites may be obscured by Jupiter or its shadow.) By following Jupiter from night to night, confirm that the satellites move around Jupiter. This proves the Copernican dictum that not all celestial objects move in Earth-centered orbits.

Two Methods for Exploring Space: Understanding Gravity and Understanding Light

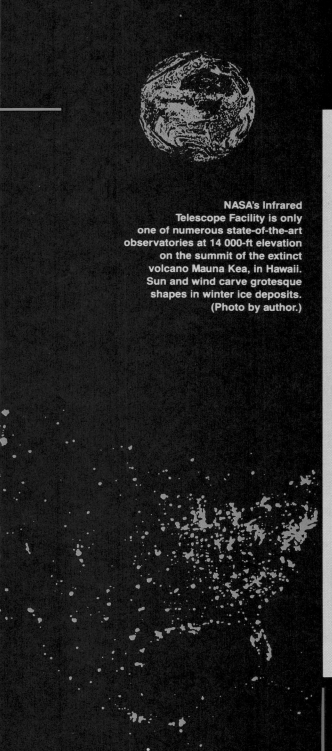

NASA's Infrared Telescope Facility is only one of numerous state-of-the-art observatories at 14 000-ft elevation on the summit of the extinct volcano Mauna Kea, in Hawaii. Sun and wind carve grotesque shapes in winter ice deposits. (Photo by author.)

CHAPTER 4

Gravity and the Conquest of Space

In Part A of this book we have seen how humans arrived at a conception of the Earth moving around the Sun among the planets. But how can we proceed from there to learn more about the physical nature of the astronomical bodies around us? To explore our space environment, we need to apply a variety of tools: both figurative tools, like physical theories of gravity and light waves, and literal tools, like telescopes and spaceships. Note that by *space environment* I mean not empty space but the rich diversity of planets, moons, gas, dust, energy fields, and stars that stretch in all directions outward from Earth.

There are two ways to explore our space environment. We can actually go there, or we can interpret light signals coming from there. In this chapter we will concentrate on the first method: how humans learned to understand, and then overcome, the bonds of gravity, so that we can send instruments and people to distant worlds in space. But some celestial bodies are too far away to visit, and the next chapter will show how we use light waves—nature's messages from space—to gain information about them.

DREAMS OF ESCAPING EARTH

As soon as humans began thinking about the arrangement of worlds in space, they began to dream of being able to leave Earth and soar through the heavens. Amazingly, fictional flights to the Moon appear in literature as far back as Greco-Roman times and recur throughout European history. The Greek satirist Lucian (c. A.D. 190) had one of his characters put on vulture and eagle wings, take off from Mt. Olympus, and fly to the Moon to learn how the stars came to be "scattered up and down the heavens carelessly." Around 1500, Leonardo da Vinci sketched devices for manned flight (Figure 4-1). In 1528 the Spaniard Eugenio Torralba

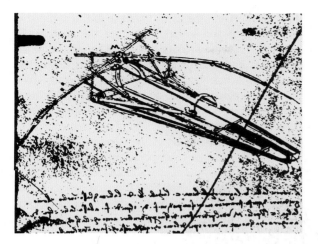

Figure 4-1 Leonardo da Vinci drew this engineering sketch for the framework of a flying machine in about 1500.

confessed to the Inquisition that he had flown near the Moon under the guidance of a demon.

Lucian's stories of lunar flight were translated into English in 1634 and since then literary trips to the Moon have occurred often. Among them are Cyrano de Bergerac's 1656 account of *Empires of the Moon* and Daniel Defoe's *Journey to the World in the Moon* in 1705. Over a hundred fictional descriptions of voyages to the Moon were published in Europe between 1493 and 1783, when the first balloon ascensions were made (Nicholson, 1949). As Figure 4-2 dramatizes, flight to the Moon was not just an idea conjured up by twentieth-century engineers upon hearing the knock of technological opportunity!

NEWTON'S LAW OF GRAVITATIONAL FORCE

The last chapter described the life of the great scientist Isaac Newton. One of his greatest accomplishments was to recognize some simple principles that describe how gravitational attraction works. This allowed him to calculate the force that one body exerts on another. This breakthrough allows all humans to master their environment in many ways: The astronomer can calculate the orbital motion of a moon around a planet; the rocket scientist can calculate the power needed to lift a cargo into orbit; the civil engineer can calculate the stresses in a bridge spanning a river.

Whether or not Newton was inspired by watching an apple fall in his garden, he started thinking about why an object falls toward Earth. He concluded that the

Figure 4-2 The rocket was still not recognized as the best mechanism for space travel by the late 1800s, when French artist Gustave Doré made this illustration of a lunar voyage.

material in Earth exerts an attraction on any material nearby, whether an apple, a stone, or a man. When he thought about the Moon, he realized that it, too, must be attracted toward Earth. Remember that **Newton's first law of motion** (Chapter 3) says:

> **A body remains stationary or moves in a straight line unless a force acts on it.**

Since the moon moves in a curve around Earth, Newton reasoned, a force must be acting on it, and that force must be coming from Earth to keep the Moon circling around Earth. Newton realized that he could calculate from the Moon's known orbital motion how fast it "fell away" from a hypothetical straight line. He then compared this acceleration rate of a falling body at the Earth's surface. After gathering accurate data, Newton showed that whereas the Moon is 60 times farther than the Earth's surface from the Earth's center, its gravitational acceleration is about 1/3600, or $1/60^2$, of the acceleration experienced at the Earth's surface. The force diminishes as the *inverse square of the distance*. Gravitational force is thus said to follow an **inverse square law.**

This result is not surprising. If a force or a substance spreads out from a point in straight lines in all directions, it must become less concentrated as it gets farther from that point. Light, radio waves, and water spray from a fast-rotating sprinkler are examples. Newton, in a leap of imagination, concluded that gravity works the same way. Earth (or any individual atom of it) acts like a source of gravity, but the farther away you go from the source, the weaker the force.

The inverse square relation is illustrated by Figure 4-3. Imagine light from a candle shining into a pyramid. If the pyramid is cut at the point where its base has an area of 1 cm², then all the candlelight entering the tip of the pyramid passes through this square centimeter.

Twice as far from the light, the base of the pyramid is twice as wide and so has four times the area, but it receives the same amount of light. The original radiation is now dispersed over 4 cm². Thus the base receives one-fourth as much light per unit area at twice any given distance. Similarly, three times as far from the light, the light is dispersed over nine times the area, and the base receives one-ninth as much light per unit area. Hence light radiation follows the inverse square law.

Newton thus knew how distance affected gravity, but he did not know what else affected it. He eventually showed that the gravitational attraction between two bodies is proportional to the amount of material in each body—that is, to its **mass**. The more mass, the more

OPTIONAL BASIC EQUATION II

Newton's Universal Law of Gravitation

Once Newton realized that the gravitational force between two objects is proportional to their masses and inversely proportional to the distance between them, he could express this more simply in math than in words:

$$F = G\frac{Mm}{R^2}$$

where

F = force of gravitational attraction between two bodies

G = gravitational constant, measured to be 6.67 × 10^{-11} N·m²/kg², in SI units

M = mass of larger body

m = mass of smaller body

R = distance between centers of M and m

The gravitational constant, G, may require some explanation. It is called a constant of proportionality. To take another example, in the familiar equation for the area of a circle, $A = \pi R^2$, the Greek letter pi is used to stand for the number 3.1416, which is the constant of proportionality between the variables A (area) and R^2 (square of radius). Similarly, using SI units of kilograms for m and meters for R, scientists measure the numerical value of $G = 6.67 \times 10^{-11}$ for all known conditions.

A striking fact is that this equation is true throughout the known universe, according to spacecraft and telescopic measurements. The equation and the value for G are true whether measuring the gravitational force between two large lead balls in the lab, the Earth and your body, the Earth and the Moon, the Sun and Mars, Saturn and its moons, or two stars orbiting around each other. Hence the equation is known as the *Universal* law of gravitation. It seemed remarkable to Newton and his contemporaries in 1687 that the reasoning of a single human could reveal the elegant simplicity, $F = GMm/R^2$, behind such a profound phenomenon of nature.

Note that since we use SI units in this book, we express mass in kilograms and distance in meters, and we get the SI unit for force, called the newton, which may be unfamiliar. Europeans and other users of the metric system commonly express amounts of material in terms of mass (kilograms); newtons usually appear only in physics and engineering problems. (Another slight complication is that Americans and English express amount of material in terms of weight or force units (pounds), instead of mass units. This will not affect problem solving in this book. You should remember the approximate rule of thumb that 1 kg has a weight of 2.2 lb on the Earth's surface. In any problem that

attraction. Thus if either mass doubles, the force between them doubles; if the distance between their centers doubles, the force drops by a factor of 4. The relationship is discussed further as Optional Basic Equation II.

Mass should not be confused with *weight*. The weight of an object is merely the gravitational force with which the Earth pulls the object against the Earth's surface. *Weight* is merely a convenient name for gravitational force. The weight of an object would be different on different planets because the gravity of different planets differs. The mass, or amount of material, of a body remains the same regardless of the body's location.

Note an aspect of the scientific method that is illustrated here. Neither Newton nor we claim to know any

"ultimate underlying cause" of gravity. We treat gravity only as a property manifested by matter, a property we can confirm by experiment and by observations of nature.

Circular Velocity: How to Launch a Satellite

Newton realized that an object could be launched into orbit around the Earth. His *Principia* contains a diagram of an Earth satellite fired from a cannon on a mountaintop, with the barrel pointed parallel to the ground (Figure 4-4). He merely applied his first law of motion. In the case of the cannonball, the force of gravity pulls the cannonball toward the ground. If the launch speed is too slow, the cannonball falls to the ground near the

starts with an amount of material expressed in pounds, convert to kilograms before proceeding with calculations using the SI units of this book.)

Sample Problem 1. Suppose you are flying your 10-ton (10 000-kg) spaceship at a distance of 10 000 km past a planet, and you detect a gravitational force of 40 000 newtons deflecting you toward the planet. Determine the mass of the planet. *Solution:* We want to solve for M, rather than F. Rearranging the equation, we have

$$M = \frac{FR^2}{Gm}$$

Substituting in values, using units of newtons, kilograms, and meters, we have

$$M = \frac{40\,000\,(10^7)^2}{6.67\,(10^{-11})(10^4)} = 6.00\,(10^{24})\ \text{kg}$$

Note from Table 8-1 that this planet is almost exactly the mass of Earth. This problem is instructive because it shows how scientists can measure the mass of planets, moons, or stars by measuring forces on objects moving near them, such as space vehicles or natural satellites.

Sample Problem 2. If the Moon were moved twice as far away, how much weaker would be the Earth's grav-

itational pull on it? *Solution:* We simply note that force is inversely proportional to the square of the distance. Thus if R increases by a factor 2, F decreases by a factor 4. The force would be one-fourth as much as for the present Moon.

Sample Problem 3. The Earth is about 3.7 times as big as the Moon and has about 81 times as much mass. Prove that an astronaut on the Moon weighs about one-sixth as much as on Earth. *Solution:* In this problem we don't have to worry about absolute weights, pounds, or kilograms of mass, because we are only asked to get the *ratio* of weights, or forces of attraction, on Earth and Moon. To get the ratio, we can simply divide the force experienced on the Moon by the force experienced on Earth. The student will find

$$\frac{f_m}{f_e} = \frac{M_m}{M_e}\left(\frac{R_e}{R_m}\right)^2$$

Note that the beauty of doing the problem this way is that the gravitational constant G, as well as the actual Earth weight of the astronaut, m, cancel out and we don't have to be bothered with these numbers. Substituting 1/81 for the mass ratio and 3.7 for the size ratio, we find that the astronaut weighs 1/5.9 as much on the Moon as on Earth, or about one-sixth as much.

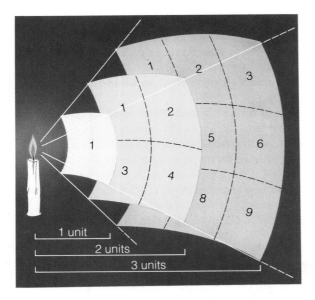

Figure 4-3 The inverse square law. At twice the distance from the source, the light is spread over four times the area. At three times the distance, the light is spread over nine times the area.

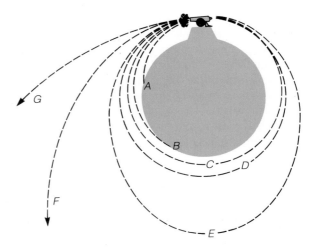

Figure 4-4 Newton realized that a vehicle launched parallel to the surface of Earth from a mountaintop could travel varying distances depending on its launch velocity. At low speed, it would drop along curve *A*, hitting nearby. At a higher speed, it could travel halfway around Earth on curve *B*. A slightly higher speed would put it in an elliptical orbit with a low point (or perigee) at *C*. A still higher speed, called circular velocity, would put it in a special type of elliptical orbit, the circular orbit *D*. A slightly higher speed would create an elliptical orbit with the farthest point (or apogee) at *E*. Escape velocity would create a parabolic orbit *F* that never returns to the Earth. A still higher speed creates a hyperbolic orbit *G*, which also never returns.

cannon (curve *A* in the figure). At a higher speed it travels farther (curve *B*). At a high enough speed, it curves toward the ground, but the surface of Earth, being round, curves away at the same rate. Thus the projectile never reaches the ground, but travels all the way around Earth and returns to the mountaintop on circular orbit *D*.[1] This speed—the speed at which an object must move parallel to the surface of a body in order to stay in circular orbit around it—is called the **circular velocity.**

The farther from Earth or other central body, the less the force of gravity that must be overcome, and therefore the lower the circular velocity. At Earth's surface, 6378 km from the Earth's center, the circular velocity is 8 km/s, or nearly 18 000 mph. The Moon, 384 000 km from that same center, moves at only about 1 km/s in its circular orbit.

In short, launching a satellite into Earth orbit is a seventeenth-century idea! One must simply get an object

above the atmosphere (so that air resistance isn't a problem), point it in a direction parallel to the ground, and accelerate it to 8 km/s.

An orbiting body's closest point to the Earth is called its **perigee**; its farthest point is its **apogee.**

Escape Velocity

As a body is launched at higher and higher speeds, the apogee point is farther and farther away. Each orbit is an **ellipse**—a type of curve describing the closed orbit of one body around a second body. A high enough launch speed would send a body from perigee near the Earth to an apogee far beyond the Moon. At a slightly higher speed, the body would travel infinitely far from the Earth (neglecting the gravitational influence of the Sun), following a curve the shape of a **parabola** (curve *F* in Figure 4-4), and it would never come back. This unique speed, which allows the object to escape the Earth forever, is called the **escape velocity**, or *parabolic velocity*. At a point near the surface of the Earth, the escape

---

[1]Newton's mountaintop cannon neglects two realities: In the length of a cannon barrel, no shell could be accelerated to the necessary speed (8 km/s) without shattering; and the Earth's atmosphere would retard the satellite, making it fall back to the ground. Thus rockets must be used, and the projectile must be launched above the atmosphere.

velocity is about 11 km/s; it is less at more distant points. Launched at a still higher speed, a body travels a similar curve called a **hyperbola** (such as curve *G*) and does not return. Thus a launch speed exceeding escape velocity is called a *hyperbolic velocity.*

In addition to the Earth, each other body in the universe, such as a planet, moon, or star, has its own gravitational field. Hence a unique escape velocity applies at the surface of each body.

ROCKETS AND SPACESHIPS

How can a body be propelled to circular velocity or escape velocity? Jules Verne, in *From the Earth to the Moon*, imagined using a 900-ft-long cannon and 400 000 lb of explosive. But as just noted, a cannon is impractical.

A more realistic technology has since been devised. A spacecraft must carry its own means of propulsion

---

OPTIONAL BASIC EQUATION III

Calculating Circular and Escape Velocities

One of Newton's achievements in systematizing physics was that his results, once derived for a specific case such as the Earth, could be generalized. The basic physical laws, such as Newton's law of gravitation, apply to bodies orbiting around other planets or around stars. From the law of gravitation, we can *derive* other equations of interest. For example, for any small satellite orbiting an object of much greater mass *M*, the circular velocity is

$$V_{\text{circ}} = \sqrt{\frac{GM}{R}}$$

where

G = Newton's gravitational constant = 6.67×10^{-11} N·m²/kg², in SI units
M = mass of central body
R = distance of orbiter from center of central body

See Figure 4-5. This is the third of nine simple but important equations that are presented in this text. This simple equation shows that the greater the mass *M*, the greater the circular velocity; and the greater the distance *R*, the less the circular velocity.

A useful fact to memorize is that the escape velocity of a small satellite is *always* $\sqrt{2}$ times the circular velocity at that distance from the planet or primary body. Thus, for example, the circular velocity of the Earth just above the atmosphere is about 8 km/s, and the escape velocity of that location is about 11 km/s.

Sample Problem 1. Show that the circular velocity not far above the Earth's surface is about 8 km/s. *Solution:* Expressing units in the SI system, we start with the mass and radius of the Earth:

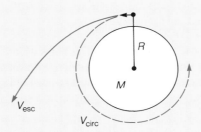

Figure 4-5 The equation for circular velocity gives the velocity V_{circ}, required to place a small object in circular orbit (dashed line) around any large mass *M* from an initial position at any distance *R*. The velocity for escape V_{esc} from this same position is always $\sqrt{2}\ V_{\text{circ}}$.

$$M = 5.98 \times 10^{24} \text{ kg}$$
$$R = 6.38 \times 10^{6} \text{ m}$$

Substituting these values into the equation, we find that a vehicle must reach a speed of 7.91×10^{3} m/s, or about 8 km/s, to stay in circular orbit, confirming the value quoted above.

Sample Problem 2. The Moon is 384 000 km from the Earth. Suppose a spaceship is following the Moon's circular orbit at a great distance from the Moon (so that it is affected primarily by the Earth's gravity, not the Moon's). By how much would it have to increase its velocity to escape from Earth altogether? *Solution:* Using the same reasoning as before, but changing the value of *R* to 3.84×10^{8} m, we find a circular velocity of only 1.02 km/s. Multiplying by $\sqrt{2}$, we find a circular velocity of 1.45 km/s. Therefore the spaceship has to increase its speed by only 0.43 km/s, or about 430 m/s. For present-day spaceships this is a relatively small change in speed!

and operate in the vacuum of space. The second point rules out propeller or jet aircraft. Around the turn of the century, several experimenters and visionaries realized that rockets were ideal. Their use was first recorded in China and Europe in the 1200s. Rockets work essentially by **Newton's third law of motion:**

> **For every action, there is an equal and opposite reaction.**

For example, if you sit on a wagon and throw a large mass (like a cinder block) out the back, the wagon coasts forward; the force needed to expel the mass causes an opposite force on the vehicle. In the same way, the force used to expel high-velocity gases out the back of a rocket nozzle pushes the rocket forward with equal force. This force is called **thrust.**

The Russian experimenter Konstantin Tsiolkovsky, beginning in 1898, and the American Robert Goddard, in the 1920s, studied and fired rockets. Both were mavericks, however, and their work was almost ignored by their contemporaries. In the 1920s in Germany, Hermann Oberth, who remarked that Verne's book was an inspiration, published several books on rocket-powered space travel. Oberth's work attracted a group of enthusiasts, including Wernher von Braun, whose astronautical experiments were converted into the V-2 guided missile program under the Nazis. At the end of World War II, about 125 German rocket experts, including von Braun, moved to the United States and continued the chain of space-travel development that stretches back to Lucian, Newton, and Verne. A remarkable footnote to our times is Oberth's survival through war and political vicissitudes to attend as a NASA guest the launch of the first successful Moon flight in 1969.

The First Satellites

After several secret postwar studies of satellites had been made, President Eisenhower announced in 1955 that the United States would launch a satellite during the International Geophysical Year (1957–1958). This was to be a civilian program using a nonmilitary rocket called Vanguard. Within days, Soviet scientists announced their plan to launch satellites larger than the American one. This plan was not taken seriously in the West: Americans viewed the Soviets as the Soviets portrayed themselves in their poster art—as unsophisticated, shirt-sleeved tractor drivers.

On October 4, 1957, the Soviet Union astonished the world by launching the first artificial satellite, the

Figure 4-6 The first artificial satellite. Russian artist A. Sokolov and cosmonaut A. Leonov collaborated on this painting of Sputnik I.

83-kg (184-lb) instrumented sphere shown in Figure 4-6. It was named Sputnik I (Russian for "satellite"). In November the half-ton Sputnik II went up, carrying a dog as a biological test. Because it was easily visible to the naked eye when it passed overhead at dusk, as shown in Figure 4-7, Sputnik II electrified the Western citizenry. In December, under hasty orders, American technicians tried to launch a small satellite in one of the Vanguard test rockets, but as millions watched on live television the rocket blew up on the launch pad.

These three months produced a crisis in Western confidence and soul searching in American education. After years of chafing at the bit, the Army team under von Braun was given the go-ahead in November; 84 days later, on January 31, 1958, the first American satellite, Explorer I, was orbited. Vanguard I, a 1½-kg (3-lb) sphere, went into orbit in March. Sputnik III went up in May. At 1½ tons,[2] Sputnik III was 56 times as

---

[2] One metric ton approximately equals the familiar English ton. Specifically, 1 metric ton = 1000 kg = 2200 lb.

Figure 4-7 This time exposure of Sputnik II shows the trail left by the satellite (lower right) as it passed a few degrees from the Moon. (Photo by Donald L. Strittmatter, January 1958.)

massive as the three American satellites combined and intensified American anxiety during that spring.

The first satellites were designed primarily to probe the nearby environment of space. Among their discoveries were the **Van Allen belts** (doughnut-shaped zones of energetic atomic particles surrounding Earth) and Earth's slight bulge in the Southern Hemisphere, which gives our planet a slight pear shape.[3]

---

[3]Ironically, it is now known that Sputnik II first detected the Van Allen radiation belts. But because the Russians did not have a global tracking network and did not tell other countries how to decode the radio signals, the Russians themselves did not get enough tracking data from Sputnik II to recognize the belts. NASA researcher A. J. Dessler (1984) has noted: "Because of their perceived need for secrecy, the Russians missed making one of the most dramatic discoveries in space science." Instead of being named the Van Allen belts (after an Iowa scientist who built the detectors), they would have been called the Vernov radiation belts (for the corresponding Russian scientist)!

The First Manned Space Flights

American engineers concentrated on miniaturizing precision instruments, while Soviet engineers, lacking the technology to miniaturize, concentrated on rocket power. The Soviets' large, powerful rockets gave them an edge during early space missions. After putting the first probe on the Moon and photographing the Moon's far side for the first time in 1959, the Russians began to test the biological possibilities of space flight with dogs, some of which they recovered from orbit. On April 12, 1961, in a 5-ton craft, a 27-year-old Russian, Yuri Gagarin, became the first person to orbit the Earth, which he did in 108 min (see Figure 4-8).

The first American manned rocket flight was Alan Shepard's 15-min suborbital flight on May 5, 1961. On August 7, Russian cosmonaut Gherman Titov made 17 orbits in a full-day flight. America's first single-orbit flight came 6 mo later on February 20, 1962, when John

Figure 4-8 The launch of Vostok I and Yuri Gagarin, the first human to circle the Earth, in April 1961.

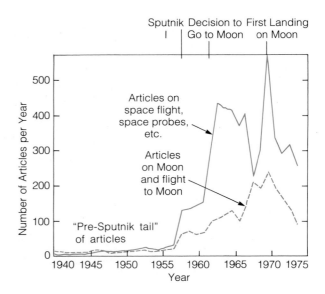

Figure 4-9 The sudden dawn of the space age. The evolution of public awareness of space exploration is reflected in the number of related articles published each year in major periodicals. The coming potential of space flight caught the media and the public unprepared when the first satellite was launched in 1957; prior to that time, the media tended to ignore the possibility or treat it as a "Buck Rogers" fantasy.

Glenn piloted a Mercury capsule. One scientific result of these flights was to allay fears that the Van Allen radiation belts or meteoroids might prevent manned space flight.

THE DECISION TO EXPLORE THE MOON: SCIENCE AND NATIONAL POLICY

Space exploration obviously has not been a purely scientific effort divorced from politics. The pacing of the whole enterprise has been determined by social judgments about national prestige and by funding decisions. The technical resources needed to attack a major scientific problem have become so complex that they demand not a single genius in a hand-built laboratory but a coordinated, well-managed program with heavy financial, political, and industrial support. The financial and political commitment must last throughout the program. Even a wealthy Isaac Newton could hardly be expected to manufacture the heavy castings, special glass, and solid-

Figure 4-10 Apollo 14 departs, carrying three astronauts on their way to the Moon in 1971. (Photo by author.)

state electronics necessary for sophisticated astronomical observations today, not to mention a fueled, 111-m (364-ft) Saturn rocket! Newton's work might have been lost in a society that did not favor inquiry, disseminate opinions, and preserve results. Similarly, the end of theoretical dreaming and the beginning of a practical effort to overcome gravity and explore space involved its own set of social conditions.

Figure 4-9 is one way of showing these social conditions. Articles on space flight and lunar exploration had a long but sparse pre-Sputnik history. Sputnik took the public by surprise. Soon an avalanche of articles on space followed, creating a strong public awareness of space by the beginning of the Apollo program.

Verne imagined a voyage to the Moon undertaken by a group of shrewd Yankees and funded by an international subscription. In the late 1930s Robert Heinlein imagined it as a commercial venture taking place in 1978, funded by an industrial tycoon. In reality it was to be a $20 billion government-inspired enterprise, conducted by an agency whose greatest problem was not the technology but the coordination of widespread resources necessary for the undertaking.

Between 1957 and 1961, planners had roughed out technical requirements and timetables for a lunar voyage. Though President Kennedy, who took office in January 1961, sought national goals that would spur creative effort, he was at first skeptical about a lunar pro-

gram, particularly because of the existing Russian lead. The president's science advisor recalls Kennedy remarking (Logsdon, 1970):

"If you had a scientific spectacular on this earth that would be more useful . . . or something that is just as dramatic and convincing as space, then we would do it." We talked about a lot of things . . . and the answer was that you couldn't make another choice.

Shepard's May 5 suborbital flight gave some reason for optimism. On May 25, 1961, the goal was set in an extraordinary presidential speech before Congress:

The dramatic achievements in space which occurred in recent weeks should have made clear to us all, as did the Sputnik in 1957, the impact of this adventure on the minds of men everywhere. . . .

I believe that this nation should commit itself to achieving the goal, before this decade is out, of landing a man on the moon and returning him safely to earth.

After Kennedy's assassination in November 1963, the NASA program, especially the Apollo Moon-landing program (Figure 4-10), became almost a memorial to the president. This prevented funding cutbacks until the program was completed. After the sixth successful landing, in 1972, a few proposed additional landings were canceled and funding for planetary exploration began a long-term decline.

AFTER APOLLO

The last remaining Apollo spacecraft was used in a regrettably short-lived effort at global cooperation in space exploration. In 1975, this spacecraft, carrying three American astronauts, linked in orbit with a Russian Soyuz spacecraft carrying two cosmonauts. The explorers shared good-humored handshakes, conducted mutual experiments, and televised pictures of each others' countries to audiences below in a gesture of goodwill. The Apollo–Soyuz project was remarkable; the challenges included an initial political agreement between Presidents Nixon and Kosygin in 1972, design of a mating tunnel to dock the two vehicles, travel of engineers and astronauts between the two countries, and learning rudiments of each others' languages on the part of astronauts and technicians. The program proved that if political leaders are willing to set challenging, cooperative goals, technical communities in the involved countries can respond with enthusiasm—a more hopeful situation than having these communities engage in bomb building!

Meanwhile, American engineers developed a fleet of four Space Shuttles designed to fulfill most of the nation's launch needs for scientific, commercial, and military payloads (Figure 4-11). Its first 24 flights, from 1981 through 1985, were highly successful. Milestones included flights of the first American woman and black astronauts, launch of satellites for Indonesia, West Germany, and other countries, flight of the European Space Agency's Spacelab, manned by a European-American team, and demonstration of an efficient drug-manufacturing technique utilizing weightless conditions. During the twenty-fifth shuttle launch, however, on January 28, 1986, an explosion destroyed one of the four shuttles, killing the racially mixed crew of two women and five men. Nearly all of America's spaceflight eggs had been put in the shuttle basket, to save money, and an inadequate number of unmanned rocket boosters were being manufactured to meet the needs for launching America's commercial and scientific payloads into space. A crisis has thus temporarily paralyzed the American space program.

SPACE EXPLORATION: COSTS AND RESULTS

The results of space exploration will be clearer to our grandchildren than to us, but we can at least compare some costs and benefits.

Figure 4-11 The Space Shuttle gave humans new capability to launch, construct, and repair payloads in space. This vertical view, looking down onto the Earth, shows the shuttle Challenger against a background of clouds during a 1983 flight. The U.S. space program was disastrously delayed when this spacecraft was demolished by an explosion during launch in 1986. (NASA.)

Costs

While we geared up for a space program from 1959 to 1969 (the year of the first lunar landing), the total budget of the United States was $1400 billion. The NASA budget (including traditional aircraft research as well as spacecraft development) was $35 billion, or about 2.5% of the total. During the American military buildup of the 1980s the NASA budget dropped to only about 1% of the total. In the 1970s and 1980s funding for all scientific research (civilian and military) averaged around 6% of the total U.S. budget. In contrast, two departments— Health, Education, and Welfare and Defense—*each* spent about one-third of the total budget.

Some writers have asked whether society might be improved by canceling space exploration and spending the money on programs to alleviate such social problems as poverty, illness, malnutrition, and the energy crisis. But as the preceding figures show, the science

Figure 4-12 Use of space photography in weather forecasting has become familiar on the evening news. This photo, made from the Applications Technology Satellite (ATS) on September 27, 1971, shows U.S. cloud cover and two tropical storms: Ginger (right) and Olivia (lower left). Advance storm warnings reduce property damage in the United States and other countries. (NASA.)

budget is too small to have much impact even if it were diverted to these areas. Most analysts believe that the few percent of the budget devoted to research more than pays for itself—and is indeed the cutting edge that leads us into the technological future that is being simultaneously shaped in Japan, the Soviet Union, and elsewhere. If the U.S. and U.S.S.R., for example, could negotiate a mutual reduction of only 10% in their military budgets, we could double the NASA budget *and* the Energy Department budget to develop technologies for the next century, and still have around $15 billion left over to help reduce the huge national debt incurred during the 1980s.

Practical Results

Seven centuries elapsed between the first use of rockets (probably in a Mongol battle in 1232) and their first scientific application (in an atmospheric research flight conducted by Goddard in 1929). Constructive benefits have come only in recent years. The first weather satellite launched in 1960 led to nearly continuous monitoring of weather conditions. TV weather reports now routinely display satellite photographs, such as the one shown in Figure 4-12. Thousands of lives and millions of dollars have been saved through the use of these photos to predict storms on land and sea.

Another important practical consequence of space flight is the communications satellite. Communications satellites, such as the Indonesian satellite launched by the shuttle, provide the first effective communications between central governments and outlying villages in some Third World countries. Events ranging from Chinese ballet and Olympic games to outbreaks of war are now broadcast between continents. In 1983, U.S. and Russian scientists collaborated on live TV hookups between Washington and Moscow to hold an international conference on nuclear disarmament. Such events

Figure 4-13 Photo made in 1973 by the Earth Resources Technology Satellite (ERTS) shows polluted waters entering the southwest corner of Lake Erie from highly urbanized areas along the Detroit (top) and Maumee (lower left) rivers. Lighter tones, brought out by contrast enhancement, reveal currents of polluted water. Similar satellite photos can be used to aid in enforcing pollution control standards. (NASA.)

Figure 4-14 View of the Texas and Louisiana Gulf Coast shows hazy air over land, contrasting with clear air over ocean (lower right foreground). Plumes of agricultural smoke drift out to sea (right center) and polluted water empties out of Galveston Bay (lower left corner). Researchers are concerned about CO_2 buildup in atmosphere from widespread deforestation by burning, revealed by satellite photos such as this. (NASA photo from 1966 Gemini manned spacecraft.)

illuminate relationships among peoples. Buckminster Fuller and other writers have predicted that this improved communication will strengthen our sense of human community on this planet, producing a "global village"—just as the bickering American colonies eventually came to accept a common identity.

Intercontinental telephone communications are made possible by a group of satellites orbiting far above the equator about 42 000 km (26 000 mi) from Earth's center. With an orbital period of 24 h, such satellites stay fixed in relation to a transmitting station. This is why backyard satellite-TV dish antennas are pointed at a fixed spot in the sky: the position of the distant satellite. A broadcaster's TV signal beamed at the satellite is thus retransmitted to other parts of Earth.[4]

Another use of satellites is in photography for mapping. Amazingly enough, in some remote areas of the world—such as the Amazon jungles of Brazil and certain parts of Ethiopia—satellite photos are better than any existing map. Ultraviolet and infrared satellite sensors are being used in agriculture, forestry, and prospecting. Satellites can also spot certain crop blights by subtle color changes before the infestations are detected on the ground. Moreover, they can detect and track pollutants in rivers, coastal waters, and the atmosphere, as seen in Figures 4-13 and 4-14, and locate

---

[4]This system was first proposed in the late 1940s by science fiction writer Arthur C. Clarke. In practice today, it causes curious pauses in transcontinental phone calls. If your call goes by normal ground links from the East Coast to the West Coast, the gap

between your question and your friend's answer is about 0.2 s, the normal response time for the brain to frame an answer. However, if your call is beamed up to a satellite about 36 000 km above the surface, then these radio waves must complete two round trips (144 000 km) at the speed of light (300 000 km/s) between question and answer. This takes roughly 0.5 s. Added to the 0.2 s mentioned above, it gives a characteristic awkward pause of 0.7 s between verbal exchanges.

mineral deposits by "reading" geologic features and subtle soil colorations. The weightless, clean, and airless environments in space vehicles also offer opportunities to test new manufacturing techniques in space. These applications are likely to expand in the next decade. The drug processing equipment tested on the shuttle is being developed commercially, and many observers believe it will become the first self-supporting space industry.

Intangible Results

There are two important intangible results of space exploration. First, space flight creates a cosmic perspective. By no coincidence, the ecological movement emerged just at the time of the first lunar flights. Astronaut Lovell, on Christmas Eve 1968, radioed from lunar orbit that the "vast loneliness . . . of the moon . . . makes you realize just what you have back there on earth." Astronaut Anders, on the same flight, noted the Earth as "the only color in the universe—very fragile . . . it reminded me of a Christmas tree ornament" (Figure 4-15). The cosmic perspective also opens the possibility of human survival even in the event of a natural or human-caused catastrophe on Earth. Rocket pioneer Wernher von Braun commented shortly after the first lunar landing (Lewis, 1969): "The ability for man to walk and actually live on other worlds has virtually assured mankind of immortality."

Second, space exploration provides a frontier and a sense of adventure that is important to human well-being (Figures 4-16 and 4-17). After several decades during which frontiers seemed to be closing all around us on the Earth, we now see a new frontier opening above us. Princeton physicist Freeman Dyson (1969) noted:

We are historically attuned to living in small exclusive groups, and we carry in us a stubborn disinclination to treat all men as brothers. On the other hand, we live on a shrinking and vulnerable planet which our lack of foresight is rapidly turning into a slum. Never again on this planet will there be unoccupied land, cultural isolation, freedom from bureaucracy, freedom for people to get lost and be on their own. Never again on this planet. But how about somewhere else?

LOOKING TO THE FUTURE

Every planet and satellite in the inner solar system has now been studied by spacecraft at close range several times. Jupiter, Saturn, Uranus, and their amazing moons (Chapters 11 and 12), not to mention Halley's comet, have been surveyed, and our space frontier is now near Neptune.

Numerous projects by various nations promise exciting results in the 1990s. In the United States, many projects were delayed by the Challenger disaster; production of unmanned launchers had been virtually halted in favor of the Space Shuttle program, and it took several years for the shuttle program to get back on track.

In 1989, the shuttle launched the first major American planetary missions in years, the Magellan probe to map Venus' surface in detail, and the Galileo robotic probe to Jupiter and its satellites. Galileo will drop a probe into Jupiter's atmosphere and map the moons in detail.

An important U.S. launch for astromers came in 1990 when the shuttle orbited the giant Hubble Space Telescope. However, this became a national embarrassment when the first tests in orbit revealed that because of design or manufacturing flaws, at least one major mirror in the HST is defective (see p. 107).

A third mission is the Mars observer, a smaller unmanned satellite designed to orbit Mars and obtain data about that planet's atmosphere and surface.

Additional space exploration missions, such as lunar orbiters and a mission to an asteroid and a comet, have been suggested but are competing with other programs for funding from Congress. In his 1984 State of the Union address, President Reagan called for a permanently inhabited U.S. space station, which will require advanced construction (Figures 4-18 and 4-19). It may involve international participation and will allow us to develop our capabilities for operating in space (Banks and Black, 1987). The space station project, however, requires a heavy commitment of the Space Shuttle fleet, and NASA managers are now proposing to build up a mixed fleet of manned shuttles and unmanned boosters to enable the launches required for a vigorous civilian space exploration program. Meanwhile the Department of Defense has more funding for space operations than all of NASA; most of it is secret and nonscientific.

During the 1980s, the United States lost some of the commanding leadership position it had acquired in

Figure 4-15 The image of our Earth hanging in space, as seen from partway to the Moon, provided humanity with a new perception of our planet as a fragile and finite globe, helping to raise environmental consciousness. In this view, much of North America is under winter cloud cover, but the Southwest and Baja California are prominent. (NASA photo by Apollo 13 astronauts.)

Figure 4-16 Astronaut working outside the U.S. Skylab space station in 1973 gave an early demonstration of our ability to work in space. (NASA.)

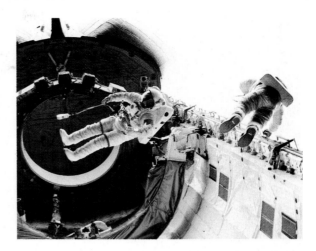

Figure 4-17 Astronauts working in the cargo bay of the Space Shuttle in 1983 symbolize our growing ability to operate in the space environment. A cloudscape on Earth's surface fills the background. (NASA.)

space with its first manned lunar landings in 1969 and its detailed unmanned missions to Mars, Jupiter, Saturn, and Uranus. Congress declined to fund an American mission to Halley's comet, an enterprise that was left to other nations. Brilliant success on Halley's comet was achieved in 1986 by probes from Europe, the Soviet Union, and Japan.

On July 20, 1989, the twentieth anniversary of the first human lunar landing, President Bush committed the United States to returning to the Moon to establish a permanent, inhabited lunar base, and then to proceed toward human exploration of Mars. Timetables, however, remain unclear, and NASA is presently preparing reports on how to achieve these exploration goals.

The Soviet Union, meanwhile, has continued an ambitious space program. (This is described in a good *Scientific American* review by Banks and Ride, 1989.) The Soviets have already established a large manned space station, Mir (Peace). They have logged more than 5600 days on board this and earlier stations (one is shown in Figure 4-20), and individual cosmonauts have lived in Mir as long as a record-breaking 1 y—long enough for a flight to Mars.

While the Soviets have made all manned flights through 1989 in vehicles that parachute back to Earth (see Figure 4-20), they launched a near duplicate of the Space Shuttle in an unmanned trip in 1988, and it may be adopted for manned flights soon. As for space telescopes, the Soviets docked a space observatory with their space station in 1987 and observed a supernova (a type of exploding star) that appeared that year.

In planetary exploration, they have landed numerous probes on Venus and constructed detailed radar maps of part of that planet. In 1988 they used a new generation of spacecraft to study Mars and one of its moons, Phobos. Of two probes launched, one was lost en route due to computer errors, and the other obtained limited data on Mars and Phobos before prematurely malfunctioning.

The Soviets used to be very secretive about their future plans in space, but perhaps because of their suc-

Figure 4-18 A 1984 test of a free-flying "manned maneuvering unit" allowed astronauts to fly away from the Space Shuttle to pursue construction and repair activities. During such a flight, the astronaut is an independent satellite of the Earth controlled by small jets of compressed gas. (NASA.)

Figure 4-19 The United States plans to construct a space station in orbit around the Earth in the 1990s, to be serviced by the Space Shuttle. The Soviet Union already has a space station in orbit, which is being expanded by add-on modules. Such stations allow low-gravity and solar energy experiments, astronomical observations, and construction facilities for space probes. In this view the Sun is hidden behind one module of the station as astronauts construct a solar panel. (Painting by Ron Miller.)

cesses and Gorbachev's policy of *glasnost* ("openness") they have recently been much more open about their future plans. Recently they have discussed a mission with French collaboration in the mid-1990s, designed to fly near several asteroids and a comet, and possibly close to Mars as well. Also, they have announced an ambitious series of Mars probes. A major unmanned mission in 1994, again with probable French participation, would drop balloons into Mars' atmosphere, along with small surface probes. A possible mission in the late 1990s would return samples from Mars. Soviet scientists in 1989 began discussing possible manned flights

to the Moon around the turn of the century. They have successfully launched a huge new rocket booster named Energia. It is the biggest operational booster in the world today—comparable to the Saturn rocket that launched American astronauts to the Moon but was then phased out of production. It is rumored that the Soviet Union may be planning manned expeditions to Mars by the turn of the century using such equipment.

Japan placed a small probe in orbit around the Moon in 1989. It had few instruments, but signals Japanese interest in space exploration.

Many observers believe that space exploration and

Figure 4-20 The Soviet manned spaceship Soyuz (Union) approaches the early 1980s Soviet space station Salyut 7 (top). Numerous flights have been made to the even larger, current Soviet space station, Mir (Peace), and new modules are constantly being added to enlarge it. (Painting by leading Soviet space artist, Andrei Sokolov.)

astronomical projects in space will become increasingly international—partly because of their expense and partly because scientists, business leaders, and politicians are becoming increasingly involved in international projects as global communications improve. For example, the United States and Europe are negotiating about collaboration on the space station and on the "Cassini" mission, which would explore Saturn and its largest moon, the cloud-shrouded satellite Titan. A pact coordinating U.S. and Soviet cooperation was allowed to lapse by President Reagan in 1982, but a new pact was signed in 1987 spelling out areas of future cooperation—extensive exchange of astronomical data, continued cooperation in studying Venus data, and coordination of Mars missions, for example. Most exciting has been discus-

sion of an unmanned, joint American–Russian sample-return mission. The Russians might build and fly a lander that would launch the sample back to Earth (they may do that anyway), and the Americans would simultaneously land a tractorlike "rover" that would roam around parts of Mars picking up different samples, which it would then put into the Russian sample-return capsule. The returned samples would be divided among international scientists for study. Such a mission in the 1990s might lead to eventual joint American–Russian manned expeditions to the Moon and Mars after 2000.

A number of scientists believe that such programs offer at least the beginnings of an alternative to the present military competition between the two countries—a chance to have their scientific communities shift from building bombs to building devices for mutual exploration of the universe (see Hartmann, Miller, and Lee, 1984).

Looking even further into the future, the president's National Commission on Space presented a bold vision of the next 50 y—a vision calling for research colonies on the Moon and Mars and a "space infrastructure" of space stations and interplanetary vehicles to support them. In their report published commercially in 1986 (National Commission on Space, 1986), they visualized this as an enterprise analogous to the opening of the American West by government survey expeditions and railroads in the 1800s.

Space operations may ultimately help solve energy, raw material, and pollution problems on Earth. For example, large satellites could collect pollution-free solar energy and beam it down to large (10-km) "antenna farms" that would replace power-generating plants on Earth. This proposal could provide energy with less environmental damage than that caused by coal-fired or nuclear generating plants, though some analysts have expressed concern over the effects of the microwave beams that would carry the energy to the receiving stations. While the cost of such a solar energy space program would be high (the Apollo program cost about $20 billion over 10 y), the United States now sends more than $50 billion each year to OPEC countries for oil. Once burned, the oil is gone for good, whereas each investment in space gives us new capabilities for the future (Arnold, 1980).

In a more visionary vein, we can imagine that industrial activity in space may reduce industrial and manufacturing pollution on Earth. Because of the availability of iron from asteroids, hydrogen from lunar soil, energy from the Sun, and other space resources, space colonies may not only be feasible but may also pay for themselves (O'Neill, 1977). Someday the Earth may be appreciated as a Hawaii in a universe of Siberias, the only place where we can enjoy the natural environment unprotected. This view may encourage us to take many industrial activities into space, where waste products cannot harm planetary ecosystems. If organized with sufficient political freedom and flexibility, space colonies might even allow experiments in new systems of government, which might defuse political tensions in the closed societies of the Earth. They might also, in the long-term future, ensure human survival in the event of an environmental or military disaster on Earth.

SUMMARY

The idea of carrying people or instruments into space existed long before the technological possibility. Newton's theory of gravitational attraction made it possible to calculate how fast objects would have to go in order to orbit around Earth or escape from Earth. The development of rocket technology early in this century provided the means to reach these speeds. Political and social factors, including strong public support for space exploration, allowed the implementation of these means.

The technology of space travel has been used in four areas: (1) improving communication, weather prediction, and manufacturing processes; (2) helping us search for new nonrenewable resources, while helping us realize that Earth and its supply of resources are finite; (3) exploring other bodies in the solar system; and (4) improving telescopic observations of stars and galaxies far beyond the solar system. Space flight has provided a sense of adventure and exploration unmatched since the Renaissance voyages to the New World.

CONCEPTS

Newton's first law of motion

inverse square law

mass

circular velocity

perigee

apogee

ellipse

parabola

escape velocity

hyperbola

Newton's third law of motion

thrust

Van Allen belts

PROBLEMS

1. If an astronaut is flying in circular orbit just above the atmosphere and wants to escape from Earth, how much additional velocity does he or she need?

2. Rocket engineers typically speak of the difference between one orbit and another in terms of velocity difference, for example, in meters per second. Why is this? (Consult Problem 1 if necessary.)

3. Because Earth rotates, the equatorial regions move at about 1600 km/h (1000 mph). Why is it easiest to launch a satellite in a west-to-east orbit over the equator?

4. You have just rowed a rowboat to a point at rest next to a dock. You step off the rowboat toward the dock.
    a. Why does the rowboat move away from the dock?
    b. In this instance, how are you analogous to the exhaust from a rocket?

5. Do you think a proposal to spend $20 billion to land a man on the Moon for the first time would be endorsed by Congress or the public if it had happened this year instead of 1961? Explain your answer, accounting for similarities or differences in public attitudes during these two periods.

6. Do you believe that in the next century manned flights to other planets, satellites, or asteroids will be common? Do you think this would have positive or negative effects on social progress, intellectual stimulation, the economy, the environment, the availability of energy and materials, and other characteristics of our civilization?

ADVANCED PROBLEMS

7. The lowest point on Earth's land surface is 392 m below sea level, at the Dead Sea in Israel and Jordan. The highest point is 8847 m above sea level, on Mt. Everest. Using a mean sea level radius for Earth of 6371 km, and assuming your average weight is measured at sea level, calculate your weight near the Dead Sea and on Mt. Everest. Use SI units as needed, but convert your answer back into pounds. Does Earth's surface make much difference in gravity from place to place?

8. Suppose you landed on a planet with a mass of 0.11 Earth-masses and a radius 0.53 times that of Earth.
    a. Compare your weight on that planet with your weight on Earth.
    b. Using data in Table 8-1 (pp. 172–173), compare this situation with that of an astronaut on Mars.

9. Suppose an astronaut is orbiting Earth at the same distance as the Moon in a circular orbit, but not near the Moon. How much faster would he or she have to move to escape Earth altogether?

CHAPTER 5

Light and the Spectrum: Messages from Space

If we send astronauts to the surface of the Moon or a robot past Neptune, we are reaching out to "touch" other parts of the universe. But most parts of the universe are too far away to reach in person or even with our space probes. To get information about these regions we have to rely on nature's messages from them reaching us in the form of light. In this chapter we will study light in all its forms.[1] We will then apply this information in this and the next few chapters to see how information can be gained about planets. In later chapters we will expand on this information and apply it to stars and galaxies.

THE NATURE OF LIGHT: WAVES VS. PARTICLES

Suppose you stand by a quiet swimming pool where a cork is floating. You disturb the cork by jiggling it. You will notice that this disturbance causes a set of waves to move out across the water. The waves have a certain spacing from one crest to the next, called the **wavelength**. They move at only one fixed speed. As they move by a given point on the water's surface, that point moves up and down; the number of these up-and-down pulsations per second at any one spot is called the **frequency** of the wave.[2]

---

[1]*Note to teachers and students:* Basic concepts useful in the next chapters are introduced here. The student will be able to develop familiarity with them as we discuss the planets. More concepts, such as Kirchhoff's laws, the Doppler effect, and the Stefan–Boltzmann law are introduced and applied to the Sun and stars in Chapters 15 and 16. This appears a useful way to introduce these physical concepts a little at a time, as needed.

[2]*Note to advanced students:* The velocity of a wave is equal to the wavelength (the distance between two crests) times the frequency (1 divided by the time interval between crests as they pass a fixed

These waves provide a useful analogy, but not a perfect description, of some **wavelike properties of light**. For instance, just as a water wave expands from its source, light spreads out in all directions from its source. **Visible light** has a tiny wavelength, around 400 to 700 nm (0.0000004–0.0000007 m—a nanometer, abbreviated nm, is a billionth of a meter; see Appendix 1). Radiation with still shorter wavelengths exists, but it is too deep-violet for our eyes to perceive; it is called **ultraviolet light**. Light with wavelengths longer than red light is too deep-red for the eye to perceive and is called **infrared light**. Extremely long wavelength infrared waves are called radio waves; radio waves are just another form of light that we cannot see. Just as the speed of the water waves is constant, the **speed of light** through empty space is constant, about 300 000 km/s (186 000 mi/s). It has been measured, for example, by the time interval necessary to communicate with a distant spacecraft.

Now suppose you shoot a BB at something. The BB has a certain energy. Unlike a wave, which takes a while to pass a fixed object and then die out, the BB delivers all its energy to its target at the moment it hits. The BB is an analog for some **particlelike properties of light**, but again it is not a perfect description of light. For instance, light has its energy concentrated in individual units, instead of having it spread out along the wave. The energy-containing unit of light is called the **photon**. It can be visualized roughly as a microscopic, BB-like particle that moves at the speed of light, yet has a certain wavelength associated with it. An important concept is that each color of light corresponds to a photon of different wavelength and energy. *The bluer the light, the shorter the wavelength and the more energetic the photon.*

From the 1600s to the 1800s, scientists argued whether light is "really" a wave or a particle. For example, Newton argued for a particle theory of light, while the Dutch astronomer-optician Huygens (1625–1695) argued for a wave theory. Arguments of Huygens and others established that light definitely has many wave-

like properties. For example, if your swimming pool has an inward-protruding corner or wall, you can observe that the wave bends slightly as it goes past the corner, so that some energy (wave motion) reaches a target slightly behind the wall as seen from the wave's source. Light does the same thing, indicating a wavelike property. A BB would move in a straight line past the wall and not hit such a target. This process of light bending its path slightly as it passes an edge is called **diffraction**; it limits the sharpness that can be achieved in the image formed by a telescope.

On the other hand, Einstein described one of the important particlelike properties of light in 1905.[3] This is the so-called photoelectric effect—an effect in which light rays can knock electrons out of metal surfaces. The effect can be explained only if the light energy arrives in individual "packets"—that is, photons—rather than being "smeared out" along the wave.

In summary, light spreads out through space something like a wave in which energy is carried by tiny, particlelike photons. Which is light: a wave or a particle? Consider a platypus. It has some ducklike and some beaverlike properties, but it is neither. Similarly, light has some wavelike and some particlelike properties, but it is neither a pure wave nor a pure particle. Considered in microscopic detail, light is a phenomenon not wholly familiar in terms of analogs in our everyday world.

THE SPECTRUM

An arrangement of all colors, in order of wavelength (or in order of photon energy), is called the **spectrum** (plural: spectra). Newton discovered he could see the spectrum of visible light by passing sunlight through a glass prism. The resulting band of colors, cast on a wall in a dark room, looks like the colored band in Figure 5-1. Water droplets in a rainstorm act like little prisms and allow us to see the spectrum—the same arrangement of colors from violet to red—in the rainbow. Newton's discovery proved that "white" light from the Sun is really made up of all the different colors. The spectrum extends to far shorter and longer wavelengths than the ones we can see. Figure 5-2 shows the names attached to different parts of the spectrum. As this figure indi-

---

point). Check the dimensions: Wavelength × frequency has dimensions of distance/time, that is, dimensions of velocity. Note that a light wave and its color can be specified by either its wavelength or its frequency. Some scientists tend to characterize waves by their frequency (as in radio parlance where stations are listed by cycles/s), but astronomers tend to refer to waves by their wavelength—a practice we will follow in this book.

---

[3]Interestingly, Einstein eventually got his Nobel Prize primarily for this work, not directly for his better-known work on relativity.

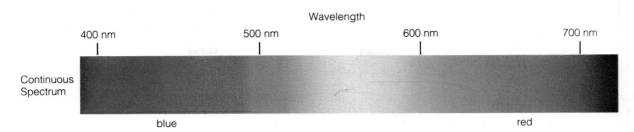

Figure 5-1 The visible part of the spectrum—an array of colors from violet to red. The wavelength scale at the top is graduated in nanometers, or billionths of a meter.

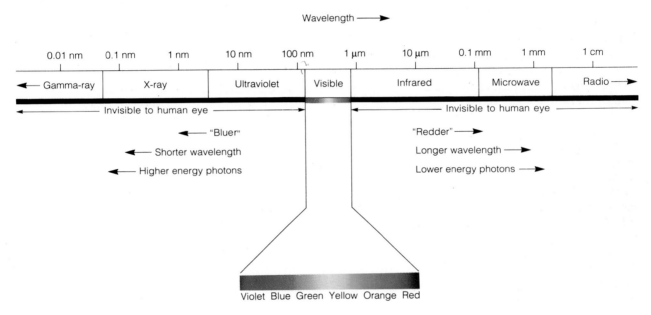

Figure 5-2 The spectrum. Top chart shows a wide range of wavelengths with names attached to different regions. Lower chart shows the order of colors in the narrow range that is visible to the eye.

cates, the radio waves we receive on our radios and TVs are simply long-wavelength versions of the light we see with our eyes.

Figure 5-3 shows an important way astronomers present this information. The bar along the bottom labels the colors. This bar, or the color band in Figure 5-1, shows the kind of picture we could get if we simply photographed a projected spectrum. But instead of photographing, suppose we used a little photocell, like a photographer's exposure meter; we could scan along the spectrum, left to right, to measure the amount of

energy (that is, light intensity) at each wavelength. Then we could plot the results as the graph in the upper part of Figure 5-3. This allows us to see the amount of light of each color. For example, Figure 5-3 shows that sunlight's strongest intensity is reached at greenish-yellow wavelengths. Astronomers often present the spectrum in this way.

Astronomers' study of the properties of the spectrum of different objects is called **spectroscopy**. *Spectroscopy is the most important method for learning about remote planets, stars, and galaxies.*

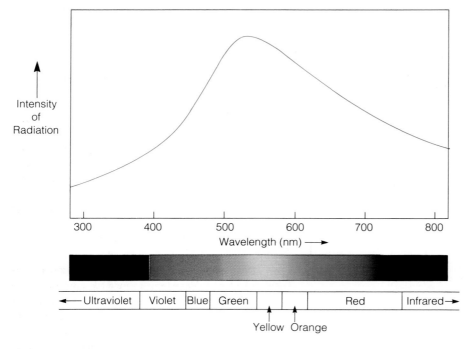

Figure 5-3 A representation of the spectrum of sunlight. The graph gives the intensity, or amount, of light at each wavelength and the names of the corresponding colors along the bottom. The Sun's dominant radiation occurs at wavelengths we perceive as greenish-yellow; we perceive the blend of *all* the Sun's radiation at different wavelengths as yellowish-white.

ORIGINS OF LIGHT: ELECTROMAGNETIC DISTURBANCES

Consider again the cork in the swimming pool. We had to disturb it to get a wave. If you put a second cork in the swimming pool and watch as you make a wave, you would see it bob up and down as the wave passed by. The cork would be a *detector* of the passing wave. In the same way, if you had a tiny enough electrically charged particle, or a tiny enough compass, you could detect light waves passing by. As the wave passes, the electric charge would vibrate with the frequency of the wave and the compass needle would oscillate rapidly with this same frequency. For this reason, physicists say that light is a disturbance of electric and magnetic **fields** in space. A field is an important concept in much of physics. A field is said to exist in a certain volume of space if some physical effect can be measured and assigned a numerical value *at every point* throughout that volume. For example, at any point near Earth, Earth attracts a given "test particle" with a measurable force; therefore a gravitational field is said to exist. Similarly, where a compass measures a certain response at any point, a magnetic field is said to exist. Also, when an electrically charged particle experiences a force at any point, an electric field exists.

The charged particle and the compass just mentioned are actual tools with which we can detect electric and magnetic fields in space or in your room. They prove that light involves pulsations in these electromagnetic fields (coexisting electric and magnetic fields). For this reason, we say that all light, of whatever wavelength—gamma-ray, X-ray, ultraviolet, visible, infrared, microwave, or radio—is **electromagnetic radiation**. The word *radiation* is used for light of any wavelength.[4]

If the second floating cork were hit by a BB or by another drifting cork, this would disturb its motion and cause it to radiate a new wave. In the same way, a drifting electron can be disturbed in such a way as to release electromagnetic radiation. Each electron has a certain amount of energy associated with it, by virtue of its motion, spin, and other properties. If an electron is disturbed in such a way as to reduce its total energy,

---

[4]However, the word must be read with caution, since physicists and engineers sometimes speak of streams of atomic particles, such as protons or electrons, as radiation. For example, charged particles escaping from radioactive material are sometimes called "radiation," or better, "particle radiation." These atomic particles are not light or electromagnetic radiation. Other examples are "radiation belts" and "cosmic rays," both of which are particles, not electromagnetic radiation.

it can give up the energy in the form of radiation—a photon that speeds away. The resulting photon has an energy equal to the energy the electron lost. In other words, the wavelength and color of the emitted radiation are exactly controlled by the amount of energy lost by the electron. As long as this electron is unattached to an atom, it can make a change in energy by any amount, and hence can radiate photons of any wavelength. Thus, free electrons, if being disturbed, could radiate an array of photons of all wavelengths. This produces a spectrum like Figure 5-1—a continuous band of colors called a **continuum**, or continuous spectrum.

Thermal Radiation

Continuum radiation is emitted all the time, all around us. Electrons (and atoms and molecules) in all gases, liquids, and solids are in constant motion, jostling each other. **Temperature** is simply a measure of the average energy of these motions, which are thus called **thermal motions** (from the prefix *thermo-*, referring to heat). The **absolute temperature scale** is a scale used by scientists, with 0 being the state of no thermal motion—the coldest temperature possible—and 273° being the freezing point of water. Each degree on the absolute scale is the same as a centigrade (also called Celsius) degree. The units are called Kelvins (after a famous physicist), abbreviated K. Thus, the freezing point of water can be written 32°F, 0°C, or 273 K, read as 32 degrees Fahrenheit, 0 degrees centigrade, or 273 Kelvins. Because electrons are being constantly disturbed by thermal motions, all objects continuously radiate a continuum spectrum of radiation. The higher the temperature, the faster the thermal motions and the more disturbed the atomic particles, and the more radiation is given off. **Thermal radiation** is the name given to the kind of radiation that depends on the temperature of the radiating material.

Wien's Law

It is important to realize that the light you see from a rock or a planet, or from this book, is *not* thermal radiation. It is **reflected radiation**—light from the Sun or some other source, bouncing off the object.

Why can't we see the thermal radiation given off by a rock, planet, or book? The basic reason is that they are too cold. Cold objects give off much less radiation than hot objects. More important, however, they give

off the wrong color of radiation for us to see. Let us explain with an important physical law discovered in 1898 by German physicist Wilhelm Wien (pronounced VEEN). It is called **Wien's law**:

> **The hotter an object, the bluer the radiation it emits.**

This effect is familiar in everyday life. Ordinarily, a nail emits no visible radiation. If you heat it over a stove, however, it begins to radiate a dull red light. Higher temperature leads to a bright orange-red glow. At a still higher temperature, the dominant color is "white hot," or yellowish-white. If the nail could be heated enough, its radiation would become distinctly bluish. This effect is shown in terms of spectra in Figure 5-4, where we see the light change from red to blue, and increase in intensity, as the nail is heated.

Despite appearances, the nail is radiating regardless of its temperature. When it is at room temperature, the light it radiates is so red (of such long wavelength) that we cannot see it because our eyes are not sensitive to infrared radiation. For example, note that room temperature is about 300 K, and the curve labeled "300 K" on Figure 5-4 is so far to the right it is almost off the diagram. Similarly, a hot enough object radiates ultraviolet light, whose wavelength is too short for us to see. For example, note that the curve labeled "8000 K" peaks at the beginning of the ultraviolet, and hotter objects would emit a still larger fraction of their light in the ultraviolet part of the spectrum. Though we cannot see infrared and ultraviolet radiation, we can build instruments to detect it.

Measuring the Temperature of a Planet (or a Star)

Now we can see one of the ways that light carries messages from space. Instruments to detect infrared radiation of planets and stars were first built around World War II and are still being improved. With such infrared detectors, we can measure the temperature of a planet by measuring the color (wavelength) at which its thermal radiation peaks. As shown in Figure 5-5, an object with peak radiation at a certain infrared wavelength has one temperature, while a star with peak radiation at visible wavelengths is hotter. The peak thermal radiation lies in the infrared parts of the spectrum for all planets and moons, from Mercury to Pluto. They are all too cold to emit appreciable visible light; we see them not by their *own* light but by the sunlight they reflect.

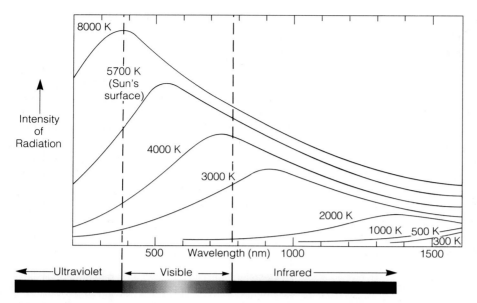

Figure 5-4 Schematic diagrams of the spectrum of a single object (a nail, lava, or a planet) as its temperature is changed. At room temperatures (~300 K), the object glows only in the infrared, with radiation invisible to the eye. As predicted by Wien's law, the dominant color becomes more blue as the temperature increases. As it heats, the object glows first deep red, then orange, and so on. At 5700 K, the material radiates light like the Sun's (cf. Figure 5-3). The total energy radiated by the object (which is proportional to the area under the curve) also increases as the temperature grows.

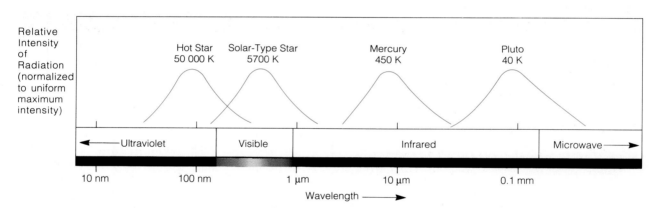

Figure 5-5 Schematic spectra of thermal radiation from four different celestial objects, adjusted to the same intensity for clarity. As predicted by Wien's law, hotter objects are bluer. The temperature can be determined by the wavelength of the dominant radiation. Planets are so cold they radiate their thermal radiation only in the infrared part of the spectrum; infrared detectors are needed to measure them. Similarly, ultraviolet detectors help study the hottest stars.

Nonthermal Radiation

The thermal radiation we have been discussing occurs when electrons change energy levels at random due to disturbances arising from the *thermal* motions of atomic particles. For completeness, we should add that other types of disturbance can create a continuum spectrum of different shape than in Figure 5-4. These types of

radiation, unrelated to thermal motions of atomic particles, are called **nonthermal radiation**.

EMISSION LINES AND BANDS

We've been describing ways to produce light as electrons are disturbed. This process, in which a particle emits a photon, is called **emission**. We started off picturing a freely drifting electron (like a cork floating in a pool), not attached to an atom. Now consider an electron orbiting around the nucleus of an atom. As soon as we consider electrons attached to atoms, extraordinary modifications occur to the emission process, and these are critical to all of astronomy.

Atoms and Emission Lines

Figure 5-6 shows the schematic structure of an atom, with its central nucleus and some orbital paths available to a specific electron. This electron may be only one of many, depending on the element: hydrogen, helium, and so on. In some ways, the atom is like a tiny solar system—that is, a relatively massive central object and tiny orbiting particles. But there is an extraordinary

difference. In the solar system we can put a rocket into any orbit we choose; each orbit would have its own velocity and hence energy. But in an atom, electrons can occupy only *certain* orbits. These are called **energy levels**, since each orbit has one specific energy. Physicists describe this by saying the atom has quantized energy levels, because only certain quantities of orbital energy are possible. This was discovered in 1913 by Danish physicist Niels Bohr. In Figure 5-6, the solid circle represents the orbit occupied by the electron, and the dashed circles are other allowed energy levels. At the top are a few of the infinite number of energy levels outside the atom, where energy levels are not discrete.

Now let us disturb the electron as before (hitting it with a photon or a neighboring atom) so that it drops to a lower energy level. (In fact, this process can also

---

OPTIONAL BASIC EQUATION IV

Measuring Temperatures of Astronomical Bodies: Wien's Law

Wien's law tells which wavelength W corresponds to the maximum amount of radiation, given the temperature T. Using SI metric units, the wavelength is given in meters and the temperature in Kelvins, abbreviated K. (These units are sometimes called "degrees Kelvin," since they define a temperature scale in degrees; but the correct SI terminology is "Kelvins.")

The law is

$$W = \frac{0.00290}{T}$$

The number 0.00290 is a constant of proportionality and remains the same in all applications of the law. Thus as T increases, W decreases, giving shorter wavelengths and hence bluer light.

Sample Problem 1. The Sun has a surface temperature of about 5700 K. Calculate the wavelength of the strongest solar radiation. *Solution:* Using SI metric units and powers of 10, we get

$$W = \frac{2.9 \times 10^{-3}}{5.7 \times 10^3} = 5.1 \times 10^{-7} \, \text{m}$$

This, of course, is exactly in the middle of the range to which the eye is sensitive. (Otherwise we could not

see sunlight!) Light of this wavelength corresponds to a greenish-yellow color. The blend of all solar wavelengths we see as yellowish-white.

Sample Problem 2. A certain planet (not unlike Earth) has a mean temperature of 290 K (around room temperature). Characterize its thermal radiation. *Solution:* Inserting $T = 290$ K in the equation, we find a wavelength $W = 10^{-5}$ m, or 10 μm. This is well out into the infrared—invisible to the eye but detectable by instruments.

Sample Problem 3. A certain star, cooler than the Sun, has a temperature of 2900 K. Characterize the color of its radiation. *Solution:* Since it is cooler than the Sun, it must be at least somewhat redder. But how much? Is its peak radiation orange, red, or what? Inserting $T = 2900$ K in the equation, we find that $W = 10^{-6}$ m, or 1 μm. This means the peak radiation is in the infrared—just redder than the eye can see. But much radiation will spill into the nearby red and orange parts of the visible spectrum (see curves in Figure 5-4), so the star will look orangish-red.

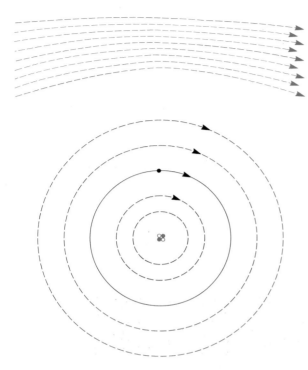

Figure 5-6 Schematic view of an atom's structure. If an electron (dot) finds itself in an orbit (solid line) around the nucleus of an atom, there are only certain other orbits, or "energy levels," that it can occupy in the atom (dashed circles). However, at large distances from the nucleus (top) virtually any orbits, or "energy levels," are possible.

happen spontaneously without outside disturbance.) As it drops, it emits a photon whose energy equals the difference in energy between the two levels. But note that being inside an atom, it can experience only certain, fixed changes in energy level, as shown in Figure 5-7. For instance, if it starts in level 3, it can drop only to level 2 or 1. Thus it can emit only certain amounts of energy corresponding to these differences in energy level; it can emit only photons of certain wavelength or color. As a result, *each element can emit only certain wavelengths of light.* Figure 5-8 shows, for example, the emissions produced by a sample of hydrogen gas containing neutral (that is, uncharged) atoms, with their electrons in different energy levels. These emissions are called **emission lines** because they appear in a projected spectrum as *lines* or narrow bars of color, as seen in Figure 5-8.

Here, then, is an astounding and useful fact! If you see a certain set of emission lines, you can match it with a certain element and infer that atoms of *that* element are present and glowing in the distant object. You can tell something about the object's composition, even without having a sample!

Note that if the electrons in an atom are in their lowest possible energy level, that atom cannot produce an emission line. This is because the electrons cannot drop to any lower energy level. An atom in which all electrons are in the lowest possible energy level is said to be in its **ground state**. An atom in which one or more electrons are in energy levels higher than the lowest available ones is said to be in an **excited state**.[5] Excited states usually last only a short time before the electrons "decay" to the lowest available energy level, producing the ground state. Atoms generally need to be disturbed to produce and maintain excited states. For this reason, hot gases are more apt to produce emission lines than cold gases, because atoms in hot gases collide faster and more often.

Molecules, Crystals, and Emission Bands

The electron structure in a molecule is more complex than in an atom. The electron's path may take it around two or more nuclei. Thus the emission line structure from a molecule is not so simple as that from an atom. For example, in a gas containing water molecules (H_2O), we get more complex emission lines than in a gas containing single H and O atoms. The molecule has various ways of responding to a disturbance in addition to having its electron change energy levels—for example, it may vibrate like two balls linked with a spring. As a result, the energy levels are vastly more numerous and the resulting emission lines blend together. Instead of sharp emission lines, we get blended clusters of barely separated emission lines over a range of wavelengths, as shown in Figure 5-9. The resulting broader emission feature from a molecule is called an **emission band**. The rest of the story is the same, though: A given molecule (such as H_2O) can produce only certain emission bands. Thus we can identify the molecules glowing in a given remote source just by detecting their emission bands.

---

[5]Each energy level can be occupied only by a certain number of electrons. For example, the lowest level can take only two.

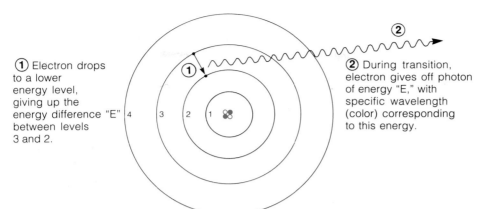

① Electron drops to a lower energy level, giving up the energy difference "E" between levels 3 and 2.

② During transition, electron gives off photon of energy "E," with specific wavelength (color) corresponding to this energy.

Figure 5-7 Schematic view of emission of a photon of light as an electron drops from a higher to a lower energy level in an atom.

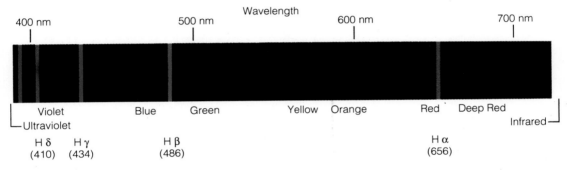

Wavelength

400 nm   500 nm   600 nm   700 nm

Violet   Blue   Green   Yellow   Orange   Red   Deep Red

└ Ultraviolet   Infrared ┘

H δ   H γ   H β   H α
(410)   (434)   (486)   (656)

Figure 5-8 A visible-light spectrum consisting only of the emission lines produced by glowing, neutral hydrogen gas. Wavelength scale is given at the top. The red line, called the hydrogen alpha line, is especially prominent in many astronomical examples of glowing gas, causing many astronomical objects to glow with reddish light.

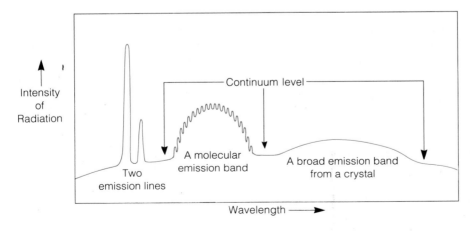

Intensity of Radiation

Continuum level

Two emission lines

A molecular emission band

A broad emission band from a crystal

Wavelength ⟶

Figure 5-9 Schematic examples of emission lines and bands as they might appear in a spectrum. The molecular emission band shows a "fine structure" associated with the structure of the molecule.

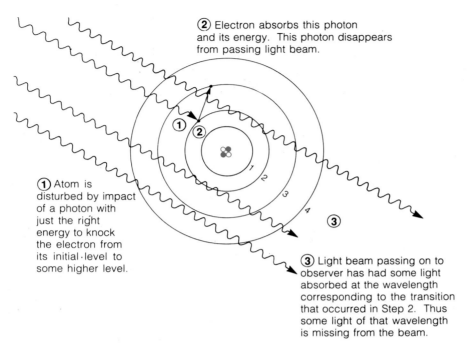

② Electron absorbs this photon and its energy. This photon disappears from passing light beam.

① Atom is disturbed by impact of a photon with just the right energy to knock the electron from its initial level to some higher level.

③ Light beam passing on to observer has had some light absorbed at the wavelength corresponding to the transition that occurred in Step 2. Thus some light of that wavelength is missing from the beam.

Figure 5-10 Schematic view of absorption of a photon of light as an electron is knocked from a lower to a higher energy level in an atom.

In order for atoms and molecules to produce clear emission lines and bands, they must be detached from one another, as in gases. If the atoms were linked together, they would form molecules, and if the molecules were linked, they would form a solid or liquid. In most solid or liquid substances, the electron structure is so complex that emissions are not confined to one wavelength, but are smeared out. Therefore, emission features of solids and liquids are barely discernible. Most emission lines and bands arise from gases.

However, an important exception occurs among rock-forming minerals, which make up most planetary surfaces and interstellar dust grains. Most rock-forming minerals are crystals, which can be thought of as giant molecules. In a crystal, atoms are joined in a fixed pattern that simply repeats as the crystal grows bigger. Light can penetrate through the outer millimeter or so of a crystal, and very broad, faint emission bands can result, as shown in Figure 5-9. Warm grains of silicate material near other stars have been identified from their emission bands, for example.

ABSORPTION LINES AND BANDS

Let us return, in Figure 5-10, to an atom with an orbiting electron. This time it gets disturbed by a passing light wave (photon) that bumps it up into a higher energy level. Energy was removed from the beam to do this. This process of energy removal from a light beam is called **absorption**.

Atoms and Absorption Lines

Only the specific energy corresponding to the specific transition (level 2 to 3 in the case of Figure 5-10) can be absorbed in an atom. Thus only a photon of specific energy or wavelength can be removed from the beam. A beam of light passing through a cloud of such atoms will have many of these photons removed, thus absorbing some light of that color. As a result, the light in a narrow interval of the spectrum is lost, and this missing interval is called an **absorption line**. Various absorption lines can result from the various possible upward transitions—for example, level 2 to 3, 2 to 4, 2 to 5, 1 to 2, 1 to 3, and so on. The visible part of the spectrum, with absorption lines due to hydrogen, is shown in color in Figure 5-11.

Since the *intervals* between energy levels in a given atom are the same whether absorption or emission is occurring, the pattern of emission and absorption lines for a given element is the same. An element can be identified from either its emission lines or its absorption lines.

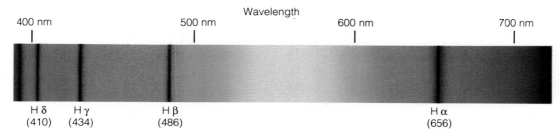

Figure 5-11 Visible portion of the spectrum showing absorption lines due to hydrogen. Such a spectrum would be seen if hydrogen gas lay between the observer and a light source with a continuous spectrum; the hydrogen absorbs only the specific "missing" colors. These absorption lines have the same positions as the emission lines in Figure 5-8.

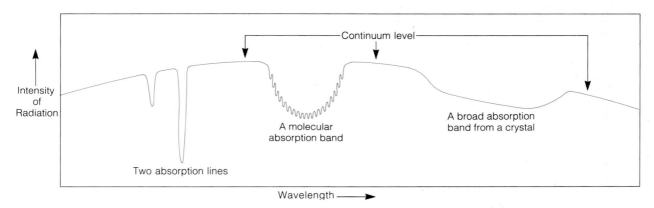

Figure 5-12 Schematic examples of absorption lines and bands as they might appear in a spectrum. The molecular absorption band shows "fine structure" associated with the structure of the molecule.

Molecules, Crystals, and Absorption Bands

Much of the preceding also applies to molecules and crystals. As light penetrates through a cloud of molecules (a gas) or through the upper millimeter of a crystal (in a rock surface), transitions of electrons among its numerous, closely spaced energy levels produce **absorption bands**. The molecules in the gas, or the composition of the crystal, can be identified if the absorption bands themselves can be detected and identified. Figure 5-12 shows schematic examples of absorption lines and absorption bands. The absorption bands of a substance have the same wavelength intervals as its emission bands.

Measuring the Composition of an Atmosphere of a Planet (or a Star)

Now we see how light carries even more messages. Suppose we look at a planet with an atmosphere. Light passes into the atmosphere and back out, as shown in Figure 5-13. As the light passes through the gas, it may acquire absorption lines and bands that allow us to identify some of the atoms and molecules in the gas. Not all atoms and molecules have equally prominent lines and bands. Some, such as carbon dioxide (CO_2), are easy to detect, and others, such as nitrogen (N_2), are hard to detect by spectroscopy. Thus while the carbon dioxide of Venus and Mars was discovered by Earth-based spectroscopy decades ago, the abundant nitrogen of

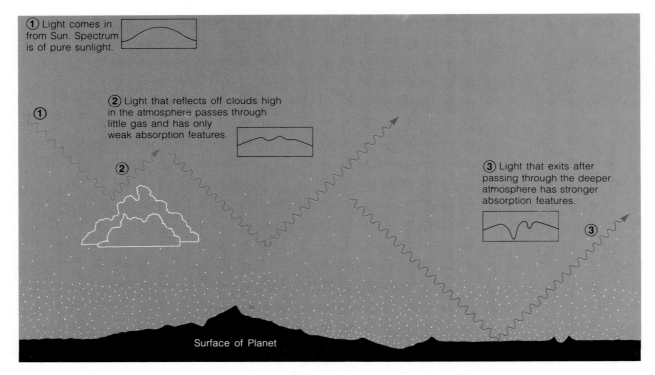

① Light comes in from Sun. Spectrum is of pure sunlight.

② Light that reflects off clouds high in the atmosphere passes through little gas and has only weak absorption features.

③ Light that exits after passing through the deeper atmosphere has stronger absorption features.

Surface of Planet

Figure 5-13 Production of absorption lines or bands in the spectrum of a planet as light passes through the atmosphere of the planet. Identification of the lines or bands allows identification of at least some constituents in the atmosphere. The same process occurs as light radiates from a star outward through the star's surrounding gaseous atmosphere.

Saturn's Moon, Titan, was not discovered until a Voyager spacecraft made a close pass by Titan.

ANALYZING SPECTRA

How to Measure Properties of a Planet's Surface from a Distance

As mentioned earlier, the absorption bands caused by mineral crystals are broad, often shallow, and hard to detect. For this reason, spectroscopy has only limited success in identifying planetary surface materials. There are two approaches: measuring colors and detecting actual absorption bands. Measuring the color of an object is essentially measuring the relative intensities of radiation in different parts of the spectrum. A red object looks red because it reflects more light at red wavelengths than at bluer wavelengths. A measurement of

color often restricts the possible materials on a surface. For instance, most satellites of Saturn have a bluish-white color similar to ice, but inconsistent with the red rocks and minerals on Mars.

If we can detect a mineral absorption band, we can probably identify the mineral. For example, the satellites of Saturn just mentioned display absorption bands caused by crystals of frozen water, in the form of ice or frost, confirming the implication of the color studies that ice is a dominant material on their surfaces.

Figure 5-14 shows some results that might be obtained if we could scan a planet's spectrum all the way from the ultraviolet to the infrared. (This diagram is somewhat idealized, since different instruments are needed in different regions. One astronomer may obtain the spectrum only in one region. Figure 5-14 might require the combined work of many astronomers using different instruments at different observatories.) In the visible region we see reflected sunlight. Superimposed on it are some absorption lines and bands that tell us the planet has a gaseous atmosphere containing certain

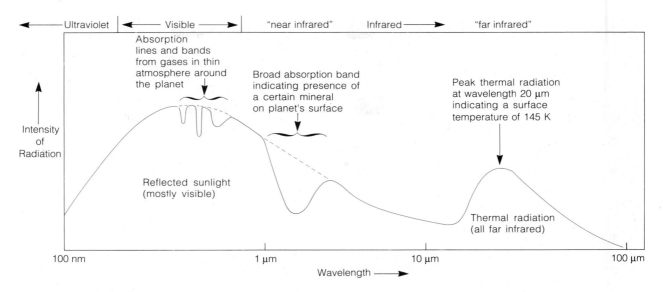

Figure 5-14 Simplified spectrum of an imaginary planet from the ultraviolet to infrared wavelengths, showing the reflected sunlight and the thermal radiation from the planet itself. The dashed curves show the shape of the reflected solar spectrum that we would see if there were no absorption lines and bands. A thin atmosphere creates some absorption lines and bands, but lets enough light through to the surface that we see a broad absorption band (at about 1.5 µm) due to minerals on the planet's surface. (See text for further description.)

elements and compounds; these arose as shown in Figure 5-13. In the near infrared, just beyond 1 µm wavelength, we see a broad absorption feature characteristic of a rock-forming mineral. This gives us some indication of the rock type on the surface, indicating at the same time that clouds are not thick enough to block most of the light from reaching the surface. Finally, in the far infrared at 20 µm wavelength, we see the thermal emission from the planet. Noting that the dominant wavelength is at 20 µm, we can use Wien's law to estimate the surface temperature on the planet. In this case, it is a cold 145 K (−198°F).

The Spectrum and Our Atmosphere: Seeing into Space from the Earth

Figure 5-15 illustrates several important points about the relationship between astronomy and the Earth's atmosphere. First, clouds block much of our view of space. One reason for putting observatories on top of mountains is to get above the clouds. Another reason is to get above the dense, shimmery air that makes

telescopic star images dance and twinkle when seen from sea level.

But a more important reason can be seen by comparing Figure 5-15a made in normal visible light, with Figure 5-15b, made in infrared light. Many atmospheric molecules, such as water molecules, strongly absorb infrared light just beyond the reddest wavelengths we can see. Figure 5-15b is an image of Earth made with light of the infrared wavelength partly absorbed by water. Almost none of this light can make it all the way to the ground and back out into space. The infrared photons that *do* make it back into space have gone only partway down to the ground, reflecting off clouds as haze. Sure enough, the image shows water vapor clouds and haze over most of our planet. Conversely, a sea level astronomer who wants to look for water vapor absorptions on, say, Mars, has a problem: The infrared light that carries this message from Mars is blocked by water vapor in our own atmosphere before it ever reaches his or her telescope on the ground, as can be seen in Figure 5-16. Even the thin, high atmosphere blocks many wavelengths. For example, the **ozone (O_3) layer**, about 20 to 50 km up, absorbs nearly all ultraviolet radiation (thus protecting us from severe sunburn).

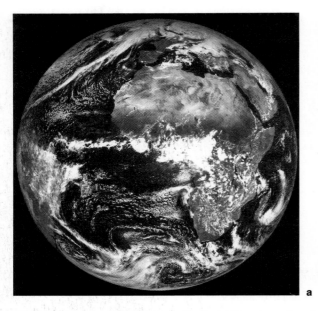 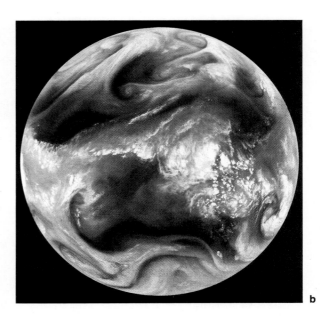

Figure 5-15 Views of Earth based on two different wavelengths of light. **a** View in visible light, showing clouds and continents. Africa dominates this image. **b** View of the same part of Earth at a wavelength of reflected solar infrared light. Much of this light is absorbed by water vapor gas in the air. Due to absorption, water vapor haze obscures most of Earth's surface. (European Meteorological Satellite images, courtesy C. R. Chapman.)

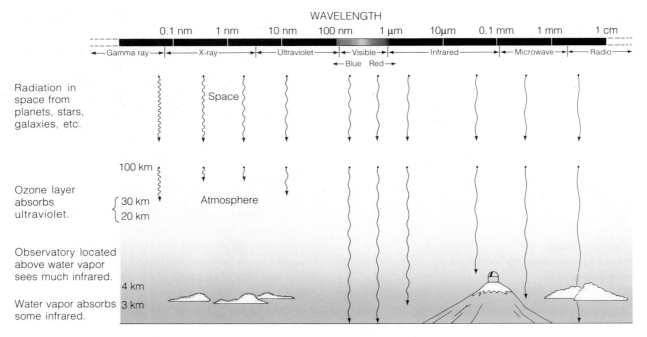

Figure 5-16 Electromagnetic spectrum is shown at top. All these types of radiation from various astronomical sources can be detected by a telescope in space (upper middle), but many of them are absorbed by various gases in our atmosphere (bottom).

These absorptions of certain wavelengths by our atmosphere are a main reason for putting modern observatories on mountains. In particular, high observatories are above most of our atmosphere's water vapor, which is concentrated in the bottom 3 km (9000 ft) of the atmosphere and which absorbs much of the infrared light from planets and stars. At the same time, high-altitude observatories escape the clouds and the shimmering, hazy air of low elevations. Thus, for example, astronomers at 14 000-ft Mauna Kea Observatory in Hawaii can measure most of the infrared spectrum almost every night. To measure most of the ultraviolet spectrum, we must get above the ozone layer, far above even the highest mountains. Therefore, various ultraviolet, X-ray, and gamma-ray telescopes, as well as infrared and optical telescopes, have been put in orbit. These have made amazing breakthroughs that we will describe in later chapters. Of special importance is the **Hubble Space Telescope,** placed into orbit in 1990 (Bahcall and Spitzer, 1982). The telescope has a mirror 2.4 m (94 in.) across, making it one of the larger telescopes in operation. In an embarassing failure of American design or manufacturing, its optics were found to be defective after it was launched. However, it is still expected to produce revolutionary astronomical discoveries because of its space-based opportunity to sample a wide range of ultraviolet to infrared wavelengths. It may be repaired by shuttle astronauts, with instruments on it being updated and replaced.

Figures 5-15 and 5-16 illustrate one of the advantages of access to a wide range of wavelengths. Photos using light of different wavelengths penetrate gaseous atmospheres of planets or stars to different depths (because of the different rate of absorption at different wavelengths). Therefore they show different features at different depths in the atmosphere. Comparison of images at different wavelengths allows us to tell something about the vertical structure of the atmosphere we are observing.

THE THREE FUNCTIONS OF TELESCOPES

A **telescope** is a device with three functions. First and most obvious, it magnifies the image of an object to a larger angular size than we perceive with our naked eye. The term **magnification** applies primarily to a telescope designed for visual observation; the magnification is the apparent angular size of a distant object seen through the eyepiece, relative to its apparent size seen by the naked eye. If a telescope makes something look 10 times larger, we say it has a magnification of 10×. The second function is **resolution**—the ability to discriminate fine detail. Whereas the eye can resolve angular details only a few minutes of arc across, a telescope might show detail a few seconds of arc across. The telescope's third function is **light-gathering power**—the ability to collect light and reveal fainter details than the naked eye can see. When light from a distant planet or star reaches Earth, a certain number of photons strike each square centimeter of Earth each second. The pupil of the eye has a diameter less than a centimeter and can receive only a limited number of photons per second. But a telescope collects *all* the photons striking a lens or mirror many centimeters across. This makes a much brighter image.

Early telescopes were designed entirely for observers to look through. In modern professional telescopes, the human eye is replaced by instruments that can make more precise measurements. Thus modern astronomers rarely look through their giant telescopes! Nevertheless, the three functions still apply. A large telescope used with film to take pictures, for example, can be compared to an ordinary camera instead of to the eye: It obtains a more magnified image, a more clearly resolved image, and a brighter image than the camera.

Two Designs for Optical Telescopes

Two basic designs have been used for telescopes. The first to be built, the **refractor**, uses a lens to bend, or refract, light rays to a focus, as in Figure 5-17. Galileo first used this type astronomically in 1609. The second type, the **reflector**, uses a curved mirror to reflect light rays to a focus, as in Figure 5-18. Isaac Newton built the first reflector in 1668. Several reflector designs have since been constructed, but the simplest is Newton's, sometimes called the Newtonian reflector. In recent years new designs have combined lenses and mirrors. Often called **compound telescopes**, these designs are often very compact and portable.

Focal Length, Aperture, and the Telescope's Functions

In a telescope, the main lens or mirror is called the **objective**, as shown in Figures 5-17 and 5-18. The

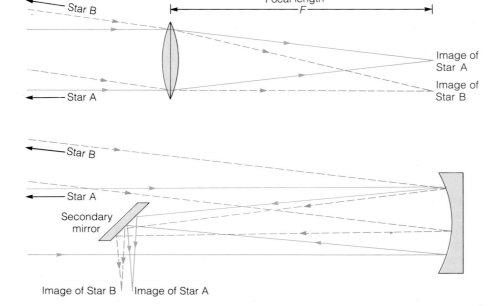

Figure 5-17 Cross section through a lens, showing image formation in a refracting telescope (and most cameras). Light rays from two stars are focused into two images.

Figure 5-18 Cross section through a mirror system, showing image formation in a reflecting telescope. Light rays strike curved mirror (right) and are reflected back toward focus. The focus would normally lie in an inconvenient position in front of the mirror, but a secondary mirror is used to beam light rays to one side for easier access to the image.

distance from the objective to the place where the image is focused is called the **focal length** of the objective (marked F in Figure 5-17). The diameter of the objective is called the **aperture**.

The three functions of a telescope are controlled by the focal length and the aperture. The longer the focal length, the more the magnification and the bigger the image. This principle also applies to camera lenses. Lenses with long focal length give big images and are called telephoto lenses. The wider the aperture, the better the resolution. Note that magnification is not the same as resolution. Magnification alone only produces a big image, not necessarily a sharp image. By increasing focal length, we get a bigger image, but it may be blurry. If we increase the aperture (an expensive process, since we have to buy a bigger lens or mirror), we can get a sharper image. The third function, light-gathering power, is connected with aperture also: The larger the aperture, the more light collected.

Radio Telescopes

Radio telescopes serve the same functions as optical telescopes. But because a 50-cm radio wave is a million times longer in wavelength than a visible light wave and carries less energy per photon, radio telescopes need larger surfaces to collect and focus enough energy to give strong signals. As shown in Figure 5-19, radio telescopes are reflectors, with a large curved surface (the "dish") of metal or wire mesh, which reflects the radio waves to a focus, where a smaller radio detector picks up the signal.

USING VISUAL TELESCOPES

An optical telescope designed to be looked through, as opposed to a radio telescope or a camera, may be called a *visual telescope*. The optical systems of Figures 5-17 and 5-18 can be converted into visual telescopes simply by adding an eyepiece (and usually tubing to keep out stray light), as shown in Figure 5-20. The eyepiece is simply a lens or system of lenses designed to magnify the image still more and allow the eye to examine it.

One of the first questions asked of backyard telescope enthusiasts is: "How far can you see with that thing?" This is not the right question to ask, because no telescope is limited by distance. Every optical system can see as far as there is an object large enough or bright enough to detect. The naked eye as well as the

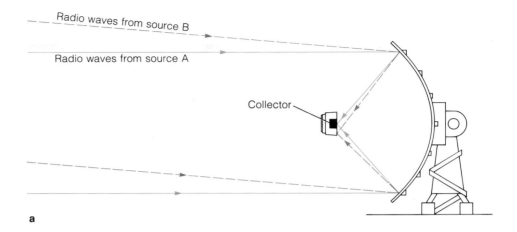

a

5-m (200-in.) Palomar telescope can see the Andromeda galaxy 19 billion billion kilometers away—but the
telescope shows more detail and fainter regions.

Magnifying Power

In any given telescope, the magnifying power, or *power*
as it is usually called, is controlled by the eyepiece.
Suppose a distant object subtends an angle of 1 minute
of arc (about the angular size of Jupiter when it is prominent). An eyepiece that makes the object appear 100
minutes across when seen through the telescope is said
to give 100 power.[6]

In theory, any telescope can be made to give any
magnifying power simply by inserting an eyepiece of
short enough focal length. You might think you'd want
to use as high a magnification as possible in any telescope: the bigger the image, the better. In practice,
however, three effects limit the magnifying power that
can be used. First, because we look into space through
turbulent air, higher powers magnify air turbulence, and
too high a power causes the image to shimmer hopelessly. (The air quality, called **seeing**, varies from night
to night, with occasional nights of good seeing occurring
when the air is still.)

---

[6]The magnifying power is given by the formula

$$\text{Power} = \frac{\text{focal length of objective}}{\text{focal length of eyepiece}}$$

For instance, if a 1-cm eyepiece is inserted in a telescope of objective focal length 100 cm, we get 100 power, often written 100×.

b

Figure 5-19 a Schematic cross section of a radio
telescope, showing similarity to the design of the reflector-
type optical telescope. **b** View of one of the radio
telescopes in the Very Large Array in central New Mexico.
This view shows the underside of the "dish" (top), or curved
reflector, that collects the radio waves. This telescope
(along with its neighbors) can be repositioned in different
arrays along the railroad tracks, giving different receiving
qualities for different observing projects. (Photo by author.)

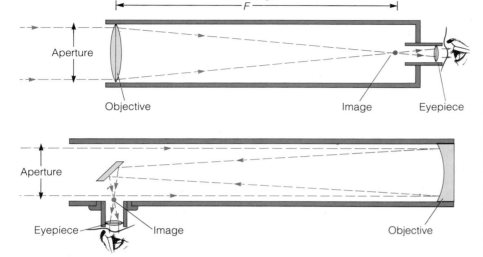

Figure 5-20 Cross sections showing how the systems of Figures 5-17 and 5-18 are converted to visual telescopes by the addition of an eyepiece. The eyepiece is a small magnifying lens (or several lenses mounted together) used to examine the image. Tubing from the objective to the eyepiece helps cut out stray light.

Second, as the magnification is increased, the image gets fainter and fainter because the light is spread out over a larger and larger area. Too high a power gives a hopelessly faint image.

Third, due to fundamental properties of light waves, a telescope of A-cm aperture cannot resolve details smaller than $12/A$ seconds of arc. Thus a 12-cm (5-in.) telescope can resolve about 1 second, but nothing smaller. Therefore, too high a power gives a hopelessly fuzzy image.

Astronomers look forward to solving the first problem by putting more telescopes into orbit in space—above the atmosphere. The last two problems can be overcome by increasing the telescope aperture, giving more light and resolution of smaller angular detail. However, the maximum useful magnifying power for most Earth-based telescopes on most nights is about $20\times$ per centimeter of aperture. For example, a 5-cm (2-in.) telescope might be used at $100\times$; a 15-cm (6-in.) at $300\times$; and a 30-cm (12-in.) at $600\times$.

Light-Gathering Power

The other important function of a telescope, gathering light, is controlled solely by the aperture size. Many astronomical objects, such as nebulae and galaxies, are very faint, and to see their details one needs light more than magnifying power. In fact, the best visual impression of faint nebulae comes from a large telescope (lots of light) used at *low* power (to concentrate light). For

this purpose, one might use only $4\times$ per centimeter of aperture; for instance, a 15-cm (6-in.) telescope at $60\times$. If you have access to a telescope, you can experiment with these recommendations while looking at different celestial objects.

Generally, the amount of detail that can be seen on astronomical objects depends ultimately on the stillness and clarity of the air and on the telescope aperture, since these control both the useful magnifying power and the light-gathering power.

Observing the Moon and Planets

Although any visual telescope can be used to observe the heavens, certain designs and techniques are best suited to certain purposes. To observe the Moon and planets, magnification is the most important function, because one wants to get a big enough image to see detail. Telescopes with rather long focal lengths, about 6 to 10 times the aperture (often stated as f 6 to f 10), are desirable. Eyepieces giving magnifications of 100 to $200\times$ are recommended. Some of the sights you can see with such equipment, in order of increasing difficulty, are the mountains and craters on the Moon, the phases of Venus, the satellites of Jupiter, the satellites of Saturn, the rings of Saturn, the cloud bands on Jupiter, divisions in the rings of Saturn, the phases of Mercury, the polar ice caps of Mars, the dusky markings of Mars, the cloud bands on Saturn, and the dusky markings on Mercury. An aperture of 5 cm (2 in.) suffices to see the first five items on the list; an aperture of around 25 cm (10 in.) is recommended for all items and is usually necessary for the last one.

Observing Star Fields, Nebulae, and Galaxies

The most prominent star clusters, nebulae, and galaxies are faint but cover larger angular areas than planets; hence light-gathering ability is the most important function in observing them. Telescopes with shorter focal lengths of only about three to six times the aperture (ƒ 3 to ƒ 6) are useful, and magnifications of 20 to 100 × are recommended. Some double stars and star clusters are interesting to study under higher power, but higher powers spread out the light too much to give good views of nebulae and galaxies, which are mostly very faint. As always, the largest possible aperture gives the best view.

Observing the Sun

A NORMAL TELESCOPE SHOULD NEVER, UNDER ANY CIRCUMSTANCES, BE POINTED AT THE SUN. An eye could be immediately burned or blinded, since the objective acts like a giant magnifying glass, concentrating solar light, heat, and ultraviolet radiation at a point near the eyepiece. Furthermore, the telescope is likely to be damaged, since the solar heat can crack the glass in eyepieces or secondary mirrors. Some telescopes can be modified for safe solar observing. One usually starts by putting an opaque card over the objective with a new aperture as small as 1 or 2 cm, then projecting the image onto a white card so it can be viewed. Instructors or telescope owners should be consulted before attempting this procedure.

PHOTOGRAPHY WITH TELESCOPES

Instead of in an eyepiece, an image can be formed on a piece of photographic film. Development of properly exposed film then provides a permanent record of the observation. Many telescopes come equipped with photographic attachments for this purpose. In the case of a bright object like the Moon or a planet, the eye usually sees more than a photograph with the same telescope, because the eye can take advantage of moments of perfect seeing, whereas the photograph averages moments of poor seeing, which blurs the image. On the other hand, faint objects like nebulae and galaxies are usually better shown in photos, because light can be accumulated in long exposures whereas the eye cannot "store" light.

Many photos of star fields and nebulae in this book were taken with small telescopes or ordinary cameras. The reader should consult the picture captions, most of which describe the equipment and exposure used. In many cases, readers can duplicate or improve the results with their own equipment.

In large, modern observatories, the recording instruments are often electronic, instead of photographic or visual, as shown in Figure 5-21. Furthermore, computer-directed machines can point the telescope almost precisely at any known celestial object. Thus an astronomer may spend nearly the whole night observing without looking through the telescope, though he may look through a small telescope called a *finder*, mounted parallel on the side, to help find his target object, or he may view television screens that monitor the image.

PHOTOMETRY

Pictures of astronomical objects are interesting, of course, but much astronomical work requires measuring an object's brightness—the amount of light coming from it at all wavelengths or at selected ranges of wavelengths (such as blue to green). This is called **photometry**. By giving a precise measure of the amount of light at various wavelengths, photometry allows astronomers to measure temperature, composition, and other properties of a remote object. Until a few decades ago, this was done by measuring the size and density of the image on a photograph. Today it is done much more precisely with electronic devices. The basic device is a photomultiplier placed at the telescope's focus. Each photon of light collected by the telescope strikes the photomultiplier surface, which is designed to release a shower of electrons with each photon impact. In some designs, each electron strikes a surface and releases more electrons, thus multiplying the effect into a weak electric current that can be measured. Measurement of the current tells the number of photons arriving from the astronomical object, thus giving a measurement of its light intensity.

In recent years, these techniques have been extended into video-imaging devices that can produce TV images of objects too faint to photograph. Most important of

a

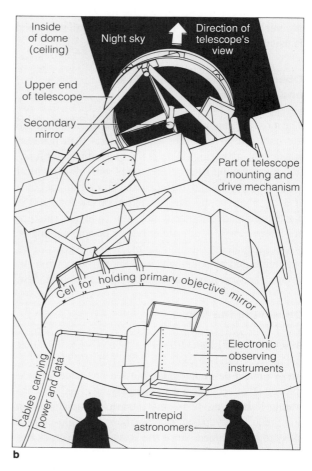

Inside of dome (ceiling)

Night sky

Direction of telescope's view

Upper end of telescope

Secondary mirror

Part of telescope mounting and drive mechanism

Cell for holding primary objective mirror

Electronic observing instruments

Cables carrying power and data

Intrepid astronomers

b

Figure 5-21 **a** A view of a modern telescope, at Mauna Kea Observatory, Hawaii. In this telescope design, light is focused through a hole in the center of the main mirror, where the astronomer mounts electronic instruments to record data. Data are fed through cables to computer equipment, which performs initial analysis and stores data for further analysis. The dome (top) is open to night air in order to give the clearest observing conditions; temperatures inside the dome often drop below freezing during observing sessions. (Photo by author.) **b** Schematic diagram of *a*.

the modern detectors are charge-coupled devices, more commonly known as CCDs. These are extremely light-sensitive detectors, made possible by microelectronic technology. They are often used in arrays of, say, 500×500 light detectors, about the size of a postage stamp. An image can be created from the resulting 250 000 "dots"; each dot corresponds to one measurement of brightness, made by one detector. Each dot is called a picture element, or *pixel*. CCD imaging arrays can record in 5 min more faint detail than photographic film exposed for 100 min in the same telescope. (For a review of CCDs see Kristian and Blouke, 1982.)

IMAGE PROCESSING

Extraordinary advances have occurred in the last decade in our ability to process images in order to gain maximum information from them. There are three main areas of interest: photographic images, digitized images, and false color images.

Photographic Images

Photographic images are composed of microscopic dots arranged in a random pattern; each dot represents a grain in the chemicals composing the film. The more light hitting a given grain, the larger and darker it appears in the final (negative) image. The negative image is then processed to make a positive print or transparency. Color films, such as Kodachrome, have several layers of light-sensitive chemicals with different color dyes to produce a color image from a single exposure. However, newer techniques allow us to combine separate black-and-white images made with different-colored filters—such as three images made with red, yellow, and blue filters in a spacecraft—to create a single color photo. Some of the color photos in this book were produced in this way. A major problem in classic photography has been that film can reproduce only a limited range of contrast. If the bright head of a comet is 10 000 times brighter than its faint tail, an exposure long enough to record the faint details of the tail overexposes the head, giving the false impression of a bright blob rather than an intense star-like point surrounded by a diffuse glow. Recent photographic techniques, such as using variable-density "masks" that screen light from the brightest regions, allow improved images of high-contrast astronomical subjects.

Digitized Images

As described earlier under photometry, some modern sensors such as CCDs have a *grid* of dots, or pixels, in which the brightness is measured for each dot. Imagine a blowup of a newspaper photo. You can see each individual pixel as a dot. Now imagine that each dot is replaced by a number from zero (blackest tones) to 100 (brightest), representing the brightness of the light at that point in the image. Such an image may come not only from a CCD; an ordinary photo can be scanned and measured and converted into such pixels, which is called *digitizing the image*. The position of each dot and its brightness is stored in a computer. Then the image can be processed in many ways. We could order the computer to make a print such that zero is the blackest tone that can be represented on a photo and 100 the brightest. Then we would preserve the entire range of brightness in the original picture, but at low contrast. Suppose we are interested in some subtle contrast features that appear only in the light gray areas between 70 and 90 on the brightness scale. Then we could ask the computer to make a new print in which the darkest tones of the print correspond to 70 in the original image and the brightest correspond to 90. Every part of the image

that was fainter than 70 now appears as black; every part that was brighter than 90 now appears as white. We now have an extreme-contrast version of all the detail between 70 and 90—which might reveal some interesting new features that we missed before.

False Color Images

Another technique is to assign different colors to each brightness level. For instance, we could use violet for the dark tones of 1 to 10, blue for 10 to 20, green for 20 to 30, and so on. Or we could divide the brightness scale even finer, with yellow for 50 to 52, orange for 52 to 54, and so on. In this way, we could produce an image in which the colors have nothing to do with the original colors but are used merely as a code to allow us to separate features of slightly different brightness. An example is seen in Figure 5-22. Such images are called *false color images*. In recent years, they have become the darling of magazine art directors who are attracted by the flashy pictorial quality of the colors. Indeed, some false color images have been published without adequate explanation, and TV journalists have been heard to exclaim over the strange, "fried-egg" appearance of Halley's comet. Such pictures had nothing to do with the actual appearance of the comet. They were merely false color images with different rings of color representing different brightness!

For such reasons, "true color" images are used as much as possible in this book, though many texts and magazines are full of false color imagery. "True color" is in quotes because at some level of detail it is difficult to reproduce all the nuances of natural color, especially in faint objects. But the colors in most images in this book do give some impression of the actual colors of the objects.

SPECTROPHOTOMETRY

As we stressed in the first part of this chapter, it is important to measure not only the total amount of light coming from an astronomical object but also the intensity at each wavelength. This process is called **spectrophotometry**. It is done by passing the light through a spectroscopic device that breaks the light into different colors before it enters the photomultiplier. Then the photomultiplier can measure the amount in each color (that is, wavelength) range. Similarly, photos or

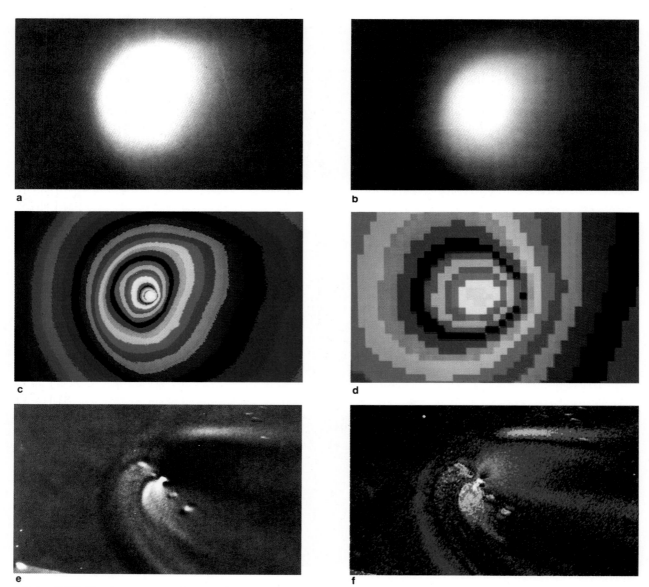

Figure 5-22 New image processing techniques are shown by these images of the head of Halley's comet. **a** A direct image of the comet's head taken by a CCD on a large telescope resembles an ordinary photo and shows an overexposed bright central condensation surrounded by glowing gas. The image has some defects, such as the streak at right, due to internal properties of the CCD.
b Addition of several images and correction of flaws makes a cleaner image. The central region is still overexposed.
c The same image in false colors, with different-colored bands used to represent different brightness levels. This image, though less realistic in terms of visual appearance, reveals the smooth increase of brightness toward the center, even within the regions overexposed on print *a*.
d An enlargement of the central part of *c* shows the individual pixels composing the image. **e** More sophisticated image enhancement has been used to exaggerate localized intensity boundaries within the seemingly uniform contours. This enhanced image reveals faint spiral jets, present in the earlier images but swamped by the comet's glow. Bright blobs are images of a star the comet passed during the several different exposures. The star is barely visible in *b* and *c*, but here it is enhanced by the processing. **f** A false color version of *e* emphasizes the spiral jet structure and the increasing brightness toward the center, but it is even further removed from the appearance the comet would have to the naked eye, since the colors are arbitrary and the overall glow around the comet's head is suppressed to reveal the curved streaks. (All images by Stephen M. Larson, Lunar and Planetary Lab, University of Arizona, from observations made at Boyden Observatory, South Africa, on the day of Giotto probe encounter; see Giotto images in Chapter 13.)

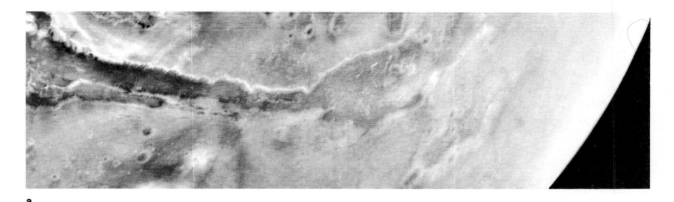

a

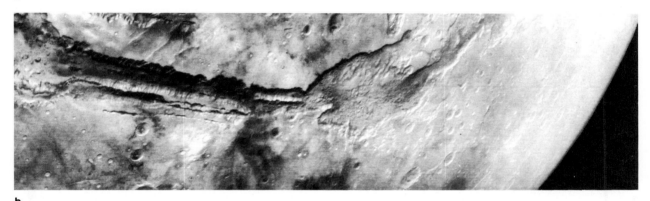

b

Figure 5-23 Visible-light and thermal infrared images of a huge canyon on Mars, photographed in 1989 by the Soviet space probe Phobos-2. **a** Image in visible light (reflected sunlight). **b** Image made at far infrared wavelengths around 11 μm, using the infrared radiation deriving from heat of the surface. Cooler areas are dark; warmer areas are bright. Note that the infrared light penetrates better through the haze at the horizon (right). Comparison of the two images reveals surface properties (see text). North is at the top. (Courtesy A. Selivanov and M. Maraeva, Glavcosmos, USSR.)

TV images can be produced showing the object as seen in different colors (i.e., different wavelengths). An interesting example comes from a 1989 Soviet probe to Mars in which two cameras took matching pictures, one using reflected visible sunlight and the other using infrared light at a wavelength of 11 μm (Figure 5-23). The first is an ordinary photo, but the second uses the infrared thermal radiation caused by the warmth of Mars' surface, so that warm spots show up as bright, while cool spots are dark. This principle is made clearer by the discussion on page 97 and by Figure 5-5 on page 98, which show that a planet radiates infrared light at a wavelength around 10 μm. Thus, the images help clarify how different Martian geologic units absorb and re-radiate the heat of the Sun. We will see other examples throughout this book.

INTERFEROMETRY

In recent years, astronomers have applied a clever technique to improve telescope performance. In effect, this trick makes it appear as if we have a bigger telescope than we really do. The technique, called **interferometry**, involves the use of widely separated telescopes in a special arrangement to increase the resolution of fine details in astronomical objects. In our earlier discussion of telescopes, we noted that the resolution of a telescope can be increased by increasing the telescope's aperture. If we could build a good-quality optical telescope or radio telescope a mile in diameter, for example, we could achieve fantastic resolution (especially if the telescope were in space, where the full resolution could be realized without being ruined by atmospheric

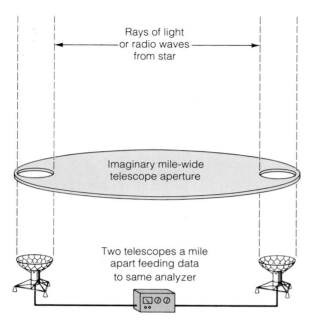

Rays of light or radio waves from star

Imaginary mile-wide telescope aperture

Two telescopes a mile apart feeding data to same analyzer

Figure 5-24 Schematic sketch of an interferometer. Two telescopes a mile apart are linked to a single detector and act like two parts of a mile-wide telescope.

shimmering). At present, a mile-wide telescope is too big to build. Cunning astronomers, however, have devised a partial solution to the problem. Imagine a mile-wide telescope covered by a mile-wide disk-shaped aluminum sheet, as in Figure 5-24. Now cut a meter-wide hole at each edge of the disk; in effect, we have a mile-wide telescope with most of its aperture blocked. It would not have the light-gathering power of a mile-wide telescope, but it would have much of the resolution.

Although it is now impossible to build a mile-wide telescope, it is easy to build two big telescopes and locate them a mile apart. The trick is that their light (or radio waves in the case of a radio telescope) must be fed into a single detector, just as all the light from a single telescope goes into one detector. The two telescopes must act like one, even though they are separated. This blending is now commonly achieved by sending the electrical signals received by the instruments on the separated telescopes to one central facility, where they are processed together using sophisticated computer techniques.

With the technique of interferometry, astronomers have been able to measure incredibly small angular details—fractions of a second of arc—in distant stars and galaxies. The technique has been especially fruitful for radio telescopes, which have made beautiful images of distant galaxies' details. One of the most advanced such observatories is the Very Large Array of the National Radio Astronomy Observatory. It includes 27 movable radio telescopes (see Figure 5-19) spread over an area up to 35 km across (Hjellming and Bignell, 1982). Astronomers in 1986 achieved the first interferometry using a radio telescope in space linked to one on the ground, thus increasing the baseline to distances wider than the diameter of the Earth (Levy and others, 1986)! This technique holds much promise for the future.

LIGHT POLLUTION: A THREAT TO ASTRONOMY

Figure 5-25 shows the pattern of urban lights across the United States as seen on a cloud-free night from space. Many observatories, especially in the West, are near rapidly sprawling cities. The glow from these cities lights up the sky, reducing the contrast between faint stars and the dark sky background. This is why it is so hard to get a good view of the sky from urban areas. Moreover, many types of fluorescent streetlights emit emission lines that interfere with astronomical work. These problems have rendered many famous observatory sites, such as Mt. Palomar in southern California, nearly worthless for the most sensitive modern instruments. A classic solution to this problem has been for astronomers to move even further toward the frontier—as when Percival Lowell established his observatory in Flagstaff, Arizona, in the 1890s or when modern astronomers send telescopes into space. Nonetheless, many cities have acted to protect the investment in nearby observatories by adopting only streetlights that emit minimally disruptive wavelengths of light and by installing hoods that block lights from shining up into the sky.

SUMMARY

Most of our knowledge of the universe comes from deciphering messages carried from extraterrestrial bodies by electromagnetic radiation. To do this we need to under-

Figure 5-25 A beautiful pattern that threatens modern astronomy. Compositing of satellite photos has produced a montage that shows North American urban lights as seen from space with no cloud cover. Skies over the Eastern megalopolis, southern California, and other regions are "polluted" with light so strongly that neither the eye nor the telescope can see the faintest stars that would otherwise be visible from these regions. (Photo by W. T. Sullivan III, courtesy National Optical Astronomy Observatories.)

Telescopes are devices that not only magnify the angular size of distant objects but also, just as importantly, enable us to gather much more light than is gathered by the eye. Thus we can see objects much fainter than visible to the unaided eye. This chapter contains advice on viewing different astronomical objects with small telescopes. Astronomers bolt various kinds of instruments on large telescopes to analyze the light from planets, stars, and so on. One of the most important instruments is the spectrophotometer, which measures the amount of light at each wavelength over a range of wavelengths, allowing precise measurement of the spectrum of the object being observed. We are just entering an era when large telescopes are being placed in orbit above the atmosphere in order to give access to unblurred images throughout all wavelengths of the spectrum.

stand the nature of the radiation and how it is generated in atomic particles. Electromagnetic radiation is released in the form of photons, each of which has its own specific wavelength and energy. The light coming from an object, arranged in order of wavelength, is called the spectrum of the object. Each wavelength corresponds to a "color," ranging from too short (too blue) to see, through visible colors such as blue and red, to wavelengths too long (too red) to see. Photons of blue light have more energy than photons of red light.

Atoms and molecules each produce absorption and emission lines and bands in the spectrum as photons interact with them. These derive from the energy level structure of their electronic orbits. Each element and compound has its own lines and bands. Materials in a distant object can be identified by obtaining the object's spectrum and identifying emission or absorption lines or bands, if present. Many gases in planetary atmospheres or stars can be identified in this way, but solid surfaces of planets can be only roughly characterized, since solid materials have only weak, poorly defined absorption bands.

CONCEPTS

wavelength

frequency

wavelike properties of light

visible light

ultraviolet light

infrared light

speed of light

particlelike properties of light

photon

diffraction

spectrum

spectroscopy

fields

electromagnetic radiation

continuum

temperature

absolute temperature scale

thermal motion

thermal radiation

reflected radiation

Wien's law

nonthermal radiation

emission

energy level

emission line

ground state

excited state

emission band

absorption

absorption line

absorption band

ozone layer

Hubble Space Telescope

telescope

magnification

resolution

light-gathering power

refractor

reflector

compound telescope

objective

focal length

aperture

seeing

photometry

spectrophotometry

interferometry

PROBLEMS

1. Photographic films are usually more efficient when measuring more energetic photons. Which would you expect to be easier to photograph: a faint blue, red, or infrared star?

2. The ground cools at night by radiating thermal infrared radiation into space. Explain why the air stays warmer on a moist, cloudy night than on a very dry, clear night.

3. A spacecraft passes close to a hitherto unknown planet. A camera system on the spacecraft photographs surface details quite clearly in all wavelengths except the absorption bands of carbon dioxide. At these wavelengths, the image is featureless. Interpret these results.

4. Why has there been more emphasis on getting ultraviolet and infrared telescopes into orbit than on getting radio telescopes into orbit?

5. You are considering buying a pair of binoculars. Two available pairs each have lenses with 50-mm aperture, but one has a magnifying power of $7 \times$ and the other, $20 \times$. Which pair:
   a. Gives the larger apparent image size?
   b. Gives a greater apparent brightness in the observed image?
   c. Is easier to hold still enough to observe? (Remember that the unsteadiness of your hands is magnified as much as the image.)

6. For your birthday, your parents offer you a choice of lenses for your 35-mm camera. One has a focal length of 24 mm and the other has a focal length of 90 mm. (Focal length numbers are marked on the inner rim of all camera lenses.)
   a. Which one would give a bigger image but cover a narrower field of view in terms of degrees?
   b. Which one would give a smaller image but cover a wider angular field of view?
   c. Which one would be called a telephoto lens?

7. a. Why can an atom in the ground state not produce emission lines? **b.** Use this fact to explain why the colorful emission-line glows of certain nebulae occur near hot stars, but not in the cold gas of interstellar space.

ADVANCED PROBLEMS

8. Confirm that the thermal infrared radiation in Figure 5-14 could come from a planetary surface at a temperature of 145 K.

9. A planetary astronomer has an instrument that scans the spectrum from 5 to 10 μm wavelength. Looking at a certain moon, he detects thermal infrared radiation that increases from 5 to 10 μm but does not peak in this range. Apparently it peaks at a longer wavelength than 10 μm. What can you say about this moon's surface temperature?

10. What wavelength of thermal radiation is being emitted with greatest intensity by this book?

11. Your friend's Newtonian reflector telescope has a focal length of 70 in. and he has an eyepiece with a focal length of ½ in.
   a. What magnification does the telescope give?
   b. With the preceding information, discuss what can be said about the brightness of the image given by this telescope.
   c. If you look through the telescope at the Moon, what angular size would the Moon appear to be?
   d. Another friend appears and asks how big a lens the telescope has. How would you answer?

12. Two telescopes each have a 4-in. aperture, but one has a short focal length and gives $30 \times$ magnification and the other has a long focal length and gives $150 \times$ magnification.
   a. Which one would give a better view of the planet Mars, which subtends an angle of about 20 seconds of arc?
   b. Which one would give a better view of the faint, wispy Andromeda galaxy, which subtends as much as 3°?

PROJECTS

1. Compare views through binoculars of different sizes. (Binoculars carry a designation such as 7 × 35, where the first number is the magnifying power and the second is the aperture.) Confirm that at fixed power, larger apertures give brighter images.

2. Observe and sketch a rainbow. Confirm that the colors appear in order of wavelength as seen in Figure 5-1.

3. Study a room light equipped with a dimmer switch. Turn the switch from bright to the faintest visible setting. Note the redder color of the light bulb at the dimmest position. Relate this effect to Wien's law.

4. Experiment with two lenses having different focal lengths, such as 20 cm and 5 cm. Measure their focal lengths by focusing sunlight on a surface. Then (repeating experiments that must have been done first by European lensmakers around 1600) hold or mount the two lenses to make a simple refracting telescope like that in Figure 5-20. Tubing is not essential. Estimate your telescope's magnification by comparing the magnified image with a naked-eye view, and then compare with the formula on page 109. REMINDER: DO NOT LOOK DIRECTLY AT THE SUN WITH ANY TELESCOPE.

Exploring the Earth–Moon System

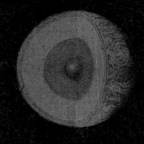

The only photo showing Earth and its moon from a great distance was made by the Voyager 1 spacecraft from a distance of 7 million miles as it left on its flight to Jupiter and the outer solar system. Background stars are too faint to be recorded in space photos. (NASA.)

CHAPTER 6

Earth as a Planet

Some readers might wonder, "Why discuss Earth in an astronomy book? Why not relegate this material to a course in geology?" The answer is dramatized by Figure 6-1. The soil in your backyard is just as much a sample of planetary material as a piece of the Moon. People have taken a long time to realize this, and many people still don't recognize the implications of the astronomical discovery that we are spinning through space on a small, finite ball.

When Renaissance astronomers realized that other planets were worlds, they assumed that they closely resembled Earth. This led to the idea of "the plurality of worlds"—the assumption that other planets had climates and life forms like Earth's. Now we realize this is only partially true. Our spaceships have brought us photos of other worlds, showing lonely lava flows, rugged craters, dusty winds, and scudding clouds. We realize now that Earth is just one planet out of many, one natural laboratory in which one set of "input" conditions has acted to produce one set of rock types, one kind of climate, and a certain set of life forms. If we can understand how Earth's conditions have evolved, we will be better equipped to understand how other input conditions have produced different environments on other planets.

Conversely, astronomical exploration of the planets has helped us evolve a new picture of Earth. We now know that a planet's surface is not a static platform but a dynamically evolving crust. It may be pummelled by meteorites and blanketed with lava flows and, in Earth's case, crumpled by forces arising in the interior and eroded by flowing water, as exemplified in Figure 6-2. Its climate may be affected by subtle variations in solar radiation and, in Earth's case, by pollutants. By studying Earth as one of the planets, we are discovering how these processes work and learning how Earth displays similarities to, as well as differences from, other worlds.

Figure 6-1 Earth as a planet. The crescent Earth was photographed by Apollo 12 astronauts on their way to the Moon in 1969. They exclaimed on the beauty of the scene as the Sun emerged from behind Earth. (NASA.)

Figure 6-2 Earth's mountain ranges are a unique feature of our planet. Crumpling of the crust creates uplifted areas, and subsequent erosion by water and wind creates types of sharp canyons and peaks not found on other worlds. Water, especially, carves valleys (right center) and deposits the eroded sediments at the valley mouth (light triangular deposit). (French Alps above Val d'Isere; photo by author.)

EARTH'S AGE

What makes Earth evolve? How did it come into being? How old is it?

A variety of astronomical and geologic evidence indicates that the Earth was formed about 4.6 billion (4.6×10^9) y ago from particles orbiting the Sun and that Earth displays many ongoing evolutionary processes caused by internal heat and external forces. Chapter 14 will detail the process that formed Earth and all the other planets; here we emphasize evolutionary processes that will help us understand the conditions we find not just on Earth but also on other planets.

Debates About Earth's Age

Our view of Earth as an evolving planet of great age has not come easily. Until a few centuries ago, Earth's age was hardly even considered, except for vague ideas that it was very old. In 1646, the English scholar Sir Thomas Brown wrote that determining Earth's age "without inspiration . . . is impossible and beyond the Arithmetick of God himself." Around 1650, several other scholars hypothesized that Earth's age and history could be deciphered from references in ancient scriptures.

Archbishop James Ussher used this method and calculated that the whole cosmos was formed on Sunday, October 23, 4004 B.C., and that humanity was created on Friday, October 28.

However, also in the seventeenth century, naturalists realized that if sediments accumulated at the rate measured at the mouths of European rivers, many more millennia would have been required to accumulate the sediment deposits actually observed.

Even as evidence for Earth's great age accumulated, the idea of geological evolution, or change, was as controversial as the later question of biological evolution. Many naturalists argued that all landforms were created by sudden catastrophes, like earthquakes, rather than by very long term processes. As late as the 1700s other people argued that fossils and strata deposits interpreted by geologists as signs of evolving climates and landforms were "devices of the devil" put in rocks to mislead us. Even today some "creationists" argue that fossils and other geological features were created and put in the ground all at once. By the same logic one might imagine that all "ancient" artifacts and history books were fabricated and placed in museums in, say, 1836. Scientists reject such views on the grounds that they have little productive consequence; they don't lead us toward new observations or possibilities for new discoveries.

Scientists who studied our planet from the 1500s to the 1800s were influenced by medieval conceptions that Earth started in an Eden-like state (Davies, 1969): "a magnificent, lush estate which God had stocked with everything necessary for human well-being before admitting man as a freehold tenant." Those who accepted Earth's evolution felt that it had been a process of shriveling, as an apple dries out. Mountains were evidence of Earth's decline in old age, "even as warts, tumours, wenns, and excrescencies are engendered in . . . men's bodies." This view even affected perceptions of the landscape. When John Dennis crossed the Alps in 1693, he reported not the majestic panoramas that impress us, but only "vast, but horrid, hideous, and ghastly ruins."

The new concept of changing geological features fascinated Victorian thinkers. Tennyson wrote, "Where the long street roars hath been the stillness of the central sea . . . the solid lands, like clouds they shape themselves and go."

Study of landscapes and rocks during the 1800s and 1900s contradicted theories of sudden catastrophes or mere passive shriveling as mechanisms forming most geological features. For example, study Figure 6-2. Several thoughts emerge. First, such a smooth fold would not result from a sudden catastrophic compression. Rocks shatter in such an event. This fold suggests a slow evolution of Earth's landscapes. Second, at measured deformation and erosion rates, Earth must be much more than a few thousand years old to allow time to produce such folds and then to allow erosion to expose them. Third, enormous forces must be at work to compress and crumple regions the size of mountain chains. Photographs of Earth taken from space enable us to sense this more clearly as seen in Figure 6-3. At the same time, observations of volcanoes, earthquakes, movements along fractures, and other features show that mountain formation continues even today. Earth is still actively evolving.

Measuring Rock Ages by Radioactivity

All questions of the Earth's age and evolution rates were clarified by the discovery of a reliable technique to determine the ages of rocks. This subject might seem unrelated to astronomy, but remember that Earth's rocks, like lunar rocks and meteorites, are cosmic materials. We live on a large space-rock.

In 1896 the French physicist Antoine Becquerel accidentally left some photographic plates in a drawer with some uranium-bearing minerals. Later he opened the drawer and found the plates fogged. Being a good physicist, he did not dismiss the event but investigated. He found that the uranium emitted "rays," which, like Roentgen's X-rays of 1895, could pass through cardboard. The rays turned out to be energetic particles emitted by unstable atoms. Radioactivity had been discovered.

A **radioactive atom** is an unstable atom that spontaneously changes into a stable form by emitting a particle from its nucleus. The original atom thus becomes either a new *element* (change in the number of protons in the nucleus) or a new *isotope* of the same element (no change in the number of protons but a change in the number of neutrons). The original atom is called the **parent isotope** and the new atom is called the **daughter isotope**.

The time required for half of the original parent isotopes to disintegrate into daughter isotopes is called the **half-life** of the radioactive element. If a billion atoms of a parent isotope were present in a certain mineral grain in a rock, a half billion would be left after one half-life, a quarter billion after the second half-life, and so on. As examples, half the rubidium-87 atoms in any given sample change into strontium-87 in 50 billion years; uranium-238 changes (after a chain of successive decays) into lead-206 with an effective half-life of 4.51 billion years; potassium-40 changes into argon-40 with a half-life of 1.30 billion years; and carbon-14 decays into nitrogen-14 with a half-life of only 5570 y.

Early in the twentieth century, physicists realized that here was a way to determine the ages of rocks—the date when a given rock formed. Suppose we could determine the *original* number of parent and daughter isotope atoms in a rock. In part, this can be done by measuring numbers of stable isotope atoms, which usually occur in certain proportions to the unstable isotope atoms in a given fresh mineral. Then, if we simply count the *present* numbers of parent and daughter isotope atoms

Figure 6-3 An earthquake zone: Los Angeles from orbit. This view from the Landsat unmanned satellite uses infrared light. Since infrared light is not visible to the eye, image processing must be used to render the scene. The processing makes vegetation appear red (because it reflects brightly in the infrared), and water is black. The San Andreas fault is the left–right linear border on the north side of the San Gabriel Mountains, north of Los Angeles. For millions of years, the lower half of the region has slid leftward (northwest) in small jumps during earthquakes. The gray V (upper right edge) is a good example of sediments eroding out of the mountains, depositing a V-shaped alluvial fan. (NASA.)

in the rock, we can tell how many parent atoms have decayed into daughter atoms and hence tell how old the rock is. If half the parent atoms have decayed, the age of the rock equals one half-life of the radioactive parent element being studied. This technique of dating rocks is called **radioisotopic dating**, since radioactive atoms that decay are called radioisotopes.

Note that the quantity being determined in radioisotopic dating of rocks is the time since the rock began to retain the daughter element. In most cases being discussed here, this is the time since the rock solidified from an earlier molten material, such as lava. Once the rock solidifies, any daughter isotope atoms are trapped in the solid mineral structure.

Measuring Earth's Age

To measure Earth's age, you might expect that we would merely search for the oldest known rock on Earth and assume that it dates from the Earth's creation. In reality

it is not so simple. Earth is geologically so active that rocks dating from Earth's formation have long since eroded away. For instance, many rocks in the Rocky Mountains formed about 60 million years ago or less; many rocks in the eastern United States formed a few hundred million years ago. Older, more stable parts of North America exist in Canada, where **Earth's oldest known rocks** have been found, dating from 3.96 billion years[1] ago. Such ancient, stable regions are called **continental shields,** after their flat, circular shapes. Several shields in different parts of the world have yielded rocks 3.5 to 3.9 billion years old. In the Australian shield,

---

[1] Expressing geological ages is a continuing editorial problem. This book uses the most common American convention: 1 billion years $= 10^9$ y $= 1\ 000\ 000\ 000$ (abbreviated "1 b.y."). This would be fine except that the English use 1 billion to mean 10^{12}! Therefore some authors have adopted 10^9 y $= 1$ aeon (or eon). Other authors have adopted the international prefix "giga-," which stands for 10^9, and they use 10^9 y $= 1$ Gy (see Appendix 1).

a few isolated crystals as old as 4.3 billion years have been found, but no complete rocks that old have been found; they have been destroyed by erosion.

These results mean that Earth must have formed more than 4.3 billion years ago. Without additional information, we would not be able to fix the age exactly. As we will see in the next few chapters, however, more information comes from outside the Earth. Lunar rocks and meteorites show that *all the planets formed within a relatively short interval (about 50 to 90 million years) about 4.6 billion years ago* (Pepin, 1976). The **age of Earth** is therefore believed to be 4.6 billion years.

EARTH'S INTERNAL STRUCTURE

The world's deepest drill hole is in the Soviet Union, 250 km north of the Arctic Circle (Kozlovsky, 1984). It reached a depth of 12.5 km in 1989, on the way to a planned 15 km. It yields interesting data on deep rocks, but is not deep enough to reveal Earth's deep internal structure. Our best clues about the interior come, instead, from earthquake waves that pass through Earth's material. If you make a small splash in a calm swimming pool, waves radiate across the surface from the disturbance. Observers at other edges of the pool can gain information about the disturbance by observing the waves—for example, they can locate the disturbance by comparing the directions from which the waves come.

In the same way, waves are generated in Earth by earthquakes. Waves traveling through the Earth (or other planets) are called **seismic waves**. Some travel along the surface and others penetrate through the Earth. The velocity and characteristics of the waves depend on the type of rock or molten material they traverse. For example, as shown in Figure 6-4, waves that travel through Earth from an earthquake on the far side traverse deep interior regions. An earthquake produces several different types of waves, and some of these are unable to pass through liquids. Since the outer part of the core is molten, these waves are blocked by the core. As Figure 6-4 indicates, such waves from an earthquake at A might be seen by the observer, but from an adjacent earthquake at B they would not be seen. The presence of the Earth's core was discovered in this way.

Such studies have revealed two important types of layering in the Earth: chemical and physical. Chemical layering refers to layers of different composition. Phys-

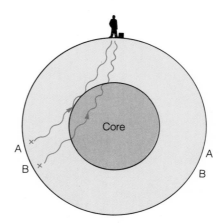

Figure 6-4 Earth's core was discovered in 1906 when observers found that they received modified seismic waves from earthquakes on the far side of the Earth (beyond B from the observer) compared to waves of nearer earthquakes (closer than A). Waves from B and beyond are altered by passing through the core, and certain types of waves from B and beyond do not even reach the observer (see text). The wave paths are curved because rock layers at different depths have different properties.

ical layering refers to layers of different mechanical properties, such as rigid layers versus fluid layers.

Chemical Layering of Earth: Core, Mantle, Crust

Chemical layering was the first type of layering recognized. Seismic and other data indicate that the Earth contains a central **core** of nickel-iron metal. The core's radius is about 3500 km, just over half the Earth's radius, as shown by the dark material sketched in Figure 6-5.

The core is surrounded most of the way out to the surface by a layer of dense rock, called the **mantle**. Near the surface, the densities of the rocks are typically lower. The **crust** is a thin outer layer of lower-density rock about 5 km thick under the oceans and about 30 km thick under the continents.

The core–mantle–crust structure gives us important clues about the history of the Earth and other planets. First, it shows the importance of **differentiation** processes—processes that separate materials of different composition from one another. Most geologists believe that the key differentiation process in the Earth was the melting of much of the inner rock material after the Earth formed. The source of the heat was

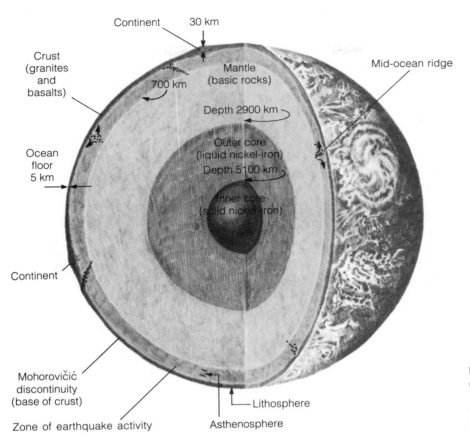

Continent   30 km

Crust
(granites
and
basalts)

Mantle
(basic rocks)

Mid-ocean ridge

700 km

Depth 2900 km

Outer core
(liquid nickel-iron)

Depth 5100 km

Inner core
(solid nickel-iron)

Ocean
floor
5 km

Continent

Mohorovičić
discontinuity
(base of crust)

Zone of earthquake activity

Lithosphere

Asthenosphere

Figure 6-5 A simplified diagram
of the Earth's interior structure as
revealed by modern seismic
studies. Dots indicate typical
earthquake positions.

radioactive minerals trapped in the Earth as it formed. Gradually those minerals released heat as radioactive atoms decayed. The interior of the Earth was so well insulated by overlying rock that the heat could not escape. The temperature rose until the rock melted. When the rock melted, heavy portions like metals were able to flow downward toward the center, while lighter, low-density minerals could float toward the surface, where they eventually solidified into a crust of low-density rock.

As a more detailed example of this theory, we note that one of the lowest-density and most common minerals to form in cooling, molten rock is called *feldspar*. If the theory is right, feldspar should have formed and floated toward the surface. Indeed, we find that surface rocks of both the Earth and Moon are extremely rich in feldspar. One of the rock types formed from mixtures of minerals rich in feldspar is called **basalt**. Basalt is the common lava that erupts from volcanoes that tap the crust and upper mantle, and basalt is also common on the Moon's surface.

A still lower-density type of rock is **granite**, the light-colored rock commonly formed from molten materials in continents. In some ways, continents seem to be a low-density granitic "scum" floating on the denser rocks of the basaltic lower crust and upper mantle.

A simple example of differentiation that can produce such layering occurs during the smelting of metal ores. When the ore is melted, the metal sinks to the bottom of the vat (core) while the bulk of the rock fills the upper part of the vat (mantle). On the surface floats a thin scum of slag, or low-density rock (crust). Note that certain amounts of metals are chemically attracted to minerals in the crust and hence tended to stay with those minerals as they formed. Thus we can have iron and other dense ores in the crustal rocks, even though much of Earth's iron is in the core.

Support for these ideas comes from meteorites, the stone and metal fragments that fall onto the surface of Earth from space. These will be discussed further in Chapter 13, but we note here that they are believed to be fragments from the surfaces and interiors of small interplanetary bodies broken during collisions. Interestingly, they show many of the rock types we have discussed: Nickel–iron meteorites are probably metal fragments from cores of these small worlds, similar to Earth's core. Basaltlike meteorites are probably fragments of mantles and crusts. Thus we see that many worlds may have gone through a differentiation process that led to structures like the Earth's.

Physical Layering of Earth: Lithosphere and Asthenosphere

A second type of layered structure has been discovered involving layers of different rigidity. This type of layering is extremely important in determining what types of features we see in the landscapes around us on Earth, and on other planets, as we will see in a moment.

Layering by rigidity was created during the cooling of Earth. As you can imagine from the example of the smelter vat, the interior stays molten for a long time because it is hard for the heat to get out, but the surface cools fast because it is exposed to the atmosphere and surrounding space, and the heat can easily radiate away. Thus if a planet were melted and left to cool, a solid layer would form on the surface, underlain by a still-molten liquid layer. This picture is complicated by the fact that the layers at different depths have different compositions (due to the differentiation discussed above) and are also at different pressures. They may thus have different solidifying temperatures, and they may actually form alternating layers of solid material, molten material, or partly molten slushy material. A partly molten layer can by visualized as a layer in which some low-melting-point minerals are melted, but high-melting-point minerals are not. (For example, if you heated a mixture of ice and sand, you would get a mixture of solid sand grains with water between them.)

The solid layer at the surface of Earth or any other such planet is called the **lithosphere** (from Greek roots meaning "rock shell"), and it is underlain by a partly melted layer that is much less rigid and less strong. As shown in Figure 6-5, the underlying partly melted layer is called the **asthenosphere** (from Greek roots for "weak shell").

The reason the lithosphere and asthenosphere are important is that their interplay determines the surface features of Earth, such as mountains, seafloors, and continents. To understand this, we need to review some principles of heat flow—to describe how heat gets out

of the hot interior of Earth or any other planet, once the planet is heated by radioactivity. Heat always flows from hot to cooler regions, and it can flow by three methods. The first is radiation; an example is the radiant heat that reaches us through space from the Sun. The second is conduction; an example is the flow of heat through a metal cooking pot, whose handle might grow too hot to touch. The third is convection, which is heat flow by movement of the heat-carrying medium; an example is the buoyant ascent of warm air in a thundercloud or the rise of a hot-air balloon. If you put an inch-deep layer of cooking oil in a pan and heat it from below on a stove, you may be able to see patterns of currents, called convection cells, set up in the oil as it convects.

The relation of these ideas to a planet is as follows. Once a planet heats up, its insides cool by all three processes. Heat radiates from the surface into space; heat is conducted outward from the hot center through rock; and convection may occur in some of the more fluid layers where currents can flow, like the convection cells in the cooking oil. As the planet cools, the surface layers solidify, forming a relatively rigid surface layer of rock—the lithosphere. In the Earth, the relatively rigid lithosphere is a layer about 100 km deep, and the relatively fluid asthenosphere is the underlying layer from about 100 to 350 km deep. Even though the asthenosphere is not totally molten, it seems to be plastic enough for sluggish convection currents to flow. These currents bring up hot mantle material, creating "hot spots" in the Earth's crust where volcanoes are likely. The currents also drag on the underside of the lithosphere, setting up stresses in it and tending to crack it into separate, large-scale pieces.

Just like arctic explorers crossing ice floes that are floating on the ocean, we are living on a rigid layer that "floats" on a more fluid base. Of course, we are not in quite so much danger of a cracking floor as the arctic explorer, but cracks do occur. Figure 6-6 shows a result. As the asthenosphere shifts, it can stretch the lithosphere only so far before the brittle lithosphere cracks. This is what we perceive as an earthquake.

LITHOSPHERES AND PLATE TECTONICS: AN EXPLANATION OF PLANETARY LANDSCAPES

Tectonics is the study of movements in a planetary lithosphere, such as the movements that cause earthquakes,

Figure 6-6 Occasionally nature reminds us that we are living on a thin, brittle crustal planetary lithosphere that is often broken by motions of underlying semifluid layers. This devastation was caused in 1906 in San Francisco by an earthquake and resultant fire. The quake involved movement on the San Andreas fault, which passes under the downtown area. (Photo courtesy S. M. Larson.)

mountain building, and so on. For more than a century, geologists studied the Earth's tectonics without recognizing the underlying nature of the forces that cause these movements, as just outlined. Only since the 1960s have geologists pieced together a real understanding of how these principles affect the Earth and how they apply to other planets (Hurley, 1968). This new understanding is called the theory of **plate tectonics**.

As the asthenosphere drags on the more brittle lithosphere, it cracks the lithosphere into large, continent-scale pieces called *plates*. Further asthenosphere movements tend to drag and jostle the floating plates, sometimes pulling them apart from each other and sometimes pushing them into each other. Cracks along the margins of plates are usually the sites of volcanoes and earthquakes, since molten magma from below can squeeze up to the surface through the cracks, and since plate collisions cause stresses as plates rub together, eventually leading to rock fracturing in the form of earthquakes. This can be seen in a map of volcanoes and earthquakes, such as Figure 6-7. The string of shallow earthquakes down the mid-Atlantic (marked also by volcanoes of Iceland and the Azores) marks a plate boundary where new lava is rising from below. On the

ocean floor it erupts and piles into a feature mapped by oceanographers in the 1960s—the Mid-Atlantic Ridge. Eruption of new lavas at these sites pushes apart the neighboring plates, causing the Americas to drift westward and Europe eastward relative to the plate boundary.

On the west coast of the Americas, the American continental plate collides with the Pacific plates, crumpling and riding up over them as the margins of the Pacific plates slide under the American plates. This crumpling is the explanation of major western American mountain chains like the Sierra Nevada and the Andes. Contorted structures at depths down to 15 km where plate boundaries collide have been mapped with new techniques (Mutter, 1986).

Such zones are common sites of earthquakes, as California residents can attest. Fractures caused by earthquakes are called **faults**. Colliding plate regions are laced with faults like the famous San Andreas fault, seen from space in Figure 6-3. Many earthquakes are caused by movements along these faults. As shown by Figure 6-6, the devastation can be extreme. Although earthquake prediction is only in its infancy, the U.S. Geological Survey has predicted a moderate to large earthquake on the central San Andreas fault by 1992

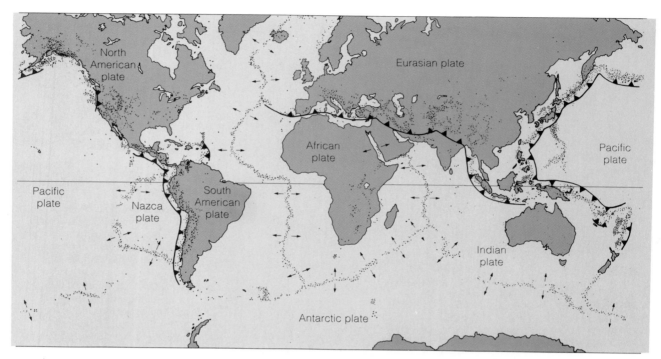

Figure 6-7 A "snapshot" of tectonic activity in Earth's present-day lithosphere. Each dot represents a recorded earthquake. Earthquakes cluster along margins of plates, where tectonic forces cause the most activity. There are two kinds of margins. The lines with arrows on each side are spreading centers, usually mid-ocean ridges, where new magma wells up from the mantle. It pushes the crustal surface in the direction of the arrows across the plate and causes continental drift, whose rates are actually measured by geophysicists. At the far margin, shown as lines with teeth, the material plunges back down into the mantle at an angle under the margin of the adjacent plate. Teeth are on descending side of line.

(near Parkfield, between Los Angeles and San Francisco) and have instrumented the area in hopes of learning more pre-earthquake warning signs. They expect a still larger earthquake in the Riverside–Palm Springs area of southern California by about 2020 (Silberner, 1987).

Other scenes of plate tectonic activity can be glimpsed in Figure 6-7. For instance, the plate containing the subcontinent of India years ago drifted northward, as shown in Figure 6-8. It collided with the Asian plate and caused the massive crumpling that created the Himalayas, marked by shallow earthquakes.

Much geological evidence supports the theory of plate tectonics. Matches of geological provinces of equal date and rock type in eastern South America and western Africa show that they were once part of the same landmass, as seen in Figure 6-8. Traces of glaciation and other evidence in southern Africa indicate that this landmass drifted north from a position once much closer to the icebound south pole. The theory of plate tectonics also explains why rock units on Earth older than 1 or 2 billion years are so rare. Most older surfaces have been crumpled beyond recognition or driven downward under other plates, only to be remelted, mixed with mantle material, and perhaps reerupted as new lavas. Some of these complex processes of Earth's active geology can be seen in cross section in Figure 6-9. They are well reviewed by Maxwell (1985).

Now we can see a connection with the landscapes on other planets, which will be discussed further in later chapters. Smaller worlds, like Mars and the Moon, do not have well-developed crumpled mountain ranges or

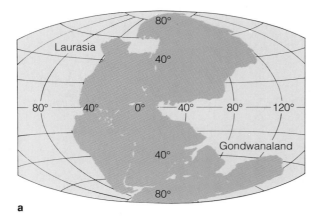
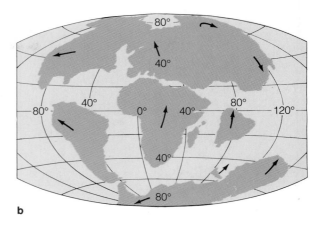

a   b

Figure 6-8 Two "frames" in a time-lapse movie of Earth's surface, showing the dramatic movement of continental land masses due to plate tectonic motions in the last 4 percent of Earth's history. Geologists are not certain of the motions in earlier times, though there is evidence of previous collisions and splittings. **a** 170 million years ago, Pangaea, composed of two major regions called Laurasia and Gondwanaland, was just beginning to split apart. **b** 70 million years ago, the Atlantic was widening and present landmasses were beginning to take shape. India had not yet collided with Asia to make the Himalayas. (After Ingmanson and Wallace, *Oceanography*, Wadsworth Publishing, 1985.)

plate boundaries shown in Figure 6-9. The reason goes back to the basic principles of heat flow and lithosphere formation. Smaller worlds cool faster than big worlds. Therefore, their lithospheres get thicker in the same amount of time. Thus their surfaces are more stable and more protected against asthenosphere currents far below. Lava does not so frequently gain access to the surface. Convection can't so easily drive plates apart or cause them to drift into each other. Therefore, the surfaces of Mars and the Moon have much more ancient structures (such as ancient impact craters) than Earth does.

OTHER IMPORTANT PROCESSES IN EARTH'S EVOLUTION

Volcanism is the eruption of molten materials from a planet's interior onto its surface. On Earth, the asthenosphere contains pockets of partly melted materials, as indicated by seismic wave analysis. This underground molten rock, called **magma**, is under pressure, often charged with gas such as steam, less dense than surrounding rock, and highly corrosive. Therefore it tends to work its way to the surface, especially in regions where fractures provide access. When it reaches the

surface, it erupts and is then called **lava**. It may shoot in a foamlike form into the air (due to pressure from dissolved gas, like the spray from an agitated can of pop when it is opened) or ooze out and flow for many kilometers across the ground. If enough lava is erupted, it may accumulate into volcanic mountains. During intervals ranging from years to millions of years, volcanism thus creates landforms ranging from flat lava flows to craters and volcanic peaks, as seen in Figures 6-10 and 6-11.

Space exploration has shown that volcanism is one of the most important processes forming landscapes on other planets. Some planets have huge lava flows and volcanoes. Study of the Earth's volcanoes helps us understand these alien landscapes. Conversely, study of the other planets' volcanic features helps us understand relations we see among terrestrial volcanoes.

Among all known planets, the Earth undergoes the most active processes of landform destruction. Largely, this activity is due to the Earth's thick atmosphere and flowing water, which other planets lack. **Erosion** includes all processes by which rock materials are broken down and transported across a planet's surface; such processes include water flow, chemical weathering, and wind-blown transport of dust. **Deposition** includes all processes by which the materials are deposited and

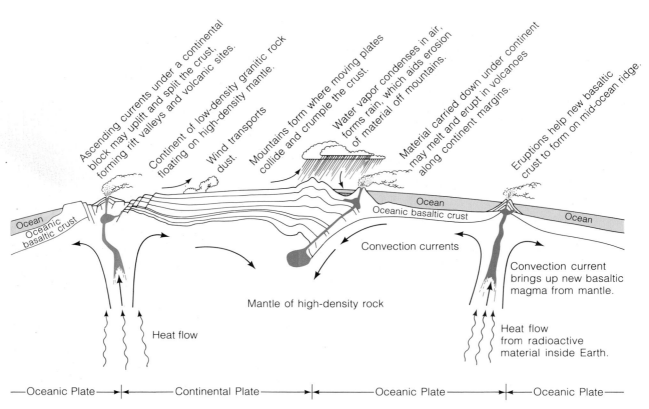

Figure 6-9 Schematic diagram of some of the processes making the Earth more geologically active than most other worlds. Convective heat flow from the interior drives convection currents in the asthenosphere. These in turn crack the more rigid surface layers and move plates. Plate movements cause earthquakes, faulting, the formation of rift valleys (left), and crumpled mountain chains (right). In addition, atmospheric processes and water flow cause rapid erosion. Vertical scale is exaggerated for clarity.

accumulated; such processes include deposition of sediments in lake and ocean bottoms and dropping of wind-blown dust in dune deposits.

On the Earth, erosion and deposition, especially by flowing water, are the most important landscape-forming processes, but this is not true on all other planets. The extreme activity of erosion and deposition on Earth is responsible for wearing away or covering up the most ancient rocks and for wearing away ancient impact craters so rapidly that they were not even recognized on Earth until the last few decades. Similarly, most familiar mountain ranges do not display their initial forms, which may have been caused by volcanism, fracturing, and folding. Instead, most mountain ranges are merely resistant rock "cores" left after uplift and erosion of more massive units of crustal strata.

EARTH'S MAGNETIC FIELD

Physicists use the term **field** to describe any property that can be measured throughout a volume of space. For instance, we know Earth's gravity field extends far out beyond the Moon because we can measure its effects on the motions of the Moon, spaceships, and other planets. Similarly, if we move a compass throughout the region around Earth, we discover that there is a **magnetic field** defined by the fact that the compass is deflected in a direction roughly toward the North Pole. The compass does not point exactly at the true North Pole (defined by the Earth's spin) but at a point called the magnetic North Pole, which moves slightly from year to year and is presently located among arctic islands off northern Canada.

Figure 6-10 Volcanic activity on Earth builds varied landforms. This crater formed as the ground collapsed around an erupting vent; its light-colored floor is filled with windblown dust. It is on the flank of a shield volcano, a type found on several planets and named for its profile, as seen on the horizon. The rest of the landscape is covered by overlapping lavas and cinders and is dotted by volcanic cones. (Pinacate volcanic field, Mexico; photo by author.)

Figure 6-11 During Earth's history, eruptions of lava have created new landmasses, pushing back the sea even as the sea eroded coastlines of old landmasses. (Photo by author; 1988 Kalapana lava flow, Hawaii.)

What causes Earth's magnetism? The interior of Earth cannot be magnetized like a giant bar magnet, because the interior regions are too hot to maintain magnetism. Rather, geophysicists believe that liquid iron in the outer core is slowly convecting, and that these motions set up electric currents in the metal. A magnetic field will exist around any electric current, and Earth's field is thus attributed to currents in the liquid metal core. The presence or absence of a magnetic field thus becomes an important indicator of the presence of an iron core inside a planet.

EARTH'S ATMOSPHERE AND OCEANS

Just as Earth's interior and surface have dramatically evolved, so have the atmosphere and oceans—the linked layers of gas and fluid on the surface. The original atmosphere, called the **primary atmosphere**, had less

oxygen than today's atmosphere. Evidence for this is that sediments deposited more than about 2.5 billion years ago are less oxidized than modern sediments, indicating less oxygen was in the air. In addition, the earliest fossil life forms, dating from about 2.5 to 3.5 billion years ago, are types of algae found today only in oxygen-poor environments, such as salt marshes along seacoasts.

Indeed, most of the Earth's present air is regarded as a **secondary atmosphere**—gas that was not originally present but was added by outgassing of the Earth's interior from volcanic vents and fissures, as shown in Figure 6-12.[2] Measurements of such volcanic gases show that they are rich in water vapor (H_2O), carbon dioxide (CO_2), and nitrogen (N_2). Today's atmosphere is 76%

---

[2] The primary atmosphere was probably rich in hydrogen (H) compounds, such as methane (CH_4), ammonia (NH_3), and water (H_2O), since hydrogen dominated the gas in the early solar system (see Chapter 14).

Figure 6-12 Gases emitted by volcanic fissures add water vapor, carbon dioxide, nitrogen, and other chemicals to the atmosphere. Volcanism was a major factor in altering the Earth's original atmospheric composition. Yellow deposits are crystals of pure sulfur condensed from the hot, sulfur-rich gas on the cool surface rocks, just as water vapor from a teakettle may condense on a nearby window. (Hawaii Volcanoes National Park, Hawaii; photo by author.)

nitrogen and 23% oxygen by weight. The proposed evolution from volcanic gases to the present atmosphere is described below.

The water vapor emitted by the earliest volcanism condensed into liquid water and eventually formed oceans. Because of Earth's unique temperature regime between 0° and 100°C, Earth is the only planet with vast expanses of liquid water, as Figure 6-13 reminds us.

The carbon dioxide that was emitted at the same time mostly dissolved in the oceans, where it formed a weak carbonic acid solution. (Most of the carbon dioxide emitted by modern industrial processes does the same thing, with only a small portion of the total remaining in the air.) The carbonic acid attacks rocks and sediments of the seafloor, eventually forming carbonate rocks. Through this direct chemical process, as well as by biological processes, most of the carbon dioxide has thus been tied up in carbonate rocks such as limestone. The nitrogen (N_2), being chemically inactive, remained in the air, forming the nitrogen observed today.

A small amount of water is always in the air as water vapor, and molecules of water are sometimes struck by powerful ultraviolet rays of sunlight and broken into hydrogen and oxygen atoms. The hydrogen, being the lightest element, tends to rise and eventually escape into space. The oxygen remains behind. Until about 2 or $2\frac{1}{2}$ billion years ago, this was the only source of oxygen, and the atmospheric oxygen content was increasing slowly. After this time, abundant plant life evolved, consuming carbon dioxide and releasing oxygen. The atmospheric oxygen content thus increased more rapidly.

Dynamics of Earth's Atmosphere

Why do winds blow? Why don't they just blow themselves out, leaving a calm atmosphere? Any turbulent motions require an energy source to keep them going. The motions of our atmosphere turn out to be another astronomical application of the same physics discussed in the case of motions of materials inside Earth: Heat

Figure 6-13 The most common "landscape" on Earth is unique in the solar system: An open ocean of liquid water covers about three-fourths of our planet. (Pacific Ocean; photo by author.)

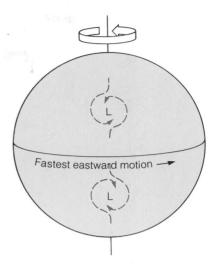

Figure 6-14 Due to the Coriolis drift, airmasses moving toward or away from the equator are deflected east or west, respectively, relative to the ground. Air moving into low-pressure areas (L) produces cyclonic storm patterns; the sense of rotation is opposite in the two hemispheres and is proof of Earth's rotation.

is transferred through the atmosphere and sets up convective motions. In this case the energy source is sunlight. Suppose we magically stopped all winds. Sunlight would beat down on the noontime side of the Earth, warming the ground and warming the air. The air near the ground would eventually become warm enough to expand and rise. New air would flow in laterally along the ground to take its place; thus winds would arise.

If the temperature gradient in the air is steep enough (that is, very warm air near the ground and considerably cooler air aloft), strong convection currents will be set up, lifting the air to form towering thunderhead clouds.

Coriolis Drift

The Earth's rotation is also involved in determining winds and weather patterns. Consider an airmass moving northward off the equator toward a low-pressure region in the atmosphere, as seen in the northern half of Figure 6-14. Since the air has to flow from higher-pressure to lower-pressure regions, it tries to reach the center of the low-pressure regions, marked L in the figure. Because Earth is turning, the equatorial airmass would make one trip around the circumference in 24 h. This means an equatorial airmass is moving eastward at nearly 1700 km/h, whereas material near the pole has no such

speed. As the equatorial mass moves north, therefore, it finds itself moving eastward faster than the ground or local air in that region. It is deflected east, as shown in Figure 6-14. Meanwhile, a polar airmass drifting south toward L finds itself moving slower than L and lags to the west.

Such eastward and westward deflections of winds, due to a planet's rotation, are called **Coriolis drift** (after a French mathematician of the 1800s). Coriolis drift sets up a spiral motion counterclockwise in the air moving into a low-pressure storm system, as shown in Figure 6-14. Conversely, in the Southern Hemisphere the circulation around a storm system produces a clockwise spiral motion, as also seen in the figure. Coriolis drift thus explains the spiral cloud patterns so prominent in space photos of Earth (as in Figures 6-15 and 6-16) and also explains the spiral whirling motions of cyclones and hurricanes. No Coriolis drifts, hence no spiral patterns, would occur on nonrotating planets.

THE ASTRONOMICAL CONNECTION

Most people think of the Earth as isolated from astronomical influences, having evolved independently as a result of its own internal forces. But recent research

Figure 6-15 A spiral, cyclonic storm system is seen near the horizon in this photo made at dusk by Apollo 11 astronauts. (NASA.)

indicates a number of ways in which *external* astronomical forces influence Earth's evolution in dramatic ways.

Meteorite impact craters have been created by impacts of interplanetary bodies up to 10 km or more in diameter. The eroded craters, reaching at least 140 km (90 mi) across, have been recognized on the Earth only in the last century. In the last two decades, dozens have been identified. Most are thousands or millions of years old and have been damaged by erosion, but Figure 6-17 shows a young, well-preserved example. A few are important for exposing oil- or mineral-bearing strata. For example, a 40-mi-wide region in Ontario, called the Sudbury Complex, is an unusual concentration of metal ores. Its origin has been a long-standing geological mystery, but recent work indicates it formed when a giant impact 1.8 billion years ago melted crustal rocks, allowing the native metals to concentrate (Faggart and others, 1985). The metal in a nickel in your pocket is likely to have come from this impact site.

Impact craters formed at a much more rapid rate in the early days of the Earth's history. Although the most ancient craters have been obliterated by erosion on the Earth (Figure 6-18), they are preserved in great numbers on the Moon and planets. The intense early bombardment of Earth may have affected Earth's early crustal structure, the early environment for forming life, and the origin of the Moon (see next chapter).

These "astronomical connections" between our terrestrial environment and our cosmic environment may seem inconsequential. After all, geologists and biologists have worked for two centuries without being much concerned about any astronomical connections. But new evidence indicates that the history of life and climate on Earth may be more controlled by astronomical factors than earlier scientists ever dreamed.

To understand this, note in Table 6-1 that geologists have divided Earth's history into periods, based on fossil life forms and on radioisotopic dates. This table is

Figure 6-16 Photos of Earth from space show a characteristic swirling cloud pattern due to Coriolis drift. A counterclockwise cloud spiral into a low-pressure storm is seen on the upper right edge in the Northern Hemisphere, and a clockwise spiral is seen on the lower right in the south. Brightest regions of Earth are barren desert areas such as the Sahara (upper left); desert expansion in recent decades concerns agriculturalists. (NASA photo by Apollo 17 astronauts.)

a

b

Figure 6-17 Meteor Crater, in Arizona, testifies to the effects of interplanetary debris hitting Earth. A meteorite's impact formed this 1.2-km crater about 20 000 y ago. **a** Road and museum buildings (left) give scale in the aerial view. **b** View from rim at sunset shows strata distorted by the explosion and boulders thrown out onto the rim. (Photos by author.)

Figure 6-18 The more ancient a surface we study on Earth, the better our chances of finding ancient meteorite impact craters because of the longer exposure time. A number of highly eroded craters have been identified in Canada. This ring shaped lake occupies the remnants of one such crater, called Manicougan, seen here from space. The crater formed about 210 million years ago and was about 70 km across, but it has been eroded nearly flat by the passage of glaciers over it during the ice ages. It gives direct evidence of gigantic impacts during ancient prehistory. (NASA photo.)

called the **geological time scale.** It chronicles slow but striking shifts in Earth's biology and environment. Radical environmental changes occurred at certain times, marked by dramatic alterations in fossil life forms. The question is: What caused them? Some may have been caused by effects arising within Earth itself, such as mountain building that changed sea levels or volcanism that destroyed large areas and changed climates with smoke fumes. But other changes may have had external, astronomical causes. During the especially dramatic change 65 million years ago, about three-fourths of all species became extinct within a few million years—or perhaps much less! This extraordinary change marks the division between the Cretaceous period (see Table 6-1) and a subperiod called the Tertiary. The radical extinctions at that time are called the **Cretaceous–Tertiary extinctions.** No one could explain the Cretaceous-Tertiary extinctions and environmental change until about 1980. Then, scientists discovered that a soil layer marking the transition contains unusually high amounts of elements common in many meteorites. Meteorites, which are stones that fall out of space onto Earth, are believed to be fragments of inter-

planetary bodies called asteroids. Within a few years the discovery was confirmed at sites around the world: Some unusual event had distributed the asteroidlike materials all over the Earth. The data thus suggested the impact of an asteroid. From the total amount of asteroidal elements, scientists estimated that an asteroid 10 km across crashed into Earth, as shown in Figure 6-19 (Alvarez and others, 1980; Ganapathy, 1980).

This idea has been controversial, but more and more data support it. Some scientists argued that the unusual soil layer was associated with volcanoes, but 1987 studies showed that the soil contains shock-fractured rock crystals, which are produced by shock waves of an energetic impact but not by volcanism (Kerr, 1987a). Asteroid statistics indicate that such impacts should occur every few hundred million years or so, supporting the idea's plausibility.

But would an asteroid's impact do the job of exterminating species and altering life on Earth? The soil layer was found to contain soot, indicating widespread fires (Wolbach and others, 1986). Calculations on the effects of a large impact indicate that dust from the impact and soot from the resulting wildfires would be thrown into the high atmosphere, blocking sunlight and making it too dark to see for a few months (Toon and others, 1982). This in turn would wipe out lower parts of the food chain, such as sunlight-dependent plankton and small plants, explaining eventual extinction of many higher species. After the large reptiles were wiped out, the hitherto obscure mammals emerged to inherit Earth!

In an example of pure research dovetailing with practical concerns, scientists have used the climate calculations to study effects of a cataclysm associated with nuclear war (Levi and Rothman, 1985; Schneider, 1987). They found that dust and soot raised by explosions and fires in all-out nuclear attacks would block the sunlight, as would the after-effects of a large impact, dropping temperatures and damaging crops in both target countries and aggressor countries. This theory of **nuclear winter** has been widely publicized and has affected military strategists' thinking about nuclear war. The detailed climate calculations applied to the theory of an asteroid impact and the theory of nuclear winter symbolize our growing understanding of how our planetary climate could be altered by terrible, unforeseen events, both in the past and the future.

The theory that an asteroid's impact caused one of the major discontinuities in evolution and climate has led scientists to look for evidence of asteroid impacts

TABLE 6-1

Geological Time Scale

Era	Age (millions of years)	Period	Life Forms	Events
Cenozoic	0	Quaternary	Humanity	Technological environmental modification; extraterrestrial travel; ice ages
	3	Tertiary		
			Mammals	Building of Rocky Mountains
	65	Cretaceous	Extinction of many species of plants and animals, including dinosaurs	Large meteorite impact (65 million years ago); continents taking present shape
Mesozoic	130	Jurassic	Dinosaurs	
	180	Triassic	Reptiles	
	240			
		Permian	Conifers; extinction of many species of plants and animals	Building of Appalachian Mountains
	280	Pennsylvanian	Ferns	Pangaea breaking apart
	310	Mississippian		
Paleozoic	340	Devonian	Fishes	
	405	Silurian	Early land plants	
	450	Ordovician		
	500	Cambrian	Trilobites	Earliest abundant fossils (trilobites, etc.)
	570			
		Ediacarian	Small soft forms	
	640			
	1000		First macroscopic life forms; sexually reproducing life forms	Growth of protocontinents; scattered fossils
Proterozoic	2000			
	2600		Oxygen-producing microbes	Oxygen increasing in atmosphere
Archeozoic	3000			Crustal and atmospheral evolution Earliest fossils (algae)
	3600		Microscopic life	
				Oldest rocks; crustal formation? Heavy meteoritic cratering
	4500			
Formative	4600			Formation of Sun and planets (see Chap. 14)
Presolar	12 000?			Formation of Milky Way galaxy (see Chap. 23)
	16 000?			Origin of universe (see Chap. 27)

Source: Data in part from Schopf (1975); Morris (1987).

a

b

c

Figure 6-19 Impacts by interplanetary bodies as much as 10 km across and even larger may explain some sudden environmental changes found in the ancient geologic record. These views show the impact of a 10-km asteroid as seen from an altitude of 100 km (60 mi) in views 10 s before impact, 1 s before, and 60 s after. Massive amounts of dust being ejected in the final picture will settle into the atmosphere and be blown around the world by high-altitude winds. Chemical evidence suggests that such an impact 65 million years ago led to extinction of dinosaurs and many other life forms, probably due to climate changes resulting from the dust pall in the atmosphere, which blocked most sunlight for many months. (See text; paintings by author.)

associated with other important discontinuities in geological history. A few other cases of meteorite element concentrations have been reported in certain layers, including possible examples about 34 and 240 million years ago at a time when there were other mass extinctions of many species (Table 6-1), but these examples have yet to be confirmed. Scientists are thus still debating the role of asteroids in shaping life on Earth (Kerr, 1987a).

Still another example of an astronomical connection to the Earth's history comes from studies of the Earth's orbit. Many studies suggest that minor orbital changes are caused by gravitational forces of nearby planets, which cause changes in the Earth's tilt, distance from the Sun, and exposure to sunlight. These effects seem to explain most of the climatic changes associated with ice ages in the past million years or more (see review by Kerr, 1987b).

As discussed in Chapter 15, certain disturbances on the surface of the Sun also appear to be correlated with disturbances in the Earth's climate. Thus modern research suggests many links between astronomy and climate effects on the Earth.

SUMMARY

Geological studies of strata and fossils have established a chronology of events in the history of the planet Earth, and radioisotopic dating of rocks has established the actual ages of these events, as summarized in Table 6-1. It is broadly divided into intervals called *eras*. Only the last 14%

of the Earth's history has yielded enough rock evidence, such as fossils and dates, to provide finer divisions, which geologists call *periods*.

Drawing from material presented in this chapter, we can now construct a thumbnail sketch of Earth's history.

Some $4\frac{1}{2}$ billion years ago Earth formed. Giant meteorites struck much more often than today, scarring the primal landscape with great impact craters. The surface was lifeless.

Life probably originated within the first few hundred million years of Earth's history, since the earliest microscopic fossils date back about $3\frac{1}{2}$ billion years ago.

The heating of Earth's interior, probably due to radioactivity, led to a prompt internal melting within the first few hundred million years. This, in turn, caused differentiation—a draining of metal toward the center to form a nickel–iron core and a floating of lighter minerals to form a low-density crust overlying a dense rock mantle.

As Earth cooled and solidified, the surface layer formed a rigid lithosphere overlying a more plastic asthenosphere. Currents in the asthenosphere broke the relatively thin lithosphere into *plates*. Continents repeatedly split and rejoined as rifts broke the lithosphere and plate motions caused crustal masses to collide.

Mountainous landscapes thus replaced the original, cratered surface, though vestiges of the old surfaces still appear in the cratered continental shields. With the coming of plants and animals, the landscape became familiar to modern eyes. Figure 6-13 reminds us that the dominant "landscape" of our planet is a vast expanse of water—a view unique in the solar system, as we will see in the next few chapters.

In the scale of planetary evolution, recent events such as the ice ages seem the merest of details. Larger climate changes have punctuated the past. They may have had various terrestrial and astronomical causes. For example, extinction of many species probably resulted from atmospheric dust caused by the large impact 65 million years ago. It is important to understand such events. Even small shifts in climate can produce droughts, affecting world food production. Scientists studying the 65-million-year-old event recently concluded that dust and smoke from nuclear warfare would trigger a devastating climate change known as "nuclear winter"—yet another argument for stopping nuclear arms proliferation!

Let us conclude our summary of Earth's history by compressing events into a single day. Life evolved sometime in the early morning, but the fossil-producing trilobites that begin the traditional geological time scale in the Cambrian period did not appear until about 9:30 in the evening. By 10 P.M. there were fishes in the sea, and by 11 P.M., dinosaurs on the land. Mammals did not appear until about 11:40. Human beings, who have been here (depending on your definition of *human*) perhaps 2 million years, made their appearance only 30 seconds before midnight. The last few thousand years—civilization—occurred in a tenth of a second—represented by the pop of a single flashbulb at midnight. The question is: What will be here a tenth of a second after midnight?

CONCEPTS

radioactive atom	plate tectonics
parent isotope	fault
daughter isotope	volcanism
half-life	magma
radioisotopic dating	lava
Earth's oldest known rocks	erosion
	deposition
continental shield	field
age of the Earth	magnetic field
seismic waves	primary atmosphere
core	secondary atmosphere
mantle	Coriolis drift
crust	meteorite impact crater
differentiation	geological time scale
basalt	Cretaceous–Tertiary extinctions
granite	
lithosphere	nuclear winter
asthenosphere	

PROBLEMS

1. When were the last major earthquakes in your region? Is your region seismically active or inactive? How is it located with respect to the boundaries of tectonic plates?

2. If a rock sample can be shown to contain one-eighth of its original amount of radioactive uranium-235, how old is it?

3. In view of the preservation of ancient craters and the lack of folded mountain ranges on the Moon, would you predict the Moon to have more or less seismic activity than the Earth? Discuss your reasoning.

4. Fossils of apelike predecessors of the genus *Homo* (such as *Australopithecus*), found in Africa, are believed to date back at least 2 to 3 million years. What percentage of the Earth's age is this? Can you accept, philosophically, that events happened during most of the history of the Earth before anyone was around to see, hear, or record them?

5. If the Earth is 4.6 billion years old, about how much more radioactive uranium-238 did it have when it formed? Would the heat-production rate from radioactivity when the Earth formed have been more, less, or the same as it is now?

6. The composition of the Earth's atmosphere is probably much changed from what it first was.

a. What happened to the abundant hydrogen atoms initially present or produced by breakup of molecules such as methane (CH_4)?

b. Given that much ammonia (NH_3) was initially present and that ammonia molecules break apart when struck by solar particles in the atmosphere, account for one source of the Earth's now abundant nitrogen.

c. What two gases were added abundantly by volcanoes, and where did these two gases finally end up?

7. Describe how you would expect conditions on the Earth to be if Earth were so close to the Sun that the mean surface temperature was above 373 K (100°C). What if the Earth were far enough from the Sun that the mean surface temperature was below 273 K (0°C)? Explain why a view like Figure 6-13 is probably unique to the Earth.

ADVANCED PROBLEMS

8. The average near-surface temperature gradient in the Earth is 20°C/km.

a. Assuming the surface temperature is about 20°C, how many kilometers would one have to drill to reach a depth where water would boil? (The temperature of boiling water is 100°C.) Compare this depth with that of the deepest mines, roughly 3.5 km. (The actual boiling temperature at depth would be somewhat higher than 100°C because of the increased air pressure, an effect we ignore in this problem.)

b. If such depths could be reached economically, dual pipes could be lowered, with water pumped down one pipe, converted to steam, and blown up the other pipe. Steam-powered plants could thus tap the planetary energy source in any part of the world. Describe the possible economic and political consequences of such a project.

9. As seen from Mars, the Sun subtends an angle of about $\frac{1}{3}°$. Suppose that the Earth passes exactly between Mars and the Sun.

a. Is the Earth big enough to cause a total eclipse of the Sun as seen from Mars?

b. Assuming that the human eye can resolve a disk as small as 2 minutes of arc (120 seconds of arc), could the Earth be detected by the unaided eye as it crossed the Sun, as seen from Mars? (*Hint:* Use the small-angle equation. The Earth's diameter is 12 756 km, and its distance from Mars is 60 000 000 km.)

PROJECT

1. Place cooking oil an inch or two deep in a flat pan over low heat and under a single strong light. Because the oil does not readily boil, heat is soon transmitted to the surface in visible convection currents. Note how the currents divide the surface into cells, or regions of ascent, lateral flow, and descent. These cells are analogous to tectonic plates in the Earth's surface layers. Sprinkle a slight skin of flour on the oil's surface to simulate floating continental rocks and watch for examples of continental drift and plate collisions. Experiment with different depths of oil and different temperature gradients (by changing the heat setting), and record the results. Be careful not to turn up the heat too high (especially on a gas stove) to avoid having the cooking oil catch fire!

CHAPTER 7

The Moon

The same Moon that we see today shone down on the breakup of the continents, gleamed in the eyes of the last dinosaurs, and illuminated the antics of the first protohumans. The same Moon was seen by all the historical figures we have mentioned in preceding chapters: Stonehenge builders, Mayan eclipse observers, Aristarchus, al-Battani, Isaac Newton. They all asked what it is, where it came from. Today the Moon is a little different: It has footprints on it and we know some of the answers.

The Moon is the Earth's only natural **satellite**—a body that orbits around a larger body. (The term *moon* is also used generically to mean a satellite.) As a world, it is respectable: It has a diameter of 3476 km, about one-fourth the diameter of the Earth. It is a rocky world splotched with dark gray flows of ancient lavas and dotted with craters formed by explosions when meteorites hit it in the ancient past. Ours is the first generation to have seen a few of the Moon's starkly beautiful landscapes. We know, not just from instruments but from the *experience* of our astronauts, that it has no air, no water, no weather, no blue sky, no clouds, no life.

In spite of the Moon's new familiarity, many people are still confused about even its simple phases. Can you recall which way the horns of the crescent point in the evening sky? Cartoonists often draw the horns pointing down toward the horizon. Not so. Since the fully illuminated edge of the crescent must face the Sun, which has just set below the horizon, the horns must point upward, away from the horizon. Scoff, too, at the novelist who describes the full Moon rising at midnight! To be fully illuminated, the Moon must be opposite the Sun and hence must rise as the Sun sets.

These relations can be seen by studying Figures 7-1 and 7-2, showing the Moon's movements along its orbit. The whole subject of its motions, though it might seem mundane, provides interesting clues about the ancient

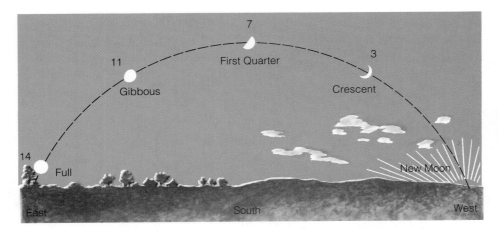

Figure 7-1 Wide-angle view of the sky, looking south, showing some of the phases of the Moon as seen by a Northern Hemisphere observer at sunset on the indicated days of the monthly lunar cycle, counting from the day of new moon.

history of the Earth–Moon pair. First we will consider these motions, then examine the Moon's surface and the astronauts' discoveries, and finally explore the puzzling problem of the Moon's origin.

THE MOON'S PHASES AND ROTATION

As can be seen by comparison of Figures 7-1 and 7-2, the Moon's **phases**, or shapes on different days, are directly caused by the Moon's motion in its monthly orbit around Earth. Let us say that on day 0 the Moon crosses the line between the Earth and the Sun. Here the Moon is nearly in front of the Sun, as seen from the Earth, and is lost in the Sun's glare. On this day the Moon generally cannot be glimpsed. This is called the date of **new moon**. A couple of days later, the Moon has moved far enough from the Sun to be glimpsed in the early evening sky; it is backlit by the Sun, giving it a crescent shape. On day 7, it is 90° from the Sun, a phase called **first quarter**, because it is a quarter of the way around its orbit. It looks half-illuminated at this time. For the next week it is more than half-illuminated, a phase called **gibbous** (hard *g*, as in *give*). On day 14 or 15, it is opposite the Sun and fully illuminated—a phase called **full moon**. This is the day on which the Moon rises at sunset and fills the night sky with its brightest possible light. For the next 2 weeks, the Moon rises after midnight and is visible primarily in the early morning sky. On day 22, the Moon is three-quarters of the way around its orbit and now in a half-lit phase called

third quarter. On day 29, the Moon is back to the new moon phase.

The term *waxing moon* refers to the first 2-week period, when the Moon is growing more illuminated each day; *waning moon* refers to the second 2-week period, when the Moon is growing slimmer.

The Moon takes 27.3 d to complete one revolution around Earth relative to the stars—an interval called the Moon's **sidereal period**. During this period, Earth moves roughly 27° around the Sun, so that the Moon has to move through this additional angle to complete its cycle of phases relative to the Sun. Therefore the cycle of lunar phases takes 29.5 d.

Whenever the Moon is visible, no matter what its phase, we can always distinguish at least some of the dark lava plains that make up the features of the so-called man in the moon. This is because of a curious characteristic of the Moon's motion: It always keeps the same side facing Earth, as shown by the stylized mountain in Figure 7-2. In 1680, the French astronomer G. D. Cassini explained this characteristic in a statement that is sometimes hard to grasp:

> **The Moon rotates on its axis with a period equal to its orbital revolution period around Earth, so that the same side keeps facing Earth at all times.**

The rotation of any satellite in this way is called **synchronous rotation** since it is synchronized with the satellite's own revolution, as shown in Figure 7-2. How can the Moon rotate at all, you might ask, if it always keeps the same side toward the Earth? The best answer

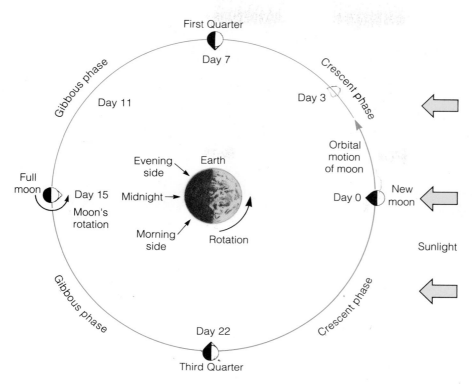

Figure 7-2 The Moon's motion around Earth. Synchronous rotation of the Moon is shown by a mountain (indicated by a triangle) that always faces Earth. The Moon completes one rotation during one complete revolution around Earth. Phases seen by an earthbound observer are indicated at different points in the orbit. First quarter is seen in the evening sky, third quarter in the early morning sky.

is provided by an example. Put a chair in the middle of a room. The chair is the Earth; the walls, the distant stars. To represent lunar orbital motion, walk around the chair. If you walk around the chair always facing it, so that an Earthly observer in the chair never sees your back, you will find that you have faced all sides of the room during one circuit. In other words, you have rotated once on your axis and made one revolution around the chair. (If you put a strong light in one corner of the room and hold up a ball as you walk around the chair, the observer in the chair can see the cycle of phases on the ball.)

Even writers who should know better sometimes speak of the Moon's "eternally dark side." This mistake comes from a popular belief that the side eternally hidden from us must always be dark. But the far side, just like the near side, has day and night—periods of sunlight and periods of darkness. This can be seen in Figure 7-2. Each period lasts about 2 weeks, since the Moon takes about 4 weeks to make a complete rotation. There is always a dark side, but it isn't necessarily the *far* side.

The Moon's synchronous rotation has another consequence. For an astronaut at any spot on the near side

of the Moon, the Earth hangs forever in the same spot in the sky. (Imagine living on the lunar mountaintop in Figure 7-2; Earth would always be overhead.) Nonetheless, Earth goes through a complete cycle of phases every 4 weeks, as seen by a lunar astronaut. Figure 7-3 shows Earth's crescent and gibbous phases as seen from the Moon.

Is there a reason why the Moon keeps one side toward Earth? When asking for a reason, a scientist normally means to ask: "Could the observation be explained by some more fundamental properties of nature, so that it becomes a special case of a more general phenomenon?" Here the answer is yes. In the 1780s and 1790s, Joseph Louis Lagrange and Pierre Simon de Laplace used Newton's law of gravity (Chapter 4) to show that if the Moon were slightly egg-shaped or football-shaped, gravitational forces would make the longest axis point toward Earth at all times. Confirming this, space vehicles in the last few decades have shown that one axis of the Moon *is* indeed about 2 or 3 km longer than the others and points steadily toward Earth.

If the Moon always kept *exactly* the same side toward Earth, earthbound observers could never see more than

a

b

Figure 7-3 Seen from the Moon, Earth goes through phases. **a** The crescent Earth rises, as seen from orbit over the cratered lunar highlands. **b** The gibbous Earth seen on another occasion from a similar position. (NASA photos by Apollo 17 and 8 astronauts, respectively, orbiting over the Moon.)

exactly 50% of it. Careful mapping, however, has shown that the Moon wobbles and that during a period of years 59% of the Moon can be seen, with first one side and then another being turned slightly farther from Earth than its average position.

TIDAL EVOLUTION OF THE EARTH–MOON SYSTEM

Gravitational forces in the Earth–Moon–Sun system cause **tides**, or bulges in the shape of Earth and the Moon. The closer two objects are, the stronger the gravitational force each exerts on the other. Thus the side of the Moon facing the Earth has a stronger force on it than the far side, because the facing side is closer. This is shown by the small arrows in Figure 7-4. As a result, the Moon stretches slightly along this line, limited by the small elasticity of its rock interior. This stretching is called a **body tide**. Similar forces acting on the Earth produce not only a body tide but also an **ocean tide** in the fluid layer of water on the Earth's surface. These tides take the form of bulges on the front and back side of both the Earth and the Moon, since each body comes

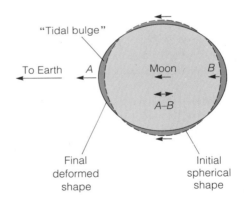

Figure 7-4 The gravitational attraction of the Earth on the Moon is greater on the near side (*A*) than on the far side (*B*). The difference (*A* − *B*) acts as a stretching force deforming the Moon from its unstressed spherical shape to a flattened shape with tidal bulges on the near and far sides.

into equilibrium with the gravitational forces by stretching along the Earth–Moon line.

The Earth's body tides are hard to detect, but its ocean tides are obvious to anyone who visits a beach for more than an hour. They range in height from about ½ m (2 ft) to over 15 m (50 ft). One might suppose that

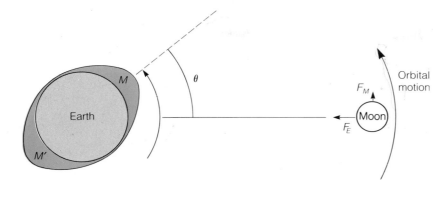

Figure 7-5 The cause of the Moon's tidal recession. Because of Earth's relatively rapid rotation, Earth's tidal bulge (MM') gets dragged off the Earth–Moon line at an angle θ. Thus, in addition to the normal gravitational force of Earth (F_E), there is a net forward force (F_M) caused by the difference in attractions between the nearer bulge (M) and the farther bulge (M'). This force (F_M) pushes the Moon ahead in its orbit, causing it to spiral slowly outward.

high tide always occurs when the Moon is overhead, with the highest tides at the new Moon or full Moon, because the tidal forces act along the Earth–Moon line (or, in the strongest case, along the Earth–Moon–Sun line). However, this is only a rough tendency; complications result from motions of water around the irregularly shaped oceans and seas and because of the Earth's rotation. Coastline geometry in some places can produce remarkable wave effects associated with tides, but so-called tidal waves are related not to tides but to earthquakes or volcanic activity at sea; they are properly called by their Japanese name, **tsunamis**.

Tidal bulges raised on the Earth and the Moon have four major effects, first described in 1898 by George Darwin, son of the famous naturalist. These effects are described in the following four subsections.

Underlying Cause of the Moon's Synchronous Rotation

The effect of tides explains why the Moon keeps one side facing Earth. As mentioned earlier, any elongation in the Moon's shape would tend to make one side always face Earth. Tides guarantee such an elongation, since they create bulges. If the Moon were initially spherical and rotating at a nonsynchronous rate, gravitational forces acting on it would create tidal bulges. Since Earth would be trying to keep the bulges aligned with it (horizontal in Figure 7-4)—but a nonsynchronous lunar rotation would be trying to drag the tidal bulges around with it— the bulges would tend to exert a frictional drag, slowing the rotation until it became synchronized and one side faced the Earth at all times. This effect explains why most other satellites in the solar system, besides ours, keep one face toward *their* planet.

Recession of the Moon

The second effect is a slow **tidal recession** of the Moon away from the Earth because of gravitational forces on tidal bulges. As shown in Figure 7-5, Earth's rapid rotation drags its tidal bulge slightly ahead of the Earth–Moon line by some angle θ. The gravitational effect of the bulge M exceeds that of M', because M is closer to the Moon. Because M is in front of the Moon, the net force from M tends to pull the Moon ahead in its orbit. This accelerates the Moon forward and makes it spiral very slowly out from the Earth. Laser beam reflectors placed on the Moon by astronauts have allowed scientists to measure this slow outward movement directly. The Moon was much closer to Earth several billion years ago. Analyses do not indicate the exact date, but most researchers believe the Moon was closest to Earth about 4.6 billion years ago during the formation of the two bodies.

Slowing Earth's Rotation

The third effect of tides is that the Moon slows Earth's rotation, just as Earth has slowed the Moon to synchronous rotation. The effect can be seen in Figure 7-5, where the Moon pulls back on the tidal bulge M, acting like a brake on Earth. Studies indicate that billions of years ago, when the Moon was closer, the Earth's day was only about 5 or 6 h long. These theoretical results have been confirmed by an unexpected finding. Certain marine creatures create daily and monthly banded structures in their shells or other hard parts, allowing biologists to count the number of day bands in a monthly cycle. Fossil evidence suggests that whereas the present month is 29.5 d long, it was only about 29.1 d long

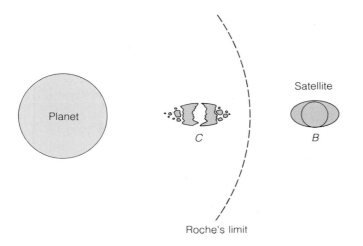

Satellite

Planet   *C*   *B*   *A*

Roche's limit

Figure 7-6 Roche's limit. If a satellite is located at *A*, a small tidal bulge develops. At closer distance *B*, a larger bulge develops. Within a critical distance, called Roche's limit, the stretching force differential between the near and far sides is so great that the satellite is torn apart (for example, at distance *C*).

45 million years ago. Older fossils show that tides existed as long as 2.8 billion years ago and that the month was as little as 17 d long (Kaula and Harris, 1975). We cannot extrapolate such data further back in time because the tidal effects depend on the configuration of terrestrial oceans and continents, and these configurations were different by unknown amounts in the past.

As Earth slows and the Moon recedes in the future, it might eventually recede so far that Earth's gravitational pull would be very weak. Before it can escape into an orbit around the Sun, however, small tides will be raised on Earth by the Sun. These tides will slow Earth's rotation so that the length of the day and the month will become equal, stopping the tidal recession. Earth and the Moon will then *both* be in synchronous rotation. Eventually, because of this small solar effect, Earth's rotation will slow so much that the day will exceed the month and the tidal process will reverse: The Moon will begin to approach Earth again.

Roche's Limit

The fourth application of tidal theory shows that a small body, if close enough to a large body, can be torn apart by tides. Calculated around 1850 by French mathematician Edward Roche, **Roche's limit** is the distance between any two different-sized bodies within which the tide-raising force exerted on the smaller body is sufficient to disrupt it. The effect occurs because the *difference* between the forces on the near and far sides of a satellite increases as the satellite moves closer to the primary body. This effect causes the satellite to

stretch into a slight egg shape as it approaches the primary, as shown in Figure 7-6. The position of Roche's limit depends on the size, density, and strength of the satellite. For instance, a small metal spacecraft is not torn apart while orbiting near Earth. But a body as big as the Moon, if similarly placed for a long period, would develop internal stresses and fractures and eventually disintegrate into a cloud of orbiting particles like Saturn's rings. The Moon could not remain an intact body much closer to Earth than about 18 000 km (11 000 mi). (Its present distance is 384 000 km, or 240 000 mi.)

SURFACE FEATURES OF THE MOON

Until the telescope was invented around 1608, no one knew much about the lunar surface features except for the gray patches that make up the face of "the man in the Moon" and "the woman in the Moon" (Figure 7-7). Some thought the Moon was a polished sphere. Thomas Harriot and Galileo Galilei, the first persons known to have seen the Moon's features through a telescope, made their early observations in 1609 and 1610 (see page 66). They both recorded rugged regions with prominent shadows cast by mountains and crater rims along the **terminator**, the line dividing lunar day from lunar night (see Figure 7-8). Details were less prominent under high lighting (full moon) or at the edge of the disk, called the **limb**. Galileo reported:

The moon certainly does not possess a smooth surface, but one rough and uneven, and just like the face of the

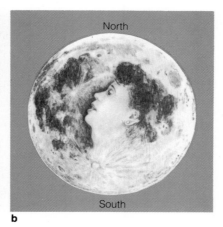

North

South

Figure 7-7 a At full moon, the rugged appearance of the terrain is minimized by the absence of shadows. Compare with Figure 7-9. Full lighting emphasizes different features, such as the bright rays emanating from the crater Tycho (bottom). **b** Several folklore figures, such as "the woman in the Moon," are formed by the pattern of dark lava plains. Squinting at *a* and *b* may help you see these features. (Photo from Lunar and Planetary Laboratory, University of Arizona.)

earth itself, is everywhere full of vast protuberances, deep chasms, and sinuosities.

These features were later recorded in better detail by telescopic photographs (Figure 7-9). Most of the roughness was caused by thousands of **impact craters**, or circular depressions caused by meteorite impacts (Figures 7-10 and 7-11), ranging up to 1200 km in diameter. Smaller ones (up to a few kilometers) are bowl-shaped, while larger ones have a more complex structure, such as central mountains or concentric rings of cliffs. Galileo found the dark gray patches that are visible to the naked eye and form "the man in the Moon" to be much smoother than the brighter, cratered areas, as seen in his sketch in Figure 7-8. He mistook these dark patches for seas. Using Latin, as scientists did in the 1600s, he called them *maria* (singular **mare** [MAH-ray]). These "seas," as Galileo himself probably eventually realized, are actually vast plains, which we now know are covered with dark lava. Mare surfaces cover much of the front side of the Moon, but only 15% of the whole Moon. The bright regions are cratered, rugged upland areas. Figure 7-12 shows a view that encompasses cratered uplands and mare plains.

Galileo's work greatly strengthened the conception that the Moon is an Earth-like, planetary body having familiar features such as mountains. He used this as a proof against those who argued that the Earth was not to be included among the planets.

The first reasonably accurate lunar map was produced by the German Johannes Hevelius in 1647. (A modern lunar map with names of certain prominent fea-

Figure 7-8 Comparison of one of Galileo's first lunar telescopic drawings (left), made in 1610, to a photograph of the Moon at the same phase. Letters show corresponding features. Galileo detected mountains, plains, and craters (bottom). (Courtesy Ewen A. Whitaker, University of Arizona.)

tures is seen in Figure 7-13.) In 1651 an Italian priest, Riccioli, started the present practice of naming craters after well-known scientists and philosophers, such as Copernicus, Tycho, and Plato. The maria were given poetic and fanciful names, such as Mare Imbrium (Sea of Rains). Lunar mountains were named for prominent terrestrial ranges, such as the Alps. These mountains, however, are unlike terrestrial folded ranges; they are the rims of vast multiringed craters, called **basins**, which

a b

Figure 7-9 Early and modern photos of the Moon at similar phase. **a** One of the first known lunar photos, an 1851 daguerreotype, probably by American astronomer J. W. Draper or his son, Henry. Smallest details are about 35 km across. (NASA.) **b** Modern view, illustrating a century's advance in astrophotography. Smallest details are about 1 km across. (Lunar and Planetary Laboratory, University of Arizona.)

in turn contain the mare plains. Figure 7-14 illustrates one of these huge impact features and its probable subsurface structure. Bright streaks called **rays**, which radiate from various craters but show no relief, are fine debris blasted out of the craters. Most maria are merely seas of lava that have flooded ancients basins.

Astronomers of the 1700s and 1800s undertook a tantalizing endeavor, searching with ever larger telescopes with ever greater magnification for diagnostic details that would show how the Moon formed. Were there any signs of changes—new craters forming or volcanic activity? Were there any artificial structures that might have been built by the inhabitants that popular writers continued to place on the Moon?[1]

Despite years of search, no changes or artificial structures were found. Lunar photographs, first made in 1849 (see Figure 7-9), showed the same structures year after year. Because there seemed to be nothing new to discover, interest in the Moon dwindled, and many astronomers saw it as a sterile nuisance that lit up the night sky and blotted out the faint objects they wanted to observe.

Nonetheless, evidence for minor geological activity on the present-day Moon came in 1963, when on two occasions observers at Lowell Observatory saw a red glow near the crater Aristarchus. This glow may have been a volcanic eruption or gas discharge; later data from Apollo flights indicated occasional gas emission in this region.

Why is the Moon's geological character different from Earth's? Why do craters dominate? Why are there no major mountain ranges? Where did the Moon come from? What would it be like to stand on the Moon? Such questions motivated actual journeys to the Moon.

---

[1] Or—an idea conceived more recently—by alien interstellar travelers who might have visited the solar system in past eras? See, for example, Arthur C. Clarke's book *2001*.

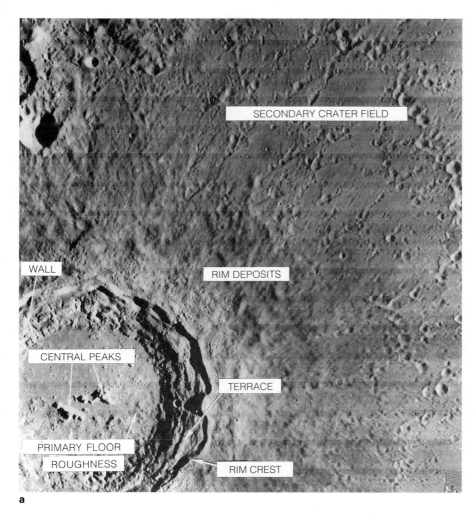

SECONDARY CRATER FIELD

WALL

RIM DEPOSITS

CENTRAL PEAKS

TERRACE

PRIMARY FLOOR

ROUGHNESS

RIM CREST

a

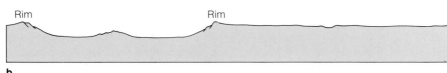

Rim   Rim

b

Figure 7-10 a The typical structures in a large lunar crater are exemplified by this Orbiter photograph of the crater Copernicus, which is about 90 km across. The main bowl was excavated by the impact of a meteorite a few kilometers in diameter. Central peaks and terraces are rebound and slump features. Rim deposits and satellite craters are caused by debris blasted out of the crater during its formation. (Courtesy James W. Head III, Brown University.) **b** An approximate cross section of such a crater.

FLIGHTS TO THE MOON

The first human device to reach the Moon was a Russian spacecraft that carried little scientific equipment and crashed into the Moon in 1959. The first close-up photos of the surface, from the American probe Ranger VII in 1964, showed that the surface was not craggy but covered by a gently rolling layer of powdery soil, scat-

tered rocks, and shallow craters of various sizes. This type of soil cover, shown well in Figure 7-15, is present nearly everywhere on the Moon and is called the **regolith** (rocky layer). The lunar regolith is typically 3 to 30 m (10 to 100 ft) deep and made primarily of debris blasted out of lunar craters as they were formed. Each well-preserved lunar crater is surrounded by a sheet of such debris, called an **ejecta blanket**. The regolith is

Figure 7-11 The sculpturing of the lunar highlands from countless impacts of meteorites of all sizes can be imagined from this nearly vertical view. This photo, looking past the Apollo 16 command module down onto the Moon's far side, was taken from the landing module just after its undocking from the command module in order to proceed to the lunar surface. (NASA.)

therefore said to be composed of overlapping ejecta blankets. The powdery surface of the regolith is due to small meteorites (some microscopic), which are so abundant that they have "sandblasted" most of the upper few meters into fine dust.

Table 7-1 lists the six Apollo lunar landings and two earlier test flights. Twelve men walked on the Moon during the Apollo program from 1969 to 1972. Figures 7-15 through 7-22 include typical scenes observed during these missions. The first two Apollo missions were cautious tests that touched down on flat, smooth plains. The subsequent four Apollo landings sampled a variety of complex and rugged sites. Astronauts collected many samples and placed various instruments in position to make measurements (Figure 7-16).

As seen in Table 7-1, these flights proved that the lunar surface features are very old. The maria—the dark regions visible to the naked eye from Earth—turned out to be vast flows of basaltic lava, 3 to 4 billion years old. One such lava sample is shown in Figure 7-17.

The light-colored bright uplands are still older, formed during the early years of the solar system about 4 to $4\frac{1}{2}$ billion years ago, although most of the oldest rocks are heavily fragmented or pulverized. Chemical evidence from the rocks shows that around $4\frac{1}{2}$ billion years ago, the surface layers of the Moon were molten, forming a vast sea of lava called a **magma ocean**. Low-density feldspar crystals then solidified in the magma ocean. Because of their low density, these floated in the magma ocean, accumulating into an ever-thickening lithosphere of basaltlike rock called anorthosite. This is the rock type that now composes most of the uplands.

Most of the ancient anorthosite rocks did not survive intact. Numbers and ages of craters prove that the rate of meteorite impact in the first half-billion years of lunar history was thousands of times higher than today. The high cratering rate pulverized most of the primordial lunar rocks. Upland sites visited by astronauts revealed overlapping layers of ejecta blankets from many craters, composed of dust, glass droplets, rock chips,

Figure 7-12 The Apollo 12 lunar landing module sails over several terrain types on the Moon. Ancient cratered terrain dominates the right side. Somewhat younger, smooth lava flows, peppered with small craters and light-toned dust, cover the lower left corner. In the distance, more pristine, dark-colored lava lies near the horizon. (NASA.)

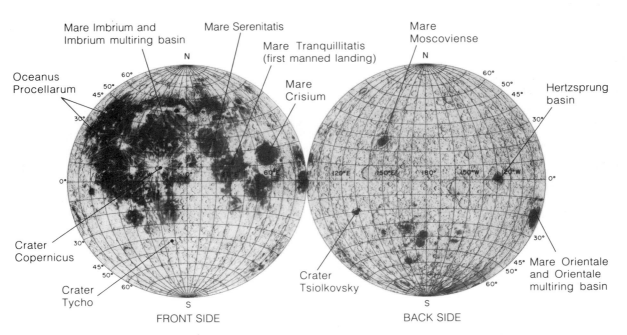

Figure 7-13 Map of the front (left) and far (right) sides of the Moon, showing some prominent features. (U.S. Geological Survey, courtesy R. M. Batson.)

a

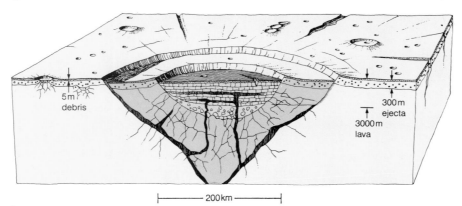

b

Figure 7-14 a The Orientale basin is the youngest and most dramatic multiring basin on the Moon. The outermost ring of cliffs is nearly 1000 km in diameter. Lava erupted to form dark "ponds" along fractures at the base of some cliffs. Map shows scale. (NASA photo from unmanned orbiter spacecraft.) **b** Hypothetical cross section.

Figure 7-15 The powdery texture of the lunar surface is seen in the foreground of this famous photograph, taken shortly after the first human footsteps on the Moon. (NASA Apollo 11 photo.)

and **breccias**, or rocks composed of cemented rock fragments. These results explain why rocks older than about 4.2 billion years are rare, or heavily pulverized, and why rocks younger than 3.0 billion years are also rare. The oldest rocks were destroyed by impacts. And after 3 billion years ago, the Moon had cooled enough, and the lithosphere was thick enough, that subsequent volcanism or rock-forming activity was infrequent. Because 3 to $4\frac{1}{2}$ billion-year-old rocks are rare on Earth, lunar rocks have given geologists a welcome insight into

conditions in the Earth–Moon system during that era. More detailed descriptions of the advances resulting from Apollo exploration are given by Taylor (1975).

LUNAR ROCKS: IMPLICATIONS FOR THE MOON AND EARTH

Rocks are cosmically significant. They are solid materials that contain many clues to their long histories of

TABLE 7·1

Manned Apollo Explorations of the Moon

Mission	Date	Landing Site	Results	Typical Rock Ages (billions of years)[b]
Orbital Missions				
Apollo 8	Dec. 24, 1968[a]	———	First lunar orbit. Orbital mapping. 115-km minimum altitude.	
Apollo 10	May 21, 1969[a]	———	Test of approach to approximately 17-km minimum altitude.	
Landing Missions				
Apollo 11	Jul. 20, 1969	Mare Tranquillitatis	First landing. Samples of mare material.	3.5–3.7
Apollo 12	Nov. 18, 1969	Oceanus Procellarum	Samples of mare material.	3.2–3.4
Apollo 14	Feb. 5, 1971	Fra Mauro (ejecta from Imbrium basin)	Samples of ejecta from Imbrium basin.	3.9
Apollo 15	Jul. 30, 1971	Edge of Mare Imbrium at foot of Apennine Mts.	Samples of material from Apennine Mts., forming rim of Imbrium basin. Samples of mare material. First use of roving vehicle.	up to 4.3 (upland) 3.3–3.4 (mare)
Apollo 16	Apr. 20, 1972	Lunar uplands near crater Descartes	First landing in uplands. Samples of upland materials.	3.8–4.2
Apollo 17	Dec. 11, 1972	Taurus Mts.; edge of Mare Serenitatis	Samples from a region suspected of recent volcanism.	3.8 (mare) 4.2–4.4 (upland fragments)

Note: In addition to samples mentioned above, lunar soil samples were returned to Earth by three unmanned Soviet probes, Luna 16, 20, and 24. Ages were 3.3–3.4 billion years for basalt lavas and about 4.4 billion years for an anorthosite upland rock chip. Additional studies of isotopes in all the rock samples indicate that the Moon as a whole formed around 4.5 billion years ago.

[a]Date lunar orbit began.

[b]Date of solidification of crystalline rocks, or formation of breccias, based primarily on rubidium–strontium results of Wasserburg, Papanasstasiou, Tera, and colleagues at the California Institute of Technology, and a summary by Taylor (1982). Additional rocks and soils were collected.

crystallization, melting, recrystallization, and so on. Rocks also tell the histories of their parent planets. Rocks can be analyzed in various ways:

1. By their elements and isotopes, which indicate the material from which the parent planet originally formed

2. By their minerals, which indicate the degree of differentiation that produced different chemical compounds inside the planet

3. By their structure, which indicates their environments through history

4. By their radioisotopic ages, which indicate when the planet-forming, differentiating, and rock-altering processes occurred

If the concepts in this list are not clear, the reader should review Chapter 6.

The elements and isotopes of lunar rocks are similar to those of Earth's, though the Moon has much less iron. No new elements or bizarre compounds were found on the Moon, and the proportions of elements suggest that the Moon formed from material similar to that of Earth's mantle. However, relative to Earth's mantle, the Moon is strongly depleted in **volatile elements** and compounds—substances that are driven off by heating, such as water. Similarly, the Moon is enriched in **refractory elements** and compounds—substances with high boiling points, such as aluminum and titanium. These findings indicate that the lunar material may have

Figure 7-16 Astronaut Harrison Schmitt collects lunar pebble samples with a "rake" during the Apollo 17 mission, final mission to the Moon. Schmitt stands here on dark mare soil; brighter hills of the lunar highlands can be seen in the distance. (NASA.)

Figure 7-17 A sample of basaltic lava from the Moon. This sample shows frothy texture caused when bubbles of gas formed inside the molten rock; similar textures are found in terrestrial lava. Other textures are also found on both the Moon and Earth. (NASA; Apollo 15 sample.)

been strongly heated before the Moon formed, driving off the water and other volatiles into space.

Although the minerals of the lunar rocks are similar to those of Earth's mantle and crustal basalts, there is a very interesting difference. The Moon's rocks are not as differentiated as Earth's, and there are no granitic continental blocks or plates in the lunar crust. These facts are among many lines of evidence suggesting that the Moon has not been nearly so geologically active as Earth. This is because the Moon is so much smaller than Earth that it cooled more quickly, formed a thicker lithosphere, and never developed the plate movements or perpetual volcanic and tectonic activity that make Earth's geology so complex.

The evidence for a lunar magma ocean has led to a new view that many or all planets may have initially had magma oceans that allowed low-density minerals to concentrate near their surfaces. This influenced the rock types, minerals, ores, and so on, that we now find on the surfaces of these bodies.

All of the huge impact basins formed as part of the intense cratering that occurred before or around 4 billion years ago. The biggest impacts created fractures that allowed basaltic lavas to erupt from depths of around 300 km (a depth determined from certain mineral properties), and lunar lava flows filled many of the impact basins between 4 and 3 billion years ago. Because the Moon is smaller than the Earth, however, it cooled much faster than the Earth and formed a thick, solid, rigid lithosphere. It has been relatively quiet since then.

It is extraordinary to realize that astronauts have walked and driven across landscapes whose mountains and craters have lain still and silent since trilobites crawled on our own seafloors and the first lizardlike creatures crawled out on our land. Perhaps the most important fact to remember is that the lunar lithosphere formed very early and was very thick, as deep as 1000 km. Thus it was too rigid to allow mountain-building activity or plate tectonics, and it blocked magmas from reaching the surface. Thus the Moon's surface preserves craters and rocks formed 3 to 4 billion years ago. In contrast, the Earth has a thin lithosphere only 100 km thick and easily broken by earthquakes and volcanic eruptions. It is so active that it has destroyed most landforms and

Figure 7-18 Apollo 16 astronauts drove their battery-powered "rover" (rear) to the rim of this 40-m-diameter crater in 1972. Note that the crater is old enough to have been smoothed by the sandblasting effect of innumerable small meteorite impacts that pock it with smaller pits and blanket it in regolith. (NASA.)

Figure 7-19 The landing module at the Apollo 15 site at the foot of the Apennine Mountains. The tracks of the astronauts' battery-powered "rover" vehicle are visible in the powdery regolith. (NASA.)

Figure 7-20 Apollo 17 astronauts obtained samples from huge boulders, apparently dislodged in the past from bedrock further up the slopes of the upland hills near their landing site. (NASA.)

rocks formed during this period. Thus the Moon supplies some heretofore missing information about the early history of the planets and the early environment of the Earth–Moon system.

THE INTERIOR OF THE MOON

We can understand the Moon better by considering its interior structure, shown in Figure 7-23.

First, the **mean density** (total mass divided by

total volume) of the Moon is much less than that of the Earth—3300 versus 5500 kg/m³ (3.3 versus 5.5 g/cm³). This proves that the Moon is mostly rocky, like Earth's mantle, and that it lacks a big iron core. Second, measurements indicate that the Moon has virtually no magnetic field. This again suggests the lack of a large molten iron core, because scientists believe that the planets' magnetic fields originate in currents in such cores. Nonetheless, magnetic measurements on lunar rocks indicate that when they solidified billions of years ago,

OPTIONAL BASIC EQUATION V

The Definition of Mean Density

When an astronomer begins to get data on a new object, such as a newly measured planet, asteroid, or star, one of the first things he or she wants to know is the mean density of the object, because this helps clarify the composition of the object. The mean density is defined as

$$\text{Mean density} = \frac{\text{total mass of object}}{\text{total volume of object}}$$

This is called the *mean* density because it gives only the average density, averaging over the whole object. Parts of the object's interior might have higher or lower densities. The Earth's mean density, for example, is about 5500 kg/m³, but the iron core has much higher values, in excess of 8000 kg/m³, while surface rocks have values closer to 2500 kg/m³. Thus, for example, if you sighted a round object in space, it would make quite a difference to your interpretation if the mean density turned out to be 8000 kg/m³ instead of 2000 kg/m³!

The value of 1000 kg/m³ is of particular interest, since this is the density of water and the approximate density of ice. (When the metric system was set up, the kilogram was defined such that 1000 kg equaled the mass of a cubic meter of water.)

Since most objects in astronomy are roughly spheroidal in shape, we can replace volume by the expression for the volume of a sphere, $4\pi R^3/3$. Thus, using R as the radius, M as the mass, and the Greek letter ρ (rho) traditionally used as the density, we get the equation

$$\text{Mean density} = \rho = \frac{3M}{4\pi R^3}$$

Thus if an astronomer or astronaut can measure the mass M of an object (perhaps by its gravitational force on another object) and its radius R, he or she can quickly calculate the mean density and make some comments about the types of materials that might compose the object.

Note that in the SI system of units, M is measured in kilograms and R in meters, so that the density is expressed as kilograms per cubic meter. While water and ice have densities of about 1000 kg/m³, rock has densities around 2500 to 3000 kg/m³, and gaseous objects have densities less than 1000 kg/m³, unless they are compressed by strong gravity. In many scientific and other books, density is more commonly expressed in units of grams per cubic centimeter, which are 1/1000 of the value in kg/m³. For example, water has a density of 1 g/cm³.

Sample Problem. Voyager spacecraft and Earth-based measurements showed that the small moon of Saturn, Enceladus, has a surface of frozen water, a radius of 250 km, and a mass of 8.4×10^{19} kg. Find the mean density of Enceladus and comment on whether it might be largely icy or rocky throughout. *Solution:* Converting R to meters and inserting the given values for R and M, we compute a density of 1283 kg/m³. Since R and M are given only to two significant figures, we should round off our answer to two significant figures, or 1300 kg/m³. Note that this is only 30% greater than the density of ice! Thus Enceladus could not be made mostly of rock, but it could be mostly icy throughout.

Figure 7-21 View down a canyonlike lunar rille (valley). Apollo 15 astronauts attempted to determine the origin of rilles by visiting this example, but found no conclusive evidence. Rilles probably are formed as channels along which lavas flowed during mare formation. (NASA; Apollo 15.)

Figure 7-22 The payoff from the Apollo missions came with the returns to Earth. This blastoff from the Moon was televised to Earth by a remote camera set up by Apollo 16 astronauts. Debris flying away from spacecraft includes material blown off the descent stage (lower) by ignition of the ascent engine. (NASA.)

there was a lunar magnetic field roughly 4% the strength of the Earth's field. Because no such field exists today, the implication is that the Moon had a molten iron core only in the past, when its interior temperatures were higher. A small core might still exist but reach no farther than about 350 km from the Moon's center, or about 20% of the radius (Lammlein and others, 1974).

Third, seismic data show much less quake activity on the Moon than on Earth. Large lunar quakes rank only 0.5 to 1.3 on the Richter scale, compared with 5 to 8 for major earthquakes. In a year, the Earth expends

about 100 million million times as much seismic energy as the Moon.

Nonetheless, moonquakes tell us something about the Moon's interior. The quakes occur mostly at depths of 700 to 1200 km. Just as Earth's quakes are concentrated at the bottom of the brittle lithosphere and the top of the sluggishly moving asthenosphere, the Moon's quakes are believed to mark the bottom of a 1000-km-thick lunar lithosphere. Many moonquakes occur in monthly cycles associated with tidal flexing as the Moon's slightly elliptical orbit brings it toward and then away from the Earth (Toksoz and others, 1974). Some quakes detected by Apollo instruments had a different source— the impacts of modest-sized meteorites (too small to make craters visible from Earth).

CRATERING OF THE MOON AND EARTH

As noted earlier, the Moon underwent an intense bombardment from about 4.6 to 4.0 billion years ago, with a cratering rate thousands of times higher than today's rate. All large impact basins, such as the one in Figure 7-14, as well as the heavily cratered uplands formed at that time. From about 4.0 to 3.0 billion years ago, the rate declined to the present level, which has been nearly constant since that time. Earth has had roughly the same cratering rate, and scientists assume that both Earth and the Moon experienced similar cratering histories. The Apollo data from the Moon thus help clarify

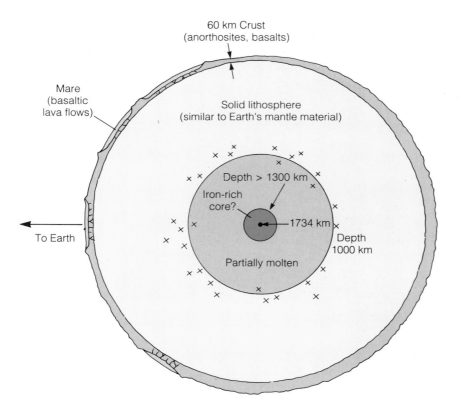

Figure 7-23 Cross section of the Moon as revealed by Apollo and other sources. The crust is thinner on the front side than on the far side. Fractures under large impact craters on the front side allowed lava to reach the surface and create more mare lava flows than on the far side. An earthquake zone (x) marks the bottom of the lithosphere. The existence of an iron-rich core is uncertain.

the impact history of Earth, paving the way for recent theories that giant impacts may have altered climates and biological evolution (see Chapter 6).

The **early intense bombardment** of the Moon represents the final stages of the sweep-up of debris left over after planet formation (see Chapter 14). The now visible lunar craters may represent only the last interplanetary bodies to be swept up, with still earlier craters obliterated by the ones we now see.

Craters are the most important landforms on many other planets and moons as well. The photo in Figure 7-24 shows how meteorite impact explosions may have looked. The meteorites ranged from abundant microscopic particles, through numerous kilometer-sized chunks, to objects over 100 km in diameter. The latter made craters about 1000 km across.

Repeated cratering by a rain of meteorites explains lunar topography. In the lava plains, there have been only enough craters in the last 3 billion years to cover a small portion of the surface with multikilometer craters and to grind up a layer of lava into the regolith some tens of meters thick. But in the uplands, the 4- to 4½-billion-year-old surface is virtually saturated with

multikilometer craters. Small impacts and overlapping ejecta blankets have smoothed the once craggy landscape, producing the gently rolling hills photographed by astronauts. Only occasionally did astronauts encounter medium-sized craters young enough to hint at the chaotic rubble created by the original explosive impacts, as seen in Figure 7-25.

Among the many lunar landscapes not yet seen by astronauts, large young craters would be the most rugged and dramatic, as shown by Figure 7-26.

WHERE DID THE MOON COME FROM?

The Moon's origin has long frustrated theorists. Prior to the Apollo missions, astronomers debated three main theories of lunar origin. The **fission theory** proposed that the Moon is made of material spun off the outer layers of the Earth and that it has receded ever since, due to tidal action. The **co-accretion theory** proposed that the Moon formed in orbit around the Earth at the same time the Earth was forming. The **capture the-**

Figure 7-24 Explosion tests on Earth simulate many features of meteorite impacts. This explosion of 100 tons of TNT produced a crater 39 m (128 ft) across and 7 m (23 ft) deep. Secondary impact craters formed as far as 110 m (360 ft) away and the furthest ejecta went 200 m (660 ft). The jets of material may simulate material that formed lunar rays; the turbulent cloud at the base of the explosion expands across the landscape and deposits much of the ejecta blanket. (Photo by author.)

Figure 7-25 A relatively young 900-m-diameter (½-mi-diameter) impact crater in the lunar uplands. Rock samples indicate that the crater formed about 50 million years ago. (NASA; Apollo 16.)

Figure 7-26 Views of the prominent, large, young crater Tycho, about 90 km in diameter. The age of the crater has been estimated at 100 million to 270 million years.
a Remote view of the Moon centered above Tycho, showing Tycho's system of bright rays. These are caused by streamers of ejecta shot out from the crater during the impact explosion that formed it. (Rectified photo, Lunar and Planetary Laboratory, University of Arizona.) **b** View of Tycho showing rugged interior and surroundings and central peaks. Box shows location of photo c. (NASA; Orbiter 5.) **c** Close-up of the extremely rugged surface just above the central peak shown in the preceding view. Smallest details are about 100 m (100 yd) across. Tycho is apparently too young for cratering to have "sandblasted" these rugged features and converted them to a smooth regolith. (NASA; Orbiter 5.)

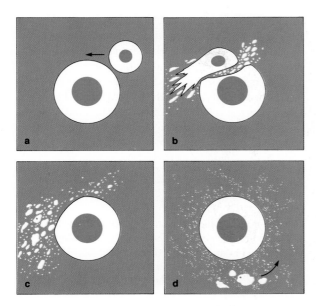

Figure 7-27 Schematic view of the giant-impact hypothesis for lunar origin. A large interplanetary body approaches **a** and collides with the primordial Earth **b**, following formation of iron cores (dark) in both bodies. Hot gas and condensing debris from the mantles of both bodies are thrown into near-Earth space **c**. Part of the debris forms an orbiting cloud around Earth, in which the Moon begins to aggregate (**d**). (Adapted from computer models by A. Cameron, W. Benz, W. Slattery, M. Kipp, and J. Melosh.)

ory proposed that the Moon formed somewhere else in the solar system, and then was later captured into an orbit around the Earth.[2] Each of these theories now seems to have fatal flaws. The fission theory does not explain how enough energy could have been provided to lift a swarm of particles off Earth. The co-accretion theory does not explain how the Moon could have formed near Earth but have received so much less iron than Earth. The capture theory does not explain where the Moon formed in order to receive its low-iron composition. Moreover, calculations show that capture of a "moon" approaching from some other region is extremely improbable because the Moon would not easily have slowed to a speed near circular velocity.

Apollo studies provided new information. Especially important was the finding that the Moon's bulk material is generally like the Earth's mantle, but with less vol-

atiles and slightly more refractories. Another important result was that the proportions of different isotopes composing lunar oxygen exactly match those in terrestrial oxygen. Studies in the 1970s showed that oxygen in rocks from other parts of the solar system (such as meteorites) has different isotopic proportions. This establishes that lunar material formed at about Earth's distance from the Sun and rules out, for example, capture of a body from far away.

In 1984, an international conference agreed on a new leading theory, called the **giant-impact hypothesis**.[3] It suggests that during the final stages of Earth's formation, but after its iron core formed, Earth was hit by a large interplanetary body—perhaps as large as Mars. (As will be clearer in Chapter 14, many sizable interplanetary bodies were growing as the planets formed.) The impact blasted out hot debris from the upper mantles of both Earth and the impactor (Figures 7-27 and 7-28). This debris, lacking iron and depleted in volatiles because of the heat of the impact, aggregated to form the Moon. This theory explains a Moon with roughly mantlelike composition, little iron, few volatiles, and terrestrial oxygen-isotope proportions. It also explains the Moon's uniqueness: Such an impact might have happened to only one of the nine planets. Continuing work on this theory includes chemical studies of lunar rocks and computer models of the giant impact (Hartmann, Phillips, and Taylor, 1986).

The firmest information so far obtained from lunar rocks about lunar origin is not about the mode but about the *timing* of the event. The Moon formed about 4.6 billion years ago, and the process was relatively fast, in astronomical terms, taking no more than about 90 million years—the time interval estimated for formation of the entire solar system.

RETURN TO THE MOON?

Ever since Apollo 17 blasted off the Moon in 1972 (Figure 7-29), the Moon has been deserted. During the 1980s, a number of scientists and NASA planners began

---

[2]With a bit of tongue-in-cheek male chauvinism, NASA geologist Bevan French has called them the daughter, sister, and pickup theories.

[3]This theory was first published in 1975 by D. R. Davis and myself and, essentially independently, in 1976 by A. Cameron and W. Ward, but little further work was done on it for several years. By the time of the 1984 conference, new data were available and sophisticated computer models of impact were possible, and it was exciting to see our idea emerge from relative obscurity into the status of the leading theory!

Figure 7-28 The giant impact that blew mantle material out of primordial Earth's mantle. This view, based on computer models of the event, shows luminous matter, some as hot as the Sun's surface, spraying outward about $\frac{1}{2}$ h after the impact. (Painting by author.)

advocating a return to the Moon, perhaps as part of a large-scale plan to establish research colonies on the Moon and Mars. An unmanned satellite to orbit the Moon has been advocated but not funded in the United States; the Soviet Union and Japan are both reportedly planning one. It would measure lunar rock compositions over the whole globe of the Moon, clarifying what resources might be available there. Permanent manned lunar stations could produce oxygen from lunar rocks for breathing and for fuel (Thomsen, 1986). Supply of liquid oxygen fuel from lunar rocks for spaceships and space stations might ultimately be cheaper than hauling it up from Earth, because so much less energy is needed to launch material off the Moon than off the Earth. Such

lunar research stations would also clarify the Moon's origin and evolution, and would conduct astronomical research more effectively than from low Earth orbit, where half the sky is blocked by the Earth.

Construction techniques for future lunar bases have already been tested on Earth. Although modules might be transported ready-made from Earth, 1986 tests on lunar soil showed that it is ideal for making concrete (Lin, 1986)!

SUMMARY

The Moon is an ancient planetary body, little disturbed since the formative days of the solar system. Much of the information in this chapter yields a chronological history of the Moon. Because the Earth shared much, if not all, of this history, we summarize the history of the Earth–Moon system in Table 7-2.

Origin Lunar rocks and meteorites reveal that the Moon and planets were formed 4.6 billion years ago. The Moon's formation may have involved a giant impact. The Earth and Moon were close together shortly thereafter.

Duration of Formative Process Differences in ages among lunar and meteorite specimens indicate that the Moon and planets reached approximately their present sizes a few million to 90 million years after the formative process began.

Magma Ocean Analyses of lunar rocks indicate that the lunar surface was initially covered with a molten magma ocean several hundred kilometers deep. An initial magma ocean may also have formed on Earth, but evidence here has been destroyed.

Early Intense Bombardment Nearly all lunar rocks that formed before about 4 billion years ago have been pulverized by repeated bombardment. According to Apollo data, the cratering rate 4 billion years ago was 1000 times the present rate, and before that it was still higher. The Earth was presumably also bombarded at that time. Many large craters formed on the Earth and the Moon during this period. The cratering rate declined to the current value by 3 billion years ago.

Tidal Movement of the Moon The Moon was once much closer to the Earth than it is now. Tidal analysis proves that from its moment of closest approach, it moved out quickly, reaching about half its present distance in only 100 million years, or 2% of the time since Earth's forma-

Figure 7-29 Apollo lunar landing module on its way back to command module and Earth, after blasting off the Moon. Such flights raise the question of what nation or nations will return to the Moon to continue lunar research. (NASA.)

tion. If the Moon formed alongside the Earth as the Earth formed, then it probably approached its present orbit between 4.5 and 4 billion years ago, and the Earth's day approached its present length at that time.

Heating and Differentiation Radioactive elements in the Earth and Moon heated their interiors, allowing heavy metal materials to sink to the center, probably within a billion years of the formation of these bodies. The Earth formed a large iron core, but the Moon had less iron and formed little or no core.

Mare Lava Flows Basaltic lavas erupted in many places on the front side of the Moon, especially where the crust was thinnest because of large-scale impacts. Successive lava flows formed the mare plains 3.8 to 3.2 billion years ago. Because the Moon is smaller than the Earth, it cooled more quickly and most volcanism died out around 3 billion years ago.

Lithosphere Evolution The Moon cooled off and formed a 1000-km-thick lithosphere that shielded its surface from internally caused changes. Convection currents continued in the Earth, breaking the 100-km-thick lithosphere into moving plates.

Sporadic Recent Cratering Occasional large impacts, involving meteorites a few kilometers across, excavated major craters, throwing bright rays of pulverized ejecta across the dark maria and old uplands. The most prominent example is shown in Figure 7-26. Smaller meteorite impacts churned the upper soil, producing the regolith. On Earth some of the large craters formed by meteorites during the last billion years have also been preserved (as described in Chapter 6).

Terrestrial Erosion The parts of terrestrial history studied by most geologists are only the tail end of Earth–Moon history. Because of plate tectonics and atmo-

TABLE 7.2

History of the Earth–Moon System

Time (billions of years ago)   **(Present)**

4   3   2   1   0

EARTH

Formation

Intense cratering[a]

Crustal formation

Shorter day

Length of day
increasing (due to tidal
interaction with Moon)[a]

Early life

Oldest
surviving rocks

Atmospheric changes
(O_2 increase due to plants)

? ←Continental evolution——→

Mammals→| |←
Man→|| ←

Earliest substantial fossils

MOON

Formation[a]

Magma ocean[a]

Intense cratering[a]

Early heating[a]

Interior cooling[a]

Sporadic cratering[a]

Rapid motion
away from Earth

←Continued slow movement
away from Earth→

Occasional moonquakes
and gas emissions[a]

Magma ocean cools[a]

Mare formation[a]

4   3   2   1   0

Time (billions of years ago)   **(Present)**

[a]Information discovered or improved through Apollo-related lunar research.

spheric erosion, most of the Earth's surface rocks represent only the last 12% of Earth–Moon history, whereas most lunar surface structures date back through 84% of it. The Moon shows more clearly than the Earth the early and middle parts of geological history and the combined results of internal geological processes (for example, partial melting) and external astronomical processes (impacts). The Moon has given us new understanding of processes that shaped our own world, and we have left our mark on it.

CONCEPTS

satellite	impact crater
phases	mare (maria)
new moon	basin
first quarter	ray
gibbous	regolith
full moon	ejecta blanket
third quarter	magma ocean
sidereal period	breccia
synchronous rotation	volatile elements
tide	refractory elements
body tide	mean density
ocean tide	early intense
tsunami	bombardment
tidal recession	fission theory
Roche's limit	co-accretion theory
terminator	capture theory
limb	giant-impact hypothesis

PROBLEMS

1. At what time of day (or night) does:
   a. The first quarter moon rise?
   b. The full moon?
   c. The last quarter moon?
   d. The new moon?

2. When a terrestrial observer is recording a new moon, what phase would the Earth appear to have to an observer on the Moon?

3. Draw a diagram like Figure 7-2 and show approximate locations from where Figures 7-3a and 7-3b could have been taken.

4. Explain why the Moon keeps one side toward the Earth.

5. Why might one expect the highest tides to occur at noon or midnight on the date of the new moon or the full moon? Why don't the highest tides always occur at these times?

6. Suppose an astronaut orbiting just above the atmosphere releases into an orbit of their own two Ping-Pong balls just in contact with each other. Would you expect them to stay in contact with each other indefinitely? Why or why not?

7. Imagine you are selecting a lunar landing site.
   a. What type of feature might offer fresh bedrock where regolith layers have been stripped away?
   b. Would you expect the landscape inside a young 100-km-diameter crater to be rougher or smoother than inside an old crater of the same size?
   c. Where might astronauts land to seek evidence of recent volcanic activity?

8. Imagine you are an astronaut exploring the Moon.
   a. Would an ordinary compass work on the Moon? Why or why not?
   b. What celestial object could serve as a directional aid (like the North Star) for astronauts hiking on the Moon's front side?
   c. What property of this object's apparent motion would make it especially useful as a navigational aid during a lunar stay of several months?

9. If the ages of the Earth and Moon are identical, as believed, why are most rocks found on the Moon so much older than Earth rocks?

10. Suppose sedimentary rocks had been discovered on the Moon. How would this affect our beliefs about the Moon's history?

11. By comparing pictures of lunar maria (3 to 4 billion years old) and uplands (4 to 4.5 billion years old), prove that the meteoritic cratering rate was much higher during the first few hundred million years of lunar history than during the last 3 billion years.

ADVANCED PROBLEMS

12. The Moon has about 0.012 of the Earth's mass and 0.27 of the Earth's radius.
   a. Using Newton's law of gravity, show that the Moon's surface gravity is about one-sixth that of the Earth's.
   b. How much would an 82-kg (180-lb) person weigh on the Moon? Give your answer in pounds.
   c. Note that pounds are a measure of weight, whereas kilograms are a measure of mass. Would an 82-kg person have a different mass on the Moon?

13. Prove that the Moon subtends an angle of about ½°. It is 3476 km in diameter and averages 384 000 km away.

14. How fast must a projectile move to escape from the Moon? (Mass of Moon = 7.35×10^{22} kg; radius = 1.74×10^{6} m.) Compare this with the escape velocity from Earth.

15. From planetary data given in Table 8-1, confirm that the Earth's mean density is roughly 5500 kg/m^3 and that the Moon's is roughly 3300 kg/m^3. Explain how this result alone proves that the Moon cannot contain as much iron as the Earth.

16. If you have a small telescope that resolves details 2 seconds of arc across, what is the smallest crater you can see on the moon?

PROJECTS

1. Observe the Moon at different phases with a telescope of at least 5-cm (2-in.) aperture. Locate and compare the texture of upland regions with maria (dark plains). Compare visibility of detail near the terminator and away from the terminator, and explain the difference. Sketch examples of craters, ray systems, and mountains.

2. With a telescope, find an example of a bright-ray crater, such as Tycho or Copernicus, and compare its appearance at full Moon (high lighting) with its appearance near the terminator (low lighting). Why do the rays disappear under low lighting? Make a simulation of this effect by scattering a thin dusting of white flour or powder on a slightly darker, textured surface with a raylike pattern. Illuminate with a bright light bulb from above (full moon) and from the side at a low angle (low lighting), and compare the appearance.

3. Prepare a box with white flour several centimeters deep and a light dusting of darker surface powder (flush with the top edge of the box). Drop different-sized pebbles into the box to make craters. Illuminate with a bright light bulb from various angles and compare with the appearance of the lunar surface. Compare the number of craters needed to simulate a mare region and an upland region. Can the surface be saturated with craters if enough stones are dropped? What physical differences exist between this experiment and lunar reality? (Example: These stones hit at a few meters per second, whereas meteorites hit the Moon at several kilometers per second and cause violent explosions.)

The Solar
System

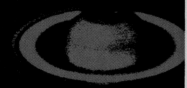

Three worlds in one photo.
Voyager 1 made this image showing giant,
cloudy Jupiter in the background and
two of its moons, Europa (white) and
Io (orangish), passing in front of it.
(NASA.)

CHAPTER 8

Introducing the Planets—Mercury

If we could journey far beyond the Moon and look back, we would see the Sun and its family of planets—the **solar system**. We would discover that the Earth is only the fifth largest of many worlds that orbit around the average-sized star we call the Sun. This is a far cry from the conception of twenty generations ago, when the Earth was viewed as a kind of imperial capital of the universe—a unique stationary scene of human activities around which the Sun, Moon, planets, and stars moved. The exciting transition from the older idea to the modern conception began what astronomer Carl Sagan has called "the cosmic connection," the realization that we are only one part of a larger system of worlds—an idea still growing in our consciousness even today.

Chapter 3 described how the arrangement of the planets' orbits was discovered, and Figure 3-10 showed our modern knowledge of that arrangement. In this chapter we begin describing the other planets, their moons, and the interplanetary bodies. We will focus here not so much on their orbits as on their properties as worlds. The idea that we live in a system of worlds suggests a new conception of a large theater in which we may travel, investigate many new examples of geological, meteorological, and biological processes, and perhaps exploit new sources of energy and materials.

A SURVEY OF THE PLANETS

The solar system is defined as the Sun, its nine orbiting planets, their own satellites, and a host of small interplanetary bodies, such as asteroids and comets. Starting in the center of the solar system, the major bodies and their symbols are:

⊙ **Sun**
☿ **Mercury**
♀ **Venus**

⊕ **Earth**
♂ **Mars**
♃ **Jupiter**
♄ **Saturn**
♂ **Uranus**
Ψ **Neptune**
♇ **Pluto**

The symbols, mostly derived from ancient astrology, are sometimes used as convenient abbreviations today. A traditional memory aid for this outward sequence is *"Men Very Early Made Jars Stand Upright Nicely, Period."* Surely today's students can do better![1] To avoid confusion about the positions of Saturn, Uranus, and Neptune, remember that the *SUN* is a member of the system, too.

Some simple facts about the bodies of the solar system are useful to remember. For example, the Sun is about 10 times the diameter of Jupiter, and Jupiter is about 10 times the diameter of Earth. Whereas the Sun is a **star**, composed of gas and emitting radiation by its own internal energy sources, **planets** are bodies at least partly solid, orbiting the Sun, and known to us primarily by reflected sunlight. **Satellites**, in turn, are solid bodies orbiting the planets.

The planets divide into two groups. **Terrestrial planets** are the four inner planets, Mercury through Mars. They most nearly resemble Earth in size and in rocky composition. The **giant planets** are the four large planets of the outer solar system, Jupiter through Neptune. Much bigger than the terrestrial planets, they also have a different composition, being rich in icy or gaseous hydrogen compounds such as methane (CH_4), ammonia (NH_3), and water (H_2O). Pluto falls into neither category, being a special, mysterious case.

Table 8-1 presents a comprehensive list of data. Notice that some satellites are bigger than some planets! The diameters of the remotest satellites are uncertain because they are hard to observe, but the largest satellites in the solar system are Jupiter's moon Ganymede and Saturn's moon Titan, with diameters over 5000 km. They are both bigger than the planets Mercury and Pluto. Planets and moons are not the only large bodies in the solar system. The large asteroid Ceres (diameter

1020 km), which orbits the Sun in a planetlike orbit between Mars and Jupiter, is bigger than half the known satellites of the solar system. As Figure 8-1 shows, there are probably 26 worlds (in addition to the Sun) in the solar system larger than 1000 km across.

Until the decade of the 1970s, even the best telescopic views of Uranus, Neptune, Pluto, and all satellites beyond the Moon revealed only poorly perceived disks, like a pinhead held at arm's length. Virtually nothing was known about the features on these worlds. The satellites were especially anonymous—assumed to be cratered globes not much different from our own Moon. Perhaps the most astonishing discovery of solar system exploration in the last decade was the unforeseen variety among these worlds. As we will see in the next chapters, distant moons include a near-featureless iceball, a sulfur-orange world with dozens of active volcanoes, a world with one blackboard-black hemisphere and an opposing snowy-white hemisphere, a cloudy world where gasolinelike compounds may rain out of the sky, and a world that may have an ocean of liquid nitrogen! The solar system is not just nine planets and a sun; it has dozens of worlds, each with its own personality.

COMPARATIVE PLANETOLOGY: AN APPROACH TO STUDYING PLANETS

Planetology is the study of individual planets and systems of planets. In the early years of planetary studies through telescopes, each planet tended to be characterized as a world unto itself: Certain markings could be glimpsed on Mars; Jupiter had a different type of markings; Saturn had rings; and so on. But in recent decades, as we have come to learn more about planets' surfaces, atmospheres, interiors, and evolution, with spacecraft and with sophisticated astronomical instruments, a new style of planetology has come into being. It is called **comparative planetology**: a systematic study of how planets compare with each other, why they are different, and why certain planets have certain similarities.

In comparative planetology, each planet and moon is regarded as an experiment that teaches us what type of environment evolves if you start with a certain mass, with a certain composition, at a certain distance from the Sun. A good example of this approach comes from

---

[1]A prizewinner in 1984 was created by a student at the University of Hawaii: *My Very Erotic Mate Joyfully Satisfies Unusual Needs Passionately.*

TABLE 8·1

Objects in the Solar System
(Including the Sun, all planets, satellites known through mid 1990, the six largest asteroids, and Chiron)

Object	Equatorial Diameter (km)	Mass (kg)[a]	Rotation Period[b] (d)	Orbital Period (days unless marked)	Distance from Primary (10³ km unless marked)	Orbit Inclination[b,c] (degrees)	Orbit Eccentricity	Escape Velocity (km/s unless marked)	Known or Probable Surface Material
Sun	1 391 400	1.99 (30)	25.4	—	0	—	—	617	Ionized gas
Mercury	4878	3.30 (23)	58.6	89	0.387 AU	7.0	0.206	4.2	Basaltic dust & rock
Venus	12 104	4.87 (24)	243R	225	0.723 AU	3.4	0.007	10.4	Basaltic & granite rock
Earth	12 756	5.98 (24)	1.00	365	1.00 AU	0.0	0.017	11.2	Water, granitic soil
Moon	3476	7.35 (22)	S	27	384	18–29	0.055	2.4	Basaltic dust & rock
Mars	6787	6.44 (23)	1.02	687	1.52 AU	1.8	0.093	5.0	Basaltic dust & rock
Phobos	27 × 19	9.6 (15)	S	0.32	9.4	1.0	0.015	11 m/s	Carbonaceous soil
Deimos	15 × 11	?	S	1.26	23	2.8	0.001	6 m/s	Carbonaceous soil
Asteroids									
1 Ceres	1020	1.2 (21)?	0.38	4.6 y	2.77 AU	10.6	0.08	0.6	Carbonaceous soil
4 Vesta	549	2.4 (20)?	0.22	3.6 y	2.36 AU	7.1	0.09	0.3	Basaltic soil
2 Pallas	538	?	0.33	4.6 y	2.77 AU	34.8	0.24	0.3	Meteoritic soil
10 Hygiea	443	?	0.75	5.6 y	3.15 AU	3.8	0.10	0.2	Carbonaceous soil
511 Davida	341	?	0.21	5.7 y	3.19 AU	15.8	0.17	0.2	Carbonaceous soil
704 Interamnia	338	?	0.36	5.4 y	3.06 AU	17.3	0.15	0.2	Unidentified soil
Jupiter	142 800	1.90 (27)	0.41	11.9 y	5.20 AU	1.3	0.048	60	Liquid hydrogen?
J16 Metis	40	?	?	0.29	128	0.0	0.0	20 m/s	Rock?
J15 Adrastea	25 × 15	?	?	0.30	129	0.0	0.0	10 m/s	Rock?
J5 Amalthea	270 × 150	?	S	0.50	181	0.4	0.003	0.13	Sulfur-coated rock?
J14 Thebe	120? × 90	?	?	0.67	222	0.0	0.0	60 m/s	Rock?
J1 Io	3630	8.94 (22)	S	1.77	422	0.0	0.000	2.6	Sulfur compounds
J2 Europa	3138	4.80 (22)	S	3.55	671	0.5	0.000	2.0	H_2O ice
J3 Ganymede	5262	1.48 (23)	S	7.16	1070	0.2	0.001	3.6	H_2O ice, dust
J4 Callisto	4800	1.08 (23)	S	16.69	1883	0.2	0.008	2.4	Dust, H_2O ice
J13 Leda	8?	?	?	239	11 094	26.7	0.146	4 m/s?	Carbonaceous rock?
J6 Himalia	180	?	0.4	251	11 480	27.6	0.158	90 m/s?	Carbonaceous rock?
J10 Lysithea	40	?	?	259	11 720	29.0	0.130	20 m/s?	Carbonaceous rock?
J7 Elara	80	?	?	260	11 737	28.0	0.207	40 m/s?	Carbonaceous rock?
J12 Ananke	30	?	?	631	21 200	147R	0.17	16 m/s?	Carbonaceous rock?
J11 Carme	44	?	?	692	22 600	163R	0.21	20 m/s?	Carbonaceous rock?
J8 Pasiphae	70	?	?	735	23 500	148R	0.38	40 m/s?	Carbonaceous rock?
J9 Sinope	40	?	?	758	23 700	153R	0.28	20 m/s?	Carbonaceous rock?
Saturn	120 660	5.69 (26)	0.43	29.5 y	9.54 AU	2.49	0.056	36	Liquid hydrogen?
S18 Pan	20	?	?	0.58	134	0	0	10 m/s?	Ice?
S15 Atlas	38 × 28	?	?	0.60	138	0.0	0.000	13 m/s?	Ice?
S16 Prometheus	140 × 74	?	?	0.61	139	0.0	0.002	50 m/s?	Ice?
S17 Pandora	110 × 66	?	?	0.63	142	0.0	0.004	35 m/s?	Ice?
S11 Epimetheus	140 × 100	?	S	0.69	151	0.3	0.009	50 m/s?	Ice?

Object	Equatorial Diameter (km)	Mass (kg)[e]		Rotation Period[b] (d)	Orbital Period (days unless marked)	Distance from Primary (10^3 km unless marked)	Orbit Inclination[b,c] (degrees)	Orbit Eccentricity	Escape Velocity (km/s unless marked)	Known or Probable Surface Material
S10 Janus	220 × 160	?		S	0.69	151	0.1	0.007	70 m/s?	Ice?
S1 Mimas	394	3.8	(19)	S	0.94	186	1.5	0.02	0.2	Mostly H_2O ice
S2 Enceladus	502	8.4	(19)	S	1.37	234	0.0	0.00	0.2	Mostly H_2O ice
S3 Tethys	1048	7.6	(20)	S	1.89	295	1.1	0.00	0.4	Mostly H_2O ice
S13 Telesto	≈25 × 11	?		?	1.89	295[d]	0	0	7 m/s?	?
S14 Calypso	30 × 16	?		?	1.89	295[d]	0	0	9 m/s?	?
S4 Dione	1118	1.0	(21)	S	2.74	377	0.0	0.00	0.5	Mostly H_2O ice
S12 Helene	36 × 20	?		?	2.74	377[d]	0.2	0.00	11 m/s?	Mostly H_2O ice
S5 Rhea	1528	2.5	(21)	S	4.52	527	0.4	0.00	0.7	Mostly H_2O ice
S6 Titan	5150	1.3	(23)	?	15.94	1222	0.3	0.03	2.7	Ices, liquid NH_3 & CH_4
S7 Hyperion	350 × 200	?		chaotic	21.28	1481	0.4	0.10	0.1	Ices?
S8 Iapetus	1436	1.9	(21)	S	79.33	3560	14.7	0.03	0.6	Ice and soil
S9 Phoebe	230 × 210	?		0.4	550.5	12,930	150R	0.16	0.1	Carbonaceous soil
Asteroid/Comet 2060 Chiron[e]	350?	?		0.25	50.7 y	13.70 AU	7.0	0.38	0.1?	Carbonaceous soil (?) and volatile ices
Uranus	50 800	8.76	(25)	0.72R	84.0 y	19.18 AU	0.8	0.05	21	?
U6 Cordelia	26	?		?	0.34	49.3	0	0	14 m/s?	Ice and soil
U7 Ophelia	32	?		?	0.38	53.3	0	0	17 m/s?	Ice and soil
U8 Bianca	44	?		?	0.44	59.1	0	0	23 m/s?	Ice and soil
U9 Cressida	66	?		?	0.46	61.75	0	0	35 m/s?	Ice and soil
U10 Desdemona	58	?		?	0.48	62.7	0	0	31 m/s?	Ice and soil
U11 Juliet	84	?		?	0.49	64.35	0	0	44 m/s?	Ice and soil
U12 Portia	110	?		?	0.52	66.09	0	0	58 m/s?	Ice and soil
U13 Rosalind	58	?		?	0.56	69.92	0	0	31 m/s?	Ice and soil
U14 Belinda	68	?		?	0.62	75.10	0	0	36 m/s?	Ice and soil
U15 Puck	160 × 150	?		?	0.76	85.89	0	0	126 m/s?	Ice and soil
U5 Miranda	484	7	(19)	S	1.41	130	3.4	0.02	0.4	H_2O ice, soil
U1 Ariel	1160	1.4	(21)	S	2.52	192	0	0.00	0.7	H_2O ice, soil
U2 Umbriel	1190	1.2	(21)	S	4.14	267	0	0.00	0.6	H_2O ice, soil
U3 Titania	1600	3.4	(21)?	S	8.71	438	0	0.00	1.1	H_2O ice, soil
U4 Oberon	1550	2.9	(21)	S	13.46	586	0	0.00	1.0	H_2O ice, soil
Neptune	48 600	1.03	(26)	0.67	164.8 y	30.07 AU	1.8	0.01	24	?
1989 N6	54	?		?	0.30	48.2	4.5	?	26 m/s?	Ice and soil
1989 N5	80	?		?	0.31	50.0	<1	?	48 m/s?	Ice and soil
1989 N3	150	?		?	0.33	52.5	<1	?	74 m/s?	Ice and soil
1989 N4	180	?		?	0.40	62.0	<1	?	84 m/s?	Ice and soil
1989 N2	190	?		?	0.55	73.6	<1	?	106 m/s?	Ice and soil
1989 N1	400	?		?	1.12	117.6	<1	?	222 m/s?	Ice and soil
N1 Triton	2705	1.3	(23)	S	5.88	354	159R	0.00	2.5	CH_4 ice
N2 Nereid	400	?		?	360.2	5515	27.6	0.75	0.2?	CH_4 ice
Pluto	2300	1	(22)?	6.4	247.7 y	39.44 AU	17.2	0.25	0.9?	CH_4 ice
P1 Charon	1190	1	(21)?	S	6.39	19	0R	0.00	0.6?	CH_4 ice?

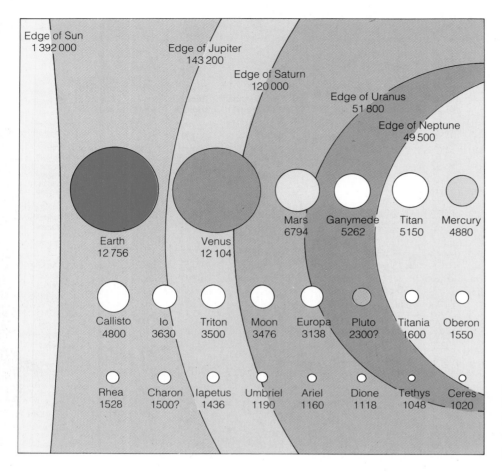

Figure 8-1 The 27 largest bodies in the solar system, shown to true relative scale. They include 1 star, 9 planets, 16 satellites, and 1 asteroid. Diameters are in kilometers.

comparing Earth and Venus. Here are two planets with nearly the same mass and size; but they have radically different atmospheres and climates, as seen in Figure 8-2. The comparative planetologist tries to understand why. Is it related to their different distances from the Sun? Or is there another factor? Such scientists then try to use this knowledge to clarify our understanding of the Earth. Since the atmosphere of Venus has much more carbon dioxide than Earth's atmosphere, for example, we may learn something about the effects of the significant increase in CO_2 in Earth's atmosphere due to pollution.

More possibilities for comparative planetology may be glimpsed from Figures 8-1 and 8-2. Figure 8-2 gives a color comparison of several selected planetary bodies to scale. We see orange deserts on Mars contrasting with the gray lavas of the Moon. Jupiter's icy-gray moon Ganymede (lower left) contrasts amazingly with Sat-

urn's same-sized moon, orange smog-covered Titan. Explaining such differences clarifies not only each world but also the whole system of worlds, including our own. Each world tells not only its own story but also a story of connections. In the next few chapters, we will explore these stories one at a time.

THE PLANET MERCURY

We will start our survey of the planets with the planet closest to the Sun—Mercury. Mercury is the second smallest planet in the solar system. It is only about 40% the size of Earth and about 40% bigger than the Moon. Figure 8-3, showing Mercury passing in front of the Sun, emphasizes its tiny size seen from Earth. For this reason, little was known about Mercury prior to the era of space exploration. A telescope with an aperture larger

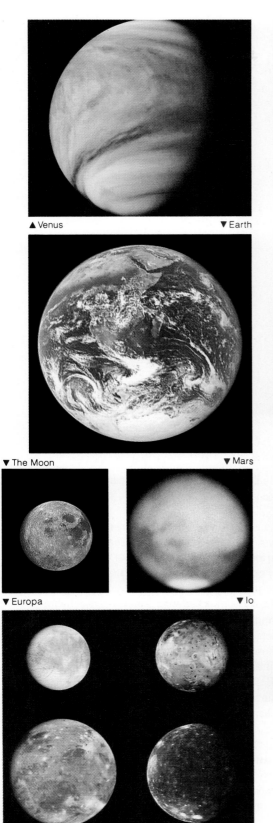

▲ Venus   ▼ Earth

▼ The Moon   ▼ Mars

▼ Europa   ▼ Io

▲ Ganymede   ▲ Callisto

▲ Jupiter   ▼ Saturn

▲ Titan

Figure 8-2 The contrasting appearances of several selected worlds at the same scale. Colors are fairly realistic, except in the cases of Venus (where ultraviolet images are used to bring out the cloud patterns of the relatively featureless yellowish-white clouds) and Saturn's rings (where false color is used to emphasize differences among the whitish rings). (NASA photos.)

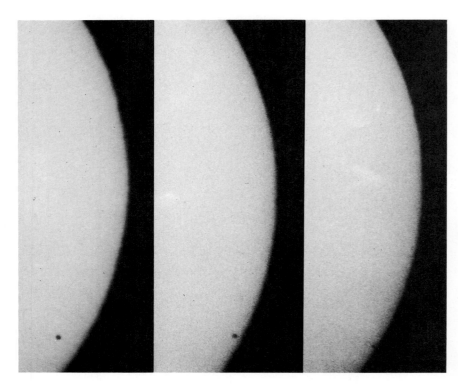

Figure 8-3 Three photos made of Mercury as it transited the Sun on November 10, 1973. Mercury (the small black disk at bottom) is moving off the disk of the Sun. These photos show the small size of Mercury's disk during its closest approach to the Earth, illustrating how poor an earthbound observer's view of the planet is. (Photos with a Questar telescope by W. A. Feibelman.)

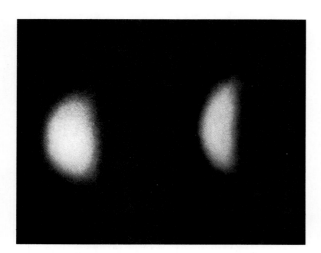

Figure 8-4 Two photographs of Mercury (June 7 and 11, 1934) showing the changing phases of the planet. Virtually no surface detail can be seen; very few photos from the Earth show reliable detail. (Lowell Observatory.)

than about 15 cm (6 in.) reveals that Mercury has phases (Figure 8-4), very faint dusky markings, and a pinkish-gray cast.

In 1974 and 1975, the American spacecraft Mariner 10 sailed past Mercury on three different occasions. Due to Mariner 10's orbit, its cameras were able to record details on only about half the surface of Mercury. But this was sufficient to reveal that Mercury is a world much like the Moon, pocked with craters, marked with giant multiring basins and lava flows, and with virtually no atmosphere. From the standpoint of comparative planetology, therefore, Mercury gives a valuable example of a planet intermediate in size between the Moon and the Earth. Before we go on to show how Mercury's size affects its surface features, we pause to consider some rather peculiar properties of Mercury's motions.

Rotation and Revolution of Mercury

There is a curious relationship between Mercury's rotation and its orbital revolution around the Sun. Due to complex tidal effects (similar to those that hold the Moon

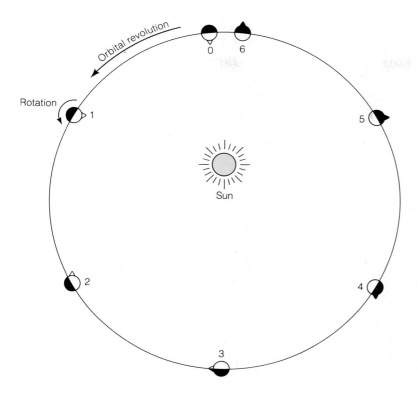

Figure 8-5 A view of Mercury's orbit showing the relationship between its rotation and revolution. The cartoons represent Mercury as a globe with one high mountain marking the rotation position. Shading shows the nighttime side. See the text for a discussion.

facing the Earth), Mercury has gotten locked into a 59-d period of rotation, which is just two-thirds of its 88-d orbital revolution period. The combination of these two rates means that the Sun moves very slowly across Mercury's sky, taking 176 d to go from noon until the next noon!

This can be seen better from Figure 8-5, in which we visualize Mercury at position 0, where the noontime sun is overhead at a certain mountain. At position 1, Mercury has completed a quarter of its rotation relative to the stars. By position 3, Mercury is halfway through its orbit, 44 d have elapsed, and the Sun is just setting on the mountain. By position 4, Mercury is two-thirds of the way around its orbit, but it has completed one rotation, since the mountain now points in the same direction as in position 0. By position 6, Mercury has gone once around the Sun in 88 d, but it is now midnight on the mountain, since the mountain faces directly away from the Sun. Another 88 d are necessary to bring the mountain back to noontime, as you can confirm by sketching in the next orbital trip, positions 7 through 12. Thus the Mercurian "day" (from sunrise to sunrise) is not the 59-d period of rotation but 176 d.

Mercury and Dr. Einstein

At first glance, Mercury's orbital motion seems to contradict the laws of Kepler and Newton: The *perihelion*—the point nearest to the Sun—shifts in position slowly around the Sun from year to year. This movement is called **orbital precession.** Some precession had been predicted from Newton's laws, but observers found an excess shift of 43 seconds of arc each century—a tiny amount, but enough to consternate orbital theorists!

The French astronomer U. J. Leverrier, around 1860, thought that the excess shift might be caused by the gravity of a small, undiscovered planet inside Mercury's orbit. He even gave the planet a name—Vulcan (Moore, 1954). Leverrier had already successfully predicted Neptune's existence from similar gravitational disturbances in the motion of Uranus, but he was wrong in the case of Mercury. Twentieth-century observations reveal no planet inside Mercury's orbit.

But how can Mercury's orbital precession be explained? The solution came in 1915, when Albert Einstein showed that the great mass of the Sun disturbs the orbits of nearby planets in a way unpredicted by

Figure 8-6 Mariner 10 spacecraft photo of Mercury shows a heavily cratered planet that resembles the Moon. Compare with Figure 7-9b. (NASA.)

Newton's laws. Einstein's theory of relativity predicted almost exactly the excess precession observed—43.03 seconds of arc per century. Einstein predicted smaller excesses for Venus and Earth, and these too were confirmed by observation. Thus Einstein's contribution to solving the puzzle of Mercury's precession played a major role in the acceptance of his theory of relativity.

Surface Properties of Mercury

Because Mercury is so close to the Sun, its daytime surface is much hotter than Earth's. Measurements by infrared detectors of the thermal radiation from both the day and night side of Mercury show that temperatures in the upper few millimeters of soil range well above 500 K (441°F) in the "early afternoon" near

perihelion to lows of about 100 K (−279°F) at night. In some areas, depending on soil type, the temperature might exceed 600 K (621°F). Though science fiction decades ago used to describe hypothetical pools of molten metal on the daylight side, this now seems unlikely for several reasons. First, the melting point of lead is about 600 K; that of aluminum is 832 K. Second, because of the insulating effect of the overlying soil, the temperature slightly below the surface is only 314 to 446 K.

Mariner 10 produced the best available data about Mercury, including photographs of the surface (Figure 8-6) and magnetic measurements. The surface resembles the Moon's in its abundance of rugged craters and in its huge, multiringed craters known as basins. A few vague features glimpsed from Earth correlate with certain bright and dark features photographed by Mariner

Figure 8-7 The Caloris basin on Mercury. The center of the basin lies in shadow out of the frame to the left. The left half of the frame is dominated by curved cliffs and fractures surrounding the impact site. The photo from top to bottom covers an area of about 1300 km; the cliffs are believed to be about 2 km (6000 ft) high. Compare with similar lunar features in Figure 7-14. (NASA.)

10. The most prominent basin, shown in Figure 8-7, is named the Caloris basin (in keeping with the high Mercurian temperatures). Its concentric ring structure, more than 1200 km in diameter, strongly resembles the Orientale ringed basin on the Moon (see Figure 7-14). Evidently the same impact and lava flow processes occurred on Mercury as on the Moon some 3.5 to 4.5 billion years ago. From this discovery, most scientists believe all planets suffered an intense bombardment by interplanetary debris at the close of the planet-forming process.

Mercury's landscape probably superficially resembles the lunar landscape: Rolling, dust-covered hills have been eroded by eons of bombarding by meteorites and covered by a regolith. Dusty lava plains and fault-cliffs remind us of violent ancient volcanism. Fresh impact craters might display rugged boulders and outcrops of craggy rocks. As shown in Figure 8-8, the sky is black, dominated during the nearly 88-d period of sunlight by a Sun looking about $2\frac{1}{2}$ times bigger in angular size than the Sun in the Earth's sky. Besides the Sun, two jewel-like planets, much brighter than any stars, occasionally dominate the sky. These are yellowish-white Venus and bluish Earth.

Mercury's Internal Properties and Tectonic Activity

Another comparison with the Moon relates to lithosphere structure. As we indicated in the last two chapters, the lithosphere's thickness is a key to a planet's surface evolution. If the planet is small, like the Moon,

Figure 8-8 Landscapes on Mercury probably resemble those on the Moon. In this imaginary view, the Sun, which would appear $2\frac{1}{2}$ times bigger than from Earth, is hidden behind a rock. The solar corona, or outer atmosphere of the Sun (see Chapter 15), dominates the sky. In the upper right, Venus and Earth (bluer) are seen in conjunction. Zodiacal light (see Chapter 13) stretches to the upper right. (Painting by the author.)

it cools rapidly and a thick, rigid lithosphere forms, preventing any internal activity from breaking through to disturb the surface. If the world is as big as the Earth, however, it takes a long time to cool and only a thin lithosphere has had time to form. In the case of Earth, it is so thin that it is readily broken by faulting and volcanism associated with plate tectonics.

Scientists were thus interested to study the surface features of Mercury, because it is in between the size of the Moon and Earth (but closer to lunar size). Using the philosophy of comparative planetology, we might predict that Mercury would have a thinner lithosphere and more signs of surface tectonic disturbance than the Moon. This turns out to be correct. Cliffs such as those in Figure 8-9 appear to mark huge faults, suggesting that the lithosphere was just thin enough to fracture when Mercury contracted as it cooled. As shown in Figure 8-9, the faults appear to be compressional faults, formed because the surface was growing smaller as Mercury contracted. Aside from these faults, which are not found on the Moon, the rest of Mercury's surface is very lunarlike, with occasional patches of smooth lava

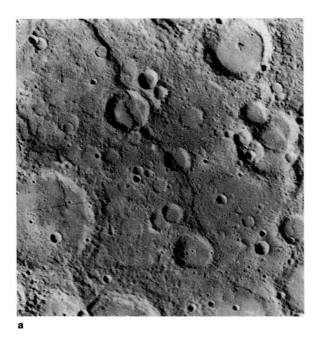

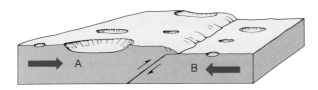

Figure 8-9 **a** A heavily cratered portion of Mercury resembles the lunar uplands. The cliff running from the upper left to the lower right, called Discovery Scarp, may be a fault caused by contraction of the planet following early heating and volcanism. The largest crater (left center) is about 125 km across. (NASA, Mariner 10.) **b** Schematic diagram showing probable geometry of Discovery Scarp.

cury's high mean density (5500 kg/m^3)—high compared to that of the Moon (3300 kg/m^3), which lacks much of an iron core.

A Trace of an Atmosphere on Mercury

Although Mercury has essentially no atmosphere, there is a local concentration of gas atoms around the planet from several sources. Most abundant are sodium atoms, discovered in 1985 by astronomers in Texas. These atoms are probably continually knocked off the surface by gases streaming from the Sun. This does not mean that sodium is the most abundant material on Mercury's surface, but only that sodium is easy to dislodge from the rock minerals because of its chemical properties. Another type of gas concentrated around Mercury is the solar gas itself, which contributes hydrogen and helium atoms to Mercury's environment, but at even lower concentrations than the sodium. This "atmosphere" of gases is so thin that it exerts less than one hundred-billionth of the pressure exerted by our atmosphere. It would not be noticeable to an astronaut, of course, but it is interesting to astronomers because of the process it reveals of solar gas interaction with Mercury's surface.

A LESSON IN COMPARATIVE PLANETOLOGY: SURFACE FEATURES VS. PLANET SIZE

We learned from studying the Moon that intense cratering occurred during the interval from 4.5 to about 4.0 billion years ago. Millions of large meteorites rained down onto the lunar surface. This **intense early bombardment** happened to all the planets—it was the great sweep-up of interplanetary debris left over after the planets formed. It allows us to explain some general features of planets' surfaces.

The basic idea here is a competition between the construction of *internally* derived features, such as volcanoes, and their destruction by the *external* bombardment. The intense early bombardment lasted for the first 500 million years of planetary history—from 4.5 to 4.0 billion years ago. Imagine all the planets starting with a magma ocean and hot interior, products of the formative process and heating by radioactive minerals

filling the larger craters. There is some evidence that Mercury's cratered uplands contain more ancient lava flows than the Moon—again confirming the idea that the larger the world, the more geological activity it has.

The magnetic field of Mercury is only about 1% that of Earth's. Nonetheless, Mercury's magnetic field has the form calculated for an internally generated field like the Earth's, with a magnetic axis tilted about 7° to the planet's rotation axis. These results suggest that this field is caused by weak currents in an iron core like the Earth's core. A large iron core also accounts for Mer-

buried in the planets. Volcanic mountains, lava flows, and faulted cliffs might form as long as a planet remains hot and active and its lithosphere is not too thick. If a planet is small, it cools rapidly and forms a thick lithosphere, say, within 300 million years. Formation of volcanoes and tectonic features stops. The intense bombardment continues for another 200 million years, saturating the surface with craters and destroying traces of the constructional features—just as bombers pounding a city might reduce its buildings to a rubble of rocks and craters. Such a world would show only craters on its surface. On a slightly larger world, like the Moon, a few "last gasp" lava flows might escape from the interior through the lithosphere after 4.0 billion years ago, leaving dark plains on the surface. Because the intense early bombardment had already ended, these features would not be destroyed. On a still larger world, like Mercury, the cooling and lithosphere formation persists after 4.0 billion years ago, creating faulted cliffs as well as lavas that survive. On Earth, of course, the interior remains hot and active to this day, continuing to break the lithosphere with faults, mountain chains, volcanic cones, and the like. Because of the small crater production rate in recent geological time, the construction processes, together with erosion from Earth's winds and rains, win out over the cratering. Relatively few meteorite craters can be found on Earth.

Thus we see a rule of thumb: The larger a world (or, more precisely, the more *massive* a world), the younger and more active its surface; the smaller a world, the older and more heavily cratered its surface. There are four intermediate-sized worlds between Mercury's size and Earth's size: Mars, Venus, Jupiter's moon Ganymede, and Saturn's moon Titan. As we will see in the next few chapters, they tend to confirm our rule.

SUMMARY

According to present data, the solar system includes about 26 planetary worlds larger than 1000 km and a host of smaller bodies. These worlds show a great deal of variety, depending primarily on their mass but also on other conditions, such as distance from the Sun. Details of surface structure and composition are becoming clear for most of these worlds only since the advent of space exploration in the 1960s. A current approach to studying these worlds, called comparative planetology, is to examine them not as isolated cases but in comparison with each other and especially in comparison with the Earth. In this way, we learn which forces determine such properties as surface geology.

Mercury is the closest planet to the Sun. It is the most Moonlike of the planets and is only 40% larger than the Moon. It has a heavily cratered surface that is broken in a few places by large, faulted cliffs produced by contraction. Mercury has virtually no atmosphere.

CONCEPTS

solar system	giant planets
star	planetology
planet	comparative planetology
satellite	orbital precession
terrestrial planets	intense early bombardment

PROBLEMS

1. Which planet can come closest to Earth? (*Hint:* See Table 8-1.)

2. How many times bigger in diameter than Earth is the largest planet? How many times more massive than Earth is the most massive planet?

3. Which planet has the largest satellite?

4. Which planet has the largest satellite measured in terms of the diameter of its planet—that is, which satellite's diameter is the largest fraction of its planet's diameter?

5. Which planet can come closer to Uranus: Earth or Pluto?

6. Using Table 8-1, list some systematic differences in physical properties between terrestrial planets and giant planets. Consider size, temperature, density, number of satellites, and orbital properties.

7. List processes of planetary evolution that have occurred on both Mercury and the Moon. List the processes, if any, that are unique to each. Is there any indication of plate tectonic activity on Mercury? What might this indicate about heat flow and mantle convection inside Mercury?

8. Some of the best telescopic observations of Mercury are made during midday instead of after sunset or before sunrise. Why? Why are none made at midnight?

9. Suppose astronauts plan to land on Mercury wearing spacesuits of the Apollo type used on the Moon. What modifications, if any, might have to be made for using the suits on either the daytime side or the nighttime side of the planet?

ADVANCED PROBLEMS

10. Calculate and compare the orbital velocities of the Earth (1.5×10^{13} cm from the Sun) and Pluto (about 5.9×10^{14} cm from the Sun). The Sun's mass is 2.0×10^{33} g.

11. Suppose the Sun were replaced by a star four times as massive, and the Earth were moved to a new orbit 2 AU from this star. Compare the gravitational force on this "new earth" to that in the present solar system.

12. When earthbound observers can see Mercury partly illuminated, it is at a distance of about 0.9 AU, or 135 million kilometers.

    a. If you used a telescope that revealed details 1 second of arc across, what would be the smallest features you could see on Mercury? (*Hint:* Use the small-angle equation.)

    b. Why can virtually no detail be seen on the surface of Mercury when it is closest to Earth, about 0.61 AU away?

    c. With the telescope in part (a), what would be the thinnest cloud layers visible on Venus at its closest approach to Earth?

13. Of the terrestrial planets, Mercury is most like the Moon in size and surface character. Using the data in Table 8-1, calculate the mean density of Mercury and use this to confirm that Mercury has an iron percentage much more like the Earth's than the Moon's. Comment on how this might affect a lunar origin theory in which the Moon is viewed as an ordinary planetary body captured into orbit around the Earth.

14. Using Mercury's high mean density of 5500 kg/m^3, cited in the text, suppose that Mercury's mantle has a density like the Moon or Earth's mantle, around 3300 kg/m^3, and that the core is made of iron alloys with density 8000 kg/m^3.

    a. Describe how you might calculate the radius of Mercury's core from this information.

    b. Do the calculation if you can. (*Hint:* Recall that the volume of a sphere is $\frac{4}{3}\pi R^3$, where R is the radius of the sphere.)

PROJECTS

1. Try to see Mercury. This is likely to be harder than it sounds, because Mercury is prominent only during its elongation period, which lasts only a few days and then only for an hour or less each day just after sundown or before sunrise. Determine from an almanac or an astronomy magazine when Mercury comes to evening elongation and find a site with a very clear western horizon. It is best to start a few days before the elongation so you can become familiar with the background stars in the appropriate region of the sky.

2. If Mercury becomes visible during your studies, observe it with a telescope of at least 20 cm (8 in.) aperture. Try to detect the phase of the planet. Observe on several successive days, and see if you can detect a change in phase as Mercury circles the Sun.

3. If a planetarium is nearby, arrange a demonstration of the motions of Mercury as seen from space or from Earth.

TABLE 8·1

Sources and Notes

Sources: Data from IAU announcements 3463–3476; Voyager team reports (*Science,* 1979, *204,* 964ff.; 1979, *206,* 934ff.; 1981, *212,* 159ff.); Reitsema, Smith, and Larson (*Icarus,* 1980, *43,* 16); Voyager press releases on Neptune satellites, 1986; Dunbar and Tedesco (on Pluto; *Astron. J.,* *92,* 1201); data tables in *Satellites,* ed. J. Burns, 1986; Thomas and others, 1989, on small Uranus satellites. S18 discovered in 1990; designation "Pan" is suggested but not yet official.

Notes: Numbers assigned to asteroids and outer planets' satellites indicate order of discovery, except for largest satellites. The tables in this book use a system of symbols to indicate data that is approximate ($\sim$), uncertain (?), and not available or not applicable ($-$).

[a]Numbers in parentheses are powers of 10.

[b]An R in this column indicates retrograde motion; S indicates that synchronous rotation has been confirmed.

[c]To ecliptic for planets; to planet equator for satellites.

[d]S12 in the leading Lagrangian point of Dione's orbit. S13 and S14 are in following and leading Lagrangian points of Tethys' orbit, respectively. Lagrangian points are stable points for small bodies in larger bodies' orbits, and are discussed further in Chapter 13.

[e]D. Tholen, W. Hartmann, and D. Cruikshank discovered anomalous brightening of this strange object in 1988, probably due to cometary activity. Karen Meech and M. Belton found a coma in 1989. Though it is catalogued as an asteroid, Chiron turns out to be the largest known comet nucleus!

CHAPTER 9

Venus

Venus is widely regarded as Earth's sister planet because its size is so similar to Earth's. Venus has about 95% of the Earth's diameter (see Figure 8-2) and about 82% of its mass. It has no moons.

From the viewpoint of comparative planetology, the similarities between Venus and Earth make us eager to compare the geological and atmospheric properties of the two planets. Since Venus comes closer to Earth than any other planet, you might think that it would be one of the best-observed worlds. Its surface remained very mysterious until space probes could be sent there, however, because Venus is completely obscured by clouds. Indeed, the brilliant, yellowish-white, nearly blank cloud layer is the first feature to impress a telescopic observer. It prevented early astronomers from observing a single surface detail or even the rotational properties of the planet. Unmanned space probes, both in orbit around the planet and on its surface, have revealed that the atmosphere is totally different from Earth's, while the surface has some similarities and some differences from the ground we walk on. Venus is an interesting challenge to comparative planetologists: Why should a planet close to Earth's orbit, and of nearly the same size, show so many differences? In this chapter we will search for answers.

THE SLOW RETROGRADE ROTATION OF VENUS

Radar signals, bounced off Venus from Earth, reveal that Venus has an unusual rotation unlike that of Mercury, Earth, or Mars. Those planets all have **prograde rotation**—spin from west to east. Venus has **retrograde rotation**—spin from east to west (Figure 9-1). Venus' spin is also unusual in being very slow, taking 243 d to make a complete turn on its axis. These properties were discovered in 1962, when radar signals were first bounced off the planet. The cause of the unusual

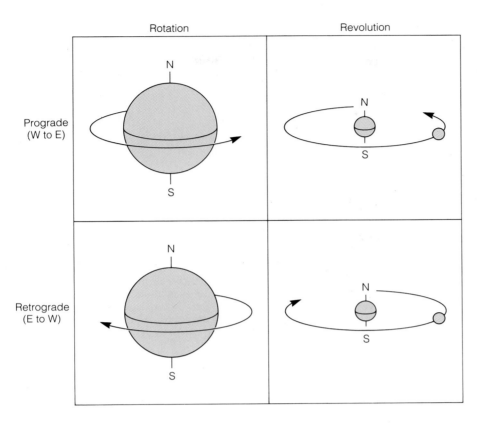

Rotation   Revolution

Prograde
(W to E)

Retrograde
(E to W)

Figure 9-1 Rotation (motion around an internal axis) and revolution (motion around an external body). Prograde motion (top) is west to east. Most planets, including Earth, have prograde rotation and revolution. The rotation of Venus is retrograde, as diagrammed in the lower left figure. Remember the difference between rotation and revolution by noting that the common handgun should be called a rotator, not a revolver!

reverse spin may involve tidal forces between the Earth and Venus or an ancient collision with a body larger than our Moon, of which there may have been many in the early solar system.

VENUS' INFERNAL ATMOSPHERE

The first proof that Venus has an atmosphere came as long ago as 1761, when the Russian scientist M. Lomonosov observed the backlit atmosphere extending around the disk, as shown in Figure 9-2. This phenomenon occurs when Venus is approximately between the Earth and the Sun, so that sunlight backlights the hazy upper atmosphere of Venus, revealing the haze layer ringing the planet.

The Venusian atmosphere intrigued scientists for two centuries after it was discovered.[1] What caused this whitish shroud? What kind of planetary surface did it hide? In 1928 the American astronomer Frank Ross photographed dusky patterns by using plates and film sensitive to ultraviolet. These patterns, shown in Figure 9-3, are formed by cloud layers differing in composition, particle size, or altitude.

In 1932 Mt. Wilson astronomers studied the spectrum of Venus and detected extraordinary amounts of carbon dioxide (CO_2, the same gas dissolved in carbonated soft drinks). Later data showed that Venus' atmosphere is about 96% CO_2, as seen in Table 9-1.

Still undiscovered was the composition of the opaque cloud layer and the air below it (Figure 9-4). In the 1940s and 1950s some writers imagined stormy clouds of water droplets, a surface swept by torrential rains, and vegetation like that of a Brazilian rain forest. Some supposed that the highly reflective clouds would shade the surface and moderate the climate in spite of Venus'

---

[1]Adjectives for planets are controversial. Some writers, claiming that "Venusian" is an ugly word, use "Cytherean" (from a name of Aphrodite), which is merely confusing. "Venereal" is already preempted by other areas of human endeavor. While "Venusian" (ve-NOO-sian) has the sanctity of science fiction tradition, many astronomers use "Venerian" or the noun "Venus" as an adjective.

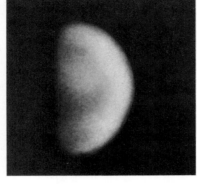

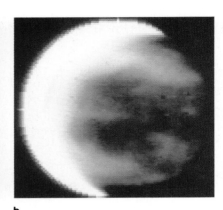

a

b

Figure 9-2 A telescope view of Venus as it passed nearly between the Earth and the Sun. Backlighting by the Sun illuminates the atmosphere of Venus all the way around the disk. This effect, first recorded in 1761, not only proves the existence of an atmosphere but gives information on its structure. (New Mexico State University.)

Figure 9-3 Earth-based images of Venus' clouds. **a** Ultraviolet photo; ultraviolet images generally show faint cloud patterns whose contrast is too low to show up in visible light. (Lunar and Planetary Laboratory, University of Arizona.) **b** Unusual infrared view taken in 1988 shows the cloud patterns on Venus' night side. Sunlit crescent (left) is overexposed. Darker regions on night side are clouds 50 km above surface; lighter regions are clear areas where infrared radiation from hot lower layers breaks through the clouds. (Courtesy W. M. Sinton, Institute for Astronomy, University of Hawaii.)

closeness to the Sun. In the 1960s, however, when the planet's thermal radiation was measured at far-infrared and radio wavelengths, Wien's law revealed that the lower atmosphere has a temperature of about 750 K (891°F)—hardly conducive to liquid water or life as we know it!

In 1970, **Venera 7**, a Soviet probe that was the first spacecraft to land successfully on another planet, transmitted data from the Venusian surface for 23 min. The data confirmed the high temperature and revealed an atmospheric pressure about 90 times as great as Earth's! Instead of our 101 000 N/m^2 (14.7 lb of force pressing on every square inch of surface) the pressure on Venus is about 9 000 000 N/m^2 (1320 lb/in.^{2}), equivalent to that endured by a diver nearly a kilometer (3000 ft) below the terrestrial ocean surface! Later Soviet spacecraft landings confirmed these results. Venus is a stranger world than most humans had imagined!

Venus was soon revealed to be stranger yet. In 1972–1973 astronomers discovered that the clouds of Venus consist not of water droplets, like the Earth's clouds, but rather of tiny droplets of sulfuric acid (H_2SO_4)! In 1978 probes dropped by the American spacecraft

Pioneer Venus showed that the clouds lie primarily in a high layer 48 to 58 km above the surface.

In 1985, two balloons (dropped by Russian probes on their way to Halley's comet) floated in the clouds for 46 h, measuring hurricane-like winds (150 mph) but relatively pleasant conditions at this altitude ($T = 95°F$ and pressure like that on Earth's surface). Venus' clouds are higher than Earth's, which are mostly less than 10 km high.

In spite of the differences, the clouds of Venus and Earth form in a similar way, in atmospheric layers where the temperature and pressure cause condensation of some relatively minor atmospheric constituent. On Earth, this constituent is H_2O, condensed either into droplets (lower clouds) or ice crystals (high cirrus clouds). On Venus, it is H_2SO_4 droplets, which begin to fall as they grow. If a droplet gets big enough, it falls out of the cloud deck, where it encounters much higher temperatures and evaporates. Thus Venus' weird H_2SO_4 rain never reaches the ground. This explains why the clouds have a well-defined bottom surface, as detected by the Pioneer probes, and why the lower atmosphere and surface are clear, as found by the Russian Venera Landers.

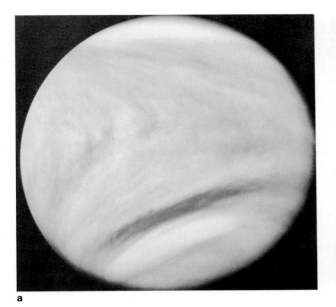 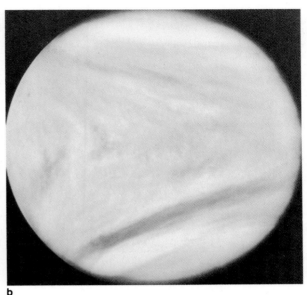

a b

Figure 9-4 Circulation of Venus' atmosphere is shown in these two views taken 5 h apart. In **b**, the features have shifted to the west (left; note *V*-shaped dusky marking in left center). While the planet itself takes 243 d to turn, 200-mph winds blow the clouds around the planet in about 4 d. This is a false-color image made in ultraviolet light. In visible light, the contrast is so low that the yellowish-white clouds look nearly featureless, although telescope observers have sometimes reported the bright polar cloud cap and adjacent dark band. (NASA, Pioneer Venus Orbiter.)

In 1978 the American and Russian Venus probes made yet another startling discovery—terrific blasts of lightning play among the clouds of Venus.

The Greenhouse Effect on Venus

Why is Venus so hot? Planets absorb sunlight and, following Wien's law, radiate infrared light. A planet's surface temperature is determined by the balance between the amount of visible sunlight it absorbs and the amount of infrared radiation it emits. If the solar energy absorbed each second is greater than the infrared energy radiated each second, the planet heats up. If the incoming amount is less than the outgoing amount, the planet cools. The mean, or equilibrium, temperature is reached when the two rates are equal.

The incoming rate is easily calculated as the total sunlight striking the planet minus the amount reflected. The total energy striking the planet per square meter in 1 s is called the **solar constant** for that planet. The outgoing radiation increases as the planet's surface temperature increases.

Suppose the planet has no atmosphere. Then the situation is easy to predict. The surface rocks heat up until their temperature is so high that outgoing infrared radiation equals the incoming sunlight. But if the planet has an atmosphere that absorbs some sunlight (as in Figures 5-13 and 5-16), the atmosphere and the surface both warm up and radiate infrared energy. Because of the nature of atmospheric gases, the outgoing infrared energy from the surface may not escape directly into space. In fact, CO_2 and H_2O gases absorb a great deal of the outgoing infrared, thus adding energy to the atmosphere and warming it even more. The warming continues until the amount of infrared escaping from the top of the atmosphere equals the amount of incoming sunlight. On Venus, the lower atmosphere has to reach about 750 K before this condition is met (Figure 9-5).

This heating is called the **greenhouse effect** because of its resemblance to the heating physics of a greenhouse. The glass panes of a greenhouse admit sunlight but block the escape of the infrared. (They also keep the warm air from escaping—an important function that, in the case of planets, is performed by grav-

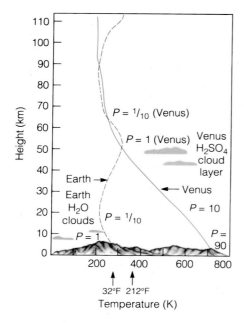

Figure 9-5 Temperature and pressure of Venus' atmosphere (solid curve) and Earth's atmosphere (dashed). The greenhouse effect greatly heats the lower atmosphere of Venus. Pressures (P) are measured in units of the Earth's surface pressure.

ity.) Hence the inside becomes warmer than the outside. The greenhouse effect explains why Venus is so hot: Its massive CO_2 atmosphere blocks outgoing infrared radiation. The greenhouse effect also explains why a cloudy night often stays warmer than a very clear night on Earth, since the water vapor in the cloud layer blocks outgoing infrared radiation from the cooling Earth.

The greenhouse effect is of great concern to environmental scientists on Earth. Burning of fossil fuels has increased Earth's atmospheric CO_2 by perhaps 10% since 1860 (Walker, 1977, pp. 127–128). Of course, there is much less CO_2 here than on Venus, but greenhouse effects associated with this and other atmospheric changes could alter climates and agricultural productivity on Earth. Venus, with its CO_2–caused greenhouse effect, thus serves as a "natural lab" for understanding environmental change on Earth.

Why Venus Has a CO_2 Atmosphere

Why should Venus' atmosphere be mostly CO_2 instead of N_2 and O_2, like the Earth's? (Compare columns in Table 9-1.) The answer is clearer if we rephrase the

question: Why does Earth *not* have a massive CO_2 atmosphere?

The groundwork for answering this question was laid in Chapter 6, where we saw that volcanic degassing of terrestrial planets produces secondary atmospheres. Volcanoes, especially those that release primordial lavas from the mantle, emit mostly H_2O and CO_2 gas, with some N_2. On the Earth, the H_2O formed oceans, and the CO_2 dissolved in the oceans and ended up in carbonate rocks. This explains why Earth was left with a nitrogen-rich atmosphere.

Thus we see that if for some reason a terrestrial planet did not form oceans of water, the CO_2 emitted by its volcanoes would not dissolve in the oceans and would be left as a dominant atmospheric gas. Then if the H_2O molecules disappeared in some fashion, an atmosphere of nearly pure CO_2 would be left. This happened on Venus in the following way.

Many scientists assume that Venus also had primordial volcanoes that emitted much water vapor from its interior. (This view is somewhat controversial pending further surface chemical studies. Other scientists argue that Venus had noticeably less water to begin with; see Grinspoon, 1987.) Because of Venus' heat, any water that found itself in liquid form eventually evaporated, forming H_2O molecules in the air. The H_2O molecules were broken into H and O atoms by energetic solar radiation. The H atoms, being very light, tended to float to the top of the atmosphere and escape into space, a process well documented by atmospheric chemists. Now the question is: What happened to Venus' leftover oxygen? Soviet and American scientists, studying rock and soil data returned by the various Venus landers, believe that most of the oxygen combined with rock minerals and disappeared from the air.

If you could estimate the total amount of O_2 bound up in the oxidized minerals of Venus' rocks, and then imagine reconstituting it into the original water molecules, you could estimate how much water Venus originally had. Interestingly, scientists have concluded Venus once had enough water to make "oceans" at least 10 m (30 ft) deep, though it is not known whether that much water ever actually collected on the surface at once. At any rate, the greenhouse effect and the resulting high temperatures destroyed any liquid oceans; Venus' CO_2 couldn't dissolve and was left as the major atmospheric gas.

If this theory is right, the sister planets, Earth and Venus, should have emitted similar total amounts of CO_2

TABLE 9-1

Atmospheres of Venus and Earth

Venus		Earth	
Gas	Percent Volume	Gas	Percent Volume
CO_2	96.5	N_2	78.1
N_2	3.5	O_2	20.9
SO_2	0.015	H_2O	0.05 to 2 (variable)
H_2O	0.01	Ar	0.9
Ar	0.007	CO_2	0.03
CO	0.002	Ne	0.0018
He	0.001	He	0.0005
O_2	≤ 0.002	CH_4	0.0002
Ne	0.0007	Kr	0.0001
H_2S	0.0003	H_2	0.00005
C_2H_6	0.0002	N_2O	0.00005
HCl	0.00004	Xe	0.000009

Note: Compositions are for near-surface conditions, with terrestrial data other than H_2O tabulated for dry conditions. CO_2 on Earth is probably increasing by 2% to 3% of the listed amount in each decade, because we are burning so much fossil fuel. This activity may be modifying Earth's climate.

Sources: Oyama and others (1979); von Zahn and others (1983).

from their volcanoes and the Earth's CO_2 should be traceable somewhere. It is. The theory received strong support from inventories of Earth's carbonate rocks (rocks like limestone, formed by actions of sea creatures and by reaction of seafloor rocks with the ocean's dissolved CO_2, or carbonic acid). The inventories showed that Earth's carbonate rocks contain about the same amount of CO_2 as Venus' atmosphere! Thus Venus and Earth *did* produce similar amounts of volcanic gases, but Earth's got trapped in its rocks through the mediation of the ocean. Earth's unusual N_2–O_2 atmosphere, which makes the higher life forms possible on Earth, thus seems to be a special consequence of our H_2O oceans, which prevented Earth from developing the "normal" CO_2 atmosphere of a terrestrial planet.

LANDSCAPES ON VENUS

Soviet space scientists have made Venus "their" planet, so to speak, by landing numerous probes on it. Their Venera 4 probe, which crashed on Venus in 1967, was the first human-built object to touch another planet.

After many attempts, the Soviets successfully landed a number of probes that lasted in the surface heat for many minutes, making measurements of surface conditions. Veneras 9 and 10, in 1975, and Veneras 13 and 14, in 1982, made panoramic photos stretching from near the spacecraft to a bit of the horizon. These photos are shown in Figure 9-6. Their clarity shows that although Venus has high clouds, the surface is free from haze.

The photos revealed stark, dramatic landscapes with angular boulders, gravel, flat outcrops, and fine soil alternating at different sites. The varied states of erosion, with angular, young-looking rocks at some sites, suggest different degrees of geological activity. Although jet-stream winds (up to 185 mph) were detected at altitudes around 40 km, four lander probes found only gentle breezes of $\frac{1}{2}$ to 3 mph at the surface (Kerzhanovich and Marov, 1983).

As seen in Figure 9-7a, the landscape has orange-brown tones because it is bathed in the orangish light that filters through the clouds. Direct sunlight never falls on Venus' rocks, and the Sun is hidden beyond a high overcast.

Studies of the color pictures reveal that the *intrinsic*

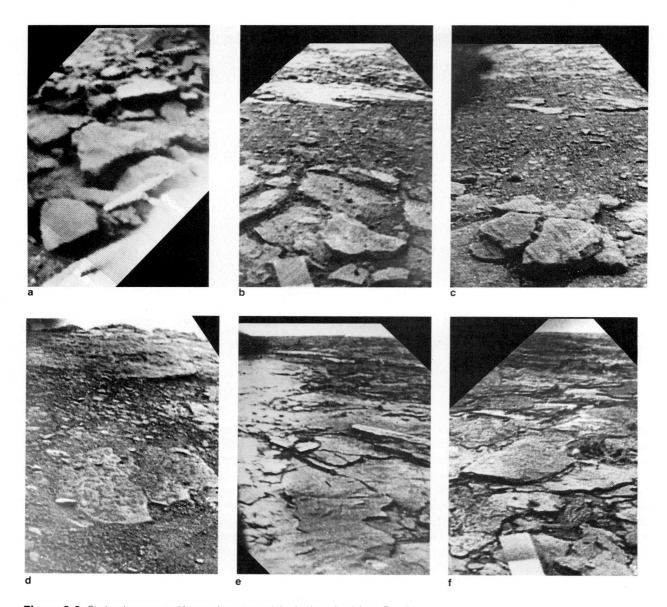

Figure 9-6 Six landscapes on Venus: views toward the horizon (top) from Russian Venera landers. Parts of landers appear in the bottom of views *a, b,* and *f*. **a** First photo from the surface of Venus, showing loose boulders near Venera 9. **b** and **c** Boulders and gravel near Venera 13. **d** Platey rock outcrops and gravel near Venera 13. **e** and **f** Platey rock surfaces near Venera 14. (Photos *b–f* courtesy C. Florensky and A. Basilevsky, Vernadsky Institute, Moscow.)

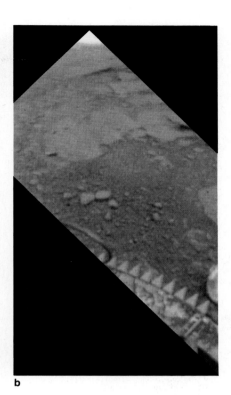

Figure 9-7 A Venus landscape in color. **a** Under the natural, mustard-colored light filtering through Venus' atmosphere, the daytime landscape displays a yellowish-orange hue. **b** If the same scene were illuminated by ordinary white light (such as sunlight on Earth or a spotlight from a spaceship), the rocks would be seen to have a neutral gray color. Part of the Russian spacecraft Venera 13 is seen at bottom. (Courtesy Vernadsky Institute, Moscow; image processing by Carlé Pieters and colleagues, Brown University.)

a   b

color of the rocks is neutral gray. Although the orangish light would make them look orangish to a (well-insulated!) astronaut looking at the daytime surface (just as your friend's gray shirt looks red in the light of a red neon sign), the rocks of Venus would look gray if the astronaut turned a normal, white-light spotlight on them at night.[2] This "white-light" appearance is shown in Figure 9-7b.

Mapping Venus

The American 1978 Pioneer Venus mission included an orbiter that mapped altitudes all over Venus by means of radar. The resulting map of Venus, shown in Figure 9-8, yields exciting scientific information that clarifies

---

[2]Curiously, the same studies revealed that if you brought the rocks inside the spaceship, they would slowly change to a reddish color as they cooled from the outdoor temperature of nearly 900°F to room temperature. This is because of optical properties of the oxidized (rusted) minerals, which are reddish at room temperature but turn gray when heated to Venus' temperatures (Pieters and others, 1986).

Venus' sisterlike relation to Earth. The planet is 60% covered with rolling lowland plains.

The radar maps also reveal Australia-sized areas raised about 2 to 5 km above the lowland plains. These are thought to be continents, or partially formed continents, indicating more evolution of Venus' lithosphere than occurred on the Moon, Mercury, or Mars.

In 1983, Russian scientists placed two radar-equipped probes in orbit around Venus and constructed much more detailed radar maps of certain areas on the border between the rolling plains and one of the "continents." The images showed volcanic mountains in the continental uplands, capped by clear examples of volcanic calderas on their summits. These were surrounded by a maze of complicated valleys and ridges. On the rolling plains are scattered, eroded meteorite impact craters, some exceeding 100 km in diameter. From the accumulated number of these craters, the Venusian plains were inferred to be perhaps a billion years old—older and less active than most regions on the Earth but younger than most regions on the Moon.

Because Venus is named for the goddess of beauty and femininity and is represented by the biological sym-

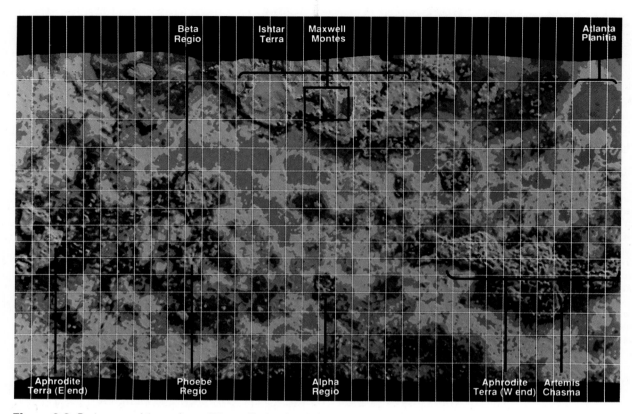

Figure 9-8 Radar map of the surface of Venus. Radar signals from the U.S. Pioneer Orbiter penetrated the clouds and measured altitudes of landforms underneath. Dark blue shows lowest terrain; red and pink, highest. Topography is reminiscent of Earth's, with the blue corresponding to seafloor interrupted by green-yellow continentlike blocks containing red mountain ranges. Latitude and longitude intervals of 10° are shown. (NASA; courtesy M. Kobrick, Jet Propulsion Laboratory.)

bol for "female," scientists have named most of its surface features after mythical or real women. The largest continent (about half the size of Africa) is called Aphrodite Terra. Other features include Ishtar Terra, with its volcanic mountains, and a crater named Eve, which marks the zero meridian on maps.

Rock and Soil Compositions on Venus

Venera probes 8–10 and 13–14, as well as VEGA probes 1 and 2 (dropped by Russian unmanned spaceships on the way to Halley's comet), carried devices to measure soil and rock composition. The results, of varied quality, indicate that Venus has much basaltic lava on its surface. Veneras 9, 10, 13, and 14, all landing on the flanks of

adjacent raised (volcanic?) areas called Beta Regio and Phoebe Regio (see Figure 9-8), measured basaltic compositions, similar to lunar basalt lavas or seafloor lavas on Earth. Venera 8, landing on plains to the east, suggested a more granitic composition. VEGAs 1 and 2, landing on the eastern flanks of the continentlike area Aphrodite, also found a generally basaltic composition. A detailed measurement by VEGA 2 indicated anorthosite rocks—the type forming the lunar uplands. The VEGA 2 and Venera 8 findings of more anorthositic or granitic rocks suggest at least partial differentiation on Venus: Low-density, feldspar-rich rocks have accumulated in low-density anorthositic or granitic "scum" that floats on a denser mantle and may make protocontinents on Venus. Recall that the Earth's continents are rich in granites and float on a denser basaltic crust.

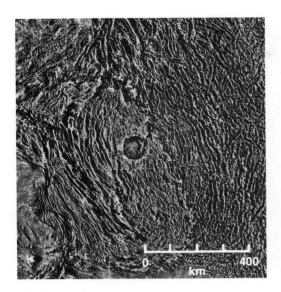

Figure 9-9 Radar image of the highest mountain summit on Venus, the 11-km-high (36 000-ft) Maxwell Montes. The image shows a 100-km-wide volcanic caldera on the summit, flanked by a complex pattern of ridges interpreted as due to tectonic compression and consequent folding. Image was constructed from radar that mapped the surface below the clouds from Russian orbiters Venera 15 and 16 in 1983–1984. (USSR Academy of Sciences.)

Scientists who have studied both the rock chemistry and the geological features of the radar maps are beginning to agree that much of Venus resembles the vast ocean-floor basaltic plains of Earth, which would be our most common terrain if you took away the water of Earth's oceans and the sediments washed in by water. On Earth, these ocean-floor basalts are erupted from volcanic rifts. Thus the range in Venus' rocks from ocean-floor-like basalts to somewhat differentiated anorthositic or granitic basalts suggests that Venus fits our rule of thumb that the more massive the planet, the greater its geological evolution. The Moon and probably Mercury have anorthosite uplands broken by basalt flows. Earth has major differentiation, with massive granitic, raised continents separated by basalt-covered ocean-floor plains. Venus, intermediate in mass but close to Earth, may have protocontinents of anorthositic or granitic rock separated by basaltic plains.

Active Volcanoes on Venus?

Spacecraft data show that volcanism has been common on Venus. The Magellan radar-mapping probe in 1990 made detailed images of lava flows, consistent with the earlier-observed basalt rocks. Also, mountain peaks as high as 11-km Mt. Everest have the profiles and summit craters of volcanoes, as shown in Figure 9-9 (Alexandrov and others, 1986).

Additional data suggest that some of the volcanoes may be active in our lifetime! First, data show a dramatic decline by a factor of 10 in sulfur dioxide (SO_2) gas in Venus' atmosphere in the 5 y after Pioneer arrived in orbit in 1978. The 1978 value was higher than had been measured earlier from Earth. Because SO_2 is a common volcanic gas, this has been interpreted as evidence of a major volcanic eruption on Venus shortly before 1978. Similar volcanic outbursts on Earth produce increases in atmospheric SO_2, which then declines due to atmospheric reactions. Second, maps of probable lightning activity suggest a concentration of lightning blasts around the peaks; lightning sometimes accompanies volcanic eruptions on Earth. (The lightning maps are controversial; see Taylor and Cloutier, 1986).

VENUS COMPARED WITH EARTH

We have seen many similarities between Venus and Earth. They are about the same size. They have probably produced similar amounts of carbon dioxide by volcanic outgassing. Both have a basaltic crust. Venus' topography may correspond to Earth's basaltic ocean-floor topography. They both have active geology and surfaces too young to be saturated with impact craters.

The meaning of these similarities is discussed in more detail in reviews by Soviet and American scientists (Basilevsky, 1989; W. Kaula, 1990). Many questions persist. Are Venus' raised areas really analogous to Earth's continents? Earth's continents are accumulations of granitic rock, whereas Venus' highlands appear to be mostly erupted volcanic rocks. Radar maps show circular structures a few hundred kilometers across, unlike any terrestrial structures (Figure 9-10). Are they eroded, ancient impact craters (with Venus' surface being somewhat older than Earth's)? Or are they a new type of tectonic structure? Seismic probes on the surface would help clarify these issues by revealing Venus' interior structure through seismic waves. Meanwhile, the first few radar images from NASA's Magellan space probe have shown the planet's rugged surface with unprecedented clarity (Figure 9-11). Their resolution, some ten times better than earlier images (such as Figure 9-10), reveals many new geologic features.

UNDISCOVERED WORLDS AMONG THE TERRESTRIAL PLANETS?

In spite of occasional suggestions to the contrary, no other major worlds or worldlets exist in the inner solar system besides Mercury, Venus, Earth, and Mars. Two regions subject to such speculation are the region inside Mercury's orbit and the region of Earth's orbit on the opposite side of the Sun from us. In the 1860s and 1870s, some observers thought they had seen a small planet, about half the Moon's size, inside Mercury's orbit. They named it Vulcan. But later observations showed it did not exist. Comets sometimes pass through this region on their journeys around the Sun, but no permanent worlds are known there. In the 1980s, American astronomers searched for even smaller bodies (nicknamed Vulcanoids) orbiting in this region, but they found none.

Some science fiction and UFO literature suggests that an Earthlike planet could exist in Earth's orbit, but on the opposite side of the Sun, where it could not be seen from Earth. Scientific study of this idea (Duncombe, 1969) showed that the hypothetical planet would be disturbed by other planets' attractions and move far enough out of the antisolar point to be detected; thus it was inferred not to exist. Subsequent spacecraft exploration has proven that no such world exists.

A LESSON IN COMPARATIVE PLANETOLOGY: WHY DO SOME PLANETS LACK ATMOSPHERES?

Our discussion of the Earth in Chapter 6 showed that the planets probably formed with an initial gas concentration called a *primitive atmosphere*, which changed to a *secondary atmosphere* as new gases were added by outgassing. Why, then, do some planets lack atmospheres while other planets have dense ones? The explanation comes from three principles that govern the motions of gas molecules in atmospheres:

1. The higher the temperature, the higher the average speed of the molecules.

2. The lighter the molecules, the higher their average speeds. Light gases like hydrogen and helium have faster average speeds than heavier gases such as oxygen, nitrogen, carbon dioxide, or water vapor.

3. The bigger the planet, the higher the speed needed for a molecule to escape into space.

Figure 9-10 Russian orbiting radar mappers obtained images of large circular features of unknown origin such as those seen at upper right, upper left, and lower center. This image is a mosaic of strips of radar data running nearly vertically; each strip comes from a different orbit of the spacecraft. Small craters—such as those between the upper two features (top center) or at lower right—are probably meteorite impact craters about 20 to 50 km across. Other features include parallel ridges that may be folded mountain belts, similar to the Appalachians. This is part of a region of Venus called Mnemosyne Regio. (Courtesy USSR Academy of Sciences.)

If you could heat a planet's atmosphere, more and more molecules would move faster than escape velocity, and fast-moving molecules moving upward near the top of the atmosphere would shoot out into space, never to return. First hydrogen, then helium, and then heavier gases would leak away into space. Cold, massive planets are most likely to retain all the gases of their primitive and secondary atmospheres; hot, small planets

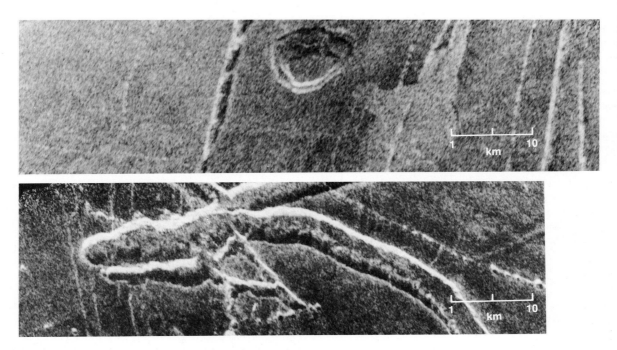

Figure 9-11 Radar images from Magellan spacecraft in 1990 show a 10-km wide volcanic caldera (top) and unusual volcanic fractures (bottom). Both images are about 70 km wide and show details as small as a few hundred meters. See back cover and its caption on page iv. (NASA.)

(with weak gravity and low escape velocity) are most likely to lose all their gases. Calculations based on these principles show that planets as small as Mercury and the Moon have lost virtually all of their gases. Venus and Earth have lost most of their hydrogen and helium but have kept heavier gases.

These principles help explain why Venus could rapidly have lost its water if water molecules broke into hydrogen and oxygen atoms: The light hydrogen atoms would quickly escape into space. (Oxygen atoms would oxidize surface rocks and thus be removed from the atmosphere as well.)

SUMMARY

A comparison of the planets can help explain planetary phenomena, particularly on Earth. Study of planets, their origins, and their development has come to be called *planetology. Comparative planetology* is the comparison of different planets to understand what makes them alike or unlike.

Only two decades ago, knowledge of phenomena on Earth, such as rock chemistry, weather circulation, and mountain building, had to be derived from knowledge found on the Earth alone. Today we can gain insights from other planets, treating them as special "laboratory examples" of other conditions. Venus, for instance, is an earth-sized planet nearer the Sun with a different rotation rate. From it we learn that an Earth without oceans could have retained much more CO_2 gas in its atmosphere, instead of in rocks. Venus confirms that atmospheric CO_2 causes atmospheric heating through a strong greenhouse effect.

From Venus we have also learned that a slower rotating Earth would have more linear weather circulation and less developed cyclonic spiral systems. The Moon, Mercury, and Venus show that smaller planets apparently do not develop enough internal energy to drive the plate tectonic activity that has broken and reformed the Earth's original, cratered crust.

From our studies of the Earth, Moon, Mercury, and Venus, we can derive three general principles that will clarify phenomena of other planets as well:

1. *The largest planets are most likely to have internal geological activity.* Internal heat is the energy source that drives geological activity such as tectonic faulting, earthquakes, and volcanism; the larger a planet, the more radioactive minerals it contains and the more radioactivity there is to release heat. Also, the larger a planet, the better insulated the interior and the harder it is for the heat to escape.

Small planets, on the other hand, cool rapidly and lose whatever heat they may have generated. The Earth, unlike Mercury and the Moon, has enough internal energy to drive plate tectonics.

2. *The larger a planet is, the younger its surface features are likely to be.* This principle follows from the one above. The more internal heat, the thinner the lithosphere and the more likely it is for the lithosphere to be broken by recent geological activity. Small planets that cooled long ago retain very ancient surface features. The Earth and probably Venus retain fewer ancient craters than Mercury and the Moon.

3. *The larger and cooler a planet is, the more likely it is to have an atmosphere, and the more likely this atmosphere is to have retained its original gases.*

CONCEPTS

prograde rotation   solar constant

retrograde rotation   greenhouse effect

Venera 7

PROBLEMS

1. Which is hotter, Mercury or Venus? Why?

2. Venus and Earth are about the same size and mass, and degassing volcanoes on each probably produced both CO_2 and H_2O gas. Why is CO_2 a major constituent of the atmosphere only on Venus, while H_2O is not a *major* constituent of the atmosphere on either planet?

3. If astronauts are to walk on Venus, what sort of space suit design might be needed? Would it need to *contain* pressure or *resist* pressure?

4. Compare the cloud patterns on Earth and Venus (see Figures 5-15, 8-2, and 9-4).
  a. What types of features are similar?
  b. These two planets have nearly the same radius, mass, and surface properties, but different rotations. What might be learned about circulation (wind patterns) of planetary atmospheres by comparing patterns of Venus and Earth?
  c. Would Coriolis drift be greater or smaller on Venus? How would this affect airflow and cloud patterns?

5. State which of the following characteristics of Venus suggest a primitive surface (little disturbed since planet formation) and which suggest an evolved surface (affected by geological processes such as erosion, differentiation, and plate tectonics):

  a. Craters
  b. A large, rifted canyon
  c. The lack of long, folded mountain ranges
  d. Basaltic surface rocks
  e. Granitic surface rocks (if any)

6. Compare the state of Venus' geological evolution with those of the Moon, Mercury, and Earth.

7. What are the chances that life as we know it exists on Venus? Why?
  a. If Venus were to have a surface temperature of about 300 K and an abundance of H_2O in its clouds, how would you rate the chances for life? Why?
  b. If Venus were exactly like the Earth, would life necessarily exist there?

8. How close is Venus during its nearest approach to Earth (see Table 8-1)? How many times farther is this than the distance to the Moon?

9. If a telescope shows Venus to be a thin crescent, where is Venus relative to the Earth and Sun?

ADVANCED PROBLEMS

10. Suppose you have a telescope that resolves $\frac{1}{2}$ second of arc. What would be the thinnest cloud layers visible on Venus in a view such as that in Figure 9-2, during its closest approach to Earth?

11. If Venus has a surface temperature of 750 K, at what wavelength is its strongest radiation emitted? Is this ultraviolet, visible, or infrared radiation?

12. Calculate the mean density of Venus and compare it with that of Earth. Would you expect Venus to have a larger or smaller amount of iron than Earth?

PROJECTS

1. Determine whether Venus will be prominent in the evening or morning sky during this semester, and observe its motions and brightness from day to day. Observe on which date it is farthest from the Sun and estimate this angle.

2. Observe Venus in a telescope of at least 5-cm (2-in.) aperture on several dates a few weeks apart. Observe and sketch the changes in phase, and explain them in terms of Venus' motion relative to the Earth and the Sun.

3. If a planetarium is convenient, arrange a demonstration of the motions of Venus.

CHAPTER 10

Mars

On a Martian summer morning in 1976, the sun came up as usual on a rock-strewn plain. The dawn temperature was around $-84°C$ ($-120°F$ or 189 K), but by afternoon the air warmed to about $-29°C$ ($-20°F$ or 244 K). The wind was light and the rust-colored rocks lay about as they had for the last few million years. The first sign of something unusual came at about 4:11 in the afternoon when a tiny, white, starlike object appeared high in the dusky red Martian sky. It was the 16-m (53-ft) diameter white parachute of the Viking 1 landing craft.

Within a few moments, the contraption would have been clearly visible from the surface. At an altitude of 1.2 km (4000 ft) the lander's three engines fired with a pale, transparent flame. Within seconds the parachute cut loose and drifted away. Slowed by its rocket engines, the spacecraft dropped for another 40 s or so. As it dropped the last few meters, reddish dust swirled into the air. The first of three lander legs hit the ground about as hard as you would if you jumped off a chair, and the jolt automatically switched off the engines. As the ungainly spacecraft bumped to rest, Viking 1 became the first human-built machine to gather data on the surface of Mars.

Until the instant that Viking's cameras clicked on, no one knew what existed on the surface of Mars; many scientists expected plants or other life forms. As might be hoped for any successful exploration, Viking's voyage confirmed some existing ideas but forced surprising revisions of other ideas.

MARS AS SEEN WITH EARTH-BASED TELESCOPES

At its closest, Mars comes within about 56 million kilometers (35 million miles) of the Earth—closer than any other planet but Venus. When it is that close, a

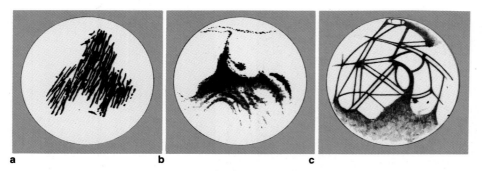

Figure 10-1 Drawings of Mars through telescopes over a three-century span. The north-extending dark triangle on all three drawings is a region known as Syrtis Major. **a** One of the earliest known sketches, by Christian Huygens in 1659. **b** Drawing by English observer W. R. Dawes during the 1864–1865 approach of Mars. Dawes recorded a streaky extension of Syrtis Major—the type of feature later called a "canal." **c** Italian observer Giovanni Schiaparelli first popularized the conception of thin, straight "canals," as in this 1888 sketch. North polar ice cap is at the top. (After Huygens, Dawes, and Schiaparelli.)

telescope of only 7 to 10 centimeters' aperture will show features on its reddish surface, including polar ice fields, clouds, and dusky markings (Figures 10-1 and 10-2)—features not unlike those of the Earth.

The Markings of Mars

The first to see Mars through a telescope was probably Galileo, who wrote in 1610 that he could see its disk and phases. These observations indicated that it was a spherical world illuminated by the Sun. When telescopes improved, observers began mapping dusky and bright patches that we can still recognize today. Dutch physicist Christian Huygens first clearly sketched these markings in 1659 (Figure 10-1a), and French–Italian observer Giovanni Domenico Cassini tracked them a few years later to determine that **Mars' rotation period** is 24^{h}37^m, only a bit longer than the Earth's.

Seasonal Changes on Mars

Mars has seasons just like Earth, though each season lasts about twice as long since the Martian year is nearly twice ours. Telescopic observations in the 1700s and 1800s (Figures 10-1b and c) revealed **seasonal changes in the features of Mars**. In the Martian hemisphere experiencing summer, the bright, white polar cap shrinks away and may disappear from view, while the dusky markings darken and grow more prominent. Some early

observers mistakenly thought the dark areas were oceans and called them **maria**, just as on the Moon. Brighter, orange areas came to be called **deserts**, a term that proved to be more appropriate.

While retaining roughly constant shapes, the markings also change slightly from year to year, as shown in Figures 10-2 and 10-3. Once these changes were established, many observers thought that the dark areas were regions of vegetation, perhaps losing their leaves in winter and turning dark and lush in summer, as on Earth. To understand the studies of Mars, it is important to realize that many observers in the late 1800s erroneously believed that Mars had climate and vegetation like the Earth. This opinion evolved even further as a result of the celebrated affair of the Martian "canals."

Canals on Mars?

In 1869 Father Angelo Secchi in Rome mapped streaky markings he called *canali*, maintaining the convention of naming dark areas after bodies of water. In 1877, Giovanni Schiaparelli, director of an observatory in Milan, popularized the term and drew the streaks much narrower and more linear than earlier observers had. He showed them forming a network of lines on Mars, as seen in Figure 10-1c. These features came to be called **canals**.

Many other observers did not see such features. The controversy grew hotter in 1895 with a vivid

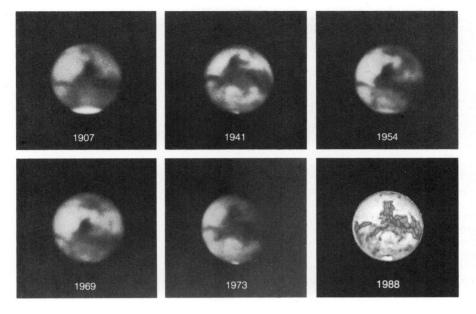

Figure 10-2 Seventy-five years in the Syrtis Major region of Mars (see Figure 10-1). Changes in the dusky markings can be seen. For example, a dark "wing" to the right of Syrtis Major and a brightening of the circular region Hellas, below it, are prominent in 1941. Note the large winter south polar cap in 1907 and the small summer cap in 1941, 1973, and 1988. (1907–1973, Lowell Observatory. 1988 view uses new CCD technology, courtesy P. Pinet, S. Chevral, C. Buil, and E. Thouvenot, Toulouse OMP Observatoire, France.)

Figure 10-3 Examples of modern Earth-based photography of Mars. **a** 1971 image showing the south polar winter ice cap and bright haze over the north pole. **b** 1973 image of nearly the same hemisphere of Mars, showing partially melted south polar cap. The light orange spot (left center) is a dust cloud marking the beginning of a large-scale dust storm. Examples can be seen of the faint wispy markings that were mapped as "canals" by some early observers. (Catalina Observatory photos; courtesy S. M. Larson, University of Arizona.)

description of the canals in a book, *Mars*, by Percival Lowell. Lowell was a wealthy Bostonian who founded his own observatory in the exceptionally clear air of Flagstaff, Arizona. Lowell said the canals were very sharp lines:

It is the systematic network of the whole that is most amazing. Each line not only goes with wonderful directness from one point to another, but at this latter spot it contrives to meet, exactly, another line which has come with like directness from quite another direction.

Lowell concluded that the features really were canals—artificial ditches built by intelligent creatures to carry water. He pointed out that spectroscopic measurements had revealed Mars to be a dry place. He hypothesized that a once moist climate was becoming desert-like as water evaporated from the thin Martian atmosphere into space, and that a Martian civilization had turned to massive irrigation canals to carry water from their polar snow fields to the dry, warm equator.

This exciting hypothesis sparked raging debate for

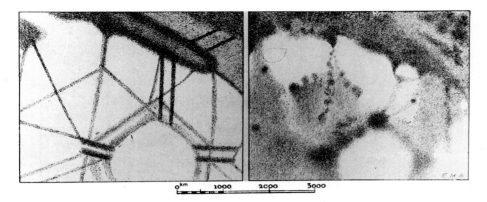

Figure 10-4 An explanation of the canals proposed by the French observer E. M. Antoniadi around 1930. Under mediocre observing conditions, canallike streaks may be seen in some regions of Mars, as Lowell described them (left). During moments of excellent atmospheric stability, observers with large telescopes find that these streaks break apart into a complex pattern of mottling and dark patches. The canals are a product of poor observing conditions combined with the tendency of the eye to connect dots into lines.

several decades. After spacecraft visits to Mars, however, we now know that a canal network does not exist as Lowell and some others drew it. What went wrong with the observations and the hypothesis of Martian civilization? Astronomer Carl Sagan has commented that drawings of geometric networks on Mars certainly do imply that intelligence is present, but the drawings alone don't tell at which end of the telescope the intelligence resides. Modern evidence indicates two reasons why canals existed only in observers' minds. First, streaky markings, including faulted canyons and dust deposits, do exist on Mars; seen through the shimmery atmosphere of Earth, these markings may resemble patterns of lines, as shown in Figure 10-4. Second, and more important, some people are more likely than others to perceive streaky patches as straight lines, especially if they already believe the lines are there. Lowell was one of these; he even drew lines on Venus.

Names of Martian Features

Schiaparelli, a classical scholar as well as astronomer, named the larger Martian dark and light regions (as well as the illusory canals) after historical, mythological, and geographic features of his native Mediterranean area. These names, such as Hades, Arabia, and Libya, are still used for the major dark and light areas. In 1971, when the Mariner 9 spacecraft revealed actual geological structures such as craters, mountains, and canyons

(Figure 10-5)—all too small to be seen from Earth—these were assigned additional names. As on the Moon, craters were named after scientists. The largest canyon complex, big enough to stretch across the United States from coast to coast, was named Valles Marineris (Valleys of Mariner). A modern map with some of the markings and geological structures is shown in Figure 10-6.

THE LURE OF MARS

On October 30, 1938, Orson Welles broadcast a realistic radio play in which listeners heard "newsmen" reporting that Martians had invaded and were laying waste to New Jersey. Thousands believed it and the resulting panic caused a national scandal. In July 1965, Mariner 4 sent back the first close-up pictures of Mars, which were widely published on front pages of newspapers; the reporting rate of UFOs shot up by a factor of 6 for several weeks. Why has Mars, of all the planets, held such a strong fascination for the public?

The answer lies in the hypotheses we have just discussed. If Copernicus knocked the Earth out of the center of the universe, Lowell and the Victorians made people realize that Earth's civilization might not be the only one around. Giordano Bruno and others in the 1600s and 1700s had already suggested that other worlds might be inhabited—an idea known as the *plurality of worlds*. And Darwin's theory of evolution, published in 1859,

Figure 10-5 A portion of Mars showing the vast canyon Valles Marineris (bottom center), whose length is comparable to the width of the entire United States. Three large volcanic mountains with summit craters are prominent at the left. Meteorite impact craters, like those of the Moon but with more erosion, dot other parts of the picture, as in right center. In this processed image the colors are nearly natural, although the clouds are somewhat whiter than the true yellowish Martian cloud colors. As on Earth, the clouds tend to form over or near mountains; they nearly surround the three volcanoes. (NASA Viking composite image; courtesy A. S. McEwen, U.S. Geological Survey.)

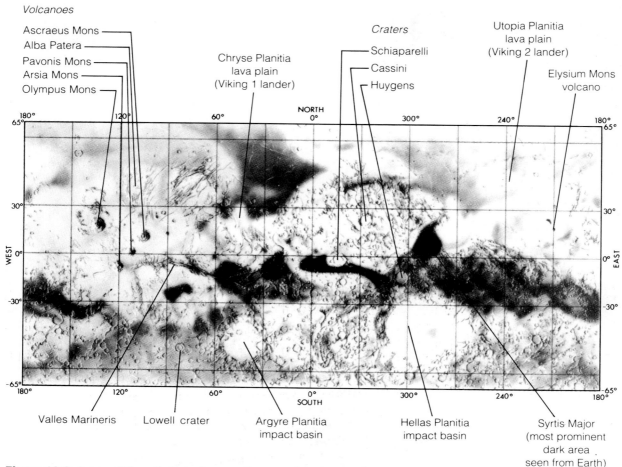

Figure 10-6 A map of Mars. Dark shadings are the somewhat changeable markings visible from Earth, probably associated with windblown dust. Topographic features such as craters and volcanoes are too small to see from Earth. The names of some features are given. (Base map courtesy R. M. Batson, U.S. Geological Survey.)

made more plausible the idea that other planets might produce totally alien species specially adapted to their environments.

The idea electrified leading thinkers. Tennyson and other poets wrote about it. In 1898, H. G. Wells published *The War of the Worlds*, in which Martians invade the Earth to escape their dying planet. Wells commented:

The Tasmanians, in spite of their human likeness, were entirely swept out of existence in a war of extermination waged by European immigrants. . . . Are we such apostles of mercy as to complain if the Martians warred in the same spirit?

There was even a UFO scare in the 1890s that produced reports of Martian spaceships (described as looking like the first dirigibles, which were then flying).

Between the days of Lowell and the first space flights to Mars, several generations of readers grew up on stories by Edgar Rice Burroughs, Ray Bradbury, and others, which pictured a Lowellian Mars with remnants of a dying civilization holding out in nearly deserted cities on a dying, drying planet. While incorrect in many details, these blends of theory and fancy fostered valid excitement about whether life might really exist on Mars and helped to encourage interest in actual voyages to the red planet. Yet, by the time of the first Martian

Figure 10-7 A close-up of the Viking spacecraft—typical of a number of space probes that have returned extraordinary pictures and data about other planets. The camera can be seen hanging on the bottom of the craft. Winglike panels are covered with solar cells to provide electricity from sunlight. Shroud with flag protects the engines. A landing probe (not shown) detached from the bottom and parachuted onto Mars, while the rest of the vehicle stayed in orbit to take pictures. (NASA.)

voyages, Earth-based measurements had already revealed that Mars had colder, thinner, drier air than had been thought. Could such a planet support advanced life forms? Microbes? Or no life at all?

VOYAGES TO THE SURFACE OF MARS

The first three human-made devices to reach the surface of Mars were unsuccessful Russian probes. Mars 2 crashed in November 1971; Mars 3 landed in December 1971 but failed after 20 s on the surface; another probe sent back data while parachuting through the

atmosphere in 1974 but failed moments before touchdown. Causes of the failures may have been design problems or hostile Martian conditions; a dust storm was raging during the 1971 landings.

The first successful landing on Mars was by the **Viking 1** spacecraft, which touched down July 20, 1976 (7 y to the day after the first human landing on the Moon). It was followed on September 3, 1976, by a duplicate spacecraft, **Viking 2** (Figure 10-7). Both landings were on plains that looked relatively smooth from orbit but turned out to be rock strewn, as seen in Figures 10-8, 10-9, and 10-10. Boulders as wide as 3 m were photographed among dunes near Lander 1. Missing were the Martians, deserted cities, canals, or the strange vegetation imagined by early writers.

Mars turned out to be a desolate, cold, yet beautiful desert. The reddish color of Mars was vividly shown by color photos from Viking landers. Though some rocks appeared dark gray, like terrestrial lavas, most rocks and soil particles were covered with a coating of rust-like, reddish iron oxide minerals. Similar iron minerals give terrestrial deserts their familiar red-to-yellow coloration, especially when moisture is present only occasionally. Viking scientists were surprised to find the daytime sky of Mars reddish-tan instead of blue (Figure 10-10). The sky color is caused by much fine red dust stirred from the surface into the air by winds and deposited even on rock tops as it settles out of the air (Figures 10-11 and 10-12).

Atmospheric Composition

The composition of the **Martian atmosphere,** shown in Table 10-1, gives several clues about the planet's history. Like Venus, Mars has a very thin atmosphere that is mostly carbon dioxide, probably generated chiefly by planetary degassing through volcanic activity. As noted in our discussions of Earth and Venus, volcanic gases, generated from melting of the interior rocky matter, are rich in carbon dioxide.

The Martian Climate

During the weeks after landing in the Martian summer, **air temperatures** at the two Viking sites ranged from nighttime lows around 187 K ($-123°$F) to afternoon highs around 244 K ($-20°$F). Temperatures of the soil,

Figure 10-8 First close-up photo of Martian soil was sent from Viking 1 lander on July 20, 1976. Fine dust and rock chips are shown. Landing leg is at right. The landing dislodged soil, which settled in the center of the concave pad at end of the leg. The larger chips are about 5 cm (2 in.) across. (NASA.)

Figure 10-9 Under late afternoon sun, rocks near the Viking 1 landing site cast picturesque shadows across the dunes of Mars. At the time of this picture, the atmosphere was particularly dusty, giving a hazy light with low-contrast shadows. Parts of the landing craft containing the camera can be seen at the edges. (NASA.)

which absorbs more sunlight than the air, exceed freezing (273 K, or 32°F) on some summer afternoons, so that any frost formed at night near the surface can melt and produce moisture or water vapor. Winds at the two sites were usually less than 17 kph, with gusts exceeding 50 kph. Much higher winds are believed to occur at certain seasons, however, raising clouds of dust that

can be observed from Earth. **Air pressure** at each site was only about 0.7% that on the Earth.

The Martian polar caps give vivid evidence that important gases freeze out of the Martian atmosphere (Figure 10-13). On the hemisphere that is having summer there is a small, permanent cap of frozen water, because even the summer temperature is below the

Figure 10-10 Rocky plain at the Viking 2 landing site, looking toward the Sun. The boulders are backlit. The pronounced orange of the Martian sky gives way to a brighter (overexposed) glow around the Sun, which is out of the picture to the upper right. (NASA.)

Figure 10-11 The largest boulder photographed by Viking landers lay among dusty dunes only 8 m (20 ft) from Viking 1. Had the lander come down on it, the vehicle probably would have been destroyed. Seen in higher light and in clearer air than in Figure 10-9, the rock shows the gray color and frothy texture of basalt lava as well as a prominent cap of ocher dust that settled onto it after the last dust storm. (NASA.)

freezing point of water. It is not known how thick the ice deposit is or how much water it contains. Many scientists believe the permanent cap at each pole is several kilometers thick. During Martian winter, temperatures plunge below the 146 K ($-197°F$) level at which carbon dioxide clouds form. Carbon dioxide snow, or "dry ice," accumulates on the polar ground to form a large polar cap. This transient winter cap is only a few meters thick and eventually shrinks during Martian spring. As seen in Figures 10-14 on page 207 and 10-15 on page 208, a thin, scenic frost layer accumulated on the ground around the Viking 2 lander at latitude 48°N during Martian winter, but not at the Viking 1 site at 22°N.

Figure 10-12 Sunrise over the plains of Utopia, 1978. The Sun sits just on the horizon, dimmed by the haze layer of dust. Light-scattering by the dust creates a soft glow just above the Sun, an effect visible in dusty twilight in terrestrial deserts. (NASA, Viking 2 photo.)

TABLE 10-1	
Composition of Martian Atmosphere	
Gas	**Percent Volume**
CO_2 (carbon dioxide)	95
N_2 (nitrogen)	2.7
Ar (argon)	1.6
CO (carbon monoxide)	0.6
O_2 (oxygen)	0.15
H_2O (water vapor)	0.03
Kr (krypton)	Trace
Xe (xenon)	Trace
O_3 (ozone)	0.000003

Note: Amounts of gases vary slightly with season and time of day. H_2O is especially variable. Some CO_2 condenses out of the atmosphere into the winter polar cap; changing cap sizes cause small changes in the total Martian atmospheric pressure.

Rock Types

Martian rocks at the two sites appear to be fragments of lava flows, as judged by their textures and colors and by the chemistry of the associated soil. Proportions of elements measured in the soil resemble those for soils derived from basaltic lavas on the Earth and Moon. A minor component in the soil is water, about 1%. This water exists not as a liquid or ice but as H_2O molecules in the crystal structure of the rock particles. When the rocks were heated to around 703 K (430°C or 800°F) during automatic experiments inside the Viking spacecraft, this water was released. It may be a remnant of more abundant water in the past, raising the possibility that future explorers could get water by heating Martian soil. Many investigators believe that ice may be present among soil particles a few meters below the surface. This would form a layer of **permafrost**, permanently frozen soil similar to that in arctic tundra regions of the Earth.

If one were to search for the most Marslike landscapes, rock types, and soil types on Earth, the best analogs would be found in extremely dry volcanic deserts. As shown in Figure 10-16, remarkably Marslike vistas can be found in such regions as the Peruvian coastal desert and high-altitude volcanic plains.

MAJOR GEOLOGICAL STRUCTURES

Data and pictures from the Viking landers have given vivid ideas of the Martian landscape. In addition, the mapping of large-scale geological structures from orbiting spacecraft has yielded information just as intriguing.

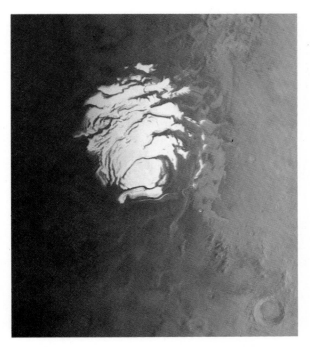

Figure 10-13 The south polar ice cap of Mars seen from orbit. This picture shows the small cap left near the south pole during the southern summer. It is roughly 360 km across and consists of CO_2 ice (possibly mixed with H_2O ice). Spiral breaks mark warmer, sun-facing ridges in the layered sedimentary deposits around the pole. In winter, condensation of CO_2 frost and snow makes the cap expand to 10 times this size. Compare with Earth-based views of cap in Figure 10-3. Night side of the planet is at the left. (NASA Viking orbiter photo; courtesy Tammy Ruck and Larry Soderblom, U.S. Geological Survey.)

Figure 10-14 Onset of Martian winter at the Viking 2 site. **a** The boulder-strewn plain, showing a trench dug by a remote-controlled sampler arm. At right lies a protective hood ejected from the sampler. **b** Later, as winter sets in, the same scene shows deposits of CO_2 frost in the cold shadows of rocks. In the meantime, the sampler has dug another trench near the first one. (NASA.)

Martian Dust: The Markings Demystified

The dark markings, once thought to be vegetation, are probably caused mostly by dust deposits. Evidence for this theory includes vast dune fields, hills, and crater rims (seen extending down from craters at the right center edge of Figure 10-5). The dark patches are dust deposits, often dropped in crater floors (Figure 10-5, right center). The streaks are dust deposits dropped when prevailing windstreams are disrupted by topography; such deposition has been simulated in lab experiments. A final evidence of the transport of dust on Mars is satellite photography of dust devils—typhoon-like "twister" columns of dust picked up by swirling winds (Figure 10-17). These giant dust columns often tower 2 to 6 km (6500 to 20 000 ft) above the Martian deserts (Thomas and Gierasch, 1985).

Seasonal changes in contrast and shapes of markings, once thought to be evidence of vegetation, are now attributed to dust transport by seasonal winds, especially summer dust storms. Even though dust storms create broad changes visible from Earth, they cause only subtle changes at the local scale. Viking 1, which worked on the surface of Mars for 3.3 full Martian years (over 6 Earth years), photographed only millimeter-scale deposition or erosion of dust in most areas, even though it weathered several dust storms (Arvidson and others, 1983). Some broad Martian markings hundreds of kilometers across, such as seen in Figures 10-3 and 10-5, may be dust deposits only millimeters thick.

Figure 10-15   Winter on Mars. At the Viking 2 site near 48°N latitude, a thin layer of mixed CO_2 frost and H_2O frost covers the ground at night. During the day the CO_2 frost sublimes, leaving frozen H_2O covering much of this frosty scene adjacent to the previous pictures. On this late winter afternoon, the Sun is low in the sky behind us and the shadow of the Viking 2 spacecraft is seen in the foreground. (NASA.)

Figure 10-16 Close terrestrial analogs to Martian landscapes occur in very arid regions with a history of weathering. **a** Glacier-dropped boulders in volcanic plains, Iceland. **b** Basaltic lava at 10 000-ft altitude on Mauna Loa volcano, Hawaii. **c** Windswept coastal desert of Peru. **d** Death Valley, California. Most of these areas are less weathered and the iron-bearing minerals are less oxidized, or rusted, than on Mars. Hence those areas are less red than Mars, but the rock structure and landscape is similar. (Photos by author.)

Martian Craters: Clues to Surface Processes

Martian craters, such as those shown in Figure 10-18, are interesting for three reasons. First, they are abundant and show that impacts have been common on Mars as on the other planets. Second, they show varied states of degradation, which indicate that erosive processes have been much more active on Mars than on the Moon or Mercury. In some regions the craters have been obliterated by young-looking lava flows; in other regions,

by massive accumulations of sediments or windblown dust hundreds of meters deep. Third, the craters give a way to estimate the age of Martian surface features. By estimating the rate at which interplanetary debris hit Mars and created the craters, analysts believe that the older, heavily cratered regions are probably a few billion years old, like the cratered surfaces of the Moon; but the youngest, sparsely cratered volcanoes, lava flows, and eroded surfaces may be as young as 100 million years or less.

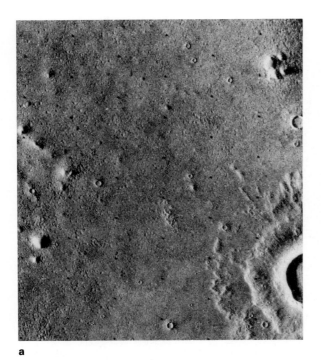

a

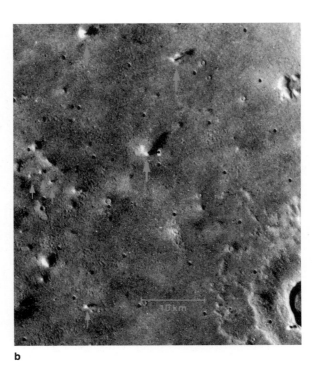

b

Figure 10-17 Dust devils on Mars. **a** On a normal afternoon in a desert region northwest of Olympus Mons volcano, an orbital view of a region slightly smaller than Rhode Island shows mostly featureless terrain with only a few hills and craters. **b** On another afternoon in summer, transient bright features cast shadows up to several kilometers long. These features are believed to be huge columns of dust like so-called dust devils on Earth. They are caused when the high midday summer sun heats the ground; air near the ground warms and rises, carrying eddies of dust into the air. In some years, when dust devils inject enough dust into the Martian air, vast dust storms occur on Mars. The crater at lower right is surrounded by splashlike lobes of ejecta; these probably formed when the meteorite impact melted permafrost layers of ice, splashing out slurrylike mud. (NASA Viking photos.)

Figure 10-18 Orbital view of one of the older, cratered portions of Mars. This view reveals Mars to be a planet falling somewhere between the Moon and the Earth in the sense of comparative planetology. On the one hand, we see lunarlike craters caused by ancient meteorite impacts; but on the other hand, the landscape shows cloudy haze and evidence of erosion unlike the Moon. (NASA Viking orbiter photo.)

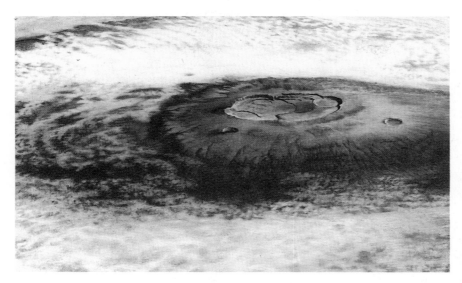

Figure 10-19 The top of the mightiest Martian volcano, Olympus Mons, protrudes through morning clouds. The lava-furrowed flanks and the 65-km summit caldera complex can be seen. (NASA hand-tinted image derived from oblique Viking orbiter photo.)

Martian Volcanoes: Clues to Lithospheric Stability

The largest known **volcanoes** in the solar system were discovered on Mars by Mariner 9 in 1971. The highest volcano, Olympus Mons (Figure 10-19), rises about 24 km (78 000 ft) above the lower Martian deserts. In contrast, Mt. Everest is only 9 km above sea level, 13 km above the mean ocean floor, and 20 km above the greatest ocean depths. The huge base of Olympus Mons is about 500 km across and would nearly cover the state of Missouri. The caldera, or volcanic crater, at the summit is about 65 km across. Olympus Mons is thus several times bigger than any similar type of volcanic cone on Earth. For example, Mauna Loa reaches about 9 km above the seafloor, is about 120 km in diameter at its seafloor base, and has a summit caldera a few kilometers wide.

Several other giant volcanic mountains dot Mars. Most are a few hundred kilometers across at the base, and one has a summit caldera 140 km across. Three are prominent in Figure 10-5. The largest Martian volcanoes are gently-sloped and clustered in a broad volcanic region up to 2500 km across (upper left quadrant of map in Figure 10-6). Geologists believe the Martian lavas were very fluid, spreading into broad flows instead of piling up in steep-sided cones like Mt. Fuji. The volcanic region is fresh-looking, perhaps 1 billion years old or less. There is some evidence that localized volcanic eruptions might persist even today (see Lucchitta, 1987), though Mars is probably not nearly as active as Earth. Apart from the giant volcanoes, many smaller volcanic cones, usually old and eroded, dot Mars, as shown in Figure 10-20.

The broad volcanic region is surrounded by swarms of radial fractures that extend thousands of kilometers from their centers. The largest fracture is a vast canyon system, the Valles Marineris, prominent in Figure 10-5. It suggests that Mars' lithosphere is thinner than the Moon's—enough so to be split by internal stresses. On the other hand, it is thicker than the Earth's—enough so to resist plate tectonic crumpling.

The large volcanoes and fracture systems all suggest that the Martian lithosphere in the volcanic region has been massively disturbed by lava upwelling from below. Uplift and volcanism may have been caused by rising hot currents similar to those suspected inside the Earth. The Martian volcanoes have probably reached such enormous size because Mars' lithosphere was too thick to form movable plates, even though it was thin enough to split. A given spot on Mars therefore stays directly over a given ascending magma current, and the lava eruptions during many years pile up into single huge mountains. In contrast, Mauna Loa and many other terrestrial volcanoes—in fact, the whole string of Hawaiian islands—are believed to be separate lava accumulations that occurred as plates in the Earth's crust drifted over a single region of intermittent mantle activ-

Figure 10-20 A highly eroded ancient Martian volcano about one-tenth the size of Olympus Mons. This orbital view is a vertical image of a gently sloped volcano whose summit caldera is at left center. Radiating from the summit are eroded channels. Some geologists believe this mountain was formed by ash flows, which would be more easily eroded than basaltic lava flows. Erosion channels may have been formed by water released during the eruption when hot ash melted icy permafrost layers in the soil. Photo covers a region about the size of Rhode Island. (NASA Viking photo; courtesy A. S. McEwen, U.S. Geological Survey.)

ity under the Pacific. A Martian volcano shows what might happen if all the Hawaiian islands were piled into one eruptive unit.

Mars confirms our general rules (Chapter 9 Summary) that large planets have more internal geological activity and younger surfaces than small planets. Small planets lose heat faster, thus developing thicker lithospheres without the hot, convecting interiors needed to drive plate tectonics or crack lithospheres. Mars, midway in size between Earth and the Moon, also falls midway between them in terms of surface geology: It has an older and less active surface than the Earth, but a younger and more active surface than the Moon.

ROCK SAMPLES FROM MARS?

Although no spacecraft has yet returned samples from Mars, it is possible that we already have some Martian rocks on Earth! Meteorites are rocks that occasionally fall from space onto Earth, and most of them are known to be fragments of interplanetary bodies called aster-

oids (see Chapters 13 and 14). Most of the hundreds of known examples are primitive rocks formed 4.5 billion years ago, at the same time the Earth and planets were forming. However, about a dozen meteorites puzzled researchers because they appear to be basaltic lavas formed only about 1.3 billion years ago. The asteroids are all too small to have had volcanic activity only 1.3 billion years ago, and researchers wondered where these objects came from.

By the 1970s, researchers knew enough about lunar rocks to rule out the Moon as the source of these meteorites, and in 1979 several researchers suggested they may be **Martian meteorites,** or rocks blown off Mars by a meteorite's impact. Studies in the 1980s show that this is almost certainly true. Among the samples that have been studied, traces of gas trapped in the rocks were found to have just the same ratios of elements as Viking measured in the air on Mars. Mineral evidence indicates they had been shocked by explosive forces consistent with ejection off Mars.

Assuming they *are* indeed from Mars, their composition tells us that Mars has a mantle similar to Earth's,

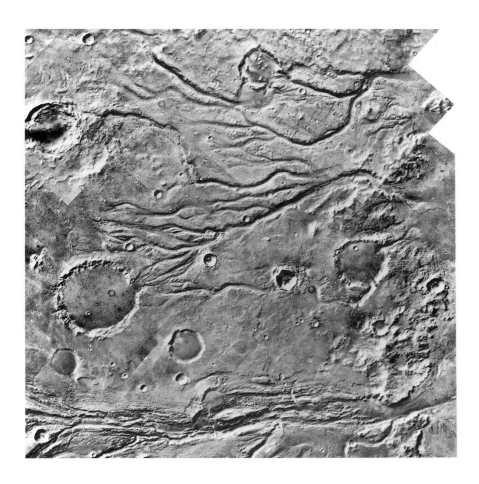

Figure 10-21 Dramatic evidence of ancient riverbeds on Mars is found in these channels and tributary systems. The area is about 180 km (110 mi) wide and drops about 3 km in the direction of flow, from the west (left) to the east. Water apparently cut into some old craters, but predated others. (NASA Viking photo from orbit.)

that Martian volcanoes erupted within the last 1.3 billion years, and that Martian subsurface materials contain large amounts of water or ice (McSween, 1985). These findings are consistent with the picture of Mars constructed from spacecraft data: Mars apparently has large regions covered by basaltic lava flows about a billion years old, and it may have large amounts of permafrost in its frozen soil.

THE MYSTERY OF THE ANCIENT MARTIAN CLIMATE

So far Mars might sound like the Moon or Mercury with a few extra volcanoes and a little air to blow the dust around. Mariner 9 shattered this conception by photographing **channels** that look like dry riverbeds, as seen in Figures 10-21 and 10-22. These channels meander in sinuous curves and often have tributaries. They get wider and deeper in the downslope direction and have sedimentary deposits on their floors. In short, they have all the features of arroyos cut by water or ice-clogged streams. Other theories of their origin—for instance, that they might have been lava flow channels—do not explain all their features.[1]

The greatest surprise of Martian exploration was to confirm that Mars is very arid, and yet at the same time to discover what appear to be riverbeds! Evidently, Mars once had flowing rivers of liquid water. The number of recent impact craters interrupting the channels indicates that the channels are perhaps 1 to 3 billion years old. The duration of flow episodes is unknown. The

---

[1] Although a few of the major riverbeds lie near reported positions of the once popular "canals," there is little correspondence in general. The channels do not explain the canals.

Figure 10-22 Along a border between lava plains (top) and ancient highlands (bottom half), riverlike channels appear to have eroded the higher ground and emptied onto the plains. The water flow appears to have mostly stopped before the lavas covered the plains, since the channels do not extend onto the lava desert. However, some flow activity from the center channel appears to have formed a small valley and delta deposits at its mouth after the lava desert formed. The largest crater has a "splash" pattern believed to result from a muddy ejecta mixture of soil and melted ice. A bright, windblown dust deposit extends to upper left from small crater in plain (upper left of center). Smallest craterlets are comparable to a football stadium in size. (NASA Viking orbiter photo, Mangala region; courtesy A. S. McEwen, U.S. Geological Survey.)

channels are younger than the most ancient cratered regions, but older than most volcanoes. Some channels emanate from chaotic collapsed areas believed to have formed when ice deposits melted and released water onto the surface. Others—a network of fine channels near the equator and large channels with tributaries— suggest that some water came from other sources, possibly rainfall from a once denser atmosphere, as visualized in Figure 10-23. Other signs of ancient erosion, such as degradation of craters and the buildup of layered sedimentary deposits, also suggest much more atmospheric and erosive activity in the past.

Consistent with this, Viking scientists found chemical evidence of a denser atmosphere in the past, based on chemically inert gases in the present-day Martian atmosphere. Gases such as argon are called inert because they don't combine readily with other elements. There-

fore, once they are injected into the atmosphere by volcanoes, they remain there. Thus, argon and other inert gases are good indicators of total volcanic activity. By measuring the present argon content, and knowing the amounts of nitrogen and carbon dioxide that are emitted by volcanoes along with the argon, Viking scientists could estimate the total N_2 and CO_2 dumped into Mars' atmosphere during Martian history. They concluded that Mars once had 10 to 100 times as much of these gases as now present in the atmosphere, making an early Martian atmosphere nearly as dense as Earth's atmosphere today (Owen and others, 1977; Haberle, 1986). Similarly, geologists studying the amount of lavas and their probable H_2O content have estimated that the total water vapor injected into Mars' atmosphere by all the volcanoes during all of Mars' history would have been enough to make a water or ice layer 50 m deep

a

b

Figure 10-23 Imaginary views of a Martian channel ("now" and "then"). **a** A typical channel under present conditions. (Painting by author.) **b** Reconstruction of same as it was being inundated by a catastrophic flow of water. This might have occurred during the several episodes under ancient climatic conditions when the atmosphere was denser. Whether such scenes really occurred on Mars is still unknown. (Painting by Ron Miller.)

over most of Mars (Greeley, 1987). Other workers studying orbital photos point to features that resemble ancient shorelines, and there actually may have been ancient temporary lakes or oceans up to hundreds of meters deep in some locations (Kerr, 1986).

This leads to exciting new questions. Although Mars' present conditions are inhospitable to evolution of life, Mars' *ancient* conditions may have been more clement.

Where have the water and atmospheric gas gone? Many Mars analysts believe that much more of the water is still on Mars. We know that much water is frozen in the permanent polar caps and that about 1% of the surface soil is chemically bound water. But much more water may still be frozen as permafrost below the surface. Some other gases may have escaped into space.

A more provocative question is: What could have caused the transition from the earlier conditions—with more liquid water, more air, and higher temperature—to the present arid, freezing conditions? Astronomers and meteorologists have combined forces to discover what factors could change planetary climates within intervals of a hundred million or a billion years.

Under present conditions, water is very unlikely to flow on Mars. It is usually frozen, and even if it warmed enough to melt, it would very rapidly evaporate into the thin Martian air. (In many regions the air pressure is so low that the water would spontaneously boil away into the air.) But calculations indicate that if Mars ever had more air, liquid water might have been more stable on the surface.

One possibility is that the early atmosphere (3 to 4 billion years ago) was denser and wetter but slowly dissipated into space. But some data suggest a more rapid mechanism of climate change. Cornell scientists have found that if solar radiation striking the Martian polar caps increased by only 10 to 15%, much of the frozen carbon dioxide in the cap would *sublime* (go from solid to gas), passing into the atmosphere. This denser atmosphere would make Mars warmer, carry more heat to the poles, and cause still more polar dry ice to sublime. Thus there would be a striking feedback effect: A small initial change in polar climate could result in a large change in global climate (Sagan, Toon, and Gierasch, 1973).

These results led scientists to look for ways in which more sunlight might have reached the Martian poles in the past. One such process was revealed by calculating the changes in the rotation of Mars caused by the accumulation of lava in the volcanoes in the domed region

around Olympus Mons. These calculations indicated that before that volcanism occurred, the poles of Mars might have been inclined toward the Sun by as much as 45°, instead of the present 25°. This dip would mean that each summer pole would have received enough warmth to melt all the water in the permanent ice cap. The water would have circulated in the atmosphere on its way toward freezing at the winter pole. Rain might have occurred and created the observed riverbeds, or at least the water that erupted from underground sources might have lasted long enough to flow and erode the surface.

A second major process for depleting the ancient Martian atmosphere was suggested by geochemists who compared Martian meteorites with Viking soil measurements. We learned from our discussion of Earth and Venus that carbon dioxide was removed from the Earth's atmosphere when it formed carbonate-rich rocks in the oceans. The same process would have operated on Mars if Mars once had oceans, forming limestone-like deposits of calcium-carbonate-rich rocks. Warren (1987) suggests this process removed vast amounts of CO_2 from Martian air. He finds that the early surface pressure of Mars' CO_2 atmosphere could have equaled the pressure of Earth's present atmosphere, consistent with the estimates by Viking scientists.

These ideas about Martian climate history show how astronomical exploration of other planets can illuminate conditions on Earth. Everyone is familiar with the ice ages that have occurred on the Earth in the last hundred thousand years, and geological records indicate even greater climate changes in the last few hundred million years. The Martian evidence forces us to realize that planetary climate changes are common and that the Earth's climate, too, may be subject to astronomically caused changes.

WHERE ARE THE MARTIANS?

Much of the motivation and excitement in exploring Mars has been in the search for extraterrestrial life. Discovery of life on Mars would be of major cultural importance. Just as the discoveries of the Copernican revolution showed that the Earth was not the center of the solar system, **life on Mars** (or elsewhere) would show that humans are not necessarily the lords of creation. On the other hand, proof that life never evolved on Mars in spite of favorable conditions would present an exciting challenge to our present concepts of the nature

Figure 10-24 Sampler arms on both Viking landers scooped up soil samples for chemical analysis inside the spacecraft. Here a Viking 2 sampler arm pushes aside a frothy-textured lava rock to gather a soil sample from under it. Protected soil in such locations would have less irradiation by solar ultraviolet light, expected to be harmful to any possible Martian life forms. This experiment occurred on October 9, 1976. (NASA.)

of life, since biological experiments suggest (but do not prove) that life should evolve whenever conditions are suitable.

The Viking mission was specifically designed to look for life on Mars.[2] Of the five Viking experiments involved, two gave strong negative results, but three gave ambiguous results. In the first experiment, the cameras showed no signs of life. The second experiment, a soil analysis, revealed no organic molecules in the soil at either Viking site, at a sensitivity of a few parts per billion. Since **organic molecules,** or massive molecules containing carbon, are essential building blocks of life as we know it, this test strongly indicates that living organisms do not now exist in Martian soil and have not existed in the recent past.

The other three experiments were designed to look for ongoing biological processes, such as metabolism and photosynthesis, by taking soil samples (Figure 10-24), putting them in special chambers (some with nutrients), and watching for chemical changes that would indicate microscopic organisms processing the material in the chamber. All three of these tests found the type of changes that indicated life in terrestrial samples!

At first, investigators believed that Viking may have actually detected microbial life on Mars, but in a quantity of less than one organic molecule per billion parts of soil. However, the magnitude and rate of the changes were unlike those for terrestrial microbes. Furthermore, because most investigators believe microbes would be unlikely, if not impossible, at the measured low abundance of organic molecules, the Viking results seem to reveal unexpected but ordinary chemical reactions in the soil—*but not life.* Tests with simulated Martian soil suggest that solar ultraviolet radiation reaching the surface of Mars causes different mineral properties than those familiar on Earth; in their altered states, Martian minerals may react in unanticipated ways in the test chambers.

Thus, Viking has shown that Mars "is not teeming with life from pole to pole" (to use the memorable understatement by astronomer Carl Sagan). It probably has no present or recent microbial life. Why? Ultraviolet light irradiates the surface because Mars lacks an ozone layer. It would break down organic molecules, preventing current life on exposed soil or rock surfaces. Conceivably, primitive life might have evolved in the past, when Mars had thicker air and more water, but its traces may have been destroyed as ultraviolet light sterilized the windblown soil. A less likely possibility is that Martian life does exist, but the Vikings missed it. Researchers since 1979 have reported microorganisms in antarctic soils where Viking-type instruments did not detect

---

[2]By *life* scientists generally mean "life as we know it"—the ability of carbon atoms to combine with other atoms and form very complex molecules, which in turn form organisms that grow and reproduce.

Figure 10-25 Manned expeditions to the surface of Mars may eventually clarify Mars' secrets. Astronauts could set up a base similar to Antarctic research stations and probe Martian geology and chemistry more effectively than unmanned probes. Such expeditions are within our technical capability; Russian cosmonauts have occupied their Salyut space station for durations equivalent to a flight to Mars. (Painting by Paul Hudson.)

organic molecules, reopening the question of whether Martian life is too subtle for Viking to find. Moreover, scientists in 1977 discovered algae and lichens in microscopic pore spaces *inside* rocks in barren, Marslike antarctic valleys where no life was previously thought to exist. Perhaps we need to seek life inside Martian rocks!

A third possibility is that life never evolved on Mars. Liquid water may never have existed long enough to allow even microbes to evolve. Future exploration may clarify whether Martian life ever existed. The dusty plains of Mars have not given up their last secrets about the origins of life in the solar system. Perhaps a Mars expedition will someday ferret out the answer (Figure 10-25). Lunar astronaut (and later U.S. senator) Harrison Schmidt has remarked that today's students will be the parents of the first Martians!

MARTIAN SATELLITES: PHOBOS AND DEIMOS

Not all of the mysteries of Mars are on its surface. In 1877, the American astronomer Asaph Hall became the first human to see a satellite of Mars. Shortly afterward, he charted the positions of two Martian moons, naming the inner satellite Phobos ("fear") and the outer one Deimos ("terror") after the chariot horses of Mars in Greek mythology. Close-up photographs by spacecraft have revealed these moons to be strange, potato-shaped, cratered chunks of rock (Figures 10-26 to 10-28). Their dimensions are given in Table 10-2.

The craters of Phobos and Deimos were caused by collisions with small bits of meteoritic debris. The largest crater on Phobos is 8 km (5 mi) across and can be

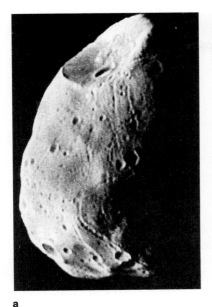

a

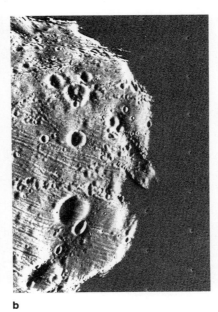

b

Figure 10-26 Mars' larger moon, 28-km-long Phobos. **a** A crescent-lit view of the entire moon, a potato-shaped rocky mass. The largest crater, Stickney (top), marks an impact that fractured and almost shattered Phobos. **b** Close-up view shows grooves running roughly radial to Stickney, perhaps marking fractures from that impact. This photo, made from 880 km (545 mi) away, shows objects as small as 40 m (130 ft) across. (NASA Viking orbiter photos.)

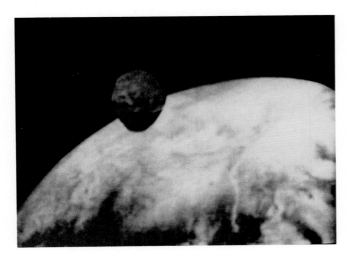

Figure 10-27 Color view of black Phobos hanging in front of ochre-colored Mars, photographed by the Soviet probe Phobos-2 in 1989. This unusual view dramatically shows Phobos' very dark tone, probably caused by carbon-rich compounds. (Courtesy B. Zhukov, IKI [Institute for Cosmic Investigations], USSR.)

seen at the top of Figure 10-26a. It is named Stickney, the maiden name of Mrs. Hall, who encouraged her husband's successful search for the satellite. A collision violent enough to create such a crater would release as much energy as 100 000 atom bombs of the Hiroshima size, or about 1000 hydrogen bombs of megaton size.

A peculiar feature of Phobos was revealed by close-up photos from Viking orbiters. Networks of grooves reach widths of around 100 m and lengths of as much as 10 km. Some are rows of adjoining craters. The exact nature of the grooves is unclear, but they radiate roughly

from the large crater Stickney and may be fractures (somewhat masked by surface dust) caused by the mighty Stickney impact, which nearly shattered Phobos.

The origins of Phobos and Deimos are unclear. Their surfaces are dark in color and apparently of a composition resembling a carbon-rich type of meteorite (called carbonaceous chondrite) known to be common in the nearby asteroid belt. For these reasons, many researchers believe Phobos and Deimos originated as asteroids and were later captured into orbit around Mars. This could have happened 4.5 billion years ago, when pri-

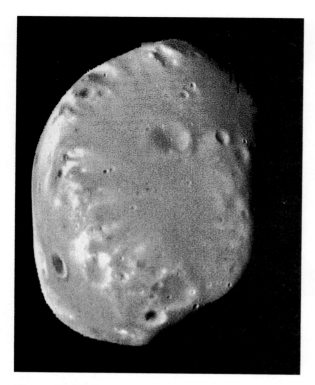

Figure 10-28 Deimos, a 16-km-long Moon of Mars. Its surface texture is smoother than that of Phobos, perhaps because of a different duration of microcratering since the most recent large impacts. Streaky markings are believed to be due to slipping of loose material in locally downhill directions in Deimos' weak, asymmetric gravity field. (NASA.)

TABLE 10-2				
Dimensions of Phobos and Deimos				
Diameter	Phobos		Deimos	
	km	mi	km	mi
Longest	28	17	16	10
Intermediate	23	14	12	$7\frac{1}{2}$
Shortest	20	12	10	6

Mars, have led to emphasis on Mars as the major target of planetary exploration, especially in the Soviet Union. In the United States, the president's National Commission on Space in 1986 urged a permanently inhabited Martian base as a long-term space goal.[3] The project could be international—serving as a model for international technical cooperation as well as an impetus for scientific and engineering growth. The National Space Commission report targets the base for the decade of the 2020s.

The Soviet Union has already announced plans for an unmanned mission to Phobos in 1989 that will maneuver within about 50 yd of that moon and then send over a lander that can hop from site to site analyzing surface chemistry.

NASA plans a "Mars Observer" that would orbit Mars and gather new data on surface mineral composition, atmospheric dust, and geological structures, but its launch has been delayed from 1990 to 1992. The Russians, however, have discussed ambitious plans to shift their focus in space exploration from Venus to Mars. They have announced plans for unmanned surface exploration featuring one or more balloons that would touch down on the surface at night and rise in the day. This mission, in 1994, might include a surface module or roving surface vehicle. The Russians have also discussed a sample-return mission in the 1990s, and there is much enthusiasm on both sides to do this as a joint U.S.–USSR undertaking, making it a test case for international technical cooperation. The samples would confirm the origin of the proposed Martian meteorites and would characterize other locales on the red planet.

mordial Mars had a more extensive atmosphere that could slow asteroids happening to pass through its outer fringes. This slowing could have caused a passing asteroid to be captured into Martian orbit. Phobos and Deimos, according to some astronomers, might be pieces of a single asteroid broken during such a capture process. Further drag from the primitive, extended atmosphere may have altered the moons' orbits, but the early atmosphere soon dissipated, leaving the moons stranded in their present orbits.

EXPEDITIONS TO MARS

Did Mars ever have ancient life? Why did its climate change? Where did Phobos and Deimos come from? Decades of interest in the comparative planetology of Mars and Earth, as well as the potential habitability of

---

[3]The commission included Neil Armstrong, astronaut Kathryn Sullivan, test pilot Chuck Yeager, U.N. ambassador Jeane Kirkpatrick, as well as several well-known scientists and engineers.

American and Soviet researchers hope to coordinate the work of all vehicles near Mars in the 1990s, but U.S. scientists are frustrated by lack of funding for new Mars missions. Meanwhile, the Soviet Union has proceeded with its manned program near Earth, where its astronauts have stayed in orbit for periods comparable in time to a Martian expedition. It is rumored they may be building toward manned expeditions around Mars or even to its surface, perhaps around the turn of the century.

A LESSON IN COMPARATIVE PLANETOLOGY: THE TOPOGRAPHY OF EARTH, VENUS, AND MARS

Orbital mapping of the planets has allowed researchers to prepare topographic maps showing the altitudes of points across the surfaces of Earth, Venus, and Mars. Figure 10-29 compares three such maps. The surface data have been digitized, with a pixel for each point of specified latitude and longitude. The color assigned to that pixel represents the altitude, shown by the color altitude scale, which is the same for all three maps.

A comparison of the maps reveals interesting differences in the geological "styles" of the planets related to their sizes. Earth, the largest, is dominated by rolling seafloor plains interrupted by continental blocks (Figure 10-29a). The same could be said for Venus, although there is not as much elevated "continental" land on Venus, perhaps because smaller Venus did not have as well-developed plate tectonic motions (Figure 10-29b). Indeed, we can see that the Earth's major mountains are arc-shaped ranges that develop when plates collide. Because a smaller planet loses its internal heat faster, Venus did not have as much internal energy to drive such plate motions.

These ideas are affirmed as we turn to the map of still-smaller Mars (Figure 10-29c). Here the map shows a quite different style. This planet was too small to generate even enough tectonic energy to destroy all its original cratered topography. Thus we see that some of the deepest depressions are well-defined circular impact basins, not rolling seafloor plains. The Hellas basin (lower right center) is a prominent example. Most of the uplands rendered in khaki green are heavily cratered. There is a link with Venus, however. The highest Martian mountain areas, such as the broad Tharsis dome (tan, left center), are simply piles of volcanic lavas surmounted by the mighty Olympus Mons volcano (Figures 10-5 and 10-19) and other volcanic peaks, not unlike Venus' Maxwell Montes (Figure 10-29b, top center; Figure 9-9).

In other words, the maps confirm our rule of thumb at the end of Chapter 9: Worlds smaller than Mars preserve ancient surfaces dominated by the *external* forces of cratering that shaped the planet. For worlds around the size of the Moon and Mars, internal energy is significant enough that volcanic forces break through the lithosphere and resurface parts of the planet. Worlds larger than Mars have surfaces dominated by these *internal* forces, including volcanism and tectonic restructuring.

SUMMARY

Mars, the planet once thought to have fields of vegetation or even a dying civilization, has been revealed by spacecraft to be a barren but beautiful desert lacking any advanced life forms. Orbital photos reveal a wide variety of landscapes including lava flows, grand canyons, landslides, polar snowfields, eroded strata, dunes, impact craters, and arroyos. Biological experiments aboard Viking landers indicate that interesting chemical reactions take place in Martian soil, and the seemingly lifeless Martian environment may yet yield interesting clues about the evolution of biochemical reactions and their dependence on climate.

Evidence about Mars' ancient climatic history seems to contradict its present-day barrenness. Nearly all water on Mars today is locked in polar ice, frozen in the soil, or chemically bound in the soil; virtually no liquid water exists. But liquid water apparently once flowed and eroded the surface, indicating different past climates. Current research indicates that the climates of both Mars and Earth may have varied significantly during their histories.

CONCEPTS

Mars' rotation period	Martian air temperature
seasonal changes in Mars' features	Martian air pressure
mare	permafrost
desert	volcanoes
canal	Martian meteorites
Viking 1	channels
Viking 2	life on Mars
Martian atmosphere	organic molecules

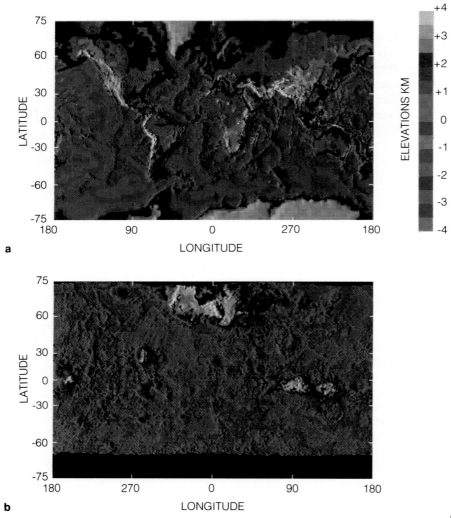

a

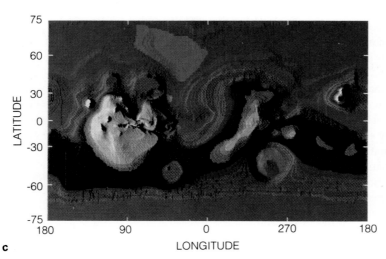

b

c

Figure 10-29 Topographic maps of the three largest terrestrial planets; altitude scale (shown by color bar) is the same in all three cases. Longitude scale at bottom of each increases in the direction of rotation of the planet. **a** Earth, showing continental blocks with arc-shaped mountain chains formed by plate collisions. **b** Venus, showing lesser development of "continental" masses. **c** Mars, showing a more primitive surface with circular impact basins and highest spots created by giant volcanic peaks. (Computer-generated topographic maps from orbital radar and other data; courtesy M. Kobrick, Jet Propulsion Lab.)

PROBLEMS

1. Describe a day on Mars as a future astronaut might experience it. Include the length of the day, the appearance of the landscape, possible clouds and winds, possible hazards, and objects visible in the sky.

2. What scientific opportunities for long-term exploration does Mars offer compared with the Moon? What qualities of the Martian environment might make operating a long-term base or colony easier on Mars than on the Moon once initial materials were delivered to the site?

3. Compare photos of craters on Mars and on the Moon.
  a. Assuming that all craters had similar sharp rims when fresh, which craters have suffered most from erosion?
  b. What does this say about lunar versus Martian environments?

4. Give examples of how the Martian environment and geology are midway between those of the smaller planet Mercury and the larger planet Earth. Comment on atmosphere, craters, volcanism, and plate tectonics.

5. What scientific knowledge might be gained from close-up investigation of Phobos and Deimos? What measurements would be of interest if rocks from Phobos and Deimos were available for study?

6. Imagine you are a visitor from outer space exploring the solar system.
  a. If two Viking-type spacecraft landed at random places on the Earth, took photos, measured the climate, and took soil samples, what might they reveal about the Earth?
  b. How many landings might be needed to characterize the Earth adequately?
  c. To characterize Mars to the same degree, how many might be needed?

ADVANCED PROBLEMS

7. What is the weight of an astronaut on Phobos if he or she weighs 130 lb on Earth?

8. In a gravitational field, objects fall through a height h in a time $t = \sqrt{2h/a}$, where a is the acceleration due to gravity. Suppose you are an astronaut exploring Deimos, where a is about 0.0042 m/s². Suppose you drop a tool from eye level.

  a. How long does it take to reach the ground?
  b. How long would it take on Earth where a is about 10 m/s²? (*Hint:* Because we are using the SI system, express h in meters.)

9. With a telescope that can resolve details as small as 1 second of arc, what is the smallest detail you can see on Mars during its closest approach to Earth at a distance of about 56 million kilometers?

10. Calculate the velocity needed to launch an object into circular orbit around Deimos from a point on the highest "mountain" on Deimos, assumed to be 8 km from the center. Assume the mass of Deimos is 4×10^{15} kg. Many people can throw an object at about 30 m/s. Would you need a rocket to launch a satellite from Deimos?

PROJECTS

1. Observe Mars with a telescope, preferably within a few weeks of an opposition and with a telescope having an aperture of at least 15 cm (6 in.). Magnification around 250 to 300 is useful. Sketch the planet. Can you see any surface details? Usually the most prominent detail is one of the polar caps, a small, brilliant white area at the north or south limb contrasting with the orangish disk. Can you see any dark regions? Compare the view on different nights and at different times of the night. (Because Mars turns about once in 24 h, the same side of Mars will be turned toward Earth on successive evenings at about the same hour.) If no markings can be seen, three explanations are possible: Observing conditions are too poor; the hemisphere of Mars with very few markings may be turned toward Earth; a major dust storm may be raging on Mars, obscuring the markings.

2. For the previous observations, determine which side of Mars you were looking at. (Your instructor may need to assist you.) First determine the date and Universal Time of your observations (UT = EST + 5 h = PST + 8 h. Thus 10 P.M. EST on April 2 = 03^h00^m on April 3, Universal Time). In *The Astronomical Almanac*, the table "Mars: Ephemeris for Physical Observations" gives the central meridian (or longitude on Mars of the center of the side facing Earth) at 0^h00^m Universal Time on each date. From these tables you can find the Martian central meridian for the time of your observation. (Mars turns about 14.7°/h.) Compare your observations with a map of Mars, locating the part of Mars that you observed.

CHAPTER 11

Jupiter and Its Moons

The terrestrial planets, which we have been studying, are huddled relatively close to the Sun. Now we leave them behind and move to *the outer solar system*—the part of the solar system beyond the asteroid belt.

INTRODUCING THE OUTER SOLAR SYSTEM

The outer solar system contains four **giant planets**—Jupiter, Saturn, Uranus, and Neptune—and a small planet, Pluto. The name of the monarch of the Roman gods is fitting for Jupiter. The biggest planet, it contains 71% of the total planetary mass—nearly 2½ times as much as all other planets combined. All four of the giant planets also have large families of satellites—at least 41 in all. The four giant planets together contain 99½% of the total planetary mass and harbor about 91% of the known satellites.

The four giant planets have much lower mean densities than the terrestrial planets—700 to 1600 kg/m^3 as compared with 3900 to 5500 kg/m^3. Saturn, at 700 kg/m^3, would float like an ice cube if we could find a big enough ocean. (Water's density is 1000 kg/m^3.) This simile is significant. The giant planets evidently *are* made largely of ices, as well as low-density liquids such as liquid hydrogen.

Jupiter's diameter measures a little over 10 times Earth's diameter, and Saturn just under 10 times. Uranus and Neptune have diameters about four times the Earth's. Placed on the face of Jupiter, Earth would look like a dime on a dinner plate (see Figure 8-2).

Perhaps the most important principle to remember about the outer solar system is that because it is further from the Sun and much colder than the inner solar sys-

a   b

Figure 11-1 Comparison of (**a**) an example of the best Earth-based telescopic imagery of Jupiter and (**b**) a close-up view from a spacecraft. Jupiter is entirely covered by multicolored clouds arranged in lacy belts swirled by wind motions. (Photo **a**: Catalina Observatory 61-in. telescope, courtesy S. M. Larson, University of Arizona; photo **b**: NASA Voyager photo, processed by A. S. McEwen, U.S. Geological Survey.)

tem, it contains much more ice than the inner solar system. Because the gases that formed the Sun and its planet-spawning surroundings were mostly hydrogen, the ices that formed in the outer solar system are frozen compounds of hydrogen, such as water (H_2O), methane (CH_3), and ammonia (NH_3). Thus, instead of forming worlds of rock like the terrestrial planets, the outer solar system formed worlds of rock *plus* ice, often in roughly 50–50 mixtures. This explains many properties of worlds in the outer solar system. The giant planets are giant for two reasons: (1) They had ices in addition to rock, and (2) once they reached a large size during their formation, their gravity was so great that they began to pull in the surrounding hydrogen-rich gases. (The gravity of Earth and smaller planets was too weak to hold these light gases.) Hence the giants are huge, hydrogen-rich worlds. Their surfaces are hidden by colored clouds. Their thick atmospheres may bear some resemblance to the primordial atmosphere of Earth.

Similarly, most of the moons in the outer solar system have icy surfaces or dirty-ice surfaces. On some of them, geological processes tended to evaporate the ice off the surface, leaving darker, soil-rich surfaces. On others, internal heating melted the ice, producing watery "lava" that erupted and formed bright ice patches.

These moons are fascinating worlds—some larger than the planets Mercury and Pluto, as seen in Figure 8-2.

THE PLANET JUPITER

Even a week's observations with a backyard telescope reveal the swirling cloud patterns of Jupiter, prominent in Figure 11-1. The most obvious pattern is the system of dark and light cloud bands parallel to Jupiter's equator, as sketched in Figure 11-2. The dark ones are called **belts**; the bright ones are **zones**. Within these bands, wispy spots and streaks arise, develop, and die out. These features look small, but some of them are larger than the Earth! Though the smaller ones evolve in days, the larger ones may last for months or years. Dark clouds may grow and darken an entire bright zone for months or years at a time (Figure 11-3). The clouds are believed to be composed primarily of ice crystals of ammonia, ammonium hydrosulfide, and frozen water.

The cloud belts and zones have distinct colors, shown vividly in Figure 11-1. Usually belts are brown, reddish, or even greenish, whereas zones are light tan, whitish, or yellowish. The variegated colors persist even among small-scale clouds, as seen in the close-up of Figure

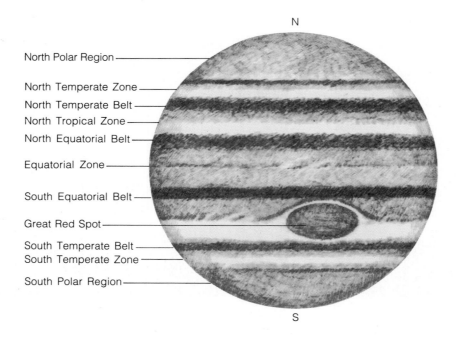

Figure 11-2 Semipermanent cloud formations of Jupiter, visible from Earth when viewed through small telescopes.

North Polar Region

North Temperate Zone
North Temperate Belt
North Tropical Zone
North Equatorial Belt

Equatorial Zone

South Equatorial Belt

Great Red Spot

South Temperate Belt
South Temperate Zone

South Polar Region

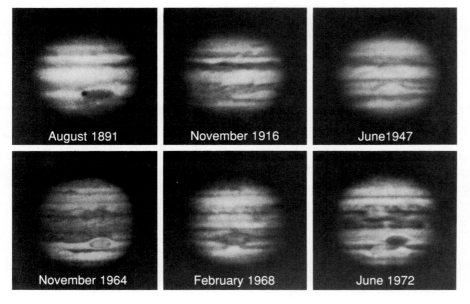

August 1891

November 1916

June 1947

November 1964

February 1968

June 1972

Figure 11-3 Photographs of Jupiter spanning 81 y, showing the changing array of Jupiter's belts and zones. The dark north temperate belt is relatively permanent, but the equatorial zone changes from bright (1891) to dark (1964). Many of these images show the Great Red Spot. (Lowell Observatory.)

11-4. The colors are in fact caused by photochemical reactions much like those that produce orangish-brown smog layers from pollutants on Earth. On Jupiter, the colorful minor constituents produced by these reactions may include hydrogen sulfide, organic particles, or metallic sodium particles.

Jupiter's Atmosphere

Just as on Earth, the cloud materials are only minor constituents of a much more extensive atmosphere of clear gas. On Earth, the gas is mostly nitrogen and oxygen, and the clouds of water droplets and ice crys-

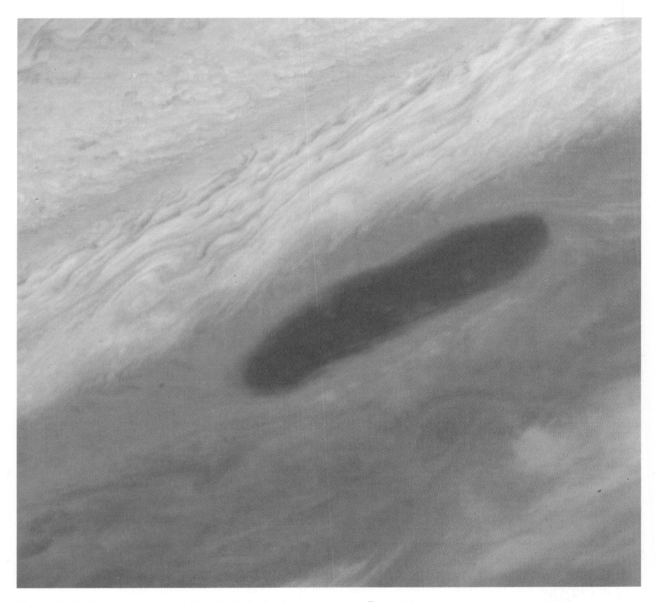

Figure 11-4 Nature's abstract painting: Jupiter's clouds at close range. The reddish-brown oval, more than half the diameter of Earth, may be a clearing in the upper cloud deck, allowing a view of deeper, redder clouds. Winds among the white clouds, which mark the north temperate zone, were measured at 260 mph. Smallest visible cloud features are about 80 km (50 mi) wide. (NASA Voyager 1 photo.)

tals condense from the small amounts of water vapor in the air. On Jupiter, the gas is mostly hydrogen and helium, while the clouds condense from ammonia, water vapor, and other minor compounds. **Jupiter's atmospheric composition** is about four-fifths hydrogen and one-fifth helium by mass. It is very different from our nitrogen/oxygen atmosphere. The hydrogen and helium are thought to be a "fossil" atmosphere of the gas that surrounded all the planets as they formed. The Sun is made from similar gas. Water, ammonia, and methane are also present in Jupiter's atmosphere. Voyagers 1 and 2 in 1979 revealed additional trace gases.

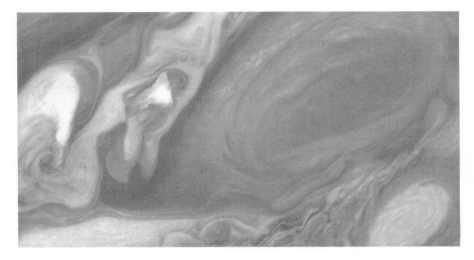

Figure 11-5 Jupiter's awesome Great Red Spot marks a turbulent storm system bigger than the whole Earth. It has been observed for several centuries. Smallest details in this view are about 100 km (60 mi) across. (NASA Voyager photo.)

Infrared radiation reveals **Jupiter's temperature** in the upper atmosphere to be very cold because of Jupiter's great distance from the Sun—about 133 K ($-220°$F), both on the sunlit and nighttime sides of the planet. At a lower level, in the poisonous upper clouds (where the air pressure equals the Earth's), the temperature is closer to 150 K ($-189°$F).

The lower atmosphere of Jupiter may be more interesting. Gaps in the highest clouds have revealed lower haze layers with temperatures of around 250 K ($-9°$F). Lower regions are even warmer. According to some scientists, the lower regions resemble the hydrogen-compound-rich primordial atmosphere of the Earth when terrestrial life originated.

A few scientists speculate that some level may even be warm enough for complex organic molecules to evolve into simple organisms that could float in the atmosphere. A recent model of Jupiter's atmosphere calls for temperatures similar to those at the Earth's surface at a level of about 60 km below the Jovian cloud tops, where the pressure would be about 10 times the Earth's surface pressure. Such conditions might be hospitable to primitive life, but strong updrafts and downdrafts would probably draw hypothetical organisms to levels that would be too cold or too hot for life. Thus most scientists doubt that any advanced life forms exist on Jupiter.

Voyagers' cameras revealed mighty blasts of lightning playing among the clouds and enormous auroral displays flickering high above the clouds in the polar regions.

For at least 300 years a storm three times bigger than the Earth has been raging on Jupiter. In the telescope it appears as an enormous reddish oval in the south tropical zone, probably first studied by G. D. Cassini in 1665. It became very prominent in 1887, when it was rediscovered and called the **Great Red Spot** (see Figure 11-5). The Red Spot has reached diameters of 40 000 km. It and other, smaller transient spots are probably vast, hurricanelike storm systems in Jupiter's atmosphere. Small clouds approaching the Red Spot get caught in a counterclockwise circulation like leaves in a great whirlpool.

The Voyagers revealed eastward jet streams blowing along the equatorial and temperate zones at wind speeds 150 m/s (338 mph) faster than in the dark equatorial belts. Figure 11-6 shows an imaginary view of the awesome cloud vistas near the top of the main cloud deck of this stormy world.

Jupiter's Rotation

Cloud belts and zones have a rotation period of about 9^h50^m near the equator, but several belts and zones at higher latitudes average 9^h56^m. The Red Spot has its own rate, sometimes lagging behind or drifting ahead of nearby clouds. The best estimate of the underlying planet's rotation is $9^h55\frac{1}{2}^m$, based on radio radiations from deep-atmosphere electrical storms. However, no one is sure whether any of these periods marks the true rotation of a well-defined solid or liquid surface beneath the clouds—or whether such a surface exists.

Figure 11-6 Enormous cloudscapes would greet a visitor to the upper atmosphere of Jupiter or the other giant planets. This imaginary view shows cloud decks at several levels. In the distance, one of the moons partially eclipses the Sun. (Painting by Ron Miller.)

Jupiter's Radiation and Magnetism

Besides reflected sunlight, Jupiter emits three other types of radiation. In order of increasing wavelength, they are infrared thermal radiation, shortwave radio radiation, and longwave radio radiation.

The first is called **Jupiter's infrared thermal radiation** because it is due to the heat of the planet itself (review Figure 5-5). From measurements of the total amount of this radiation, scientists know the total amount of energy being radiated by Jupiter. Surprisingly, this figure turns out to be about twice as much energy as Jupiter absorbs from the Sun! This is very different from the case of Earth or other terrestrial

planets, where the heat radiated from inside is negligible compared to the heat received from the Sun. Jupiter's extra internal heat must be coming from somewhere, but where?

Theorists believe Jupiter is slowly contracting, releasing gravitational energy as heat and radiation. This radiation was most intense when Jupiter formed and has declined ever since to the low level observed today.

Although Jupiter is radiating its own energy, it is not a true star, because its energy is not produced by thermonuclear fusion, the way a star's energy is. Jupiter's mass is not great enough to create the central pressure and heat necessary for starlike fusion reactions in its interior.

The second kind of radiation emitted by Jupiter is shortwave radio radiation, often called Jupiter's decimeter radiation because its wavelength is around $\frac{1}{10}$ m (1 dm). The distribution of this radiation with respect to wavelength shows that it is **synchrotron radiation**, a type of radio wave emitted when very fast electrons move through a magnetic field. (These electrons move at nearly the maximum possible speed for any object: the speed of light. The same radiation is found in synchrotron atom smashers on the Earth.)

Synchrotron radiation implies that Jupiter must have a magnetic field. The field was confirmed in 1973 and 1974 when Pioneers 10 and 11 flew through it and measured it. As shown in Table 11-1, it is the strongest planetary field known. Like the Earth's field, it is not perfectly aligned with the rotation axis but is about 15° off. The polarity is opposite the Earth's, with the north *magnetic* pole on the south side of the planet; a compass would thus point south on Jupiter. Because such fields are believed to result from the circulation of conductive material deep inside the planets, Jupiter's field indicates such a core. Further studies of Jupiter's field will help us understand the Earth's magnetic field.

Another similarity to Earth is that charged particles emitted by the Sun have been trapped by the magnetic field in doughnut-shaped rings around the planet, like the Earth's Van Allen belts. The particles are 100 000 to a million times more concentrated than near the Earth, making a zone of quite hazardous radiation near Jupiter's inner satellites.

The third kind of radiation, the strong, erratic decameter radio radiation (wavelength ~10 m), was discovered in the mid-1950s. Its sources are located on Jupiter itself rather than in nearby space. They may be localized storm areas. The strangest property was discovered by radio astronomers in 1964: The Earth receives the

TABLE 11·1

Magnetic Fields in the Solar System

Location	Approximate Field Strength Near Planet (nT)[a]	Source of Data
Sun (average surface)	200 000	Spectra
Sun (sunspot)	100 000 000	Spectra
Interplanetary space	4	Various spacecraft
Mercury	600	Mariner 10
Venus	<1[b]	Mariner 10
Earth (surface)	40 000	Field measures
Moon	≤2	Apollo data
Mars	≤30[b]	Mariner 9
Jupiter	400 000	Pioneer 10, 11
Saturn	20 000	Voyager 1
Uranus	25 000	Voyager 2

[a]The tesla (T) is a unit of magnetic field strength. Think of it as measuring the strength of response of a compass placed in the field. $1 \text{ nT} = 10^{-9} \text{ T}$. Some scientists continue to use older units of measure, the "gamma" and the "gauss"; $1 \text{ nT} = 1 \text{ gamma} = 10^{-5} \text{ gauss}$.

[b]Spacecraft near Venus and Mars detected little or no magnetic field increases over the interplanetary field strengths.

bursts of radio noise primarily when Jupiter's satellite Io is in certain positions *with respect to Earth.*

Why should radiation emitted from Jupiter depend on the orbital position of *one* of its moons—particularly on its position relative to Earth? Io moves through the magnetic field of Jupiter, disturbing this field and its particle motions in such as way as to produce beams of radiation that, like searchlight beams, sweep through space as Io moves along. Intermittently they point toward the Earth.

Jupiter's Internal Structure and Surface

What is Jupiter like under its clouds? Its mean density is too low for it to be a rock-and-iron planet like Earth. Instead, **Jupiter's interior** is believed to consist of 60% hydrogen, the rest being helium with small amounts of silicates and other "impurities." The heavier elements have sunk, so that a core of silicate or iron–silicate material—resembling terrestrial planets—may exist near the center. According to some theoretical models, the core may be over twice the Earth's size and have a temperature around 30 000 K.

Under the high pressure within Jupiter's interior,

hydrogen takes on an unfamiliar form called **metallic hydrogen**, which can conduct electric currents. As shown in Figure 11-7, much of Jupiter's interior may be taken up by a mantle of liquid metallic hydrogen, in which convection currents carry heat to the surface and electric currents generate Jupiter's magnetic field. Much of the outer half may be a sea of ordinary liquid molecular hydrogen, H_2. There may be no solid surface at all, only a slushy mixture of liquid hydrogen and crystals of various compounds.

The gravity at this ill-defined "surface" is stronger than that of any other planet. A person weighing 150 lb on Earth would weigh about 400 lb on Jupiter! Similarly, the pressure of the thick atmosphere is roughly 100 times the air pressure on Earth. Various data suggest that the cloud-obscured, high-pressure ocean of liquid hydrogen begins as much as 1000 km or more below the cloud tops.

JUPITER'S SATELLITES

Jupiter has an impressive array of at least 16 moons ranging from 8-km bodies to four large worlds, including two slightly larger than the planet Mercury. The moons

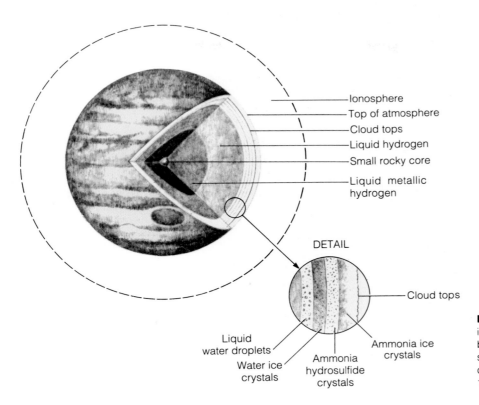

Ionosphere
Top of atmosphere
Cloud tops
Liquid hydrogen
Small rocky core
Liquid metallic hydrogen

DETAIL

Cloud tops

Liquid water droplets

Ammonia ice crystals

Ammonia hydrosulfide crystals

Water ice crystals

Figure 11-7 A view of the interior and atmosphere of Jupiter based on telescopic and spacecraft data. The atmospheric depth to the liquid zone is perhaps 100 km. (NASA)

are numbered in order of discovery. The four largest, shown together in Figure 8-2, were discovered independently by Galileo and the German astronomer Marius on two consecutive nights in 1610 as they viewed Jupiter through the newly invented telescope.[1] These bodies are called the **Galilean satellites.** You will recall their historical importance: When Galileo discovered that they orbit around Jupiter, he cited this as proof that not all bodies orbit around Earth—thus refuting the Ptolemaic theory and supporting the Copernican theory. Now we know that those moons are geologically, as well as historically, interesting.

Until the 1979 flights of Voyagers 1 and 2 through the satellite system, we knew the large satellites only

as pinhead disks in even the largest telescopes. Some observers noted vague dusky markings. Since they are in the size range of the Moon and Mercury, these moons were assumed to be cratered, dead worlds with little geological activity. Spectra showed some differences among the surface materials of these worlds, but no one realized how unique each of these moons is.

Let us start by looking at the whole system from a distance, as shown in Figure 11-8. The four large Galilean moons are in the middle. Closer to Jupiter are several quite small moons and a thin ring composed of debris: tiny particles orbiting Jupiter. Exterior to the four Galilean moons are eight more small moons. Curiously, they are in two groups of four each.

One of the exciting aspects of the systems of Jupiter's moons and the outer giant planets is that they give us an additional chance to practice comparative planetology on a grand scale. Not only can we try to learn about planetary evolution by comparing one moon to another; we can think of each as a miniature solar system that may tell us something about the planet-forming process in general. Like the planets in the solar system,

---

[1]Keen-sighted observers can sometimes see at least one of Jupiter's large moons with the naked eye; hence it may have been spotted long before 1610. In 1982, Chinese historians noted such a report of a faint Jupiter companion by ancient Chinese astronomer Gan De in 364 B.C. It will remain hard to prove such sightings; in any case, the significance—moons orbiting *around* Jupiter—was not known until 1610.

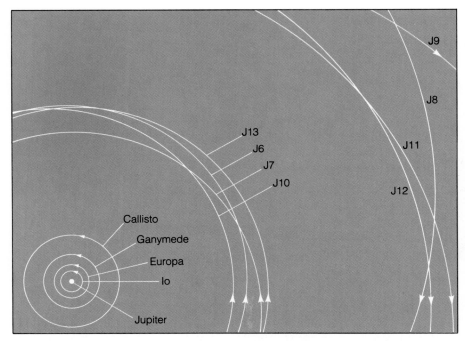

Figure 11-8 The "miniature solar system" of Jupiter and its satellites. Orbits of the four large Galilean moons are shown close to Jupiter (lower left). Still closer, smaller moons and the rings are too close to show on this scale. Two groups of outer moons lie in closely clustered orbits. Note that the outer group moves in retrograde direction—opposite to the usual sense of orbital motion in the solar system. See text for further discussion.

for example, the moons of Jupiter, Saturn, and Uranus follow nearly circular orbits and lie close to the plane of the equator of the central body. In the case of Jupiter's system, the inner small moonlets might be thought of as analogous to the terrestrial planets, while the four giant Galilean moons, at their greater distance from Jupiter, resemble the giant planets. In Chapter 14 we will examine this approach more carefully as we consider the origin of the solar system.

Two Basic Materials: Bright Ice and Black Soot

When we shift our attention from Jupiter to its moons, we are no longer dealing with matter in unfamiliar forms, such as the metallic, high-pressure hydrogen of Jupiter's interior. Moons of Jupiter and the other outer planets are made of material that is more ordinary. However, it is not quite the same composition as the familiar rocks and minerals that make up most of the Earth, Moon, and Mars.

Because the outer solar system is so cold, the dominant materials that form small worlds are not very common on Earth—our planet is too warm for many of them to have formed. A simple but useful picture of these moons can be obtained by thinking of *two* main types of materials that form at very low temperatures. The

first is ice, or, more accurately, a mixture of various types of ice. Much of the ice is like familiar ice in a cold drink, frozen H_2O, but at a much lower temperature. Other ices may include frozen CO_2 (what we call "dry ice"), frozen methane, and other frozen material. All of this ice mixture has a bright, whitish color.

The second main type of material in bright moons is rocky matter, but colored very black by dark minerals believed to be carbon and carbon compounds. The black, **carbonaceous material** is much like soot. Sometimes it is very dark chocolaty brown, probably due to colored organic compounds formed from the carbon. When it is not strongly diluted by being mixed with the bright ice, it has an extremely dark coloration, as black as black velvet.

Thus we can think of a sort of "salt and pepper" model of material on Jupiter's moons: a mixture of white stuff and black stuff. The white material is ice, and the black is sooty soil. As we will see, on some bodies' surfaces the ice has sublimed off into space, leaving the sooty material behind and giving a very black color. On other bodies, water has erupted and coated the surface with fresh ice, making a cleaner, whiter color than the Moon or Mars. Still other bodies have a grayish-tan coloration from a mixture of the two materials.

These concepts help explain why each of Jupiter's major moons has a distinct "personality" and how the

Figure 11-9 Callisto—the outermost of Jupiter's four, large Galilean moons. The surface is a dark soil with some ice mixed in. Bright spots are craters where surface soil has apparently been blasted away, exposing cleaner ice. This picture uses a UV image for the blue-color component, enhancing color contrast between ice and the background soil. (NASA Voyager 2 photo.)

personalities shift as we move from one moon to another. Because certain heating effects were strongest near Jupiter, the materials of moons near Jupiter were most modified from their initial cold, ice-and-soot compositions. Therefore, we will start our survey of the satellites with the outer moons and work our way inward toward more and more altered moons. As we will see, these altered inner moons are truly strange, with properties hardly imagined until the flight of the Voyager probes. During our discussion, you can refer to Table 8-1 (page 172) for specific physical properties of these bodies.

The Outermost Moons: Captured Black Asteroids

The eight outermost moons are so far from Jupiter that if they were much further, the Sun's gravity would be more important than Jupiter in controlling their motions. In other words, they are barely part of Jupiter's satellite system. Physically they have black surfaces that resemble surfaces of many asteroids. Any ice that may have been originally exposed on these surfaces may have sublimed into space during exposure to the Sun, leaving black, sooty surfaces. For these reasons, astronomers believe these moons are not native to Jupiter but are asteroids that originally orbited around the Sun, came close to Jupiter, and were captured into orbit. This could have happened in the early history of the solar system when Jupiter may have had a more extensive atmosphere; asteroids passing through the fringes of that atmosphere could have been slowed in a way that resulted in capture.

Satellites J6, J7, J10, and J13 have orbits around 12 million kilometers from Jupiter and inclined 27° to Jupiter's equator. All these moons, as well as the other moons mentioned so far, orbit in a **prograde** direction like our own Moon. (This is the "normal" direction of orbital motion in the solar system, counterclockwise as seen from the north celestial pole. See Figure 11-8.) But satellites J8, J9, J11, and J12 all orbit in a **retrograde** direction (clockwise as seen from the north celestial pole), 23 million kilometers from Jupiter with inclinations of around 52°. How did the outer satellites come to be clumped into two such groups? Analysts have suggested that during the capture, when the asteroids passed through the fringes of an early atmosphere, they broke into pieces as they were decelerated. Thus there may have been only two capture events, each contributing four large pieces. One object approached Jupiter in a prograde sense and fragmented to produce prograde moons in similar orbits; the other approached in a retrograde sense and produced retrograde moons. Perhaps there are smaller fragments still undiscovered in each of the two orbital groups.

Because there are no close-up photos of any of these moons, we don't know their surface features. Probably they are cratered like Mars' moons, Phobos and Deimos. Indeed, they may be quite similar to Phobos and Deimos in appearance, size, and origin: Recall that Phobos and Deimos are also believed to be dark asteroids captured into satellite orbits.

Callisto: The Cratered Moon

Moving in toward Jupiter, we come to the four Galilean moons, which must have formed as part of Jupiter's system. The outermost of these is **Callisto** (Figure 11-9). Second-biggest of Jupiter's moons, it is 4800 km across, just 2% smaller than the planet Mercury.

In terms of surface geology, Callisto is a primitive world with little sign of internal geological activity or surface heating. Craters caused by meteorite impacts

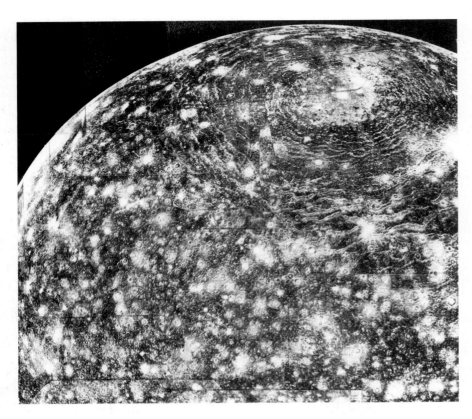

Figure 11-10 Heavily cratered surface of Callisto. Multiringed structure (top) is believed to be a remnant of a huge impact feature, the largest impact feature known in the solar system. (NASA Voyager photo.)

are the main landform, as emphasized in Figure 11-10). Compared with water's density of 1000 kg/m^3 and rock densities of about 2500 kg/m^3, Callisto's average density of 1800 kg/m^3 indicates a composition of about half ice and half rocky material. The surface is mostly a tannish-gray material that is probably a mixture of ice and carbonaceous matter. There may have been some concentration of dark soil on the surface as repeated impacts melted and vaporized some of the ice. As seen in Figure 11-10, each crater appears to blow away the surface soil and expose brighter, cleaner ice underneath. Judging from the high number of craters, the surface has changed little since its formative era 4 to 4.5 billion years ago.

Ganymede: A Giant Moon

Ganymede, with a 5262-km diameter, is the largest moon of Jupiter and probably the largest moon in the solar system. It is shown in Figure 11-11. It is 10% bigger than Callisto, 8% bigger than the planet Mercury, and nearly twice as big as Pluto. Its average density of 1900 kg/m^3 is barely higher than Callisto's, implying a similar ice–rock composition. A thin polar cap, visible in Figure 11-11, is believed to be water frost. This would indicate that water vapor has been emitted from the interior, circulated poleward in an almost negligible atmosphere, and frozen onto the ground at the poles.

The ancient parts of Ganymede's surface are very similar to Callisto's surface, being heavily cratered and covered with dark, dusty soil (Figure 11-12). Contrary to Callisto, though, these areas are divided by swaths of light-toned, grooved terrain. We can tell that the bright terrain is younger than the dark areas because it has not accumulated as many impact craters. The ancient dark crust seems to have split, allowing fresher icy material to erupt.

Such splitting may imply internal heating that melted at least the icy component and caused expansion and cracking. Water erupted, forming ice flows—the Galilean satellites' equivalent of the Earth's lava flows. The bright swaths appear to mark fractures along which water erupted (Figures 11-13 and 11-14). Scattered impact

Figure 11-11 Jupiter's largest moon, Ganymede. Part of the surface consists of old, dark, cratered terrain like the surface of Callisto. The dark terrain is broken, however, by swaths of brighter, younger, more ice-rich material. At the upper left corner we can see part of a white polar cap of frost. (NASA Voyager 2 photo.)

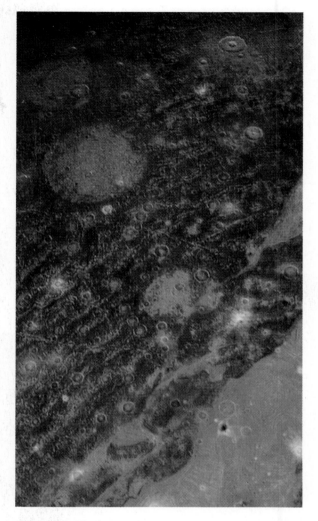

Figure 11-12 A closer view of the central region of Figure 11-11 reveals that the dark region on Ganymede is cut by concentric arc-shaped fractures—probably part of an ancient multiring system (like that in Figure 11-10 on Callisto) that predated the breakup of the crust. Impact craters dot the area but are scarcer on the younger, bright icy areas. Large bright ovals appear to be ancient craters so large that they penetrated into liquid water beneath the dark crust, filled with water, and thus left nearly rimless icy white scars. (NASA Voyager 2 photo.)

craters on the grooved terrain suggest that it formed 2 to 4 billion years ago, after which Ganymede's geological activity declined. The source of heat that drove Ganymede's geological activity may have been radiant heat released by primordial Jupiter as it formed, or radioactivity in the rocky component of Ganymede's interior.

Ganymede is telling us an important story that may help us understand the Earth better. Ganymede offers a case of incipient plate tectonics, since its brittle "lithosphere" of ice split into platelike blocks as it floated on a warmer, watery subsurface. Although Ganymede's interior was not active enough (hot enough?) to cause full-scale continental drift among the plates, offsets do show that the plates have moved slightly. Ganymede thus provides a transitional example from cold, inactive planets to larger, geologically active worlds.

Europa: The Ice-Surfaced Moon

Europa, the next inner moon, is smaller than Callisto and Ganymede—and much different. It is about as big as our own Moon. As seen in Figure 11-15, it is an icy billiard ball. There are no mountains and scarcely any craters. The surface is creamy white, as bright as the paper of this book. The only markings are delicately shaded, shallow grooves crisscrossing the surface. Spectra show that the surface is mostly frozen water, but the relatively high mean density of 2970 kg/m³ for Europa as a whole shows that Europa's interior consists

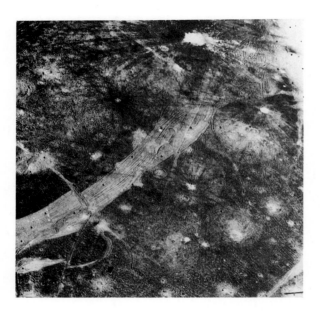

Figure 11-13 Oblique view of part of Ganymede. Lower right half is part of darker surface resembling Callisto, where each impact crater produces a bright scar (possibly exposing cleaner, underlying ice). Through left center runs a bright swath of icy material with parallel ridges and grooves a few kilometers wide. It is apparently offset by a fault at lower left. Such bright swaths appear to be fractures in original crust, in which watery "magma" rose up and froze. The area of this view would enclose Texas. (NASA Voyager photo.)

Figure 11-14 Imaginary panorama of Jupiter and the icy surface of its moon Ganymede. Ganymede's surface is split by parallel fissures, where water may have erupted and refrozen. This view, from such terrain, shows a ridge of jagged ice blocks thrust up during movements of Ganymede's crust. In the sky are Jupiter's inner moons, Europa (left) and Io (in front of Jupiter). Jupiter would look about 15 times as big as our Moon in our sky, and Europa would look slightly smaller than our Moon. (Painting by Ron Miller.)

mostly of rock. Perhaps the rocky components' radio-activity, together with tidal flexing, heated the interior, melted the ices, evaporated much of the water, and allowed the remaining water to erupt onto the surface, obliterating the original crust. That water froze into the clean, smooth ice layer we see, which has been cracked by subsequent heating or cooling. The ice layer may be 100 km (60 mi) thick, perhaps overlying a liquid water layer (see Figures 11-16 and 11-17). The near absence of impact craters shows this ice surface is younger than Callisto's or Ganymede's. It may be continuing to form by intermittent eruptions of water, even in modern times. Some Voyager team members believe they have identified a bright transient cloud marking a possible erup-

Figure 11-15 Two views of Jupiter's icy billiard ball moon, Europa. This world is about the size of our Moon but is covered by nearly smooth, whitish ice. Pale dark streaks appear to mark fractures. (NASA Voyager photo.)

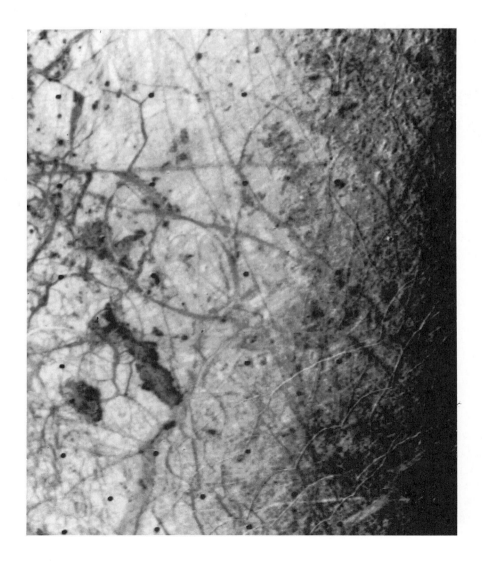

Figure 11-16 This contrast-enhanced close-up of Europa under low lighting shows a filigree of fine grooves and ridges, possibly related to fractures in the icy surface. The area is 600 × 800 km. (NASA Voyager photo.)

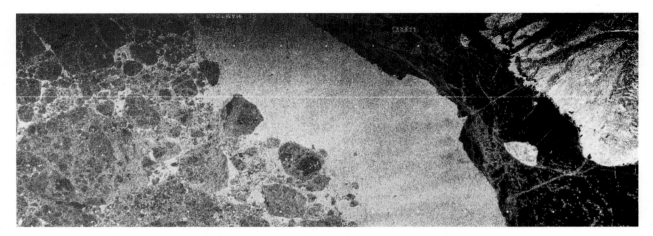

Figure 11-17 An orbital view of the fractured Arctic Ocean ice pack (left) reveals a fracture pattern similar to those seen on Ganymede and Europa, which may also involve ice slabs floating on the water. The dark lane (right) is open sea, which is bordered by an island at right. The length of the image is 120 km. (NASA Seasat satellite radar image.)

Figure 11-18 The strange volcanic moon Io. The surface is mottled by yellow, orangish, and white volcanic sulfur flows. Dark spots are believed to be hot sulfur, some actually erupting. The doughnut at the center is a vertical view through an erupting volcano's plume of debris coming out of the central dark vent and falling in a ring around it. This volcano is named Prometheus after the Greek firegod. (NASA Voyager 1 photo; courtesy A. S. McEwen, U.S. Geological Survey.)

Figure 11-19 The photo on which Io's volcanoes were first recognized. While analyzing the position of Io in order to help navigate the Voyager spacecraft, researcher Linda Morabito recognized the plume extending off the limb at right. A second eruption creates a bright plume near the terminator. (NASA Voyager photo.)

tion on the final Voyager 2 photo of Europa. Europa may occasionally have giant geysers! In his novel *2010*, science fiction writer Arthur C. Clarke speculated that life might have evolved in Europa's buried oceans. Confirmation of such ideas will have to await new spacecraft flights.

Io: The Volcanic Moon

The inner Galilean moon **Io** is probably the most bizarre world in the solar system. Like Europa, it is about the size of our Moon. But what a difference! Voyagers 1 and 2 discovered not only active volcanoes but also a fantastic mottled surface of pale yellow, orange, tan, and white sulfur deposits (Figure 11-18). When the pictures first arrived from Voyager, one researcher commented that it looked like the kind of world we used to laugh at when it appeared outside spaceship windows in grade B science fiction movies with cheap special effects.[2] But here it was, real!

---

[2]Another analyst commented that he didn't know what was wrong with Io, but it looked as though it might be fixed with a shot of penicillin. Still another wag called it the Pizza Moon, complete with cheese, tomato, and pepperoni strata.

In a realm of ice and cold carbonaceous compounds, what source of heat could possibly explain the erupting volcanoes and their sulfur flows? This would have been a total mystery when Voyager arrived, except for some brilliant detective work by California dynamicists S. J. Peale, P. Cassen, and R. Reynolds (1979). While Voyager was on its way, they calculated that although Io's orbit is nearly circular, gravitational pulls of neighboring satellites have caused Io's distance from Jupiter to vary slightly throughout its history. As the distance varies, the tidal stretching of Io changes. This flexing heats Io's interior, just as rapid flexing of a tennis ball makes it heat up from friction. The calculations showed that this heating effect is stronger in Io than in any other satellite and causes Io's interior to be molten. In a beautiful example of the scientific method at work, Peale and his colleagues published their prediction that volcanoes might be found on Io in a journal that came out just a few days *before* Voyager 1 reached Io. A few days later, as shown in Figures 11-19 and 11-20, Voyager discovered the solar system's most active volcanoes, spewing 100-km-high plumes of debris.

Another heating effect on Jupiter's moons was radiant heat from primordial Jupiter. As Jupiter formed 4½ billion years ago, it acted like a miniature sun in its own

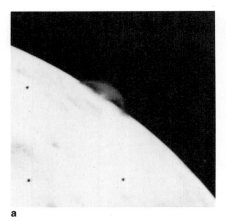

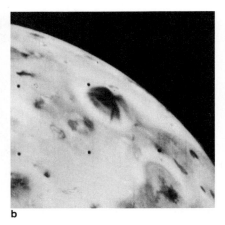

a   b

Figure 11-20 An erupting volcano on Io is seen silhouetted on the horizon (**a**) and against the background landscape (**b**). Because there is no air, the cloud does not billow as a terrestrial eruption does; rather, the debris follows curved ballistic paths and falls back to the ground. (NASA Voyager photo.)

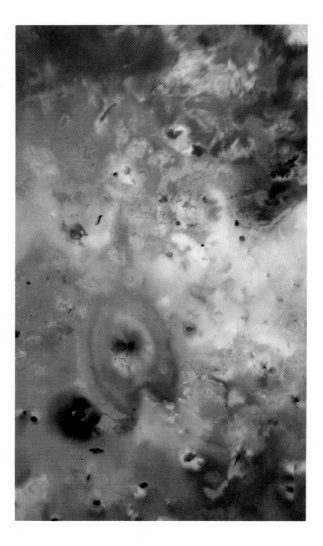

Figure 11-21 A different processing of the Voyager images renders a large part of Io in map-like form. Large black mass at upper right is in upper left of Figure 11-18. The upside-down heart-shaped feature (lower left) is another vertical view down through a volcanic plume. Its central dark vent is named Pele after the Hawaiian volcano goddess. The heart-shaped plume is about 1000 km across; because of the low gravity and furious volcanism, the debris plume spreads over an area about the size of California and Nevada! (NASA Voyager mosaic, processed by A. S. McEwen, U.S. Geological Survey.)

miniature solar system of moons, heating the surfaces of the inner moons most and the outer moons least.

Both the tidal heating and the radiant heating tended to be strongest on the inner moons. Although the tidal heating effect is strong only for Io, the combined effects may explain why the outer Galilean moon, Callisto, shows least effects of heating or ice melting, while the inner moons show more. In any case, the effects on Io were apparently so strong that any initial water (liquid or ice) was completely melted and evaporated or sublimed off the moon. The "salt-and-pepper" ice–soot mixture of other moons was destroyed. Even the carbonaceous rocky component was at least partly melted, producing sulfur-dominated lavas that spread out in overlapping layers to form a colored, sulfurous surface (Figures 11-21 and 11-22).

Although Jupiter's satellites generally have very cold surfaces (measured dawn and afternoon temperatures range between 80 and 155 K, or -315 and $-180°F$), Io's volcanoes are local hot spots. Their measured lava temperatures are as high as 600 to 700 K (621 to 800°F).

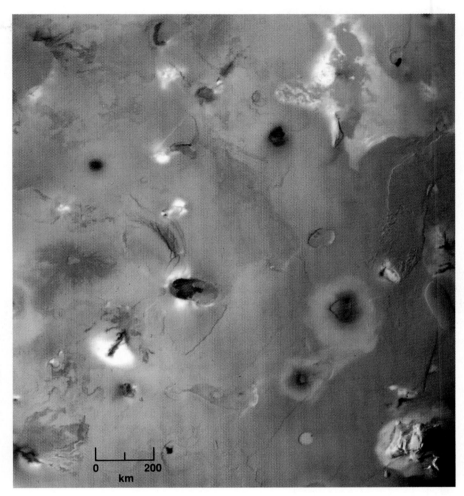

Figure 11-22 Volcanic terrain near the south pole of Io. Low Sun angle reveals several depressed calderas and a volcanic peak (lower right corner). Blackest areas may be hot or molten sulfur, whose tones grow darker at higher temperatures. Several vents (left center) show radiating spiderlike patterns of sulfurous lava flows. Plateau-like thin layers of lava have built up the surface in several areas. (NASA Voyager 1 photo.)

Infrared observations from Earth have detected explosive outbursts of Io's volcanoes, confirming the continuous volcanic activity there.

If we could survive the earthquakes, volcanic outbursts, and intense irradiation by ions trapped in Jupiter's nearby magnetic field, a day on Io would be an interesting experience. This experience is depicted in Figure 11-23. All Galilean moons display synchronous rotation—they keep one side toward Jupiter just as tidal locking keeps one side of the Moon toward the Earth (see Chapter 7). As a result, Jupiter hangs in one spot in Io's sky. It covers an angle of 20°, looking 40 times bigger than our Moon in our sky. In Figure 11-23a, we see the day starting with an eclipse as the Sun emerges from behind Jupiter above a dimly lit landscape. About

15 min later (Figure 11-23b), the Sun has emerged and the ground is covered by a whitish frost of sulfur dioxide (SO_2), condensed from volcanic gases during the cold darkness of the eclipse. This occasional phenomenon has been glimpsed from Earth.

Io's day—one rotation and one trip around Jupiter—is about 42 h long. In Figure 11-23c, 10½ h after Figure 11-23b, we have traveled a quarter of the way around Jupiter. The sulfur dioxide frost has burned off and we see the colored sulfur deposits. A strange phenomenon has happened in the sky. Sulfur and sodium atoms knocked off Io form a permanent cloud around it, and when Io moves toward or away from the Sun, these atoms give off a visible yellow glow due to a process involving the absorption of Doppler-shifted sunlight. (The

a b c

d e f

Figure 11-23 A day on Io. Sequence shows events as seen from one spot on Io as the moon moves around Jupiter in 42 h. See the text for further description. (Paintings by author.)

Doppler shift is explained in Chapter 16.) Thus a faint yellow aurora lights the sky in Figure 11-23c. This yellow glow around Io has actually been photographed from Earth, as seen in Figure 11-24. In Figure 11-23d, we have moved another quarter of the way around. The Sun is behind us, shadows stretch toward the horizon, and Jupiter is full. In Figure 11-23e, the Sun has set, the landscape is lit by Jupiter, and the faint yellow glow is back in the sky. Each time Io moves one-quarter of the way around Jupiter, incidentally, Jupiter turns once with respect to the stars, and so the same cloud features can be seen in Figure 11-23c–e from different directions. In Figure 11-23f, we witness a predawn volcanic eruption. The distant volcano spews curtains of

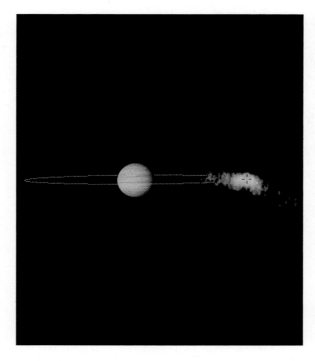

Figure 11-24 The yellow-glowing cloud of sodium gas surrounding Io and spread along its orbit. In this composite photo Jupiter is seen surrounded by Io's orbit drawn by computer. Io is at the right end of the orbit, at the position marked by the cross. Superimposed is an image of the yellow glow from sodium atoms knocked off Io's surface. (Courtesy Bruce A. Goldberg, Jet Propulsion Lab/NASA.)

ash up into the sunlight (just as high clouds may be lit by Sun just before dawn on Earth). The shadow of Io darkens the lower part of the eruptive cloud.

Moonlets on the Outskirts of the Ring

At least four additional moonlets exist inside the orbit of Io, near the edge of a faint ring discovered by the Voyagers. All are much smaller than the Galilean moons. The largest, designated J5 Amalthea, was discovered in 1892. It is a cratered lump 270 km long and 155 km wide (roughly the scale of New Jersey, which is not as cratered but might as well be, according to some observers). Amalthea has an orangish color attributed to sulfur atoms ejected from Io and coating Amalthea. The other three moons, all discovered by Voyagers, are

around 40 km across. One orbits between Amalthea and Io; the other two are nearly on the outer edge of Jupiter's ring.

THE RING OF JUPITER

The Voyagers' discovery of Jupiter's ring came as a surprise. It is too faint and too nearly edge-on to be seen from Earth (Figure 11-25). It probably consists mostly of microscopic dark particles being knocked off the inner moonlets near its edge. Forces acting on these particles make them spiral inward toward Jupiter, thus creating a ring through which material flows slowly over millions of years.

FUTURE STUDIES OF JUPITER

NASA's **Galileo program** is the only planned mission to Jupiter and its satellites announced by any of the spacefaring nations. Galileo is a large and complex unmanned spacecraft designed to eject a small probe into Jupiter's atmosphere. The mother ship will go into orbit around Jupiter in order to visit and map one satellite after another. Galileo was the last of the large, state-of-the art space vehicles to be built before funding cutbacks caused NASA to reduce the sophistication of other planned unmanned missions. It was originally to be launched by the Space Shuttle in the mid-1980s. The Challenger disaster seriously delayed it, however, and although the spacecraft is built and ready to go, it is unlikely to reach Jupiter before the mid-1990s. If all parts are still in working order, Galileo should give us much new detail about the Jupiter system.

WHY GIANT PLANETS HAVE MASSIVE ATMOSPHERES

At the end of Chapter 9 we gave three principles that explain why some planets lack atmospheres. Let us apply them here to explain in more detail why giant planets have massive atmospheres rich in hydrogen gas. The first principle says that the higher the temperature, the higher the molecules' speed. (For more detail, see accompanying Optional Basic Equation VI.) This means

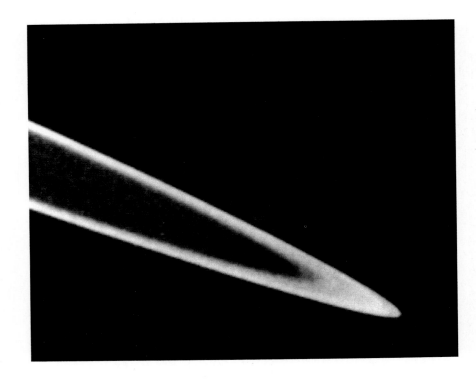

Figure 11-25 Jupiter's ring is believed to be fine material spiraling in toward Jupiter from small moons near the outer edge of the ring. Jupiter is out of the picture at left. (NASA Voyager photo.)

OPTIONAL BASIC EQUATION VI

Typical Velocities of Atoms and Molecules in a Gas

A useful result about the motions of atoms or molecules in a gas was worked out around 1860 by physicists, particularly the Scottish researcher James Clerk Maxwell and the Austrian researcher Ludwig Boltzmann. Like our other Basic Equations, it is a simple law that gives us many insights throughout astronomy. It comes from the recognition that any gas is made of particles (atoms or molecules or even microscopic dust grains), all moving at various speeds determined only by the temperature T. In fact, T is defined as a measurement of the average velocity of energy of a particle in the gas. The higher the T, the higher the mean velocity (or energy) of any representative particle, as noted at the end of Chapter 9. Now recall from basic physics that the kinetic energy of any moving particle is $\frac{1}{2}mv^2$, where m is the particle mass and v is its velocity. The discovery of Maxwell and Boltzmann is that the mean kinetic energy is proportional to T. This statement is usually written

$$\frac{1}{2}mv^2 = \frac{3}{2}kT$$

where

$$k = \text{Boltzmann constant}$$
$$= 1.38(10^{-23}) \text{ joules/degree}$$

The joule (abbreviated J) is the SI unit of energy. Solving for v, we have a value that may be thought of as the typical velocity of an average particle:

$$\text{Typical velocity} = v = \sqrt{3kT/m}$$

(To be more precise, this value is the square root of the mean squared velocity and is about 10% higher than the true mean velocity. Physicists widely use this "root mean square" velocity as an all-purpose indicator of gas atom speeds, however.)

Note that the mass of an atom or molecule is conveniently figured from

that the cold, outer planets will generally have low molecular speeds in their atmospheres. The second principle says that the lighter the molecule, the higher its speed. This means that on any given planet, hydrogen will be the fastest-moving molecule. The third principle says that the bigger the planet, the higher the speed needed for a molecule to escape the planet's gravity and shoot off into space. This means that giant planets will hold whatever gases they have much better than smaller planets like Earth.

Combining these principles, we can compare the Earth and a giant planet. On small, warm Earth, the fastest-moving hydrogen molecules will move more than fast enough to escape into space. Thus Earth has lost virtually all its original hydrogen, which was by far the most abundant primordial gas. The slower nitrogen and oxygen did not escape. Conversely, on a cold giant like Jupiter, the fast-moving hydrogen and all other gases have been retained. Since the original hydrogen was more abundant than all other gases put together, today Jupiter's hydrogen-rich atmosphere is very massive and very deep.

SUMMARY

Jupiter is the largest planet in the solar system. Its massive atmosphere is about four-fifths hydrogen and one-fifth helium by mass and is probably a remnant of gas from which the Sun and planets formed. The planet is covered by dense colored clouds arranged in bands parallel to the equator.

Jupiter's system of moons provides us with a "miniature solar system" to compare with the system of planets. The four largest moons are called the Galilean moons. Each of these satellites is unique. Callisto is cratered and has a gray ice/soil surface. Ganymede has a similar surface, except that it is broken by swaths of younger, brighter ice. Europa is completely surfaced by clean, bright, barely cratered ice. Strangest of all is Io, which has been heated by tidal forces, creating the only known satellite with active volcanoes. Indeed, Io is the most volcanically active body in the solar system and is covered with yellowish sulfur-rich lavas. Inward from the large moons are a number of small moonlets on the outskirts of a thin ring. The situation is reminiscent of the solar system's four giant planets circling beyond the small terrestrial planets. On the outskirts of Jupiter's system are eight small, black moons that seem to be captured asteroids.

m = (atomic weight of atom or molecule) × (mass of a hydrogen atom)
= (atomic weight) × $1.67(10^{-27})$ kg

Sample Problem 1. What is the typical velocity of a molecule of air in your room? *Solution:* Assume T = 68°F = 20°C = 293 K. (We have to convert to SI units, which for T are kelvin degrees.) Assume the air molecule is the nitrogen molecule, N_2. A check in a chemistry book shows that these two nitrogen atoms have total atomic weight 28 (28 × the weight of a hydrogen atom), so that m = $4.68(10^{-26})$. Therefore

$$v = \sqrt{\frac{3(1.38)(10^{-23})(293)}{4.68(10^{-26})}}$$

$$= 509 \text{ m/s}$$

Thus the atoms and molecules around you are typically hitting your skin at about ½ km/s! You don't feel the

individual hits because they are too small, but you do feel the sustained pressure exerted by the ensemble of them.

Sample Problem 2. Prove that a hydrogen atom would move much faster than a nitrogen atom in the same atmosphere. *Solution:* Since v is proportional to $\sqrt{1/m}$, we confirm that the smaller the mass, the greater the velocity. Since an N atom is 14 times more massive than an H atom, it would move $1/\sqrt{14}$ as fast. Or, to say the same thing, the H atom moves 3.7 times faster. You should also show that an H *atom* would move 5.3 times faster than an N_2 *molecule* in the same gas.

As illustrated in the advanced problems at the end of this chapter, this basic equation can easily be used to show why hydrogen has escaped from the Earth and not from Jupiter.

CONCEPTS

giant planet

belts

zones

Jupiter's atmospheric
composition

Jupiter's temperature

Great Red Spot

Jupiter's infrared thermal
radiation

synchrotron radiation

Jupiter's interior

metallic hydrogen

Galilean satellites

carbonaceous material

prograde and retrograde
satellite orbits

Callisto

Ganymede

Europa

Io

Galileo program

PROBLEMS

1. Answer the following problems:

a. How might studying cloud patterns on Jupiter, Mars, and Venus help us understand terrestrial meteorological theory?

b. What planetary or environmental characteristics might figure in such a theory?

c. How do these factors vary among these planets?

d. Why are other planets not included on the list?

2. How would gravity on the Galilean satellites compare with gravity on the Moon (see data in Table 8-1)? Which three bodies in the solar system would you expect to have general environments most like the Moon's?

3. Describe a landscape on Io.

4. Two large planets have the same size and mass but different orbits.

a. Which would you expect to have more hydrogen? Why?

b. If they have had different amounts of volcanism, which would you expect to have more carbon dioxide? Why?

ADVANCED PROBLEMS

5. Jupiter's four Galilean satellites revolve in orbits about 400 000 to 2 million kilometers from the planet.

a. What would be their maximum angular separation from the planet when Jupiter is at its closest distance of about 4 AU from Earth?

b. These satellites have a brightness of about the sixth magnitude, equal to the faintest stars visible to the unaided eye. The eye can normally distinguish details as little as 2 minutes of arc apart. Comment on the feasibility of seeing the Galilean satellites with the naked eye.

6. If you had a telescope that could reveal details as small as ½ second of arc, would you be able to see dark markings on Ganymede (diameter of 5270 km)?

7. Calculate the weight of a 140-lb person on the "surface" of Jupiter. Use Table 8-1.

8. What is the typical velocity of a hydrogen atom:

a. In the 1500-K gas at the top of Earth's atmosphere?

b. In a thin 1500-K gas at the surface of the Moon?

c. In a 200-K gas at the top of Jupiter's atmosphere?

d. The fastest atoms in a gas may move three or more times faster than the typical velocity calculated above, due to random collisions in the gas. Calculate the escape velocities of the Earth, Moon, and Jupiter, and use these facts to explain the absence of hydrogen on the Earth or Moon but its retention on Jupiter.

PROJECTS

1. Observe Jupiter with a telescope with at least an 8-cm (3-in.) aperture. Sketch the pattern of belts and zones. Which are the most prominent belts? Which zones are brightest? Compare these results with photos in this book. Is the Red Spot or other dark or bright spots visible on the side of Jupiter being observed?

2. Observe Jupiter with large binoculars or a telescope with at least a 2½-cm (1-in.) aperture. How many satellites are visible? Observe the satellite system at different hours over a period of several days and try to identify the satellites. (This could be done as a class project, with different students making sketches at different hours.)

3. With a telescope having at least a 15-cm (6-in.) aperture, determine the rotation period of Jupiter by recording the time when the Red Spot is centered on the disk. Note that intervals between appearances must be an integral number (1, 2, 3, and so on) of rotation periods.

CHAPTER 12

The Outermost Planets and Their Moons

Jupiter is only the first of four giant planets that occupy the outer solar system. Beyond Jupiter are the giants Saturn, Uranus, and Neptune. Each of the four giants has a massive gaseous atmosphere, an extensive system of interesting satellites, and probably a unique ring system. (The status of a ring, or segments of rings, around Neptune is unclear, as we will see.) The existence of four such systems offers us good opportunities for comparative planetology. By comparing the giants, the systematics of the different moon systems, and the rings, we can gain insights into the evolution of worlds, the common themes of planetary evolution, and the circumstances that produce unique properties, such as the volcanoes of Io.

Overlapping the orbit of Neptune is the orbit of the tiny world Pluto. Pluto is conventionally listed as the ninth planet, but it has distinctive properties, and is smaller than our own Moon. As we will see in this chapter and the next, labeling Pluto as an independent ninth planet may not be the best way to look at it.

THE PLANET SATURN

Saturn is most famous for its rings, the only ring system that can be easily seen in Earth-based telescopes (Figure 12-1). The globe itself is less interesting than Jupiter and lacks colorful features such as the Great Red Spot. Saturn's clouds are organized into bright zones and dark belts, sketched in Figure 12-2. The colors are yellowish and tan, and, as indicated in Figure 12-3, the contrast is less than on Jupiter. Because Saturn is nearly twice as far from the Sun as Jupiter, it is colder. This difference explains the lower degree of color, because at the cloudtop temperatures, around 100 K ($-173°C$ = $-279°F$), there is less formation of colored organic compounds. In spite of the blander colors, the overall composition of Saturn's atmosphere is similar to Jupiter's—mostly hydrogen (H_2) and helium (He), with minor amounts of methane (CH_4) and other gases.

Just as with Earth and other planets, Saturn radiates energy from its interior, in the form of infrared radiation. As in the case of Jupiter, this energy output from the interior has been found to exceed the input from the Sun. Both amounts of heat are feeble, but the finding means that the lower atmosphere of each planet receives around two to three times more warmth from below than from above—a fact that affects atmospheric circulation and structure.

As on Jupiter, atmospheric circulation varies with latitude. Rotation periods near the equator are about 10^h14^m, but cloud formations near the poles show longer periods, such as 10^h40^m.

That the mean density of Saturn is the lowest of any planet and less than that of water indicates that Saturn has only a small core of rocky material. Its layer of metallic hydrogen is smaller than Jupiter's. Most of its bulk is molecular hydrogen in either liquid or highly compressed gaseous form. A core of rocky compounds, something like a terrestrial planet but with 15 times the mass of Earth, exists at the center of Saturn.

Saturn's atmosphere resembles Jupiter's in composition and structure (Table 12-1). Figure 12-4 dramatizes the strong absorption of certain light wavelengths by methane. (Compare with the absorption by water vapor in the Earth's atmosphere, pp. 132–133.) High layers of such absorbing haze help reduce the contrast among Saturn's cloud markings. Nonetheless, small-scale cloud features may be seen in contrast-enhanced Voyager photos, such as Figure 12-5. Such photos revealed turbulent structures and stronger eastward jet streams than on Jupiter, with wind speeds up to 450 m/s (1000 mph) eastward along the equatorial zone. Only the larger disturbances can be seen from Earth; they can be tracked for many weeks.

SATURN'S RINGS

Galileo's first sight of Saturn, in 1610, was a strange one: a blurry disk with blurry objects on either side. He drew it as a triple planet. Not until 1655 did Christian Huygens discover that a ring system encircled the planet.[1]

Figure 12-1 An example of the best photography of Saturn from Earth-based telescopes. The yellowish color of the clouds contrasts with the whitish tones of the ice composing the ring particles. (Catalina Observatory 61-in. telescope; courtesy S. M. Larson, University of Arizona.)

A small modern telescope easily shows the astonishingly beautiful rings, largest in the solar system. The rings' proportions are surprising. They stretch about 274 000 km (171 000 mi) tip to tip, but they are probably less than 100 m thick! The system is so thin that terrestrial observers lose sight of it altogether when it appears edge-on for a few days once every 15 y as the Earth passes through the ring system's plane (see Figure 12-6).

Saturn's rings are composed of billions of separate particles, too small to be seen individually from Earth or in Voyager photos. This was proved by a stepwise combination of theory and observation. In 1895, Scottish physicist James Clerk Maxwell showed that the rings could not be a solid disk because they are inside Roche's limit, where a disk would be pulled apart into myriads of particles. In 1895, American astronomer James Keeler confirmed spectroscopically that different

---

[1]Huygens announced this discovery in the form of a famous anagram (succession of apparently meaningless letters). Huygens later explained that the anagram was to represent the sentence

"Annulo cingitur, tenui, plano, nusquam cohaerente, ad eclipticam inclinato" ("It is surrounded by a thin, flat ring, nowhere touching, inclined to the ecliptic"). In the days before copyrights, such anagrams were a common form in which scientists published initial results, thus establishing priority while giving themselves time for more observations. Not until about 1661 were most observers convinced of Huygens's discovery (Alexander, 1962).

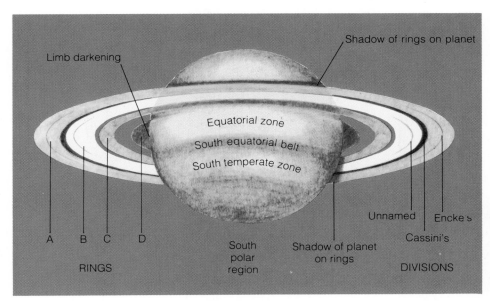

Limb darkening

Shadow of rings on planet

Equatorial zone

South equatorial belt

South temperate zone

A   B   C   D

RINGS

South polar region

Shadow of planet on rings

Unnamed

Cassini's

Encke's

DIVISIONS

Figure 12-2 Features of Saturn. Telescopes with apertures as small as 5 cm (2 in.) will show the rings; telescopes larger than about 25 cm (10 in.) will sometimes show all these features.

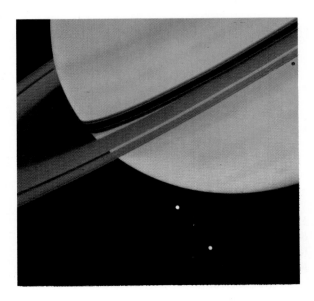

Figure 12-3 Saturn and two of its moons, Tethys (above) and Dione, photographed from close range by Voyager 1. The moons are about 1000 km across, less than one-third the size of our Moon. One of the moons casts a shadow on Saturn's surface (upper right). The photo dramatically shows the gap in the rings called Cassini's division (see Figure 12-2). The rings are not bright because they are lit at only a grazing angle by the Sun; their shadows are visible on the surface of the globe. Only faint cloud bands are visible in Saturn's orangish atmosphere. (NASA.)

TABLE 12·1

Atmospheric Compositions of Jupiter and Saturn

Gas	Estimated Percentage by Mass	
	Jupiter	Saturn
H_2 (Hydrogen)	79	88
He (helium)	19	11
Ne (neon)	1?	?
H_2O (water)	Trace	?
NH_3 (ammonia)	0.5?	0.2
Ar (argon)	0.3?	?
CH_4 (methane)	0.2?	0.6
C_2H_6 (ethane)	Trace	0.02
PH_3 (phosphine)	Trace	Trace
C_2H_2 (acetylene)	Trace	Trace

Note: Data derived mainly from Voyagers 1 and 2 infrared observations (*Science 204*:972; *206*:952; *212*:192). "Trace" indicates a small fraction detected; "?" indicates a likely constituent that has not been measured. The hydrogen and helium abundances are very close to those in the Sun, 78% and 20%, respectively.

Figure 12-4 Photographs of Saturn with films and filters sensitive to different colors of light. Top = blue (450 nm), the most common view; bottom = wavelength absorbed by methane gas (898 nm). Abundant methane in the atmosphere absorbs most of the light, while the rings remain bright, being covered with ice that reflects sunlight at this wavelength. (Lunar and Planetary Laboratory, University of Arizona.)

Figure 12-5 Contrast-enhanced close-up of Saturn's cloud belts reveals structure similar to that in Jupiter's clouds. Two dark ovals are visible at upper right; each is about half the size of Jupiter's Great Red Spot. They may have origins similar to the Red Spot. The north equatorial belt is at the bottom (dark); north temperate zone is in the middle (bright). (NASA Voyager 1 photo.)

parts of the rings move at different speeds, each in its own orbit around Saturn. Modern proof that the rings are not solid comes from spacecraft photos showing that we can see through them, as in Figure 12-7. In 1970, American astronomers proved spectroscopically that the ring particles are composed of, or covered by, frozen water (Lebofsky, Johnson, and McCord, 1970). Various measurements from Earth and from the Voyagers show that common particles in the rings range from ping pong ball to house sizes. Some particles of larger or smaller size may also exist. The remarkable view from within the rings is suggested by Figure 12-8.

The rings are divided by several gaps. The biggest is called **Cassini's division**, after G. D. Cassini who discovered it in 1675 (see Figures 12-1, 12-2, 12-3, and

12-7). A dusky ring region outside Cassini's division is called Ring A; the brighter region just inside it is called Ring B. Fainter inner ring regions are C and D (compare Figures 12-2 and 12-7). Ring A is cut by Encke's division; other finer divisions have been mapped by telescopic observers.

Scientists were astonished when Voyager 1 and 2 photos showed not just a few major divisions but thousands of fine-scale divisions and ringlets, such as seen in Figure 12-9. Wavy ring-edges, gentle twists in ringlets, and transient radial shadings were also found, as seen in Figure 12-10. What maintains such intricate structure? One theory was that unseen moonlets clear swaths among the ring particles, create wavy edges, and maintain ringlets by their gravitational effects. Sup-

Figure 12-6 Varying aspects of Saturn during its 29-y orbit around the Sun, as photographed with Earth-based telescopes. Because the rings maintain a fixed relation to the ecliptic, the Earth-based observer sometimes sees from "above" and sometimes "below" the ring plane. This series shows half the 29-y cycle, including two views that show the disappearance of the rings as Earth passes through the ring plane. (Lowell Observatory.)

porting this theory was the Voyager 1 discovery of two 200-km moonlets straddling the narrow F ring just outside Ring A, confining Ring F to a narrow zone. Moons of this type, which confine rings into narrow zones, are called **shepherd satellites**. Voyager 2 searches failed to turn up other shepherd moons among the rings, down to a diameter of a few kilometers. Nonetheless, ring structure may result from a combination of poorly understood gravitational effects caused by known moons and still-unseen moonlets.

Voyager radios picked up distinct static signals caused by electrical discharges in the rings. The discovery suggests that electrostatic forces may be more important in the ring particle interactions than hitherto suspected.

How did such a remarkable ring system form? A clue is that any crowded swarm of particles surrounding Saturn (or any planet) would have to form a flat ring system. Mutual collisions plus gravitational forces from Saturn's oblate mass would force the particles into Saturn's equatorial plane. Since they are inside Roche's limit, tidal forces prevent them from coalescing into a large satellite.

Figure 12-7 This extraordinary view of Saturn in crescent phase was obtained by Voyager 2 as it left Saturn's system. Such backlighting of Saturn can never be seen from Earth. The rings' shadow makes a black band on the globe, and the shadow of the globe cuts a black swath through the rings. Note that the globe of Saturn can be seen *through* most parts of the rings where they pass in front of the globe, proving that the rings are not solid. (NASA.)

Figure 12-8 In this imaginary view we are floating among the icy particles within Saturn's rings. Saturn's globe is at left, and the distant Sun (right) backlights the scene. The rings, perhaps less than 100 m thick, stretch off into the distance, lower center. (Painting by author.)

The problem remains of where the particles came from in the first place. They may have been:

1. Particles condensed from gas as Saturn formed, like hailstones condensing in a cooling cloud

2. Part of a satellite blown apart by collision with a comet or asteroid

3. Part of a comet or asteroid broken apart by tidal forces after approaching too close to Saturn

Some theorists have calculated that the rate of erosion of the ice particles in the present rings—due to collisions of the particles with one another and with small meteorites—is too fast for the rings to have lasted

Figure 12-9 Close-up photos of Saturn's rings show unexpectedly intricate structure—including thousands of individual ringlets. Overexposed Saturn can be seen in the background, along with the shadow of the rings (top). The dimmest ring (top) is Ring C; Ring A is in the foreground. (NASA Voyager photo.)

Figure 12-10 Orbital motion of ring material around Saturn is shown in this sequence of photos, taken half an hour apart. Note the movement of the dusky "spokes" from one position to another. (NASA Voyager 1 sequence.)

since Saturn formed 4.5 billion years ago. Other theorists claim that the rate of impact of larger meteorites falling toward Saturn is great enough to break up some of the innermost small moons sporadically during the solar system's history. The meteorites would concentrate in number and increase in speed as they were attracted toward Saturn due to Saturn's gravity. Thus an inner moon near Saturn would have a greater chance of breakup than a similar-sized outer moon. Some fragments from such a breakup would reassemble into a new moon or moons, but others would be added to the ring system. Thus while many astronomers argue that the rings are permanent, immutable, and as old as Saturn, a growing number believe that the rings of Saturn and other giant planets are constantly evolving and changing due to fragmentation of moonlets and input of new ring particles. Hence we may be living in an era in which the rings are unusually dramatic! Future spacecraft voyages toward the rings may clarify this ring evolution issue.

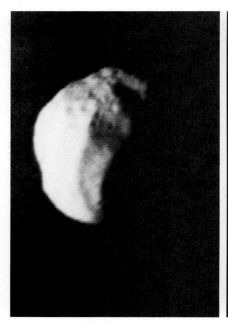 

Figure 12-11 Saturn's satellite S11 is a 70 × 135 km object, probably eroded into its irregular shape by intense cratering. These two views, taken 13 min apart, . show the shadow of Saturn's nearby rings moving across the satellite. (NASA Voyager 1 photo.)

SATURN'S SATELLITES

Saturn's satellite system, like Jupiter's, is extensive and includes exotic worlds. There are at least 17 moons (see Table 8-1, page 172), including a host of small moons near the rings, a small outer moon, and a grouping of larger moons at intermediate distance, including the giant moon Titan.

If we start on the outskirts of the rings, we find five small moons with diameters between 20 and 220 km, mostly discovered in 1979 and 1980 by Voyager 1 and by intense Earth-based observation during a brief period when the rings were seen edge-on and thus did not obscure the satellites by their glare. Voyager photos have shown some of these small moons to be lumpy and cratered (Figure 12-11). Bright white tones of at least some of them suggest icy surfaces, but their interior composition is uncertain. They seem to be battered iceballs.

Mimas: A Cratered Iceball

The next moon is **Mimas**, 394 km across, round, icy, and heavily cratered with one very large, fresh-looking crater (Figure 12-12). Mima's bright surface (60% reflectivity) and low bulk density (1200 kg/m^3 compared to 1000 for pure ice and around 3000 for rock) indicate that Mimas is composed mostly of ice.

Scientists who have studied rate of impacts of large meteorites onto the inner satellites believe that Mimas may have been fragmented one or more times. If this is true, the Mimas we see today has been reassembled from fragments. (The fragments did not have to refit into the perfect sphere we see today; gravitational forces would have caused the ice to deform into a spherical shape in much the way that a glacier's ice flows.) The other small moonlets on the outskirts of Saturn's rings might also be fragments from such a breakup of such a moon. This theory thus neatly explains the appearance of numerous small moonlets on the outskirts of the rings.

Enceladus: Signs of Geological Activity

The next satellite, **Enceladus**, is especially fascinating as a "missing link" between Jupiter's moons Europa and Ganymede: Ençeladus (Figure 12-13) has the linear grooves and bright, sparsely cratered, icy surface of Europa, but it also has some moderately cratered, fractured regions that resemble Ganymede. It has the brightest surface known in the solar system. Enceladus is 502 km across, only one-seventh as big as our Moon.

a

b

Figure 12-12 Saturn's small satellite Mimas, 394 km (245 mi) across, is an example of a worldlet whose topography is totally dominated by craters. **a** Color view by Voyager 1. **b** Another view by Voyager 1 reveals a crater unusually large relative to the satellite itself. It is about 130 km across and has a large central peak that rises some 6 km (20 000 ft) from its floor. The impact that made it was almost large enough to fragment Mimas. Voyager scientists nicknamed this photo the "Darth Vader Deathstar picture" for its fancied resemblance to the *Star Wars* villain's space station with its large, circular landing port. (NASA.)

It surprised scientists, for its surface appears much younger and more geologically active than expected for such a tiny, icy world. Its surface seems to have been created by eruptions of water and ice crystals from an interior heated by tidal interactions with other Saturnian moons. Calculations like those used to predict Io's volcanoes show that Enceladus would be the Saturnian moon most strongly heated by this mechanism, although the calculations suggest tidal heating is barely adequate to produce such activity. Its small size keeps it from building up enough internal heat to power major geological activity as on Io, which is seven times as big. The mean density of Enceladus, 1100 kg/m³, indicates that Enceladus is mostly icy throughout, so that radioactivity from rock minerals is an unlikely heat source. Nonetheless, researchers have discovered a so-called E ring of tiny particles extending far beyond the main rings of Saturn and having a concentration in Enceladus' orbit (see Figure 12-14). Together with the cracks, brightness of the ice, and presence of sparsely cratered young regions, the particles suggest that occasional eruptions of water may recoat the surface with fresh ice and blow some ice grains clear off the surface. Enceladus bears further study!

Tethys, Dione, and Rhea

The next three moons—**Tethys, Dione,** and **Rhea**—are around 1000 to 1500 km across, about one-third to one-half size of our Moon. They have bright, icy surfaces. Tethys has a bulk density of only 1000 kg/m³ and hence must be nearly pure ice. Dione and Rhea have densities of around 1300–1400 kg/m³ and probably have some rocky materials mixed with their ice. These worlds are moderately to heavily cratered, as seen in Figures 12-14 to 12-17. Generally, they confirm our rule of thumb that worlds smaller than our Moon are relatively inactive, preserving ancient cratered surfaces, at least in the absence of special heating mechanisms. However, Tethys' surface is broken by a huge canyon system and

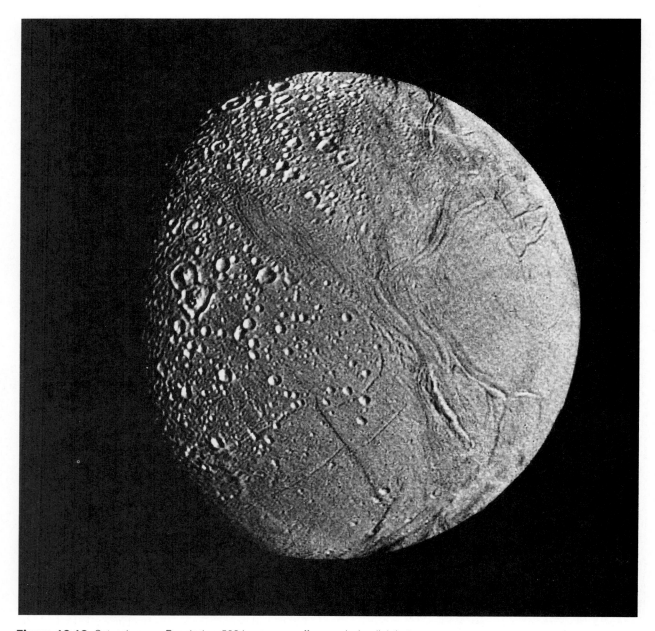

Figure 12-13 Saturn's moon Enceladus, 502 km across, offers a missing link between the "cracked billiard ball" surface of Jupiter's moon Europa and the cratered, fissured surface of Jupiter's moon Ganymede—though Enceladus is only one-sixth the size of Europa. Portions of the ancient cratered surface seem to have been cracked as water erupted and froze into smooth ice plains, obliterating older craters—for example, bottom central region. This suggests that Enceladus may have been heated enough to melt portions of its icy interior at some time in its history. (NASA Voyager 2 photo.)

Figure 12-14 A landscape during the beginning of an eclipse on Saturn's icy moon, Tethys. Sunlight passing through Saturn's atmosphere colors the foreground. The fuzzy extension of the rings (upper left) is the E ring, backlit by the Sun. It is composed of tiny particles believed to have been ejected from the neighboring moon, Enceladus, visible at upper left. (Painting by author.)

Figure 12-15 Saturn's satellite Dione, 1100 km across, has a surface marked mostly by craters but showing a few bright swaths (lower right edge) and fracture-like canyons (upper left edge) along the sunset line. (NASA Voyager 1 photo.)

Dione's by swaths of light-toned material (visible in Figure 12-16) reminiscent again of Ganymede.

In 1980, Earth-based observers discovered interesting additional small moons roughly 20 to 50 km across in the orbits of Tethys and Dione. One orbits 60° behind Tethys; one 60° ahead of Tethys; and the third 60° ahead of Dione. Gravitational forces make this 60° point a stable location for small objects. Certain asteroids are known in similar 60° points ahead of and behind Jupiter. These discoveries expand the list of similarities between satellite families and the solar system as a whole—supporting the concept of satellite families as miniature solar systems.

Titan: The Moon with a Thick Atmosphere

We called Io the most bizarre moon, but Saturn's largest moon, **Titan**, is a close runner-up. With a diameter of 5150 km (3200 mi) it is probably the second-largest moon in the solar system and the only moon with a thick atmosphere. Methane (Ch_4) was discovered from spectra of Titan in 1944. Observations in 1973 showed that Titan's sky is not clear but filled with reddish haze, seen in Figure 12-18 (bottom right). Later observations showed that this haze is photochemical smog produced by reactions of the methane and other compounds when they are exposed to sunlight—like the smog produced by the action of sunlight on hydrocarbons over Los Angeles. Titan is the smoggiest world in the solar system.

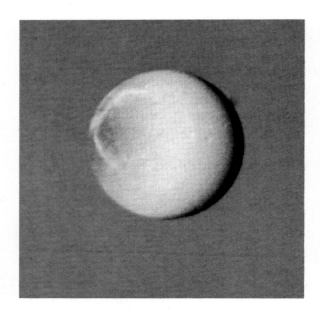

Figure 12-16 An unusual view of Dione silhouetted not against space, as in most views, but against the orangish clouds of part of Saturn in the background. The color contrast emphasizes Dione's whitish, icy surface. The trailing hemisphere (left) has more dusky mottling and bright swaths of uncertain origin. (NASA Voyager photo.)

The Voyagers showed that the methane and smog are not more than 10% of the atmosphere. The main constituent is nitrogen, meaning that the atmosphere is denser than realized before the Voyager flights. Voyager 1 discovered that the surface air pressure is 1.6 times that on Earth! Since the Earth's air is also mostly nitrogen, Titan's atmosphere offers fascinating comparisons to present or primeval conditions on Earth. The main difference is that Titan is very cold, around 93 K ($-292°F$). Minor constituents detected in Titan's air include ethane, acetylene, ethylene, and hydrogen cyanide. The abundance of these organic molecules suggests that Titan offers a good natural laboratory for research on the origins of life.

Based on the measurements of temperature and pressure at Titan's surface, researchers visualize dramatic weather conditions, illustrated in Figure 12-19. Methane may exist not only as a gas but may also be able to rain out of the clouds and exist as snow or ice, playing the same triple role of gas, liquid, and icy solid as water does on Earth. In addition to methane (CH_4), meteorologists conclude that ethane (C_2H_6) would form in the atmosphere, slowly drizzle out, and accumulate

Figure 12-17 A close-up look at the cratered surface of Saturn's largest icy moon, 1528-km Rhea, nearly half as big as our own Moon. In this view under high sun, the craters' rims look especially bright, apparently exposing regions of clean ice. (NASA Voyager 1 photo.)

on the ground. Thus the surface of Titan may be largely covered by a cold ocean of liquid methane and liquid ethane estimated to be a kilometer deep. Still more complicated, gasolinelike compounds may form in the smog and rain out of the hazy clouds. As these facts unfolded during the Voyager mission, one Voyager scientist characterized Titan as "a bizarre murky swamp," shown in Figure 12-19b. More interesting to scientists are the possible biochemical reactions among these organic molecules. Life is unlikely in the cold temperatures of Titan, but biochemists would be very excited to learn what complex organic compounds have formed on Titan. Indeed, Titan may be a natural laboratory for studying primordial biochemical evolution on Earth! Titan remains a fascinating target for future exploration, and NASA has considered long-range plans to parachute probes beneath the mysterious clouds to see what is really on the surface.

Hyperion: A Tumbling Biscuit

As if to reemphasize the rule that each world is unique, the next satellite is completely different. **Hyperion** is a biscuit-shaped chunk roughly 350 km across and 200 km thick. As seen in Figure 12-20, Voyager 2 photos show an irregular, cratered shape. Hyperion's long axis is not quite lined up with Saturn, as would be expected if tidal forces had acted over a long period. Due to complex forces acting on its irregular shape, its rotation rate varies irregularly. This type of irregular rotation, where the rotation rate and the pole orientation change continuously over periods of some months, is called **chaotic rotation**. Hyperion is the best-known example of chaotic rotation. Some scientists believe Hyperion may be an oddly shaped, wobbling piece of a former moon broken by the impact of a giant meteorite.

Iapetus: One Black Side and One White Side

The next outer moon, 1436-km (900-mi) **Iapetus** (I-ăp′-e-tus), is unique in being two-faced! The hemisphere on the leading side in Iapetus' orbital motion around Saturn is covered with dark, frosty soil the color of reddish tar, but the trailing hemisphere is covered with bright, white ice. As seen in Figure 12-21, the

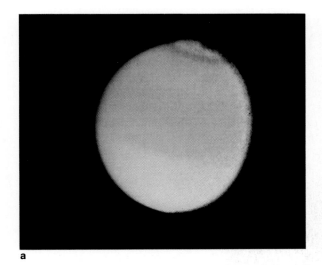

a

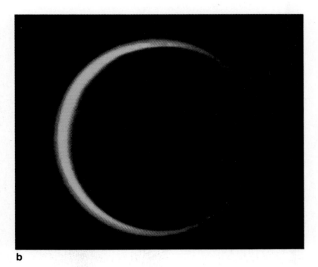

b

Figure 12-18 Saturn's giant moon, Titan, about as big as Mercury. Titan has a smoggy atmosphere of nitrogen, colored orangish by organic compounds. **a** In this gibbous lighting view, a dark cloud band can be seen around the north polar region, and the southern hemisphere haze is brighter than the northern hemisphere. **b** Under crescent lighting, the illuminated atmosphere can be traced all the way around the moon. The color is bluer where sunlight has been scattered farther through the atmosphere—for the same reason that scattered sunlight colors our own sky blue. (NASA Voyager photos.)

a

b

Figure 12-19 Scenes on Saturn's giant moon, Titan. **a** From above the smoggy layer of orangish haze, visitors would see Saturn hanging in a blue sky. (Painting by author.) **b** Below the clouds it is quite dark. Measurements of surface temperature and pressure suggest snowstorms and rainstorms of solid and liquid methane compounds—with possible rivers, "waterfalls," and oceans of liquid methane and liquid ethane. (Painting by Ron Miller.)

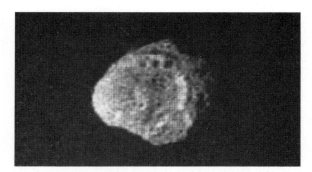

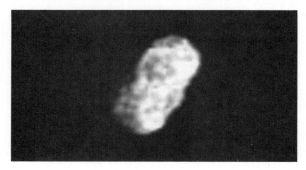

Figure 12-20 Top view and side view of Saturn's biscuit-shaped satellite, Hyperion. The irregular shape may have been created by a major meteorite impact. The side view (bottom) was made from a greater distance and is less sharp. (NASA Voyager 2 photos.)

difference between the two sides is striking. What could create this black and white moon? The answer almost certainly involves Iapetus' neighbor moon—Saturn's outermost satellite Phoebe.

Phoebe: Another Black, Captured, Outer Moon

Phoebe is a 220-km (140-mi) black-surfaced moon moving around the outskirts of Saturn's gravitational influence in a retrograde orbit. Since its color is unlike the other moons' and its orbit is so unusual, many scientists believe Phoebe did not form as one of Saturn's original moons but instead is an asteroid captured into orbit by Saturn's gravity.

How does Phoebe relate to two-faced, black and white Iapetus? Black dust knocked off Phoebe by meteorites would spiral inward toward Saturn and hit the leading side of Iapetus at high speed. Darkening of Iapetus' leading-side ice layer by black dust from Phoebe may partly explain Iapetus' dark leading side. But Iapetus' dark side is more reddish-black than Phoebe; some scientists think the story must be more complex. They theorize that reddish-black organic minerals form from methane in regions of fresh Iapetus ice disturbed by dust from Phoebe. Moons can be complicated places!

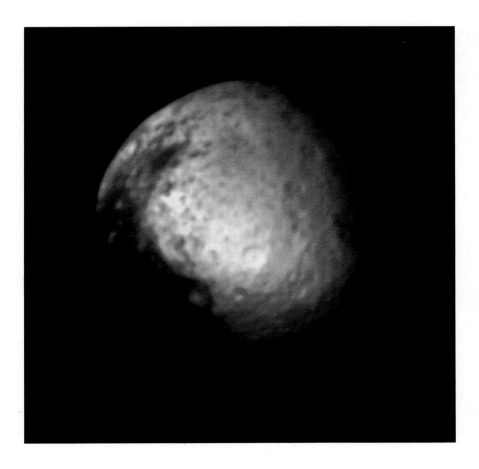

Figure 12-21 Saturn's 1440-km moon Iapetus has one face (right) surfaced with clean bright ice and the opposing hemisphere covered with dark material. The dark matter is probably a mixture of ice and dark carbon-rich minerals. The difference between the two hemispheres is puzzling but may involve the neighbor moon, Phoebe (see text). (NASA Voyager 2 photo.)

THE PLANET URANUS

In 1781 William Herschel became the first known person to discover a new planet when he detected **Uranus** during an ambitious star-mapping project. Uranus is so far from Earth that an ordinary telescope reveals almost no markings on it. But in 1986, Voyager 2 flew past Uranus, giving us our first close-ups of its bluish haze (Figure 12-22), its dark rings, and its unexpectedly varied moons.

The planet's name was suggested by J. Bode (of Bode's law) because in mythology Uranus was the father of Saturn, who in turn was the father of Jupiter. Because of Uranus' long 84-y period of revolution around the Sun, it has made only two orbital circuits since its discovery.

Herschel found several satellites of Uranus as well. The orbits of these satellites lie in the plane of the planet's equator but are steeply inclined to the planet's *orbital plane* by 98°. As the French dynamicist Laplace noted in 1829, steep inclination to the orbital plane can arise only if the planet has a substantial equatorial bulge, whose gravitational attraction holds the satellites over the planet's equator. Thus Uranus itself must have an equatorial bulge, and its equator must be inclined 98° to its orbital plane, an unusual planetary configuration shown in Figure 12-23. This **obliquity** (deviation of the equator from the plane of orbit) being greater than 90° means that the rotation of Uranus is *retrograde*, as illustrated in Figure 12-23. The rotation period of the planet, under the clouds, was measured by Voyager to be 17.2 h.

Thus Uranus has a unique seasonal sequence. When the north pole points almost directly toward the Sun, the southern hemisphere is plunged into a long, dark winter, lasting for about a quarter of the planet's 84-y revolution. After this 21-y south polar winter and north polar summer, the Sun shines on the equatorial regions. Each point on the planet now goes from day to night during the planet's 17-h rotation. After 21 more years the south pole points approximately toward the Sun and

a

Figure 12-22 The hazy sky-blue planet, Uranus. **a** As Voyager 2 approached Uranus from the inner solar system and the Sun's direction, it took this photo with Uranus at a gibbous phase under nearly full sun. The view looks down on the pole, which is pointed nearly toward the Sun due to Uranus' unusual axial tilt. Cloud features are nearly invisible, though a faint dark band can be seen along the upper right edge. **b** As Voyager left, it captured this view of crescent Uranus looking back toward the Sun. (NASA Voyager 2 photos.)

b

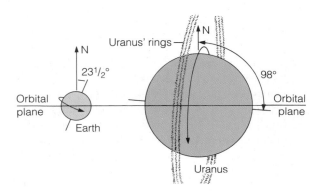

Figure 12-23 Comparison of the sizes and rotations of Earth and Uranus. The Earth has an obliquity (or axial tilt) of $23\frac{1}{2}°$ and a prograde (west-to-east) rotation. Uranus has much steeper obliquity and retrograde rotation.

the southern hemisphere experiences a 21-y summer.

The Voyager flyby revealed that Uranus' atmosphere has a deep layer of almost featureless hazy gas overlying the deeper cloud. The composition of the atmosphere is similar to those of Jupiter and Saturn—about three-fourths hydrogen (H_2) and one-fourth helium (He) by mass, with small amounts of methane (CH_4) and other gases. The haze layer is very cold: Voyager recorded a temperature minimum of 51 K ($-368°F$). The temperature increases, however, down into the clouds. Perhaps because of the cold, Uranus' visible haze layer lacks the orange and tan-colored compounds that give rich color to the clouds of Jupiter and Saturn. In the absence of such colored clouds, the traces of methane in the high haze absorb red light and scatter blue light, lending an ethereal sky-blue color to the disk as seen in Figure 12-22. Only extreme processing of the relatively featureless images reveals faint cloud patterns showing through the high haze.

Calculations suggest that the inner quarter of the planet's diameter is occupied by a central rocky core, perhaps molten, similar to Earth in size. This core is surrounded by a deep ocean of hot, high-pressure, liquid water. Above this is the deep atmosphere comprising the outer half of the planet.

Uranus' Rings

In 1977 Uranus passed in front of a relatively bright star. A number of astronomers watched, expecting to see the star dim as it passed behind Uranus' upper atmosphere and thus to learn about the haze layer's structure. They were astonished to see, in addition, the star dim several times at some distance from the Uranian disk, but symmetrically on each side of the disk. These observations marked the discovery of narrow rings of dark material around the planet. **Uranus' rings** blocked the starlight as Uranus moved. Voyager 2 obtained the first clear pictures of these rings (Figure 12-24), which are much narrower than the rings of Jupiter or Saturn. Scientists do not fully understand what confined Uranus' rings into such narrow bands, although shepherd satellites, like those on each side of Saturn's narrow F ring, are believed to be involved. Voyager cameras discovered two shepherd moons straddling the thickest Uranian ring (Figure 12-24). Gravitational forces of other undiscovered moonlets may help confine the rings. Further research is under way to clarify these curious "shepherding" effects.

URANUS' SATELLITES

Before Voyager, five moons of Uranus were known and named after characters in Shakespeare's *Midsummer Night's Dream* (see Table 8-1, page 172). But ten smaller moons were discovered in 1985 and 1986 as Voyager 2 sped past the planet!

Scientists had assumed that at such great solar distance, with such low temperatures and small sizes, the Uranian moons would be little more interesting than inert iceballs. But once again, as in the cases of Jupiter and Saturn, Voyager revealed that the whole system of **Uranus' satellites** is much more interesting than had been predicted from Earth.

The ten new moonlets are small, dark bodies on the edges of the rings, similar to the innermost moonlets in Saturn's system. Of the five larger moons, the outer

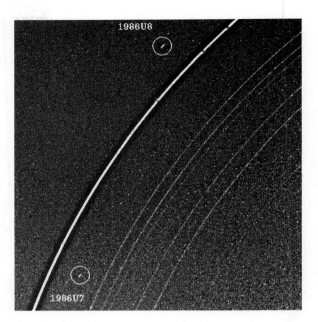

Figure 12-24 Portion of the rings of Uranus and two "shepherd moons" (circled), whose gravitational pulls help keep ring particles confined to narrow widths. Width of widest, outer ring is only 100 km (60 mi). All nine rings appear; inner three may be glimpsed by holding up book and sighting along the line of the rings. (NASA Voyager 2 photo.)

four are the largest, all about a third the size of our Moon. True to our rule of thumb that small worlds show primarily external cratering without internal disturbance, these moons are fairly heavily cratered and somewhat resemble Saturn's moons of similar size, as seen in Figures 12-25 and 12-26. But there are complications. Several of the outer four are strongly fractured by canyons, as seen in the figures. These fissures may indicate internal heating from unknown sources or dynamic forces acting on each satellite from gravitational interaction with other moons.

Miranda

By far the greatest surprise came from the close-up photos of the inner of the five large moons—modest-sized **Miranda**. At a mere 484-km diameter, it is the smallest of the five main satellites. In accord with our rule of thumb that smaller size signifies less internal

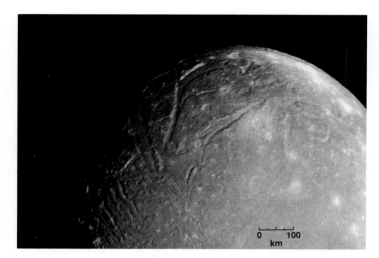

Figure 12-25 Uranus' 1160-km-diameter moon, Ariel. This view shows the prominent system of fractures and canyons that have broken the surface of this moon. Brightest spots are impact craters that have apparently blown away the darker soil and exposed cleaner ice. (NASA Voyager 2 photo.)

a

b

Figure 12-26 Uranus' strange, 484-km-diameter moon, Miranda. Old crater surfaces (bottom) are broken by swaths of striped, younger terrain, which have accumulated fewer impacts. Deep fractured canyons can be seen at the top. **a** Mosaic made by computer-processed combination of high-resolution photos. **b** Lower-resolution color view of part of the same hemisphere. (NASA Voyager 2 photos; mosaic processed by U.S. Geological Survey; courtesy Jet Propulsion Laboratory.)

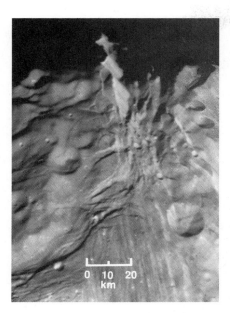

Figure 12-27 Detailed view of a system of fractures, faults, and valleys on Uranus' moon Miranda. Compare with the full-disk view of Miranda (Figure 12-26), where this valley appears near the top. At the top, extending onto the night side, is an enormous cliff with an estimated height of 5 km (16 000 ft) and a slope of about 30°. Much of the region to its right, and in the foreground, has slumped along parallel fractures. (NASA Voyager 2 photo; upper right corner, outside original frame, printed as uniform gray here.)

activity, it might have been predicted to show only craters. But unexpectedly it is the most fractured and resurfaced of all, as seen in Figure 12-26. As shown in Figure 12-27, one cliff is a sheer, sculpted fault face sloping to a height of 5 km (16 000 ft)! This cliff is only part of the system of grooves and fractures that swathe Miranda. How could such a small, asteroid-sized moon have generated enough internal energy to create these impressive features? Voyager scientists suggested, as with Saturn's inner large moon Mimas, that it might have been smashed by impact and reassembled, with the fractures forming as the new moon adjusted its shape while the fragments settled. Another suggestion is that Miranda was forced into a period of chaotic rotation during its past history, and stresses might have caused fractures (Marcialis and Greenberg, 1987). Whatever its history, it rotates smoothly and synchronously now, and its fractured face is the cause of much bewilderment.

THE PLANET NEPTUNE

The discovery of Uranus led to the discovery of Neptune. After 1800, theorists tried unsuccessfully to fit observations of Uranus' position into Kepler's laws of planetary motion. Uranus seemed to have its own somewhat irregular motions. A few scientists thought this might signal a breakdown of Newton's law of gravity at great distances from the Sun. Others correctly suggested that Uranus was being attracted by a still more distant planet. In the 1840s an English astronomer and a French mathematician set out independently to predict where the new planet should be.

Both men tried to shorten their laborious calculations by following the Bode's rule, assuming that the new planet would be 38 AU from the sun.[2] But Neptune lies at 30 AU—a serious exception to Bode's rule. The English astronomer J.C. Adams finished his prediction in 1845 after two years of calculations, but he had trouble getting his senior professors at Cambridge interested in searching for the new planet with a telescope. In 1846 English astronomers began a desultory search. Several times they actually charted the new planet but failed to realize what they had seen.

Meanwhile, the French mathematician Leverrier, who had finished his work soon after Adams, interested two young German astronomers in the search. Armed with Leverrier's predictions, they located the new planet within half an hour of starting their search on September 12, 1846. It was given the name **Neptune**. Although Adams and Leverrier are both now credited with the discovery, the incident became an international scandal at the time because of the Britons' failure to grasp their opportunity.

Neptune's Atmosphere and Interior

Voyager 2 gave us our first really close look at this distant planet. Like Uranus, Neptune turned out to have an atmosphere that is dominantly composed of hydrogen gas (H_2), with helium composing most of the rest. Traces of methane and other gases were found. Ultraviolet sunlight converts the methane to hydrocarbon smog compounds, such as ethane and acetylene. As on Uranus, the blue color is caused by methane's absorption of red

---

[2]Today the necessary calculations could be done quickly on computers, but in the 1800s they took months or years.

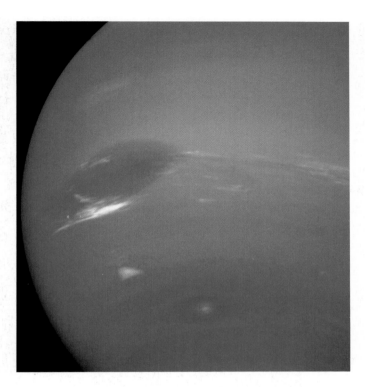

Figure 12-28 Color view of a portion of Neptune shows the bluish color of its atmosphere. Dark oval clouds, such as the prominent "Great Dark Oval," mark circulating storm systems, capped by white condensed clouds higher in the atmosphere. (NASA Voyager 2 photo.)

Figure 12-29 The rings of Neptune. Two rings exhibit denser arclike segments (bottom), but are continuous around Neptune. Fainter broad rings were also found inside each of these rings. (NASA Voyager 2 photo.)

light, so that as sunlight goes into the atmosphere and reflects back to us it loses its red component.

Because it is so far from the sun, Neptune is extremely cold, with an effective mean radiating temperature of about 59 K (− 353°F).

Although it is roughly the same size as Uranus, Neptune shows two major differences. First, it has much more dramatic cloud structures, including an Earth-sized large dark cloud called the **Great Dark Spot** (Figure 12-28). It is a large storm system circulating in an anti-cyclonic (counterclockwise) direction. Generally, the atmosphere seems more turbulent than Uranus'. The second difference may explain the first. While Uranus has very little internal heat of its own and radiates about as much heat as it receives from the Sun, Neptune has a strong, mysterious inner heat source and radiates 2.7 times as much heat as it gets from the Sun. This may be the result of radioactivity or of large impacts that stirred up Neptune's interior. In any case, the internal heating of the atmosphere from below may explain why Neptune's atmosphere is turbulent, with the Great Dark Spot and other storm systems visible in Figure 12-28.

Neptune's Arc-Rings

Prior to Voyager 2's flight past Neptune, Earth-based observers attempted to repeat the successful discovery of Uranus' rings by watching Neptune pass in front of stars. They got puzzling results. Sometimes they saw the star dimming before or after the planet passed it (indicating a ring), but other times they saw nothing. They concluded that the rings of Neptune were not uniform, but contained thicker arc-like segments extending only part of the way around the planet, with the rest of the ring being thin or nonexistent.

As shown in Figure 12-29, Voyager 2 confirmed that the rings are not uniform. The photo shows brighter sections, including strong arc-like segments. The gravitational factors that concentrate Neptune's ring particles in certain parts of the rings are not understood, but as in the Uranus system the odd effects are probably associated with the presence in the rings of small satellites, which may produce concentrations of particles.

Neptune's Satellites

Neptune has a satellite system that is peculiar: The largest satellite is moving in a circular but highly inclined retrograde orbit, instead of the low-inclination prograde

Figure 12-30 Triton shows a strangely mottled surface of bright ice and frost deposits. Few impact craters are found; fractures cross the central region. The top, brighter area is around the south pole and is believed to mark a polar frost cap. (NASA Voyager 2 mosaic.)

orbit common for large moons. This satellite, **Triton**, has a diameter of 2705 km, making it the seventh largest moon in the solar system. It was discovered in 1846.

Triton is strange in another respect. Due to tidal forces it is slowly spiraling in toward Neptune. Analysts predict it will eventually crash into Neptune. Neptune's outermost moon, Nereid, discovered in 1949, is only one-eighth as large as Triton and moves in a highly elliptical prograde orbit, contrasting with Triton's circular, retrograde orbit. The two orbits seem so unusual that many theorists think they may have been disturbed at some time in the past by passage of a large body (a giant comet or Chiron-like body?) through the system. Nereid's properties, including a fairly dark surface, suggest it might be a captured comet.

Figure 12-31 Portions of the photographs on which Pluto (arrow) was discovered. The planet was distinguished from stars by its motion. (Lowell Observatory.)

Voyager 2 found six additional moons inside Triton's orbit, ranging from 400 km (slightly bigger than Nereid) down to 54 km diameter. Thus, Neptune has eight known satellites in all, one large and seven small.

Voyager 2's close-up look at Triton revealed an unexpectedly unique world. With a surface temperature of 38 K ($-$391°F), Triton is the coldest body yet observed at close range. It has an extremely thin atmosphere, with only 0.016 millibar pressure, compared with Mars' 6 millibars and Earth's 1000 millibars. The atmosphere is predominantly nitrogen gas (N_2) as on Earth and on Saturn's moon, Titan. Creamy-colored seasonal ice deposits cover the poles (top part of Figure 12-30). A thin bluish-white frost layer, possibly nitrogen frost, covers the equator (bottom of Fig. 12-30). The surface is geologically young, with few impact craters, but what can be the source of surface renewal on such a cold world? Voyager photos show geyser-like eruptions sending dark columns of smoke 8 km straight up into the atmosphere, where they are sheared flat by high-altitude winds. Other geyser vents have dark, streaky deposits, apparently dark debris blown downwind from the vents (dark streaks, top half of Figure 12-30). Other parts of the surface show peculiar hillocks and fractures.

The geysers and surface variety were surprises for such a cold world. Triton's "backward" orbit indicates an unusual history. Voyager scientists think Triton may have experienced violent collisions or tidal heating, which kept its interior warm enough to drive the eruptions and disrupt the surface (Goldreich and others, 1989; Voyager Team, 1989).

THE PLANET PLUTO AND ITS MOON

Scientists had expected that gravitational pulls by Neptune would explain irregularities in Uranus' motions. But Neptune's gravity did not account for all of them, and Neptune itself has some unexplained irregular motions. These irregularities suggested still another planet beyond Neptune. For this reason, Percival Lowell in 1905 began a search for a ninth planet. As shown in Figure 12-31, this search led to the discovery of **Pluto** in 1930, when Clyde Tombaugh found it on Lowell Observatory photos. After it was named for a god of the underworld, a new planetary symbol (℗) was created from the first two letters, which were also Lowell's initials.

Though Pluto was hailed as the ninth and final planet, its status as a planet is currently being questioned. There are at least five reasons:

1. It is much smaller than any other full-fledged planet. At diameter 2300 km, it is smaller than our Moon and only 47% as big as Mercury.

2. It fails to continue the trend of the giant planets in size; among them, it is an anomaly.

3. Its orbit is not close to twice the size of Neptune's, in keeping with other planet spacings; instead, it overlaps Neptune's orbit.

4. An interplanetary body was discovered in 1977 between Saturn and Uranus, raising the possibility that the outermost solar system has many interplanetary bodies of various sizes. Moreover, we know that many comet nuclei exist beyond Neptune's orbit (see next chapter). Pluto might just be the largest of such interplanetary bodies.

5. Pluto's moon is much larger with respect to Pluto than the moons of most planets. It is about 57% as big.

For these reasons astronomers are beginning to think of Pluto (and its moon) as more like a giant double asteroid rather than a normal planet. The distinction may be more than just semantic: Pluto may indeed be more closely related to asteroids and comets than to full-fledged planets. In a few years, children may be taught that the solar system has only eight planets, not nine! But for the present, most writers list Pluto as the ninth planet.

Pluto's light variations and eclipses of its moon establish that it rotates every 6.39 d on an axis tipped about 120° to its orbit. Spectroscopic observations indicate water ice on the surface and perhaps a thin atmosphere of methane gas, probably derived from methane ice deposits on the surface. The surface reflects some 60% of the light falling on it, characteristic of bright, icy material.

Pluto's satellite, named **Charon**, was discovered in 1978. The name comes from a mythological figure associated with Pluto. Charon (pronounced KEHR-on) circles Pluto in 6.39 d, keeping one face toward Pluto. Charon appears somewhat darker than Pluto, reflecting about 40% of the incident light—a figure characteristic of slightly dirty ice. The densities of Pluto and Charon are probably similar. The mean value of the two has been measured at 2030 ± 35 kg/m³ (Tholen and Buie, 1989), suggesting a composition of somewhat more than half rock and the rest ice.

Although Pluto and Charon are too close together in angular separation for either one to be studied separately in a telescope, the series of eclipses of Charon by Pluto in the 1980s permitted astronomers to study the spectrum of Pluto alone, when Charon was hidden behind it. Not only did this opportunity give the spectrum of Pluto; the difference between that spectrum and the normal spectrum of Pluto plus Charon (that is, the brightness of both minus the brightness of Pluto alone) gave the spectrum of Charon alone. These studies show interesting differences between the two bodies. Charon is not only darker than Pluto, but it has less methane ice and more water ice. Charon is grayish-white in color, but Pluto is somewhat pinker, perhaps due to its methane compounds. It is a mystery, therefore, whether these two bodies formed together with the same initial composition or whether they are two independent bodies that somehow came into orbit around each other.

Our general rules about planetary evolution suggest one possible scenario. Methane ice sublimes more readily (at lower temperature) than water ice. Now suppose that Pluto and Charon formed together of similar materials. If Charon by chance started out with a slightly darker color, it would absorb more sunlight and get slightly warmer. Its methane ice would begin to sublime, turning into gas. Because Charon is so small and has such slight gravity, the atoms of methane gas could escape into space. The subliming ice would leave behind dark soil particles, causing the surface to grow still darker, resulting in a feedback effect. Eventually, only water ice and dark soil would be left on Charon's surface, explaining its present appearance (Marcialis and others, 1987).

On cold, remote Pluto and Charon, the surface temperature is only about 50 K (−369°F) and the Sun would be too far away to be perceived as a disk. From this outpost of the solar system, the Sun would look like an intensely bright streetlight across the street, reminding us that it is one star out of many.

PLANET X?

Several astronomers have sought dynamical or photographic evidence of a planet beyond Pluto, sometimes called "Planet X." Clyde Tombaugh, the discoverer of Pluto, conducted a long search of the region beyond Pluto and ruled out any planet as large as Neptune near the plane of the solar system out to a distance of around 100 AU.

However, the 1978 discovery of Pluto's satellite allowed a better mass determination than had been possible earlier for Pluto, and the mass turned out to be too small to account for all of the perturbations on the motions of Neptune. This finding has prompted a number of astronomers to suspect that other objects may exist, perhaps well out of the plane of the solar system. Naval Observatory astronomer Thomas Van Flandern (and others, 1981) has proposed a search for an object with as much as three Earth masses in a highly inclined eccentric orbit. There is growing suspicion that the outer solar system may harbor additional small bodies. Incomplete searches reported in 1979 by Charles Kowal and in 1988 by Jane Luu and David Jewitt netted no new objects beyond Pluto, but the quest continues.

COMPARISON OF GIANT PLANET ATMOSPHERES—FROM COLD TO COLDER

Each giant planet has a massive atmosphere that is roughly three-quarters hydrogen and one-quarter helium by mass. In this regard, they are similar. But each giant planet is nearly twice as far from the Sun as its inward neighbor, so their temperatures drop dramatically as we go outward from the Sun. Thus we can understand certain trends we observe as we go from Jupiter to Neptune. For instance, the dropping temperature causes significant differences in some of the minor compounds that freeze from gas to ice crystals. Thus, for example, ammonia exists as an abundant gas prominent in the spectrum of Jupiter, but it declines in prominence as a gas in the outer planets because it freezes into ice crystals that form clouds lower in the atmosphere. Methane is therefore the most important trace gas in the atmospheres of Uranus and Neptune.

The same effect explains color trends among the giants. Jupiter is quite colorful due to the red, orange, and brown organic compounds formed in its clouds. These clouds are just barely visible on colder Saturn. In the coldness of Uranus' and Neptune's atmosphere, these compounds are evidently less readily formed. Here the color is dominated by a high-altitude methane-rich haze that absorbs the redder tones of sunlight and leaves the blue colors we observe.

A LESSON IN COMPARATIVE PLANETOLOGY: COMPARISONS AMONG MOON SYSTEMS

With all the processes that make unique moons—erupting volcanoes, featureless ice worlds, fractured plains, intense cratering, thick atmospheres, geysers—are there any systematic trends in common among satellite systems? Neptune's system may have been seriously disrupted in the past. If so, its commonality with other systems may have been destroyed.

Nonetheless, among Jupiter's, Saturn's, Uranus', and Neptune's systems, there are common features. All four have a ring. All four have tiny moonlets on the outskirts of the rings, perhaps because of fragmentation of original moons by impacts. All four have their largest moons in the middle of their satellite systems—just as Jupiter

and Saturn, the largest planets, are in the middle of the sequence of planets as we go out from the Sun. Thus the world-building process may be most efficient at an intermediate distance from the central body. (We will take up the details of planet and satellite formation in Chapter 14.) Jupiter, Saturn, Mars, and possibly Neptune all seem to have captured black asteroids, suggesting that there was an abundant supply of these dark objects in the early solar system and that capture events were not uncommon. Ice and black sooty material containing carbon and possibly colored organic molecules were common matter for forming all the moons of the outer planets. But local processes—most notably the heating of Io—have depleted ice or altered surface layers of some of the moons. Thus we find some variety in surface appearance as we go from moon to moon, even though the starting materials were similar.

Despite a variety that might at first seem bewildering, we do see basic similarities in the satellite systems (and with the solar system as a whole). These suggest that the processes by which the satellite systems and solar system were formed were similar and more or less orderly, but with occasional statistical "fluke" events, such as large-scale collisions.

SUMMARY

The four giant planets—Jupiter, Saturn, Uranus, Neptune—and their satellite systems show some systematic similarities as well as differences. All four of the giant planets have dense, cloudy atmospheres, averaging around four-fifths hydrogen and one-fifth helium by mass. All have rings composed of billions of small particles ranging in scale from microscopic to house-sized. All have systems of satellites. And most of the moons contain abundant ice as well as black soil, because this part of the solar system is so cold that ices were a main constituent of the material available for building moons and planets.

Among differences, we find that the colder, more remote planets have more hazy, less colorful cloud patterns. Jupiter has reddish and tan clouds; Uranus and Neptune have greenish-blue clouds from methane-rich haze. The ring systems have different amounts of material; Saturn's is most prominent, and Neptune's is concentrated in unusual arcs. Uranus and its satellite system are tipped so that the equator and satellite orbits lie nearly perpendicular to the plane of the solar system.

Pluto is an "oddball"—much smaller than the other planets and having a moon half as big as itself. It may be more closely related to other small bodies (such as asteroids and comets—see the next chapter) than to the true planets.

CONCEPTS

Saturn	Phoebe
Saturn's rings	Uranus
Cassini's division	obliquity
shepherd satellites	Uranus' rings
Saturn's satellite system	Uranus' satellites
Mimas	Miranda
Enceladus	Neptune
Tethys, Dione, and Rhea	Great Dark Spot
Titan	Triton
Hyperion	Pluto
chaotic rotation	Charon
Iapetus	"Planet X"

PROBLEMS

1. Why do Uranus and Neptune have more Earthlike colors (bluish) than Jupiter and Saturn?

2. Discuss how rings may have formed around giant planets.

3. Why doesn't Earth have a hydrogen/helium atmosphere like the giant planets do?

4. Describe the appearance of the sky to an observer flying above the clouds in the upper atmosphere of Saturn at dusk at low latitudes. Do the same for Saturn's regions.

5. Discuss why a probe to the surface of Titan might be interesting to biologists trying to understand how life started on Earth.

6. What difficulties might be met in sending a spacecraft through the rings of Saturn to examine the ring environment? Consider a pass *perpendicular* to the ring plane versus a pass *in* the ring plane. (Note that without elaborate retrorockets, such a spacecraft would probably travel about 20 to 30 km/s relative to the rings.)

7. Describe the seasons and other effects that would occur if Earth had the same obliquity as Uranus.

8. Do orbits of planets ever cross, as seen from far north or south of the plane of the solar system?

9. Where would Bode's rule, if extended, predict a tenth planet? Would it be easier to detect if covered with snow or with rock? Why?

10. Explain why observers on Earth can never see Saturn appearing as it does in Figure 12-7.

ADVANCED PROBLEMS

11. If a spacecraft with an infrared sensing device flew by Jupiter or Saturn and measured a cloud formation whose strongest radiation came at a wavelength of 2×10^{-5}m, what would be the temperature of the material in the cloud?

12. Assume the mass of Saturn is 5.7×10^{26} kg.
a. What is the orbital velocity of a particle orbiting around Saturn in Saturn's rings, about 250 000 km from the center of Saturn?
b. Suppose this particle hits another one that orbits 1 km farther away from Saturn. Estimate how fast they come together, using whatever mathematical techniques you know.

13. Calculate the weight of a 140-lb person on the "surface" of Titan. Use Table 8-1.

14. Calculate the mean density of Saturn (lowest of any known planetary body), and discuss its implications for a dense core of rock or metal inside Saturn. Use Table 8-1.

PROJECTS

1. With a telescope having at least an 8-cm (3-in.) aperture, observe Saturn. Sketch the rings. Estimate the angle by which the rings are tilted toward Earth during your observation. Does this angle change much from day to day? From year to year? (Your teacher might save drawings by students from past years for comparison.) Can you see any belts or zones? (Usually they are less prominent than on Jupiter.) Can you see Cassini's divisions? Sketch nearby starlike objects that may be satellites and track them from night to night. Identify Titan, the brightest satellite.

2. With a telescope having at least a 15-cm (6-in.) aperture, observe Uranus and Neptune. What color are they? Can you see any detail? Any satellites?

CHAPTER 13

Comets, Meteors, Asteroids, and Meteorites

On June 30, 1908, a mysterious explosion occurred in Siberia. English observatories, 3600 km (2200 mi) away, noted unusual air pressure waves. Seismic vibrations were recorded 1000 km (600 mi) away. At 500 km, observers reported "deafening bangs" and a fiery cloud. The explosion was caused by an unknown object that struck the Earth's atmosphere from space.

Some 200 km (120 mi) from the explosion, the object was seen as "an irregularly shaped, brilliantly white, somewhat elongated mass . . . with [angular] diameter far greater than the moon's." Carpenters were thrown from a building and crockery knocked off shelves. An eyewitness 110 km from the blast reported that

the whole northern part of the sky appeared to be covered with fire. . . . I felt great heat as if my shirt had caught fire . . . there was a . . . mighty crash. . . . I was thrown onto the ground about [7 m] from the porch. . . . A hot wind, as from a cannon, blew past the huts from the north. . . . Many panes in the windows were blown out, and the iron hasp in the door of the barn was broken.

Probably the closest observers were some reindeer herders asleep in their tents about 80 km (49 mi) from the site. They and their tents were blown into the air and several of the herders lost consciousness momentarily. "Everything around was shrouded in smoke and fog from the burning fallen trees."

The cause of this remarkable explosion was a collision between the Earth and a relatively modest bit of interplanetary debris. Many such objects circle the Sun. The Siberian event was merely the fall-in of the largest object to hit the Earth in the last century or so. Even larger objects have hit the Earth in earlier eras, as proved by the 20 000-y-old Arizona crater shown in Figure 6-17 and by the evidence for a giant impact 65 million years ago that ended the reign of dinosaurs, as discussed in Chapter 6.

Smaller impacts have been recorded more often; interplanetary stones fall from the sky in various locations every year (sometimes remote and sometimes inhabited—several houses have been hit in recent decades). Tiny dust grains that burn up in the atmosphere before hitting the ground can be seen every night; these are sometimes called "shooting stars."

Two conclusions are apparent. First, events ranging from small meteorite impacts to large, rare impact explosions may have occurred in the ancient past, perhaps influencing the beliefs of ancient people about forces in the sky.

The second conclusion is that the solar system contains more than just the planets and satellites we have already discussed. Floating between the planets are debris of many different sizes. They have been divided into different types depending on observed or inferred properties.

Comets are icy worlds a few kilometers across. **Asteroids** are rocky-metallic worlds ranging from a few hundred meters or less up to 1000 km in diameter. **Meteoroids** are still smaller bits of debris, probably dislodged from comets and asteroids, floating in their own orbits in space. Table 13-1 shows orbital and other data on selected examples of all these classes.

Some of these bodies have planetlike orbits that are nearly circular, prograde, and lying in the plane of the solar system. Others have very different orbits—highly elliptical, retrograde, or inclined at steep angles to the solar system's plane. Some of these orbits are shown in Figure 13-1, which emphasizes the diversity of interplanetary objects.

Objects whose orbits cross the orbits of the Earth and other planets can collide with the planets. When they collide with Earth, they are heated by friction with the atmosphere (just as returning space vehicles are). **Meteors** are the smaller ones (usually up to a few centimeters in diameter) that burn out completely before striking the ground. **Meteorites** are the larger ones that survive and fall to the ground: They are made of rock or metal. They are free samples of other planetary worlds. Their properties testify that they broke off of larger bodies, called **parent bodies,** in the remote past. One unsolved mystery is the question of what objects were the parent bodies of different meteorite groups.

Comets, asteroids, and meteoroids of every size are all debris left over from the origin of the solar system 4.6 billion years ago. As the present planets formed, the solar system was filled with innumerable small, pre-

planetary bodies, ranging up to 1000 km across, known as **planetesimals.**

Comets, asteroids, and meteorites are all leftover planetesimals and planetesimal fragments. Thus they provide a direct link to the conditions under which the planets and the Sun itself formed. By implication, they provide clues to the formative conditions of other stars and hence provide a link with the most distant stellar regions, which will be a topic of much of the rest of this book.

COMETS

Comets are the most spectacular of the small bodies in the solar system. When they pass through the inner solar system near the Earth, they can be seen drifting slowly from night to night among the stars. (Writers sometimes incorrectly describe comets as "flashing cross the sky" like shooting stars. They do not. They seem to hang motionless and ghostly among the stars. Their motion relative to the stars can be detected by the naked eye only after a few hours.)

Comets have several parts, as seen in Figures 13-2 and 13-3. The brightest part is the **comet head.** The **comet tail** is a fainter glow extending out of the head, usually pointing away from the Sun. Although a typical comet tail can be traced for only a few degrees by the naked eye, binoculars or long-exposure photos may reveal fainter extensions of the tail extending tens of degrees or even extending clear across the night sky. A telescope reveals a brilliant, starlike point at the center of the comet head. At the center of this bright point is the **comet nucleus,** which is the only substantial, solid part of the comet, but is too small to be resolved by telescopes on Earth. Studies reveal that a typical comet nucleus is a worldlet of dirty ice only about 1 to 20 km (a few miles) across—tiny compared to most planets and moons! The gas and dust that make up the rest of the comet's head and tail are material emitted from the nucleus. As the comet nucleus moves through the inner solar system, the sunlight warms it and causes the ice to evaporate[1] into the form of gas. This gas, together with dislodged dust grains from the dirt in the nucleus, is then carried away from the nucleus by the pressure

---

[1]A more correct technical word is *sublime,* which means to change from solid form directly into gaseous form. *Evaporate* technically means to change from liquid form into gaseous form.

TABLE 13·1

Properties of Selected Small Bodies in the Solar System

Class	Example	Diameter (km)	Orbit Semimajor Axis (AU)	Eccentricity	Inclination	Remarks
Comets						
Short-period	Encke	1–8	2.2	0.85	12°	Probably dirty ice
	Halley	Few?	18	0.97	162°	Probably dirty ice
Long-period	Kohoutek	8 × 15	Very large	1.0	14°	Probably dirty ice
Meteors						
Shower	Perseid	10^{-6}	40	0.97	114°	Cometary debris
	Taurid	10^{-6}	2.2	0.80	2°	
Fireballs	July 31, 1966	?	32	0.98	42°	May be related to comets or asteroids
	May 31, 1966	?	3.0	0.80	9°	
Asteroids						
Belt	1 Ceres	1020	2.8	0.08	11°	Carbonaceous rock surface
	2 Pallas	538	2.8	0.23	35°	Rocky surface
	3 Juno	248	2.7	0.26	13°	Rocky surface
	4 Vesta	549	2.4	0.09	7°	Lava-like surface
	14 Irene	170	2.6	0.16	9°	Rocky surface
Trojan	624 Hektor	100 × 300	5.1	0.02	18°	Unusual shape
Apollo	433 Eros	7 × 19 × 30	1.5	0.22	11°	Elongated
	Apollo	?	1.5	0.56	6°	Lost before orbit precisely measured
Unusual "asteroid"	2060 Chiron	100–320?	13.7	0.38	7°	Erupted in 1988 Possibly a comet
Meteorites (chondrites)						
	Pribram	10^{-4}	2.5	0.68	10°	
	Lost City	10^{-4}	1.7	0.42	12°	
	Leutkirch[a]	10^{-4}?	1.6	0.40	2.5°	

[a]An object photographed over Europe in 1974. No fragments recovered as of late 1974, but believed to be a stone meteorite.
Source: Data from Chapman and Morrison (1974); Hartmann (1975, 1983); Gehrels (1972). The diameter of Comet Encke was measured with radar by Kamoun and others (1982).

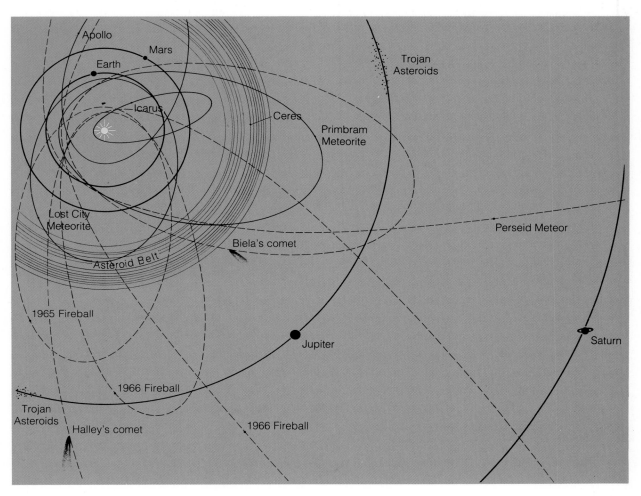

Figure 13-1 A portion of the solar system showing orbits of a few selected interplanetary bodies including comets, main-belt asteroids, Trojan asteroids, Apollo asteroids, meteorites, and large meteors, or fireballs.

of radiation and thin gas rushing outward from the Sun (Brandt and Niedner, 1986). This outrushing solar gas is called the **solar wind.** Only microscopic grains get blown outward by it; larger grains are too heavy to be caught in the solar wind. (Similarly, a handful of dust gets blown by the wind but stones do not.) The gas and dust from the nucleus form the comet's tail, often stretching more than an astronomical unit. Caught in the solar wind, the tail streams out behind the comet as the comet approaches the Sun, but leads as the comet recedes from the Sun (Figure 13-4). It is like the long hair of a woman streaming out behind her as she walks into the wind, but in front of her if she reverses direction.

Comets as Omens

Before comets' true natures were known, they were often regarded as evil omens. For example, the appearance of Halley's comet in A.D. 66 was said to have heralded the destruction of Jerusalem in A.D. 70. Five circuits later it was said to mark the defeat of Attila the Hun in 451. In 1066 it presided over the Norman conquest of England. In 1456, its appearance coincided with a threatened invasion of Europe by the Turks, who had already taken Constantinople three years before. Pope Calixtus III ordered prayers for deliverance "from the devil, the Turk, and the comet." Of course, we now

a

b

Figure 13-2 The motion of Comet Ikeya-Seki in the dawn sky during an interval of a few days in 1965. The comet's head is the sharply defined tip nearest the horizon; the tail extends diffusely upward. **a** The comet at dawn. Movement relative to *b* can be seen by comparing stars: The lower tail overlaps a bright star in *a*, but in *b* the comet's head has moved to the right of that star (35-mm camera; 15-min exposure guided on comet). **b** The comet in the predawn sky, photographed with an ordinary, stationary 35-mm camera (20-s exposure at f1.9 on Tri-X film). (Both photos by S. M. Larson, University of Arizona.)

recognize that some noteworthy events are likely in most cultures every decade or so, and comets are no longer regarded as ominous portents by most people.

Discovery of Comets' Orbits

In 1704 the English astronomer Edmond Halley applied newly developed methods of computing orbits and discovered that comets travel on *long, elliptical orbits* around the Sun and that certain comets reappear. Calculating the orbits of 24 well-recorded comets by Newton's methods, Halley found that four comets (seen in 1456, 1531, 1607, and 1682) had the same orbit and a periodicity near 75 y. Halley correctly inferred that these appearances were by a single comet and that the slight irregularities in periodicity were caused by gravitational disturbances from the planets, especially Jupiter. Halley

predicted that the comet would return about 1758. It did so on Christmas night. It was named **Halley's comet**[2]. The discovery that comets are visitors on ordinary elliptical orbits, with predictable motions, helped dispel the superstition that comets are evil omens!

Halley's comet became very famous due to its prominence in 1910, when the Earth passed through its tail. It was not so prominent in 1986 during its most recent approach, because it did not pass as close to the Earth (Figure 13-5). Even so, the 1986 approach was historic because it marked the first close-up study of a comet by space probes.

---

[2]Comets discovered today are first named for their discoverers but are later given scientific names in order of their passage around the Sun—for example, Comet 1982 I.

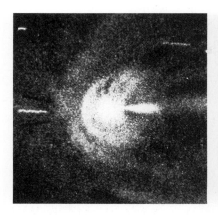

Figure 13-3 A new look at old photos of Halley's comet. The head of Comet Halley was photographed on different days as it passed unusually close to the Earth in June 1910, but photographic prints made by standard techniques of that era showed only vague structure. In 1984, astronomers Z. Sekanina and S. M. Larson reprocessed some of the old plates with modern equipment that digitizes the image and enhances low-contrast detail. The nucleus is in the brightest region at the center of each image. These images clearly show individual curved jets of dusty debris blown off the nucleus by individual outbursts. Sun is on left; tail is to right. Streaks are star images trailed by long-exposure tracking on comet. (Photos courtesy S. M. Larson, University of Arizona.)

Where Comets Come From: The Oort Cloud

In ancient times no one knew how far away comets were, and many people thought comets were phenomena in our own atmosphere. Seneca, the Roman contemporary of Jesus, wrote: "Some day there will arise a man who will demonstrate in what regions of the heavens the comets take their way." That man was Tycho Brahe, who in 1577 arranged observations of a bright comet from two different locations. Finding no parallactic shift in the comet's angular position relative to the stars (see page 64), Tycho correctly concluded that it was more distant than the Moon and thus not terrestrial.

Comet orbits have several important characteristics. First, they are inclined at nearly random angles to the plane of the solar system. Thus comets populate a roughly spherical volume of space around the Sun, rather than being confined to a disklike volume as the planets are.

From statistics of comets' orbits, Dutch astronomer Jan Oort discovered around 1950 that this spherical swarm of comets is very extensive and that most comets spend most of their time about 50 000 AU from the Sun. This swarm, containing millions of comets, is called the **Oort cloud.** Here comets are so far from the Sun

that gravitational attractions of nearby stars, as well as the Sun, are important. Following Kepler's and Newton's laws, comet nuclei drift slowly at these great distances, being disturbed by stellar forces, until they eventually fall back into the inner solar system. Still following Kepler's laws, they then move rapidly through the inner solar system on elliptical, nearly parabolic orbits (see page 65).

When a comet moves through the inner solar system, it usually spends only a few months near enough to the Sun to be visible from Earth. Then it swings slowly back toward the Oort cloud, where it may spend hundreds or thousands of years. Occasionally some comets pass near Jupiter or other planets and are redirected by gravitational forces into orbits that remain in the inner solar system. Many of the best-known comets, which return every few decades, are of this type. These comets are called **short-period comets.** In Figure 13-1, Biela's comet can be seen to be such an object, while Halley's comet can be seen to have a much larger orbit reaching toward the outer solar system.

The fact that the Oort cloud is part of the solar system (albeit remote) indicates that comets are part of the Sun's family. Rarely, if ever, do any comets come from, or escape to, regions as remote as the stars.

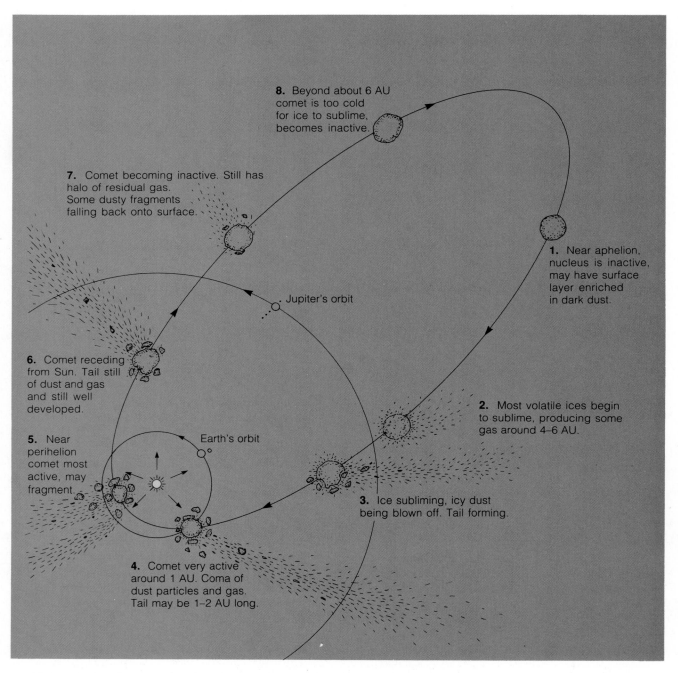

8. Beyond about 6 AU comet is too cold for ice to sublime, becomes inactive.

7. Comet becoming inactive. Still has halo of residual gas. Some dusty fragments falling back onto surface.

1. Near aphelion, nucleus is inactive, may have surface layer enriched in dark dust.

Jupiter's orbit

6. Comet receding from Sun. Tail still of dust and gas and still well developed.

2. Most volatile ices begin to sublime, producing some gas around 4–6 AU.

Earth's orbit

5. Near perihelion comet most active, may fragment.

3. Ice subliming, icy dust being blown off. Tail forming.

4. Comet very active around 1 AU. Coma of dust particles and gas. Tail may be 1–2 AU long.

Figure 13-4 Stages in the development of a typical comet as it travels from the outer solar system into the inner solar system, where it loops rapidly around the Sun and moves outward again. The tail develops only within a few astronomical units of the Sun, where sunlight is warm enough to sublime the ice in the nucleus. In order to show comet detail, the figure is not drawn to scale.

a

b

Figure 13-5 Two views of Halley's comet during its approach in 1986. In that year, it was in Southern Hemisphere skies and not well placed for northern observers. Both photos show the pinkish color of the diffuse dust tail and the bluish color of the narrower, straighter gaseous tail. **a** On the morning of March 8, the tails were well displayed. **b** By April 14, the comet had moved to a position where our line of sight was almost directly down the dust tail, making it look much more diffuse and fan-shaped. Two other interesting astronomical objects are in the extreme distant background: The very distant galaxy Centaurus A, split by a dark dust lane, is right of the comet; the star cluster Omega Centauri, in our own galaxy, dominates the bottom. Compare more detailed picture of Centaurus A in Figure 25-7. (Photos by William Liller from Easter Island for NASA's International Halley Watch; 4- and 5-min exposures on Fujichrome 400 film with 8-in. Schmidt telescopic camera at f1.5.)

Comets Near the Sun

As comets loop into the inner solar system, they pass the Sun at various distances. Some that pass very close actually break apart as a result of stresses during the furious degassing of the ice (see Figure 13-6). In 1979 a rare but interesting subclass of comets was found: comets whose point of nearest approach to the Sun is less than one solar radius from the Sun's center. That is, these comets collide with the Sun. An example is seen in Figure 13-7, a series of photos taken from a satellite. At least two additional Sun-colliding comets were discovered by space telescopes in the following three years. One of the satellite telescopes, which had finished its designated mission, was "shot down" by the Defense Department in an antisatellite weapon test, ending these discoveries.

Comets' Nuclei: Dirty Icebergs in Space

By the 1700s, when people discovered that comets are not wispy atmospheric phenomena or evil visitors but objects orbiting the Sun, the next question became:

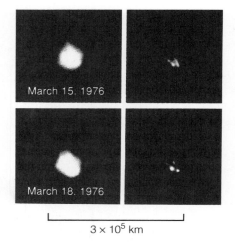

March 15. 1976

March 18. 1976

3×10^5 km

Figure 13-6 Several comets have been seen to break apart. These pictures show the breakup of Comet West (1975n) into four pieces in March 1976. The left column photos are long exposures; the right column photos are shorter exposures, revealing four nucleus fragments, each with its own stubby tail. The pieces separated at relative speeds of about 1 to 5 m/s (2 to 11 mph). One piece was seen for only two weeks; others were followed for over six months. (Lunar and Planetary Laboratory 1.5-m telescope photo, University of Arizona, courtesy S. M. Larson.)

What is the physical nature of the comet? Progress on this question came in 1868 when English astronomer William Huggins first studied comets through a spectroscope. As discussed in Chapter 5, the spectroscope allows astronomers to measure properties of the material in a celestial body. Huggins found three bright emission bands, which he could identify with gaseous carbon. Modern astronomers have added many other kinds of atoms, molecules, and ions to the list, including H (hydrogen), O (oxygen), CN, CH, OH, H_2O^+, CN^+, CH^+, OH^+, N_2^+, CO^+, and CO_2^+. Recent studies have made exciting additions to the list: more complex organic molecules such as CH_3CN, H_2CO, and HCO. This discovery proves that organic molecules—the building blocks of life—form elsewhere in the universe besides Earth.

Note that all these atoms (H, O, C, N) and molecules are just the ones that would be expected if the gas were formed by the sublimation of common solar system ices such as H_2O (water), CH_4 (methane), NH_3 (ammonia), and CO_2 (carbon dioxide).

Apart from revealing gases, the spectroscope shows that some particles in comet tails are much bigger than

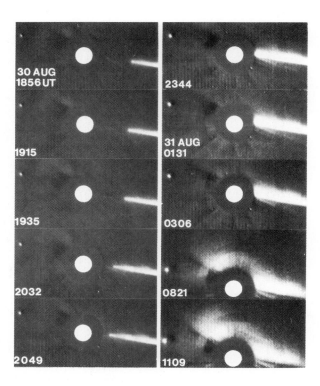

30 AUG 1856 UT

2344

1915

31 AUG 0131

1935

0306

2032

0821

2049

1109

Figure 13-7 The collision of Comet 1979 XI with the Sun on August 30, 1979, photographed with an orbiting telescope. The white disk shows the size of the Sun, which was covered by a larger disk (making a black ring around the Sun). Tracking data show that the comet hit the outer atmosphere of the Sun. The nucleus did not survive. In the postimpact series (right) the contrast has been enhanced; rapid sublimation of ices in the nucleus has added material to the tail. Material from impact or breakup of the nucleus streams outward in several directions. The comet did not reemerge on the left side of the Sun. Venus is at left. (LOLWIND satellite photos from Naval Research Laboratory, courtesy D. J. Michels.)

molecules. They are dust grains. Thus a comet tail contains both gas and dust coming off the nucleus. Indeed, these two materials can easily be distinguished in many color comet photos. The gas tends to blow straight back from the Sun in the solar wind, but the heavier dust grains move partly under the influence of the solar wind and partly under the influence of Keplerian motions, like microplanets. Thus two slightly separated but distinct tails often develop. The gas tail tends to scatter blue light just as the gas in our atmosphere. The dust tail is composed of dark, reddish-brown dust and tends to have a pinkish cast. Thus comets can often be seen to have a straight, bluish gas tail and a slightly curved,

Figure 13-8 The head of Comet Bennett, in 1970, displayed spiral jets of dust and gas, indicating that the nucleus is rotating. Similar jets in Halley's comet are shown in Figure 5-22. (S.M. Larson, Lunar and Planetary Laboratory, University of Arizona.)

pinkish dust tail as in Figure 13-5. These colors can't be sensed with the eye because comet tails are so faint, but they do show up in color time-exposure photos.

Evidence of this kind proved that the nucleus, whatever it is, can slough off both gas and dust when warmed by the Sun. Around 1950, Harvard astronomer Fred Whipple first put together all this evidence to give essentially a correct prediction of the nature of a comet nucleus. He argued that the gases do indeed come from subliming ice and that the ice must contain dust grains. As the ice sublimes, the dust grains are dislodged and blown away by the solar wind along with the gas. The ice was predicted to be mostly frozen water with certain amounts of frozen methane (CH_4), carbon dioxide (CO_2), frozen ammonia (NH_3), and other materials to explain the various C, H, O, and N atoms in the gas. Thus Whipple's theoretical picture of a comet nucleus came to be called the **dirty iceberg model,** and it has been confirmed by modern data.

Observations in the 1970s and 1980s made a great forward leap in our understanding of the nucleus. There are four lines of evidence. First, certain comets have been observed to break apart spontaneously (Figure 13-6). Thus the nucleus must be a distinct, solid body but fairly weak, perhaps divided by fractures. Second, irregular movements of comets were attributed to jet-like outbursts of gas from individual spots on the nucleus—perhaps individual fissures or vents that expose fresh ice which sublimes rapidly, shooting out jets of gas. The jets often show spiral curvature, like the spray

from a rotating garden sprinkler, as seen in Figures 13-8 and 5-22. This pattern indicates that the nucleus is rotating. Third, comet dust has colors similar to certain very dark brownish-black asteroids that are found only in the outer solar system. Thus it was predicted that comets may be planetesimals colored by the same dark dust found in outer solar system asteroids. And fourth, space probes returned data from near Halley's comet in 1986.

A Comet Nucleus at Close Range

An international fleet of five spacecraft (two Japanese, two Russian, and one European) probed Halley in 1986. The first close-up data and pictures supported the earlier work.[3] After the first four craft tested the environment at moderate distance, the European probe flew by closest, only about 600 km from the nucleus. It was named Giotto, after the Italian artist whose 1304 painting of the comet may have been the first to show it.

The heart of Halley's comet was a hazy environment where the probes found more fine dust than expected. Giotto was hit by a dust grain about 960 km from the nucleus, knocking it partially out of commission for 32 min during the closest approach! As seen in Figures 13-9 to 13-11, the photos revealed bright jets of illuminated gas and dust coming off "active" spots on the black, peanut- or potato-shaped nucleus, about 15 km long and 7 to 10 km wide. The haze made the pictures fuzzy. The coma gas was found to be about 80% water vapor by volume. The dust particles were found to be rich in C, H, O, and N and somewhat resembling carbon-rich meteoritic particles collected on Earth. There was so much carbon, hydrogen, oxygen, and nitrogen in these particles that they came to be called CHON particles. They are different from familiar terrestrial dust, which is richer in silicon, iron, other metals, and their oxides.

---

[3]The encounter dramatized that while our news media may be adept at reporting tragedies, they fail in dealing with the adventure of human exploration. During the night of the first Russian encounter, a TV hookup with the USSR allowed live coverage of humanity's first close look at a comet; but my local affiliate carried instead an old Evil Knievel movie. The next day the papers ran a front-page picture of the comet, reportedly from the Russian spacecraft. A few days later, however, on an inner page they reported sheepishly that this was not the new Russian picture at all but a computer-processed 1910 photo that had been misidentified by a news agency.

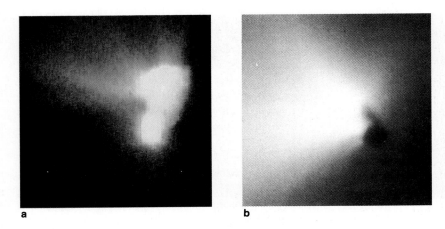

a   b

Figure 13-9 The nucleus of Halley's comet at moderate distance from two different space probes in March 1986. Both views are hazy because of the dusty environment, but they show luminous jets issuing from a peanut-shaped nucleus. Sunlight is from left in both views, and jets come from sunlit side, where gas is subliming rapidly. **a** View from Russian Vega 2 probe at 8000-km distance. Lighting at gibbous phase illuminates much of nucleus surface in this view. (USSR Academy of Sciences.) **b** View from European Giotto probe at about 20 000 km distance. Crescent lighting silhouettes dark side of the nucleus against the luminous head. (Copyright Max-Planck-Institut für Astronomie.)

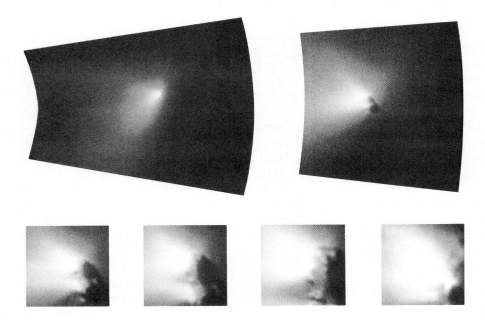

Figure 13-10 The flight of Europe's Giotto probe past the nucleus of Halley's comet. This sequence of images shows the telephoto camera's field of view as it flew by the nucleus. Camera was designed to aim at the brightest object in the field of view, which turned out to be the largest jet of gas and dust spraying off the sunlit side. (Copyright Max-Planck-Institut für Astronomie, courtesy H. U. Keller.)

Figure 13-11 In the weeks after Giotto's flight past the nucleus of Halley's comet, the hazy images were highly processed to improve clarity and combine the best parts of different images at different scale. This resulting picture is the best available. It shows bright jets of gas and dust shooting from localized active vents on the sunlit side of the nucleus. Smallest details are about the size of a football field. There is little color to be seen in the nucleus, which is surprisingly irregular in shape. The unlit "night" side is silhouetted against the softly glowing head on the right. Because of the journalistic tradition of reporting only "today's news," indistinct pictures of the nucleus were widely reproduced in newspapers at the time of encounter, but this much improved image has rarely been published in the popular media. (Giotto photo courtesy Harold Reisema and Alan Delamere, Ball Aerospace; copyright 1986 MPAE.)

Although some astronomers had predicted dark, asteroidlike dust on comets, Halley's nucleus turned out to be darker than many astronomers expected, reflecting only 4% of the light that strikes it. (Compare with 60 to 80% for clean ice.) It is as dark as black velvet! Earth-based observers found a similar result for the nucleus of another comet, Neujmin 1, in 1987 (Campins and others, 1987). Since abundant carbon was observed in the dust shed from Halley, the black material is likely to be very carbonaceous. Scientific interest is thus growing in this cometary dust, which is apparently rich in the building blocks needed for life. No life is expected to have evolved in the frozen interiors or surface layers of comets, but perhaps comets contributed organic materials as they struck primordial planets.

The Origin of Comets

From the evidence described so far, astronomers have theorized a four-stage history that explains the origin of comets. First, comets must have formed by aggregations of ice crystals and dust grains in the cold, outer regions of the solar system while the planets were forming. We know from studying the gases in the atmospheres of Jupiter and other giant planets that hydrogen, carbon, oxygen, and other elements were available as the outermost planets formed. In the environment of the outer solar system these would have condensed into ice grains (such as H_2O) that could accumulate into icy planetesimals—the comet nuclei. Second, many of the nuclei that were still orbiting around the Sun in the outer solar system by the time the giant planets had formed must have approached close to these planets and been flung by the planets' gravity into highly elliptical orbits that took them out of the solar system. Due to Kepler's laws, they move slowest and spend most of their time far beyond Pluto, near aphelion. By a similar process, Voyager and Pioneer space vehicles were flung out of the solar system by close encounters with Jupiter. This process randomized many comet orbits and formed the Oort cloud of objects circling the Sun beyond

Figure 13-12 A proposed future comet mission would park a probe in orbit alongside a comet and watch it "turn on" as it approached the Sun. This painting was made two months before the Halley encounters and was a final attempt to predict some features of the nucleus. It shows jetting similar to that seen in spacecraft photos such as Figure 13-9. (Painting by author.)

Pluto. Third, comets were "stored" in the Oort cloud for long periods of time. Fourth, comets in the Oort cloud occasionally found themselves on trajectories back into the inner solar system, where they formed the bright heads and tails of gas and dust for which comets are known.

Future Work on Comets

Scientists would like to learn more about the elements and organic compounds in comets in order to compare them with terrestrial data on the origins of life. They would also like to understand the physical processes that occur during comet activity. Is the black dirt mixed through the nucleus' ice? Or does a layer of black surface gravel mask cleaner ice below? Could future astronauts use comet ices for water supplies?

American scientists have proposed a "CRAF" (Comet Rendezvous, Asteroid Flyby) mission. As shown in Figure 13-12, this probe would "park" in a position near a comet far from the Sun and then observe the turn-on processes as the comet became active and began to blow off surface materials. A torpedolike "penetrator" would be fired into the comet to measure the compo-

sition of its dirty ice. On the way to the comet, the probe would pass by one or more asteroids, thus getting data for comparing asteroids and comets. A group of scientists has been selected to build the equipment, but as of 1987 the mission had not yet been funded for actual construction. Meanwhile, the European Space Agency, which successfully flew Giotto to Halley's comet, is considering a probe to fly through a comet's head and return a sample of the gases and particles to Earth.

METEORS AND METEOR SHOWERS

Meteors that flash momentarily across the sky might seem totally unrelated to comets. However, a study of their frequency on different nights reveals a direct connection. On an average night you may see about 3 meteors per hour before midnight and about 15 meteors per hour after midnight. (You see more meteors after midnight because you are then located on the leading edge of the Earth as it moves forward in its orbit, sweeping up interplanetary debris.) But on certain dates each year you may see **meteor showers** of 60 meteors or more per hour, all radiating from one direction in the sky (Table 13-2).

TABLE 13·2

Dates of Prominent Meteor Showers

Shower Name (After Source Constellation)	Date of Maximum Activity[a]	Associated Comet
Lyrid	April 21, morning	1861 I
Perseid	August 12, morning	1862 III
Draconid[b]	October 10, evening	Giacobini–Zinner
Orionid	October 21, morning	Halley
Taurid	November 7, midnight	Encke
Leonid	November 16, morning	1866 I
Geminid	December 12, morning	"Asteroid" 1983 TB[c]

[a]Showers can last several days before and after the peak activity on the listed date. Observations are best when the constellation in question is high above the horizon, usually just before dawn.

[b]The Draconids are now weak because their orbits have been disturbed by the gravity of planets, but further disturbances may again strengthen the shower in the future.

[c]This object was discovered in 1983 by the IRAS satellite and is probably a "burnt-out" comet nucleus.

Figure 13-13 A rare meteor shower; the Leonids of November 17, 1966. The rate of meteors visible to the naked eye was estimated to exceed 2000 per minute. The brightest star (upper left) is Rigel, in the constellation Orion. This exposure of a few minutes' duration was made with a 35-mm camera. (D. R. McLean.)

The best-known example is the Perseid shower, which occurs every year around August 12, when bright meteors streak across the sky every few minutes from the direction of the constellation Perseus. (A shower is named for the constellation most prominent in the area of the sky from which the shower radiates.) Occasionally the showers are so intense that meteors fall too fast to count. During the Leonid shower of November 17, 1966, meteors fell like snowflakes in a blizzard for some minutes, at a rate estimated to be more than 2000 meteors per minute (see Figure 13-13).

What is the connection between the showers and the comets? In 1866 G. V. Schiaparelli (of Martian canal fame) discovered that the Perseid meteor shower occurred whenever the Earth crossed the orbit of Comet 1862 III. The Perseids, then, must be spread out along the orbit of that comet. Other relationships were soon found between specific meteor showers and specific

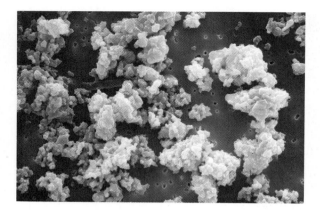

Figure 13-14 Clumps of microscopic meteoroidal particles collected by high-flying aircraft. They may have originated in comets. The largest clump is about 60 μm across. (Courtesy D. E. Brownlee, University of Washington, Seattle.)

comets, as Table 13-2 shows. In 1983, an infrared astronomical telescope in orbit (called IRAS) discovered the thermal infrared emission from a swarm of meteor dust spread along the orbit of Comet Tempel 2. Here, then, was a direct detection of the dust scattered by a comet. Even the orbits of individual sporadic meteors, tracked photographically, often resemble long- or short-period cometary orbits. *Therefore, most meteors must be small bits of debris scattered from comets.*

Since the 1960s, rockets and balloons have collected microscopic fragments believed to be meteoroids, shown in Figure 13-14. These fragments are irregular glassy silicate and metallic particles, supporting the theory that comet nuclei are dirty icebergs with bits of entrapped grit. Their compositions are similar to those of certain types of meteorites. Most meteors are far too small to reach the ground, "burning" at altitudes around 75 to 100 km. Occasional large ones, called **fireballs,** are very bright and spectacular. They generally explode in the air instead of hitting the ground, again indicating that they are too fragile to survive atmospheric entry.

In 1986, scientists discovered the "missing link" between comets and meteor showers. The IRAS satellite had mapped the sky in infrared light coming from all sorts of celestial materials. Among the asteroids, nebulae, and other sources, researchers found trails of dust spread along the orbits of active comets (Sykes and others, 1986). The solar system is ringed with these tenuous **dust trails** of microscopic particles blown off

the comets' surfaces when they are active. Visible meteors may be just the larger particles in these trails. As we will see in later chapters, these rings of dust may relate to rings of dust found around other stars.

ASTEROIDS

Asteroids are the largest of the interplanetary bodies, ranging from the biggest, Ceres (about 1000 km in diameter), down to a few hundred meters across and probably less. Spectral studies show that they are rocky, often with metal embedded in them. They appear in various parts of the solar system but are most abundant in the **asteroid belt**—a region between Mars and Jupiter.

Discovery of Asteroids

Asteroids have played an interesting role in the mapping of the solar system. Normally too small to be seen by the naked eye, they were unknown before 1800. Bode's rule, confirmed by the discovery of Uranus in 1781, called for a planet at 2.8 AU from the Sun in the large space between Mars and Jupiter. Therefore, astronomers set out to find the "missing planet" in 1800; success came on the first night of 1801 when Ceres was discovered—just at 2.8 AU.[4]

Between 1802 and 1807, three more small, planetlike bodies turned up between 2.3 and 2.8 AU from the Sun. Because of their small size, they came to be called minor planets, or *asteroids,* a name that now applies to any interplanetary body that is not a comet (that is, is not known to emit gas). By 1890, some 300 asteroids were known. In 1891, German astronomer Max Wolf began searching for them photographically by time exposures, detecting many new asteroids by their telltale motion among the stars.

Asteroids are known by numbers (assigned in order of discovery, but only after the orbit has been accurately identified) and a name (chosen by the discoverer)— 1 Ceres and 2 Pallas, for example. The names cover a wide range of human interests, including cities, mythol-

[4]As noted by Arthur C. Clarke, this discovery came just when the philosopher Hegel had "proved" philosophically that there could be no more than the seven then-known planetary bodies, Mercury through Uranus—which suggests that going out and looking is worth more than sitting at home speculating.

a   b   c   d   e   f

g

Figure 13-15 a–f Sequence of false color images of asteroid "1989 PB" about 27 min apart, showing extraordinary dumbbell shape. Blue indicates the weakest radar reflection, grading through yellow and red to white as the strongest reflection off the center of each lobe. (Images by Steven Ostro and colleagues at Arecibo Observatory radio telescope, Puerto Rico, in 1989.) **g** Possible closeup appearance of the asteroid, perhaps formed by low-speed collision of two rounded bodies. (1978 painting by author.)

ogy (1915 Quetzalcoatl), politicians (1932 Hooveria!), spouses, and other lovers.[5] About 4000 are now cataloged, and perhaps 100 000 observable ones remain uncharted.

Asteroids Outside the Belt

Several interesting subgroups of asteroids have orbits outside the main asteroid belt. **Apollo asteroids,** for example, come into the inner solar system and cross the Earth's orbit. They are named after Apollo, first of the group to be discovered. Dozens are known, including some only a few hundred meters across. Apollos can come very close to Earth, and some even hit Earth! They generally survive only a hundred million years or so before hitting Earth or some other planet, creating impact craters. The supply is replenished as other asteroid fragments are thrown out of the belt by the gravitational force of massive Jupiter. Still other Apollos may be remnants of burnt-out comets that have lost their ices and are no longer active.

Most asteroids are too distant for telescopes to show their shapes, but astronomers use a clever technique to estimate their shapes. As it rotates, the asteroid brightens and fades as it presents first a broadside view, then an end-on view, and so on. Measurements of these changes show that larger asteroids like 1 Ceres are relatively spherical, but many small asteroids, including many Apollos, are elongated or irregular. They are probably splinter-like fragments of larger parent bodies. In 1989, an Apollo asteroid came so close to Earth that it could be well measured by the technique of bouncing radar waves off it. This method showed it to be an amazing dumbbell shape, possibly caused by coalescence of two rounded fragments (Figure 13-15).

Trojan asteroids[6] lie in two swarms in Jupiter's orbit, 60° ahead of and 60° behind the planet, called **Lagrangian points.** The astronomer Joseph Louis Lagrange discovered that particles can be held in the two swarms by Jovian and solar gravitational forces. At least 45 Trojans have been found at one Lagrangian point, and observers estimate about 700 observable Trojans. The largest Trojan is Hektor, estimated to be 100 km wide and 300 km long, tumbling end over end in Jupiter's orbit. Trojan asteroids and their neighbors, the outer satellites of Jupiter (J6 through J13), are similar in having a very dark color and a different composition (more carbon minerals) than that of belt asteroids. They probably formed near Jupiter's orbit. Jupiter's outer satellites may even be Trojan-like asteroids that were captured by Jupiter's strong gravity into orbits around the giant planet.

A unique asteroid even further from the Sun is 2060 Chiron, discovered in 1977. It orbits the Sun, ranging from a point just inside Saturn's orbit out to a point just inside Uranus' orbit. It is estimated to be 200 km across and is the only asteroid known beyond Saturn's orbit. In 1987, as it was moving closer to the Sun, it surprised astronomers by brightening suddenly over a period of months and developing a coma. It had turned into a

---

[5]Recently I learned of a new astronomical amusement—making sentences using only asteroid names. My favorite (not my own creation, I'm sorry to say) is "Rockefellia Neva Edda McDonalda Hamburga."

---

[6]The name *Trojan* comes from the tradition of naming these particular asteroids after heroes in the Homeric epics.

Figure 13-16 Beyond the Moon, the closest planetary bodies are Apollo asteroids. Small Apollos, a few hundred meters across, may approach close enough to the Earth–Moon system (lower right) to be visited relatively easily by spaceships. **a** Astronauts transfer from a "parked" ship to the asteroid surface. Some researchers believe their materials could be economically exploited for use on or near Earth. (Painting by author.) **b** The small asteroid's gravity is so weak that astronauts could float nearby as they take geological samples. (Painting by Pamela Lee.)

a

b

comet! This shows it contains ice, which warmed as it approached the sun. Possibly Chiron and Pluto are only the biggest examples of a host of ice-rich worldlets waiting to be discovered on the outskirts of the solar system.

The Rocky Material of Asteroids

As indicated in Figure 5-14, the spectra of asteroids can be used to estimate their mineral properties. The asteroids divide into several classes with different spectra, and the spectra of these classes have been compared with spectra of minerals and rocks from Earth, the Moon, and many meteorites. The spectra of some asteroid classes closely match those of some types of meteorites. This is strong evidence that certain meteorites are fragments from certain classes of asteroids.

Astronomers discovered in the 1980s that these different asteroid compositional classes are concentrated at different distances from the Sun, so that the asteroid belt has a rough zonal structure. Thus, in the inner and central parts of the belt,[7] we find many asteroids with a rock or mixed rock-and-iron composition similar to two important classes of meteorites called chondrites and stony-irons (see the next section). In some aster-

---

[7]"Inner" asteroid belt is the portion nearest Mars.

oids, such as 1 Ceres, the largest (1020 km across), the surface materials include water-bearing minerals; the water molecules are chemically bound in the mineral grains (Lebofsky, 1981). Also in this region are some asteroids (such as 4 Vesta, the second largest, 549 km across) that did melt—they have surface rock similar to basaltic lavas of the Earth and Moon. In the outer belt are many asteroids with very black surface rock material composed of carbon-rich and water-rich minerals. These materials (also found in certain meteorites) formed in a region of the solar system colder than the Earth's environs. Still more remote asteroids, such as some of the Trojans, have reddish-black surfaces probably consisting of carbonaceous minerals colored by organic compounds. Some of the remote asteroids may also contain ices, mostly hidden by the black soil. Near or beyond Jupiter may be stony-icy asteroids that are transitions between normal asteroids and comet nuclei of the outermost solar system.

Asteroids that come near the Earth, such as Apollo asteroids, are a mixture of these various types, apparently perturbed onto Earth-crossing orbits from various regions of the solar system.

Although no asteroid has been photographed at close range, asteroids are known to have rounded and irregular shapes; the close-up appearance may resemble that of Phobos and Deimos (Figures 10-26 and 10-28).

Mines in the Sky?

Because some asteroids probably contain metals and other useful minerals, and because some come very near the Earth, asteroids may someday be economically exploited. Studies are already under way to examine the feasibility of flying to Apollo asteroids for economic and scientific exploration, as suggested by Figure 13-16. In terms of energy expenditure, some close asteroids are actually easier to reach and return from than the Moon! Asteroids of pure nickel–iron might be mined in space, brought to space stations for processing, or shaped into crude entry vehicles and landed on the Earth in remote areas. Based on meteorite samples, other Apollo asteroids are believed to contain good ores of economically important platinum-group metals. Masses worth billions of dollars—enough to supply the Earth's needs of certain metals for decades—might be obtainable (Gaffey and McCord, 1977; Hartmann, 1982; O'Leary, 1983). Other asteroid materials such as rock, ice, and hydrogen may be used for constructing and maintaining space stations.

Figure 13-17 Imaginary view of the collision of two asteroids. Fragments created by the collision may ultimately fall on the Earth and planets as meteorites. Since two asteroids are of different mineralogical types, some of the meteorites produced will be mixtures of different rock types. The larger body is about 30 km across. (Painting by author.)

Origin of Asteroids

Asteroids are probably planetesimals that never finished accumulating into a planet. Possibly disturbed by nearby Jupiter, the planetesimals in the belt region collided too fast to coalesce into a planet. Instead, they collided and broke into thousands of fragments, as depicted in Figure 13-17. Most smaller asteroids in the main belt are such fragments. In 1983, the IRAS satellite discovered distinct rings of dust circling the Sun in the asteroid belt, each interpreted as debris from a relatively recent, individual asteroid collision. Thus the collisions continue, sporadically, even today.

METEORITES

Meteorites are stony and metallic objects that fall from the sky. Modern evidence suggests that they are fragments of Apollo-type asteroids that cross the Earth's orbit and eventually collide with Earth.

Meteorites as Venerated Objects

Stones from the sky have long been a source of awe. Meteorites have been buried with American Indian artifacts (Figure 13-18), wrapped in "mummy cloth" (Casas Grandes, Mexico), and kept as a sacred possession of an Alaskan tribe. A stone venerated in the Temple of Diana at Ephesis (one of the seven wonders of the ancient world) reportedly fell from the sky and was probably a meteorite. A black stone, enshrined around A.D. 600 or earlier in the sacred Moslem shrine at the Ka'aba in Mecca is also believed to be a meteorite (Sagan, 1975). When a meteorite fell at Ensisheim, France, in 1492 (Figure 13-19), the emperor Maximilian, in residence nearby, decided he should go on a Crusade. The main mass of the meteorite, suspended in the Ensisheim church, was regarded with reverent awe by the peasants.

Scientific Discovery of Meteorites

In 1794 a German physicist, E.F.F. Chladni, reported that the Ensisheim and other supposed celestial stones seemed similar to each other and different from normal terrestrial stones. He concluded that these "meteorites did indeed fall from the sky." This conclusion started a controversy among naturalists. Upon hearing that a meteorite had fallen in Connecticut, Thomas Jefferson, himself an accomplished naturalist, is supposed to have said, "It is easier to believe that Yankee professors would lie than that stones would fall from heaven."

The French Academy, the scientific establishment of the day, dismissed as superstition the notion of stones from the sky. As luck would have it, a meteorite exploded over a French town in 1803, pelting the area with stones. Cautiously, the academy sent the noted physicist J. B. Biot to investigate. His report, one of the historic documents of science, methodically constructed an irrefutable chain of evidence from eyewitness accounts, measurements of the 2 × 6 km area of impacts, and specimens of the meteorites themselves. This report established that stones can indeed fall from the sky.

Figure 13-18 Burial of the Winona stony meteorite in a crypt constructed by prehistoric Indians in northern Arizona. The Indians probably saw it fall and thus attached special significance to it. (Museum of Northern Arizona, Anthropological Collections; photo by author.)

Meteorite Impacts on Earth

Interplanetary debris collides with the Earth at very high speeds, usually 11 to 60 km/s (24 000 to 134 000 mph). At such speeds, material is heated by friction with the air. Dust grains and pea-sized pieces burn up before striking the ground. Larger pieces are usually slowed by drag, although at least two grapefruit-scale specimens have punched through house roofs, one bruising a woman occupant! Since they pass through the atmosphere too fast for their interiors to be strongly heated, stories of meteorites remaining red-hot for hours after falling are untrue.

Large meteorites are rare. Only a few brick-sized meteorites are recovered each year. In 1972 an object weighing perhaps 1000 tons just missed the Earth, skipping off the outer atmosphere; it was filmed from the ground and detected by Air Force reconnaissance satellites. Objects weighing 10 000 tons, like the Siberian object of 1908, which are large enough to cause nuclear-scale blasts, fall every few centuries. Larger blasts, thousands of years apart, may form multikilometer-scale craters, such as that in Figure 6-17.

Since meteorites are "free samples" of planetary matter, recovering and reporting a meteorite is a rare

Figure 13-19 Halley's comet on Bayeux tapestry, events of 1066; the words say "They marvel at the star." (The Bettmann Archive.)

honor. Meteorite discoveries should be presented, or at least lent, to research institutions—for example, the Smithsonian Institution in Cambridge, Massachusetts, and Washington, D.C., and the Center for Meteorite Studies, Arizona State University, Tempe.

Types and Origin of Meteorites

Meteorites are among the most complex rocks studied by geologists. There are many different types, and sometimes they appear as **brecciated meteorites**—meteorites made of mixed fragments of different meteorite types all jumbled into one rock! After decades of study, scientists have begun to understand how meteorites formed and what they have to tell us about the ancient history of the solar system. The key to the story is that meteorites are fragments of asteroids (and sometimes mixtures of fragments of dissimilar asteroids), created when asteroids collided and blew apart at various times during the history of the solar system. Their fragments were ejected in various directions, often getting thrown into orbits that eventually intersected the Earth's.

The different types of meteorites thus tell us about conditions inside the asteroidal parent bodies. Some are samples of surface rock layers, and others are samples of deep cores inside these bodies. Table 13-3 summarizes the major types and their abundances. The first two types are called **chondrites** (KON-drites) because they contain peculiar BB-like glassy spherules called **chondrules** (from the Greek term for seedgrains), as shown in Figure 13-20. These chondrules are believed to be spherules that solidified from molten droplets sprayed out during planetesimal collisions as the planets formed. The preservation of the chondrules in these meteorites proves that they were never melted or severely altered since their formation long ago. This is also shown in the case of **carbonaceous chondrites,** blackish meteorites with water chemically bound in their minerals; this water can be driven off by mild heating to only a few hundred degrees. The carbonaceous chondrites were probably formed in the cold, outermost asteroid belt or outer solar system where ices were abundant. They are undoubtedly fragments of the blackish carbonaceous types of asteroids found in these regions.

Achondrite meteorites, on the other hand, are rocky meteorites with no chondrules. They resemble

TABLE 13·3

Types of Meteorites

Meteorite Type	Percentage of All Falls[a]	Remarks
Stony		
Carbonaceous chondrite	5.7	Most primitive, least altered material available from early solar system.
Chondrite	80.0	Commonest type. Defined by millimeter-scale spherical silicate inclusions, sometimes glassy, called *chondrules*.
Achondrite	7.1	Most nearly like terrestrial rocks. Defined by lack of chondrules, which have been destroyed by a heating process. Some resemblance to terrestrial lunar igneous rocks of basaltic type.
Stony–iron	1.5	Contain stony and metallic sections in contact with each other.
Iron	5.7	Nickel–iron material. Museum specimens are often cut, polished, and etched to show interlocking crystal structure.
Total	100	—

[a]This table is based on meteorites called *falls*—those actually seen to fall. Meteorites found by chance in the soil, called *finds,* are more numerous but less valuable statistically because they are biased toward iron meteorites, which attract attention whenever found in the ground, while stony meteorites eventually weather to resemble ordinary stones.

basaltic lavas found on the Earth, Moon, and Mars and are fragments of regions that were once melted and then resolidified inside or on parent asteroids.

Stony–iron meteorites (mixtures of stone and iron alloy) and **iron meteorites** (pure iron alloy metal) are probably samples from the central regions of asteroids that once melted, probably due to heat-producing radioactive minerals incorporated in them when they formed (see Figures 13-21 and 13-22). As they melted, the heavy molten metals sank to the center while the lighter silicate magmas floated. As the asteroids later cooled, the central regions formed nickel–iron cores (like that believed to exist in the Earth) while the surface regions formed rocky lithospheres. When the asteroids were smashed in collisions, fragments of the metal regions, or mixed fragments of metal and achondrite rock, were released.

Studies of minerals in several types of meteorites show that they formed inside parent bodies with diameters of tens to hundreds of kilometers—fitting the theory that they formed inside asteroids.

Ages of Meteorites

One of the most important aspects of meteorites is that they can be dated by the techniques discussed in Chapter 6, and this gives us a history of events in the early solar system. Dating by various techniques shows that the most primitive meteorites formed during a "brief" interval of only 20 million years, 4.6 billion years ago. "Formed" means that they made the transition from dispersed dust grain and gas to solid, rocklike objects. This interval, then, marked the birth of planetary material. Other types of dating show that many major colli-

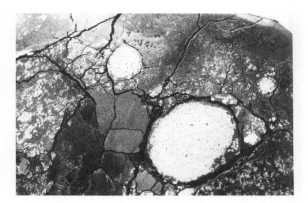

Figure 13-21 Cut and polished section of a stony–iron meteorite from the Bondoc Peninsula, Philippines. The dark fractured matrix is silicate rock, possibly fractured in collisions. The bright nodules are nickel–iron nuggets a few centimeters across. (Photo by author.)

Figure 13-20 A 3-cm portion of the chondrite meteorite Bjurbole showing many chondrules of various sizes, including a large one protruding at lower left. (Center for Meteorite Studies, Arizona State University.)

Figure 13-22 A cut and polished section of the Campo del Cielo (Argentina) iron meteorite. Intersecting metal crystals can be studied to reveal environmental conditions inside the parent body. The dark inclusions are stony silicate bodies; the largest is about 4.5 cm across. This pattern of metal crystals is called the Widmanstätten pattern, after its discoverer. (Photo by author.)

sions happened at this time, 4.6 billion years ago, and sporadic additional collisions smashed some meteorites' parent bodies at other times scattered through solar system history. We will discuss some of these important events in more detail in the next chapter.

ZODIACAL LIGHT

The smallest interplanetary particles are microscopic dust grains and individual molecules and atoms spread out along the plane of the solar system and concentrated toward the Sun. If you look west in a very clear rural sky as the last glow of evening twilight disappears (or look east before sunrise), you can detect the diffuse glow of sunlight reflecting off the cloud of these particles. It appears as the **zodiacal light**—a faint glowing band of light extending up from the horizon and along

the ecliptic plane, shown in Figure 13-23. It is brightest at the horizon. Measurements made by astronauts indicate that it merges with the bright glow of the Sun's atmosphere.

The zodiacal light can be thought of as the visible effect of countless dust grains—once distributed in trails along comets' orbits but eventually spread out into a uniform disk-shaped cloud of dust around the Sun.

Figure 13-23 Sunset and the emergence of the zodiacal light. **a** Sunset. **b** Blue color
still dominates the sky $\frac{1}{2}$ h later and obliterates the faint zodiacal light. The crescent moon,
Venus, and Jupiter, in order up from the horizon, are emerging and define the position of
the zodiac. **c** The zodiacal light begins to be visible along the zodiac 1 h after sunset.
d The zodiacal light reaches its best visibility about $1\frac{1}{2}$ h after sunset, appearing as a
diffuse band of light rising nearly vertically from the horizon and running near Venus and
Jupiter. **e** and **f** The zodiacal light slowly sets 2 h and $2\frac{1}{2}$ h after sunset, respectively.
Features in these photos can be seen only away from cities under dark, clear skies.
(Exposures *a–b* on Kodachrome 64, *c–f* on 3M ASA 1000 film; 28-mm wide-angle lens
with nearly 90° vertical field of view; exposures *c–f* 5 min at f 2.8. (Photos by author from
Mauna Kea Observatory.)

Figure 13-24 Devastation caused by the great Tunguska meteorite fall of 1908. This view, more than a decade later, shows trees some miles from "ground zero" blown over by the force of the explosion. (Photo from E. L. Krinov, 1966.)

SIBERIA REVISITED: ASTEROID OR COMET IMPACT?

We now find the mysterious Siberian explosion of 1908, which we described at the beginning of this chapter, easier to understand. Interplanetary space contains debris of many sizes. Rocky and icy chunks up to a few kilometers in size cross the Earth's orbit and are likely to collide with the Earth from time to time. One of these objects apparently struck the atmosphere over Siberia on June 30, 1908.

Other dramatic events in Russia at this time kept Russian scientists from visiting the site until 1927. During the 1927 expedition and later expeditions, scientists found that, surprisingly, the object did not reach the ground and form a crater. It apparently exploded in the atmosphere. Trees at "ground zero" were still standing but had their branches stripped by the blast forces in a downward direction. Trees had been knocked over by the blast out to 30 km from ground zero, as shown in Figure 13-24. A forest fire was started and trees were scorched by the blast out to about 14 km. Researchers found carbonaceous chondrite dust but no meteorites.

Most investigators believe the object was composed of weak material that disintegrated in the atmosphere. It may have been a small asteroid resembling a stony meteorite some 90 to 200 m across (Sekanina, 1983) or a similar-sized icy comet fragment. It injected much dust high into the stratosphere. On June 30 and July 1, sunlight shining over the North Pole illuminated this dust during the night, and newspapers could be read at midnight in western Siberia and Europe.

Such was the effect of a relatively small bit of interplanetary debris striking the Earth. To put this event in perspective, if the same object had exploded over New York City, the scorched area would have reached nearly to Newark, New Jersey. Trees would have been felled beyond Newark and over a third of Long Island. The man knocked off his porch could have been in suburban Philadelphia. "Deafening bangs" might have been heard in Pittsburgh, Washington, D.C., and Montreal.

New reconnaissance satellites may provide early warning against such objects, which may hit the Earth every century or so, allowing us to take action or even intercept such objects to use their materials for more constructive purposes.

An interesting sidelight on large ancient impacts is that the distortions of geological strata on the Earth have been economically valuable in some cases. Rich iron ore deposits at Sudbury, Ontario, occur in a probable impact crater more than 100 km across and about 1.8 billion years old; the impact exposed native iron-bearing strata. Several oil deposits in North America and the Soviet Union occur in sediments collected in ancient craters.

SUMMARY

All of the solar system's small bodies—comets, meteors, asteroids, and meteorites—are examples of the planetesimals or their fragments. Planetesimals were the preplanetary bodies that formed in the solar system 4.6 billion years ago, during an interval of less than 100 million years.

Most asteroids are now in a belt between Mars and Jupiter; the planetesimals in other regions interacted with planets and either were ejected from the solar system after near misses or crashed onto planet and satellite surfaces, making craters. Asteroids in the inner belt are dominated by metal-rich rock materials; asteroids from the outer belt to Jupiter have carbon- and water-rich minerals. Planetesimals formed farther out were ice rich and evolved to what we know as comets. The relation between the various asteroids, comets, meteors, meteorites, and craters that we see today is summarized in Figure 13-25.

The most detailed information about these bodies comes from meteorites, whose chemistry reveals that they came

Origin   Evolution   Today

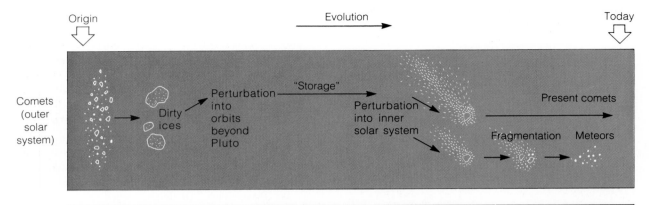

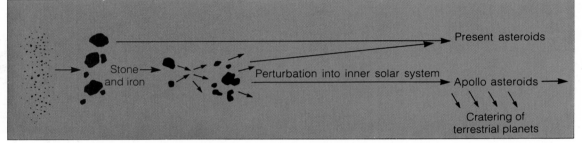

4.6 billion years ago

Figure 13-25 Schematic histories of cometary and asteroidal material, showing condensation into multikilometer bodies (left) and subsequent perturbation and fragmentation to explain phenomena now observable.

from asteroidlike parent bodies a few hundred kilometers across. Many of these were melted, causing minerals to separate into iron and silicate phases. Others survive with only minimal heating, providing samples nearly unaltered since the solar system's earliest days.

Thus the small bodies of the solar system provide some of the best clues about the origin of planets, both in our solar system and perhaps near other stars. Recent studies also indicate that they may be exploited as a source of raw materials for use on the Earth and for space exploration.

Oort cloud	Lagrangian points
short-period comet	brecciated meteorite
dirty iceberg model	chondrite
meteor shower	chondrule
fireball	carbonaceous chondrite
dust trails	achondrite
asteroid belt	stony–iron meteorite
Apollo asteroids	iron meteorite
Trojan asteroids	zodiacal light

CONCEPTS

comet	planetesimal
asteroid	comet head
meteoroid	comet tail
meteor	comet nucleus
meteorite	solar wind
parent body	Halley's comet

PROBLEMS

1. If a comet should happen to pass through Saturn's satellite system, why would it probably not be detected from Earth?

2. Kepler's third law states that $a^3 = P^2$, where a is the semimajor axis of a body orbiting the Sun (expressed in astronomical units) and P is the period (expressed in years). Should this result apply to comets? If a typical comet in Oort's cloud has a semimajor axis of about 100 000 AU (10^5 AU), how often would it return to the inner solar system?

3. In terms of measuring and reporting useful scientific information, what actions would be appropriate if you observed an extraordinarily bright meteor or fireball? In two columns, list examples of useful and nonuseful descriptions of the fireball's speed, brightness, and apparent size. What actions would be appropriate if you saw a meteorite strike the ground?

4. Summarize relations between the bodies and particles responsible for meteors, comets, the zodiacal light, asteroids, and meteorites.

5. Suppose future astronauts could match orbits with a comet and reach its nucleus. Describe the possible surface appearance of a comet. Consider gravity, surface materials, sky appearance, and so on. Which would be easier to match orbits with: a long-period or short-period comet?

6. Typical interplanetary material may move at about 15 km/s relative to the Earth–Moon system.

    a. If a kilometer-scale asteroid was discovered on a collision course with the Earth when it was 15 million kilometers away, how much warning time would we have?

    b. What would be the potential dangers?

    c. If a much smaller asteroid was similarly discovered at the distance of the Moon, how much warning time would we have?

    d. How long would the objects take to pass through the 100-km thickness of the atmosphere?

7. Based on everyday experience, what is the danger of an event such as that in Problem 6 compared with the danger of other natural disasters, such as earthquakes? Is a large-scale meteorite disaster a plausible source of myths during the 10 000-year history of humanity? Defend your answer.

ADVANCED PROBLEMS

8. Suppose a comet on a parabolic orbit is closest to the Sun at a point very near the Earth and moves in the same direction as the Earth. How fast does it move (in km/s) relative to the Earth?

9. Use the small-angle equation to calculate how close a 2-km-diameter asteroid would have to come to Earth to allow its shape to be resolved by a telescope that can reveal angular details 1 second of arc across. How does this distance compare with the Moon's distance?

10. Suppose a small asteroid of pure nickel–iron, with radius $r = 100$ m, could be located and exploited. If the value of the alloy were 90¢ per kilogram, what would be the potential economic value of the asteroid? Compare this with the $20 billion dollar cost of the Apollo program. Assume that the density of the material is 8000 kg/m^3.

11. How many asteroids would be required to provide enough material to make one planet the size of Earth? Assume a typical asteroid is 120 km across and the Earth is about 12 000 km across.

PROJECTS

1. Use a large piece of cardboard to make a model of the inner solar system out to the orbit of Jupiter. Assume that the planets travel approximately in the plane of the cardboard. Use orbital properties listed in Table 13-1 to cut out scale models of the orbits of various interplanetary bodies. (A slit through the first cardboard could be used to show how comet or asteroid orbits penetrate through the ecliptic plane. Note that the Sun must always occupy one focus of each orbit.)

    a. Show how the geometry of passage of a comet (or other body) through the ecliptic plane, especially for highly eccentric orbits, depends on the angle between the perihelion point and the ecliptic plane, measured in the orbit. (This angle is fixed for each body but is omitted from Table 13-1 for simplicity.)

    b. Show how the prominence of a given comet may depend strongly on where the Earth is in its orbit as the comet passes through the inner solar system.

2. Visit Meteor Crater, Arizona. Why is this feature misnamed? Observe the blocks of ejecta and deformation of rock strata, as explained in museum signs and tapes. What would prehistoric observers, if any (estimated impact date was 20 000 y ago), have witnessed at various distances from the blast that formed this crater nearly a mile across? (Other impact sites are known in various states, but they are eroded or undeveloped.)

3. Examine meteorite specimens in a local museum. Compare the appearance of stones and irons. Are chondrules visible in any stones? Heat damage? In iron samples that have been cut, etched, and polished, look for the crystal patterns that give information about cooling rate and environment when the meteorite formed inside its parent body.

CHAPTER 14

The Origin of the Solar System

The planets, satellites, meteoritic material, and Sun did not exist before about 4.6 billion years ago. Samples from the Earth, Moon, and meteorites suggest that the solar system formed from preexisting material within an interval of about 20 to 100 million years. Prior to that time, the atoms in the earth, in this book, and in your own body were floating in clouds of thin gas in interstellar space. It is remarkable to realize that this interstellar material somehow aggregated not only into a dazzling star but also into a family of surrounding planets, as symbolized in Figure 14-1. How did the Sun and its planetary system form?[1]

FACTS TO BE EXPLAINED BY A THEORY OF ORIGIN

Nobel laureate Hannes Alfvén, who has spent years researching the solar system's origin, once said: "To trace the origin of the solar system is archaeology, not physics." He meant that our ignorance of the initial conditions forces us to work backward through time, reasoning from whatever clues we can find. The most important clues are *facts about the solar system that have no obvious explanation from present-day conditions* but must have arisen from initial conditions as the solar system formed. Table 14-1 lists some of these clues. In this chapter, we will account for them one by one. This process leads to an interesting realization. As we sift through clues found in meteorites, lunar rocks, and orbits in our own solar system, we will learn about the formation of not only our own system but also the stars themselves. Thus this chapter makes a link to the ensuing chapters.

[1]Note that this question is not addressing the origin of the whole universe, which probably occurred around 14 billion years ago, or the origin of our galaxy, which probably occured around 12–14 billion years ago. These topics will be taken up in later chapters.

Figure 14-1 Earth, almost lost in the glare of its star, symbolizes the question of how planets formed around the Sun. (NASA photo from Apollo 14 on the way to the Moon.)

TABLE 14-1

Solar System Characteristics to Be Explained by a Theory of Origin

1. All the planets' orbits lie roughly in a single plane.
2. The Sun's rotational equator lies nearly in this plane.
3. Planetary orbits are nearly circular.
4. The planets and the Sun all revolve in the same west-to-east direction, called prograde (or direct) revolution.
5. Planets differ in composition.
6. The composition of planets varies roughly with distance from the Sun: Dense, metal-rich planets lie in the inner system, whereas giant, hydrogen-rich planets lie in the outer system.
7. Meteorites differ in chemical and geological properties from all known planetary and lunar rocks.
8. The Sun and all the planets except Venus and Uranus rotate on their axis in the same direction (prograde rotation) as well. Obliquity (tilt between equatorial and orbital planes) is generally small.
9. Planets and most asteroids rotate with rather similar periods, about 5 to 10 h, unless obvious tidal forces slow them (as in the Earth's case).
10. Distances between planets usually obey the simple Bode's rule.
11. Planet–satellite systems resemble the solar system.
12. As a group, comets' orbits define a large, almost spherical cloud around the solar system.
13. The planets have much more angular momentum (a measure relating orbital speed, size, and mass) than the Sun. (Failure to explain this was the great flaw of the early evolutionary theories.)

CATASTROPHIC VERSUS EVOLUTIONARY THEORIES

Throughout history, theories about the origins of things have ranged from so-called catastrophic theories to evolutionary ones. **Catastrophic theories** tend to explain things by sudden events, rare accidents, unusual circumstances, or even the intervention of a Creator whose actions happened once and then ceased. **Evolutionary theories** tend to explain things as relatively gradual developments from relatively normal earlier conditions. For example, catastrophists (especially in the 1700s) argued that the Earth's mountains were created in violent, sudden upheavals and that all biological species were created at once. Most modern naturalists, however, refute this view with overwhelming evidence derived from rock ages, fossils, and observations of ongoing changes in mountains and species today. This evidence indicates that mountains and species evolved over very long periods by processes that included many slow changes as well as occasional rapid changes. Modern data strongly indicate that many natural systems, such as clouds, ocean basins, planets, stars, galaxies, and biological species evolved.

This viewpoint does not deny sudden catastrophic events, however. The physical matter of the universe seems to have come into being in such an event—the "big bang" discussed in Chapter 27. On a more provincial scale, the amazing change in the Earth's climate that resulted in the extinction of many terrestrial species 65 million years ago (see Chapter 6) probably involved a sudden asteroidal impact.

Because of the controversy accompanying the acceptance of Darwin's theory of the evolution of species, many people think that science defends only evolutionary theories. Many state legislators, scientists, and theologians have become embroiled in controversies over this subject. Even today proposals have been made that any presentation of evolutionary theories (at least in biology) be accompanied by an "equal-time" presentation of a catastrophist (sometimes called creationist) theory. Usually the creationist alternative is drawn from a fundamentalist interpretation of Judeo-Christian tradition with 3000-y-old roots in the Middle East, but other alternatives could be proposed. Advocates of such proposals don't seem to realize that science doesn't work by the advocacy system, such as that used by two lawyers paid to defend two prechosen positions. Instead, it works by laying all available *evidence* on the table and then seeing if any picture emerges on which all observers can agree. Scientists gain the regard of their colleagues not so much by "winning" arguments as by presenting new pieces of evidence. The evidence, in a sense, argues its own case.

In this chapter we will show how *evidence,* much of it from the preceding chapters, is used to reconstruct the ancient processes that formed planets and explain the facts listed in Table 14-1.

Historically, both catastrophic and evolutionary theories of the solar system's origin have been proposed. The first evolutionary theory was advanced by the French philosopher René Descartes in 1644. He assumed that regardless of how matter was created "in the beginning," it was free to evolve according to its incorporated laws of nature. He proposed that space was initially filled with swirling gas in which local eddies, or dense regions, evolved into individual stars. Smaller eddies around them made planets.

The first catastrophic theory was French, as well. The naturalist Georges Buffon suggested in about 1745 that the Sun had been accidentally hit by a passing star and the debris formed planets. Calculations today show that such collisions would be rare. As we will see, the evidence favors an evolutionary view for formation of planetary systems around stars.

THE PROTOSUN

Before the Sun existed, its material must have been distributed in interstellar space in a large cloud like the interstellar clouds we see today. The Sun began to form a little more than 4.6 billion years ago when this cloud (or part of it) became so dense that its own inward-pulling gravitational forces became stronger than the outward force of the pressure generated by the motions of its atoms and molecules. The cloud thus began to shrink.

Early Contraction and Flattening

Even if gas in the cloud had been randomly circulating initially, the contracting cloud would have developed a net rotation in one direction or another. To show this, an experiment can be done with a cup of coffee or a pan of water. If you stir the liquid vigorously but as randomly as you can and then wait a moment and put a drop of cream in it, the cream will usually reveal a smooth rotation in one direction or the other, because the sum of the random motions will usually be a small net angular motion in one direction or the other. Because the interstellar cloud was relatively isolated from the rest of the universe (as such clouds are today), this rotating motion, or angular momentum, could not be easily transferred anywhere and thus was *conserved* as the cloud shrank. (The term *angular momentum* refers to a quantity used by physicists to measure the total rotary motion of a system—it increases with the amount of rotating material and with the speed and diameter of the rotating system.)

This principle of **conservation of angular momentum** predicts that the cloud would have rotated faster as its mass contracted toward its center, just as a figure skater spins faster when she pulls in her arms. Mathematical studies show that centrifugal effects associated with the faster spin caused the outer parts of the cloud to flatten into a disk, while material in the center contracted fastest, forming the Sun, as shown in Figure 14-2. The planets formed in this disk, accounting for fact 1 in Table 14-1: *the single plane of all planets' orbits.* Because the Sun itself was an integral part of this disk, fact 2, *the Sun's rotation in the same plane,* is also explained.

At first, when the cloud was large, its atoms were far apart and could have fallen more or less freely toward the center of gravity in the middle of the cloud. This state is called **free-fall contraction.** If free-fall persisted, the cloud could have completely contracted into the Sun in only a few thousand years. But eventually, because of their growing concentration and random motions, the atoms in the cloud began to collide with each other. Atoms interacting in a gas cause outward *pressure.* A buildup of pressure in the cloud would have slowed the collapse. Thus the inward force of gravity and the outward force of gas pressure competed. Because the properties of gases are well understood, astrophysicists can analyze the later contraction and evolution of the cloud under these competing forces.

Helmholtz Contraction

Gravitational contraction in which the shrinkage is slowed by outward pressure is called **Helmholtz contraction** (after Hermann von Helmholtz, the German astrophysicist who first studied it). In 1871, Helmholtz showed

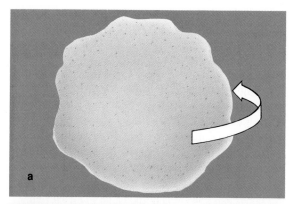
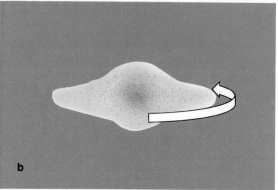
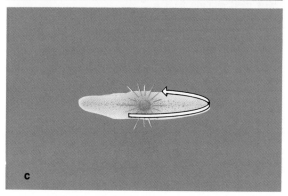

Figure 14-2 Three stages in the evolution of the protosun. **a** A slowly rotating interstellar gas cloud begins to contract because of its own gravity. **b** A central condensation forms and the cloud rotates faster and flattens. **c** The Sun forms in the cloud center, surrounded by a rotating disk of gas.

how contraction would have caused heat to accumulate in the contracting protosolar cloud. Helmholtz noted the following (quoted in Shapley and Howarth, 1929):

If a weight falls from a height and strikes the ground, its mass loses . . . the visible motion which it had as a

*whole—in fact, however, this is not lost; it is trans-
ferred to the smallest elementary particles of the mass,
and this invisible vibration of the molecules is [what we
call] heat.*

In the contracting protosun, atoms or swarms of atoms
would have fallen toward the center until they collided
with other parts of the gas cloud. Temperature would
have increased inside the cloud. Wien's law and other
physical laws guarantee that the protosun would have
radiated energy and warmed the surrounding cloud.

Calculations based on the Helmholtz theory indicate
the conditions: Eventually the cloud's central temperature rose to 10 million Kelvin or more, starting the
nuclear reactions that made it a star and not just a ball
of inert gas. Meanwhile, and more important for the
formation of planets, the outer parts of the cloud, shown
in Figure 14-2, formed a disk of gas as big as the solar
system, at temperatures of a few thousand Kelvin. *These
theorized steps have been confirmed by actual observations
of newly formed stars surrounded by clouds of gas and
dust.*

THE SOLAR NEBULA

A cloud of gas and dust in space is called a **nebula**
(plural: *nebulae*) from the Latin term for mist. The disk-
shaped nebula that surrounded the contracting Sun is
called the **solar nebula.** Molecules of gas or grains of
dust must have moved in circular orbits, because noncircular orbits would have crossed the paths of other
particles, leading to collisions that would have damped
out the noncircular motions—in the same way that noncircular eddies get damped, or canceled, in the cup of
coffee mentioned earlier. Thus, neglecting small-scale
eddies in the gas, broad-scale motions in the cloud were
in parallel circular orbits, accounting for facts 3 and 4
in Table 14-1. As the nebula stabilized, its gas began to
cool.

Condensation of Dust in
the Solar Nebula

The solar nebula was initially gas that Helmholtz contraction heated to at least 2000 K. At such a temperature, virtually all elements were in gaseous form. As is
the case with other cosmic gas, most solar nebula atoms
were hydrogen, but a few percent were heavier atoms

such as silicon, iron, and other planet-forming material.

How did solid particles form in this gas? The answer
can be seen on the Earth. When air masses cool, their
condensable constituents form particles: snowflakes,
raindrops, hailstones, or the ice crystals in cirrus clouds.
Similarly, as the solar nebula cooled, condensable constituents formed tiny solid particles of dust. Various
mineral compounds appeared in a sequence known as
the **condensation sequence.**

Chemical studies show that as the temperature in
any part of the nebula dropped toward 1600 K, certain
metallic elements such as aluminum and titanium condensed to form metallic oxides in the form of **grains,**
or microscopic solid particles, as shown in Table 14-2.
At about 1400 K a more important constituent, iron,
condensed. Microscopic bits of nickel–iron alloy formed
as grains or perhaps coated existing grains. Still more
important, at about 1300 K abundant silicates began to
appear in solid form. For instance, a magnesium silicate
mineral, enstatite ($MgSiO_3$), formed at about 1200 K
(Lewis, 1974). These silicate minerals are the common
rock-forming materials, so the solar nebula at this point
was acquiring a large quantity of fine dust with rocky
composition.

In the outer, colder regions of the nebula, some
grains that formed in interstellar space before the solar
nebula existed may have survived. These could have
been mixed with the newly condensing grains. This may
explain some unusual grains that chemists have found
in ancient meteorites.

Complex mixtures of magnesium-, calcium-, and iron-
rich silicates condensed, depending on the temperature, pressure, and composition of the gas at various
points in the solar nebula. Since local conditions were
determined by the distance from the newly formed Sun,
different compositions of mineral particles may have
dominated at different locations, explaining facts 5 and
6 in Table 14-1. Exploration of the planets has helped
clarify conditions in the nebula. For example, comparison of Pioneer probe results on the atmosphere of Venus
in 1979 with earlier Viking results on Mars shows that
the amount of primordial argon gas incorporated from
the solar nebula into the planets decreased by a few
hundred times from Venus, past the Earth, to Mars.
This suggests that the nebular gas further out from the
Sun had lower pressure (Pollack and Black, 1979). Similarly, the outer nebula must have been cooler than the
inner nebula, being further from the Sun and more shaded
by the inner dusty nebula (Cameron, 1975). In these

Figure 14-3 A piece of the Allende carbonaceous chondrite, showing white inclusions. These demonstrate how material that formed in one environment was later trapped in other material. The black matrix is composed of microscopic dust grains condensed at a few hundred Kelvin. The white inclusions contain aluminum-rich minerals condensed at high temperatures. As recognized in the 1970s, the inclusions shed light on the earliest formative conditions in the solar system. (Courtesy R. S. Clarke, Smithsonian Institution.)

1   5
cm

TABLE 14-2

Condensation Sequence in the Solar Nebula

Approximate Temperature (K)	Element Condensing	Form of Condensate (with Examples)	Comments
2000	None		Gaseous nebula
1600	Al, Ti, Ca	Oxides (Al_2O_3, CaO)	
1400	Fe, Ni	Nickel–iron grains	Parent material of planetary cores, iron meteorites?
1300	Si	Silicate and ferrosilicate minerals [enstatite, $MgSiO_3$; pyroxene, $CaMgSi_2O_6$; olivine, $(Mg, Fe)_2 SiO_4$] in form of microscopic grains	First stony material, combined to form meteorites; some still preserved in primitive meteorites
300 ↓ 100	H, N, C	Ice particles (water, H_2O; ammonia, NH_3; methane, CH_4)	Large amounts of ice; still preserved in outer planets and comets

Source: Data from Lewis (1974), Grossman (1975), and others.

regions, at about 300 K, water was trapped in many minerals. Beyond the asteroid belt, snowflakes of H_2O ice condensed. At 100 to 200 K, in the outermost nebula, ammonia and methane ices also condensed. In the outer solar system, these ices survive even direct sunlight; they survive today on comets and icy satellites of the giant planets.

Meteorites as Evidence

The compositions of meteorites strongly support this theory of condensation. Carbonaceous chondrites contain inclusions of material believed to be among the earliest solid particles in the solar system. These inclusions, shown in Figure 14-3, are rich in elements that

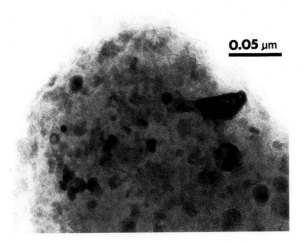

0.05 μm

Figure 14-4 A bit of interplanetary dust at much higher magnification than in Figure 13-14. In this transmission electron microscope image, larger dark blobs are mineral grains, probably magnetite (Fe_3O_4) or iron–nickel sulfide. They are surrounded by an extremely fine-grained carbon-rich matrix, probably consisting of amorphous carbon particles. Researchers believe this fine-grained material may be the original condensates formed in the solar nebula or formed in interstellar clouds and then trapped in the solar nebula. This microscopic structure supports the theory that tiny mineral grains condensed and then accreted into larger bodies as planetary material formed in the dusty nebula around the primordial Sun. (Courtesy Roy Christoffersen and Peter Buseck, Arizona State University.)

would have condensed first (at the highest temperatures), such as osmium and tungsten. Their minerals formed at temperatures of about 1450–1840 K, consistent with fact 7 in Table 14-1. Yet the bulk of carbonaceous chondrites contain microscopic grains formed at low temperatures, showing that different kinds of primeval dust aggregated into individual bodies (Figure 14-4). Water is a common constituent in minerals in parts of carbonaceous chondrites. Also as predicted by the theory, enstatite is a common mineral in meteorites.

A PRESOLAR EXPLOSION?

As meteorite compositions were being studied in the 1970s, an interesting mystery emerged. Light-colored inclusions of high-temperature minerals in certain carbonaceous chondrites, pictured in Figure 14-3, were found to have queer abundances of certain isotopes.

Most elements occur in one main, stable isotopic form and several other forms, both unstable (radioactive) and stable. In the light-colored inclusions, researchers found certain isotopes that could only have been created very shortly before the solar system itself.

For instance, they found xenon-129, a form of xenon that arises from the decay of radioactive iodine-129. This decay process is very fast, geologically speaking, once the iodine is created—iodine-129's half-life is only 17 million years. Therefore, the parent iodine must have been trapped in the meteorite inclusions in the brief interval (1 to 20 million years?) between the creation of the iodine and its decay into xenon. Remember that the inclusions were among the first minerals condensed in the solar nebula, and they are the main ones that got a dose of the mysterious iodine.

The mystery is how a batch of radioactive iodine and other isotopes was created just *before* the first planetary material formed and how it got trapped in that material. The types of isotopes involved are now believed to be created by nuclear reactions inside certain massive stars, which burn their nuclear fuel quickly and then explode. Thus the evidence indicates that such a star was located near the presolar nebula and exploded, spewing short-lived radioactive isotopes and other debris into the cloud that was becoming the solar nebula. Indeed, the blast from the explosion probably helped compress the presolar cloud, helping to initiate the collapse that produced the Sun!

This may sound like a very unlikely catastrophic theory—a giant stellar explosion close enough to play a chance role in forming the Sun. Not so. As will become clearer in Chapter 18, stars form in close groups, that usually include massive exploding stars. Explosions initiating new sequences of gas cloud compression, star formation, and silicate dust condensation now seem to be a normal aspect of the slow, somewhat sporadic evolution of matter in our galaxy over time spans of billions of years (Herbst and Assousa, 1979).

FROM PLANETESIMALS TO PLANETS

Although the preplanetary particles may have formed as microscopic grains, they clearly grew bigger. (Otherwise there would be no planets!) The hypothetical intermediate bodies, from millimeters to many kilometers in size, are usually called **planetesimals,** as noted in

Chapter 13. Evidence that they existed includes the following:

1. Craters on planets and satellites indicate impacts of planetesimals with diameters of at least 100 km (Figure 14-5).

2. Meteorites are their surviving fragments, and their microstructures reveal how the dust grains clumped together.

3. Asteroids and comet nuclei reaching more than 100 km in diameter survived until today throughout the solar system.

But exactly how did microscopic grains aggregate to produce 100-km-sized planetesimals? If they had circled the Sun in paths comparable to present asteroidal and cometary orbits, they would have collided with one another at speeds much faster than rifle bullets. They would have shattered (as seen in Figure 13-17) and the solar system would still be a nebula of dust and grit.

But did dust particles have such high speeds in the early solar nebula? According to dynamical analyses, they collected in a swarm of particles with nearly parallel, circular orbits in the central plane of the disk. Because the orbits were nearly parallel, planetesimals approached each other gently and collision velocities were low. In low-velocity collisions, some dust grains simply stuck together, held by weak adhesive forces such as gravity and electrostatic attraction. Mutual gravity probably caused groups of planetesimals to clump together, growing to multikilometer size in only a few thousand years—a mere moment in cosmic time. At this point, the solar nebula would have resembled the scene in Figure 14-6, with rocky planetesimals orbiting in dusty clouds that dimmed the central Sun.

What did the newly forming planetesimals look like? As they grew from diameters of meters to diameters of hundreds of kilometers, they probably resembled asteroids—perhaps with irregular shapes due to mergers of bodies and fractures due to collisions, as seen in Figure 14-7.

Dynamical studies indicate that coalescing particles tend to form bodies rotating in a prograde motion (fact 8, Table 14-1) with similar rotation periods (fact 9). No one is sure why the planetary spacings are regular (fact 10), but studies suggest that gravitational forces tended to divide the solar nebula into ring-shaped zones, each favoring the formation of one planet.

The larger planetesimals collided with smaller ones,

knocking off debris. Some small particles fell back on their surfaces, forming a powdery soil layer that, like the lunar regolith, was effective in trapping other small fragments. Thus the largest planetesimals grew fastest, sweeping up the others. These bodies may have grown to 100-km size in a few million years. Some planetesimals, the *parent bodies* of meteorites, were heated, melted, and differentiated into metal and rock portions. Some were shattered by collisions with other large or fast neighbors, freeing iron, stony–iron, and achondrite meteorites from their interiors, as shown in Figure 13-17. As some shattered, others grew to replace them.

This scenario of gradual growth through collisions allows us to explain the major classes of bodies in the solar system—the terrestrial planets, giant planets, asteroids, and comets—as follows.

Terrestrial Planets

In the inner part of the solar system, this process simply continued until Mercury-sized to Earth-sized planets were formed. Most of the planetesimals in this region were composed of silicate rocky material, familiar to us from the rocks of the Earth and Moon. Thus all the terrestrial planets grew from this material until most of it was swept up; some remaining gas and dust was blown away by outrushing gas and radiation from the newly heated sun.

Giant Planets

The giant planets started out forming in the same way as the terrestrial planets, accreting planetesimals. More material was apparently available in the giant planet zone, however, perhaps because ices as well as rocky material had condensed in this cold region, augmenting local planetesimal masses. Thus the "embryo planets" that were to become Jupiter, Saturn, Uranus, and Neptune grew to Earth's size and beyond.

By the time they reached about 15 times the mass of present-day Earth, an interesting thing happened. At 15 Earth masses, they had such strong gravity that they began to pull in gas from the surrounding solar nebula. Thus they accreted not only planetesimals to make a solid/liquid planet, but they also accreted massive atmospheres of gas whose composition approximately equaled that of the nebular gas. Hence the giant planets can be thought of as two-phase planets. Their cores are "giant terrestrial planets" averaging around 15 times more

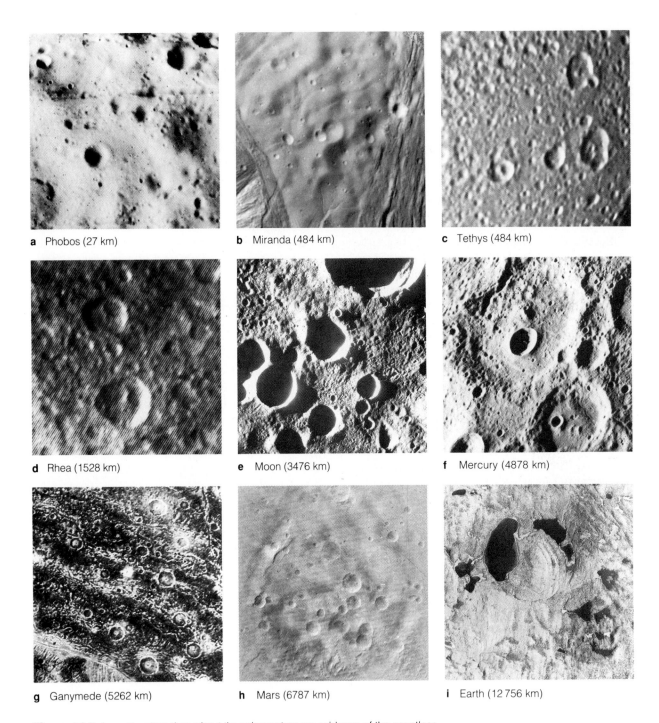

a Phobos (27 km)

b Miranda (484 km)

c Tethys (484 km)

d Rhea (1528 km)

e Moon (3476 km)

f Mercury (4878 km)

g Ganymede (5262 km)

h Mars (6787 km)

i Earth (12 756 km)

Figure 14-5 Impact craters throughout the solar system are evidence of the countless planetesimals that crashed into the planets and their moons. These photos show craters on nine different-sized worlds. The diameter of each world is given. The photos are at different scales. The arrangement in order of planetary size shows that the larger planets have enough erosional activity to degrade the more ancient craters; notice evidence of noncratering geological processes in the bottom row. (Photos *a–g:* NASA; photo *h:* Soviet Mars 5 orbiter photo, courtesy C. Florensky, Vernadsky Institute, Moscow; photo *i:* 45-million-year-old eroded Canadian crater, Earth Physics Branch of the Department of Energy, Mines, and Resources, Ottawa, Canada.)

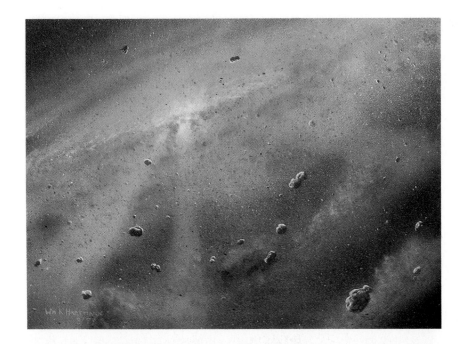

Figure 14-6 A scene in the early solar nebula. Planetesimals of rocky and icy materials orbit in the foreground. The Sun is partially obscured and reddened by dust and gas in the inner nebula. (Painting by author.)

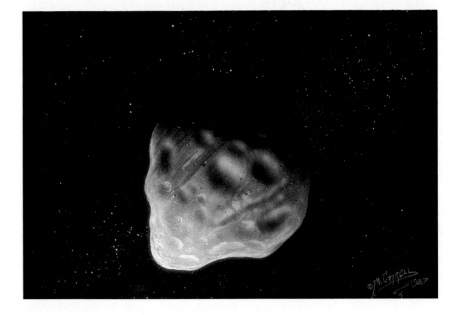

Figure 14-7 Artist's conception of one of the planetesimals in the primeval solar system. This rocky body has grown by aggregation of innumerable smaller grains and fragments, has been fractured and cratered by larger collisions, and has been rounded by sandblasting effects of smaller collisions. (Painting by Michael Carroll.)

massive than Earth; and these cores are surrounded by giant atmospheres. The terrestrial planets never accumulated these giant atmospheres of hydrogen-rich nebular gas because they never attained enough mass to pull in the nebular gas.

To support this scenario, we can compare the compositions of the terrestrial planets with those of the giants and with that of the nebular gas. Although we don't have samples of the nebular gas, we *can* measure the Sun's present-day composition, which is believed to be about the same as that of the nebular gas from which the Sun formed. Table 14-3 makes this comparison. Here we see that the gas from which the Sun and nebula formed were mostly hydrogen and about one-fifth helium.

TABLE 14·3

Comparison of Solar Nebula and Giant Planet Atmospheres

	Composition Percentages by Mass			
Gas	Solar Nebula (Sun's Present Composition)	Jupiter	Saturn	Uranus
H_2 (hydrogen)	76	79	88	76
He (helium)	22	19	11	23

Source: Anders and Ebihara (1982, solar); Voyager reports (Stone and others, 1979, 1982, 1986); Hubbard and Stevenson (1986).

Note: Uncertainties are a few percent. Results are basically similar; scientists debate significance of lower helium content for Saturn.

In the giant planets we see a similar situation. Recall that the terrestrial planets' atmospheres are mostly carbon dioxide degassed from inside the planet; they have virtually no trace of massive hydrogen–helium concentrations. Given that the uncertainties in the Table 14-3 data are a few percent, we see that the hydrogen–helium concentrations of the three measured giant planets are remarkably like the composition of the solar nebula. These measurements support the theory of giant planet formation with nebular atmospheres added to the planets by their gravity.

As the nebular gas was attracted into each of the four giant planets, it formed a miniature "solar nebula" around each giant. And in this disk-shaped cloud of gas and dust, the accretionary growth process was repeated all over again. That is, each giant planet became an analog of the Sun, and moons grew around it analogous to planets. In these miniature planetary systems, the most abundant building materials were ice and the black carbonaceous dirt common to the outer solar system. Thus giant planets spawned systems of dirty-ice satellites in prograde, circular, coplanar orbits around them, accounting for fact 11 in Table 14-1.

Asteroids

The asteroids are easily understood in this scenario. They are planetesimals that never made it all the way to planethood. There is probably a specific reason why an asteroid belt was left stranded between Mars and Jupiter: This zone would normally have been a location where a planet should have grown. Ceres, the largest asteroid, had already grown to 1000-km diameter by the time the growth process stopped in that zone. Probably the reason it stopped was that Jupiter had grown so huge that its gravity disturbed the motions of the asteroids in the zone we now know as the asteroid belt. As mentioned in the last chapter, this disturbance increased the collision velocities of the asteroids, causing them to smash into innumerable fragments during each collision instead of coalescing into an even-larger body.

Comets

Our general picture of planet formation also allows us to explain comets. There must have been asteroidlike planetesimals that formed from dirty ice in the present region of the giant planets. Most of them were consumed in growing the rocky-icy cores of the giant planets. But millions were left in interplanetary space as the giant planets grew. The gravitational forces of the giant planets were so strong that whenever an icy planetesimal experienced a near-miss with a giant planet, the planetesimal was thrown off into a new orbit. Such orbits typically had aphelia thousands of astronomical units from the Sun. That is, the icy planetesimals were thrown into the Oort cloud! Thus we can account for fact 12 in Table 14-1. The Oort cloud of comet nuclei is nothing but a storehouse of icy planetesimals that were ejected from the outer solar system during the final stages of giant planet growth.

What Became of the Remaining Planetesimals and the Solar Nebula?

Most of the planetesimals were eventually accumulated by the planets, scattered into the Oort cloud, or left stranded in the asteroid belt, probably within 100 million years of the solar system's beginning. Each time a planetesimal crashed into a planet to add to the planet's mass, a crater was formed. For another 500 million years, as lunar rocks have taught us, the last remaining interplanetary planetesimals rained down onto planetary surfaces, creating still more craters on all planets and satellites, as shown in Figure 14-5. A few planetesimals were captured into orbits around the newly formed planets, explaining the dozen or so satellites with irregular orbits that seem to have been captured instead of growing in a nebular disk around the planet.

EVOLUTION PLUS A FEW CATASTROPHES

We have stressed that this growth of planets by innumerable small planetesimals produced certain regularities of the solar system: planetary orbits that lie in the plane of the Sun's equator, the regular Bode law spacings of the planets' orbits, prograde orbital revolutions, prograde rotations, the mostly small obliquities of planets, and the regular systems of prograde satellites lying in equatorial planes of their planets' equators. These regularities of the solar system require smooth evolutionary growth from a system of many small planetesimals, not a catastrophic creation in some chaotic system.

Nonetheless, one of the beauties of the modern picture of planet formation is that it requires a few catastrophic events that can explain some of the nonregularities of the solar system (Wetherill, 1985). These catastrophic events would involve the collisions or near-misses of the planets with the largest of their nearby planetesimal neighbors as the planetesimals were being swept up. As each planet grew, it experienced collisions of different sizes. If a collision was large enough, it could have had important effects.

We have already seen one example of this type of thinking: the theory that a Mars-sized planetesimal hit Earth late in its growth and blew off material from which the Moon formed. Another example is that Uranus was probably hit by a relatively large planetesimal, tipping its rotation axis to lie nearly *in* the plane of the solar system instead of nearly perpendicular to it, as is the case for most other planets. Other properties of the solar system—such as the different styles of ring systems and the geological differences between hemispheres of some planets, such as Mars—may trace back to large impacts. The largest impacts were not big enough to randomize the characteristics of planets and their orbits, but they were large enough to give planets individuality of character!

THE CHEMICAL COMPOSITIONS OF PLANETS

Why do the planets vary in their chemical composition? Why do meteorites differ from lunar and terrestrial materials? The answers to these questions now become quite clear. The condensation sequence shows that different groups of minerals existed at different *temperatures* in the nebula. In the cold outer parts of the nebula, hydrogen-rich ices formed, but in the hot inner parts, only metals and silicates. Massive bodies such as the Sun and Jupiter retained their original gases; studies of these bodies show that the nebula was originally mostly hydrogen and helium with only a small fraction of heavier atoms such as silicon, oxygen, and iron, as shown in Table 14-4.

Table 14-4 shows what happened after the gas condensed into solid planetary materials. Virtually none of the hydrogen or helium was trapped in the solid materials of the inner solar system. Primitive meteorites of the carbonaceous chondrite and chondrite types show that primitive silicates were composed mostly of oxygen, iron, silicon, and magnesium. As soon as these materials melted, however, probably due to heat from radioactivity in certain minerals, the iron drained to form central cores resembling iron meteorites. This differentiation process meant that the surface materials of planets became more iron-poor and silicon-rich than the primitive materials. This sequence can be seen by following Table 14-4 from left to right. Note that *surface rocks of lunar and terrestrial uplands have only 2%–4% iron, compared with 24% in the original solid planetary material. On the other hand, they have increased percentages of elements like silicon and oxygen, which combine into lightweight minerals. (Review also the discussion of differentiation on page 124.)

TABLE 14.4

Elemental Abundances in Cosmic Matter (Selected Elements)

Element	Designation[a]	Sun ("Cosmic" Composition)[b]	Carbonaceous Chondrite (Most Primitive) Meteorite	Chondrite Meteorite	Earth: Ultra-basic Rock	Mars: Basaltic Rock	Moon: Mare Basalt	Earth: Basalt	Moon: Upland	Earth: Granite (Continental Crust)
					Percentage by Weight					
Hydrogen (H)	V	78	—	—	—	—	—	—	—	—
Helium (He)	V	20	—	—	—	—	—	—	—	—
Oxygen (O)	L	0.8	34	35	43	—	43	44	45	48
Iron (Fe)	S	0.04	24	26	9.4	13	11	8.6	4.6	2.1
Silicon (Si)	L	—	15	18.5	20	21	21	23	23	33
Magnesium (Mg)	L	—	13	15	20	5	5.5	4.6	4.4	0.6
Sulfur (S)	—	—	8.6	2.3	0.03	3	0.2	0.3	—	0.3
Calcium (Ca)	L	—	1.5	1.3	2.5	4	8	7.6	7	1.5
Nickel (Ni)	S	—	1.4	1.4	0.2	—	0.0001	0.01	0.0001–0.01	0.001
Aluminum (Al)	L	—	1.2	1.1	2	3	8	7.8	13	8
Sodium (Na)	L	—	0.46	0.70	0.4	—	0.4	1.8	0.4	2.7
Titanium (Ti)	L	—	0.08	0.08	0.003	0.5	2.1	1.4	0.3	0.2
Potassium (K)	L	—	0.05	0.09	0.004	0.25	0.1	0.83	0.01–0.1	3

Source: Fairbridge (1972); Taylor (1973); Mason and Melson (1970); Page (1973); Clark and others (1977).

[a] V = volatile (driven off by heating); L = lithophile (concentrated in silicaceous rocks; sulfur usually classified separately); S = siderophile (iron-affinity elements; concentrated in basic rocks and metallic minerals, as in the Earth's core and mantle).

[b] This composition is called cosmic because it is believed to be the same gas that formed not only the Sun but all nearby stars.

MAGNETIC EFFECTS AND THE SUN'S SPIN

As noted before, the early Sun would have been spinning fast enough after its contraction to deform into a disk, but today it is rotating slowly and is spherical. It has evidently slowed down. What happened to the missing angular momentum? For many years this unanswered question was the fatal flaw in evolutionary theories of planet formation. However, in the 1950s and 1960s Swedish–American astronomer H. Alfvén and others realized that the early Sun probably had a strong magnetic field. The hot surface of the early Sun shot out ionized gas, probably at a greater rate than it does today. Because ions must move with any local magnetic field, the rotating magnetic field of the Sun tried to drag the ions of the inner solar nebula with it. The gas would thus have exerted a braking force on the rotating Sun, just as water does on a spinning tennis ball dropped into a pool. The ball's rotation slows.

This **magnetic braking** probably slowed the Sun's rotation, transferring angular momentum to the planetary material in the solar nebula. This theory solves the angular momentum problem of item 13 in Table 14-1.

STELLAR EVIDENCE FOR OTHER PLANETARY SYSTEMS

Since no unusually rare processes were necessary to form the solar system, other stars may also have spawned planetary systems. All stars should form from gravitationally unstable, collapsing nebulae. All stars should have some angular momentum and thus produce rotating, flattened clouds around themselves as they form. Such clouds should cool as their heat radiates away into space, and dust grains of varied compositions should appear. Many observations of stars indicate that these processes occur elsewhere.

But would the subordinate material always accumulate into planet-sized bodies? Probably not, according to simple observations of stars. Many if not most stars are double or triple star systems. In such a system the second biggest object is not a planet, like Jupiter, but a full-fledged star with nuclear reactions and an incandescent gaseous surface. Such massive companions may interfere with planets forming in the system.

However, one of the exciting areas of astronomy in the 1980s has been the discovery of star systems with features related to those of our planetary system. First, there have been numerous detections of ever-smaller companions to apparently single stars. None may be quite as small as Jupiter, but smaller ones are reported each year, and the new discoveries are approaching Jupiter in size. Second, astronomers were surprised to discover with infrared satellites that many seemingly single, sunlike stars have thin clouds of dust around them. These clouds may be analogs of the dust trails from comets or the zodiacal light. They may be indicators that planetary objects—comets, asteroids, or even full-fledged planets—lurk still unseen near these stars.

Thus there is a stimulating new merger between studies of our own planetary system and the study of other stars. A new interdisciplinary field is emerging: the study of planetary material around other stars. If only 0.1% of all stars had planets on which liquid water could exist, then 100 million habitable planets could plausibly exist in our galaxy! We will come back to this subject—and its provocative consequences for possible alien life—in later chapters.

SUMMARY

Information about the origin of the solar system has been culled from meteorites, lunar samples, the oldest terrestrial rocks, and chemical and dynamical analysis of planets and satellites. This information indicates that the Sun formed from a contracting cloud of gas about 4.6 billion years ago. As outer parts of this cloud cooled, solid grains of various minerals and ices condensed and accumulated into planetesimals during a relatively brief interval, lasting from a few million to 100 million years.

During this interval, neighboring planetesimals collided, often gently enough to allow them to hold together by gravity. In this way small planetesimals aggregated into a few larger bodies. Sometimes these bodies collided at high enough speeds to shatter each other, producing meteoritelike fragments. The largest bodies survived and grew into planets. Moons formed near planets by a similar process. Many planetesimals crashed into planets and moons, forming craters. When icy planetesimals in the outer solar system made near-miss encounters with giant planets, the planets' gravity flung many of them almost out of the solar system, forming the Oort cloud reservoir of comets (as described in Chapter 13).

The Sun and the outer planets, with their high masses and strong gravity, retained the light, hydrogen-rich gases of the original cloud. Lower-gravity terrestrial planets lost these gases.

No data about the early solar system restrict these processes to the solar system. Thus many scientists suspect that planetary debris or full-fledged planets may have formed near some other stars besides our Sun.

CONCEPTS

catastrophic theories

evolutionary theories

conservation of angular momentum

free-fall contraction

Helmholtz contraction

nebula

solar nebula

condensation sequence

grain

planetesimal

magnetic braking

PROBLEMS

1. The gas in the early solar system was about 76% hydrogen (Table 14-3). Considering the theory of thermal escape of gases from planetary atmospheres (see Chapter 9), explain the absence of abundant hydrogen in the terrestrial planets' atmospheres. Why does this not need to be listed as a fact to be explained by theories of solar system *origin* in Table 14-1?

2. Since all planetary material condensed from the same nebula, why do meteorites have different chemical and geological properties than rocks you might find in your own yard?

3. Because of heating by the Sun and by the contraction process, gases in the inner solar system were probably warmer than gases in the outer solar system when planetary solid matter formed. In terms of the condensation sequence, relate this to the estimated or observed composition of the planets.

4. Judging from planetary composition, where was the inner boundary of the part of the solar nebula where ices condensed?

5. If planets orbited the Sun in randomly inclined orbits, both prograde and retrograde, how might theories of such

a solar system's origin differ from the theory described in this chapter?

6. List some observations that support the theory of solar system origin described in this chapter.

7. How are theories of solar system origin different in principle from theories of the origin of the universe?

8. What reasons do we have for believing that the entire solar system formed during a single, relatively short interval?

9. If a 5-year-old member of your family asked where the world came from, how would you answer?

ADVANCED PROBLEMS

10. If a planetary system was forming around a nearby star and the star was obscured by a "solar nebula" of dust grains at a temperature of 1000 K, explain how such a dust nebula might be detected. (*Hint:* Apply Wien's law and assume that a large telescope is available with infrared detectors.)

11. If dust grains in circular orbits in the early solar nebula collided at a speed equaling 0.1% of their orbital velocity, how fast would they have collided near the present orbit of Earth? How fast near the present orbit of Pluto?

12. A rule of thumb is that a planet can retain a gaseous atmosphere for a geologically long time if the typical velocities of the molecules are no more than about one-fourth the escape velocity of the planet. Assume that icy bodies with density 1000 kg/m^3 were growing in the outer solar system in a nebula of hydrogen at temperature 500 K.

a. Use this rule of thumb to show that if any primitive planets grew to around 38 000 km across they would form "cores" that would begin to retain hydrogen atoms and form massive gaseous envelopes, thus beginning to form giant planets.

b. Analysts have shown that the giant planets all have cores of ice and rock that have about 8 to 30 times the mass of Earth. Prove that the mass of the "cores" described in Problem 12a would fall between Earth's mass and the mass of these giant planet cores.

c. Comment on how these simple calculations help explain how the giant planets may have formed.

PART E

Stars and Their Evolution

Star cycles. The bright reddish object on the left is a cloud of gas and dust where stars form. The brightest star on the right is a supernova, or exploding star, that appeared in 1987. Such an explosion marks a massive star's death. (National Optical Astronomy Observatories.)

CHAPTER 15

The Sun:
The Nature of
the Nearest Star

Much of modern astronomy deals with stars—how they generate their energy, what kinds of light they radiate, how they form, and how they evolve. How can we study such remote objects as the stars? We can study the closest one, which is nearer to the Earth than five of the nine planets. Light from its surface reaches us in only eight minutes. Our eyes are dazzled by it. The Earth is bathed in its flow of radiation, washed by the winds of its outer atmosphere, blasted by seething swarms of atoms blown out of it, bombarded by bursts of X rays and radio waves emitted by it. It is our Sun, a million-kilometer ball of hydrogen and helium in the center of the solar system.

The Sun's million-km diameter is roughly 10 times Jupiter's size and roughly 100 times the Earth's size.

Humans pondered the stars for many centuries before they realized that the Sun is just another star, and the stars are suns. The Sun is the only star on which we can see surface details. It is so close that we can watch storms develop on its surface and track them as they are carried around it by **solar rotation.** As shown in Figure 15-1, this rotation averages about 25.4 d relative to the stars (and 27.3 d relative to the Earth, since the Earth's orbital motion is in the same direction as the solar rotation and must be added in). As in Jupiter's atmosphere, the equatorial region rotates faster than the polar regions—proof that the Sun has a gaseous, not solid, surface.

SPECTROSCOPIC DISCOVERIES

We cannot send space vehicles to land on the Sun, as we have done for planets. To study the Sun or stars we must rely on interpreting their light, sampled by means of our telescopes.

Chapter 5 describes certain properties of light, such as the radiation maximum occurring at certain colors,

314

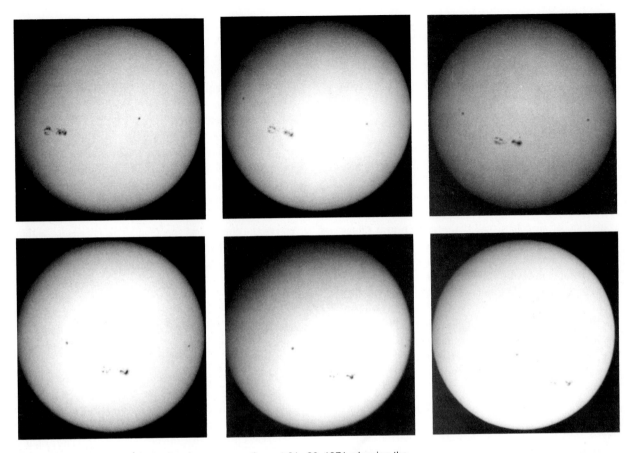

Figure 15-1 Six photos of the Sun in sequence, August 21–26, 1971, showing the rotation of the Sun by the movement of sunspot groups. (W. A. Feibelman.)

or wavelengths, depending on the temperature of the radiating body (Wien's law, page 97). Here we will touch on a few of these basic concepts again as we explain how they apply to the Sun.

The Solar Spectrum

As Newton showed, sunlight is a mixture of all colors. Light with this specific mixture of colors is called **white light** because it looks white to our eyes. Sunlight passing through Newton's prism revealed a **spectrum,** the array of these colors in order of wavelength; see Figure 5-1. Review pages 102–103 and Figure 5-13 to recall how parts of the ultraviolet and infrared spectrum of sunlight and starlight are absorbed by our atmosphere before reaching the ground.

In 1817, German physicist Joseph Fraunhofer found that certain wavelengths were missing from the Sun's spectrum, so that the spectrum appeared to be crossed by narrow, dark lines.[1] As explained in Chapter 5, these are called **absorption lines** (Figure 5-11). Fraunhofer named them A, B, C, and so on, from red to blue.

What were these lines? Let us review and expand on what we learned in Chapter 5. By the mid-1800s, scientists discovered that when a given element is burned, it emits glows of certain colors and no others. These very narrow wavelength intervals, unique to each element and as unmistakable as a set of fingerprints, are called **emission lines.** Researchers soon found that some of the emission lines exactly matched the position

---

[1]For a good review of the early discoveries about solar radiation and its effects on Earth, see Meadows (1984).

of Fraunhofer's solar absorption lines. For instance, Fraunhofer's D absorption line (actually a close pair) matched an emission line from sodium; his H and K lines, calcium. Did this mean that the D line in the solar spectrum was caused by sodium in the Sun, and the H and K lines by calcium?

Kirchhoff's Laws of Radiation

The answer proved to be yes. The proof? In the 1850s, German physicist Gustav Kirchhoff discovered in the laboratory the conditions that produce the three different kinds of spectra described in Chapter 5: the glow consisting of all colors, called the **continuum** (illustrated in Figure 5-1); absorption lines; and emission lines. For instance, when Kirchhoff looked through his spectroscope toward a sodium flame against a dark background, he saw the sodium D line in emission, like the emission lines in Figure 5-8. But when he changed the background to a brilliant beam of sunlight passing through the same flame, he saw a strong D absorption line similar to the absorption lines shown in Figure 5-11. In each case, the D lines came from the gaseous sodium atoms in the flame.

Thus, in simple terms, the absorption lines found by Fraunhofer in the solar spectrum occur when photons of solar white light pass outward through the cooler gas of the solar atmosphere. Photons of certain wavelengths (such as the D line), when striking certain atoms (such as sodium), are absorbed from the outgoing beam, causing an absence of that color in the beam.

Kirchhoff reduced such observations to three statements called **Kirchhoff's laws of radiation:**

> **1. A gas at high pressure, a liquid, or a solid, if heated to incandescence, will glow with a continuous spectrum, or continuum.**
>
> **2. A hot gas under low pressure will produce only certain bright colors, called emission lines.**
>
> **3. A cool gas at low pressure, if placed between the observer and a hot continuous-spectrum source, absorbs certain colors, causing absorption lines in the observed spectrum.**

Chapter 5 explained why these laws work. The continuum arises when free electrons are available; this occurs in high-pressure gases, liquids, or solids. Emission lines arise from electrons inside atoms in an excited

state, as in a hot gas. Absorption lines arise when atoms are in, or near, the ground state, as in a cooler gas. If these principles seem unclear, you should review Chapter 5, especially pages 96–97.

Kirchhoff himself found that the absorption lines and emission lines of a given gas have identical wavelengths, as seen by comparing Figures 5-8 and 5-11. What is seen depends on the temperature and density of the gas relative to the radiation coming from behind it, as indicated in the third law. Later an important modification was made to Kirchhoff's laws: An absorption spectrum need not originate *in front of,* or in a cooler gas than, the continuous spectrum. It can arise within the same gas as the continuous spectrum. This is because within a single gas, electrons may be jumping upward in some atoms (making absorption lines) and downward from the free state into other atoms (forming the continuum). Indeed, in the Sun some absorption lines originate in the same surface layers that produce the continuous spectrum we call sunlight. These layers form the well-defined visible surface of the Sun, called the **photosphere.**

In summary, spectroscopy allows us to determine many things about the properties of the Sun. Most important, *identification of the spectral lines allows identification of the elements in the Sun.* Application of Kirchhoff's laws proves that the photosphere is a layer of hot gas. Application of Wien's law (page 97) allows us to use the wavelength of the strongest solar radiation to determine the photosphere temperature. Other spectroscopic principles enable astronomers to measure temperatures and pressures at different depths in the gas near the Sun's surface.

Images of the Sun

Normal photographs of the Sun (or any other scene) are made from light of many colors, since normal photographic films are sensitive to many wavelengths. However, color filters and other devices can restrict the wavelength range. By using a narrow enough range, we can see the Sun in only the light emitted by a chosen gas (hydrogen, for example) in a specific atomic state, thus tracing the distribution of that gas. The **spectroheliograph** is an ingenious instrument that allows this to be done. It spreads the light into different colors, from which one color alone is selected to form an image. When hydrogen is in certain states of temperature and pressure, it emits an especially useful red color—the well-known **hydrogen alpha line,** or *Hα emission,*

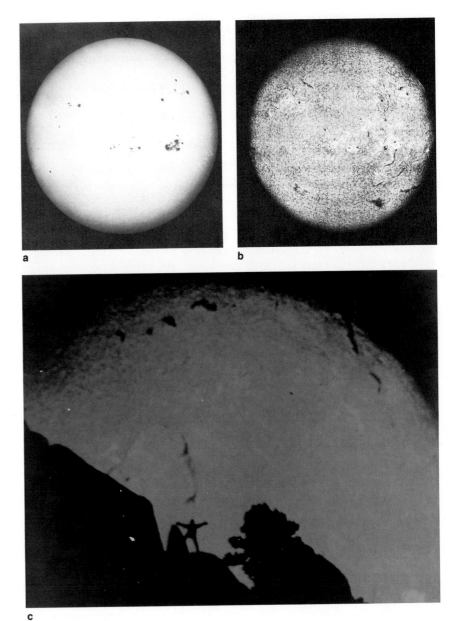

Figure 15-2 a The Sun in normal visible light, with sunspots. Shading around edge, called limb darkening, is caused by solar atmosphere's absorption of light. **b** The Sun on the same date, photographed by a spectroheliograph in red light emitted by hydrogen. Bright areas involve intense hydrogen emission. **c** An unusual spectroheliograph photo. Researchers and hiker cooperated to silhouette the hiker against the setting Sun. (Images *a* and *b* from Hale Observatories; *c* from National Oceanic and Atmospheric Administration.)

shown in Figure 5-7. In an image using Hα light, as in Figure 15-2 (right side), the only bright regions are those where hydrogen of a certain temperature is emitting this specific red glow. Dark parts of the image are regions where there is still much hydrogen but little of it is in this Hα-emitting condition. Thus images in Hα light, or light emitted by other gaseous atoms, are very useful in mapping the conditions of that particular gas in various regions of the surface. Special telescopes with such equipment are devoted to solar studies (Figure 15-3).

COMPOSITION OF THE SUN

After more than a century of spectroscopic study, the **Sun's composition** is accurately known. It is about three-quarters hydrogen and one-fifth helium by mass—roughly the same H/He proportions we found in the giant planets' atmospheres. The heavy elements common in the Earth comprise only 2% of the Sun by mass. The most abundant elements are listed in Table 15-1, which is believed to reflect the bulk composition of material from which the solar system formed.

Figure 15-3 The setting Sun silhouettes a solar observatory on Kitt Peak, Arizona. The diagonal structure at left is the Kitt Peak National Observatory solar telescope. At the top of the vertical tower is a mirror that tracks the Sun and reflects its image down the diagonal tunnel to an underground lab. Conventional observatory domes are also seen. The picture was taken during three years of trial photos from positions in the adjacent desert calculated for each date to position the Sun behind the domes. This view was from a site 23 mi from Kitt Peak; the camera had to be positioned within 10 ft to give the desired view. (Photo by Don Strittmatter with 2300-mm telephoto lens.)

TABLE 15-1

Composition of the Sun

Element	Percent Mass of the Sun	Atomic Number
Hydrogen (H)	76.4	1
Helium (He)	21.8	2
Oxygen (O)	0.8	8
Carbon (C)	0.4	6
Neon (Ne)	0.2	10
Iron (Fe)	0.1	26
Nitrogen (N)	0.1	7
Silicon (Si)	0.08	14
Magnesium (Mg)	0.07	12
Sulfur (S)	0.05	16
Nickel (Ni)	0.01	28

Source: Adapted from Anders and Ebihara (1982).
Note: Based on spectroscopic measurements of the Sun and measurements of meteorites and other samples.

An interesting episode occurred in 1868, when the French astronomer Pierre Janssen and the English astronomer Norman Lockyer independently found solar spectral lines corresponding to an unknown element. This element, named *helium* (from the Greek *helios*, "sun"), was the first to be discovered in space instead of on the Earth (where it was not observed until 1891). After hydrogen, it is the second most abundant element in the Sun.

SOLAR ENERGY FROM NUCLEAR REACTIONS

Hermann von Helmholtz showed in 1871 that the energy output of the Sun corresponds to the burning of 1500 lb of coal every hour on every *square foot* of the sun's surface. No ordinary chemical reactions can produce energy at this rate! Thus, Helmholtz realized, the Sun is not "burning" in the normal sense.

Then what *is* the source of the Sun's heat and light? In the 1920s, astrophysicists realized that the energy of the Sun and other stars comes from **nuclear reactions**—interactions of atomic nuclei near the star's center. Normally, nuclei are protected from interacting by their surrounding clouds of electrons. Familiar **chemical reactions,** such as coal burning, involve interactions only between *electrons* of different atoms, far outside the central nucleus. If the temperature and pressure are high enough, however, atoms collide fast enough to knock away electrons, allowing the nuclei to interact. This happens in the Sun's central core.

States of Matter in Stars and in the Universe

When we deal with nuclear reactions and the material inside stars, we are no longer dealing with the familiar forms of matter that make up the solids, liquids, and gases of our planet and our bodies. We are about to make the leap from cold planetary matter to hot stellar matter. This is a good moment to pause and describe the way that matter behaves under a variety of conditions.

There is a simple way to arrange the states of matter—by temperature as in Table 15-2. Although this arrangement oversimplifies certain effects of pressure and other variables, it is very useful in understanding the forms of material we are dealing with in this book. The table is arranged from the bottom up in order of increasing temperature.

Starting at the bottom, at the lowest temperatures, we find that matter is "frozen." It exists as a solid. This means that the atoms are bonded together, often in a lattice pattern. The bonds are formed by the sharing of electrons (tiny dots) between nuclei. The nuclei consist of protons (blue) and neutrons (white). Most of the mass of each planet is in this solid state—indeed, the crystals that form rocks are good examples of atoms bound together in lattice patterns.

Now recall from page 97 that our conception of temperature is merely a way of measuring the rate of motions of the atoms. The faster the motions, the higher the temperature and vice versa. At absolute zero temperature, 0 K, atomic particles would have no motion. But as we raise the temperature of a solid, its atoms vibrate faster and faster.

By the time room temperature is reached, around 300 K, the typical atoms in typical substances are moving at around $\frac{1}{2}$ km/s. This activity is sufficient to break many of the atoms loose from their lattice. We perceive this as the substance melting, or turning into a liquid. The liquid oceans of planet Earth are a good example. In the liquid state, chains or groups of atoms may move

among each other. By the time we reach another 100 or 200 K higher, the atoms are moving at around 1 km/s and the chains are broken. Now we have created a gas.

In a gas, the individual atoms (or molecules, such as H_2O in water vapor) are moving freely. This is the state of matter in the air we breathe. But all these forms of matter are cooler than matter in the Sun or in most stars. If we keep heating the gas, the speeds of the atoms increase. They hit harder and harder.

At a temperature of a few thousand degrees, the atoms are hitting each other so hard that they break the electrons free from their orbits. The Sun's surface, for instance, has a temperature a little over 5000 K. There the hydrogen nuclei, many of them stripped of their electrons, move at speeds around 10 km/s. Gas that has had its electrons knocked off is called **ionized gas,** or *plasma.* This is the form of matter not only in the Sun's surface layers but also in a familiar flame. Because the electrons are constantly being bumped off atoms and rejoining them, they are constantly changing energy states and giving off light. Following Kirchhoff's second law, the light of a candle flame is in the form of emission lines. But as we noted above, in accordance with Kirchhoff's first law, the light given off from the high-pressure gas in the surface of the Sun or a star is in the form of continuum.

The insides of stars are even hotter than their surfaces. The center of the Sun, for instance, is at about 15 million Kelvin. What happens if we confine the gas (in a container or inside a star) and keep raising the temperature toward such a value? The electrons have all been stripped off and the bare nuclei of atoms are colliding. As always, when the temperature increases, the particles collide harder and harder. Atomic nuclei might be compared to spitballs: If they just brush together at low speed, they merely bound apart; but if they hit hard enough they fuse together. Among nuclei, it takes temperatures of the order 10 million Kelvin before the fastest nuclei begin to fuse. Typical hydrogen nuclei at this temperature move at around 500 km/s, and the fastest ones in a sample of the ionized gas will be moving even faster. This merger of nuclei at high temperatures is called *fusion.* The fusion may be among individual protons (blue in Table 15-2) and neutrons (white), or it may be between nuclei to build even larger nuclei with many protons and neutrons.

Fusion of atoms inside stars is one of the most important processes in the universe. We will be dealing

TABLE 15·2

States of Matter in the Universe

Schematic Chart

Approximate Temperature Scale	Velocity of Typical Particles		State of Matter	Typical Location	Typical Radiation Emitted
60 billion K	Relativistic (appreciable fraction of speed of light)		NUCLEAR FRAGMENTS (nuclear particles collide hard enough to shatter)	Accretion disk near a black hole	Gamma rays X rays
10 million K	500 km/s		BARE NUCLEI (nuclei collide and fuse, causing nuclear fusion reactions)	Core of a star	X rays Ultraviolet light
5000 K	10 km/s		IONIZED GAS (electrons knocked free)	Atmosphere of a star	Visible light
500 K	1 km/s		GAS (separate atoms)	Atmosphere of a planet	Infrared light
300 K	$\frac{1}{2}$ km/s		LIQUID (some atoms linked in chains)	Water	Far infrared light
0 K	0		SOLID (atoms linked in lattice)	Rock	Radio waves

TABLE 15·3

The Proton–Proton Chain: Energy Source of the Sun

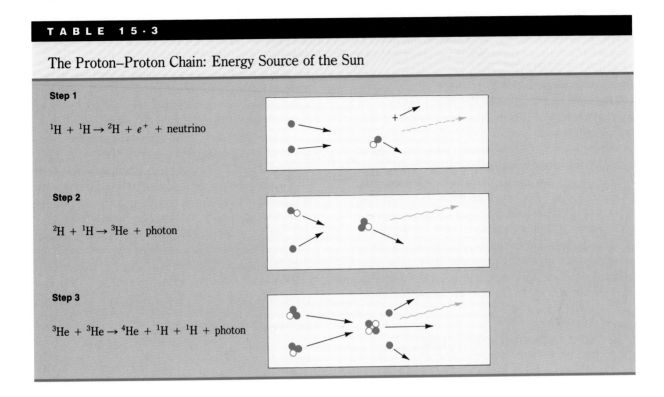

Step 1

$$^1H + {}^1H \rightarrow {}^2H + e^+ + \text{neutrino}$$

Step 2

$$^2H + {}^1H \rightarrow {}^3He + \text{photon}$$

Step 3

$$^3He + {}^3He \rightarrow {}^4He + {}^1H + {}^1H + \text{photon}$$

with its consequences throughout the rest of this book. One of its most important effects is that the fusion process gives off energy, shown in Table 15-2 as wavy lines representing radiation. This energy heats the gas, maintains the temperature inside the Sun at around 15 million Kelvin, and thus keeps the reactions going.

But what happens if we force the temperature still higher in a sample of ionized gas? Eventually, at temperatures of billions of degrees, the particles are moving at nearly the speed of light. Then the nuclei hit so hard that they shatter in a shower of radiation and tiny subatomic particles. In later chapters, we will encounter such environments—for example, near black holes.

The Nuclear Reactions Inside the Sun

Nuclear reactions inside stars do two important things: They generate energy and they gradually change the star's composition because they build up more and more heavy nuclei. The principal reactions inside the Sun are believed to be a three-part sequence that fuses four hydrogen atoms into a helium atom. Because this chain of reactions starts with two hydrogen nuclei—that is,

two single protons—it is called the **proton–proton chain.** To see how it works, let us follow it in detail in Table 15-3. The reaction occurs in three steps, shown both in diagrammatic form and in the form of the nuclear reaction equations written by physicists. In step 1, two protons collide and fuse. The fusion produces a form of hydrogen nucleus designated ^2H. A tiny positive particle called a positron is given off, together with a massless (or virtually massless) particle called a **neutrino.** In step 2, the hydrogen nucleus hits another proton and fuses into a form of helium known as helium-3, designated ^3He. A photon of radiation is emitted. In step 3, two of the ^3He nuclei collide and fuse into the normal form of helium, helium-4, designated ^4He. Two protons are left over and another photon is emitted.

Note that the reaction has done two important things: It has given off energy in the form of radiation of photons, and it has created helium out of the lighter element hydrogen.

The total amount of mass left at the end of the three-step chain is slightly less than the mass of the initial hydrogen atoms. During the fusion, a small amount of mass m is converted to an amount of energy E, according to Einstein's famous equation $E = mc^2$. (The constant c is the velocity of light, 3×10^8 m/s. When units are in the SI metric system, m is in kilograms and E is in joules.) In the Sun's fusion sequence, about 0.007 kg of matter is converted into energy for each kilogram of hydrogen processed. This liberates 4×10^{26} J/s inside the Sun, and the Sun radiates this much every second to maintain its equilibrium. This corresponds to 400 trillion trillion watts—which equals a lot of light bulbs!

Every second, the Sun converts 4 million tons of hydrogen into energy and radiates it into space. Long before the Sun can use up all its mass, the solar core will have converted so much hydrogen to helium that there will not be enough hydrogen left in the core to fuel further reactions and the reactions will stop. According to a recent calculation, the Sun won't run out of hydrogen for about 4 billion years. The consumption of hydrogen inside stars proves the important point that stars are not permanent, but must evolve and run down.

THE SUN'S INTERIOR STRUCTURE

As a result of Helmholtz contraction (see Chapter 14), the temperature at the Sun's center reached 15 million Kelvin as the proton–proton reactions became established. This remains the current temperature in the **solar core,** where nuclear energy is generated.

Approximate conditions in the layers between the core and the surface can be calculated by using equations that describe the pressure at any depth, the properties of gas under different pressures, and energy generation rates at points inside the Sun. The results appear in Figure 15-4.

These calculations indicate that the gas pressure at the Sun's core is about 250 billion times the air pressure at the Earth's surface. This high pressure compresses the gas in the core to a density of about 158 000 kg/m^3—158 times denser than water and about 20 times denser than iron. One cubic inch of this gas would weigh nearly 6 lb! The core of the Sun occupies about the inner quarter of the Sun's radius. This 1/64 of the Sun's volume contains about half the solar mass and generates 99% of the solar energy.

How Energy Gets from the Core to the Surface

As heat energy always flows from hot to cool regions, solar energy travels outward from the hot core, through a cooler zone of mixed hydrogen and helium, toward the surface. Throughout most of the Sun's volume, this energy moves primarily by **radiation.** That is, the energy radiates through the gas in the form of light, just as light travels through our atmosphere. Very little moves by **conduction,** the mechanism by which a pan on a stove becomes hot.

In the outer part of the Sun we find a third mechanism of energy transport—**convection** (page 126). Convection occurs when the temperature difference per unit length between the hot and cold regions is so great that neither radiation nor conduction can carry off the outward-bound energy fast enough. So-called "cells" of gas, having become heated enough to expand, become less dense than their surroundings and rise toward the surface, move across the Sun at about 20 m/s, cool by radiating their energy into space (sunlight!), and sink (Thomsen, 1985).

The Solar Neutrino Puzzle

Although solar astronomers believe they understand the basic structure and energy sources of the Sun, a frustrating problem has arisen in recent years. Step 1 of the proton–proton chain (Table 15-3) produces a neutrino. From the Sun's total energy production and the rate of reaction of the proton–proton chain, astronomers predicted the amount of neutrinos that should be released by the Sun. Even though neutrinos are hard to detect (they pass through huge amounts of material without being absorbed), detectors to measure neutrino numbers were constructed in the 1970s. To scientists' amazement, the experiments detect only about a third as many solar neutrinos as predicted.

Could our understanding of solar processes or atomic physics be fatally flawed? Three interesting new theories advanced in the mid-1980s suggest ways out of the puzzle. First, a certain hitherto unmeasured type of subatomic particle might exist, charmingly called WIMP, for *w*eakly *i*nteracting *m*assive *p*article. WIMPs would transport energy out of the solar core, cool it, and thus retard production rates of neutrinos. A second theory proposes that instead of zero mass, neutrinos have a tiny, unmeasured mass, which would affect the experiments and possibly solve the puzzle. A third theory, from Russian work, suggests that certain types of neutrinos transmute into undetectable particles, thus explaining the deficiency. Each theory is being tested by new experiments. The results will improve knowl-

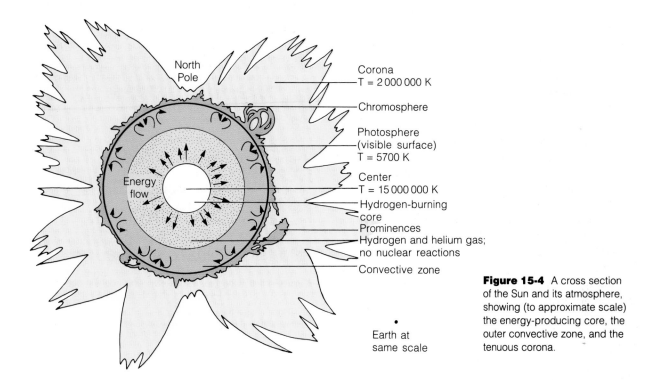

Figure 15-4 A cross section of the Sun and its atmosphere, showing (to approximate scale) the energy-producing core, the outer convective zone, and the tenuous corona.

edge of solar and subnuclear processes. This illustrates the close tie between modern astronomy and basic physics (Weneser and Friedlander, 1987; Bahcall, 1990).

THE PHOTOSPHERE: THE SOLAR SURFACE

Energy ascending from inside the Sun heats the photosphere—the bright surface layer of gas that radiates the visible light of the Sun. The photosphere's temperature can be found quickly from Wien's law. In the solar spectrum the maximum solar radiation is yellow, with a wavelength of about 510 nm. Application of Wien's law shows that the temperature of the photosphere is about 5700 K.

If the Sun is a giant ball of gas, why does it appear to have a sharply defined surface? The answer involves the **opacity** of the gas—its ability to obscure light passing through it. Air, for example, has low opacity. Gas in the photosphere has many **negative hydrogen ions** (hydrogen atoms with an extra electron, designated H^-), which obstruct light and cause high opacity. They produce an opaque layer beyond which we cannot see. Only

a few hundred kilometers above this layer, at the top of the photosphere, there are few H^- ions and the gas is clear. Most of the Sun's light comes from a layer about 400 km thick, giving the appearance of a sharply defined surface. There is also hydrogen in the solar atmosphere above the photosphere, but no abrupt change in the gas density. A solid probe, if it could survive the 5700-K temperature, could drop directly through the photospheric "surface" and plunge into the Sun, like an airplane passing through the surface of a cloud.

The convective motions that bring mass and energy from the interior disturb the photospheric surface. A photograph of the Sun shows pronounced **granules** in the photosphere, as shown in Figure 15-5. (Compare with Figure 15-6.) Each bright granule is a convection cell 1000 to 2000 km across, rising from the subphotospheric layers. Each granule rises at a speed of 2 to 3 km/s and lasts for a few minutes. Slightly darker regions between granules mark areas where cooled gas descends again into the Sun.

Surging wave motions have also been observed in the gas, with wavelengths of about 5000 km and periods of about 5 min. Such sizes may seem abstract, but keep in mind that many moving masses being churned on the surface of the Sun are as large as the entire Earth!

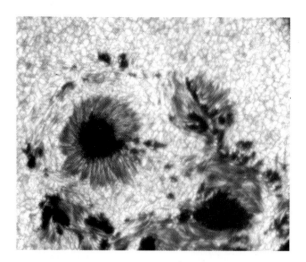

Figure 15-5 A detailed photo of the solar surface in the region of a sunspot. Outside the sunspot, the normal solar surface is mottled by granules believed to be convection cells in the solar gas. The main sunspots are comparable to the Earth in size, and the large granules are comparable to continents. (Balloon-borne telescope photo, Princeton University, Project Stratoscope, supported by NSF, ONR, and NASA.)

Figure 15-6 Oblique downward view of convection cells in cumulus clouds from about 10 km over Indiana, showing a convective pattern similar to that found on a larger scale in the solar gas (see upper right part of Figure 15-5). (Photo by author.)

CHROMOSPHERE AND CORONA: THE SOLAR ATMOSPHERE

Shooting up into the solar atmosphere are **spicules**: columns of incandescent gas 10 000 km long. Essentially giant flames, they fade in 2 to 5 min and are a transition from the photosphere to the overlying layer called the chromosphere. They are shown in Figures 15-7 and 15-8.

The **chromosphere** (which means "color layer") is a pink-glowing region of gas just above the photosphere. Its light is mainly the red Hα emission line described in Figure 5-8. E. G. Gibson (1973) has called the chromosphere "froth on top of the turbulent and relatively dense photosphere." The chromosphere can be seen by the naked eye during a solar eclipse. When the Moon covers the rest of the solar disk, this thin outer layer is visible as a ring of small, intense red flames, just visible in Figure 15-9 (top edge). As drawn in Figure 15-4, the chromosphere is a thin layer about 2500 km thick. In its upper regions the temperature exceeds 10 000 K.

Above the chromosphere is the rarefied, hot gas of the **corona**. Gas in the corona reaches the amazing temperature of 2 million Kelvin, due to heating by violent convective motion in the photosphere and chromosphere. As a result of the extreme heat, it expands rapidly into space. Whereas the gas density is about 0.001 kg/m^3 within a few hundred kilometers of the photosphere, it drops to 10^{-7} kg/m^3 in the middle chromosphere and to less than 10^{-11} kg/m^3 in the lower corona. Clearly the corona is only the outermost, tenuous atmosphere of the Sun.

During eclipses, the corona is visible to the naked eye as a pearly, glowing gas around the Sun, as seen in Figures 15-9 and 1-21. Although Plutarch recorded this, some early astronomers thought it was an optical illusion. Early photographs finally proved its existence. Spectra were obtained in 1869, eventually revealing that the coronal gas has extremely high temperatures of 1 to 2 million Kelvin! In 1930 French astronomer Bernard Lyot built an instrument called the **coronograph,** which artificially eclipses the Sun and allows the intriguing gases of the corona and chromosphere to be studied at will (Figure 15-10).

Why are both the chromosphere and the corona hotter than the photosphere? After all, they are farther from the Sun's internal energy source. The answer is that magnetic effects and shock waves from the violent subsurface convection transfer a lot of energy to this gas. The Sun's magnetic field controls motions of gas in the corona, creating delicate streamers. (See review

Figure 15-7 Spicules, or clustered jets of solar gas, photographed in red light of hydrogen Hα emission. Though the spicules are luminous gas, they appear darker than the bright solar background when silhouetted against the photosphere. (Hale Observatories.)

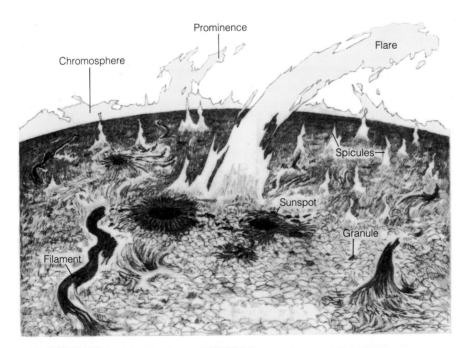

Figure 15-8 A schematic, oblique view of the solar surface showing features mentioned in the text.

Figure 15-9 The Sun's inner atmosphere during an eclipse. Pink flames of hot hydrogen, colored by the red Hα emission of excited hydrogen atoms, protrude from the Sun's surface at the top and several other points around the Moon's black silhouette. The pearly diffuse glow is the inner corona. (NASA photo by astronauts using solar telescope in Skylab space station.)

by Wolfson, 1983.) The coronal gas expands and merges with the interplanetary gas and dust. While the inner glow of the corona comes from solar gas, the outer glow is really the inner zodiacal light—sunlight reflected from interplanetary dust.

SUNSPOTS AND SUNSPOT ACTIVITY

Although sunspots sometimes can be seen with the naked eye when the Sun is dimmed by fog or a dark glass,[2] their nature was not realized until 1613, when Galileo studied them and concluded that they are located on the solar surface and are carried around the Sun by solar rotation, as seen in Figure 15-1.

The Nature of Sunspots

A **sunspot** is a magnetically disturbed region that is cooler than its surroundings. A sunspot looks dark only

---

[2]**Never point a telescope or binoculars at the Sun without professional aid!** Unmodified telescope optics collect enough solar light and heat to blind the observer. Even pointing a telescope *near* the Sun may concentrate light inside the instrument in such a way as to damage it, as well as risking an observer's eyesight.

because its gases, at 4000–4500 K, radiate less than the surrounding gas at about 5700 K. Motions of solar gas near sunspots are controlled not by atmospheric forces, as with terrestrial storms, but by magnetic fields of the Sun. *Ions* (charged atoms or molecules), which are common in the Sun, cannot move freely in a magnetic field but must stream in the direction of the field—for example, from the north magnetic pole to the south.

A gas with many ions is called a **plasma,** and unlike a neutral gas, its motions are strongly influenced by magnetic fields. For this reason, plasmas in the sunspots and elsewhere in the solar atmosphere move in peculiar patterns that indicate the twisted patterns of the solar magnetic field.

Huge clouds of gas, larger than the whole Earth, erupt from the disturbed regions of sunspots. These **prominences** can be seen when silhouetted above the solar limb, or edge. *Eruptive prominences,* as seen in Figure 15-10, shoot out at speeds averaging about 1000 km/s. *Quiescent prominences,* as seen in Figure 15-11, are masses of flowing gas held in relatively fixed positions above the Sun for hours or days by magnetic fields. The largest blasts of material and their very active sunspot sites are called **flares.** Figure 15-12 shows a vertical view of a flare near a sunspot pair.

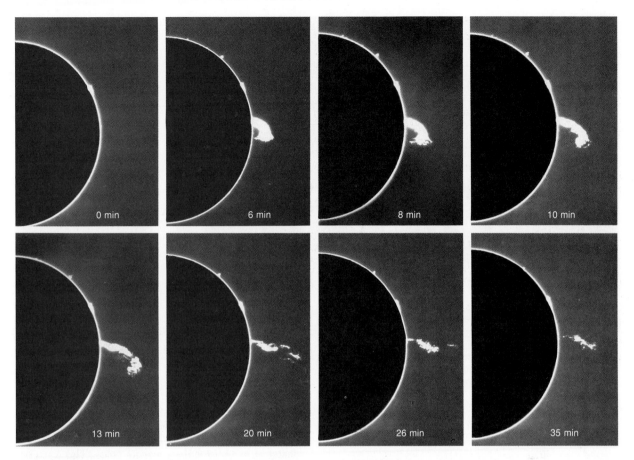

Figure 15-10 This sequence of photos shows a jet of gas blasting off the Sun over a period of 35 min. The photos were made with a coronograph, which obscures the bright solar disk and allows solar atmospheric activity to be monitored. (National Center for Atmospheric Research.)

The 22-y Solar Cycle

Around 1830 an obscure German amateur astronomer, H. Schwabe, began observing sunspots as a hobby. After years of tabulating his counts, he announced in 1851 a **solar cycle:** The *number* and *positions* of sunspots vary in a cycle, as shown in Figure 15-13. This discovery, followed a year later by the discovery that terrestrial magnetic compass deviations exactly follow the same cycle, was a key step in understanding the Sun and its effects on the Earth.

The cycle's duration averages 22 y and consists of two 11-y subcycles, as shown in Figure 15-14. At a time of "sunspot minimum" (a minimum number of sunspots), the few visible spots are grouped within about 10° of the solar equator. When a new cycle begins in a year or so, groups of new spots appear at high latitudes,

about 30° from the solar equator. The spots often appear in pairs, and the eastern spot in each northern hemisphere pair is of specific polarity—for example, a north magnetic pole. In the southern hemisphere the polarity is reversed. After a few years, the sunspot number reaches a maximum and the spots are at intermediate latitudes, about 20° from the solar equator. After about 11 y, the spots appear mostly about 10° from the equator, and a sunspot minimum occurs again.

Now the cycle begins to repeat, except for a noticeable difference. The new spots forming at ±30° have reversed polarity. The eastern spots in the northern hemisphere are now south magnetic poles! Thus it takes another 11 y to complete the full cycle, when all features resume their initial pattern. A sunspot maximum should occur around 1990.

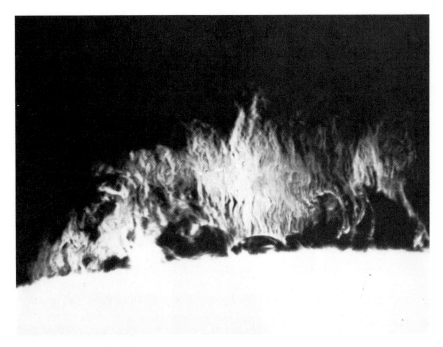

Figure 15-11 A quiescent prominence, about 50 000 km high, hanging like a sheet of flame above the solar surface. The Earth's diameter is less than a quarter the height of this plasma cloud. (Sacramento Peak Observatory.)

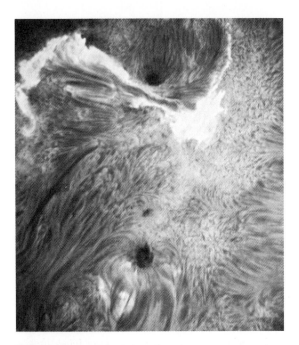

Figure 15-12 Vertical view of gaseous streamers and solar flare activity (bright region, top) around a pair of sunspots (dark areas at top and bottom). (Air Force Cambridge Research Laboratories.)

The sunspot cycle occurs because the Sun rotates faster at its equator than near its poles. This causes a shearing and twisting of the magnetic field that controls motions of the ionized solar gas, as shown in Figure 15-15. The field is further distorted by convection, until the twisted pattern breaks and a new pattern forms, starting a new cycle. (See reviews by Newkirk and Frazier, 1982; Parker, 1983.)

Still more remarkable is the fact that the magnetic field of *the entire Sun* reverses during each 11-y subcycle; thus the entire Sun participates in the full 22-y cycle. Imaginary observers on the Sun would find their compasses pointing north in one direction for 11 y (subject to disturbances by frequent magnetic storms) and in exactly the opposite direction for the next 11 y. This behavior is not entirely unknown: The Earth's field reverses every few hundred thousand to few million years. Both patterns of reversal may involve cyclic flow patterns in the deep fluid cores of the two bodies. The sunspot cycle is important to us because during years of maximum sunspot activity, solar particles shooting off the Sun affect the magnetic field and upper atmosphere of Earth, disturbing radio communications and causing aurorae.

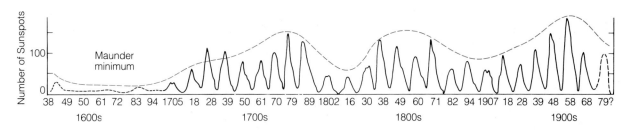

Figure 15-13 Sunspot counts since the 1600s show the cycle averaging 11 y for sunspot numbers (half the 22-y magnetic cycle), with evidence for a longer 80-y cycle (blue dashed line). The extensive period of low sunspot activity in 1600s is believed to correlate with climate changes at that time. (After data of M. Waldmeier in Gibson, 1973; Pasachoff, 1980.)

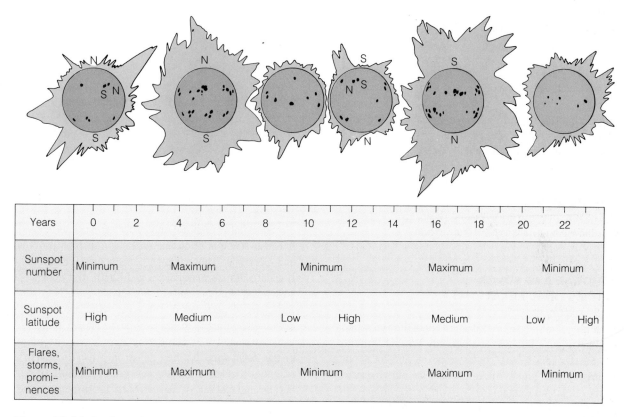

Years	0	2	4	6	8	10	12	14	16	18	20	22
Sunspot number	Minimum		Maximum			Minimum			Maximum			Minimum
Sunspot latitude	High		Medium			Low		High	Medium		Low	High
Flares, storms, prominences	Minimum		Maximum			Minimum			Maximum			Minimum

Figure 15-14 A schematic sequence of solar changes during the 22-y cycle. Typical coronal appearances are shown for the minimum and maximum of the cycle in the two drawings at left. (See text.)

SOLAR WIND

Particles blasted out of flares and spots rush outward through interplanetary space. The solar coronal plasma, having been heated to nearly 2 million Kelvin by the violence of photospheric convection, also expands rapidly into space (limited only by magnetic forces acting on charged particles). Together these effects cause the **solar wind:** an outrush of gas past the Earth and beyond the outer planets. Near the Earth, the solar wind trav-

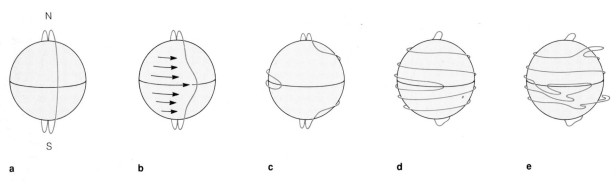

a   b   c   d   e

Figure 15-15 Some features of the solar cycle are explained by this simple model following the history of a "magnetic field line," which defines the direction of the solar magnetic field **a**. Because of differential rotation of the Sun **b**, and because the field lines must move with the ionized gas, field lines tend to distort **c**, becoming wrapped around the Sun **d**. Convection and eruptions further distort the field **e**. After about 11 y the field becomes unrecognizable and neutralized, and a new field forms.

els at velocities near 600 km/s, and sometimes reaches 1000 km/s. The gas has cooled only to 200 000 K, but it is so thin that it transmits no appreciable heat to the Earth. According to spacecraft data, the solar wind extends at least as far as Saturn's orbit.

In addition to the solar wind, solar radiation itself exerts an outward force on small dust particles. This effect, which is greater on small particles, is called **radiation pressure.** Together these are the forces that blow comet tails away from the Sun.

AURORAE AND SOLAR–TERRESTRIAL RELATIONS

Solar radiation and particles blasted out of the Sun strike Earth. Because 99.98% of all energy passing through the Earth's atmosphere comes from the Sun,[3] it is not surprising that small "flickers" in solar output cause major effects on Earth.

Effects come from changes in both solar radiation and solar particles. During a solar flare, the total visible radiation from the Sun changes by much less than 1%, but the X-ray radiation may increase by a hundredfold. The X-ray photons are energetic because they have very short wavelength. When they strike Earth's upper atmosphere, they change its distribution of ions and affect radio transmission on the ground. Imaging of the Sun at X-ray wavelengths from above the atmosphere has permitted us to visualize the remarkable appearance of the X-ray Sun, as shown in Figure 15-16, which vividly shows flares and active areas emitting X rays. The flares shoot material upward into the corona, affecting the coronal structure, as seen in the sequence of Figures 15-17 through 15-19.

Solar flares emit not only radiation, such as X rays, but also streams of atomic particles, such as protons and electrons. These join and enhance the solar wind. If a flare directs material toward Earth, the enhanced solar wind hits Earth after a few days' travel. Normally the charged particles of the solar wind are strongly deflected by Earth's magnetic field, as shown in the left side of Figure 15-20. But during solar flares, the surge in the solar wind is often so strong that the Earth's magnetic field is seriously distorted, affecting distributions and motions of charged particles throughout the Earth's vicinity.

The ions near Earth are concentrated into doughnut-shaped regions around Earth, about over the equator. These **Van Allen belts** of radiation were discovered in 1958 by the first artificial satellites. They are shown by the *x*'s in Figure 15-20. They are particularly affected by disturbances in the solar wind. Under normal conditions, as the solar wind sweeps around the outer limits of the ionosphere, it builds up voltages of 100 000 volts or more between the outer regions of the magnetic field and the atmosphere, driving some charged particles along

---

[3]Based on a tabulation by Hubbert (1971); the rest is in the form of tidal energy, starlight, cosmic rays (atomic particles from space), and subterranean geothermal energy.

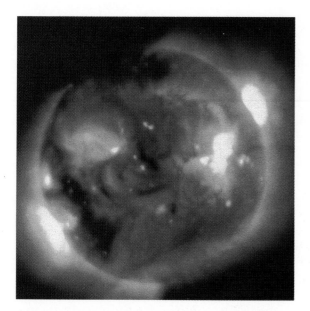

Figure 15-16 A view of the Sun in X-ray wavelengths showing flares as sites of intense X radiation. X rays are invisible, and image processors here chose false-color orange tones usually associated with sunlight. As with the ultraviolet helium emissions (see Figure 16-15, page 358), the polar regions are relatively inactive in X-ray emissions. (NASA photo by Skylab astronauts.)

the magnetic field lines toward the poles (dashed lines in the middle of Figure 15-20). These particles crash into the upper atmosphere, excite the gas atoms there, and cause them to glow. This glow can be seen from the ground; it is called the **aurora** (plural: *aurorae*). The ions crash into the atmosphere only near the magnetic poles, causing intermittent aurorae in northern Canada and Antarctica, as seen from space in Figure 15-21. These aurorae are also called the northern lights and southern lights (or aurora borealis and aurora australis, respectively).

Although the northern lights are normally visible only in Alaska and Canada, large solar flares disrupt the situation so much that the ions may be dumped into the atmosphere over lower latitudes, including much of the United States. Under these conditions, many people may see an awesome auroral display (Figure 15-22; see reference by Akasofu, 1989). The sky may be filled with a stationary red glow, moving bands of red, white, and greenish light, or huge displays like colorful draperies rippling silently in the wind. The common drapery form is really a curtainlike display of glowing atoms controlled by the form of Earth's magnetic field; this is well seen in a photo taken by astronauts orbiting in Skylab 3

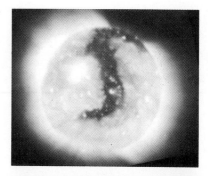

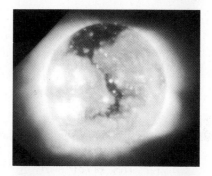

Figure 15-17 The coronal structure of the Sun's atmosphere, photographed by Skylab astronauts using an X-ray-sensitive telescope. This technique revealed rifts in the coronal structure, through which the cooler (darker) surface gases can be seen. The evolution of such a rift is shown in the sequence. (NASA; American Science and Engineering, Inc.)

Figure 15-18 Surface and atmospheric activity of the Sun. During the June 30, 1973, solar eclipse in Kenya, the outer corona of the Sun was photographed in white light (outer image). About an hour earlier, astronauts in the orbiting Skylab photographed the solar surface (circular inset) using X radiation, showing centers of flare activity. Major streamers in the outer corona are shown to be aligned with X-ray flares on the surface. (National Center for Atmospheric Research; American Science and Engineering, Inc.; NASA.)

Figure 15-19 The same eclipse shown in Figure 15-18 was photographed with special techniques to reveal the great length of glowing gas streamers extending into the outer corona. (Los Alamos Scientific Laboratory, J-Division, courtesy Charles F. Keller.)

who witnessed such a display from above (Figure 15-23). During the greatest flare-related disturbances, aurorae may be seen from latitudes as low as Mexico and India, compass needles may deviate by a degree from normal, and electromagnetic effects may overload electric power lines, burn out transformers, and disrupt telephone, telegraph, radio, and power transmission. Because these effects are related to mighty solar flares, they are most common around the years of sunspot maximum, such as around 1990.

Discovering these causes of the aurora took 250 y. As early as the 1700s, Edmond Halley, of Halley's comet fame, discovered that auroral structures lie along lines of Earth's magnetic field. Around 1920, Norwegian observer F. C. Störmer showed that aurorae occur at altitudes usually exceeding 100 km (65 mi). But the connection with the solar wind and the Van Allen belts was not clarified until artificial satellites began mapping ions in Earth's vicinity.

Even on days or nights when there is no aurora, reactions among molecules in the high atmosphere cause a faint glow at various wavelengths. This glow is called **airglow.** An example is the ultraviolet emission on the daylight side of the Earth in Figure 15-21.

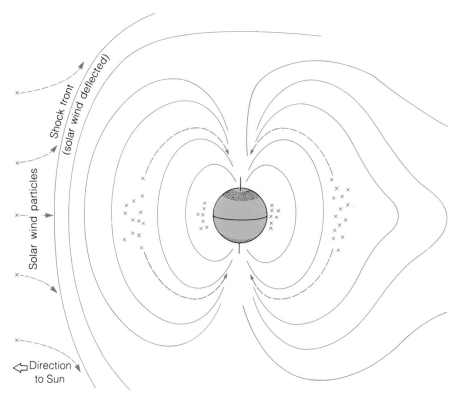

Figure 15-20 Interaction of solar wind ions (from left) with Earth's magnetic field. The shock front is analogous to the bow wave cut by a moving boat. Dashed lines show the typical paths of solar ions (x's). Those ions that penetrate Earth's field accumulate in doughnut-shaped Van Allen radiation belts around Earth. Concentrations of x's mark their positions. Ions in the belts eventually empty into Earth's polar atmosphere, colliding with air molecules, and forming auroral zones near the north and south magnetic poles.

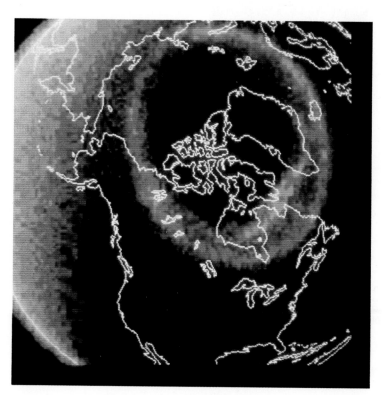

Figure 15-21 Auroral activity on Earth. This ultraviolet image (wavelength 130 nm) shows airglow from excited oxygen, both on the illuminated crescent of Earth (left) and in a ring of auroral glow centered on Earth's north magnetic pole. This glowing ring, corresponding to the stippled ring on the globe in Figure 15-20, is where ions tend to be dumped into the atmosphere from the Van Allen belts. In this false-color image, the computer has added continental outlines. (Dynamics Explorer 1 satellite image; courtesy L. A. Frank, University of Iowa/NASA.)

a

b

Figure 15-22 Various forms of the aurora. **a** Part of a typical curtain display similar to that in Figure 15-23, but seen from Earth's surface in the Alaska Range, Alaska. (Photo by Nancy Simmerman, Alaska Photo.) **b** Unusual ring of vertical rays over Alaska. (Photo by Michio Hoshino, Alaska Photo.)

IS THE SUN CONSTANT?

Solar radiation reaches the Earth at a rate of 1.37 kW/m^2, called the **solar constant.** That is, 1370 J of energy reach every sunward-facing square meter of Earth every second.

This rate is called the "solar constant" because it was once assumed not to vary. But is it truly constant?

Apparently not. During the few days of a single flare, X radiation may change dramatically, though the total solar output changes by less than a percent. During the 22-y solar cycle, the number of these flares changes. Moreover, the 22-y cycle itself is erratic and solar activity varies from one cycle to another. During the period 1645–1715, for example, sunspot numbers were probably unusually low (see Figure 15-13), solar flares were

Figure 15-23 An auroral display as seen from *above*. This spectacular scene was photographed by astronauts in Skylab 3 orbiting in a space shuttle halfway between Antarctica and Australia. Faint moonlit clouds can be seen on Earth, as bright vertical curtains of aurora play across the upper atmosphere. (NASA.)

virtually absent, and coronal light was pale during eclipses (Eddy, 1976). This period is called the **Maunder minimum** after the nineteenth-century astronomer who discovered it in old sunspot records.

Tree-ring patterns in the western U.S. indicate 22-y cycles in droughts and isotope chemistry of trees, suggesting solar cycle effects on weather and plant growth over a 1000-y period (Dicke, 1979). Similarly, Figure 15-24 shows a correlation between the sunspot cycle and murderous Ethiopian droughts during a 434-y period. Monitoring of solar changes themselves is growing more sophisticated. Measurements in the 1980s by the Solar Maximum Mission satellite showed a correlation between the Sun's *total* luminosity at all wavelengths and the sunspot cycle. The total luminosity was about 0.1% brighter during periods of maximum sunspot activity (Foukal, 1990).

These changes may seem small, but small changes in the heat budget of Earth can have drastic effects. Changes in mean annual temperature by only a degree can cause dramatic changes in climate and food production. This is why signs of global warming are causing such concern. Any solar changes, of course, are compounded by human changes to the environment, such as addition of CO_2, which traps the incoming solar heat and raises the temperature of the Earth. The whole question of solar influence on climate and agriculture is

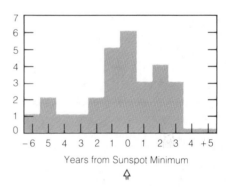

Years from Sunspot Minimum

Figure 15-24 A graph of suspected correlation between terrestrial weather and the solar sunspot cycle. Major droughts in Ethiopia, recorded from 1540 to 1974, appear to correlate with sunspot minima. Such studies may benefit agriculture by helping to predict weather cycles. (Data from Wood and Lovett, 1974.)

attracting new research. A Soviet–American agreement increased international cooperation on these studies (Wilcox, 1976). Solar studies may have growing importance for astronomy, meteorology, agriculture, and world economics. Astronomers trying to understand solar variations have focused on processes in the Sun's outer convective envelope, where instabilities distort the Sun's magnetic field, create flares, and alter the total luminosity (Gilliland, 1982; Sofia and others, 1985).

SOLAR ENERGY AND OTHER COSMIC FUELS

Solar energy has accumulated in the ground as coal and petroleum, originally stored in organisms. Industrial civilization has been built with this fossil solar energy, and a region's standard of living is correlated with its rate of consumption of such energy. But while it took about 300 million years to accumulate the fossil fuels, we have burned through a significant fraction of them in only 100 years. If today's trends continue, fossil fuels will be largely gone in a few decades.

A cosmic perspective helps us recognize a hurdle we face: As a planetary culture, we must be sure that the dwindling reserves of fossil fuel are used to ensure the production of alternative fuels—because *after* we run out of fossil fuels, it won't be easy to develop the technology to produce alternative sources of energy.

With the Sun providing every daylit square meter in the Earth's vicinity with 1.37 kW, solar energy seems a likely alternative to fossil fuel. A 30% efficient 1-m^2 solar collector could run a 400-W appliance.

A giant test of such technology is the power station Solar One, near Barstow, California. It uses 1818 mirrors to focus sunlight on a central collector, generating 10 000 kW of electricity. (See review by Kreith and Meyer, 1983.) The *total* energy needs of the United States (at the use rate of the 1970s) could be met by collectors operating at only 10% efficiency, spread in giant solar farms, perhaps in desert areas covering only a tenth the area of Arizona. Greater collection efficiency would be possible in the permanent sunlight of space. Engineering studies (Arnold, 1980; Lewis, 1977) have indicated the economic feasibility of building orbiting space colonies from lunar or asteroidal materials, with solar collectors that convert solar energy to radio waves and beam it back to Earth (Figure 15-25).

More down-to-Earth is the fact that the roof area of a typical home intercepts roughly 1000 to 2000 kW·h of energy per day, equaling the average daily energy consumption of an American household (Snell and others, 1976). Rooftop solar collectors are a means of using this energy to create heat and electricity. Many people are just realizing that American homes built during the 1950s and 1960s were designed when energy costs were being kept artificially low through government policies. Insulation was skimpy because it was more expensive than the electricity to heat or cool the building. Today, as energy runs out and energy costs rise correspond-ingly, energy costs in many American homes exceed $1000 per year and may rise further. Thus self-sufficient, solar energized homes are increasingly attractive.

In contrast to fossil fuels, energy from basic planetary or astronomical sources might be called **cosmic fuels.** In addition to solar energy, cosmic fuels include nuclear fission energy (used in present-day reactors), nuclear fusion energy (use of solar-type fusion reactions, possibly using water as a source of hydrogen; still in the experimental stage), geothermal energy (heat from the interior of the Earth), and energy from tides and winds. Although nuclear fission energy is the current favorite to get us over the short-term energy hurdle, it has the overwhelming disadvantage of involving and creating deadly radioactive wastes that could wreak health havoc if they escaped into the environment, either by accident or terrorism. New designs of nuclear plants might lessen these dangers (Lester, 1986).

The other cosmic fuels are cleaner than nuclear or fossil fuels. They have an additional subtle advantage: Since they can be collected anywhere in the world, they could make countries energy independent, allowing a more stable planetary culture. As long as the planet depends on fossil fuels, it can be dominated by the economic power of whatever nation controls the cheapest reserves. (The United States once did, but has now burned through them.) The United States and USSR might reduce some of the tensions that come from a "Persian Gulf mentality" by collaborating to develop solar or other cosmic-fueled energy generating systems for Third World as well as First World countries, thus reducing power struggles over fossil fuel deposits. In any case, our generation must respond to the coming exhaustion of fossil fuels not by inaction or a panic-stricken rush to exploit the last reserves, but by a systematic conversion from fossil fuels to safe cosmic fuels.

S U M M A R Y

The Sun can be studied both observationally and theoretically. Observational studies have yielded information about solar gas revealed by spectral absorption and emission lines created in the solar spectrum as sunlight from the Sun's interior passes outward through the layers of solar gas. These studies reveal, among other things, that the Sun is about three-quarters hydrogen; most of the rest is helium, and a few percent is composed of heavier elements.

Theoretical studies reveal that as the Sun formed by contraction of an interstellar gas cloud, it got so hot that

Figure 15-25 In the near future, astronauts may construct giant arrays of solar panels in space to collect "free" solar energy for applications in space stations. Researchers in the United States and USSR have also proposed beaming power from such panels to Earth as a nonpolluting alternative to dwindling supplies of fossil fuels. (Painting by Pamela Lee.)

atoms at the center collided at high speeds. These collisions cause nuclear reactions in which hydrogen atoms are fused into helium atoms, releasing energy. This nuclear fusion is the source of the Sun's light and heat. The most important reactions in the present-day Sun are a series called the proton–proton cycle.

Transport of this energy from the Sun's center to the outer layers violently disturbs the surface, producing phenomena such as granules, prominences, flares, and sunspots. Particles are shot off the Sun in outward-moving gas called the solar wind, which interacts with the Earth, causing aurorae and other phenomena.

Comparing the Sun with other energy sources used on the Earth shows that we are rapidly consuming our planetary budget of fossil fuels and will soon have to convert to cosmic energy sources, such as solar energy, to maintain our present rates of energy consumption.

CONCEPTS

solar rotation

white light

spectrum

absorption line

emission line

continuum

Kirchhoff's laws of radiation

photosphere

spectroheliograph

hydrogen alpha line

Sun's composition

opacity

negative hydrogen ions

nuclear reaction

chemical reaction

ionized gas

proton–proton chain

neutrino

solar core

radiation

conduction

convection

granule

spicule

chromosphere

corona

coronograph

sunspot

plasma

prominence

flare

solar cycle

solar wind

radiation pressure

Van Allen belts

aurora

airglow

solar constant

Maunder minimum

cosmic fuels

PROBLEMS

1. Why do astronomers infer that the Sun's energy comes from nuclear fusion reactions of the proton–proton cycle? How do we know it does not come from chemical burning?

2. If you see a cluster of sunspots in the center of the Sun's disk, how long would the spots take to reach the limb, carried by the Sun's rotation? How long would it take for the cluster to appear again at the center of the disk?

3. Why is the solar cycle said to be 22 y long, even though the number of sunspots rises and falls every 11 y?

4. Suppose you could make detailed comparisons of the appearance of the Sun at different moments of time. What variations in appearance would you see if the intervals were:
   a. 10 min?
   b. 1 wk?
   c. 5 y?
   d. 10 y?
   e. 100 million years?

5. Why is a radio disturbance on the Earth likely to occur within minutes of a solar flare near the center of the Sun's disk, whereas an aurora occurs a day or two later, if at all?

6. How much more massive is the Sun than the total of all planetary mass (see Table 8-1 for data)?

7. What is the most abundant element in the solar system? The second most abundant element?

8. If the Earth formed in the same gas cloud as the Sun, why is the Earth made from different material than the Sun?

9. Why will the Sun change drastically in several billion years?

10. Why is the Sun's energy generated mostly at its center, and not near its surface?

ADVANCED PROBLEMS

11. Using the small-angle equation, calculate the following:
   a. The angular size of a sunspot that has the same diameter as Earth.
   b. The angular size of Earth as seen by an imaginary observer on the Sun.

12. At what velocity must particles move to escape from a region of the corona 2 solar radii out from the center of the Sun?

13. Use Wien's law to confirm the temperature estimate for the Sun's surface, based on a maximum energy emission at wavelength 510 nm, as quoted in the text.

14. a. Calculate the speed of a typical hydrogen ion in a 2-million-degree portion of the solar corona.
   b. Compare this with the escape velocity in the corona (from Problem 12). Using the rule of thumb that the fastest atoms move more than three times faster than the typical thermal velocity, comment on the possibility of solar atoms escaping from the corona and thus supplying gas for the solar wind.

PROJECTS

1. According to the principle of the pinhole camera, light passing through a small hole will cast an image if projected onto a screen many hole diameters away. Confirm this by cutting a 1-cm hole of any shape in a large cardboard sheet and allowing sunlight to pass through the hole onto a white sheet in a dark room or enclosure several feet away (3 m— more if possible). Confirm that the projected image is round, an actual image of the Sun's disk. Are any sunspots visible?

2. Cut a 1-cm hole in a sheet of cardboard and use it as a mask over the end of a small telescope. *After* masking the telescope, point it toward the Sun and project an image of the Sun through an eyepiece onto a white card. **Under no circumstances should anyone ever look through the eyepiece at the Sun, since all the light entering the telescope is concentrated at that point and can burn the retina!** Professional supervision of the telescope is suggested. Note also that an unmasked large telescope may concentrate enough light and heat to crack eyepiece lenses.

Are sunspots visible? If so, trace the image on a piece of paper and reobserve on the next day. Confirm the rotation of the Sun by following sunspot positions for several days. Class records kept from year to year can be used to record the cyclic variations of the numbers of sunspots.

CHAPTER 16

Measuring the Basic Properties of Stars

Now we are ready to take the great leap out of the solar system into the realm of the stars. It is quite a leap—the next star beyond the Sun is 260 000 times as far as the Sun and 6800 times as far as Pluto. If we make a model solar system the size of a half dollar, the neighbor stars would be dots smaller than a period in this book, scattered about half a block apart.

Because of optical limitations and turbulence in the Earth's atmosphere, the world's largest telescope cannot distinguish details smaller than a few hundredths of a second of arc (about 10 millionths of a degree). But the stars with the largest apparent angular size are no larger than about 0.04 seconds. So the disks and surface details of nearly all distant stars are hidden from us, in contrast to the great detail we can study on the Sun.

In spite of this, we know there are giant stars bigger than the whole orbit of Mars, stars the size of the Earth, and stars the size of an asteroid. There are red stars and blue stars. There are stars of gas so thin you can see through parts of them, and stars with rocklike crusts that may contain diamonds. There are stars that are isolated spheres, stars with disklike rings around them, and stars that are exploding.

All of these bodies fit the definition of stars: **Stars** are objects with so much central heat and pressure that energy is (or has been) generated in their interiors by nuclear reactions. The most familiar stars visible in the night sky are balls of gas with solar composition and sizes usually a few times smaller or larger than the Sun.

How can we know all these details about stars if we cannot even see the disks of stars in telescopes? In the next chapters we will describe the details of familiar and unfamiliar types of stars, but first we will describe *how* we learn about them.

NAMES OF STARS

Stars are named and cataloged by several systems. Because Ptolemy's *Almagest* was passed on by Arab astronomers, many of the brightest stars ended up with Arabic names. Since *al-* is the common Arabic article, many star names start with *al:* Algol, Aldebaran, Altair, Alcor. Other scientific *"al words"* also have Arab origins: *algebra, alchemy, alkali,* and *almanac.* Stars in constellations are cataloged in approximate order of brightness using Greek letters. Thus the brightest star in the constellation of the Centaur is called Alpha Centauri (α Centauri). Fainter stars or stars with unusual properties are often known by English letters followed by constellation names or by catalog numbers, such as T Tauri or B.D. 4° + 4048.

IMAGES OF STARS

Figure 16-1 shows a typical telescopic photograph of a bright star. The strange rays or circle have nothing to do with the actual shapes of the stars, which would be tiny pinpoints buried in these overexposed images. The rays and circle are caused by optical effects in the telescope and film system.

Visual examination of a bright star in a modest telescope shows what appears to be a tiny disk surrounded by faint rings, but the disk and rings are caused by an optical effect called **diffraction.** The disk is sometimes called the **Airy disk** after the English astronomer who explained it. It again has nothing to do with the real star and in fact prevents the much smaller real star image from being seen.

An additional problem arises in color photos of stars and other celestial light sources such as galaxies and nebulae. The length of time of the exposure (that is, the amount of light collected) has to be just right to render the color correctly. If the exposure is too short, the image is too faint to be clear. But if the exposure is too long, the image is overexposed and the film (or other recording medium) renders it as bright as possible—that is, white. Thus a red star's or blue star's overexposed image shows up as white (as in Figure 16-1). The true color may show only in an aura around it where fewer photons have struck the film. Thus in a long exposure designed to show faint details of stars and nebulae with widely different intensity, the brightest stars and brightest parts of the nebulae may look white because

they have been exposed too long to preserve their true colors.

In 1975 astronomers at Kitt Peak National Observatory picked one of the stars with the largest apparent angular size, the bright star Betelgeuse in the constellation Orion, and used computer techniques to remove the blurring due to atmospheric turbulence and diffraction. Their analysis produced an image of a fuzzy disk, shown in Figure 16-2, which is believed to be the actual disk of Betelgeuse. Analysis showed that this star's actual linear diameter is 500 to 750 times larger than the Sun's. If Betelgeuse replaced the Sun, it would engulf the Sun and the terrestrial planets and extend out into the asteroid belt! This alone proves that not all stars are alike.

DEFINING A STELLAR DISTANCE SCALE: LIGHT-YEARS AND PARSECS

The vast distances that separate stars and make them so hard to observe are awkward to express in ordinary units. Astronomers use units appropriate to these distances. The easiest to understand is the **light-year** (abbreviated "ly"), the distance light travels in 1 y, which is about 6 million million miles, or 10^{16} m.

Remember that the light-year is a unit of *distance,* not time. The common mistake of using light-year as if it were a unit of time is like saying that the ball game lasted for 2 miles.[1]

The nearest star beyond the Sun, Proxima Centauri (which is in orbit around Alpha Centauri), is about 4.3 light-years away. The Sun could be said to be 8 light-minutes away. The North Star, Polaris, is about 650 light-years away. Polaris' light takes about 650 y to reach us, so we are seeing it now as it was in the 1300s! If Polaris had suddenly exploded in 1950, we would not know it until about A.D. 2600.

Astronomers more commonly use a still larger unit of distance called the **parsec** (abbreviated "pc"):[2]

$$1 \text{ parsec} = 3.26 \text{ light-years}$$
$$= 3 \times 10^{16} \text{ m}$$

In this book we will use the parsec and its multiples (kiloparsecs and megaparsecs) to express cosmic dis-

---

[1] Even noted space pilot Han Solo confused distance and time units when he claimed in *Star Wars* that the Millennium Falcon could "make the Kessel run" in less than 12 parsecs.

[2] The reason for defining this unit will become clearer on page 353.

Figure 16-1 Telescopic image of the bluish star Alcyone—brightest star in the Pleiades—and adjacent stars. The true disk of the star would be a microscopic point in the center of the overexposed image. The four spikes and the surrounding circle are not stellar phenomena, but only optical effects produced inside the reflecting telescope and photographic plate used to make the picture. Optical effects distribute a small percentage of each star's light in this pattern, but only for overexposed stars like Alcyone is there enough light to record the spikes or ring. The spikes are called *diffraction spikes*. The fainter stars in the picture illustrate the great range of apparent brightness among stars. Faint blue wisps around Alcyone are portions of a gas cloud associated with the star—a feature rarely visible among stars. (Copyright by California Institute of Technology and Carnegie Institution of Washington, by permission from the Hale Observatories.)

Figure 16-2 Special optical techniques were used to create this red-light image (wavelength 710 nm) of the disk of the red supergiant star Betelgeuse in the constellation Orion. This star, only 0.06 second of arc across, has one of the largest angular sizes of any star seen from Earth. The brighter region is probably hotter gas upwelling from the star's interior. (False color image with 4.2-m telescope, courtesy David Bushcer, Mullard Radio Astronomy Observatory, Cambridge UK.)

tances as we move to more remote parts of the universe. You can convert parsecs to light-years approximately by multiplying by 3. Another convenient fact to remember is that near the Sun, stars are roughly a parsec apart. For instance, Alpha and Proxima Centauri, the closest stars to the solar system, are about 1.3 parsecs away.

DEFINING A BRIGHTNESS SCALE: APPARENT MAGNITUDE

A nearby candle may *appear* to be brighter than a distant streetlight, but in *absolute* terms the candle is much dimmer. This statement contains the essence of the problem of stellar brightness. A casual glance at a star does not reveal whether it is a nearby glowing ember or a distant great beacon. Hence astronomers distinguish between **apparent brightness** (the brightness perceived by an observer on Earth) and **absolute brightness** (the brightness that would be perceived if all stars were magically placed at a standard distance).

Let us look first at apparent brightness. Because scientists generally use the decimal system, we might expect astronomers to use a brightness scale in which each unit of brightness difference corresponds to a factor of 10. No such luck. Astronomers, too, are victims of history. When Hipparchus cataloged 1000 stars in about 130 B.C., he ranked their **apparent magnitude,** or apparent brightness, on a scale of 1 to 6, with 1st-magnitude stars the brightest and 6th-magnitude stars the faintest visible to the naked eye. This system stuck. As the famous astrophysicist Arthur Eddington remarked: "You have to remember that stellar magnitude is like a golfer's handicap—the bigger the number, the worse the performance."

For more precision, nineteenth- and twentieth-century astronomers extended that scale to cover objects brighter than 1st magnitude and dimmer than 6th mag-

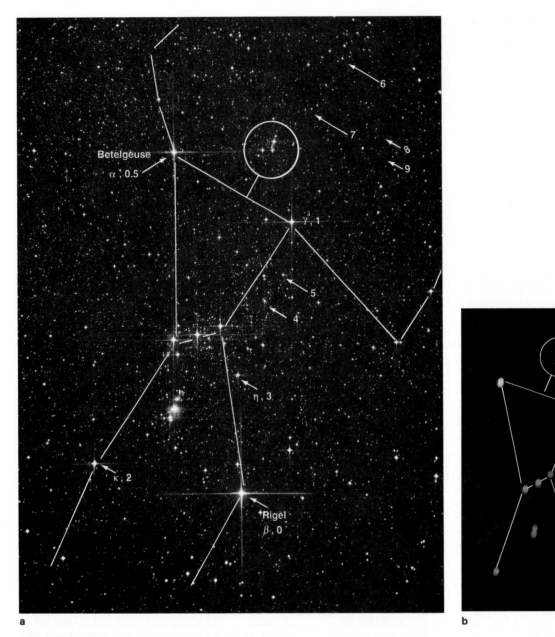

a

b

Figure 16-3 a The constellation of Orion the Hunter (stick figure) showing Greek letter designations and approximate visual apparent magnitudes of certain stars. This exposure records stars down to about 10th magnitude. Different colors of stars are evident; most stars in this region are hot, blue stars, but Betelgeuse is a prominent cool, orangish-red star. (These color differences can be seen by the naked eye.) The three stars below Orion's Belt mark the sword. The middle "star" of the sword is the Orion Nebula (see Chapter 20), marked by red-glowing hydrogen gas. (Photo by author; 11-min guided exposure on 3M ISO 1000 film, 55-mm lens, f2.8, "star filter" emphasize bright stars with diffraction spikes.) **b** An image of the same region, deliberately out of focus, expands the images of the stars into blobs that emphasize their color. Note the blue colors of most stars, due to their high temperatures, and note the difference in reddish hue between the orange thermal radiation of Betelgeuse and the red nonthermal Hα emission from the Orion Nebula. (Photo by author, 12-min guided exposure on Ektachrome ISO 1600 55-mm lens, f2.8.)

TABLE 16·1

Apparent Magnitudes of Selected Objects

Object	Apparent Magnitude
Sun	− 26.5
Full moon	− 12.5
Venus (at brightest)	− 4.4
Mars (at brightest)	− 2.7
Jupiter (at brightest)	− 2.6
Sirius (brightest star)	− 1.4
Canopus (second brightest star)[a]	− 0.7
Vega	0.0
Spica	1.0
Naked eye limit in urban areas	3–4
Uranus	5.5
Naked eye limit in rural areas	6–6.5
Bright asteroid	6
Neptune	7.8
Limit for typical binoculars	9–10
Limit for 15-cm (6-in.) telescope	13
Pluto	15
Limit for visual observation with largest telescopes	19.5
Limit for photographs with largest telescopes	23.5
Expected limit for Hubble Space Telescope	28±

Source: Data from Allen (1973)

[a]This lesser-known Southern Hemisphere star is used as a prime orientation point for spacecraft. A small light detector on spacecraft is called the "Canopus sensor."

nitude. Because Hipparchus' 1st-magnitude stars are about 100 times brighter than his 6th-magnitude stars, they used this as the definition of the brightness system. A difference of five magnitudes is *defined* as a factor of 100 in brightness. Therefore, any given star

is about $2\frac{1}{2}$ times brighter than a star of the next fainter magnitude.[3]

The system is illustrated in Figure 16-3, which shows a photo of the constellation Orion the Hunter, with some stars labeled with their magnitudes (0 to 9) on the right and, in the case of brighter stars, their Greek letter names on the left.

Table 16-1 shows how the apparent magnitude scale has been extended, including fractional magnitudes. Extending the scale beyond the brightest objects included by Hipparchus meant using negative numbers. Thus the planet Venus at its brightest ranks more than minus 4th magnitude (− 4.4), whereas the sun reaches beyond minus 26th (− 26.5).

Our accuracy in measuring stellar brightness is only 30 times better than Ptolemy's was 1800 y ago! One problem is that stars have different colors, and light detectors (the eye, photographic films, and photoelectric devices) have different sensitivities to different colors. For this reason, astronomers specify exactly what color any set of measurements refers to. Standards have been derived to express apparent magnitudes measured in blue light, red light, infrared light, and so on. Here we will usually be refering to a system having the same color sensitivity as the human eye, sometimes called **visual apparent magnitude** (often abbreviated m_v).

A MAGNITUDE SCALE FOR EXPRESSING "TRUE" BRIGHTNESS OF STARS

The apparent magnitudes that we have just discussed describe only the *apparent* brightness of each star as seen from Earth. This depends on the star's distance and thus doesn't express the star's true energy output. This true, or intrinsic, energy output can be expressed using a magnitude system called the star's **absolute magnitude,** usually expressed by the apparent magnitude the star would have if it were moved to a standard reference distance of 10 pc. In this system, a star with the same energy output as the Sun has an absolute magnitude of about + 5. A star with 100 times that energy output is said to have an absolute magnitude of 0. A star 100 times fainter has an absolute magnitude of + 10.

---

[3]The actual factor is $\sqrt[5]{100}$, or 2.512. This unusual factor as a unit of brightness difference is used in no other science. Physicists are heard to mutter obscenities when they first try to analyze astronomical data expressed in magnitudes. But the 2100-y-old magnitude system is so ingrained in astronomy that we continue to use it here.

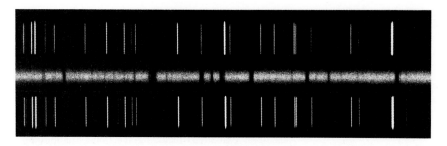

Figure 16-4 High-resolution spectra photographed on black and white film with a large telescope usually take this form. The stellar spectrum is the fuzzy horizontal luminous band in the center, crossed by vertical dark absorption lines. It is flanked at top and bottom by emission lines of a known element (usually iron) produced artificially by a spark inside the instrument. These lines, whose positions are known, form a reference scale that permits measurement of line wavelengths in the stellar spectrum. (Lick Observatory.)

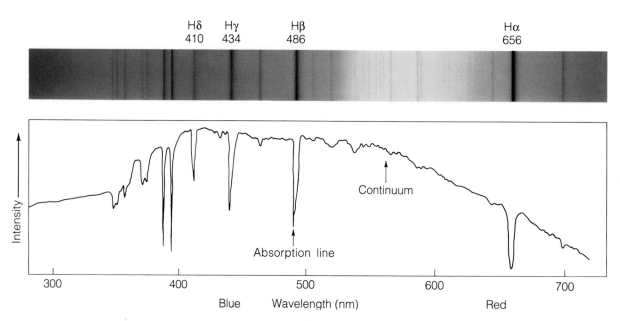

Figure 16-5 The visual portion of a typical stellar spectrogram is shown here in schematic form as it would appear on a color photographic plate (top) and in a scan by a spectrophotometer (bottom). The spectral lines appear as dark vertical streaks in the photographic spectrum and as deep valleys in the scan.

BASIC PRINCIPLES OF STELLAR SPECTRA

The last chapter showed that the Sun's **spectrum**, or distribution of light into different colors (**wavelengths),** gives information about the Sun's atmo-

sphere, surface layers, and interior. The same is true of other stars. So critical is **spectroscopy**—the study of spectra—to astronomy that many astronomers devote their entire careers to it. Chapter 5 gives more details on the principles of spectra, and you may wish to review that material while studying this chapter. Let us review

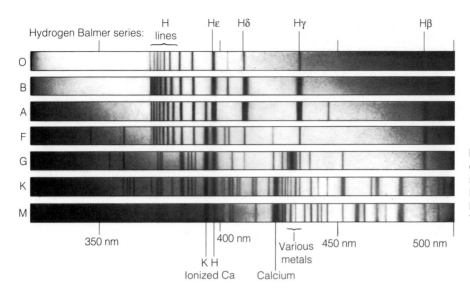

Figure 16-6 Representations of spectra of stars in the seven spectral classes. Classes A and B show especially well the Balmer hydrogen lines (compare with Table 16-2).

the nature of spectra to see how they can be used to reveal the nature of stars.

As mentioned briefly in Chapter 5, spectra can be presented in two ways—photographically (Figure 16-4) or as a chart (Figure 16-5, bottom), depending on the instrument used. As seen in Figure 16-4, a **spectrograph** produces a photographic image of the spectrum. An idealized example of a spectrum in color appears in Figure 16-5 (top). The intensities of the photographic print represent intensities of light, or radiant energy. The vertical width of the spectrum is arbitrary; sometimes such spectra are artificially enlarged vertically to make an image more readily measurable. Usually the blue end is to the left, as in the figures.

A **spectrometer** produces a spectrum in the form of a chart or graph, as in the lower portion of Figure 16-5. Wavelength is shown along the bottom, again with red usually to the right; the vertical direction represents the intensity of the light. **Absorption lines** appear as dark vertical lines on a photographic spectrum and as notches or valleys on a graph. **Emission lines** appear as bright vertical lines or as sharp peaks (review page 100). The general level of brightness between absorption or emission lines is the **continuum**.

Spectra of Stars

In 1872, Henry Draper, a pioneer in astronomical photography, first photographed stellar spectra. This represented a tremendous advance. Instead of sketching or verbally describing spectra, astronomers could directly record, compare, and measure them. Spectra of thousands of stars became available for precise analysis.

Such massive amounts of data required a classification scheme. This was begun in the 1880s by Harvard astronomer Edward C. Pickering and completed by Annie J. Cannon and a group of young women assistants who invented a system of **spectral classes** based on the number and appearance of spectral lines. The classes—A, B, C, and so on—started with spectra with strong hydrogen lines. When Annie Cannon published the Henry Draper Catalog (1918–1924), it contained spectral data on 225 320 stars and became the basis for all modern astronomical spectroscopy.

Further work showed that the classes had to be rearranged to bring them into a true physical sequence based on temperature, as shown in Figure 16-6. The sequence finally adopted begins with the hottest stars—class O—which show ionized helium lines in their spectra. The sequence of classes is O, B, A, F, G, K, and M.[4] The M stars are the coolest. About 99% of all stars can be classified into these groups. For finer discrimination the classes are sometimes divided from 0 to 9; the Sun, for example, is said to be a G2 star.

---

[4] As Annie Cannon noted, the rearrangement of letters made the sequence harder to remember. In the 1920s, American astronomer Henry Norris Russell solved this problem with Yankee ingenuity by proposing a sentence as a memory aid: Oh, Be A Fine Girl, Kiss Me! Two later classes, R and N, were easily accommodated by adding the words *Right Now*. However, when class S was added, a controversy broke out between Harvard and Cal Tech over whether it signified the word *Sweetie* or *Smack*.

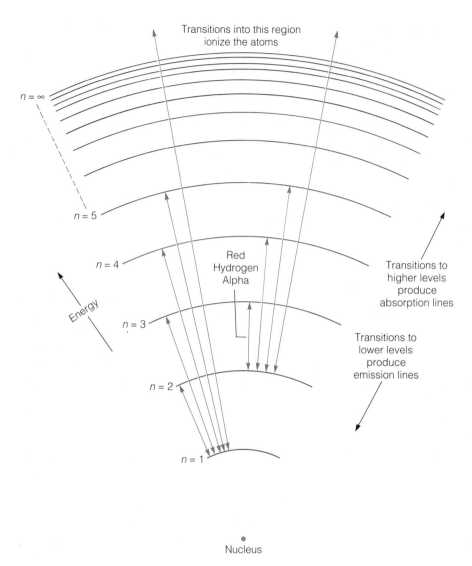

Figure 16-7 Schematic diagram of a hydrogen atom showing different possible electron orbits or energy levels (n = 1, 2, 3, . . .). Each change of orbit by an electron produces a spectral line, because energy is removed or added to the light beam. The series of lines starting from or ending on n = 1 occur in the ultraviolet part of the spectrum and are called the Lyman series. Lines starting from or ending on n = 2 occur in the visible part of the spectrum and are called the Balmer series. The well-known red Hα line of the Balmer series involves transitions between n = 2 and n = 3.

Spectral Lines: Indicators of Atomic Structure

In Chapter 5 we gave a simplified description of how emission and absorption lines depend on the orbital structure of electrons in atoms. Now we look at the process in more detail by considering the simplest and most abundant type of atom in the universe: hydrogen. As shown in Figure 16-7, hydrogen atoms' orbits can be numbered. Since a neutral hydrogen atom has just one electron, an atom in the ground state would have one electron in the orbit n = 1. If the atom had been bumped by other atoms or if it had absorbed radiation, the electron might be in the orbit n = 2, 3, and so on.

Further absorption of energy might cause it to jump from n = 3 to n = 4, creating an absorption line, or it might spontaneously revert from n = 3 to n = 2, creating an emission line. Each possible transition (1 to 2, 2 to 3, 4 to 2, and so on) creates a different line. As Figure 16-7 shows, transitions between the n = 2 level and higher levels create the lines prominent in the visible part of the spectrum, called the **hydrogen Balmer series** of lines. The famous red line called **hydrogen alpha** (Hα, shown in Figure 16-5, top) or **Balmer alpha**, involves transitions between n = 2 and n = 3 in the hydrogen atom. This line lends its brilliant red color to many astronomical gases, including the solar chromosphere and many nebulae.

A hydrogen atom needs to absorb only about 10^{-19} to 10^{-18} joules to bump its electrons from one level to another. About 2×10^{-18} joules is needed to ionize a hydrogen atom in its ground state. These are small amounts of energy, but sources of such energy are important in determining the state of the gas. In a hydrogen-rich environment where either the photons from surrounding radiation or jostling by neighboring atoms provides this much energy, many hydrogen atoms will be ionized.

The astute reader will have noticed that the strengths of different hydrogen lines from a specific gas sample will depend on the fraction of atoms with the $n = 1$ level occupied by an electron, the fraction with the $n = 2$ level occupied, and so on. This in turn is controlled by the temperature, density, and pressure in the gas. The higher the temperature, the harder the atoms bump together and the higher the fraction with electrons excited into higher energy levels. For any given element, how can we tell exactly what fractions of the atoms are in the ground state and each excited state and thus what the *relative strengths of different spectral lines will be?*

This crucial question began to be answered by the work of the Indian physicist Megh Nad Saha soon after the stellar spectral classes were recognized. In 1920 he derived an important result, now called the **Saha equation,** which allows us to tell the relative numbers of atoms in each excited state given the conditions in the gas or vice versa.

Note then the power of the knowledge we have discussed so far. *Identification* of the spectral lines tells us what elements are present in a star. *Measurement of the relative line strengths,* interpreted with the Saha equation, tells us what fraction of the atoms is in each excited state—and hence tells us something about the temperature, density, and pressure conditions in the photospheric layer where the spectrum arises in the surface of the star. This method supplements the temperature information gained from other techniques, such as Wien's law, discussed in Chapter 5.

Spectral Classes as a Temperature Sequence

Each spectral class corresponds to a different *temperature* in the gases in the light-emitting layers of the stars. **Temperature** is merely a measure of the average velocity of the atoms or molecules of a substance.

The hotter a gas, the faster its atoms and molecules move. If the temperature quadruples, the velocities of the atoms double. The faster the atoms collide, the more they disturb or dislodge each other's electrons. Furthermore, the hotter the gas, the more radiation it emits; the resulting photons also disturb electron structures of atoms. Because spectral lines depend on electron structures, and because spectral classes depend on spectral lines, spectral classes therefore form a temperature sequence.

Consider what happens as we reduce the temperature of matter. The hottest stars, the O stars, have temperatures of 40 000 K or more, as shown in Table 16-2. Here the atoms and radiation are so energetic that even the most tightly bound atoms, such as helium, have had their electrons knocked off, as the table indicates. By class B, the temperature has dropped to around 18 000 K, and helium atoms keep their electrons. By classes A and F, even relatively weakly bound hydrogen keeps its electrons. Many metals have at least one electron that can be very easily knocked off the outer part of the atom, and in G-type stars like the Sun, at around 5500 K, lines of ionized metals are prominent along with Balmer lines of hydrogen. By class K, at 4000 K, even many metals are neutral. By class M, at 3000 K, energies are so low that different atoms can stick together into molecules; even water molecules have been identified in spectra of such cool stars, as shown in Figure 16-8.

Stars cooler than M stars are faint and rarely seen, but the concept may be extended to lower temperatures. In Chapter 14, we discussed how grains of minerals such as silicates can form at low temperatures of around 2000 to 1300 K; such grains probably exist in the atmospheres of the coolest stars.

Planets, of course, have even cooler temperatures, such as 300 K for the Earth, where liquid water coexists with solid rocks. At cool enough temperatures, near absolute zero, nearly all matter would assume solid form.

Two Important Laws of Spectroscopy

We now introduce two important effects that will help us understand many astronomical phenomena. Both are described in mathematical detail in the accompanying optional boxes, but they need to be briefly described to all readers, because they are important in interpreting distant stars.

TABLE 16·2

Principal Spectral Classes of Stars

Type	Spectral Class	Typical Temperature (K)	Source of Prominent Spectral Lines	Representative Stars
Hottest, bluest	O	40 000	Ionized helium atoms	Alnitak (ζ Orionis)
Bluish	B	18 000	Neutral helium atoms	Spica (α Virginis)
Bluish-white	A	10 000	Neutral hydrogen atoms	Sirius (α Canis Majoris)
White	F	7000	Neutral hydrogen atoms	Procyon (α Canis Minoris)
Yellowish-white	G	5500	Neutral hydrogen, ionized calcium	Sun
Orangish	K	4000	Neutral metal atoms	Arcturus (α Bootes)
Coolest, reddest	M	3000	Molecules and neutral metals	Antares (α Scorpii)

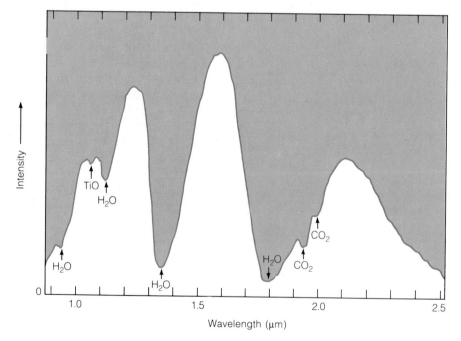

Figure 16-8 A portion of the infrared spectrum of the giant star o Ceti (Mira), showing identification of absorptions by various molecules, including the first identification of water vapor in a star. (After a diagram by G. P. Kuiper.)

The *Doppler effect* is a shift in wavelength of a spectral absorption or emission line away from its normal wavelength, caused by motion of the light source toward or away from the observer.

If the source approaches, there is a **blue shift** toward shorter wavelengths, or bluer light. If the source recedes, a **red shift** occurs toward longer wavelengths, or redder light. The amount of the shift is just proportional to the approach or recession speed of the source (as long

---

OPTIONAL BASIC EQUATION VII

The Doppler Effect: Approach and Recession Velocities

Probably the most important physical phenomenon in astronomical spectroscopy is the Doppler effect (named after the Austrian physicist Christian Doppler, who discovered it in 1842). *The Doppler effect is a change in wavelength proportional to any line-of-sight velocity between observer and source.* (That is, the wavelength of a spectral line from a fast-approaching body will differ from that of a fast-receding body *and* that of one slowly approaching.)

Doppler discussed the effect in the light from distant stars orbiting around each other, but his effect also applies to any signal transmitted by waves, including familiar sound waves on Earth. The most familiar example of the Doppler effect is the shift in pitch of sound from an approaching or receding source. As a car or train passes by and recedes in the distance, the pitch of the sound dramatically decreases. As the source approaches, the sound waves rush past the observer apparently more closely spaced than normal, decreasing the observed wavelength. As the source recedes, the waves are perceived to be more spread out, increasing the wavelength. The effect modifies whatever property of the signal is determined by wavelength—the color of light or the pitch of sound.

The equation that describes the Doppler effect is the seventh of the nine major equations we will use in this book. Since λ is used to designate wavelength, Δ to designate change, v to designate velocity, and c to designate the speed of light, the equation cited in the text is commonly written $\Delta\lambda/\lambda = v/c$. Read this as "change in wavelength divided by the original wavelength equals v divided by c." Thus if the source of light is not approaching or receding, the Doppler shift is zero. The faster the source approaches or recedes, the greater the Doppler shift in wavelength of the light.

Suppose a certain infrared spectral line normally has a wavelength of 1000 nm. If the light source is receding from the observer at 1/1000 the speed of light, the line would appear at the slightly longer wavelength of 1001 nm, a little toward the red end of the spectrum. If the source approached at 1/1000 the speed of light, the line would appear at 999 nm.

When source and observer are getting closer together, the wavelengths of all light coming from the source are *shifted toward the blue* (shorter wavelengths). But when source and observer are getting farther apart, the spectrum of light from the source is *shifted toward the red* (longer wavelengths).

The two kinds of shifts are loosely referred to as red shifts and blue shifts, and their sources are called red-shifted or blue-shifted. As a mnemonic aid, remember that *re*cession produces *re*d shifts.

Note that the observer of a Doppler shift cannot say whether it is the source or the observer who is moving relative to any absolute external frame of reference. According to the principle of relativity, only recession or approach of one body with respect to another can be measured.

Example. The hydrogen alpha absorption line in a certain star is carefully measured and found to lie at 656.4 nm instead of the normal 656.3. Describe the motion of the star in the line-of-sight direction toward or away from the observer. Does this observation tell us anything about motion perpendicular to this direction? *Solution:* Solving the equation for v, and looking up c in Appendix 2 (Table A2-2), we have

$$v = \frac{\Delta\lambda}{\lambda}c = \frac{0.1}{656.3} \times 3(10^8) = 4.6\,(10^4)\,\text{m/s}$$

The Doppler shift is toward longer wavelength, or redward, so the star must be moving away. The Doppler shift is sensitive only to motion toward or away from us, so we cannot say anything about a possible additional component of motion laterally. We can only say it is moving away from us at 46 km/s in the line of sight.

as that speed is well below the speed of light). In the shorthand of mathematics,

$$\frac{\text{Shift in wavelength}}{\text{Normal wavelength}} = \frac{\text{approach or recession speed}}{\text{velocity of light}}$$

If the light source is receding at 10% the speed of light, the light will be red-shifted by 10% of its normal wavelength; a line normally found at wavelength 500 nm would appear at 550 nm. By measuring such wavelengths, velocities of stars toward us or away from us can be studied.[5]

> The *Stefan–Boltzmann law,* discovered by two Austrian physicists about 1880, states that the higher the temperature of a surface, the more energy radiated by each square centimeter in each second.

---

[5]The same shift occurs for sound. This explains why the noise of a rapidly traveling automobile, truck, or train changes from higher pitch (shorter wavelength) to lower pitch (longer wavelength) as the vehicle rushes by. The sound a child makes to imitate a race car rushing by is a representation of the Doppler shift.

OPTIONAL BASIC EQUATION VIII

The Stefan–Boltzmann Law: Rate of Energy Radiation

Before describing how to measure specific properties of stars, we need the eighth of the nine basic equations to be used in this book. This is the Stefan–Boltzmann law, which describes the total amount of energy radiated in each second from any hot surface. The total energy radiated per second by a star is called its luminosity, abbreviated L.

In studying Wien's law, we saw that every warm body radiates. The higher the temperature, the bluer the radiation. About a century ago the Austrian physicists Josef Stefan and Ludwig Boltzmann discovered another characteristic: *The higher the temperature, the more energy is radiated each second,* as shown in Figure 16-9. Although the subjects of their study were radiating objects in the laboratory, the law applies to stars and all other bodies in the universe. The law gives the total energy L radiated per second from a body with temperature T and surface area A. For bodies that radiate efficiently, like stars, the law is:

$$L = \sigma T^4 A$$

Sigma (σ), called the Stefan–Boltzmann constant, equals 5.67×10^{-8} W/m^2·K^4, T is given in Kelvin, A in square meters, and L in the SI metric units of watts.

This equation shows that if the temperature or area of a star increases, the total energy radiated every second will increase. Suppose the star has some radius R; then its area is $A = 4\pi R^2$. Thus

$$L = 4\pi\sigma T^4 R^2$$

This variation of the Stefan–Boltzmann law allows us to calculate a star's radius if the luminosity and temperature are known.

Example. Suppose a certain B-type star is found to have a temperature of 18 000 K and a luminosity of 10 000 times that of the Sun. What is its radius relative to the Sun's radius? *Solution:* The straightforward way to do the problem is to solve the equation for R:

$$R = \sqrt{\frac{L}{4\pi\sigma T^4}}$$

We then look up the luminosity of the Sun, $L_\odot$, in Appendix 2 (Table A2-2), multiply it by 10 000 to get L, and plug in the appropriate numbers. The answer, in meters, can be compared to the Sun's radius, using the diameter listed in solar system data (Table 8-1). Solve the problem this way for experience. However, also try the following simpler way. Since we asked for R relative to the Sun, write the equation again for the Sun, with solar subscripts on R, L, and T. Then divide it into the preceding equation to get the ratio of stellar R to solar R. Many factors cancel and we get:

$$\frac{R}{R_\odot} = \sqrt{\frac{L}{L_\odot}\left(\frac{T_\odot}{T}\right)^4} = \sqrt{10\,000\left(\frac{5700}{18\,000}\right)^4} = 10.0$$

(Ask your instructor for help if this is not clear.) In this method the bothersome constants simply cancel out, since we are solving for the ratio, and we quickly see that the star is 10 times bigger than the Sun.

The mathematical form of the law gives the total energy E or luminosity L (in joules) radiated per second by a surface of area A (square meters) and temperature T (Kelvin). The law states that L is proportional to AT^4. If the temperature of a source doubles, the amount of energy radiated increases by 2^4, or 16. Thus while doubling the area of a star would increase its output twice, doubling its temperature would increase its output 16 times! Therefore, as shown in Figure 16-9, the hotter stars not only radiate bluer light than cooler stars (a result that was predicted by Wien's law), but also radiate *more* light per unit area. The Stefan–Boltzmann law thus provides us with a way to learn about the areas of stars once we measure their temperatures and total radiation output.

MEASURING 12 IMPORTANT STELLAR PROPERTIES

We now can apply the Stefan–Boltzmann law, the Doppler effect, and other principles to explain how 12

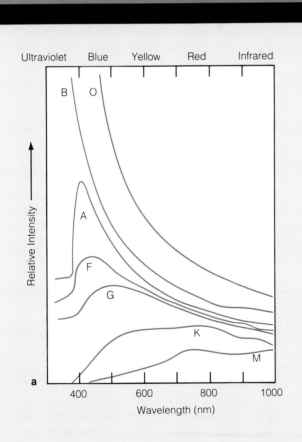

Figure 16-9 A schematic comparison of the energies emitted by various stars. **a** Curves show intensity of radiation in the continuous spectrum of each star, neglecting absorption lines. The hottest and brightest are O stars, which emit most of their energy as blue and ultraviolet light. G stars, like the Sun, emit most of their radiation in the visible part of the spectrum, especially as yellow light. The faintest stars, class M, emit most light in the red and infrared part of the spectrum. Absorptions cause irregular structure in the curves. Compare similar diagram for idealized radiator at different temperatures, Figure 5-3. **b** Schematic continuum spectrum of the O star, strongly enhanced toward blue. **c** Schematic continuum spectrum of the M star, strongly red. This spectrum would require a longer exposure than *b*, because the M star is intrinsically much fainter.

12 Basic Properties of Stars and How They Are Measured

Property	Method of Measurement
Distance	Trigonometric parallax
Luminosity (absolute magnitude)	Distance combined with apparent brightness
Temperature	Color or spectra
Diameter	Luminosity and temperature
Mass	Measures of binary stars and use of Kepler's laws
Composition	Spectra
Magnetic field	Spectra, using Zeeman effect
Rotation	Spectra, using Doppler effect
Atmospheric motions	Spectra, using Doppler effect
Atmospheric structure	Spectra, using opacity effects
Circumstellar material	Spectra, using absorption lines and Doppler effect
Motion	Astrometry or spectra, using Doppler effect

important physical properties of remote stars can actually be measured. These applications, to be discussed in the following sections, are summarized in Table 16-3.

How to Measure a Star's Distance

From what you have read so far, can you think of a way to measure the distance of a star? The question is not far-fetched, since star distances were first measured more than a century ago. A crude method may be applied to the information in Table 16-1. The brightest stars have *apparent* magnitudes about 25 magnitudes fainter than the Sun. In 1829, the English scientist William Wollaston used this simple fact to estimate that most typical stars must be at least 100 000 times more distant than the Sun, since dimming by 25 magnitudes corresponds to increasing distance 100 000 times.

Because this gives only a typical distance, we need a better technique to measure distances of individual

stars. The most important such technique is **parallax** measurement. You will recall that parallax is the apparent shift in the position of an object caused by a shift in the observer's position (p. 48).

The principle is easy to understand, as shown in Figure 16-10. Hold your finger in front of your face and look past it toward some distant objects. Your finger represents a nearby star; the objects, distant stars. Your right eye represents the view on one side of the Sun. Your left eye represents the view 3 mo later, after the Earth has traveled to a point lined up with the Sun (a shift in the Earth's position by 1 AU as seen from the star). First wink one eye and then the other. Your finger (the nearby star) seems to shift back and forth. Hold your finger only a few centimeters from your eyes; the shift is large. Hold your finger at arm's length; the shift is smaller. Likewise, the farther away the nearby star, the smaller the parallax. The parallax in this experiment may be measured in degrees, but the parallaxes of actual stars are all less than a second of arc.

Chapter 2 describes how Aristotle correctly realized, around 350 B.C., that if the Earth moved through space, nearby stars ought to show parallax. Seeing none, he concluded that the Earth stood still. Actually, the shift was there, but too small for him to measure. Likewise, around 1600, Tycho looked for stellar parallax with the naked eye and found none. He concluded (correctly) that annual movements of the Earth were of small scale compared with stellar distances.

In 1837 the German–Russian astronomer Friedrich Struve published an analysis of stars that were expected to be closest to the Earth (and therefore those most likely to have the largest and most easily detected parallaxes). He chose the brightest stars with the fastest angular motions across the sky, just as on a dark night one might sensibly infer that the closest fireflies are the ones that are brightest and move with the highest angular speeds. (These angular motions, though not apparent to the naked eye, can easily be measured from year to year with the telescope.)

Now the race was on to see who could measure the first true parallax. Credit is usually given to the German astronomer Friedrich Bessel, who in 1838 published the parallax and distance of 61 Cygni (star 61 in the constellation of Cygnus the Swan). Bessel found the parallax of 61 Cygni to be $\frac{1}{3}$ second of arc and the distance to be about 3 pc. Most stars are farther than 3 pc away, and their parallaxes are correspondingly smaller. Parallaxes as small as $\frac{1}{20}$ second of arc have

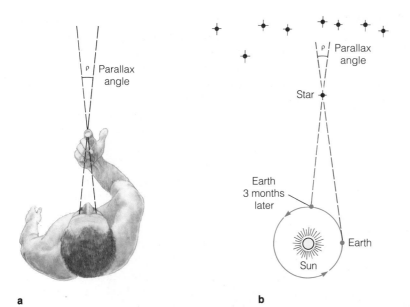

Figure 16-10 The principle of parallax determination applied to one's finger **a** and to stars **b**. Nearby objects observed from two positions appear to shift by a parallactic angle P compared with background objects; measurement of P gives the distance of the object, once the separation of the two positions is known. (By astronomical convention, the angle cataloged as a star's parallax corresponds to the angle diagrammed here; but astronomers actually observe from two positions 6 mo apart. This yields twice as large a parallactic shift, which is easier to measure.)

a   **b**

been reliably measured, and the distance of such stars is 20 pc.

As the preceding figures show, the distance of a star in parsecs is simply the inverse of its parallax angle in seconds of arc. This explains the origin of the term parsec: It is the distance corresponding to a PARallax of one SECond of arc.

The more distant a star, the smaller its parallax. Thus if a star is too distant, its parallax is too small to be measured. Parallaxes smaller than about $\frac{1}{20}$ second of arc are difficult to measure accurately. *Therefore stars farther away than about 20 pc are beyond the* **distance limit for reliable parallaxes,** although distances from 20 pc to 100 pc can be roughly estimated by the parallax technique.

Knowing accurate distances is the prime requirement of measuring most other properties of stars. Thus the estimated 1000 to 2000 stars that lie within 20 pc are our main statistical sample for measuring stellar properties. Other techniques have been devised for estimating distances to more remote stars, but they all depend ultimately on the accuracy of the parallax measures of nearby stars. (See Figure 16-11, for example, which applies the concept of absolute magnitude, or luminosity, discussed next.)

How to Measure a Star's Luminosity

The **luminosity** of a star is its *absolute brightness*— the total amount of energy it radiates each second. Unlike apparent brightness (discussed earlier in this chapter), luminosity is intrinsic to a star. The most useful concept of luminosity is **bolometric luminosity**—*the total amount of energy radiated each second in all forms at all wavelengths.*

A second concept sometimes used is *visual luminosity*—the energy radiated each second in the visible part of the spectrum (the wavelengths to which the human eye is sensitive). Since many stars radiate primarily visible light, the two kinds of luminosity, bolometric and visual, are often roughly the same. In this book *luminosity* will generally mean *bolometric luminosity,* abbreviated L. Similarly, our use of the term *absolute magnitude* will generally refer to *absolute bolometric magnitude,* a measure of the bolometric luminosity expressed in the magnitude system.

The bolometric luminosity of the Sun ($L_\odot$) is 4×10^{26} watts. (*Note:* one watt = 1 joule of energy radiated each second. The subscript is a symbol for the Sun.) Many stars, of course, are much brighter than the Sun. A star twice as luminous as the Sun, radiating 8×10^{26} watts, would be said to have a luminosity of $2 L_\odot$.

The most basic method of estimating luminosity derives from the measurement of distance. A faint light in the night may be a candle a hundred meters away, a streetlight a few kilometers away, or a brilliant lighthouse beacon 100 km away. Once we know the distance to an object, we can determine its absolute brightness. For example, Figure 16-11 shows that a 1st-magnitude star (that is, apparent magnitude $m = 1$) 10 pc away

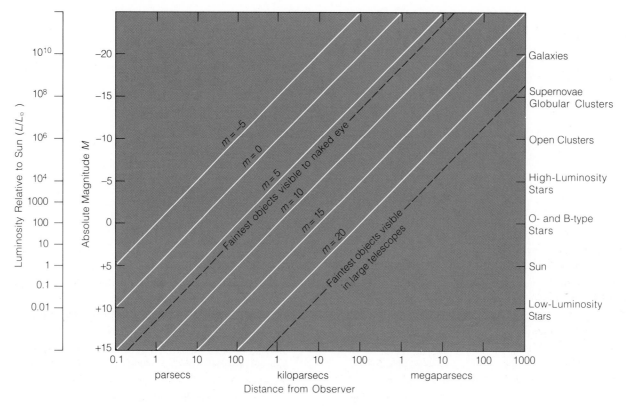

Figure 16-11 If a certain star has a known absolute magnitude M (vertical scale, possibly determined from spectra or other means), astronomers can measure its apparent magnitude m and estimate its distance from a chart like this one. For example, a high-luminosity star of $M = -5$ could be visible to the naked eye with visual magnitude $m = 5$ at a distance of around 1000 pc. Note that galaxies (see right end of graph) are so luminous that some can be visible to the naked eye at distances of about 1 Mpc, or 1 000 000 pc. The values shown here assume clear space; see Chapter 20 for a discussion of dimming due to interstellar "smog" in many regions of space.

would have an absolute magnitude of $+1$ and a luminosity of about 40 $L_\odot$.

This method becomes less accurate for very distant stars. For one thing, as mentioned earlier, parallax distance measures become unreliable beyond about 20 pc. More important, there is some interstellar haze that dims starlight over larger distances. Just as a fog bank might keep you from telling whether a distant light was a streetlight or a more distant lighthouse, variable interstellar haze in uncertain amounts complicates estimates of stars' luminosities.

However, other methods are available. For example, very luminous stars have slightly different spectral characteristics than faint stars of the same spectral class. Spectra can therefore be used to measure luminosities as shown in Figure 16-12.

Note that the luminosity, the distance, and the apparent magnitude of an object are all interrelated. As shown in Figure 16-11, if we know any two of these quantities, we can estimate the third.

How to Measure a Star's Temperature

Here again spectra come to our aid. Stellar temperatures are found by measuring colors and applying Wien's law. Interestingly, the color differences between hot, bluish stars and cooler, reddish stars are easily seen by the naked eye, especially in the constellation Orion, as shown in Figure 16-3. By studying the stars' spectra, we can measure which color is the most strongly radiated, and then use Wien's law to calculate the temperature.

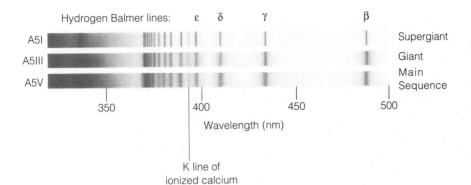

Figure 16-12 Comparison of spectra of a supergiant, giant, and main-sequence star of constant spectral class. In the larger stars, pressures are less and atoms less disturbed by collisions, resulting in slightly narrower, better-defined absorption lines.

The temperature found in this way is often called the **effective temperature,** because it is the effective average temperature of the various hotter and cooler layers of gas from which light reaches us.

As mentioned earlier, the temperature is also indicated by the relative numbers of atoms in different excited states, derived by the relative strengths of lines and use of the Saha equation.

Temperatures range all the way from about 2500 K for the cool M-type stars to 40 000 K or more for hot O-type stars.

How to Measure a Star's Diameter

The diameters of only a handful of very large stars can be measured directly from images such as Figure 16-2. Most diameters are measured from temperature and luminosity by applying the Stefan–Boltzmann law. By this law, a star radiating a certain number of watts and having a certain temperature must have a certain area A. The values can be inserted in the Stefan–Boltzmann equation and the equation solved for A. The star's diameter can be found from its area. Such measures indicate a vast range of stellar diameters, from less than Earth size, through Sun size, to huge stars bigger than the diameter of Mars' orbit!

How to Measure a Star's Mass

According to Kepler's laws as modified by Newton, when any two cosmic objects are in orbit around each other, the period of revolution (the time it takes to complete an orbit) increases as the distance between the objects increases and as the sum of the masses of the two objects decreases. Many pairs of stars orbit around each other. For many of these *binary stars* we can measure both the period of revolution and the distance between the two stars. (Binary stars are discussed in further detail in Chapter 21). Thus we can calculate the sum of the masses, which is sometimes designated $m_A + m_B$, where A and B designate the two stars.

But we want to know each individual mass, not the sum of the two. Newton showed that in a system of orbiting bodies, each body orbits around an imaginary point called the **center of mass** and that by measuring the distance of each star from the center of mass, we can measure the ratio of the masses.

Now we have both the sum of the masses and the ratio of the masses. From this information, we can get each individual mass. For example, take the double star Sirius, consisting of the bright star Sirius A (which has the greatest apparent brightness of any star in the night sky) and the faint star Sirius B. The preceding procedure shows that the sum of the masses of A and B is 3 solar masses ($3\ M_\odot$). The ratio indicates that A is twice as massive as B. From these facts, the only possible solution is that Sirius A has a mass of $2\ M_\odot$ and Sirius B, $1\ M_\odot$.

Study of many stars by this technique reveals an important fact: *Stars with nearly identical spectra usually have nearly identical masses.* This fact allows the masses of many stars to be estimated roughly from their spectral properties alone.

How to Measure a Star's Composition

As described earlier, compositions are revealed by spectra. There are two problems: detecting the *presence* of an element and measuring its *amount*. The presence of an element is detected by identifying at least one—or preferably several—of its absorption lines or emission lines in the spectrum of the star.

The amount of the element is indicated by the appearance of the spectral line. Generally, the wider and darker the absorption lines, the more atoms of the element are present. Likewise, the wider and brighter the emission lines, the more atoms. These properties are said to measure **spectral line strength**: The stronger the line, the more of the element is present. The *relative* strengths of different lines also depend on temperature, pressure, and other conditions in the photospheres of stars, as described by the Saha equation.

Astrophysicists have developed techniques to untangle the effects of composition, pressure, and so on. These techniques allow astronomers to gain information from spectra not only about composition but also about the temperature, pressure, and other properties of stars. Studies of this type around 1925 by Cecilia Payne-Gaposchkin and others showed that most nearby stars are approximately sunlike in composition.

How to Measure a Star's Magnetic Field

Spectral absorption and emission lines divide into two or more close lines in the presence of a magnetic field—a phenomenon called the *Zeeman effect*. The stronger the field, the greater the splitting. Measurements of this effect reveal that stars like the Sun typically have fields about as strong as the Sun's, but many stars have fields thousands of times stronger.

How to Measure a Star's Rotation

If a rotating star is seen from any direction except along the rotation axis, one edge will be approaching the observer and one edge will be receding (Figure 16-13). Light emitted or absorbed at the approaching edge will be blue-shifted. Light from the other edge will be red-shifted. Consider an absorption line being formed in all parts of the star's atmosphere. If the star were not rotating, neither shift would occur and the line would be very narrow. Because the star rotates, the line is broadened. The faster the rotation and the closer our line of sight to the equatorial plane, the more line broadening occurs.

The rotations of stars can thus be inferred to some extent from measurements of the broadening of spectral lines. In general, the fastest rotators are the hot, bluish, massive stars of class O, which have equatorial rotation speeds of around 300 km/s. Slightly cooler stars rotate more slowly, with a sharp break in the trend occurring at stars a little hotter and more massive than the Sun (class F). Stars cooler than F stars rotate much more slowly. The Sun's equatorial speed is only 2 km/s.

How to Measure a Star's Atmospheric Motions

If masses of stellar gas rise and fall in convection cells (as they do in the Sun), the various cells will display a range of approach and recession velocities. Therefore, blue and red Doppler shifts (respectively) would occur, broadening the stars' spectral lines, as shown in Figure 16-14. In practice, astronomers have trouble distinguishing this effect from rotational line broadening, though some discrimination is possible. Especially strong turbulence has been found in the atmospheres of certain cool, large-diameter stars known as red giants. In these stars, masses of gas rise and subside with speeds as high as 40 km/s.

How to Measure a Star's Atmospheric Structure

As noted in the last chapter, the **opacity** of an atmosphere determines how far light of a specified wavelength can penetrate into that atmosphere. Generally, a stellar atmosphere has a different opacity at each wavelength. If the opacity at a certain wavelength is low, we can see deep into the star's atmosphere at that wavelength. If the opacity at another wavelength is high, we can see only into the uppermost atmosphere. (An example of this principle is Figure 5-15b, where we look at Earth in the high-opacity wavelengths absorbed by water vapor and hence see mostly haze layers well above the surface.)

We have seen how the Saha equation allows us to derive conditions in a gas producing a given set of spectral lines. Now suppose that we could find different sets of lines arising in different layers at different depths in the star. Then we could analyze the conditions at these different levels. Applying this principle, astronomers gain information on pressure and other conditions at all levels of stellar atmospheres.

How to Detect Circumstellar Material

Unstable stars, such as newly forming or dying stars, may throw out clouds of gas and dust (see Figure 16-15). These surround the stars and are called **cir-**

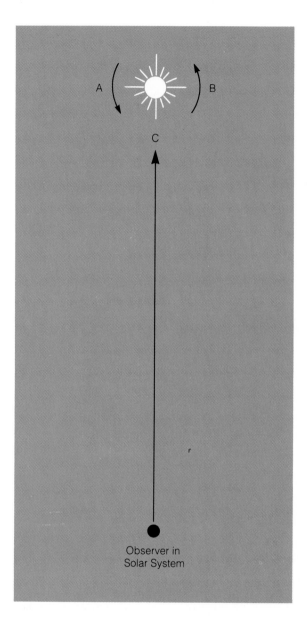

Figure 16-13 Light from a rotating star would be blue-shifted if from side *A*, red-shifted if from side *B*, or unshifted if from the central region *C*. The net result is a broadening of spectral lines.

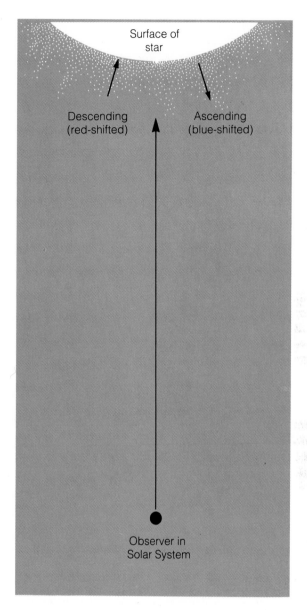

Figure 16-14 Currents of descending and ascending gas in a star's atmosphere cause Doppler shifts that broaden spectral lines.

cumstellar nebulae. Because many of these clouds are nearly transparent (as are pale flames), we can see light arising from both the near side and far side of the cloud, as shown in Figure 16-16. As such a cloud expands away from the star, absorption lines arising on the near side (where the starlight passes through the gas) are blue-shifted compared with lines in the star because the gas is approaching the observer. Emission lines from

glowing gas on the near side are also blue-shifted, but emission lines from the far side of the cloud are red-shifted.[6] These shifted lines are telltale clues to expanding circumstellar nebulae.

---

[6]For a discussion of how such spectral lines are produced, see Kirchhoff's third law (Chapter 15).

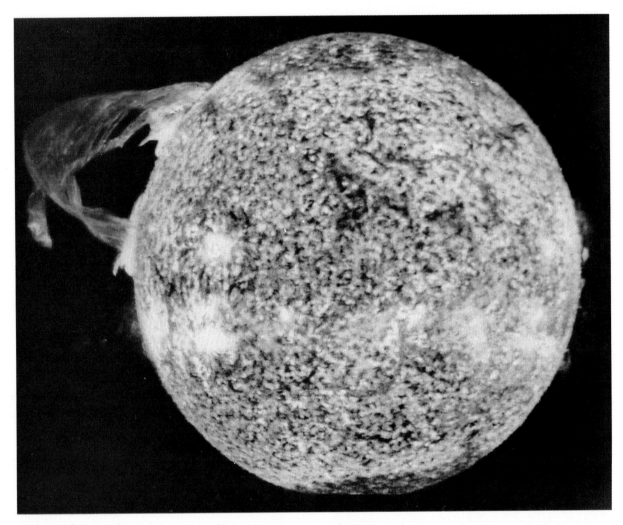

Figure 16-15 A star blowing off a cloud of gas: One of the largest solar flares ever observed shoots glowing gas off the Sun. This is a false color image of a photo made at extreme ultraviolet wavelengths (30 nm)—light emitted by excited helium atoms but invisible to the human eye. Interestingly, the photo shows that the Sun's polar regions emit little light at this wavelength, causing them to appear darker than the rest of the disk. (NASA photo by Skylab astronauts, 1973.)

In gas around certain dying stars, expansion at speeds of 1000 to 3000 km/s has been observed. This gas is believed to have been blown off the star in explosions related to the exhaustion of the star's energy supplies.

In certain newly formed stars circumstellar material has been found not in the form of glowing gas, but in the form of dust grains. These are warmed by the star and detected by the thermal infrared radiation they give off (similar to that shown in Figures 5-4 and 5-5).

How to Detect a Star's Motion

The Doppler shift is a very simple way to detect *part* of a star's motion—its **radial velocity,** or *motion along the line of sight.* A consistent blue or red shift of all of a star's spectral lines proves that the star is moving toward or away from us. Among the 50 nearest stars, about half the radial velocities measured are more than 20 km/s toward or away from us.

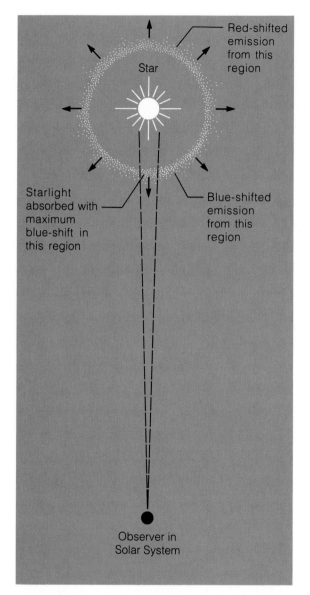

Figure 16-16 Gaseous envelopes expanding (as shown here) or rotating around stars produce characteristic spectral effects, allowing their detection even if they cannot be seen separately.

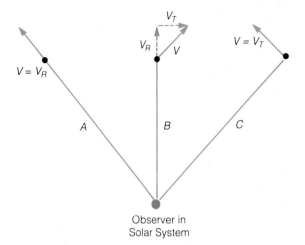

Figure 16-17 Three examples of a star's space velocity (V) compared with its radial velocity (V_R) and tangential velocity (V_T). In A, the star's space velocity is aligned with the line of sight; no proper motion would be seen. B is the most common case, in which both V_R and V_T are appreciable. In C, the space velocity is perpendicular to the line of sight; no radial velocity or Doppler shift would be seen.

The other component of a star's motion is its **tangential velocity**—the *motion perpendicular to the line of sight.* It cannot be measured as simply as radial velocity. In order to measure tangential velocity, we must measure the distance of the star and its rate of angular motion across the sky, called **proper motion.** (Values of a few seconds of arc per year are common for nearby stars.) These measurements are often lumped together in a special branch of astronomy called **astrometry.**

If both radial and tangential velocities are known, they can be combined to give the star's **space velocity:** its true speed and direction of motion in three-dimensional space *relative to the Sun* (Figure 16-17). Space velocities of most stars near the Sun are a few tens of kilometers per second and are nearly random in direction.

THE IMPORTANCE OF TELESCOPES IN SPACE

Now we can see more clearly the tremendous advantage of spaceborne telescopes, as opposed to ground-based telescopes. Figure 16-18 is a schematic diagram of several stars' spectra. The top box, which portrays the solar spectrum, shows that although most solar radiation falls in the part of the spectrum that can reach the ground, parts lie in the ultraviolet and infrared regions that can be studied only from above the atmosphere. But now look at the case of the extremely hot and cool stars in the middle box. Most of their radiation, especially the important regions of most intense emission, can be studied only by a telescope above the atmosphere. The bottom box represents an exciting 1983 discovery about the star Vega, made with the orbiting IRAS

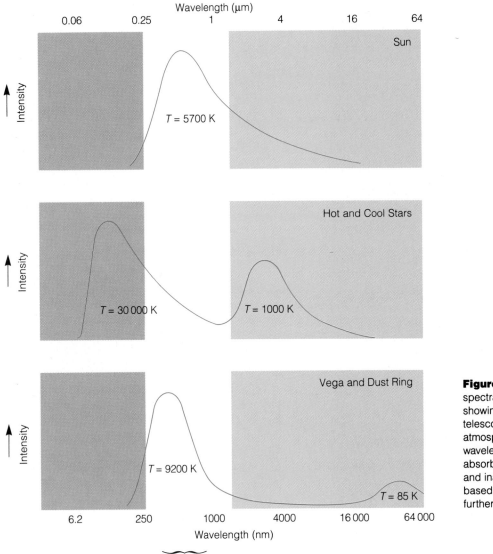

Figure 16-18 Schematic spectra of several objects, showing advantages of telescopes above the atmosphere. Shaded wavelengths are largely absorbed by the atmosphere and inaccessible to ground-based telescopes. See text for further discussion.

(infrared astronomical satellite) telescope. Before 1983, Vega seemed a solitary star somewhat hotter and more massive than the Sun. The IRAS telescope revealed far infrared radiation that (according to Wien's law) must be emitted by cool material at only 85 K (− 306°F). This is too cold for a star and is believed to be a ring of dust about 80 AU from Vega (Neugebauer and others, 1984). From the intensity of the infrared radiation, the total mass of dust is believed comparable to the mass of asteroids or comets in our own solar system. Some

believe this discovery is a new line of evidence of planetary material around other stars (Weissman, 1984).

These examples show why space telescopes (Figure 16-19) will improve our understanding of hot material, cold material, extremely energetic events, planet formation, and a host of other cosmic phenomena associated with other stars. As mentioned in Chapter 5, the Hubble Space Telescope is the largest such instrument, a 2.4-m (94-in.) telescope. Its launch, delayed by the Challenger disaster, occurred in 1990. Shockingly,

Figure 16-19 The Hubble Space Telescope (**inset**) in orbit over Lake Titicaca, South America, after launch from the Space Shuttle. (NASA Photo.) HST was designed so that astronauts could repair and replace some modules, as shown in the painting, where astronaut gives scale. Corrector lenses in replaced modules may permit salvaging of some HST observations otherwise prevented by its defective optics. (Painting by Paul DiMare.)

one mirror was found to be defective soon after launch, due to embarrassing failures of the design manufacturing/quality control process. HST will perform better than ground-based telescopes for many tasks, but not as well as had been hoped.

SUMMARY

This chapter emphasizes methods of studying stars, rather than the nature of the stars themselves. The chapter has two basic sections. The first section describes concepts:

the system of naming stars; the photographic images of stars; the definitions of *light-year* and *parsec;* the system of magnitudes for measuring brightness; stellar spectra; the Doppler effect; and the Stefan–Boltzmann law, which describes the amount of stellar radiation. The second section describes methods for determining 12 basic stellar properties, as summarized in Table 16-3.

Clearly, our journey outward from the Earth has taken us far beyond the places that we can investigate in the near future by manned or instrumented flights. Even at the speed of light (which is 10 000 times greater than speeds our instruments have achieved), it would take years to reach the nearest stars. Thus, instead of sending instruments to the stars, we must rely on the messages in the starlight coming to us. Our abilities to interpret starlight are rapidly increasing, not only with construction of new large telescopes but also with the launching of such telescopes into orbit. In the coming chapters, we will interpret what those messages to date have told us about the character and histories of the stars that surround us.

C O N C E P T S

star	Saha equation
diffraction	temperature
Airy disk	Doppler effect
light-year	blue shift
parsec	red shift
apparent brightness	Stefan–Boltzmann law
absolute brightness	parallax
apparent magnitude	distance limit for reliable parallaxes
visual apparent magnitude	luminosity
absolute magnitude	bolometric luminosity
spectrum	effective temperature
wavelength	center of mass
spectroscopy	spectral line strength
spectrograph	opacity
spectrometer	circumstellar nebula
absorption line	radial velocity
emission line	tangential velocity
continuum	proper motion
spectral class	astrometry
hydrogen Balmer series	space velocity
hydrogen alpha (Balmer alpha)	

P R O B L E M S

1. If a telescope of 25-m (1000-in.) aperture could be put in orbit or on the Moon, it could resolve an angle of only about 0.005 second of arc. Compare this resolution with the maximum angular size known for stars (0.03″). Why would this telescope perform better on the Moon than on Earth? Why do astronomers never use the highest magnifications theoretically possible with earthbound telescopes?

2. A star is 20 pc away. How many years has its light taken to reach us?

3. What is the chemical composition of most stars?

4. A reddish star and a bluish star have the same radius. Which is hotter? Which has higher luminosity? Describe your reasoning.

5. A reddish star and a bluish star have the same luminosity. Which is bigger? Describe your reasoning.

6. How does the magnitude scale differ from the scales by which we measure lengths (centimeters or inches), temperature (degrees Celsius or Kelvin), weight (kilograms or pounds), or other familiar properties?

7. A certain star has exactly the same spectrum as the Sun but is 30 magnitudes fainter in apparent magnitude. List some conclusions you would draw about it and describe your reasoning.

8. A star has a parallax of 0.05 second of arc. How far away is it?

9. In a binary pair of stars, the sum of the masses is found to be 5 $M_\odot$; the stars are equidistant from their center of mass. What are their individual masses?

10. The spectral lines of metals, such as calcium, are prominent in the solar spectrum. Hydrogen lines are less prominent. Why does this not indicate that the Sun consists mostly of these elements instead of hydrogen?

11. Each of a certain star's spectral lines is found to be spread out over a wide range of wavelength. What might cause this?

A D V A N C E D P R O B L E M S

12. How many times brighter is a daylit landscape than the same landscape lit at night by the full Moon? (*Hint:* See Table 16-1; note that 5 magnitudes equals a factor of 100 in brightness; 1 magnitude, a factor of 2½.) If your camera needs a 1/100-s exposure in daylight, what exposure might record the moonlit scene? (Actually about twice the calculated exposure will be better, because film sensitivity falls off in weak light.)

13. A certain star has a spectrum similar to the Sun's, but all the spectral lines are shifted 0.1% of their wavelength toward the red.
   a. What do you conclude about this star's motion toward or away from the observer?
   b. Is its speed in this direction unusually fast, average, or slow?

14. A star 10 pc away is observed to have a proper motion of 1 second of arc per year. Use the small-angle equation to derive its tangential velocity.

15. If the stars in Problems 13 and 14 were the same star, what would its space motion be like?

16. A star has the Sun's luminosity (4×10^{26} J/s) but a temperature of only 2850 K. How big is it?

17. a. Using optional Basic Equation VI, prove that the mean energy of atoms and ions in a star will increase as the temperature increases. (*Hint:* See discussion of Optional Basic Equation VI.)

b. Discuss how this explains the pattern of increasing ionization as temperature increases in the spectral sequence of stars.

c. In previous problems we have used a rule of thumb that the fastest atoms of any element move more than three times faster than the typical velocity of the same element in the same gas. Note that in this case they would have nine times as much kinetic energy. Assume that these fastest atoms dominate in ionizing a mass of hydrogen gas and calculate the temperature at which these fastest atoms would have enough energy to ionize adjacent hydrogen atoms. As mentioned in the text, the ionization energy for ground-state hydrogen is about 2×10^{-18} J.

d. Comment on where a star of such temperature would fall in the sequence of spectral classes, and show that hydrogen is indeed ionized in hotter stars. (*Note:* The model for ionization in problems *c* and *d* is overly simplistic. First, ionization in a star is also caused by photons of ultraviolet radiation, as well as thermal jostling by neighboring atoms. Second, the percentage of atoms ionized depends on gas density and pressure, as well as temperature, as worked out by Saha in 1920. Nonetheless, these problems give a rough idea of how ionization sets in as temperature increases in a gas, due to increasing energy of the gas particles.)

PROJECT

1. Using a telescope of fairly high magnification (such as $300\times$ or $400\times$), examine the image of a bright star on several different nights and sketch it. Are there differences from night to night? Can you identify the Airy disk, diffraction rings, or diffraction spikes? If so, label them on your sketches. Note any shimmering due to atmospheric turbulence or air currents in and around the telescope. Run the eyepiece inside or outside the focus point, making the image a round blob. Shimmering and other turbulent effects are often more evident in this way.

CHAPTER 17

The Systematics of Nearby Stars: The H–R Diagram

As soon as astronomers could begin to measure different properties of stars, they began to categorize the stars. This chapter will outline three interesting discoveries that resulted. (1) Stars come in a wide variety of forms: massive and not so massive, large and small, bright and faint. (2) One reason for the different forms is that stars evolve from one form to another. (3) Another reason is that stars of different initial mass evolve at different rates and into different final forms.

In this sense, the population of stars can be likened to that of people. If you travel from country to country and observe people, you tend to see mostly ordinary people—youngsters, middle-aged people, and elderly people. The creation of new people—the birth of babies—is secluded in special places such as hospitals and is not readily visible to the tourist. In addition, the tourist may see tombstones, but the actual deaths of people are usually brief events not commonly encountered. This metaphor explains the approach of this chapter and the next two chapters. In this chapter we will tour the local population of stars. We will find that the great majority of these stars are ordinary stars—that is, youthful, middle-aged, or growing elderly. The next chapter will show how stars come into existence and cite some observed examples of star birth. Chapter 19 will examine some of the extraordinary phenomena of stellar old age.

CLASSIFYING STAR TYPES: THE H–R DIAGRAM

When astronomers began to determine properties of stars by using the techniques described in the last chapter, they found a puzzling variety of star forms. Masses range from about 0.1 to 60 $M_\odot$. One star in a billion is even more massive than 60 $M_\odot$; a few dazzling stars hundreds of times more massive than the Sun are believed to exist (Humphreys and Davidson, 1984). Luminosities range from around one-millionth to a million times that of the Sun. Surface temperatures range from about one-third to nearly 10 times the solar temperature. Diameters range from less than 1/100 to more than 100 times the Sun's size.

When astronomers realized that there is such a variety of star forms, their first step was to devise a sensible way to arrange and study the data about them, in hopes of finding relationships among the various forms of stars. This could have been done in various ways, but one method has become traditional. This method was introduced around 1905 to 1915 by Danish astronomer Ejnar Hertzsprung and American astronomer Henry Norris Russell. They constructed diagrams plotting the spectral class of stars versus their luminosity, as shown in Russell's version of the diagram (Figure 17-1), published in 1914. This type of plot is usually called the **H–R diagram** in honor of Hertzsprung and Russell.[1]

Even the earliest H–R diagram revealed an important discovery about stars: *Among most stars there is a smooth relation between spectral class and luminosity,* indicated by the dashed lines on Figure 17-1. This band came to be called the *main sequence* of stars. Hertzsprung called stars obeying this relation (falling on the diagonal band of the H–R diagram) **main-sequence stars.** Russell correctly argued that the smooth relation between spectral type and luminosity meant that *the principal differences in stellar spectra arise mainly from variations in a single condition in the stellar atmosphere.*

As we saw in the last chapter, *temperature* is the principal factor that governs the differences between spectra of O, B, A, F, G, K, and M stars. For this reason, many astrophysicists make H–R diagrams by using a temperature scale instead of a scale showing spectral class. In this book, we identify temperature, spectral class, luminosity, and absolute magnitude along the edges of the diagram as an aid to interpretation. (Following Russell's original version, the spectral classes are arranged with *cooler* stars to the right.)

As shown in Figure 17-2, different locations on the H–R diagram correspond to different types of stars. It

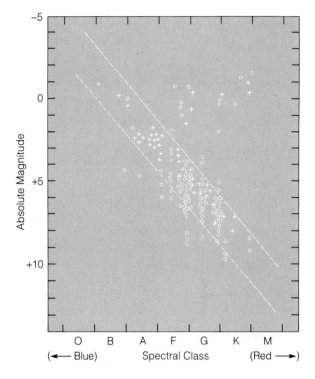

Figure 17-1 The first H–R diagram, redrawn from Russell's original 1914 publication. Dots represent stars with directly measured parallaxes; crosses represent estimated data from stars in four clusters. Dashed lines enclose Russell's identification of the main sequence—a discovery made from this diagram.

is important to understand why this is true. First, remember from Wien's law that stars of different temperature have different color. Thus, as shown schematically in Figure 17-2, the right part of the diagram corresponds to redder, cooler stars and the left part to bluer, hotter stars.

Next note that the Sun is near the middle of the diagram (luminosity 1 $L_\odot$, spectral type G). If we move upward from the Sun's position, we encounter stars more luminous than the Sun, even though they have the same temperature. Thus these stars must be bigger than the Sun. (The reasoning here is as follows: The Stefan–Boltzmann law, page 350, assures us that each square meter on a star at the same temperature as the Sun will radiate as much energy as a square meter on the Sun. Thus, to get more total luminosity, we must have more square meters of surface area and hence larger size.) Similarly, a star below the Sun's position on the diagram must be smaller than the Sun.

---

[1]Sometimes this graph is called the spectrum–luminosity diagram or color–magnitude diagram, because plots of spectral type versus luminosity or color versus absolute magnitude produce equivalent information.

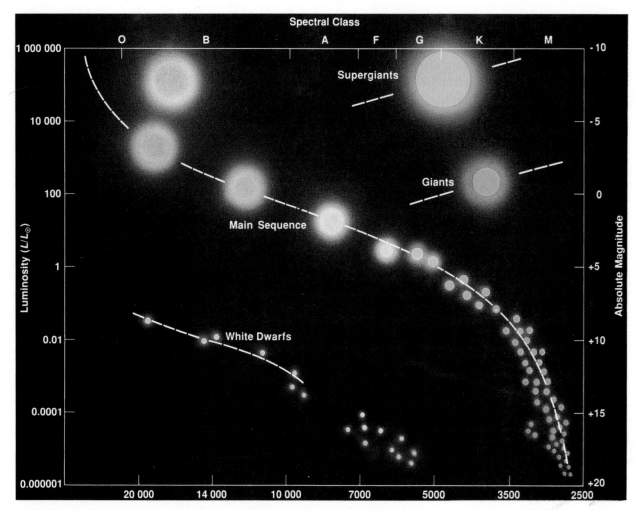

Figure 17-2 An H–R diagram for a selection of different types of stars. Schematic drawings give colors and relative sizes, but not to true scale (which would require much larger red giants). Main sequence, running diagonally across the diagram, comprises the "normal" hydrogen-burning stars. See text for further explanation.

Do stars appear *throughout* the diagram? Or are there only certain types such as main-sequence stars, giants, and dwarfs? The answer is that only certain parts of the H–R diagram are crowded with stars. Other types of stars, which would fall in other parts of the diagram, are rare or nonexistent. Using the H–R diagram, astronomers began to investigate why this is so.

THE NEARBY STARS AS A REPRESENTATIVE SAMPLE OF ALL STARS

In keeping with the pattern of expanding horizons in the

cosmic journey of this book, let's start our discussion of stars with the stars nearest the Sun. Figure 17-3 shows a modern H–R diagram for the 100 stars closest to the Sun. Stars at this range are near enough for us to measure accurate distances and to detect even very faint examples. They are also numerous enough to provide a good sample. Stars in such a random sample are believed to be **representative stars**—that is, a representative sample of all stars in our general neighborhood of the galaxy. Figure 17-3 indicates that about 93% of these stars fall on the main sequence. This shows the importance of the main sequence: *It includes the great majority of stars*. The Sun symbol plotted at spec-

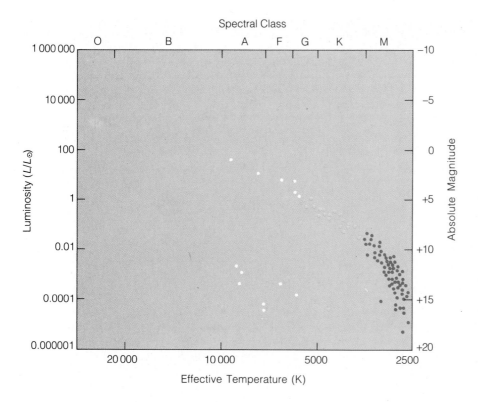

Figure 17-3 An H–R diagram for the 100 nearest stars—a representative sample from a volume of space around the Sun. This diagram shows that most stars are on the main sequence. Note also that most are fainter and cooler than the Sun. Therefore they fall in the lower right corner.

tral class G, luminosity 1 $L_\odot$, indicates that the Sun is brighter than most representative stars.

Table 17-1 lists more detailed properties of the 27 stellar systems within 4 pc of the Sun. This table reveals another interesting fact about representative stars: Most of them are inhabitants of systems in which two or three stars orbit around one another. Stars without stellar companions, such as the Sun, are in the minority.

The closest star system to the Sun is the triple system of **Alpha Centauri,** brightest star in the southern constellation of Centaurus. Shown in Figure 17-4, Alpha Centauri is easily visible from Hawaii, but it is too far south to be prominent from most of the rest of the United States. Because Alpha Centauri is so close and so bright, we can accurately measure the distances, spectral classes, temperatures, luminosities, and sizes of its components. By measuring the orbital size and period, we can use Kepler's laws of orbital motion and Newton's universal law of gravitation to deduce the masses of the components. The main two components are Alpha Centauri A and B, which are very similar to the Sun and located 1.33 ± 0.01 pc away. These two components orbit around each other in 80 y and are easily visible as a pair—yellowish and orange—in a small

telescope. In the 1800s, telescopic observer Sir John Herschel called them "beyond all comparison the most striking object of the kind in the heavens." They reached greatest separation as seen from the solar system (22 seconds of arc) in 1980 but will be only 4 seconds of arc apart in 2015.

Orbiting around A and B once every 1½ million years is a small faint star with only one-tenth the Sun's mass. It happens to be on our side of the system at a distance of only 1.29 pc (Kamper and Wesselink, 1978). Because it is the Sun's closest neighboring star, it is called Proxima Centauri (from the same root as *proximity*).

REPRESENTATIVE STARS VERSUS PROMINENT STARS

If we try to expand our statistical sample of stars by tabulating more distant ones, we run into a problem. At great distances the less luminous stars, like those in Figure 17-3, are too faint to see. The only distant stars we see are unusual, luminous ones. In fact, the **prominent stars** in our night sky are mostly distant, unusually luminous stars. As we learned from Figure 17-3,

TABLE 17·1

The Nearest 27 Stellar Systems (Stars Within 4 Parsecs)

Distance (pc)	Star Name	Component	Apparent Magnitude (m_v)	Absolute Magnitude (M_v)	Spectral Type	Mass ($M_\odot$)	Radius ($R_\odot$)	Semimajor Axis in Multiple Systems (AU)
0.0	Sun (and Jupiter	A	−27	5	G2	1.0	1.0	5.2
		B	—	—	—	0.001	0.1	
1.3	Alpha Centauri	A	0	4	G2	1.1	1.2	23.6 (AB)
		B	1	6	K0	0.9	0.9	
		C[a]	11	15	M5	0.1	?	13 000 (AC)
1.8	Barnard's Star	—	10	13	M5	?		1.3
2.3	Wolf 359	—	14	17	M8	?	?	—
2.5	BD + 36°2147	A	8	10	M2	0.35	?	0.07
		B	?	?	?	0.02	?	
2.7	L726-8 (= UV Ceti)	A	12	15	M6	0.044	?	10.9
		B	13	16	M6	0.035	?	
2.9	Sirius	A	−2	1	A1	2.3	1.8	19.9
		B	8	11	Wh. dw., A5	1.0	0.02	
2.9	Ross 154 (= Gliese 729)	—	11	13	M5	?	?	—
3.1	Ross 248	—	12	15	M6	?	?	—
3.2	L789-6	—	12	15	M6	?	?	—
3.3	ε Eridani	—	4	6	K2	0.9	?	—
3.3	Ross 128	—	11	14	M5	?	?	—
3.3	61 Cygni	A	5	8	K5	0.63	?	85 (AB)
		B	6	8	K7	0.6	?	
		C	?	?	?	0.008	?	
3.4	ε Indi	—	5	7	K5	?	?	—

these stars are quite rare in three-dimensional space. Hertzsprung aptly called them the "whales among the fishes." Thus the statistics of prominent stars do not give us a sample of representative stars. Rather, they are biased toward the "whales." Nonetheless, a tabulation of such stars is important because it reveals that some of the "whales" don't belong to the main sequence.

Non-Main-Sequence Stars

If we make an H–R diagram by adding a sampling of the most prominent stars in our sky (as ranked by apparent magnitude) to an equal-sized sampling of the nearby stars, we get an H–R diagram similar to Figure 17.-2. While the nearby stars are the small red stars at

The Nearest 27 Stellar Systems (Stars Within 4 Parsecs), *continued*

Distance (pc)	Star Name	Component	Apparent Magnitude (m_v)	Absolute Magnitude (M_v)	Spectral Type	Mass ($M_\odot$)	Radius ($R_\odot$)	Semimajor Axis in Multiple Systems (AU)
3.5	Procyon	A	0	3	F5	1.8	1.7	15.7
		B	11	13	Wh. dw.	0.6	0.01	
3.5	Σ 2398 (= Gliese 725)	A	9	11	M4	0.4	?	60
		B	10	12	M5	0.4	?	
3.5	BD + 43°44 (= Gliese 15)	A	8	10	M1	?	?	156 (AB)
		B	11	13	M6	?		
		C	?	?	K?			
3.6	τ Ceti	—	4	6	G8	?	1.0	—
3.6	CD − 36°15693	—	7	10	M2	?	?	—
3.7	BD + 5°1668 (= Luyten's Star)	A	10	12	M4	?	?	?
		B	?	?	?	?	?	
3.7	G51-15	—	15	17	?	?	?	—
3.8	L725-32	—	12	14	M5	?	?	—
3.8	CD − 39°14192	—	7	9	M0	?	?	—
3.9	Kapteyn's Star	—	9	11	M0	?	?	—
4.0	Krüger 60 (= DO Cep = Gliese 860)	A	10	12	M4	0.27	0.51	9.5 (AB)
		B	11	13	M6	0.16	?	
		C	?	?	?	0.01	?	
4.0	Ross 614	A	11	13	M5	0.14	?	3.9
		B	15	17	?	0.08	?	
4.0	BD − 12°4523 (= Gliese 628)	—	10	12	M5	?	?	—

Source: Data from van de Kamp, (1981) and Baum (1986). A companion to Bernard's Star, listed by van de Kamp, is regarded as dubious and deleted here.

[a]Proxima Centauri, the closest known star beyond the Sun.

the bottom of the main sequence, the prominent stars turn out to be mostly "whales" in the upper part of the diagram.

As shown in Figure 17-2, certain of the red K and M stars have much higher luminosity than the faint, main-sequence K and M stars. That is, they radiate more light than the fainter stars of the same temperature. According to the Stefan–Boltzmann law, they must

have more area. Hertzsprung named them **giant stars.** Most are called *red giants,* since they are reddish K and M stars. Figure 17-2 makes the situation clearer by dramatizing the relative brightnesses and sizes of these stars.

As more data came in, other stars appeared on the H–R diagram below the main sequence. By the same logic used for giants, these stars were recognized as

a

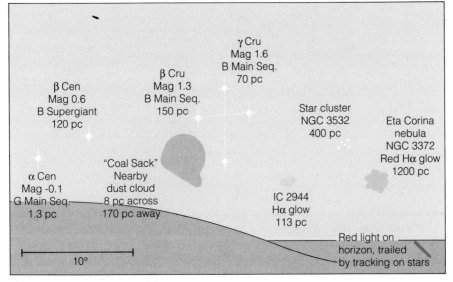

b

Figure 17-4 The nearest star system, Alpha Centauri, and the adjacent region of the sky. Neither of two companion stars orbiting around Alpha is seen here—one (Alpha B) because it is too close to Alpha A, and the other (Proxima) because it is too faint. Sketch map in **b** identifies prominent features, including the famous constellation called the Southern Cross; a dust cloud known as the Coal Sack, which blots out part of the background Milky Way; and nebulae glowing with the red color of Hα emission. The Coal Sack is easily visible to the naked eye, but the red glows are too faint for the eye to perceive as colored. (Photo by author; 18-min exposure, 55-mm lens, f2.8, Ektachrome ISO 1600 film, Mauna Kea, Hawaii.)

unusually small. They came to be called **white dwarf stars,**[2] because their colors were intermediate, like the Sun's, being neither strongly blue nor red. Still other stars were found to be brighter than the giants or the main-sequence O stars; they came to be called **supergiant stars.**

The work of Hertzsprung, Russell, and their followers was important because it showed systematic groupings of stars: one group fitting on the main sequence, a distinctly different group of giants, a group of dwarfs, and a group of supergiants. Further work showed that these groups occur in all known regions of space. This meant that universal physical processes could be sought to explain the groups.

Table 17-2 tabulates the 17 most prominent stars in our sky and reveals that nearly half are giants or supergiants. Although the whales among the fishes are rare, they account for many of the familiar stars, as illustrated by Figure 17-5.

Occasionally we find groups of stars at about the same distance from us and formed at about the same time. These "star clusters," such as the one in Figure 17-6, offer us, so to speak, living H–R diagrams in the sky. Just by studying the colors and brightness of stars in Figure 17-6, we can see the pattern of the H–R diagram. Most of the stars are on the main sequence; the brightest ones are blue-white, and the faintest ones are red. This particular scene, known as the Jewel Box cluster, is all the more attractive because the fourth brightest star is an orangish-red giant or supergiant, offering a startling color contrast.

Radii on the H–R Diagram

To emphasize the distinctness of these stellar groups, Figure 17-5 shows the same H–R diagram as Figure 17-2 but with lines added to show positions of constant radius. The important concept is this: *Any point on the H–R diagram corresponds to a star of a certain radius.* The reasoning should be clear to you already. Any point on the diagram corresponds to a certain temperature T. According to the Stefan–Boltzmann law, any surface at temperature T must radiate each second a certain amount of energy per square meter. But any point on the diagram also corresponds to a certain luminosity L,

which is the total amount of energy radiated each second. This luminosity fixes the number of square meters involved; hence the total area; hence the radius.

The diagram confirms that the names of the types are apt. Giants and supergiants can range up to 1000 times the size of the Sun, while dwarfs may be only 1/100 its size. Some well-known examples of these various types are marked on Figure 17-5. We can see that the Sun and the brightest star, Sirius, are main-sequence stars; the North Star, Polaris, is a supergiant; and the faint companion to Sirius is a white dwarf. The enormous size of the red supergiant Antares can be visualized in the imaginary close-up shown in Figure 17-7.

EXPLAINING THE TYPES OF STARS: DIFFERENT MASSES AND DIFFERENT AGES

What causes the distinctive types of stars found on the H–R diagram, such as main-sequence stars, the giants, and the white dwarfs? First, stars have different masses, as shown in Figure 17-8. *High-mass* main-sequence stars are hotter and brighter and bigger than *low-mass* main-sequence stars. Second, stars evolve from one type to another, and since we see stars with different ages, we see stars with different forms.

How did we discover the principles that control the structure and evolution of stars? The English astrophysicist Arthur Eddington (1926) once commented:

At first sight it would seem that the deep interior of the sun and stars is less accessible to scientific investigation than any other region of the universe. . . . What appliance can pierce through the outer layers of a star and test the conditions within?

The problem does not appear so hopeless when misleading metaphor is discarded. It is not our task to "probe"; we learn what we do learn by awaiting and interpreting the messages dispatched to us by the objects of nature. And the interior of a star is not wholly cut off from such communication. A gravitational field emanates from it. . . . Radiant energy from the hot interior after many deflections and transformations manages to struggle to the surface and begins its journey across space. From these two clues alone a chain of deduction can start which is perhaps the more trustworthy because it [employs] only the most universal rules of nature— the conservation of energy and momentum, the laws of

---

[2]Unfortunately, main-sequence stars are sometimes called dwarf stars to distinguish them from giants, but for clarity we reserve the term *dwarf* for white dwarfs.

TABLE 17·2					
The 17 Stars of Greatest Apparent Brightness					
Star Name	**Apparent Magnitude** (m_v)	**Bolometric Luminosity** $(L_\odot)$	**Type**[a]	**Radius**[b] $(R_\odot)$	**Distance (pc)**
Sun	−26.7	1.0	Main sequence	1.0	0.0
Sirius (α Canis Majoris)	−1.4	23	Main sequence (primary)	1.8	2.7
Canopus (α Carinae)	−0.7	(1400)	Supergiant	30	34
Arcturus (α Boötis)	−0.1	115	Red giant	(25)	11
Rigel Kent (α Centauri)	0.0	1.5	Main sequence (primary)	1.1	1.33
Vega (α Lyrae)	0.0	(58)	Main sequence	(3)	8.3
Capella (α Aurigae)	0.1	(90)	Red giant (primary)	13	14
Rigel (β Orionis)	0.1	(60 000)	Supergiant (primary)	(40)	(280)
Procyon (α Canis Minoris)	0.4	6	Main sequence (primary)	2.2	3.5

chance and averages, the second law of thermodynamics, the fundamental properties of the atom, and so on.

Eddington showed why stars are the way they are. The same two major, opposing influences at work in the formation of the Sun are competing in any star: Gravity pulls *inward* on stellar gas while gas pressure and radiation pressure push *outward*.[3] In a stable star, these forces are just balanced.

As we have already seen in the discussion of Helmholtz contraction (page 301), heat is produced during gravitational contraction of a star. A stable main-sequence star is one that has contracted until the inside is hot enough to start nuclear reactions among hydrogen atoms. At this point the interior becomes a stable heat source, radiating light and creating enough outward pressure to counterbalance the inward force of gravity. What had been a contracting ball of gas now becomes a star with a stable size, governed by the interior energy-generating conditions.

Explaining the Main Sequence

The main source of energy in a main-sequence star's interior is nuclear reactions in which hydrogen is consumed. Because stars form with a huge supply of hydrogen, stars remain stable on the main sequence for a relatively long time at a fixed size. If a stable star were

[3]As mentioned in Chapter 15, radiation exerts a distinct pressure on material through which it passes. In stars, especially massive ones, this pressure is important.

The 17 Stars of Greatest Apparent Brightness, *continued*

Star Name	Apparent Magnitude (m_v)	Bolometric Luminosity ($L_\odot$)	Type[a]	Radius[b] ($R_\odot$)	Distance (pc)
Achernar (α Eridani)	0.5	(650)	Main sequence	(7)	37
Hadar (β Centauri)	0.7	(10 000)	Giant (primary)	(10)	150
Betelgeuse (α Orionis)	0.7	10 000	Supergiant	800	160
Altair (α Aquilae)	0.8	(9)	Main sequence	1.5	5
Aldebaran (α Tauri)	0.9	125	Red giant (primary)	(40)	21
Acrux (α Crucis)	0.9	(2500)	Main sequence (primary)	(3)	(110)
Antares (α Scorpii)	0.9	(9000)	Supergiant (primary)	(600)	(160)
Spica (α Virginis)	1.0	(2300)	Main sequence (primary)	8	84

Source: Data from Burnham (1978).
Note: These are stars brighter than the first apparent visual magnitude.
[a]The designation "primary" indicates data for the brighter companion in binary pairs.
[b]Parentheses indicate estimates.

magically expanded, its gas would cool and the reactions would decline, reducing the outward pressure, and the outer layers would fall back to their original state. If it were magically compressed, the inside would get denser and the reactions would increase, raising the outward pressure and expanding the star. The star has to stay in its stable state *as long as its internal chemistry and energy production rate stay the same.*

Although Eddington didn't know the details of the hydrogen-burning process,[4] he was able to explain the main sequence by calculating the structures of stars with different masses but with fixed compositions and energy-producing processes. His calculations revealed a **mass–luminosity relation:** A hydrogen-burning star more massive than the Sun has higher luminosity and surface temperature than the Sun. Hence, on the H–R diagram it would lie to the upper left of the Sun. Similarly, a star of lower mass would lie to the lower right of the Sun. These stars therefore fall along a line in the H–R diagram—the main sequence. In other words: *The main sequence is explained as the group of stars of different masses that have reached stable configurations and are generating energy by consuming hydrogen in nuclear reactions.* Any hydrogen-burning star with $1\,M_\odot$ and solar composition must look like the Sun. Any hydrogen-burning star with $2\,M_\odot$ and solar composition must be brighter and bigger than the Sun.

---

[4]Burning technically refers to chemical reactions involving only electrons. But astrophysicists use the term informally for nuclear reactions that convert small fractions of the mass of atomic nuclei into energy.

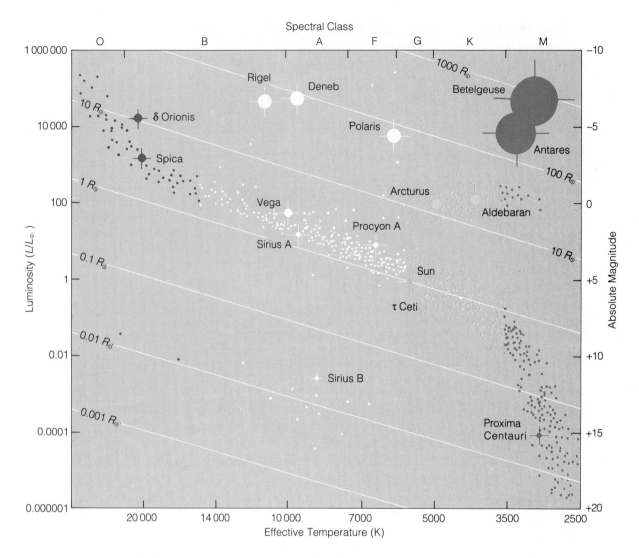

Figure 17-5 An H–R diagram with lines of constant radius showing the dimensions of stars in different parts of the diagram. Selected well-known stars are marked. Their relative sizes are schematically indicated, but there is not room to show them to true scale.

Figure 17-8 shows this trend more clearly by charting masses of some stars on the H–R diagram. Note the smooth trend of increasing mass and luminosity as we ascend the main sequence. The O-type supergiants may have 40 or even 100 times the Sun's mass.

Eddington's explanation of the main sequence helped explain why some stars lie off the main sequence. The same assumptions simply do not apply to them. In other words, giants, supergiants, and dwarfs are stars that do *not* have the same energy generation process, composition, or energy transport processes as the Sun.

The Russell–Vogt Theorem

As Eddington reached these conclusions, the astrophysicists H. N. Russell and H. Vogt independently derived a related result in 1926 known as the **Russell–Vogt theorem:**

> **The equilibrium structure of an ordinary star is determined uniquely by its mass and chemical composition.**

Figure 17-6 The Jewel Box cluster, cataloged as NGC 4755, displays different-colored stars that demonstrate the different star types charted on the H–R diagram. The cluster is named from Sir John Herschel's remark that it resembles a superb piece of jewelry. It is about 8 pc across and 2400 pc away. The brightest stars are mostly hot, bluish-white, young B stars; the faintest stars are red, low-mass stars. (Copyright Association of Universities for Research in Astronomy, Cerro Tololo Interamerican Observatory.)

Figure 17-7 Imaginary view of the red M-type supergiant star Antares as seen from a hypothetical planet $9\frac{1}{2}$ AU away—the same distance Saturn is from our Sun. Although the Sun would resemble only a brilliant dot from such a planet, Antares is so big that it would cover an angle of almost 40°. It would look 80 times bigger than our Sun in Earth's sky and have a reddish-orange color. The fainter star in the upper right is a bluish B-type star that orbits around Antares at a distance of several hundred AU. (Painting by author.)

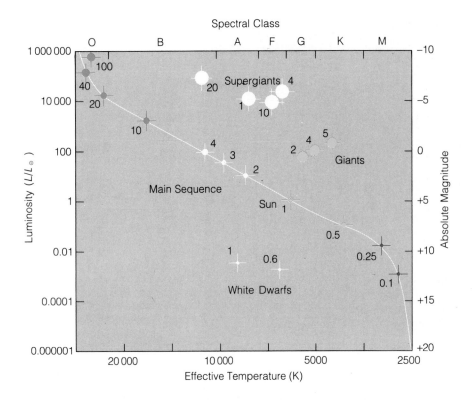

Figure 17-8 An H–R diagram with measured masses of stars (in solar masses) marked in different parts of the diagram. A smooth progression of masses is found along the main sequence, but masses in other parts of the diagram are mixed because stars of different mass evolve into the same regions—for example, the giant branch. Schematic sizes are shown (not to true scale).

This says that a certain mass of material with fixed composition—for example, one solar mass of solar composition—can reach only one stable configuration. This is the point on the H–R diagram actually occupied by the Sun. But if the composition were altered to ½ hydrogen and ½ helium, the configuration of the star and its location on the H–R diagram would be different. To put this in another way, the Russell–Vogt theorem says: Give me the mass and composition of any star, and I can tell you what the star looks like. Using these principles, astronomers have actually calculated what the stars look like—they have made tabulations of the pressures, temperatures, and other characteristics of the interiors, surfaces, and atmospheres of stars to accurately classify them.

The Russell–Vogt theorem's effect is indicated in Figure 17-9; main-sequence stars of a given mass tend to have the same radius. But giants, supergiants, and white dwarfs of the same mass, because they have evolved and have different compositions, have different equilibrium structures and radii.

Why Main-Sequence Stars Must Evolve Off the Main Sequence

The Russell–Vogt theorem explains the cause of **stellar evolution** from one form to another form: A star burning hydrogen converts it to helium and changes its composition; therefore, the star must change to a new equilibrium structure. All nuclear reactions cause changes in composition, and all changes in composition cause evolution to new structure.

PHILOSOPHICAL IMPLICATIONS OF THEORETICAL ASTROPHYSICS

Philosophically, these achievements were profound. Eddington pointed out that humans (or other intelligent

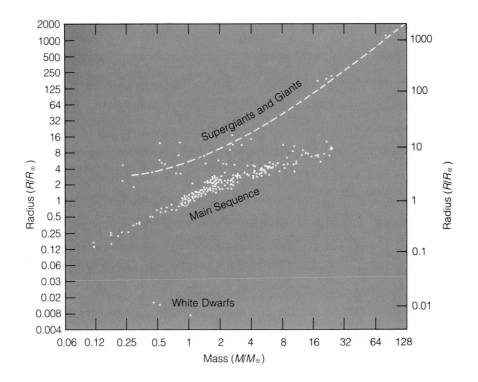

Figure 17-9 This figure reveals that distinctly different radii exist among stars of a given mass. This is a result of evolutionary changes. Most stars in the mass range shown here remain at nearly constant mass during most of their life (main sequence). However, later evolution takes them to giant size, then to dwarf size. (Data are derived from double stars' orbits from Popper, 1980; Aller; 1971; and Liebert, 1980.)

creatures), reasoning purely from elementary principles even without telescopic observation, could show that the universe must be populated by objects like stars because gravity causes gas to "clump" into star-sized masses. Smaller or larger objects have too little gravity to assemble themselves from interstellar gas, and larger objects would produce so much radiation that they would blow themselves apart. These suppositions by Eddington have been proved correct by further observation and theory: The most common size of star in the universe has a mass around 0.1 to 1.0 $M_\odot$, as shown in Figure 17-10. Stars smaller than 0.1 $M_\odot$ are less common, and stars bigger than 60 $M_\odot$ are very rare.

Eddington (1926) wrote:

We can imagine physicists working on a cloud-bound planet such as Jupiter who have never seen the stars. They should be able to deduce by [these methods] that if there is a universe existing beyond the clouds, it is likely to aggregate primarily into masses of the order of [10^{27}]

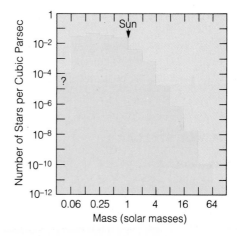

Figure 17-10 A histogram showing the frequency of occurrence of different stellar masses in the solar neighborhood. The most common stars have about $\frac{1}{4}$ $M_\odot$. Abundance of lowest-mass objects is uncertain.

tons. They could then predict that these aggregations will be globes pouring out light and heat and that their brightness will depend on the mass in the way given by the [mass–luminosity relation].

Perhaps Eddington was too optimistic; he was not giving enough credit to the interplay between observers and theorists. Nonetheless, he became fascinated by the idea that intelligent creatures could mentally derive the appearance of the universe without actually observing more than a few fundamental relations. He asked whether a superscientist, locked forever in a closed physics lab with certain minimal equipment, could discover the nature and arrangement of the universe. In other words, is the arrangement we observe the only possible arrangement? No one knows, but we can be sure that if some single physical constant—such as Newton's constant of gravitation—suddenly changed, the structure of the universe would dramatically change. If gravity suddenly decreased, for example, the Earth would expand, causing earthquakes. The Earth's rotation and orbit would change, and the Moon's orbit would change. The Sun would expand, causing a change in luminosity and a change in the Earth's temperature.

It is thus plausible that a single complex theory, using the measured values of various physical constants, would allow prediction of everything that can be measured about the universe. Such a theory would allow treatment of matter, gravity fields, electromagnetic fields, and radiation from a single set of equations. Toward the end of his life, Eddington believed that he was approaching a theory that could predict the values of certain physical constants that had previously been thought to be underivable, intrinsic properties of the universe. His last work has not been widely accepted.

AN OVERVIEW OF STELLAR EVOLUTION

Because of the work of Eddington, Russell, Vogt, and others, modern astrophysicists can calculate models of stars of any given mass and chemical composition. Thus we can calculate not only what the Sun's interior structure is like now, but what its future structure will be when 1% of the present hydrogen is gone, 2% is gone, and so forth. Each different composition corresponds to a certain (temporary) structure and to a certain position in the H–R diagram. Knowing the rates of nuclear

reactions, we can calculate the time scale associated with this evolution from one state to another. Using computer modeling, we can estimate what happens during unstable transition states, such as the transition from main-sequence to giant and post-giant stages.

Because a star's temperature and luminosity, as well as other properties, can be calculated for each state of its evolution, the stages of that evolution can be plotted on an H–R diagram. The sequence of such points is called an **evolutionary track**—the set of "footprints" a star leaves on an H–R diagram as it evolves.

Evolution of a One-Solar-Mass Star (Such as the Sun)

Figure 17-11 shows the approximate evolutionary track across the H–R diagram followed by the Sun or any star of solar mass. Some parts of the track are more certain than other parts. Consideration of different parts of the track shows clearly why different groups of stars appear on the H–R diagram. As we saw in Chapters 14 and 15, the Sun began its life as a cool, dim interstellar cloud. Therefore, a sunlike star first appears on the H–R diagram at the right, with low temperature, low luminosity, and large radius, perhaps exceeding $100\,R_{\odot}$. It contracts rapidly, becoming quite bright, and then reaches a few solar radii in about a million years (only an instant of cosmic time). At this point it approaches the main sequence. Nuclear reactions begin in the interior, because the central temperature reaches some 10 million degrees or more (see Table 15-2 and discussion on page 319). This whole process of star formation is treated in more detail in the next chapter.

Once the nuclear reactions start, the huge reservoir of hydrogen begins to be converted into helium, and a star reaches a relatively stable, main-sequence configuration. A star takes only about 100 million years to reach the main sequence, but it is destined to sit on the main sequence for roughly 9 billion years.

Eventually, the supply of hydrogen begins to be so exhausted in the central core that changes become more rapid. The core is now mostly helium. The star's composition having changed, its structure must change, according to the Russell–Vogt theorem. Because there is less hydrogen in the core to fuel the reactions, not enough energy is produced to maintain the outward pressure and keep the star in its main-sequence state. The interior contracts, keeping the temperature high enough to maintain hydrogen burning in a shell around

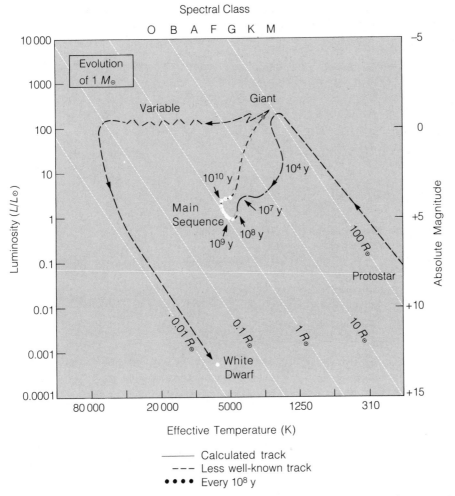

Figure 17-11 An H–R diagram showing the calculated evolutionary track for the Sun (or any star of 1 M☉). Dots are placed about 100 million years apart. The track shows the evolution from protostar state to main sequence, then to giant, variable, and white dwarf states. (Based on calculations by Westbrook and Tarter, 1975; Iben, 1967; Larson, 1969; and others.)

the helium core. Meanwhile, the outer layers of the star adjust by expanding. The star rapidly moves off the main sequence to a higher luminosity and much larger radius. It becomes a giant.

Theoretical calculations can trace a star quite accurately to the giant state, but subsequent states are less well understood, as indicated by the dashed line in Figure 17-11. Nuclear reactions begin to "burn" heavier elements, such as helium, fusing them into still heavier nuclei, as indicated in Table 17-3. Generally, after one element is consumed, slight core contractions raise the temperature to a point where still heavier elements begin

to burn in fusion reactions. Evolution now takes the star somewhere to the left. Stars in this region are unstable, often variable, and rapidly evolving. Some very massive stars in this region explode, throwing off masses of gas.

After each "combustible" element is used up by nuclear reactions in the core, the whole star contracts. Ultimately, it gets so dense it cannot contract any further. It has now reached the size of a planet—very tiny compared to its original size. A white dwarf has been formed. It is very hot and because of certain peculiarities of its nuclear structure, it cannot radiate its heat or cool very rapidly, even though it no longer produces

TABLE 17·3

Key Elements in Nuclear Reactions at Different Stages in Stellar Evolution

	Elements Being "Burned"	Comments
	Lithium and other light elements	Relatively unimportant source of energy during formation of star before main-sequence stages (see Chap. 18)
	Hydrogen	Primary energy source; defines main sequence of stars; proton–proton chain; carbon cycle (see Chap. 18)
Time	Helium	Important after hydrogen is depleted; involved in reactions during rapid evolution off main sequence and into giant branch of H–R diagram (see Chap. 19)
	Heavy elements, including carbon and metals	Energy sources during final period on giant branch and during rapid evolution off giant branch, including supergiant states (see Chap. 19)

any heat by nuclear reactions. The process of creating white dwarfs and other superdense types of stellar corpses is described in more detail in Chapter 19.

The following sequence summarizes the evolution of stars according to the groups found on the H–R diagrams (multiple arrows indicate faster evolution):

Protostar
↘ ↘ ↘
Pre-main sequence
↘ ↘
Main sequence
↘
Giant
↘ ↘
Variable or unstable
↘ ↘ ↘
White dwarf or other small star

You will get a better idea of how these stages relate if you trace the dashed line in Figure 17-11 with your pencil. Start with the pencil point in the lower right and count seconds, moving forward one round dot each second. Each second equals 100 million years. As you can see, the star almost reaches the main sequence in just one of these "cosmic seconds" but then spends nearly 100 cosmic seconds on the main sequence. Then there

is a rapid evolution to the giant stage, followed by a rapid transformation (in another cosmic second or so) to a small, faint, defunct state below the main sequence. No wonder we see so many main-sequence stars: Most stars spend nearly their whole lives in this state.

Massive Stars Evolve Fastest

Stars evolve at different rates. The more massive a star, the higher its interior temperatures and the faster it uses its nuclear fuel. A sunlike star of 1 $M_\odot$ stays on the main sequence about 9 billion years (9×10^9 y), but a star of 10 $M_\odot$ stays there only about 20 million years (2×10^7 y). The entire history of a star of 1 $M_\odot$ (protostar to white dwarf) takes about 11 billion years, whereas a 10-$M_\odot$ star lasts only about 24 million years. In spite of the differences in total time, the largest fraction of each star's life is still spent on or near the main sequence.

Since a star's hydrogen-burning lifetime depends mostly on its mass and luminosity, a simple formula gives the time the star will spend on the main sequence—which is most of the star's lifetime. It is:

Hydrogen-burning lifetime = $M/L \times 9$ billion years

where M and L are the star's mass and luminosity in solar units.

The most massive stars stay on the main sequence for only the twinkling of a cosmic eye. Some of them evolve into the supergiant region, and some less massive ones become ordinary giants. All of them quickly evolve to unstable configurations; many may explode; and all disappear from visual prominence.

Determining the Ages of Stars

Consider a group of stars that formed at the same time. After about 10 million years, stars larger than 20 $M_\odot$ will have disappeared from the main sequence. That is, the H–R diagram of the group will contain no main-sequence O stars. After about 100 million years, stars more massive than 4 $M_\odot$ will have evolved off the main sequence and the H–R diagram will contain scarcely any main-sequence B-class stars. The older the cluster, the more of the main-sequence stars will be gone. The missing stars will have been transformed into giants, white dwarfs, or even fainter terminal objects.

Thus we reach an important conclusion: *The H–R diagram can serve as a tool for dating groups of stars that formed together.* This principle will be applied often in later chapters as we probe the **ages of stars** in our galaxy.

Isolated individual stars are harder to date. The Sun's age was measured at 4.6 billion years by dating planetary matter—unavailable to us in the case of other stars. Certain indicators, such as the amount of "unburned" light elements (lithium, for example) in a star's atmosphere, can also be used to estimate a star's age. By this latter method the main-sequence stars closest to the solar system—the Alpha Centauri system (Figure 17-4)—are estimated to be at least 3 billion years old (Boesgaard and Hagen, 1974).

SUMMARY

The nearby stars, within about a hundred parsecs of the Sun, display a range of masses, luminosities, temperatures, and compositions. These properties are not randomly distributed but grouped into distinct types. Some stars are on the main sequence, some are giants, some supergiants, and some dwarfs. Differences among these types are conveniently displayed on the H–R diagram, which plots stellar luminosity against temperature.

To explain these different types of stars, this chapter developed three important principles, mentioned in the introduction. First, stars form with a variety of masses covering a range from as low as 0.1 to 60 $M_\odot$ and more. As the stars form and begin to burn hydrogen in nuclear reactions, they settle onto the main sequence, but the more massive the star, the more luminous its main-sequence position in the H–R diagram.

Second, stars evolve from one form to another. They begin with pre-main-sequence configurations, settle onto the main sequence for a relatively long time to burn their hydrogen, and then evolve off the main sequence. Post-main-sequence forms include giants, variables, and dwarfs.

Third, stars evolve at different rates. The more massive a star, the faster it goes through its sequence of life stages. This is because more massive stars reach higher central temperatures and therefore consume their hydrogen faster.

The great majority of stars, including most of those in the Sun's neighborhood, have masses in the range of 0.1 to 1 $M_\odot$. They are relatively faint and cool; they lie in the lower right part of the H–R diagram. The most massive stars are much brighter and can be seen from much further away. They lie in the upper part of the H–R diagram. Thus many of the stars prominent in the night sky are prominent not because they are close but because they are the very luminous "whales among the fishes," far away among myriads of distant fainter stars.

CONCEPTS

H–R diagram

main-sequence star

representative star

Alpha Centauri

prominent star

giant star

white dwarf star

supergiant star

mass–luminosity relation

Russell–Vogt theorem

stellar evolution

evolutionary track

ages of stars

PROBLEMS

1. Suppose you could fly around interstellar space, encountering stars at random.

 a. Describe the stars you would encounter most often. Mention masses and H–R diagram positions.

 b. About what percentage of stars would be as massive as or more massive than the Sun? (*Hint:* Use statistics from either Table 17-1 or 17-2, as appropriate.)

 c. Would the stars encountered be similar to bright stars picked at random in our night sky?

2. Do giant stars necessarily have more mass than main-sequence stars? Why are they called giants?

3. Four stars occupy the four corners of an H–R diagram. Which one has:
  a. The highest temperature?
  b. The greatest luminosity?
  c. The greatest radius?

4. Two stars lie on the main sequence in different parts of the H–R diagram. Which is:
  a. Larger?
  b. More luminous?
  c. More massive?
  d. Hotter?

5. Two stars form at the same time from the same cloud in interstellar space, but one is more massive than the other. Describe differences or similarities at later moments in time.

6. What would an imaginary terrestrial observer see as the Sun runs out of hydrogen in the future? If life is confined to Earth when this happens, would life perish from heat or from cold?

7. Which part of Problem 4 could not be answered just from location in the H–R diagram if both stars were not on the main sequence?

8. Consider newly forming cosmic objects of stellar composition. What would you expect to happen in terms of stars' life cycles as the masses decrease from $1\,M_\odot$ toward values comparable to Jupiter's mass?

9. Show that astrophysical estimates of the Sun's lifetime are consistent with meteoritic and lunar data on the age of the solar system. Why would the data be inconsistent if astrophysicists calculated that solar-type stars stay on the main sequence only 1 billion years?

10. Why don't normal stars all collapse at once to the size of asteroids or tennis balls under their own weight, since gravity always pulls their material toward their center?

11. How was the chemical composition of the Sun 3 billion years ago different from what it is now?

12. Red giants have proceeded further in the evolutionary sequence than main-sequence stars, but is it correct to make the blanket statement that red giants are older than main-sequence stars? Explain. Which star is older, the Sun or a red giant of $10\,M_\odot$?

ADVANCED PROBLEMS

13. Use Wien's law to show that an M star with temperature about 2900 K is strongly red in color and that an O star with temperature about 29 000 K would be strongly blue. Would the strongest radiation emitted by each star be visible to the human eye?

14. For a group of stars of equal radius but different temperatures T, the Stefan–Boltzmann law shows that the total luminosity will be proportional to T^4.
  a. Among such stars, how many times brighter would a 20 000-K star be than a 2000-K star?
  b. Using Figure 17-5, confirm your result by reading luminosities on one of the lines of constant radius.

15. a. As shown in Table 17-1, a typical white dwarf could have a radius of $0.015\,R_\odot$ and a mass of $0.8\,M_\odot$. Calculate the mean density of matter in such a star, and compare it with the density of familiar material such as water (1000 kg/m³), rock (about 3000 kg/m³), and lead (11 350 kg/m³).
  b. How many pounds would a matchbox full of such material weigh?

PROJECT

1. Locate or construct spherical objects that have the same proportional sizes as selected different types of stars, such as an Antares-type supergiant, the Sun, and a white dwarf.

CHAPTER 18

Stellar Evolution I: Birth and Middle Age

Just as new living individuals are being formed from chemical materials against a background of slowly changing genetic pools, new stars are being formed from interstellar gas and dust against a background of slowly changing elemental abundances in the cosmic gas. Locally—and by this we now mean a substantial part of our galaxy—there is a constantly changing environment, with new stars forming and old stars blasting material into interstellar space. From a philosophical point of view, identifying and understanding newly forming stars are challenges in our long quest to comprehend the role of stars in our universe.

THREE PROOFS OF "PRESENT-DAY" STAR FORMATION

Here are three lines of **evidence for "present-day" star formation**—that is, star formation in the current part of astronomical history.

Youth of the Solar System The solar system is only about 4.6 billion years old, whereas the entire Milky Way galaxy—our system of 100 billion stars—is at least 10 or 12 billion years old. Thus our own Sun formed much more recently than the galaxy, and stars did not all form in one burst at the beginning.

Short-Lived Star Clusters Many young, massive stars are grouped in *open star clusters*. The stars in the cluster formed at about the same time. Because of tidal disruptive forces and the tendency of each star in the cluster to follow its own orbit around the galactic center, most clusters dissociate into isolated stars in only a few hundred million years. This is confirmed by analyses of H–R diagrams of clusters, which show that few of these clusters are much older than this. The famous Pleiades cluster, for example, is only about 50 million years old. The existence of such clusters shows that star clusters did not all form at the beginning, but continue to form today.

Short-Lived Massive Stars Chapter 17 showed that massive stars evolve fastest. Calculations show that stars of 20 to 40 $M_\odot$ can last only a few million years in a visible state, yet we see them shining. They must have formed less than a few million years ago.

Thus star formation has been a continuing process during the whole history of the galaxy, including the last million years. There are stars in the sky younger than the species *Homo sapiens*.

THE PROTOSTAR STAGE

Stars form from tenuous material between other stars. The average conditions between stars are a near vacuum with very thin, cold gas. Atoms and molecules number only about 1 to 5 per cubic centimeter. About three-quarters of them are hydrogen (H atoms or H_2 molecules), and most of the rest are helium—as is the case with the atoms in most nearby stars. There are scattered dust particles. The temperature is typically around 100 K ($-279°F$).

Molecular Clouds

In certain regions of space, the density of gas and dust may be 10 000 times greater. With some 10 000 atoms per cubic centimeter, the atoms are close enough together to collide and form molecules, rather than single atoms. These regions are called *molecular clouds*. They are rich in atoms, dust grains, and diverse molecules such as molecular hydrogen (H_2), water (H_2O), and more complex forms such as ethyl alcohol (C_2H_5OH) and the amino acid glycine (NH_2CH_2COOH). Such complex molecules are virtually absent in most parts of interstellar space. Although a molecular cloud is denser than most interstellar gas, it is still a near vacuum compared with ordinary room air, where there are about 20 billion billion (2×10^{19}) molecules per cubic centimeter.

Molecular clouds are important because they are the most active regions of star formation. In these regions the atoms and molecules and dust grains are crowded close enough to begin to attract each other gravitationally—a key step in forming stars from gas (Scoville and

Young, 1984). An example of the dramatic and chaotic forms of gas and dust in such regions is shown in Figure 18-1. Many such nebulae are known. Most are cataloged in the "New General Catalog" of celestial objects and hence are usually designated by "NGC" numbers. Many also have popular names based on their appearance in telescopes; the "Eagle Nebula" in Figure 18-1b is an example.

Toward an Astrophysical Theory of Star Formation

Chapter 14 explained how the Sun formed when one of these diffuse interstellar clouds contracted and produced a central star surrounded by a dusty nebula. Now we need a more general theory of this process that can explain other stars with different masses.

When scientists use the word *theory*, they usually mean a well-tested body of related ideas often with a mathematical formulation that can be applied to a variety of cases and often backed up by observations.[1] Such a theory of star formation has been developed. This theory can be approached by asking: What causes some clouds to contract and not others? Why don't all interstellar clouds contract into stars and be done with it, once and for all? The answer involves the same two opposing forces we have considered before: gravity versus thermal pressure. Gravity pulls all the atoms in a cloud inward. But even at only 100 K the atoms are dancing, striking each other, and creating an outward pressure that opposes the tendency to collapse.

Gravity increases if the density of the cloud increases, since this gets more mass into the same amount of space. Outward pressure increases if the gas heats up, since this makes the atoms and molecules move faster. We can already say, then, that the reason not all clouds contract is that not all clouds are dense enough or cool enough. Furthermore, a noncontracting cloud can be turned into a contracting cloud if it suddenly experiences some turbulence (caused by a nearby stellar explosion, for example) that compresses it.

Astronomers want a more quantitative **theory of star formation** than this description. For interstellar material of any given density, they have calculated what

---

[1] A less complete or less tested idea is called a *hypothesis*, or a *working hypothesis*. An untested idea is often called *speculation*. Unfortunately, newspapers and magazines often publish speculations but label them with the more imposing term *theory*.

Figure 18-1 The spectacular Eagle Nebula and the associated star cluster NGC 7711 dominate a chaotic region of star formation about 2000 pc away. The nebula is a cloud of gas and dust about 6 pc across. The red color comes from hydrogen atoms excited to glow with the red "hydrogen alpha" emission line. Dark clots are concentrations of dust. Hot, bluish-white, newly formed O stars are the brightest stars in the region. The complex is only about 3 million years old. **a** General view. (Anglo-Australian Telescope Board, courtesy David Malin.) **b** The heart of the Eagle Nebula with its star cluster and silhouetted dust clouds. (Copyright Association of Universities for Research in Astronomy, NOAO.)

combination of mass and temperature a cloud would need to begin its **gravitational contraction,** or shrinkage toward stellar dimensions. The calculations are based on principles sometimes called the *Jeans theory* after English astrophysicist Sir James Jeans, who first made the calculations in 1902. (It is also sometimes called the *virial theorem* after certain principles in mechanics.)

The results are shown in Figure 18-2, where the two diagonal lines give the typical mass that would form at 10 K or 1000 K from a gas of specified density. Along the bottom, the figure shows densities ranging all the way from 10^{-32} kg/m^3 (perhaps equal to densities between galaxies) to 1 kg/m^3 (about the density of air). This brackets the interstellar density within enormous

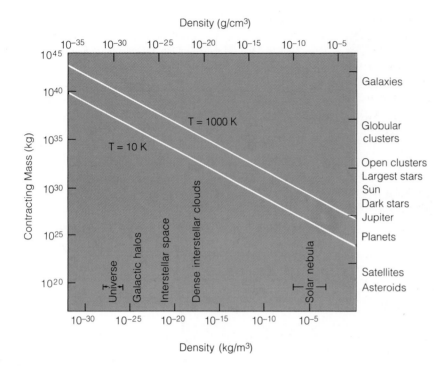

Density (g/cm³)

Contracting Mass (kg)

Density (kg/m³)

Figure 18-2 This diagram, based on the theory of star formation, shows the mass likely to be involved in self-gravitating contraction for different gas densities and two gas temperatures. Comparison of different environmental densities (bottom) with observed types of masses (right) shows that many features of the universe can be roughly explained with this simple theory. (See text.)

margins, but the theory need not apply only to interstellar gas.

First, let us apply it to the interstellar medium in star-forming regions. What masses will contract there? The figure shows that in the denser interstellar medium, at about 10^{-17} kg/m³ and 100 K, masses contracting would be around 10^{34} kg, or several thousand solar masses, and about 100 times the mass of the largest stars. But our theory was supposed to predict the formation of stars! Is something wrong? In fact, the theory agrees with observations, because, as we noted in the second of the three proofs of present-day star formation, stars apparently form in star clusters that have masses thousands of times greater than the Sun. In other words, conditions in the interstellar medium are such that *the gas subdivides into enormous concentrations containing enough mass for thousands of stars.* These huge clouds can be thought of as protoclusters.

But how do *individual* stars form? Figure 18-2 shows that as a protocluster cloud continues to contract to still higher densities, smaller, star-sized condensations can form inside it. In other words, shrinking protoclusters divide into individual stars in a process called **subfragmentation.** The Canadian researcher A. Wright (1970), for example, followed the contraction of a 6 000 000-

$M_{\odot}$, 120-pc-wide cloud and a 500-$M_{\odot}$, 12-pc-wide cloud. He found that

both large and small density perturbations will grow in size, given the conditions found throughout most of a gas cloud. It would appear that fragmentation is an almost inevitable consequence of the cloud collapse process.

The new fragments become stars, and the whole mass turns into a star cluster. It is not hard to see why the interstellar gas in the disk of the galaxy might begin to contract. The material is not uniform; clots of dust and gas exist and are constantly being stirred by the galaxy's rotation, the expansion of hot nebulae, and other influences. Naturally, some clouds accumulate material or become compressed until their density exceeds the critical density that allows contraction to begin.

Comments on the Formation of Just About Everything

It is provocative to consider the evolution of the whole universe from this point of view. For example, if all the material in the universe were spread out uniformly, a very low density of roughly 10^{-26} to 10^{-28} kg/m³ would be obtained. Figure 18-2 shows that in a more or less uniform, relatively hot gas of this density, the condensations would have the mass of galaxies, about $10^{11} M_\odot$. In other words, if there was ever a time when no galaxies existed but hot gas filled space, galaxies should have formed. We'll return to this intriguing clue about galaxy formation in Chapter 27.

Galaxies' outer regions, or halos, have densities of around 10^{-24} kg/m^3. At these densities and at temperatures of 100 K, according to the diagram, masses of $10^6 M_\odot$ could condense by their own gravity. Such objects actually exist in halolike regions around galaxies. They are called *globular star clusters*. They consist of hundreds of thousands of stars. They are second in size only to galaxies, and our theory makes it quite plausible that they formed where they did through the subfragmentation of contracting galactic masses.

We have already explained that inside galaxies, where densities are higher, open star clusters form from clouds of about $10^3 M_\odot$. We also noted that inside those clouds, where it is still denser, individual stars form by further subfragmentation involving masses around 1 $M_\odot$.

The same hierarchy of subfragmentation occurs with smaller masses. Inside the disk-shaped solar nebula, the dust settled toward the plane of the disk. So much dust concentrated there that densities in this region may have become even higher than in Figure 18-2. Chapter 14 described the theory that concentrations of dust grains in such nebulae led to even higher densities, where small masses resembling asteroids formed by self-gravitation of dust grains and then accumulated into larger planets.

Thus we see how many loose ends from earlier chapters come together into one simplified theory that accounts for many objects in the universe. This illustrates how science can help us perceive a unity in the phenomena around us. The theory as described here is simplified in that it ignores such complications as magnetism, rotation, and the effects of dust opacity. But by so simplifying, astrophysicists can concentrate on the dominant influences to rough out the origin of various objects.

The Protostar's Collapse

The preceding discussion allows us to define a **protostar** as a cloud of interstellar dust and gas that is dense and cool enough to begin contracting gravitationally into a star. An interstellar cloud may hover on the verge of this state for millions of years. A nearby exploding star

or some other disturbance might compress a gas cloud and thus trigger the collapse, which then proceeds very rapidly. For instance, typical protostars contract to stellar dimensions in a hundred thousand years—a wink of the cosmic eye. This explains why astronomers use the term **collapse:** The initial contraction is very rapid in terms of astronomical time. The collapse may not be smooth; the inner core of the cloud may collapse first, later absorbing the surrounding material; and under certain conditions a single rotating cloud may break up into two or more stars (Boss, 1985). Of course, once the cloud begins to reach stellar dimensions (a few astronomical units, or 0.00001 pc), the atoms of gas bump into each other frequently enough to produce substantial outward pressure, so that the collapse is slowed. At this point we can speak of the object as a *pre-main-sequence star.*

THE PRE-MAIN-SEQUENCE STAGE

What are the forms of pre-main-sequence stars? Where do they lie on the H–R diagram? Can they actually be observed among the stars in space? Can we actually see stars being born?

The **pre-main-sequence star** stage covers the evolution from the end of the protostar stage to the main sequence. Astrophysicists such as the Japanese theorist C. Hayashi (1961) and his American colleague L. Henyey (Henyey, LeLevier, and Levée, 1955) pioneered in calculating the evolutionary tracks and appearance of stars contracting toward their main-sequence configurations. Results of such calculations are shown in Figure 18-3, which clearly shows that stars of all masses approach the main sequence from the right in the H–R diagram. Newly formed stars would be cooler than main-sequence stars.

The energy during most of the pre-main-sequence period does *not* come from nuclear reactions, which have not yet started inside the protostar; instead, heat is generated by the Helmholtz contraction (page 301). After only a few thousand years of protostellar Helmholtz collapse, surface temperatures reach a few thousand Kelvin, thus causing visible radiation. Hayashi's work showed that convection would transport large amounts of energy from the interiors of most newly forming stars, making them very bright for short periods known as **high-luminosity phases** or *Hayashi phases*. A star of solar mass, for example, contracts in less than 1000 y from

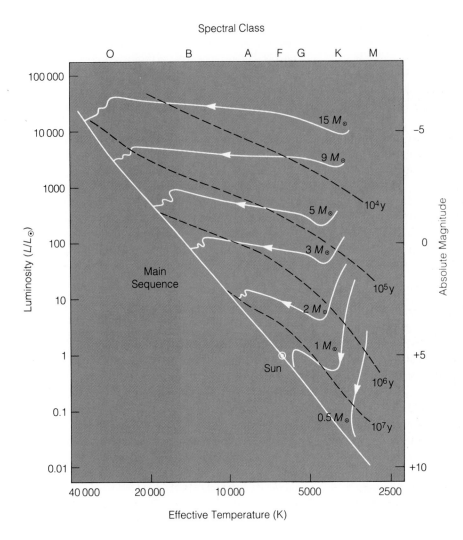

Spectral Class

Figure 18-3 H–R diagram showing simplified pre-main-sequence tracks for seven stars of different mass. Note that the high-mass stars settle on the upper end of the main sequence, while low-mass stars are on the lower end. Dashed lines show the states reached after the indicated number of years. More detailed recent calculations, taking into account the nebula around the star and other details, show more complex "squiggles" in the evolutionary tracks, but these details are not firmly established. (Data from Iben, 1965.)

a huge cloud to a size about 20 times bigger than the Sun with a luminosity about 100 times greater than the present Sun's (see Figure 17-11). Figure 18-3 shows such a star entering the H–R diagram in its pre-main-sequence stage on a rapidly descending track as it fades from its high-luminosity phase.

These calculations have three important consequences. First, they show that stars have complicated, if short-lived, evolutionary histories even before nuclear reactions start. Second, they show that *newly forming stars must lie above and to the right of the main sequence.* (This is important in identifying new stars by direct observation.) Third, the calculations indicate that stars spend only a small fraction of their lifetimes in the pre-main-sequence stage.

More massive stars evolve faster to the main sequence

than less massive stars. Note in Figure 18-3 that a 15-$M_\odot$ star reaches the main sequence in only 100 000 y. A 5-$M_\odot$ star takes about 1 million years, while a sunlike star takes several million years.

Prediction of Cocoon Nebulae and Infrared Stars

While some theorists were calculating evolving conditions in newly forming stars, others, such as the Mexican astronomer A. Poveda, hypothesized that as young stars form from collapsing clouds, remnants of the clouds might surround and obscure these stars. As in the early solar nebula, dust would form in the cooling cloud and block the outgoing starlight. The radiation from the new star would heat the nebular dust grains to a few hundred

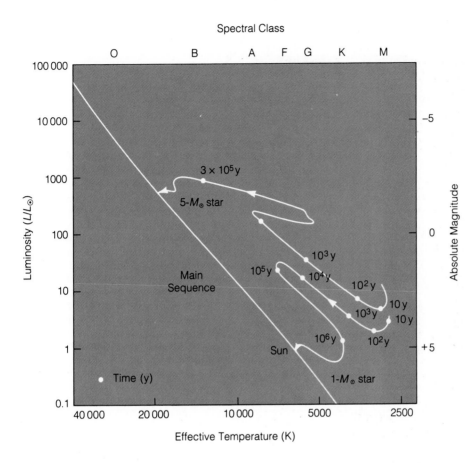

Figure 18-4 H–R diagram showing pre-main-sequence evolutionary tracks and complex luminosity changes based on the theory illustrated by Figure 18-3, modified to take into account a surrounding nebula much larger than the star. Tracks represent only the central contracting star, or "core," of the nebula. (Data from Larson, 1969.)

Kelvin, and, according to Wien's law, the dust grains would then radiate infrared light. The nebula would eventually dissipate, revealing the star.

The term **cocoon nebula** was coined in 1967 to describe the nebula that hides a star from view during its earliest formative period (perhaps a few million years for a solar-sized star). The nebula is cast off later, just as a cocoon is cast off by an emerging butterfly.

These concepts showed the need for new theoretical models, because the earlier calculations had not included the effects of surrounding clouds.

Astronomers have calculated the evolutionary tracks of stars forming as central condensations inside larger clouds. The results are shown in Figure 18-4. The track is more complicated than before, but it still approaches the main sequence from the right.

What would the newly forming star and cocoon nebula look like to an observer before the nebula dissipated? Could examples ever be observed? The outermost dust of the opaque cocoon would have a temperature of only

a few hundred Kelvin. Therefore the nebula would radiate not visible light but rather infrared light (review Wien's law, page 97), and the star–nebula complex, as seen from outside, would lie far to the right of the main sequence. Such an object could be detected as an **infrared star** at wavelengths of a few micrometers. As shown in Figure 18-5, it would evolve across this region of the H–R diagram in less than a million years and thus be detectable as an infrared star for a relatively brief period of its life. Stars of this type should therefore represent only a small fraction of all stars.

What Happens If the Mass Is Too Small?

If a cloud is dense and cool enough to contract but has less than about eight percent of a solar mass, it will contract but never develop a high enough central pressure and temperature to reach a full-fledged main-sequence state. Protostars from about 1 to 8% $M_{\odot}$ may warm up temporarily due to their Helmholtz contrac-

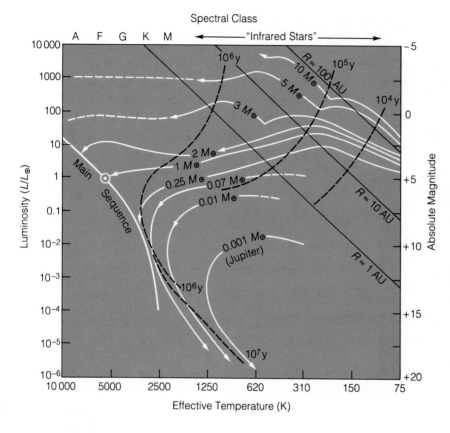

Figure 18-5 H–R diagram showing apparent tracks followed by pre-main-sequence protostars of various masses and their opaque cocoon nebulae, as viewed from outside. Unlike the central core stars (Figures 18-3 and 18-4) the configurations as a whole would be visible only as infrared stars for their first 10^5 y or so, as viewed from the Earth. Black dotted lines show time scales; black solid lines show radius of contracting nebulae. Stars smaller than 0.08 $M_\odot$ never settle onto the main sequence but continue to fade (bottom center).

tion, and may even develop a few feeble nuclear reactions, but eventually they fade without settling on the main sequence for any appreciable time. Smaller objects, including even Jupiter-sized planets, also warm up temporarily due to their Helmholtz contraction; they may glow for a while in the infrared, but they fade before any nuclear reactions can start. Even these nonstars therefore have evolutionary tracks that can be plotted on an H–R diagram, as shown at the bottom of Figure 18-5.

The smallest known luminous object in the sky, cataloged as RG 0050-2722, may be a temporarily glowing object of this type. It has an estimated mass of about 2.3% $M_\odot$ (only 23 × Jupiter's mass!) and a temperature of 2600 K. It is about 25 pc from us.

The Transition from Planet to Star: Brown Dwarfs

Objects with less than 8% $M_\odot$ (80 × Jupiter's mass), but bigger than planets, are called **brown dwarfs**, or

substellar objects, because, lacking nuclear reactions, they are not true stars. The name *brown dwarf* comes from the theory that the larger ones get red hot and glow with a dim, brownish red light.

What would brown dwarfs really be like, if we could see them up close? Suppose we could do this experiment: Start with Jupiter, keep adding solar-composition gas of hydrogen and helium, and go from planets through the brown dwarfs' size range to true stars. The result of this imaginary experiment is shown in Figure 18-6. As we add mass to Jupiter, it gets only a little bigger until it reaches roughly 2 M_{Jupiter}, where it levels off. As we add even more mass, the object actually gets a little smaller, because it has so much gravity that it can compress the gas to a denser state. Then, as we approach 80 M_{Jupiter} where nuclear reactions begin, it "turns on" as a star and again gets much bigger as we add more hydrogen fuel. Thus, conceptually, there are fairly natural dividing lines. We will use the word planet for anything smaller than 2 M_{Jupiter}. And we will use the term brown dwarf or substellar object for objects from 2 to 80 M_{Jupiter}. They would get hot, and the larger ones

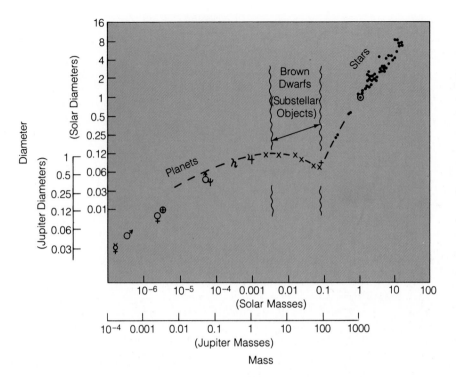

Figure 18-6 The transition from planets through brown dwarfs to stars. A combination of theoretical and observed data shows the changing size of objects as we add more mass. If we start with planetary masses (lower left) and add material of solar composition, the size levels off at about the size of Jupiter and then declines slightly due to compression by gravity. As additional mass pushes the total mass over 0.08 $M_\odot$, where nuclear reactions begin, the size again expands. (After data by D'Antona [1987], Lunine and others [1986], and Hubbard [1987, private communication].)

would look red hot, as suggested by the imaginary view in Figure 18-7. But they would lack the nuclear reactions of a true star. Only objects above 80 M_{Jupiter} have nuclear reactions and are classed as stars.

Astronomers have been searching for good examples of brown dwarfs, which would help us understand the relations between planets and stars. Brown dwarfs, of course, are hard to detect because of their faintness. One possible example was reported in 1987 by B. Zuckerman and E. Becklin of UCLA and Hawaii. Using infrared detectors they found an object with temperature 1200 K close to a white dwarf star known as Giclas 29-38. The object itself was not photographable in visible light, but analysis of its dynamics suggests that it is probably not a 1200-K dust cloud but a brown dwarf orbiting around the white dwarf. A few other brown dwarf candidates were reported in 1989 and 1990, including one with a luminosity only 0.004 that of the Sun—the lowest-luminosity object yet found outside the solar system. But it is hard to prove these objects are in the 2–80 M_{Jupiter} mass range needed to call them true brown dwarfs.

Astronomers have been surprised at the difficulty of finding brown dwarfs. Some astronomers infer that an unknown effect inhibits formation of objects less massive than about 0.1 $M_\odot$ (100 M_{Jupiter}; see Figure 17-10). This, in turn, makes the search for the elusive brown dwarfs more exciting. Not only may they be strange objects, halfway between planets and stars; they may also clarify the star-forming process.

What Happens If the Mass Is Too Big?

If the cloud is dense enough to contract but is more than about 100 $M_\odot$, the contraction is violent and produces an extremely high central temperature and pressure. Under these conditions, so much energy is generated inside the new star that the star is very luminous and may blow itself apart almost immediately without spending much time on the main sequence. This rapid destruction is more appropriate to the topic of the next chapter, and will be taken up again there. We will see that explosions of massive stars explain many features of our starry surroundings.

Figure 18-7   Imaginary view of a brown dwarf star from a hypothetical nearby planet. The brown dwarf emits a dull glow from its barely red-hot surface. In the distance is a true star, the "sun" of this imaginary system. (Painting by Ron Miller.)

EXAMPLES OF PRE-MAIN-SEQUENCE OBSERVED OBJECTS

Several types of observed objects are believed to relate to late protostellar stages or pre-main-sequence stages of star formation.

Bok Globules

Bok globules are dense clouds of dust and gas that may be contracting protostars. They are made visible by being silhouetted against brighter nebulosity in the background; examples are the Coal Sack Nebula, silhouetted against the southern Milky Way as shown in Figure 17-4, and the small, dark blobs seen in Figure 18-1.

Infrared Stars and Cocoon Nebulae

Still more interesting are infrared stars, mentioned earlier as predicted from theoretical work. Many are actually observed and may be examples of dusty cocoon nebulae around new stars—the same environments in which the Earth and the other planets formed around the early Sun, as described in Chapter 14.

The first detection of a cocoon nebula was announced in 1966, only a year after some of the first predictions of such objects. American astronomers Frank Low and Bruce J. Smith made this announcement based on their measures of R Monocerotis (star R in the constellation of Monoceros the Unicorn), a starlike object about 700 pc from the solar system (Figure 18-8). Low and Smith found that this object showed a large amount of infrared radiation and a faint amount of visible light. They hypothesized that the visible light came from a new star in the center of a cocoon nebula 200 AU in diameter, which absorbed most of the starlight. The dust grains in the nebula are heated to a temperature of about 850 K and radiate infrared light. Observations reported in 1984 suggest the nebula is disk-shaped and contains perhaps five Earth-masses of dust. These data raise the possibility that planets are forming around R Monocer-

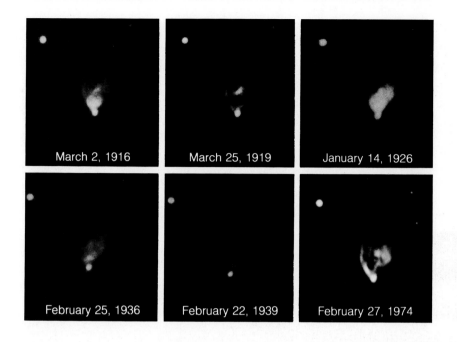

March 2, 1916

March 25, 1919

January 14, 1926

February 25, 1936

February 22, 1939

February 27, 1974

Figure 18-8 Exposures of the newly forming star R Monocerotis over a 58-y period. The star is the bright object at the bottom tip of the irregular nebulosity. The nebula, called Hubble's variable nebula, changes shape from year to year, possibly due to changes in the illuminating radiation escaping from the cocoon nebula around the associated star. The changes indicate instability in the configuration. (Lowell Observatory; 1974 photo courtesy Alan Stockton, Mauna Kea Observatory, University of Hawaii.)

otis today, just as they did in our Sun's dust-laden cocoon nebula 4½ billion years ago! What planets and landscapes will exist there 100 million years from now, when our civilization on Earth is only a dim memory?

R Monocerotis lies at the tip of a triangular nebula about 50 000 AU long, which reflects the light of the young star and its cocoon. As revealed in photos over more than six decades (Figure 18-8), the light, shape, and spectrum of the system change, again indicating that the system may be unstable and rapidly evolving, in accordance with the theory of star formation.

Many other infrared cocoon nebulae have been studied, ranging from examples that totally obscure any interior star to examples where most of the young star's light escapes.

The best examples of infrared stars and other probable newly formed stars are found in the **Orion star-forming region,** a large area of the sky around the constellation Orion. This region, which includes the R Monocerotis infrared star, is about 400 to 700 pc away and seems to be a hotbed of dense clouds and star-forming activity. When we look in this direction on a starry night, many of the prominent features of the sky are the results of recent star-forming activity, as seen in Figures 18-9 and 18-10. Many others lie hidden to our eyes inside dust clouds of interstellar space. But these are now being mapped by detectors sensitive to infrared light that passes through the clouds, even though

the clouds block visible light. Figure 18-11 shows a 1987 image of such a dark dust cloud, revealing the newly born infrared stars within it.

These and other objects exhibit many features that tie in well not only with the theory of star formation but also with evidence about conditions in the nebula that surrounded our own Sun as it formed. Among these features are:

1. Spectra indicating the presence of dust, including olivine-rich silicate and ices

2. Variable brightnesses, indicating instability

3. Detection of H_2O and OH molecules in some nebulae

4. Temperatures around 350 to 1800 K in the dust of the nebulae.

Note that these properties are very similar to the properties of the solar nebula in which the planets formed (Chapter 14). One of the most exciting implications of the observational work on newly formed stars is that we may be seeing dusty systems in which planets are forming now or might form in the "near" future—that is, in the next few million years.

T Tauri Stars

The most important pre-main-sequence stars are the **T Tauri stars,** named after the variable star T in the

a

Figure 18-9 a A panorama of star formation in the midwinter evening sky, covering about 80° around the constellation Orion. Many of the features are unusually young. Note that most of the brightest stars, such as Sirius and Orion's Belt, are recently formed bluish-white O and B stars. However, two prominent stars, Betelgeuse and Aldebaran, are red giants easily distinguished by their color (not only on this photo but also by naked eye). (11-min exposure with guided 35-mm camera, wide-angle 24-mm lens at f2.8. Star-pattern filter emphasizes bright stars and their colors by spilling some light into spike-shaped diffraction pattern; photo by author.) **b** (*Opposite*) Key map. Distances (parsecs) and ages (millions of years) are marked. Features in the constellation Monoceros (left) are shown in Figures 18-8, 18-10, and 18-17. Rings of radiating dashes mark associations of T Tauri stars, not prominent to the eye or on the photo. Compare with the enlarged view of Orion shown in Figure 16-3.

constellation Taurus. Although a great many of them are found in the region of Taurus and the neighboring constellation Orion, they can be found in many other parts of the sky. They were first recognized as a group in 1943–1944 in work almost unnoticed because the author, K. Himpel, was publishing inside Nazi Germany.

T Tauri stars may represent a transitional stage between infrared stars surrounded by opaque nebulae and stable stars that have lost their cocoons and settled on the main sequence. There are many signs of their youth and instability. The number of T Tauri stars alone in a star-forming cubic parsec may exceed the number of *all* stars per cubic parsec near the Sun by more than a factor of 10! (See Cohen and Kuhi, 1979.) T Tauri stars vary irregularly in brightness. Penn State astronomer Joan Schmeltz (1984) finds that the variations stem from changes both in the star's surface layers and in the enshrouding dust clouds. T Tauri stars lie to the right of the main sequence in the H–R diagram, where young stars are supposed to lie. Many have the infrared radia-

tion characteristic of cocoon nebulae, as pointed out by the Mexican astronomer E. Mendoza (1968).

T Tauri stars are typically 20 000 to a million years old, judging from all available evidence (Cohen and Kuhi, 1979). This is younger than the human species! Stars that have evolved beyond the T Tauri stage would probably be indistinguishable from main-sequence stars.

Disks of Dust Around Young and Middle-Aged Stars

Although there were many signs that cocoon nebulae might bear a connection to the proposed solar nebula in which the planets formed, there were no direct observations of disk-shaped nebulae around newly formed stars until the 1980s. There was, however, indirect evidence of disk shapes among some of the cocoons. Astronomers assumed that the cocoons were irregular or spheroidal in the earliest stages of protostar collapse, flattening later as the cloud collapsed further and the

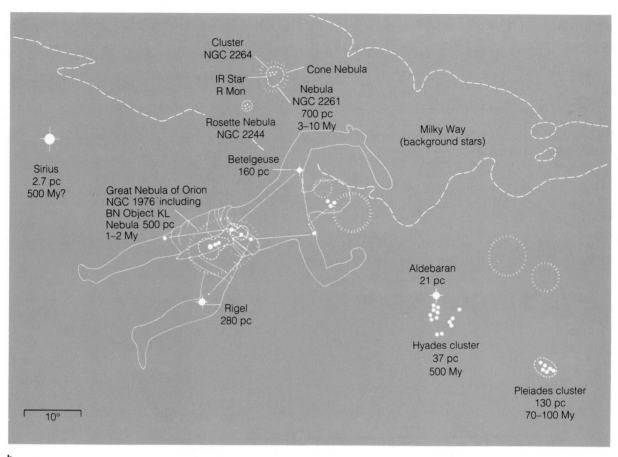

b

star formed. Researchers could only speculate that some of the dusty cocoons evolved into the thin, flat disk shape that must have spawned our system of planets.

In 1983, however, the Infrared Astronomical Satellite (IRAS) was launched into orbit with the capability of mapping stars at the far-infrared wavelengths emitted by cool dust (review pages 359–360). IRAS promptly made a new discovery: some two dozen stars that had clouds of infrared-emitting dust extending hundreds of astronomical units from the star. The discovery led ground-based astronomers to use sophisticated techniques to make detailed images of these systems in order to discover their properties. Starting in 1984, astronomers began discovering flat, disk-shaped nebulae of dust near these stars. These discoveries have been hailed as providing a "missing link" between the early stage of a ragged cocoon nebula and the later stage of a dusty disk of the type required to form planets. Among the stars studied have been HL Tauri, Beta Pictoris, and other main-sequence stars.

HL Tauri This is a classic T Tauri star probably less than 20 million years old. Observations some years ago showed that it had a surrounding swarm of microscopic dust particles. Later observations revealed that this swarm is a rotating disk about 1000 AU in radius, lying nearly edge-on to Earth, and containing about a tenth of a solar mass of material—10 times the mass of all our planets. Because the dust is in a thin disk, it doesn't completely surround the star or block its light.

Beta Pictoris This is another star with a disk of dust—in this case reaching at least 400 AU from the star and lying nearly edge-on as shown in Figure 18-12. A provocative fact about this system is that Beta Pictoris is not a classic newly forming star, such as a T Tauri star, but further evolved, like the Sun. It is classified as a normal main-sequence star of spectral class A, several times as massive as the Sun, and probably less than a billion years old. The discovery of the disk means that even main-sequence stars can have dust systems around

Figure 18-10 The Rosette Nebula and star cluster NGC 2244 are notable features of the skyscape in Figure 18-9. About 800 pc from Earth, the Rosette Nebula is a spheroidal cloud about 17 pc across. In its center is a cluster of young stars. **a** General view of the region. **b** Enlarged view showing blue colors of the brightest, hottest stars in the cluster; the dark dust globules may be contracting to form still more stars. (Copyright Anglo-Australian Telescope Board, courtesy D. Malin.)

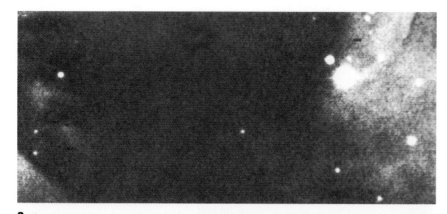

a

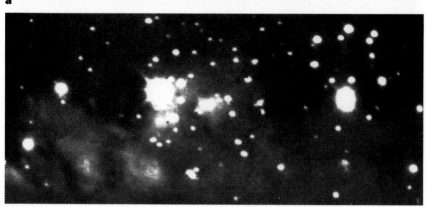

b

Figure 18-11 Piercing a dust cloud. The top picture shows part of a seemingly featureless, dark dust cloud in which few stars can be seen. The bottom picture is an image of the same region with an infrared detector sensitive to light of wavelength 2 μm. This image reveals a cluster of newly formed infrared stars whose light can escape through the dust to our telescopes. The region is a small part of the star-spawning nebula, NGC 2024, in Orion. (National Optical Astronomy Observatories.)

them—quite a step toward the possibility of planetary systems around other main-sequence stars. According to 1986 observations, the dust may be further evolved than in the HL Tauri system. There is a relatively dust-free zone close to the star, and some researchers speculate that the dust there might have aggregated into planets. Moreover, the dust that *is* seen scatters all colors of light, not just blue light, indicating that the particles are larger than light's wavelength. (If this concept is unclear, review the discussion of Rayleigh scattering on pages 437–438.) Astronomers think that the particles may be a few micrometers across—about the size of dust particles spewed out of comets.

Other Main-Sequence Stars The IRAS data also revealed a number of stars that were cataloged as perfectly ordinary main-sequence stars but had a dust system that was too tenuous to detect. If such dust were placed in orbit around the star, stellar winds would blow the dust away, just as comet tails are blown outward by the solar wind. This proves that the dust must be replenished continuously in order to exist at all. Astron-

omers have therefore inferred that there may be comets or some other unseen bodies of planetary matter orbiting the star and emitting the dust. Such dust may be analogous to the dust trails in our own solar system, lying along comets' orbits, as discovered also by IRAS.

The Mystery of Mass Ejection from New Stars

What is the process by which the thick obscuring dust around a T Tauri star thins or disappears to reveal the star? Data show that as much as $0.4\ M_\odot$ of gas and dust is blown away from T Tauri stars as they evolve. Does the material get blown outward, fall into the central star, or what? Observations show chaotic gas motions around T Tauri stars. Doppler shifts revealed gas sometimes moving inward toward the star, but more often blowing outward as fast as 50 to 200 km/s. Astronomers in the 1970s visualized T Tauri stars as having strong "stellar winds" blowing materials out from them, analogous to the solar wind, but this did not explain occasional inward motions.

Figure 18-12 The best pictorial evidence yet available of planetary dust around a nearby star, Beta Pictoris, 16 pc away. Thermal infrared radiation from the dust had been detected in 1983 by the IRAS infrared satellite telescope in space, leading astronomers to make this 1984 visual-light image from a ground-based telescope in Chile. The black central disk and crosshairs represent a system used to block the light of the star itself. With the star glare blocked, sensitive imaging equipment recorded bright material extending on either side of the star. This is an edge-on view of a dust disk extending to about the distance of the inner Oort swarm of comets in our own planetary system. The Beta Pictoris dust nebula is thicker and brighter than the zodiacal dust of our own system. Researchers suggest that planets may have recently formed in the Beta Pictoris system, though none have yet been detected. (Photo by S. Larson and S. Tapia, 1987, at Cerro Tololo International Observatory, Chile.)

Observations in the 1980s revealed a mysterious phenomenon that we will encounter again and again, not only in T Tauri stars but in certain other kinds of stars and even in galaxies. It is called **bipolar jetting**. Remember that gas and dust are organized in a thin disk in these systems. During bipolar jetting, the gas shoots out in opposite directions, perpendicular to the disk, as seen in Figure 18-13 (Lada, 1982; Bailly and Lada, 1983). Stellar winds may also blow some material away from some T Tauri stars in all directions, but in many T Tauri stars, some gas and dust apparently spiral inward toward the star, get caught in magnetic fields around the star, and then get accelerated, squirting "upward" and "downward" in two diffuse jets away from the disk, as seen in Figure 18-13. The forces that accelerate these jets are probably magnetic, but are poorly understood. They are the subject of intense current research.

In 1986 a team of astronomers from Arizona and Missouri announced another "missing link" observation. They studied an infrared star named IRAS 1629A, first detected by the IRAS satellite. Bipolar jets had already been found soon after the discovery of the star, but the 1986 observations revealed a gaseous disk lying perpendicular to the jets and reaching 800 AU from the star. The outer parts of the disk were found to be orbiting the star, but the inner parts were collapsing inward and feeding the jets. This observation strongly supports Figure 18-13's theoretical model of young stars' disks and bipolar jets.

An infrared photograph of a probable bipolar jetting system, similar in appearance to Figure 18-13, is shown in Figure 18-14. It shows warm clouds of material fanning out above and below a probable edge-on accretion disk. Figure 18-15 shows an imaginary close-up view of a better developed system; some of these are believed to have tightly defined bipolar jets.

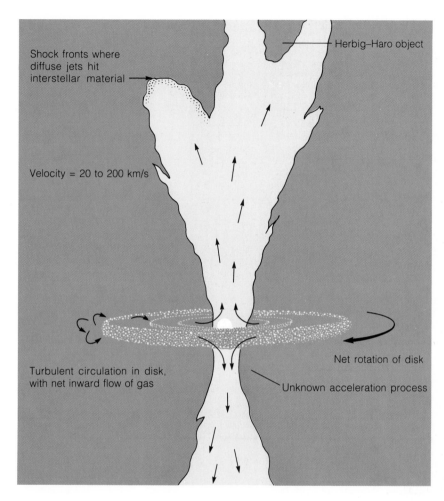

Herbig–Haro object

Shock fronts where diffuse jets hit interstellar material →

Velocity = 20 to 200 km/s

Turbulent circulation in disk, with net inward flow of gas

Net rotation of disk

Unknown acceleration process

Figure 18-13 Recent evidence suggests that as a new star turns on in the center of its turbulent disk-shaped cocoon nebula, gas is blown outward along the disk's axes by an unknown mechanism. Outflowing clouds move "upward" and "downward," at speeds up to 200 km/s, ramming into surrounding interstellar gas.

This theoretical picture explains why Doppler shifts reveal inward motions of gas around some T Tauri stars (the ones viewed from a direction along the disk), while revealing an outward motion among the majority (viewed through gas blown outward by the jets). The model also fits beautifully with evidence about the dusty, disk-shaped solar nebula in which the Earth and other planets were born near the newly formed Sun. Further work may clarify whether planets are forming from cocoon dust near other new stars, and how the jets of gas are expelled, eventually dissipating the cocoon.

The model also clarifies certain puzzling objects, such as *Herbig–Haro objects*. Herbig–Haro objects are small, variable nebulae, sometimes appearing or disappearing during a period of a few years, as shown in Figure 18-16. Astronomers once thought they might be new stars, but recent mapping of bipolar flows of gas near infrared stars indicates that Herbig–Haro objects

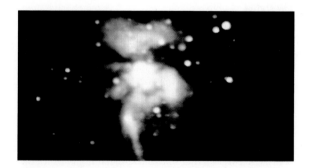

Figure 18-14 False-color infrared image of bipolar outflow system, Sharpless-106. The orientation resembles that of Figure 18-13, but the jets are more fan-shaped. The dark reddish horizontal bar across the center is believed to be the outlying parts of an edge-on dusty disk, obscuring a very young central O or B star. Abundant molecular material has been detected in the dark dust. (Image from Mauna Kea Observatory, by Klaus-Werner Hodapp and John Rayner, Institute for Astronomy, University of Hawaii.)

Figure 18-15 Probable appearance of a disk-shaped cocoon nebula of dust and gas around a newly formed star. Dust grains may be aggregating into asteroids or planetary bodies within the disk. High speed jets of gas are being expelled along the polar axes of the disk. In the background is a star-forming region and a dense cloud of dark dust. (Painting by the author.)

January 24, 1946   January 20, 1947   December 20, 1954   November 9, 1958   January 5, 1968

Figure 18-16 This cluster of Herbig–Haro objects occurs near a site of star-forming activity. The bottom grouping appears to have changed from two to three objects between 1947 and 1954, while the left-hand grouping changed from one to two objects. Such changes probably involve variable glows from supersonic gas masses near young stars. (G. Herbig, Lick Observatory.)

Herbig–Haro objects are masses of excited, glowing gas associated with the material jetted out of cocoon nebulae (Bailly and Lada, 1983). Their luminosity derives from shock waves that excite the ejected gas as it collides at high speed (100 + km/s) with nearby interstellar gas. (A shock wave is a violent heating and compression of gas as it collides with neighboring gas at faster than the local speed of sound; the sonic boom around a supersonic jet derives from a shock wave in gas compressed in front of the plane.)

In summary, gas shedding by infrared stars and T Tauri stars, and the associated creation of bipolar jets

and Herbig–Haro objects, is a challenging new area of astronomy unknown just a few years ago (Schwartz, 1983; Hartmann and Raymond, 1984). It may be teaching us not only about formation of stars per se but also about formation of planetary materials and perhaps even planet systems (Welsh and others, 1985).

Young Clusters and Associated Young Stars

Because stars form in groups, T Tauri and infrared stars are often found in groupings associated with star-form-

Figure 18-17 This swirling mass of glowing gas, young stars, and dark dust clouds, called NGC 2264, is a star-forming region. The brightest stars are all newly formed, massive, hot, blue stars. Molecules of formaldehyde (H_2CO) and other gases have been found in the region. Blue star at bottom left marks top of Cone Nebula seen on page 433. The region is mapped in Figure 18-9b. (Copyright Anglo-Australian Telescope Board.)

ing regions. In the beginning of this chapter, we described clusters of stars proven to be very young by their dynamical properties and H–R diagram position. Because low-mass stars take the longest to evolve to the main sequence, low-mass stars in these clusters ought to be found in the T Tauri stage. This has been abundantly confirmed by observations.

One of the best-known examples is the young star cluster NGC 2264, shown in Figure 18-17. Most of the low-mass stars (spectral classes A to K) in this cluster lie distinctly to the right of the main sequence, as shown in Figure 18-18, and many of these are identified as T Tauri stars. The calculated lines of constant age (dashed lines in Figure 18-18) show that the T Tauri stars match

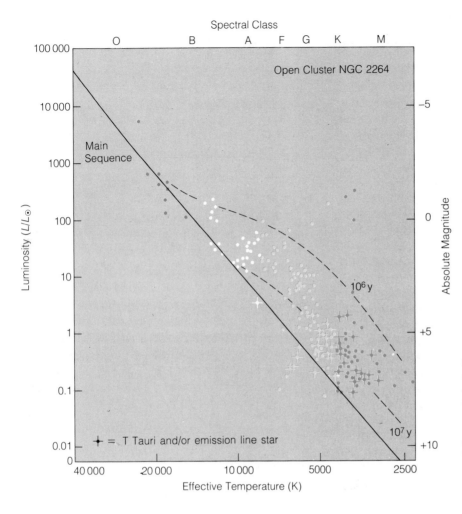

Figure 18-18 This H–R diagram of stars in the region around Figure 18-17 reveals stars lying to the right of the main sequence along age lines (dashed), suggesting ages of only a few million years. (Data from M. Walker, 1972, with isochrons from I. Iben, 1965.)

the positions predicted for an age of 3 to 30 million years. *This demonstrates how the H–R diagram can be used, together with theoretical data, to determine the **ages of star clusters.*** A study of 300 pre-main-sequence stars in this cluster indicates that many of them have dusty cocoons and that the low-mass stars formed first, at least 20 million years ago (Adams, Strom, and Strom, 1983).

Recall our rule of thumb from Chapter 17 that massive stars evolve fastest. Brilliant, massive, bluish, O-type supergiants last only a few million years. For this reason they are found only in star-forming regions a few million years old. By the time stars have dispersed from such regions, the O supergiants have already burnt out. Thus they are rare among representative field stars but

dominate pictures of star-forming areas such as Figure 18-17.

DEBRIS OF STAR FORMATION

Many star-forming regions include clouds of gas and dust that may be remnants of the star-forming process or clouds that did not get quite dense enough to fully collapse. Often, spectacular foreground clouds are seen in black silhouette against bright star groups or clouds illuminated by bright stars. Long, tubular, dark nebulae called *elephant-trunk nebulae* are sometimes found pointing toward the star-forming region as seen in Figure 18-1b and in the Cone Nebula (page 433). These may be formed as gas rushing out from the star-forming center sweeps past dense clots of dusty gas.

THE MAIN-SEQUENCE STAGE

The **main-sequence stage** is reached when the star begins to generate its energy by consuming hydrogen in nuclear reactions deep in the star's central regions. Prior to that time, the star generates energy primarily by its Helmholtz contraction (see page 301), which raises the temperature and pressure in the central regions. As a star enters the T Tauri pre-main-sequence stage, it still radiates by this means and still contracts slowly, as seen by comparing the evolutionary tracks and lines of constant radius in Figure 18-5. As the pressure near the star's center increases, the atoms jam together more tightly, and as the temperature increases, the atoms collide at greater speeds. Eventually individual nuclei hit each other so hard that they begin to fuse together. These nuclear fusion reactions generate new heat and pressure that stop the contraction. Toward the end of the T Tauri stage, as the star is just settling onto the main sequence, there may be some minor nuclear reactions involving elements that react at lower temperatures than hydrogen, but hydrogen fusion soon becomes the dominant energy source as the main sequence is reached.

In the H–R diagram, an imaginary array of stars with different masses, which have all just reached the main sequence and just begun to consume hydrogen, is called the **zero-age main sequence.** Calculations show that it would be a very narrow band of stars on the H–R diagram. In reality, the main sequence on the H–R diagram is a bit broader because it contains stars of different ages, which have converted different fractions of their hydrogen to helium. Thus they have slightly different structures and slightly different positions on the H–R diagram.

Once the star reaches the zero-age main sequence, it begins its life as a true star and commences a long sequence of various nuclear reactions. During each such reaction, tiny amounts of matter are converted to energy, providing the heat and light of the star. We can't over-emphasize the basic cause of further stellar evolution: *The nuclear reactions convert certain elements into other elements, changing the star's composition and thus the energy generation rates; this in turn causes the star to change to new forms.* This is the essence of the Rus-

sell–Vogt theorem, which you should recall from Chapter 17 (pages 374–376). To clarify these changes, we will now review the nuclear reactions in stars in more detail.

Nuclear Reactions in the Smaller Main-Sequence Stars

As described in Chapter 15, studies of our own main-sequence star, the Sun, reveal that its energy comes from a series of nuclear reactions called the **proton–proton chain.** We list it again for emphasis; you can review its details on pages 321–322.

$$^1H + {}^1H \rightarrow {}^2H + e^+ + \text{neutrino}$$
$$^2H + {}^1H \rightarrow {}^3He + \text{photon}$$
$$^3He + {}^3He \rightarrow {}^4He + {}^1H + {}^1H + \text{photon}$$

The proton–proton chain is the primary energy-producing process not only inside the Sun but also inside all main-sequence stars smaller than F stars of about $1\frac{1}{2}$ $M_\odot$. It dominates if the central temperatures are less than about 15 million Kelvin.

Nuclear Reactions in the Larger Main-Sequence Stars

In main-sequence stars larger than about $1\frac{1}{2}$ $M_\odot$ where interior temperatures are higher than about 15 million Kelvin, another reaction series dominates in producing energy. This is the **carbon cycle,** sometimes called the *CNO cycle* to reflect the involvement of carbon, nitrogen, and oxygen. The reactions are as follows:

$$^{12}C + {}^1H \rightarrow {}^{13}N + \text{photon}$$
$$^{13}N \rightarrow {}^{13}C + e^+ + \text{neutrino}$$
$$^{13}C + {}^1H \rightarrow {}^{14}N + \text{photon}$$
$$^{14}N + {}^1H \rightarrow {}^{15}O + \text{photon}$$
$$^{15}O \rightarrow {}^{15}N + e^+ + \text{neutrino}$$
$$^{15}N + {}^1H \rightarrow {}^{12}C + {}^4He$$

Again the net result is that hydrogen atoms are used up to produce helium-4 atoms with an associated release of energy. In a sense, carbon acts as a catalyst (a stimulant of change), because carbon-12 reappears at the end of the cycle, to be used again in the first reaction of a subsequent cycle.

In main-sequence stars, one or the other of these reactions consumes hydrogen in the core. Finally, major structural changes happen as the hydrogen is used up. In the next chapter we will see what dramatic events ensue.

SUMMARY

Stars are forming even today within a few hundred parsecs of the Sun and in more distant regions of space. The starry sky is not a static scene but the site of continual births of new stars out of interstellar dust and gas. Many stars and star systems are less than a few million years old—much less than 1% as old as our galaxy. Some have become visible since humanity evolved, though prominent newly formed stars have probably not appeared in our sky during recorded history.

Star formation begins with protostars, which are clouds of dust and gas that begin to contract due to their own gravity. They collapse fairly rapidly to stellar dimensions and become pre-main-sequence starlike objects. Theoretical calculations in the last two decades have revealed what features these objects probably have. Many of these features have been confirmed by observation, especially by infrared equipment, which detects the radiation from relatively low-temperature dust in nebulae around the newly formed stars.

Among the objects revealed in this way are groups of stars evolving toward the main sequence. Many groups of stars and individual stars are surrounded and obscured by cocoon nebulae consisting of dust particles (probably silicates and ices similar to those that formed the first planetary material in our own solar system). More evolved objects, such as the T Tauri stars, appear to be shedding their cocoon nebulae and have almost reached the main sequence.

Once the main sequence is reached, energy is generated in the star as hydrogen converts into helium by means of the proton–proton cycle for smaller stars and the carbon cycle for larger stars.

CONCEPTS

evidence for present-day star formation

theory of star formation

gravitational contraction

subfragmentation

protostar

collapse

pre-main-sequence star

high-luminosity phase

cocoon nebula

infrared star

brown dwarf

Bok globule

Orion star-forming region

T Tauri star

bipolar jetting

age of star clusters

main-sequence star

zero-age main sequence

proton–proton chain

carbon cycle

PROBLEMS

1. Did the Sun and solar system form in the first half or last half of our galaxy's history?

2. Suppose that you magically smoothed out all inhomogeneities in the interstellar gas so that it was all uniform.
   a. Would this help or hinder star formation?
   b. What processes would keep the gas from staying uniform indefinitely?
   c. Would star-forming conditions tend to return to normal?

3. Why do stars form in groups instead of alone?

4. How do theories of solar system formation and theories of star formation support each other? Contrast the sources of information on these two subjects.

5. Compare the time scale for significant evolution of massive pre-main-sequence stars with the time during which astronomers have recorded observations of such stars (see Figures 18-4 and 18-5). Is it reasonable that some young, pre-main-sequence stars might show evolution-related fluctuations in their properties within the time that they have been observed?

6. Suppose a cluster of stars formed 3 million years ago. Why would you expect the H–R diagram of the cluster to show no stars on either the very high mass end of the main sequence or its very low mass end?

7. Suppose a new 1-$M_\odot$ star began forming about 10 pc from the Sun. What would Earth-based observers see during the next few million years? Consider infrared observers as well as naked-eye observers.

8. Why doesn't gravity immediately cause the collapse of all interstellar clouds?

9. Why does the structure of a star stabilize when it reaches the main sequence?

ADVANCED PROBLEM

10. Suppose you observe an infrared nebula whose strongest radiation comes at wavelength 10 μm and which has absorption lines of solid silicates in its spectrum, as well as faint lines indicating a G-type star. Suppose the silicate lines are blue-shifted by about 10^{-3} of their wavelength compared with the star's wavelength. What conclusions can you draw about this system?

PROJECT

1. On a clear night (an early evening in February or a late evening in December is ideal) scan the region of Orion with your naked eyes and compare it to other regions of the sky. Note the concentration of bright blue O-, B-, and A-type stars (such as Sirius and Rigel) and star clusters (such as the Pleiades, Hyades, and the Orion Belt region) in this broad area. How do these features indicate that star formation has been going on in this general direction from the Sun in the last few percent of cosmic time?

CHAPTER 19

Stellar Evolution II: Death and Transfiguration

Stars are born and stars die. By this we mean that the nuclear reactions converting mass to energy in stars have a beginning and an end. The Sun, for example, spent some 10^7 y in its formative stages; it will spend about 10^{10} y on the main sequence, roughly 10^9 y in the giant state, and a shorter time in later unstable states consuming heavier elements until energy generation stops (see Figure 17-11). This evolutionary sequence might be likened to the periods of human life: 9 months in the womb, 65 years of normal life, 6 years of rapid aging, and perhaps a year of terminal illness. As we will soon see, stars do evolve to pathological terminal states, involving incredible forms of dense matter and processes unimagined until a few years ago.

Writer Ben Bova (1973) summed up star deaths by quoting one of Ernest Hemingway's characters, who is asked how he went bankrupt. "Two ways," he says. "Gradually and then suddenly." The gradual part is the slow expenditure of a star's hydrogen, which is scarcely noticeable to an observer. The sudden part—the subject of this chapter—comes as the star runs out of hydrogen. It goes into fits of activity, searching for new sources of fuel until, as Bova says, "gravity forecloses all the loans."

Gravity forecloses the loans by causing a continual tendency toward contraction. When the core of a main-sequence star runs out of hydrogen, several things happen. At first, the core may burn its way into outer layers that still have hydrogen. This can cause an expansion of the outer layers, so that the star temporarily balloons to a giant size. But eventually this process of hydrogen burning spreads outward into too cool a layer to sustain the reactions. Thus the hydrogen-burning reactions wind down. This means there is less outward pressure generated by reactions. Thus the core must contract due to gravitational forces pulling inward. Although some of the upper layers may be blown outward, most of the

star's mass contracts to a small size. This contraction is what we mean by "gravity forecloses the loans."

What happens after the hydrogen is exhausted and the core starts to shrink? Remember from the discussion of Helmholtz contraction (page 301) that as a mass of gas contracts, it must grow hotter. Thus the atoms in the core move faster and collide even harder than before. Soon they reach a condition where the helium atoms in the core collide hard enough to start fusion reactions. We say that the core has stopped burning hydrogen and is now burning helium. After the helium has burned, more contraction occurs and still other elements may become fuels. Eventually, there are no more fuels to react and the star contracts to a very dense state.

This chapter discusses four main stages of stellar old age: the giant stage, variable stages, explosive stages of several types, and terminal high-density stages. Table 19-1 lists examples of stars in each of these stages, which are described throughout the chapter.

Note that the most common stars—those about one solar mass—go through the giant phase and then contract rather smoothly to a small, high-density stage. But the rare, high-mass stars beyond about $8\,M_\odot$ go through explosive instabilities. Eventually they reach even higher-density stages than sunlike stars, because they have more mass and gravity pulling inward. They produce some of the most exotic objects yet discovered by astronomers, such as pulsars and black holes.

THE GIANT STAGE

As described briefly in Chapter 17, a star evolves off the main sequence as hydrogen fuels run out. At this time a thin shell, in which hydrogen is still burning, surrounds the core. Outside the shell, temperatures have never risen high enough to "ignite" the hydrogen nuclei (that is, to fuse them into helium) so there is still plenty of hydrogen fuel. This allows the hydrogen-burning shell to migrate further toward the surface. The increased temperatures cause great expansion of the outer layers, making the star evolve rapidly toward the giant state. The outermost atmosphere becomes huge, thin, and cool, even though the inner core is smaller, hotter, and denser than ever. From the outside, the outer atmospheric layers are seen to glow with a dull red color, and the star is perceived as an enormous **red giant,** such as shown in Figure 17-7.

The largest giants are truly huge—approaching 1000 times the size of the Sun. If the Sun were replaced by one of these stars, its thin, red-glowing outer atmosphere would reach beyond the orbit of Mars! A typical red giant's scale is shown by Figure 19-1.

As charted on the H–R diagram (Figure 19-2), stars from all points along the main sequence display a **funneling effect:** As they move off the main sequence, their evolutionary tracks funnel into the red giant region. They resemble patients from all walks of life crowding into the same hospital because they are victims of the same malady—hydrogen exhaustion.

Meanwhile, the star cores contract until they reach temperatures near 200 million Kelvin. This is hot enough to begin to fuse helium nuclei in the central regions of most stars, primarily by the **triple-alpha process** (named after the alpha particle, another name for the helium-4 nucleus):

$$^4\text{He} + {}^4\text{He} \leftrightharpoons {}^8\text{Be} + \text{photon}$$
$$^8\text{Be} + {}^4\text{He} \rightarrow {}^{12}\text{C} + \text{photon}$$

In this process, three helium-4 nuclei combine to produce a carbon-12 nucleus. (Because beryllium-8 is unstable, some beryllium atoms may break up before completing the process—but this merely reduces the efficiency of the process.) Notice that we are now burning helium to make an even heavier element, carbon.

The triple-alpha process produces prodigious energy, which at once heats the rest of the core, creating a sudden burst of helium burning often called the **helium flash.** In some stars, the helium flash may consume the central core's helium in only a few seconds. Since the energy from the flash diffuses through the star slowly, the heating effects seen at the surface may last thousands of years. The flash occurs as the star enters the giant region. In a Sun-sized star, it may occur 300 million years after evolution off the main sequence.

After the helium flash, "squiggles" may develop in the evolutionary track. Reduction of pressure in the outer atmosphere may allow the star to contract, temporarily moving it back toward the main sequence.

Note that the core's evolution begins to be independent of the evolution of the outer atmosphere. The characteristics of the star as perceived by an astronomer on Earth are those of the outer atmosphere. These characteristics determine the star's position on the H–R diagram. When you go outside at night and look at Betelgeuse or Antares, you are seeing the cool, red, outer

TABLE 19·1

Examples of Stars in Middle and Late Evolution

Stage of Evolution[a]	Distance from Earth (pc)	Mass $(M/M_\odot)$	Radius $(R/R_\odot)$	Luminosity $(L/L_\odot)$	Spectral Type	Density[b] (kg/m³)
Main Sequence						
Sun	<1	1	1	1	G2	1420
α Centauri B	1.3	0.85	1.2	0.36	K4	700
Procyon A	3.5	1.7	2.3	6	F5	200
Algol A	31	4	3.0	100	B8	210
Giant						
Arcturus	11	~4	25	115	K1	0.36
Aldebaran A	21	~4	45	125	K5	0.07
β Pegasi	64	~5	140	400	M2	0.003
Supergiant						
Antares A	160	~12	700	9000	M1	0.000 05
Betelgeuse	170	~18	~700	11 000	M2	0.000 07
VV Cephei A	200–1200?	20–80?	400–1600?	200–10 000?	M2	0.000 03?

Source: Data from Burnham (1978), Liebert (1980), Baize (1980), and Bonneau and others. (1982).
[a]Listed in order of increasing age.
[b]Compare these densities of familiar materials: atmosphere at sea level, 1.2 kg/m³; water, 1000 kg/m³; lead, 11 300 kg/m³.

atmosphere of a star. Hidden inside is a very hot, dense core. Thus although we may say "the star" is cooling and getting redder, it is really its atmosphere that is cooling. Instabilities after the helium flash may cause temporary heatings and coolings of the outer layers and temporary expansions to greater size; these appear as squiggles in the evolutionary track on the H–R diagram. But all this time, the core may be shrinking and growing hotter.

As the core contracts and gets hotter, it initiates reactions involving even heavier elements. Carbon burning may occur. Some of the reactions create floods of neutrons that strike nearby atoms. These initiate the so-called **s-process reactions,** in which neutrons are *slowly* added to nuclei to build up still heavier elements, and the **r-process reactions,** in which *rapid* neutron

addition builds up very heavy elements. The s-process and r-process reactions are very important because of the production of the heavier elements, such as metals. Without such reactions, the universe would still consist of almost pure hydrogen and helium and would lack sufficient heavy elements to make planets such as the Earth. Table 17-3 reviewed this sequence of element evolution in stars.

THE VARIABLE STAGE

A **variable star** is a star that varies in brightness on a time scale of hours to years. Most variable stars apparently represent *postgiant stages* of evolution

Examples of Stars in Middle and Late Evolution, *continued*

Stage of Evolution[a]	Distance from Earth (pc)	Mass $(M/M_\odot)$	Radius $(R/R_\odot)$	Luminosity $(L/L_\odot)$	Spectral Type	Density[b] (kg/m^3)
Variable and Explosive Stars						
Mira A (long-period variable)	77	2	230	Up to 1100	M6	~0.000 2
Polaris A (Cepheid variable)	120?	~6	~25	830	F8	~0.5
HD 193576B (Wolf–Rayet)	?	12	~7		"WR"	~50
DQ Herculis B (nova)	?	0.2	~0.1	<0.1	M?	300 000
White Dwarf						
Sirius B	2.7	1.0	0.022	0.002	A5	130 000 000
Procyon B	3.5	0.65	0.02	0.000 5	F	120 000 000
Pulsar (neutron star)						
Crab Nebula pulsar	1100	~2??	<0.000 02	High in UV, X ray	—	~10^{17}?
Possible Black Hole						
Cygnus X-1	≥2500	6–15?	<0.000 004	0	—	~10^{22}?

(although, as we saw in the last chapter, some variables are pre-main-sequence stars). Some pulse with a constant period; others flare up sporadically, often brightening by many magnitudes and then fading again. Such stars were recorded as long ago as 134 B.C. when Hipparchus recorded the flare-up of a "new star," proving that the heavens were not eternally constant. The star Omicron Ceti was known to the ancients as Mira (Wonderful) for its brightening from invisibility (magnitude 8–10) to the second magnitude every 331 d.

Shakespeare used the simile "constant as the northern star." He might have been interested to know that Polaris is constant neither in brightness nor in marking the north celestial pole. Polaris is a variable star that changes from magnitude 2.5 to magnitude 2.6 in just less than 4 d. Because of precession (Chapters 1 and 2), it is the North Star for only a few centuries every 26 000 y.

Some 22 650 variable stars have been cataloged and divided into as many as 28 types. The type most important to astronomers is the **Cepheid variable,** which has regular variations in brightness, usually with periods from 5 to 30 d. In 1784, the 19-year-old English astronomer John Goodricke[1] discovered the first Cepheid var-

---

[1] Goodricke offers an inspirational example of accomplishment during a short, difficult life. Deaf from infancy, he lived during the first European generation to recognize that deafness was not the same as idiocy. He attended the first English school for deaf children, took up astronomy, and made numerous interesting observations, but died at age 21 from pneumonia, possibly contracted from nighttime exposure during prolonged observing efforts.

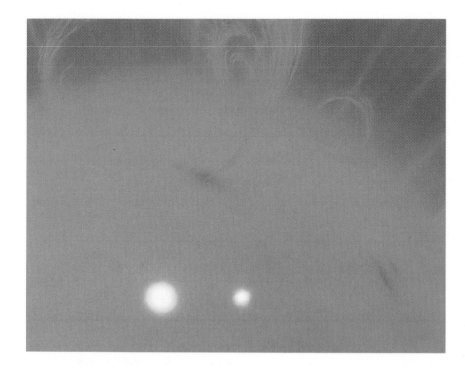

Figure 19-1 A schematic representation of the size of a typical red giant in comparison with more familiar stars. The larger of the two stars at the bottom is Sirius; the smaller is the Sun. (Painting by Tom Miller.)

iable, Delta Cephei. It varies from magnitude 4.4 to 3.7 with a period of 5.4 d; Cepheids were named after that star.

Cepheids are important for two reasons. First, because their variations are regular, they are somewhat better understood than stars whose brightness changes are unpredictable (called **irregular variables**). Second, and more important, the period of variation of each Cepheid directly correlates with its average luminosity. This relation was discovered in 1912 by one of the most famous women astronomers, Henrietta S. Leavitt, who measured hundreds of photos of Cepheid variables in the first years of this century at Harvard Observatory. Two types of Cepheids and a related type of variable called an RR Lyrae star were eventually found to have distinct period–luminosity relationships. These relationships allow astronomers to determine the luminosity of any Cepheid at any distance simply by measuring its period. This, in turn, leads to a new way to measure the distances of stars:[2] Find a Cepheid; measure its period and hence its luminosity, or absolute magnitude; then measure its apparent magnitude and derive its distance from the relations shown in Figure 16-11.

---

[2]For discussion of this very important technique in mapping distances of star clusters and similar galactic features, see Chapter 22.

Cause of Brightness Variations

Most variables change brightness because of changes in their interior nuclear reactions or in the flow of radiation from their centers. Variables vary in physical size and other properties as well as in brightness. Delta Cephei itself varies in diameter (from roughly 38 million to 41 million kilometers) and surface temperature (from 5600 to 6500 K) during its cycle.

Astrophysicists have long sought the precise cause of the internal physical disturbances that make Cepheid variable stars pulsate. Various researchers, especially the Soviet astronomer S. Zhevakin, have established that the H–R diagram contains a nearly vertical *instability strip,* shown in Figure 19-3. Stars become unstable where their evolutionary tracks cross this strip. Inside such stars ionized helium absorbs outgoing radiation and thus becomes doubly ionized. In the instability strip, this absorption occurs in the outer atmosphere, not far below the star's visible surface, where it temporarily "dams up" the radiation. Thus, again, we are seeing a phenomenon of the star's *outer* layers; the inner core may still be very hot, small, and dense.

Just like a bell (or any other object), a star has a certain frequency or time period in which it tends to vibrate in response to a disturbance. If the time required to dam up the radiation is close to the star's natural

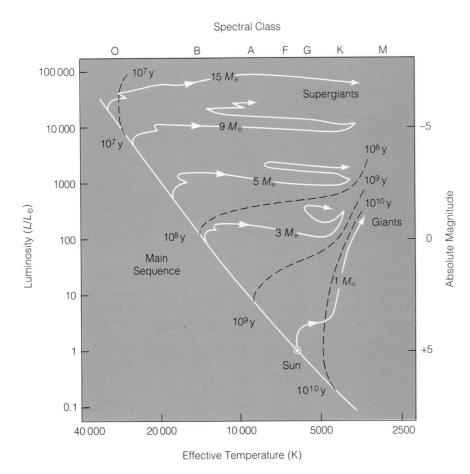

Spectral Class

Figure 19-2 Evolutionary tracks off the main sequence toward the giant region, plotted on the H–R diagram for stars of different mass. Dashed lines give the length of time since star formation; massive stars evolve fastest. All the stars evolve toward the upper right corner of the H–R diagram, the region of giant stars. (After calculations by Iben.)

oscillation period, the star will begin to oscillate repeatedly. This is what happens in a Cepheid. As the radiation is blocked, pressure builds up and the outer layers of the star expand. But once the expansion gets going, momentum and the star's natural tendency to oscillate at this speed carry the expansion too far. The atmosphere expands and becomes thinner, allowing radiation to escape easily. Then the star subsides, and the radiation once again begins to be dammed, restarting the cycle. In other types of stars, where the time scales of radiation damming and oscillation are not synchronized, the variation occurs irregularly.

The Hertzsprung Gap

Just to the left of the instability strip in the H–R diagram (Figure 19-3) is the *Hertzsprung gap,* a vertical band nearly empty of stars. Why do no stars have the combination of luminosity and temperature that would put them in this gap? Apparently the gap represents an evolutionary stage so unstable that stars evolve very quickly across it. This rapid evolution may involve a final blowoff of any remaining cool, outer atmospheric layers of gas, as well as exposure and final collapse of the dense inner core.

MASS LOSS AMONG EVOLVED STARS

As the outer atmospheres of evolved stars expand, some of the gas may be blown entirely free of the star, as shown in Figure 19-4. Some examples of mass loss follow.

Mass Loss Among Red Giants

After expansion of a red giant, some of the gas of the outer atmosphere may continue to move outward and be blown out of the star altogether. Many red giants are losing mass. This process can be observed from

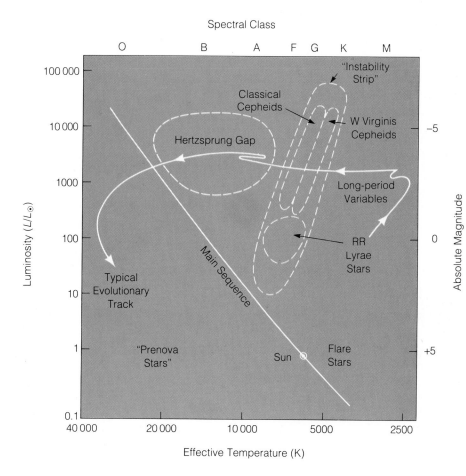

Figure 19-3 An H–R diagram showing the later stages of evolution, indicated by a typical evolutionary track (in white). The star evolves across an "instability strip" that contains Cepheid variable stars and RR Lyrae stars and then rapidly across the Hertzsprung gap. The diagram also shows the probable position of prenova stars.

Figure 19-4 A star undergoing mass loss has produced a shell of diffuse gas around itself. The star is associated with a star grouping cataloged as IC 2220. (Copyright Anglo-Australian Telescope Board, courtesy D. Malin.)

blue-shifted absorption lines arising in gas layers moving out of the star toward the observer. Probably this process grades from a smooth shedding of gas by means of a strong stellar wind, in stars of about a solar mass, to explosive shedding in stars of a few solar masses, which are less stable.

Supergiants and Mass Loss Among Massive Hot Stars

The more massive a star, the more violent and unstable its evolution. Stars of more than a few solar masses are very hot and evolve rapidly. When their atmospheres expand, these stars leave the main sequence at a luminosity already brighter than most giants. They are thus called **supergiants**—not because they are bigger than

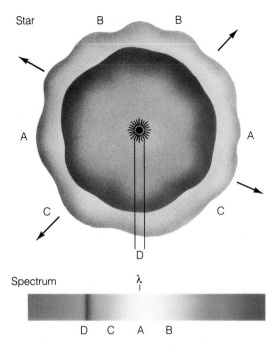

Figure 19-5 Expanding shells of thin gas, such as those around Wolf–Rayet stars or novae, can be detected spectroscopically, even when unresolved photographically. Glowing gas produces spectral emission lines of the shape shown at the bottom. In region A, light with no Doppler shift (wavelength λ) is added to the continuum spectrum of the star. In region B, red-shifted light is added, and in region C, blue-shifted light. In region D, the maximum blue shift occurs (especially in Wolf–Rayet stars), and starlight is absorbed by the nebula, producing a dark absorption band at the blue end of the emission line.

giants (some are and some aren't) but because they are brighter. They are often brighter than absolute magnitude -6 and are spread across the top of the H–R diagram, and some are hot stars classified as spectral type O and B.

O and B Stars These very massive stars have only a short lifetime on the main sequence. Ultraviolet observations from rockets and satellites, beginning in 1967, have revealed that massive O stars (greater than 20 $M_{\odot}$) have strong stellar winds from the time they leave the main sequence, blowing away as much as 10^{-5} $M_{\odot}$ of gas per year at speeds of 1500 to 1900 km/s (Conti and McCray, 1980). They could blow off several solar masses in less than a million years as they evolve toward the giant state. Plione, a B star in the Pleiades

star cluster, blew off glowing clouds revealed by spectral emission lines in 1888, 1938, and 1972.

FG Sagittae This star, a supergiant 2500 to 4000 pc away, has the fastest known evolutionary rate. It has been moving toward the giant region in the H–R diagram for decades, brightening by 4 to 6 magnitudes since 1894. From about 1960 to 1978, it has cooled from spectral class B4 to G and expanded from 24 to 70 solar radii ($R_{\odot}$). A glowing nebula around it is expanding at 34 km/s, indicating that gas was blown out about 6000 to 9600 y ago. From 1969 to 1972, certain heavy elements in the star's atmosphere, such as zirconium, have increased three to four times in abundance, indicating that gas from the interior (where the heavy elements are created) has erupted. This star is interpreted as a star that originally had 3 to 5 $M_{\odot}$ but is losing mass as it evolves from the main sequence to the giant region.

Wolf–Rayet Stars The **Wolf–Rayet stars,** named after their discoverers, are very hot stars resembling O stars except for their expanding gaseous shells and compositions. Their spectra show that their surface gases are much richer in helium, carbon, and nitrogen than most stars. As indicated in Figure 19-5, Doppler shifts show that they have thrown off surrounding clouds of gas at velocities as fast as 1000 km/s! Wolf–Rayet stars are believed to be stars that began with very high mass and eventually blew away most of their outer layers of gas, exposing their cores in which helium and heavier elements had accumulated due to hydrogen burning (Humphreys and Davidson, 1984). The ejected gas shells sometimes make a visible nebula around the star, as in Figure 19-6. More than 200 have been found since their discovery in 1867, and many are members of binary systems (pairs of coorbiting stars).

Planetary Nebulae

The gas being blown out of mass-shedding stars expands into space. It may cool enough for grains of dust, such as carbon grains, to condense in it. It may collide with other nearby gas clouds at high speed, creating glowing shock waves. Often, however, it is shed in relatively spherical bubbles of gas that surround the central star. Ultraviolet light from the star excites and ionizes the gas atoms, causing them to glow. Clouds of gas in space are called nebulae, and decades ago these particular nebulae came to be called **planetary nebulae** because

Figure 19-6 A Wolf–Rayet star is hidden in the center of several expanding shells of gas. The central bluish shell maintains the spheroidal shape of an expanding bubble, but outlying, red-glowing hydrogen is distorted into ragged loops as it collides with thin, surrounding, interstellar gas. This example is located in the star grouping NGC 2359. (Copyright Anglo-Australian Telescope Board.)

the palely glowing bubbles looked like disks of planets in small telescopes. This term is a misnomer, however, because they have nothing to do with planets. As shown in Figures 19-7, 19-8, and 19-9, they are among the most beautiful celestial features, with wispy symmetry and delicate colors (Kaler, 1986).

THE DEMISE OF SUNLIKE STARS: WHITE DWARFS

Let us now consider the further evolution of sunlike stars, by which we mean stars with initial mass similar to that of the Sun, say from 0.1 to a few solar masses. The outer atmosphere is very expanded and may be shedding its outer layers, but most of the mass is still in a dense hot core. Eventually the core runs out of fuel and gravity causes a final collapse of the whole star, including the parts of the atmosphere that have not blown off. This final collapse creates a very small, very hot, very dense star the size of Earth or other planets. Such a star is known as a **white dwarf**. About 200 are cataloged.

The Discovery and Nature of White Dwarfs

In 1844 the German astronomer Friedrich Bessel studied the motions of the brightest star in the sky, Sirius,

and found that it is being perturbed back and forth by a faint, unseen star orbiting around it. This star was not glimpsed until 1862, when American telescope maker Alvan Clark detected it. It is almost lost in the glare of Sirius, as shown in Figure 19-10. In 1915, Mt. Wilson observer W. S. Adams discovered that it was a strange, hitherto-unknown type. It is hot, bluish-white, and lies below the main sequence on the H–R diagram. It has about the mass of the Sun, but it is so faint that its total radiating surface cannot be much more than that of the Earth.

A star the size of Earth? If a solar mass were packed into an Earth-sized ball, it would be astonishingly dense. A cubic inch would weigh a ton! A thimbleful would collapse a table!

Dutch–American astronomer W. J. Luyten, who discovered a number of these stars, described what happened when astronomers first recognized Adams' amazing results (quoted by Shapley, 1960):

The figures were too staggering to be believed without further evidence. Could there be an error somewhere? What had happened? But while we pondered these things, two more stars of the same kind were found: small, extremely faint, and white, and all three were named white dwarfs.

Theoretical astrophysicists such as Sir Arthur Eddington and the Indian–American theoretician S. Chandrasekhar explained the evolution of white dwarfs. Once no more energy is available to generate outward pressure, a star collapses until all its atoms are jammed together to make a very dense material. But what does "jammed together" really mean? At these densities, some electrons move at almost the speed of light and matter loses its familiar properties. Stellar matter stops behaving like a perfect gas. In fact, it is no longer either gas, liquid, or solid, but a new form known as **degenerate matter**. By a physical law called the **Pauli exclusion principle,** only a certain number of electrons can be jammed into a given small volume of space. Thus the atomic nuclei are held apart by a sea of electrons—a so-called degenerate electron gas—at densities of about 10^8 to 10^{11} kg/m^3. White dwarfs are sometimes called *degenerate stars*. Just as there is empty space between the atoms of a quartz crystal, there is still some empty space between the electrons and other atomic particles of a white dwarf—but repulsive forces hold the electrons apart and thus they resist denser packing.

Figure 19-7 The Helix Nebula (NGC 7293) in Aquarius is a glowing spheroidal shell of gas blown off the surface layers of an evolved star by strong stellar winds. The remainder of the star is visible at the center—a very hot, blue, exposed stellar core. Strong ultraviolet radiation from this star core excites atoms of different elements at different distances in the gas around the star, causing the atoms to emit colorful glows. Dominant color is the red glow of hydrogen alpha emission from ionized hydrogen atoms. Additional blue-green comes from doubly ionized oxygen atoms. Delicate radial filaments may be related to outrushing motions of the gas. The distance and diameter are probably around 100 pc and 0.4 pc, respectively. (Copyright Anglo-Australian Telescope Board, courtesy D. Malin.)

Astronomical research on white dwarfs spurred more general physical research in the 1930s on the problem of high-density forms of matter. Luyten has gone so far as to say that analyses of white dwarfs "forged another important link in the chain of events which eventually led to the atomic bomb."

In the 1970s astronomical satellites above the atmosphere began to record UV and other shortwave radiation from sources such as white dwarfs, allowing better determinations of their temperatures, which are often high. Sirius B, for example, has a temperature of 27 000 to 32 000 K (Liebert, 1980). The United States–USSR

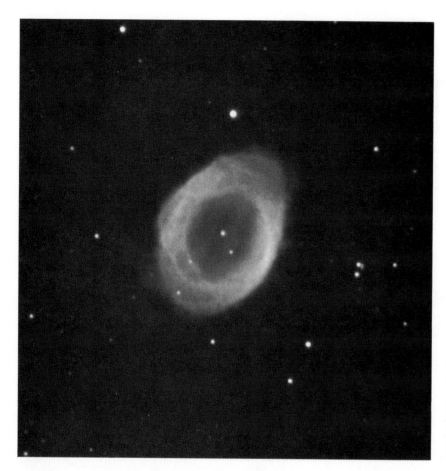

Figure 19-8 The Ring Nebula, about 700 pc away and 0.2 pc across, is similar to the Helix Nebula shown in Figure 19-7. It is a bubble of glowing gas, expanding at some 19 km/s, blown off the central star. That star is blue and extremely hot, about 100 000 K. Red Hα glow is prominent, and the inner parts have a faint blue-green glow from ionized oxygen atoms. (Copyright Association of Universities for Research in Astronomy, NOAO.)

Apollo–Soyuz joint astronaut flight in 1975 yielded the highest reported white dwarf temperature, 150 000 K, for white dwarf HZ 43. This star is a possible "missing link" between prenova explosive dwarfs and ordinary white dwarfs. BE Ursae Majoris is close behind at 100 000 K.

Because white dwarfs have low luminosities and large amounts of stored internal energy, they take a long time to radiate enough energy to cool significantly. White dwarfs thus last a long time. The coolest white dwarfs have temperatures around 5000 K, and may have cooled for at least 7 billion years to reach these temperatures (Helfand, 1983). Indeed, some white dwarfs may be among the oldest stars we can observe.

Although the interiors of white dwarfs may have densities of billions of kilograms per cubic meter, the outer layers of some may consist of ordinary matter—possibly hot gas at the surface and crystalline rocklike or glasslike solids in a crust 20 to 75 km deep. At the base of these crusts, densities may be as high as 3 million kilograms per cubic meter! Theoreticians predicted pulsations with periods of a few minutes in white dwarfs of certain temperatures, and these pulsations have been confirmed observationally (Helfand, 1983).

The Most Massive Possible White Dwarf: The Chandrasekhar Limit

White dwarfs cannot have more mass than 1.4 $M_\odot$ because the white dwarf structure becomes unstable at this point. If you tried to dump more mass on the surface of a 1.4-$M_\odot$ white dwarf, its gravity would become so strong that it would overcome the resistance of the electrons to denser packing. A still denser state of matter would arise.

The critical mass, 1.4 $M_\odot$, is called the **Chandrasekhar limit** after its Indian–American astrophysicist discoverer. Stars with masses from about 0.08 to 1.4

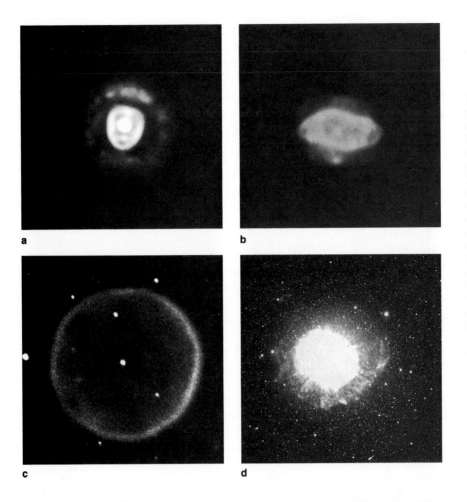

Figure 19-9 Four examples of planetary nebulae showing various shapes resulting from stars shedding their mass smoothly or sporadically in late stages of evolution. **a** NGC 2392, showing two rings, possibly indicating two distinct eruptions. About 0.2 pc in diameter and 1000 pc distant. (Mauna Kea Observatory, courtesy A. Stockton.) **b** Saturn Nebula, showing unusual lateral extensions. About 0.1 pc in diameter and 700 pc distant. (Kitt Peak National Observatory.) **c** In the absence of turbulent explosive activity, a central star may smoothly blow off a nearly perfect spheroidal shell, or "bubble," of gas. This example is Abell 39. (Kitt Peak National Observatory, courtesy George Jacoby.) **d** Long exposure of the Dumbbell Nebula (see Figure 20-13) shows a dense inner cloud and a wispy outer cloud created by two episodes of mass ejection from the central star. (Kitt Peak National Observatory.)

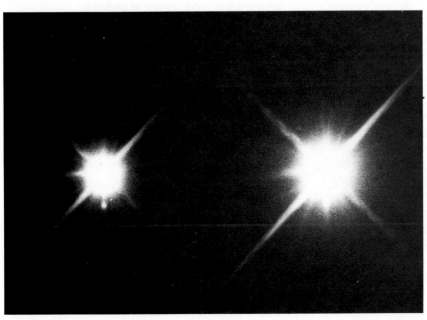

Figure 19-10 Two exposures of Sirius with the Lick 3-m telescope. Short exposure (left) reveals the faint white dwarf (below Sirius) that is the binary companion roughly 25 AU from Sirius; in the long exposure (right) the white dwarf is lost in the overexposed image. "Rays of light" are artifacts of the telescope optics. (Lick Observatory.)

$M_\odot$ apparently evolve smoothly to the white dwarf state. Current data suggest that stars from 1.4 to about 6 $M_\odot$ also evolve into white dwarfs by developing strong stellar winds (like the solar wind) or eruptive explosions that blow off mass until they are below the Chandrasekhar limit (Liebert, 1980).

White Dwarfs of Different Mass and Different Composition

According to calculations, the Sun will probably blow off about 40% of its mass when it goes through its red giant stage about 5 billion years from now. Thus it will collapse into a white dwarf of about 0.6 $M_\odot$ (Burrows, 1987).

The story gets more interesting as we consider stars of other masses. Stars in the mass range from the smallest possible star, 0.08 $M_\odot$, to about 0.25 $M_\odot$ never get hot enough or high enough pressure in their cores to initiate the burning of helium. Thus stars below 0.25 $M_\odot$ leave white dwarfs consisting mostly of helium. Larger stars, such as the Sun, do get hot enough to burn helium—for example, by the triple-alpha process mentioned earlier. Some of the helium may fuse into carbon and oxygen. Such processes happen in the most common stars—those up to a few solar masses. Thus most stars, including the Sun, will shed some of their atmosphere and leave intensely hot white dwarfs of degenerate carbon or oxygen. The larger the initial star, the larger the final white dwarf. A star of 8 $M_\odot$, which starts off as a bright, blue O or B star and then evolves into a red giant, is predicted to blow off about 82% of its mass, leaving behind a white dwarf of about 1.2 to 1.4 $M_\odot$ (Burrows, 1987).

IN THE CORES OF MASSIVE RED GIANTS: ONION-SKIN LAYERS

Now let us consider the terminal evolution of the rarer, more massive stars. Imagine a row of unusually massive red giants with different masses: 4, 6, 8, 10 $M_\odot$, and so on. Those of about 4 or 6 $M_\odot$ will have helium-rich cores hot enough to ignite the degenerate helium nuclei and fuse them into carbon. Those of around 8 $M_\odot$ will have hot enough cores to ignite the carbon and fuse it into heavier elements, such as oxygen, neon, and magnesium. A stage could arise where the outer layer is hydrogen that never burned; below that would be the outer part of the core, containing helium; the inner core would be carbon, with the carbon starting to fuse into heavier elements at the very center.

Similarly, in stars of around 10–12 $M_\odot$, a long series of reactions will fuse nuclei into elements as heavy as iron. The process has an interesting limitation at this point. Nuclear structure is such that no more energy can be produced by burning iron. Energy is consumed, not produced, as iron atoms fuse into heavier elements. Thus the iron cores of stars do not continue to ignite if they contract and get hotter. As astrophysicist Frank Shu (1982) comments: "Iron . . . is the ultimate slag heap of the universe."

Thus if the star is massive enough, the core-building process leads to a core consisting of shells of different elements, surrounding an inner core of iron, as illustrated in Figure 19-11. The shell structure can be so complex that there can be different fusion reactions happening at the surfaces of different shells, all at the same time. Hydrogen might be burning at the base of the hydrogen-rich atmosphere, while helium might be fusing into carbon at the hot base of the helium layer. Eventually, however, this situation becomes unstable, because one or more layers of the onion skin is running out of fuel, and the fusion reactions start to decline. Now gravity gets the upper hand, and the collapse of the core begins.

THE DEMISE OF VERY MASSIVE STARS: SUPERNOVAE

At the moment of collapse, the core of a massive red giant has consumed as much fuel as possible, burning its way out into the outer parts of the star. The core collapse starts because there is no point in the core where there is potential fuel *and* the possibility to create a hot enough temperature to ignite it. In a 6-$M_\odot$ star, the core may include the central 1.1 $M_\odot$ and collapse into a white dwarf of that mass. In an 8-$M_\odot$ star, the core may encompass 1.4 $M_\odot$. In a larger star, the core reaches more than 1.4 $M_\odot$. It exceeds the Chandrasekhar limit. What will happen when it tries to collapse into a white dwarf?

During the 1940s, 1950s, and 1960s, it was assumed that these stars threw off such a large fraction of their mass that they always produced white dwarfs of about 1.4 $M_\odot$. Then astronomers began to realize that denser states of matter than white dwarfs could exist. Indeed,

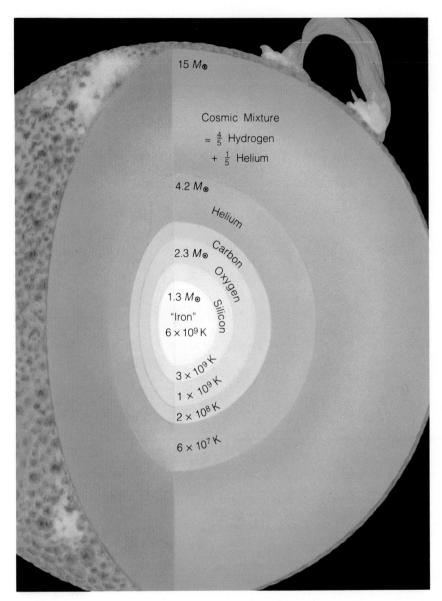

15 $M_\odot$

Cosmic Mixture

$\approx \frac{4}{5}$ Hydrogen

$+ \frac{1}{5}$ Helium

4.2 $M_\odot$

Helium

2.3 $M_\odot$

Carbon

Oxygen

1.3 $M_\odot$

Silicon

"Iron"
6×10^9 K

3×10^9 K

1×10^9 K

2×10^8 K

6×10^7 K

Figure 19-11 Schematic diagram of onion-skin layering of different elements as calculated for the precollapse core of a 15-$M_\odot$ star. Dominant element in each layer is given at top; each layer's mean temperature is given at bottom. Figures also give mass of material out to designated layer. Nuclear reactions in successively deeper layers fuse nuclei into successively heavier elements. The core and inner layers are greatly enlarged to show their structure. (After diagram by Burrows, 1987.)

collapsed stars with mass greater than 1.4 $M_\odot$ not only exist but include the strangest objects known in the universe: neutron stars and black holes. These are much smaller than white dwarfs and involve an extraordinary collapse. The collapse of the core to tiny size is aided by the fact that the iron "slag-heap" core will not ignite, even though it gets extremely hot and dense. The supercollapse releases a tremendous outburst of energy inside the already barely stable star. The outburst of energy blows off all the outer layers of the star in a titanic explosion called a **supernova** (plural: *supernovae*).

What happens inside the star during the collapse and explosion? Although there are several types of supernovae,[3] which may have somewhat different processes, the main process appears to be as follows.

---

[3]One type of supernova (called Type I) occurs in double-star pairs when a white dwarf is orbiting a star that is shedding mass. The mass crashes onto the dwarf, pushing it over the Chandrasekhar limit, making it unstable, and causing it to collapse and then explode. We will discuss this process of mass transfer further in Chapter 21. The type of supernova described above, caused by a star's own internal evolution, is called Type II.

Figure 19-12 The Crab Nebula is the remnant of a supernova explosion recorded in A.D. 1054. Red outer filaments are splatters of excited hydrogen glowing with Hα emission. Milky-green glow comes from synchrotron radiation from high-speed electrons moving in the nebula's magnetic field. The nebula is roughly 2000 pc away, 3 pc across, and expanding at a speed of about 1000 km/s. At the center is a pulsar that flashes 30 times per second, as shown in Figure 19-18. (Copyright California Institute of Technology and Carnegie Institution of Washington; by permission of Hale Observatories.)

When the incredible temperature of several billion degrees Kelvin is reached, the iron nuclei are hitting each other so hard that while some fuse into still heavier nuclei, many simply break apart into a shower of protons, neutrons, neutrinos, and other subatomic particles. Calculations suggest that once it starts, the collapse takes only a few seconds! The core material moves so fast toward the center that it overshoots, finds itself too compressed to be stable, and rebounds. This starts an outward-moving shock wave that blows away the outer layers in what astrophysicist A. Burrows calls "the fabulous explosion that we call a supernova" (Shu, 1982; Reddy, 1983; Wheeler and Nomoto, 1985; Burrows, 1987).

A supernova explosion takes the star to the amazing luminosity of 10 billion times that of the Sun! For a few days the exploding star blazes with the light of an entire *galaxy*. Then, after months, the star itself fades. During the explosion an expanding cloud of gas is launched. Initially it is too close to the star to be seen from Earth, but years later the site will be marked by a colossal, expanding nebula, called a supernova remnant.

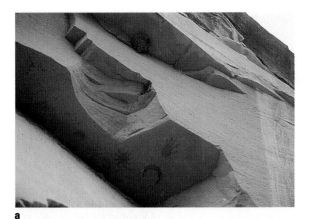

a

b

c

Figure 19-13 American Indians may have recorded the appearance of the Crab Nebula supernova on the morning of July 5, 1054. **a** Vertical view up cliff near Chaco Canyon cliff dwelling ruins in New Mexico, showing pictograph painted on overhead ledge. **b** Close-up of painting, showing orientation perceived by artist facing across canyon and leaning over backward to paint on ledge. **c** Sketch based on computer reconstruction of appearance of rising crescent moon and supernova as seen to east across the canyon from the site shortly before dawn on July 5, 1054. (Photos by author.)

Examples of Supernovae

Many supernovae (and some novae) have been close enough to the solar system to produce temporarily prominent "new stars" that were recorded by ancient people. The ancient Chinese called them "guest stars." (Some of these are listed in Table 21-3, page 472.) Some astronomers have suggested such an explanation for the Star of Bethlehem.

The most famous supernova was the explosion that produced the Crab Nebula (Figure 19-12). It was visible in broad daylight for 23 days in July 1054 and at night for the subsequent 6 months. It was recorded in Chinese, Japanese, and Islamic documents, and perhaps in American Indian rock art, as seen in Figure 19-13. The nebula is the expanding, colorful gas shot out of this supernova. Other supernova remnants are scattered throughout our galaxy (see Figures 19-14 and 19-15). Records indicate about 14 supernovae in our galaxy dur-

ing the last 2000 y, an average of one every 140 y. Any of us might live to see one bright enough to be visible in daylight.

Supernovae occur in other galaxies, too. In 1885, a supernovae in the nearby Andromeda galaxy briefly doubled that galaxy's brightness. More than 300 have been recorded in other galaxies, and their remnant clouds can sometimes be identified (Figure 19-16).

The Supernova of 1987

At 7:35 A.M. Universal Time on the morning of February 23, 1987,[4] the core inside a 20-$M_\odot$ star in the

---

[4]The date and universal time (clock time at the 0° longitude meridian in Greenwich, England) are the time when the light corresponding to the event reached Earth. Since the star's location was nearly 170 000 light-years away, the light took some 170 000 y to reach us. The explosion actually occurred around 168 000 B.C.!

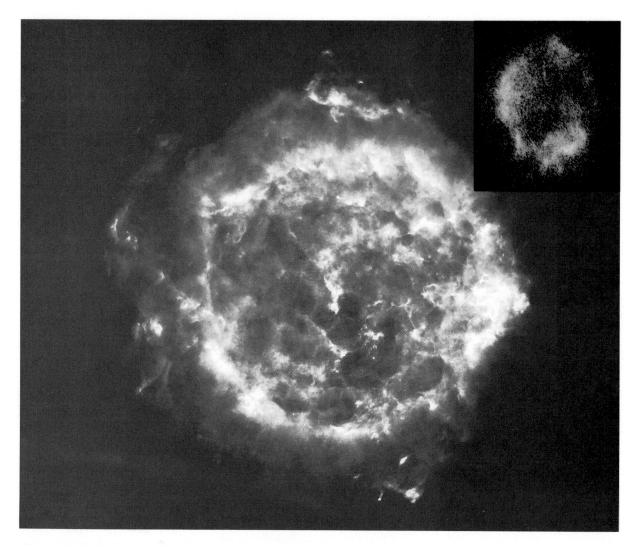

Figure 19-14 Cassiopeia A, a spherical, expanding shell of hot gas blown out of a supernova visible around A.D. 1680. This nebula is not prominent visually but has been imaged by its radio and X-ray radiation. Observations indicate that the gas has a temperature of a few million Kelvin and is rich in silicon, sulfur, and other heavy elements blown out of the inside of the supernova star, which had an original mass around 10–30 $M_{\odot}$. The cloud diameter is probably a few parsecs. Radio image at wavelength 6 cm. (National Radio Astronomy Observatory, R. Tuffs, R. Perley, M. Brown, and S. Gull.) **Inset** X-ray image from unmanned orbiting Einstein Observatory. (Courtesy S. S. Murray, Harvard/Smithsonian Center for Astrophysics.)

neighboring galaxy called the Large Magellanic Cloud became unstable. Within a second, the central iron plasma core (see Figure 19-11 on page 419), about the size of Mars, collapsed at a quarter of the speed of light down to a size of about 100 km. As temperatures reached 30 billion K, iron nuclei were fragmented (see Table 15-2 on page 320), and the star exploded in a blast of neutrinos and a flash of ultraviolet light brighter than any

other stars in the whole galaxy. Although this event was some 52 000 pc away (further than any stars we've discussed so far) it was easily detected from Earth.

Within hours, the newly brightening star was sighted by astronomers, as seen in Figure 19-17, and word was flashed to observatories around the world. For many days it was bright enough to see with the naked eye, but only at equatorial and Southern Hemisphere lati-

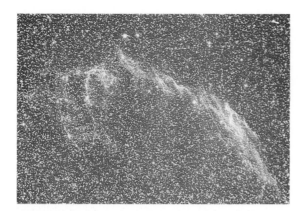

Figure 19-15 A portion of the Veil Nebula in the constellation Cygnus. This segment forms part of a giant ring-shaped arc about 2½° across—five times the apparent diameter of the Moon! The ring is expanding at about 70 km/s and is believed to be the gaseous shell blown out of a supernova roughly 40 000 y ago. It is about 460 pc away, and the shell is about 20 pc in diameter. Compare with the similar appearance of a complete supernova ring in Figure 19-16. (48-in. Schmidt telescope, Hale Observatories.)

Figure 19-16 This glowing bubble of gas is believed to be an expanded remnant cloud from a supernova explosion. This nebula lies on the outskirts of a neighbor galaxy known as the Large Magellanic Cloud. It is about 110 pc in diameter. The star that exploded to form it may have been a massive member of the central cluster of stars. (Copyright Anglo-Australian Telescope Board, courtesy D. Malin.)

a   b

Figure 19-17 The 1987 Supernova. **a** The region of the supernova before the explosion. Red wisps are gas excited by ultraviolet light of massive young stars in the region and glowing by the light of hydrogen alpha emission. **b** A few days after the explosion, the region is dominated by the brilliance of the supernova. (Copyright Anglo-Australian Telescope Board.)

tudes. Scientists checked their instrumental records from previous hours and found evidence of the neutrino burst and other early stages of the explosion. Telescopes on Earth, in the Soviet Mir space station, and on board robotic satellites were pointed at the supernova in a coordinated observing effort (Helfand, 1987).

The most exciting finding was the detection of the burst of neutrinos (by Japanese and U.S. instruments) in the first seconds of the explosion. Until then, a neutrino burst from a supernova had been merely a theoretical prediction, but now it was confirmed fact! The importance of this is that it shows how scientific theory can predict features of the universe. The neutrino detection opened a new field of neutrino astronomy, and confirmed details of the supernova processes. The outer layers (see Figure 19-11), of course, also started to collapse, but were blown into space by the outrushing blast from the core.

In the following months the supernova confirmed other basic aspects of stellar theory. Astronomers detected gamma rays emitted by the heavy element cobalt, for example, confirming that the massive parent star did synthesize heavy elements in its heart. In detail, the supernova affirmed a theoretical model in which silicon atoms in the original core (see Figure 19-11) are converted into radioactive nickel-56 atoms in the supernova shock, which in turn decay to the observed cobalt-56. This in turn decays to stable iron-56 at a rate consistent with the observed rate of brightness decline in the supernova. Reviews of the discoveries from the supernova are given by Woosley and Phillips (1988) and Woosley and Weaver (1989).

As the first nearby, readily observable supernova in three centuries, the 1987 supernova's main importance was to give astronomers their first look at a stellar explosion with the full array of modern observational techniques.

NEUTRON STARS (PULSARS): NEW LIGHT ON OLD STARS

What is left after a supernova explosion? As early as 1934, American astronomers W. Baade and F. Zwicky made a correct guess:

With all reserve we advance the view that a supernova represents the transition of an ordinary star into a neutron star, consisting mainly of neutrons. Such a star may possess a very small radius and an extremely high density.

Nuclear physicists were already working on the problem of very dense forms of matter, and they quickly took up the challenge of dense star forms. By 1938, J. Robert Oppenheimer and R. Serber showed how stars, after exhausting their nuclear fuel, could collapse to states much denser than white dwarfs.[5]

---

[5]Physicist Oppenheimer later became famous for additional nuclear research and his work as leader of the Los Alamos team that developed the atomic bomb in 1945. His physics research was halted when he lost his security clearance in the anticommunist political wrangles of the 1950s.

The concept of a **neutron star** was merely an extension of the concept of a white dwarf. In an ordinary gas, atoms are held apart by their motions, and when they collide, their nuclei are held apart by their surrounding electron swarms. In an ordinary star's core, the electrons are stripped off and nuclei begin to hit each other, reacting and releasing nuclear energy. In a white dwarf, the nuclei that are left can no longer react and are crowded randomly among a dense sea of electrons. A white dwarf, therefore, might be called an electron star since electrons control the spacings of particles.

But if a burnt-out star core is still more massive than a white dwarf, the gravity is so strong that even the repulsion between electrons is overcome. Theory indicates that this could happen in objects between 1.4 $M_\odot$ and about 3 to 5 $M_\odot$. In such objects, the nuclei themselves begin to be jammed into contact. Since whatever reactions might occur have already run their course, there are no new energy-generating reactions. Instead, the nuclei are broken into their constituent neutrons and protons. The positively charged protons coalesce with negatively charged electrons to make more of the neutrally charged neutrons. The important particles are now the neutrons, and the star is a neutron star, smaller than the white dwarf. Its density approaches the incredible densities of atomic nuclei themselves: roughly 10^{17} kg/m^3. A thimbleful would weigh 100 million tons! A skyscraper-full could contain all the mass of the Moon! A whole neutron star could contain the mass of the Sun but be no larger than a small asteroid—perhaps 20 kilometers across!

Because a neutron star is a mass of nuclear matter comparable in density to the nucleus of an atom, some astronomers have pictured a neutron star as a giant atomic nucleus with atomic mass around 10^{57}!

The Discovery of Neutron Stars

For decades, astrophysicists talked about neutron stars, but, like the weather, nobody did anything about them, because nobody *could* do anything. No known observational technique could detect them and no one could prove they existed.

But in November 1967, a 4.5-acre array of radio telescopes in England detected a strange new type of radio source in the sky. Analyzing the surveys (each equaling a 120-m roll of a paper chart), a sharp-eyed graduate student, Jocelyn Bell, was astonished to find that one celestial radio source (about a centimeter of data on the chart) emitted "beeps" every 1.33733 s!

At first, project scientists speculated that they might have actually discovered an artificial radio beacon placed in space by some alien civilization! But further work soon led away from this speculation. By January, another source was found, pulsing at a different frequency, arguing against the beacon hypothesis. Analysis showed that the first source was less than 4800 km across, much smaller than ordinary stars.

These pulsing radio sources came to be called **pulsars.** In February, Anthony Hewish and his colleagues published an analysis suggesting that the pulsars might be superdense vibrating stars that could "throw valuable light on the behavior of compact stars and also on the properties of matter at high density."

The mysterious pulsars turned out to be the long-sought neutron stars! In an exciting burst of research, the number of scientific papers on pulsars jumped from zero in 1967 to 140 in 1968. By 1973, about 100 pulsars had been discovered. The codirectors of the original discovery project, Anthony Hewish and Martin Ryle, shared the 1974 Nobel Prize in physics.

Why do neutron stars pulse? After collapsing to a small size, supernova remnant stars have very strong magnetic fields and very fast spins, rotating once every second or so. Recall from Chapter 14 and Figure 14-2 that any spinning object—even a figure skater—spins faster as it contracts. Ions trapped in the magnetic fields spin around with velocities near the speed of light. In 1968, Cornell researcher Thomas Gold showed how ions trapped in magnetic fields of spinning neutron stars produce strongly beamed radio radiation, so that the pulsar acts like a lighthouse with a beam sweeping around every second. Most pulsars have periods of ¼ to 13s. The fastest-spinning pulsar found thus far was discovered in 1982 and flashes with a 0.0016-s period. It is a mountain-sized ball of nuclear matter that spins 642 times every second (Waldrop, 1983)!

The discovery of a pulsar in the center of the Crab Nebula (Figure 19-18) and in other supernova remnants proves that pulsars are related to supernovae. Careful studies of such pulsars show that they pulse not only in radio waves but also in X rays and visible light.

Neutron stars may have solid crusts (1 km thick?) overlying a fluid "neutron soup." Tiny abrupt changes in the spin rates of pulsars have been detected and attributed to "starquakes," or fractures in the crust, which change the mass distribution and hence the rotation rate. It is strange to think of earthquake-like events happening in the solid crusts of star corpses!

Bursters

A curious type of celestial object related to neutron star pulsars has been called the **burster.** In the 1970s, satellites designed to monitor the nuclear test ban treaty by detecting gamma-ray flashes from nuclear bombs discovered gamma-ray flashes coming from the sky! In at least one case, the gamma-ray burst came from the position of a known neutron star. Researchers believe that bursts of gamma rays from the bursters may also be accompanied by flashes of visible light, X rays, and other radiation. One such visible flash may have been photographed, as shown in Figure 19-19. Such stellar flashes (lasting perhaps as much as a second) may be visible to the naked eye in the night sky, but would be very rare and difficult to confirm. Theorists have suggested that bursters may be caused by buildup of matter from a binary companion onto a neutron star until an explosion occurs (X-ray burster 4U1519-05 in Aquila seems to be in a binary system), or perhaps by sudden infall of clouds of debris or individual planetesimals onto the surface of a neutron star (Schorn, 1982).

REMNANTS OF EXTREMELY MASSIVE STARS: BLACK HOLES

As soon as pulsars were discovered, the search for still denser objects began. From this search came evidence for the strangest of all astrophysical concepts—**black holes.** Black holes are bodies so dense that their gravitational fields can keep most light (or other forms of energy and matter) from escaping.

Black holes can be visualized through an analogy with Newtonian physics made as early as 1798 by the French astronomer–mathematician Pierre Laplace. He reasoned that some bodies might be dense enough to have an escape velocity (at their surfaces) faster than the speed of light. Laplace thus assumed light could never escape from such bodies and that they would be permanently black and opaque. Although this basic idea is nearly correct, we now know that the situation is more complex because of Einstein's theory of relativity, which more correctly predicts phenomena involving speeds near that of light. But theorists using Einstein's theory have concluded that black holes probably do exist.

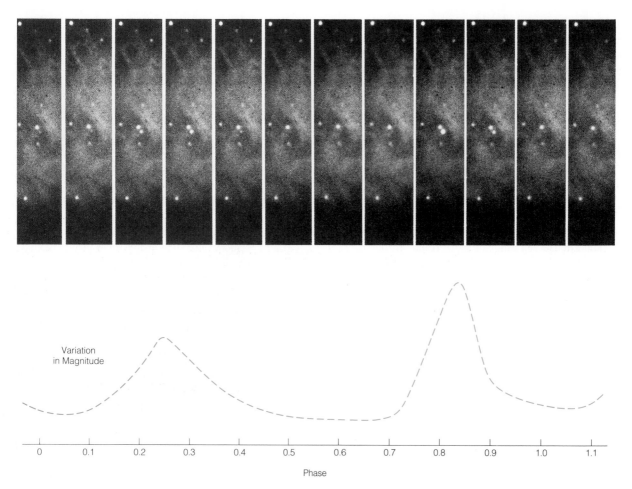

Figure 19-18 A complete sequence of flashes from the pulsar NP 0532 in the center of the Crab Nebula. Inner portions of the surrounding nebula can be seen. The entire cycle, including two flashes, lasts about 1/30 s, equaling one rotation of the pulsar. The pulsar and nebula are about 1100 to 2200 pc away. Figure 19-12 shows the full nebula. (Kitt Peak National Observatory.)

Physical Nature of Black Holes

Theorists believe that a black hole would be a very dense mass surrounded by a so-called **event horizon,** or imaginary surface from which no radiation or matter could escape.

The event horizon concept is illustrated in Figure 19-20, which shows the escape velocity at various distances from a hypothetical black hole of 1 $M_\odot$. A rocket passing at a great distance would experience the same gravity field and motions as a rocket at a great distance from the Sun. At 1 AU from the object, for example, the velocity needed to escape into interstellar space would be 42 km/s, the same as the speed needed to leave Earth's orbit. But the black hole itself might be only a kilometer or less across! As we get within a few thousand kilometers of it, the escape velocity would be thousands of kilometers per second. At a distance of around 3 km, the speed needed to escape would be the speed of light. From *within* the event horizon, an object on a ballistic trajectory (such as an atomic particle or a meteoritic dust grain) would have to move outward faster than light in order to keep from falling back—hence its escape would be impossible. This imaginary boundary region where the escape velocity equals the velocity of light is the event horizon. Calculations indicate that black

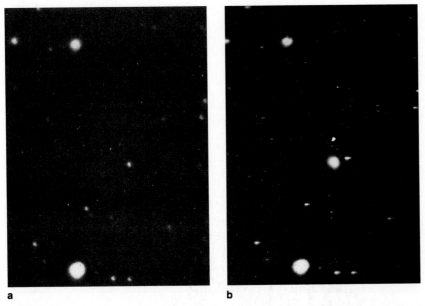

a   b

Figure 19-19 An accidental photo of a gamma-ray burster. During a sequence of 45-min exposures made in South Africa in 1928, one of the images *(b)* apparently caught the optical flash of a gamma-ray burster in action. After a gamma-ray burster was discovered in this area of the sky in 1978, astronomers made a check of old plates in search of possible evidence of optical flashes. Photo **a**, along with other photos in the series, shows no star in the position of the gamma-ray burster, but photo **b** shows a bright image there (center). The image is not trailed left-to-right like the other images (due to imperfect telescope guiding during the 45-min exposure), proving that it must have been a brief flash of light. (Courtesy Bradley E. Schaefer, Massachusetts Institute of Technology.)

holes would form from the collapse of only the most massive stars (25 $M_\odot$ or more?) or the most massive exhausted post-supernova remnants (5$M_\odot$ or more).

Two theoretical results on black holes came in the mid-1970s, especially from the English physicist Stephen W. Hawking. First, he pointed out that since matter was more densely packed in the earliest days of the universe, objects much smaller than stars could have gravitationally contracted to form black holes as small as subatomic particles, like protons and neutrons. Some of these tiny, primitive black holes might still exist in space.

Second, Hawking (1977) applied the theory of quantum mechanics and showed that *black holes need not be entirely black*, as was once thought. Quantum mechanics shows that subatomic particles often act in ways unpredicted by older theories. Thus proton-sized black holes (containing up to 10^{12} kg—microscopic motes with the mass of mountains) could radiate energy, often as an explosion emitting gamma rays. Large, kilometer-scale black holes (containing as much mass as a star, or

about 10^{30} kg) would radiate virtually nothing, just as in the original black hole theory.

If black holes exist, they must have truly unfamiliar properties. If an instrumented probe were dropped toward it, we would stop receiving signals after the probe fell through the event horizon, beyond which virtually no radiation or matter can escape. The probe, then, would seem to disappear from the observable universe. Where would it go?

Pondering this strange question, some scientists speculate that black holes amount to separate universes. Others feel that when mass "pops out of existence" by collapsing to superdensity, new mass or energy emerges somewhere else in the universe. But perhaps the probe would merely "squish" to very high density and collide with the black hole, adding a bit to its mass. Could our own universe have begun as a black hole in some other "universe"? Do black holes make fact out of the "space warps" invented decades ago by science fiction writers to allow their spaceships to wink out of sight in one place and reemerge instantaneously at some

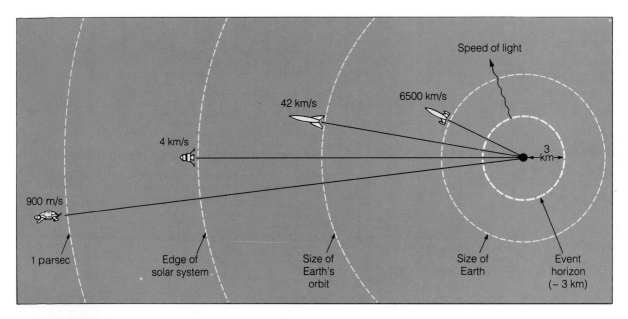

Figure 19-20 Exploring the gravitational field of a black hole of one solar mass (right). The schematic diagram (not to scale) shows the velocity needed to escape from the black hole at different distances from it. At a parsec away (left) a mere 900 m/s would suffice, but this speed increases as we get closer. At a distance of about 3 km, a body would have to be moving at the speed of light to escape; this distance is called the event horizon. No object on a ballistic trajectory could escape from inside this distance.

distant point? No one knows. As the British geneticist John B. S. Haldane put it, "My suspicion is that the universe is not only queerer than we suppose, but queerer than we *can* suppose."

Detecting a Black Hole

If a black hole itself cannot emit radiation or mass, can we ever hope to detect one? Yes. Outside their event horizons, black holes have gravity fields indistinguishable from those of ordinary stars of the same mass. Thus they can orbit around stars just like planets or binary star companions. If we observed such a star from a distance, we would not see the black hole, but we could see the star's orbital motion and calculate the mass of the unseen companion, just as astronomers routinely do in the case of ordinary faint companions. The result would indicate an unusually high-mass companion for an unseen star—maybe 5 or 10 $M_\odot$—which should tip us off that we are dealing with a black hole candidate. Furthermore, such strong gravity exists close to black holes that any matter falling into them undergoes terrific acceleration.

Suppose the black hole is orbiting around an evolved

star that has expanded into the giant state and is shedding mass.[6] Some of the expanding gas would fall toward the black hole at terrific speed. Because this gas would, on the average, have some angular momentum around the star, rather than falling directly toward it, it would form a disk of gas spiraling inward toward the black hole (Figure 19-21). This disk is called an **accretion disk.** Its gas would be extremely hot, because it would be constantly hit by new gas streaming in from the other star. Because of the high temperature, the disk would radiate very short-wave radiation, such as UV, X-ray, and gamma-ray radiation. Therefore X-ray and gamma-ray telescopes launched into orbit around the Earth have played an important role in searching for black holes. The possible appearance of the inner accretion disk and the black hole is indicated in Figure 19-22.

The best evidence for a black hole, then, would be a massive, high-temperature X-ray or gamma-ray source orbiting around another normal star. Two such candidates have been discovered. Cygnus X-1 (so named

---

[6]Many pairs of coorbiting stars of this type are known. They are discussed in more detail in Chapter 21.

Figure 19-21 An imaginary view of a star system hypothesized to explain the X rays and gamma rays from sources such as Cygnus X-1. Gas shed from the surface of a hot supergiant star (left) flows toward a very dense star, possibly a black hole, orbiting around it. It spirals around the dense object and accumulates into a disk-shaped nebula. New gas is dumped onto the nebula at very high speed, causing extremely high temperatures and emission of blue and ultraviolet light, X rays, and gamma rays. (Painting by Adolf Schaller.)

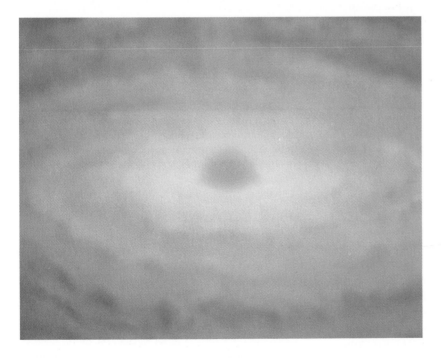

Figure 19-22 Possible appearance of the heart of an accretion disk around a black hole. The gas spiraling around the center is extremely hot and radiating blue and ultraviolet light, X rays, and gamma rays. The event horizon of the black hole appears as a reddish disk because we see red-shifted light from material falling into it but no radiation from the hot material inside it. (Painting by author.)

because it is the first X-ray source discovered in the constellation Cygnus) is a binary system about 2500 parsecs away. The visible star is a supergiant O or B star with a mass of 15 to 30 $M_\odot$. Orbiting around it is a small, hot, X-ray source with a mass of 10 to 15 $M_\odot$. The X-ray emissions suggest that the gas temperatures reach a billion Kelvin in the X-ray source (Nolan and Matteson, 1983). The second candidate is much more distant—55 000 parsecs away in a nearby galaxy called the Large Magellanic Cloud. It is called LMC X-3 and was first detected by the Uhuru X-ray satellite in the early 1970s. The visible star is a type B main-sequence star, which orbits in 1.7 d around a small, hot, X-ray source with a mass of 6 to 14 $M_\odot$.

Many astronomers believe that the hot X-ray source in each of these systems is the accretion disk surrounding a black hole. Continuing observations will document these and other black hole candidates in more detail.

Black holes and stellar explosions may also account for such unexplained phenomena as strong infrared radiation from the center of certain galaxies, unexpectedly large red shifts of some galaxies, and reported gravitational disturbances in our galaxy. An explosion powerful enough to blow outer layers off massive stars, converting the star to a black hole, could produce enormous amounts of energy. As astrophysicist R. Penrose (1972) notes, "There is no shortage of unexplained phenomena in astronomy today that might conceivably be relevant" to black holes.

THE DEMISE OF THE MOST MASSIVE STARS

Since stars of mass more than 25 $M_\odot$ are extremely rare and go through their evolution much faster than other stars, few are observed. But some of them have formed. Theorists believe that at very large masses, around 100 $M_\odot$, the star becomes so unstable that it does not even fuse nuclei all the way to iron. It may explode first. Such explosions may be so violent that it is questionable whether a black hole core even forms as a remnant; perhaps some stars are massive enough to blow themselves to smithereens. Figure 19-23 summarizes the life histories of stars of different mass, showing the trend toward more violent demises and shorter lifetimes as the mass gets greater.

PRACTICAL APPLICATIONS?

People often ask, "What good is astronomy? How can it be of use to know about places too far away to see, to touch, or almost to imagine?" We've argued that we attain firmer philosophical footing if we know about our cosmic surroundings and that exploration of the planets helps us understand the climate and resources of the Earth. This chapter provides another reason. Astronomers' understanding of dense star forms like white dwarfs, neutron stars, and black holes has developed hand in hand with physicists' understanding of dense forms of matter in the lab. But there are limits to the experiments that can be conducted with such weird forms of matter. The discovery of actual examples of such

matter in space, together with the measurement of masses, temperatures, magnetic fields, and so on, has provided physicists with examples impossible to create in their labs. Physicist D. Pines (1980) comments that such discoveries "may constitute an almost unique probe of . . . high-temperature plasma in super-strong magnetic fields, . . . neutron solids," and other superdense materials.

If research on such material still sounds esoteric to you, recall that one of the most promising long-range solutions to the energy shortage on Earth is to control nuclear fusion in the lab by jamming nuclei together in dense plasmas constrained by strong magnetic fields. Fuel for hydrogen fusion could be the cheapest and most accessible material on Earth: seawater! Experimental work in this direction is already under way in the Soviet Union, the United States, and Europe. Thus a direct line stretches from studies of neutron stars, to our understanding of matter on Earth, and on to our manipulation of matter to create new forms of usable energy.

SUMMARY

The search for the forms of aging stars has yielded some of the most fascinating objects now being studied both by physicists and by astronomers. Stellar old age leads to two basic phenomena: high-energy nuclear reactions, as ever-more-massive elements interact, and inexorable contraction, as energy sources are eventually exhausted. Some evolutionary histories of stars of different mass are summarized in Figure 19-23.

During the first stages of old age, as stars evolve off the main sequence, high energy production causes expansion of stars' outer atmospheres, producing giants and supergiants. As heavier elements go through quick reaction sequences, various kinds of instability may produce variable stars and slow mass loss. After energy generation declines to a rate too low to resist contraction, low-mass stars contract to a dense state known as a white dwarf, with final mass less than 1.4 $M_\odot$. Rarer massive stars, which start out with as much as 8 $M_\odot$ or more, undergo supernova explosions and blow off much of their initial material.

If the remnant cores of stars end up between 1.4 $M_\odot$ and about 5 $M_\odot$, they form dense, rapidly rotating neutron stars known as pulsars. If the remnant cores have more than about 5 $M_\odot$, they may form black holes. Although virtually no radiation escapes from these strange objects, they may be detectable by orbital motions of their companion stars and by high-energy radiation from material

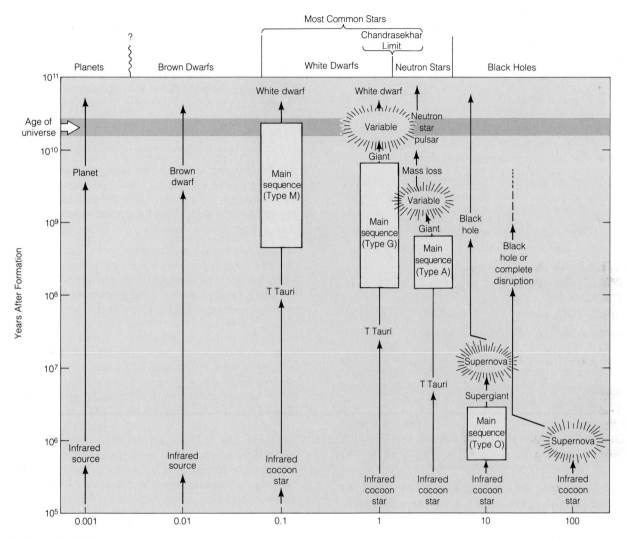

Figure 19-23 A schematic summary of stellar evolution showing the inexorable march toward high density. Evolutionary histories are shown for objects with different initial masses and the associated terms described in this and preceding chapters. Some states are hypothetical and are still being intensively researched by theoreticians and observers.

falling into them. The detailed physics of these dense, small star forms is an area of intense current research.

Complications may arise if the evolving star has a nearby coorbiting companion. Mass may be blown off the one and fall onto accretion disks around the other, changing the second star's mass and causing sudden instabilities. This probably accounts for some types of supernovae and some sources emitting X rays and gamma rays. Ultraviolet, X-ray, and gamma-ray astronomy conducted on orbiting satellites is an exciting area of new research on accretion disks, neutron stars, and black holes.

CONCEPTS

red giant	Cepheid variable
funneling effect	irregular variable
triple-alpha process	supergiant
helium flash	Wolf–Rayet star
s-process reaction	planetary nebula
r-process reaction	white dwarf
variable star	degenerate matter

Pauli exclusion principle burster

Chandrasekhar limit black hole

supernova event horizon

neutron star accretion disk

pulsar

PROBLEMS

1. Why do stars just moving off the main sequence expand to become giants instead of starting to contract at once? Why does contraction ultimately win out?

2. Many red giants are visible in the sky, even though the red giant phase of stellar evolution is relatively short-lived. Why are so many red giants visible?

3. List examples of evidence that certain stars can lose mass.

4. Which stars will eventually become:
  a. white dwarfs?
  b. neutron stars?
  c. black holes?

What will be the ultimate fate of the sun?

5. A main-sequence B3 star has about 10 times the mass of the Sun and therefore has about 10 times as much potential nuclear fuel. Why then does it have a main-sequence lifetime only 1/200 as long as that of the Sun?

6. According to the law of conservation of angular momentum, a figure skater spins faster as she pulls in her arms. How does this principle help explain why neutron stars spin much faster than main-sequence stars?

7. Comment on the roles and relations of theorists and observers in the three decades of work on white dwarfs, pulsars, and black holes. Are black holes fully understood today?

ADVANCED PROBLEMS

8. Suppose a 1-$M_\odot$ star has reached a terminal evolutionary state where it has the same diameter as the Earth.
  a. What would be the velocity of any possible material (such as captured meteoritic debris) in a circular orbit just above the star's surface?
  b. What velocity would be needed to blow material off its surface?
  c. Compare these values with values for the Earth.
  d. What type of star would this object be?

9. Use the Stefan–Boltzmann law (p. 350) to prove that the surface of a star such as FG Sagittae, evolving to the right on the H–R diagram (keeping constant total luminosity but decreasing in temperature), must be expanding.

10. Many white dwarfs have spectral types and surface temperatures similar to A or F main-sequence stars, but are much smaller. Use the Stefan–Boltzmann law to prove that this statement requires white dwarfs to lie below the main sequence on the H–R diagram.

11. Suppose a supernova occurred in the nearby star-forming region of Orion, 500 pc away. If the cloud expanded at an average velocity of 1000 km/s, how long would terrestrial observers have to wait before they could see details of the cloud's shape with telescopes resolving ½ s of arc? (*Hint:* Use small-angle equation; 1 y is about $\pi \times 10^7$ s.)

12. Prove that a gamma-ray telescope (sensitive to wavelengths less than about 0.1 nm) would be best suited to observing radiation from a billion-Kelvin accretion disk surrounding a black hole. Could such observations be made from a ground-based observatory?

13. The discussion of supernovae noted that the nuclei in stellar cores begin to break apart into protons and neutrons and other subatomic particles at temperatures of a few billion Kelvin. Laboratory experiments show that simple nuclei will fragment in this way when struck by particles moving fast enough to carry about 3.5×10^{-13} joules of energy. Prove that a gas would have to have a temperature of several billion degrees before its particles collided hard enough to begin to fragment nuclei of atoms. (*Hint:* In deriving Optional Basic Equation VI, we saw that the mean kinetic energy of a particle in an ordinary gas is ½ kT. Gases in supernovae, accretion disks, and so on, at billions of degrees are unlikely to behave as perfect gases, but assume that they do for the purpose of this estimate. Assume also that the faster particles in the gas have four times the energy of the average particle mentioned above. Then you can estimate that the energies mentioned would be reached by the faster atoms in a gas at some 4 billion Kelvin.)

14. Suppose the Sun were replaced with a black hole that had one solar mass. How would the Earth's orbital velocity change if the Earth remained in a circular orbit?

PROJECTS

1. Locate the star Mira (R.A. = 2^h14^m; Dec. = $-3°.4$) with a small telescope and determine whether it is in its faint or bright stage. If it is bright enough to see with the naked eye, record its brightness nightly by comparing it with other nearby stars of similar brightness. By checking brightnesses of these stars with star maps showing magnitudes, plot a curve of Mira's brightness over time.

2. With binoculars or a small telescope locate the star Delta Cephei (R.A. = 22^h26^m; Dec. = $+58°.1$) and compare it from night to night with other nearby stars of similar brightness. Can you detect its variations from about 4.4 to 3.7 magnitude in a period of 5.4 d?

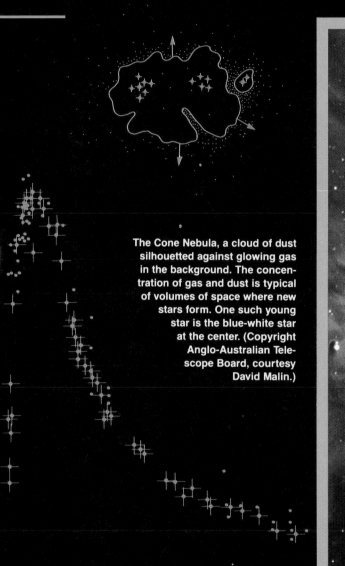

The Cone Nebula, a cloud of dust silhouetted against glowing gas in the background. The concentration of gas and dust is typical of volumes of space where new stars form. One such young star is the blue-white star at the center. (Copyright Anglo-Australian Telescope Board, courtesy David Malin.)

CHAPTER 20

Interstellar Atoms, Dust, and Nebulae

The preceding chapters discussed stars as if they were isolated individual objects. But they are not isolated. They are engulfed in a thin but chaotic medium of gas, dust, and radiation. They form from this thin material, interact with it, recycle it, and expel it to form new interstellar material.

Among the individual clouds of gas and dusts, particularly vivid nebulae were cataloged as early as 1781 by the French astronomer Charles Messier. They are thus known by their **Messier numbers,** or M numbers. The well-known Orion nebula, for example, is M 42. Others are known by **NGC numbers** or **IC numbers,** based on the more recent New General Catalog and Index Catalog, respectively. Tradition also names most bright nebulae according to their appearance in small telescopes; examples include the Crab, Dumbbell, and Ring nebulae. Some examples are listed in Table 20-1.

We sometimes casually say "space is a vacuum," but this is not quite true. While space is a better vacuum than can be achieved in labs (Table 20-2), its material cannot be neglected. What significance can this thin material have for us? For one thing, it dots our sky with **nebulae,** or vast clouds of dust and gas, some twisted into beautiful wispy forms, some dark, and some glowing with different colors. For another thing, as Joni Mitchell said in her song "Woodstock," "We are stardust." As Chapter 14 pointed out, our solar system, our Earth, and we ourselves are formed from atoms that were once part of the interstellar gas and dust. More provocatively, recent discoveries have demonstrated that interstellar material does contain complex organic molecules, and some scientists have speculated that primitive biochemical processes, perhaps related to the origin of life, may have occurred in nebulae.

TABLE 20·1

Characteristics of Selected Nebulae

Name	Constellation	Approx. Distance from Earth (pc)	Approx. Diameter (pc)	Estimated Atoms per m^3	Mass ($M_\odot$)	Spectral Type of Associated Star
Nebulae Probably Associated with Young Objects						
NGC 2261 Hubble's (R Mon)	Monoceros	700	10^{-5}	10^{18}	10^{-1}	F
Kleinmann–Low IR	Orion	500	0.1	10^{12}	100	Protostar?
Dark Nebulae						
Coal Sack	Crux	170	8	2×10^6	15	None
IC 434 Horsehead	Orion	350	3	2×10^7	0.6	B
Emission Nebulae						
M 42 Orion (central)	Orion	460	5	6×10^8	300	O
Eta Carinae	Carina	2 400	80	2×10^8	1 000	Peculiar
M 8 Lagoon	Sagittarius	1 200	9	8×10^7	1 000	O
M 20 Trifid	Sagittarius	1 000	4	10^8	1 000	O
Reflection Nebulae						
M 45 Pleiades	Taurus	126	1.5	?	?	B
Cocoon	Cygnus	1 600	2	7×10^7	7	B
Planetary Nebulae						
M 57 Ring	Lyra	700	0.2	10^9	0.2	White dwarf?
M 27 Dumbbell	Vulpecula	220	0.3	2×10^8	0.2	White dwarf?
NGC 7293 Helix	Aquarius	140	0.5	4×10^9	0.2	White dwarf?
Supernova Remnants						
M 1 Crab	Taurus	2 200	3	10^9	0.1	Pulsar
NGC 6960/2 Veil (Loop)	Cygnus	500	22	?	?	?
Gum	Puppis–Vela	460	360	10^5	100 000	Pulsar?

Source: Data from Allen (1973); Maran, Brandt, and Stecher (1973); and other sources.

TABLE 20·2

Gas Densities in Different Environments

Locale	Density^a (kg/m^3)	Particles per m^3	Typical Distance Between Particles
Air at sea level	1.2	10^{25}	1 nm
Circumstellar cocoon nebula	10^{-5}	10^{22}	50 nm
"Hard vacuum" in terrestrial	10^{-9}	10^{18}	1 μm
Orion Nebula	10^{-18}	10^{9}	0.1 cm
Typical interplanetary space	10^{-20}	10^{7}	0.5 cm
Typical interstellar space	10^{-21}	10^{6}	1 cm
Interstellar space near edge of galaxy	10^{-25}	10^{2}	20 cm
Typical intergalactic space	10^{-28}	10^{-1}	2 m

^{a}Average density of observable matter in the whole universe is estimated to be about 3×10^{-28} kg/m^3 (Shu, 1982), but many astronomers believe the average density may be somewhat higher (3×10^{-27} kg/m^3?) due to nonluminous unseen matter.

THE EFFECTS OF INTERSTELLAR MATERIAL ON STARLIGHT

Dispersed particles, whether floating in space or in the atmosphere, interact with radiation. When light from a distant star passes through clouds of interstellar atoms, molecules, and dust grains, the interaction changes the light's properties, such as intensity and color. Several complex physical laws describe these changes in some detail, but the changes can be grouped under two main principles:

> 1. When radiation (ultraviolet light, visible light, infrared, radio waves, or any other type) interacts with particles, the type of interaction depends on the types of particles and their sizes relative to the wavelength of the light.
>
> 2. The appearance of the light and the particles may depend on the direction from which the observer looks.

Three important types of interaction between radiation and matter involve atoms, molecules, and dust grains.

In reality, the interstellar material is always a mixture of gas and dust, but it is easier to understand the effects if we imagine separate interactions of light with atoms, molecules, and dust grains.

Interaction of Light with Interstellar Atoms

As shown in Figure 20-1, several things can happen if starlight passes through a cloud of interstellar gas atoms. Suppose a star radiates light of all wavelengths, and the photons of light enter a cloud of gas. Photons corresponding to certain wavelengths will have just enough energy to **excite** this gas, or knock electrons from lower to higher energy levels. They may even **ionize** the gas, or knock electrons clear out of the atoms. Each time a photon excites or ionizes an atom, that photon is consumed and disappears from the light beam. Thus an observer looking at the star through the cloud would see absorption lines created by the interstellar material, as shown in Figure 20-1.

But the energy absorbed by the cloud must be re-radiated, assuming that the cloud is in equilibrium. This

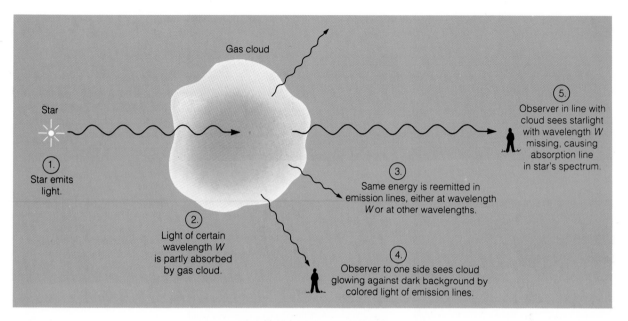

Figure 20-1 A cloud of gas (atoms and molecules) illuminated by a star and seen by an observer to one side (item 4) and an observer in line with the cloud and the star (item 5).

reradiation occurs as the electrons cascade back down through the energy levels of the atoms, creating emission lines. The photons in these emission lines leave the cloud in all directions, as shown in Figure 20-1, so that an observer off to one side would see the cloud glowing in the various colors corresponding to the emissions. Colors of nebulae are hard to see with the eye, even with large telescopes, because the light's intensity is low and the eye's color sensitivity is poor at low light levels (the reason why a moonlit scene looks less colorful than in daylight). Sensitive films and other detectors can record the extraordinary colors quite accurately, however. If a red emission line is especially strong, the cloud looks red. In another cloud, struck by photons of different wavelengths or containing atoms at different levels of excitation, a green emission line might be strongest, and the cloud would glow with green light. Photos in different colors reveal different patterns (see Figure 20-2). Since hydrogen is the most abundant gas, and since the red Hα emission line is one of its strongest emissions, many clouds of excited gas glow with a beautiful deep red color, as seen in Figure 20-3. Other colors also appear in nebulae (see Figure 20-4).

Interaction of Light with Interstellar Molecules

Molecules have similar interactions with starlight to those of atoms except that each absorption or emission affects a range of wavelengths, creating an absorption or emission *band* instead of a narrow line. Thus a gas cloud containing atoms and molecules involving different elements produces a variety of absorption lines and bands and might emit light with emission lines and bands of different wavelengths. Many molecular absorption and emission features are concentrated in the infrared part of the spectrum.

Interaction of Light with Interstellar Dust Grains

Whereas atoms are as small as 0.001 times the wavelength of visible light, and molecules are a few percent of this wavelength, many interstellar dust grains are comparable in size to light waves, and their interactions are quite different. They absorb some starlight, dimming distant stars. They also affect colors over a much

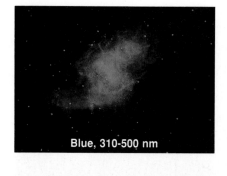

Blue, 310-500 nm

Yellow, 520-660 nm

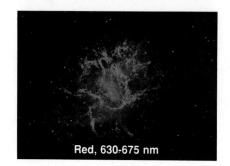

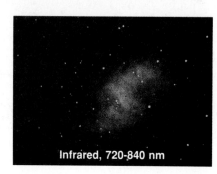

Red, 630-675 nm

Infrared, 720-840 nm

Figure 20-2 Four views of the Crab Nebula (Messier catalog number M 1), photographed in light of different colors. Red and yellow views show the outer filamentary structure corresponding to clouds of excited gas giving off spectral emission lines (especially red lines from hydrogen). Infrared view emphasizes inner amorphous clouds. Although these individual images are in black and white, many of the color figures in this book are made by combining such images; the blue image is printed in blue ink, the yellow in yellow, and so on. (Hale Observatory.)

broader range of wavelengths than individual spectral lines or bands. The most important effect is that redder light (longer wavelengths) passes through clouds of dust, whereas bluer light (shorter wavelengths) is scattered out to the side of the beam (see Figure 20-5.)[1]

The scattering of blue light results from a property of all particles smaller than the wavelength of light: They scatter more blue light than red light out of the beam. This preferential scattering of blue light is called **Rayleigh scattering** after its discoverer. It occurs in interstellar dust and gas because many grains and all atoms are smaller than the wavelengths of visible light.

Thus the observer who looks through the dust cloud at a distant star sees most of its red light, but not much of its blue light. In this way, interstellar dust makes distant stars look redder than they really are—an effect called **interstellar reddening**. This effect is seen dra-

Figure 20-3 Nebula NGC 6357 is a region where part of the hydrogen is ionized and part is excited. The result is a strong red glow from the hydrogen alpha emission line. Clouds of dust obscure parts of the region. (Copyright Anglo-Australian Telescope Board.)

matically in the dust cloud in Figure 20-6. An observer who looks at the dust cloud from the side will see the blue light scattered out of the beam, however, so that a nebula illuminated in this way will have a bluish color. Prominent examples of these blue nebulae are seen in Figure 20-7.

---

[1]Scattering of light is different from absorption, mentioned a few lines earlier, but we will not emphasize the difference. Think of absorption as a disappearance of a photon into an atom (or ion or molecule), making the atom more energetic. Think of scattering as a bouncing of a photon off an atom out of the light beam into a new direction. Either process reduces the amount of light in the beam.

Figure 20-4 The nebula NGC 6559 and surrounding clouds of dust and gas in the constellation of Sagittarius display a variety of delicate colors. (Copyright Anglo-Australian Telescope Board.)

Why Is the Sunset Red and the Sky Blue?

These same rules also apply to material in our own atmosphere. The lower few kilometers of the atmosphere are full of dust and large molecules, including many that are slightly smaller than the wavelength of light. When we look at the nearest star, our Sun, through these particles, the same effects can be seen. If the Sun is high in the sky, we look through the minimum amount of dust, as shown in Figure 20-8a; thus the reddening is minimal, and the Sun is perceived as white.

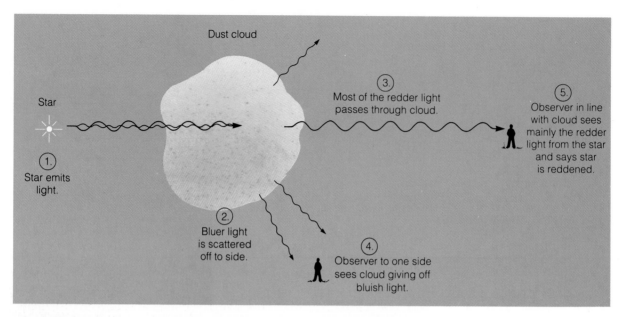

Figure 20-5 A cloud of dust grains illuminated by a star and seen by an observer to one side (item 4) and an observer in line with the cloud and the star (item 5). Compare with Figure 20-1 (p. 437), showing effects for a *gas* cloud.

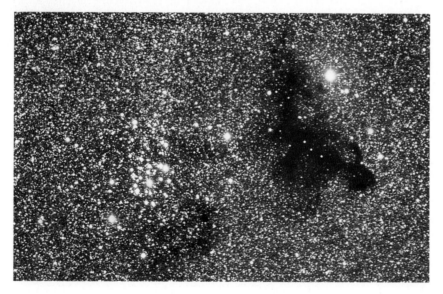

Figure 20-6 An example of interstellar reddening. A black dust cloud is silhouetted against a densely crowded field of stars in the constellation Sagittarius. Close examination of the dust cloud reveals that dimmed stars seen through it are strongly reddened. Similar examples can be found in the dust clouds of Figure 20-3. Near the cloud is the star cluster NGC 6520. (Copyright Anglo-Australian Telescope Board, courtesy D. Malin.)

At sunset, as shown in Figure 20-8b, the sunlight passes through much more gas and dust, and much of the blue light is lost from the beam by Rayleigh scattering. This strongly reddens the Sun and adjacent parts of the sky. At any time of day, if we look at some other part of the sky, as in Figure 20-8c, the light we see is the blue light scattered out of the beam of sunlight and then scattered by air molecules and dust back toward our eyes. In contrast, as seen in Figure 10-9 (page 204), the sky on Mars is reddish because many of the reddish Martian dust particles are bigger than the wavelength of light; because of that there is no Rayleigh scattering of blue light, but simply the reflection of their own red color.

Figure 20-7 The beautiful nebula NGC 1977, in the constellation Orion, shines mostly by reflected light of nearby stars. Scattering of light in the nebula's gas makes it appear blue by the same principle that makes our sky blue. (Copyright Anglo-Australian Telescope Board, courtesy D. Malin.)

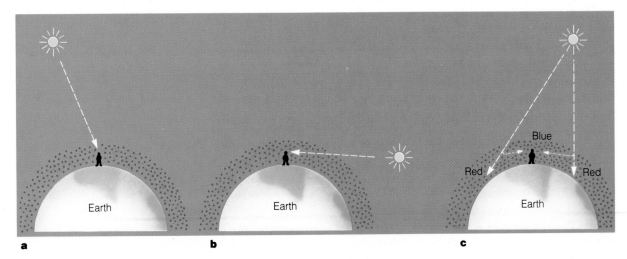

Figure 20-8 Explaining colors in the sky. **a** Light from the sun high in the sky passes through minimal dust and is minimally reddened. **b** Light from the Sun at sunset passes through the maximum amount of dust and is strongly reddened. **c** Light from the sky at any time of day is the blue light scattered out of the sunlight beam.

OBSERVED TYPES OF
INTERSTELLAR MATERIAL

Interstellar material includes gas (that is, atoms and molecules), microscopic dust grains, and possibly larger objects.

Interstellar Atoms

Atoms of interstellar gas were discovered in 1904, when German astronomer Johannes Hartmann detected their absorption lines. While studying the spectra of a binary star, he accidentally discovered absorption lines caused by interstellar calcium atoms. Certain other **interstellar atoms,** such as sodium, were soon found to produce additional prominent interstellar absorption lines. These lines are identified as interstellar by the fact that they have different Doppler shifts than the stars in whose spectra they appear.

Further studies have convinced astronomers that, although atoms such as calcium and sodium have prominent absorptions, the most common interstellar gas is the ubiquitous hydrogen. Like the Sun and stars, interstellar gas is about three-fourths hydrogen and nearly one-fourth helium.

Forbidden Lines from Interstellar Atoms A scientific mystery soon arose from interstellar gas. All experience prior to the 1920s had indicated that the distant universe was composed of the same elements as found in the solar system. However, as early as 1864, English observer William Huggins had found three greenish and bluish emission lines from the light of certain nebulae. Further work showed that these did not match the spectra of any elements known on Earth, and they were assigned to a hypothetical new element named "nebulium." By the 1920s, it was clear that there was no room in the periodic table of elements for a new element like nebulium. What then was creating the strange greenish light of nebulae?

In 1927, California astronomer Ira S. Bowen showed that in deep interstellar space, atoms could exist in so-called **metastable states,**—states with unusual electron distributions that last for several minutes. These states are unknown on Earth, because in our dense atmosphere the atoms are struck by air molecules too fast to remain in these states. The spectral lines arising from these unusual states are called **forbidden lines,** because they cannot occur in the Earth's atmospheric

gas. The nebular light turned out to be coming from strong forbidden lines of very rarefied, ionized forms of oxygen—a gas familiar in our own air. (The astronomer Henry Norris Russell thereupon quipped that the mysterious nebulium had vanished into thin air.)

Radio Radiation from Interstellar Atoms In 1944, the Dutch astronomer H. C. van de Hulst predicted that the most important type of interstellar radiation would be an emission line with wavelength 21 cm, caused by a change in the spin of hydrogen atoms' electrons. These electrons can have only certain spin rates and directions, and they emit the 21-cm radiation when they change from one state to another. Such a long wavelength is not visible light, but radio radiation. The predicted emission was confirmed in 1951 when Harvard astronomers, using radio equipment, detected this **21-cm emission line** of atomic hydrogen.

This discovery not only confirms the importance of hydrogen as a main constituent of interstellar gas, but gives radio astronomers a tool to detect where clouds of interstellar gas are concentrated. Because of the long wavelength of this radiation, it can penetrate much greater distances through the interstellar gas and dust than ordinary light. The 21-cm line of interstellar hydrogen is thus the most important emission line in radio astronomy.

Interstellar Molecules

By 1940, astronomers at Mt. Wilson Observatory in California had built spectrographs that detected absorptions due not only to interstellar atoms but also to **interstellar molecules,** such as CH and CN. Figure 20-9 shows spectral absorption lines caused by interstellar CH.

Because of molecular structure, a great many of the molecular absorptions lie in the infrared or radio parts of the spectrum. Not until after World War II did astronomers have available the technology of infrared and radio detectors to search for interstellar molecules. After being developed in the 1960s, some of the new detectors were put in satellites and large telescopes, sparking an explosion of interstellar discovery in the late 1960s and 1970s. For a time each new issue of astrophysics journals seemed to carry news of another molecule identified in interstellar space by spectral studies.

For example, the hydroxyl (OH) molecule was the only one found from 1963 to 1967, but water (H_2O),

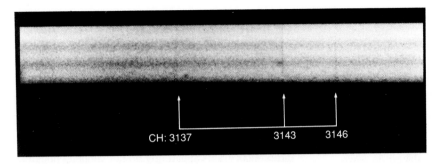

Figure 20-9 Spectral absorption lines caused by interstellar molecules of CH gas. Background light is part of the blue portion of the spectrum of the star Zeta Ophiuchi. (G. Herbig, Lick Observatory.)

ammonia (NH_3), and formaldehyde (HCHO) were found in 1968–1969, while 18 more molecules were found in 1970–1971. Some 56 varieties were cataloged by 1983. These include molecular hydrogen (H_2) and carbon monoxide (CO). There is a trend toward discovering more and more complex forms, such as the nine-atom molecule ethyl alcohol (C_2H_5OH), found in 1974.

The atoms recurring again and again in these large molecules comprise the quartet carbon, hydrogen, oxygen, and nitrogen—the "building blocks of life"! Repeatedly, these building blocks have been found in space in complex large molecules. These elements are common, and hydrogen is the most common of all, so the surprise is not that they exist in space, but that some process brought these elements together in their tenuous environment to make these complex forms.

Still more provocative is the fact that two of the detected molecules, methylamine (CH_3NH_2) and formic acid (HCOOH), can react to form glycine (NH_2CH_2COOH), one of the **amino acids.** These large molecules can join to form the huge protein molecules that occur in living cells.

Thus two exciting questions have come from research on interstellar molecules. First, does the existence of complex, carbon-rich molecules in space suggest that life could have originated elsewhere in space? The answer may be yes, and we will discuss this possibility in more detail in Chapter 28. Second, how do so many atoms come together to form these molecules? They cannot form in ordinary interstellar gas, because collisions there between atoms are extremely rare. Instead, the molecules are believed to form in the denser clouds, such as the cocoon nebulae around young stars.

Interstellar Grains

Still bigger than interstellar molecules are **interstellar grains.** They cause two observational effects: the reddening already explained and a general dimming of starlight at all wavelengths, called **interstellar obscuration.** Some grains are clumped in distinct clouds (as in Figure 20-6), but others are widely distributed throughout the interstellar gas, producing a general haze or "interstellar smog." All distant stars are harder to observe because of the haze caused by interstellar grains.

Importance of Interstellar Obscuration Interstellar obscuration played a curious role in humanity's recognition that we do not necessarily live at the center of the universe. After the Copernican revolution, astronomers assumed that the solar system did not occupy a central position. Then studies from the 1700s to the early 1900s of fainter and fainter stars seemed to indicate that the farther from the solar system we probe, the fewer stars we find. By 1922 astronomers began to wonder: Was the solar system in the *center* of a swarm of stars that was surrounded by empty space?

The answer turned out to be no. In 1931, American astronomer R. J. Trumpler published a very important paper proving that vast quantities of interstellar dust were merely obscuring the more distant stars. Trumpler proved this by studying distant star clusters. Their apparent sizes and other properties enabled him to show that the more distant a cluster, the more it is dimmed by dust in the line of sight.

Further research indicates that if we look along the Milky Way plane, dust grains dim stars by an average of about 1.9 magnitudes for every 1000 pc traversed by the beam of light. Of this amount, about 1.6 magnitudes of dimming are caused by grains concentrated in clouds and about 0.3 magnitude by grains dispersed between clouds. Over a distance of 10 000 pc, stars would be dimmed by 19 magnitudes! No wonder early observers found fewer and fewer stars the farther out they looked!

Nature and Origin of the Grains Astronomical studies reveal various properties of the interstellar grains: (1) They range in size from one-sixth to twice the wavelength of visible light; (2) they are concentrated in clouds; (3) compositions may vary somewhat from cloud to cloud; (4) many grains are elongated; and (5) many elongated grains are aligned parallel to other nearby grains, possibly because they are iron-rich and aligned with magnetic fields in space.

Grain composition has long been debated. In 1967, Indian astrophysicist N. Wickramasinghe proposed that grains are a form of carbon condensed in cooling gas blown off carbon-rich giant stars. Others suggested silicates and ices, likening grains to the dust in our primordial solar system. All these materials probably exist in interstellar grains. Carbon compounds, silicates, and ices have been identified spectroscopically, especially in star-forming regions. Reactions initiated by ultraviolet light in these materials apparently create complex organic molecules on many grains' surfaces (Greenberg, 1984).

In the early 1980s, studies of interstellar grains came much closer to home. Geochemists found that primitive carbonaceous meteorites contain grains of carbon (in various forms) and spinel (Al_2MgO_4), about 0.4 to 100 μm in size, whose isotopes indicate that they are not solar system material (Whittaker and others, 1980; Lewis and others, 1980). Further studies of the isotope compositions show that many of these grains condensed in gas blown out of red giants, novae, and supernovae, confirming carbon condensates as one grain source (Lewis and Anders, 1983; Swart and others, 1983). Magnetite grains similar to those in meteorites have been thought to contain 16% of all the iron atoms in the galaxy and may account for the magnetically controlled alignment. Only a few years ago, astronomers would have judged it implausible to study interstellar grains at firsthand. Now we seem to have them in our labs!

Interstellar Snowballs?

Astronomers are now looking for **interstellar snowballs**—hypothetical bodies much larger than grains. These could be BB-sized, baseball-sized, or even kilometer-scale bodies in interstellar space. If light interacted with such particles, it would be blocked equally at all wavelengths much less than the particle size. While microscopic grains are revealed by reddening, large particles would be difficult to detect because virtually no color effects would occur. Nonetheless, if single atoms join into molecules, and molecules into dust grains, why not expect still larger particles?

There are some indications that interstellar snowballs exist (Greenberg, 1974; Herbig, 1974). For example, the interstellar gas is strangely lacking in certain elements. Aluminum has not been found at all. Calcium is 400 times rarer than we would predict from its abundance elsewhere in the universe. Titanium, iron, and magnesium are also rarer than expected. Because the interstellar grains apparently do not contain enough of these materials to make up for the atoms missing in the gas, another class of interstellar material—snowballs—may contain the missing atoms. They might look like our present conception of comet nuclei—icy bodies with silicates and other "dirt" mixed in—justifying the term "snowball" coined by Greenberg.

FOUR TYPES OF INTERSTELLAR REGIONS

Interstellar gas and dust are far from uniform. There are thick clouds of material (nebulae) and regions of different temperature. Generally, astronomers recognize four different types of regions, defined by the condition of the gas: (1) Cold regions of dark clouds of dust and gas at about 10 K are called **molecular clouds.** Here much gas exists as molecules, including not only molecular hydrogen (H_2) but many of the more complex molecules discussed earlier. (2) Broad regions at around 100 K contain neutral hydrogen atoms and are called **HI regions** after the symbol for neutral hydrogen (HI). (3) Hot regions at around 10 000 K surround hot O and B stars, whose ultraviolet photons excite the surrounding H atoms or even knock electrons clear out of some of them, ionizing them. These are called **HII regions** after the symbol for ionized hydrogen (HII). An example was seen in Figure 20-3. As the electrons try to cascade back down to the ground state, they emit hydrogen emission lines, including the prominent Hα glow, giving many HII regions a predominantly red color. (4) Very hot regions, at about 10^6 K, exist where gas has been superheated by expanding blasts from supernovae. These are called **superbubbles.** We will discuss these four types of regions in the following sections. In very hot regions, atoms other than hydrogen may be excited or ionized, creating other spectral emission lines and additional colors besides red.

Radio telescopes and satellite X-ray telescopes have helped reveal molecular clouds and superbubbles, respectively, only in recent years. Hence it is not yet clear exactly what fractions of space are occupied by the four types of regions, although 10 to 50% of our galaxy's gas may be molecular (Blitz, 1982) and perhaps 10% in superbubbles.

Why has interstellar material not reached some uniform distribution? Suppose we magically smoothed the interstellar material to uniform density. Supernova explosions, gas ejected from giants, radiation pressure, and differential rotation of the galaxy would quickly create local dense regions. Some of these would be dense enough to contract gravitationally, making still denser clouds. Within a brief 100 million years or so the material would have reformed into a new distribution of nebulae.

Molecular Clouds

In localized regions where gas gets compressed to higher-than-average densities (perhaps by winds from neighboring expanding clouds), atoms collide more often and molecules may tend to grow. Dust grains get concentrated along with the gas. They often make the cloud opaque and dark, shielding its inner parts from nearby hot stars. More important, the dust grains can radiate heat more easily than the gas, so a region rich in dust grains gets cooler than surrounding clouds.

As we saw in Chapter 18 (especially Figure 18-2), the denser and colder a region of gas, the easier it is for stars to form in it by gravitational collapse. Thus molecular clouds are extremely important as regions of star formation. They may persist for 30 million years or so, with new stars arising here and there within them. Eventually, however, enough bright hot O and B stars arise that they heat portions of the molecular cloud, creating glowing HII regions. This explains the apparent paradox that cold molecular clouds are found in close association with hot HII regions. The most massive, short-lived stars may soon explode, blowing out million-degree gas and creating expanding superbubbles that blow the molecular cloud apart. Molecular clouds and their histories are thus intimately bound up with star births and star deaths (Blitz, 1982; Scoville and Young, 1984).

HI and HII Regions: Effects of Starlight on Interstellar Gas

One of the most important processes in creating different kinds of regions is the action of light from hot stars on the neighboring gas. The appearance and spectrum of an interstellar gas cloud depend on the states of its atoms. These states depend on the source of illumination and the range of excitations and ionizations it causes in the cloud. Two important physical laws, **Planck's law** and **Wien's law,** control the effect.

Planck's Law
The bluer the light (that is, the shorter the wavelength),the more energy each photon contains. Thus photons of ultraviolet or blue light excite or ionize atoms to a greater degree than photons of red light.

Wien's Law
Hotter sources radiate more blue light than cooler sources. (See also page 97.)

These two laws predict that the most energetic photons will be encountered near the hottest stars. Thus excitation and ionization in interstellar gas will generally be greatest near the hot, young stars of type O and least in interstellar space, far from any stars.

Photons of ultraviolet light with wavelength shorter than 91.2 nm are energetic enough to ionize hydrogen. By Wien's law, stars hotter than about 30 000 K produce most of their radiation at these wavelengths, and even stars hotter than about 20 000 K produce many such photons. Therefore hot, blue stars of spectral classes O and B are surrounded by large regions in which the interstellar hydrogen is ionized, as shown in Figures 20-10 and 20-11. If the gas were uniform, these regions would be spherical. Since the gas is nonuniform, these regions have ragged shapes.

In 1939, the Danish–American astronomer Bengt Stromgren calculated the average radii of the ionized regions around various types of stars, showing that only the hottest stars create large ionized regions. According to Stromgren's calculations, the diameters of the ionized HII regions in typical interstellar gas are:

Class	Diameter
O5	280 pc
B0	52 pc
B5	7.4 pc
A0	1.0 pc

The gas in HII regions is heated by stars to temperatures around 8000 to 10 000 K. In HI regions, farther from hot stars, temperatures are around 100 K.

a

b

Figure 20-10 The Trifid Nebula, about 1600 pc away in the constellation Sagittarius, beautifully combines red and blue. The hot gas within the cloud glows with the red light of Hα emission. It is roughly 5 pc across. Dark lanes of colder dust superimposed in front create a flowerlike pattern. Associated cool clouds have a soft blue coloration from scattering of starlight. **a** (*Opposite page*) Shorter exposure reveals details of the red nebula. The bright star in its center is a very luminous, hot O star with a blue-white light; it may be the star that excites the nebula to glow. (Copyright Association of Universities for Research in Astronomy; NOAO.) **b** Longer exposure reveals the large extent of surrounding blue-scattering gas and dust. (Copyright Anglo-Australian Telescope Board.)

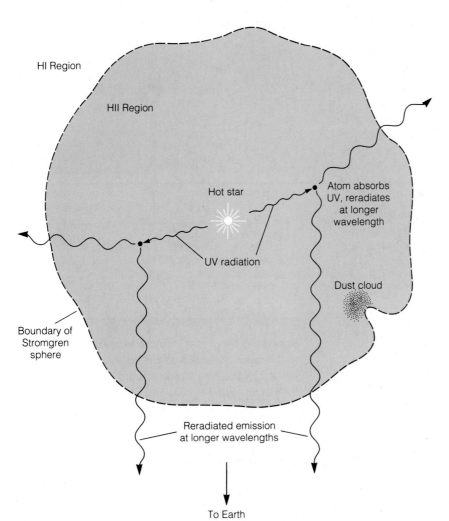

Figure 20-11 Creation of an HII region. Ultraviolet radiation from the central hot star is absorbed by hydrogen atoms, ionizing all of them out to a certain distance (dashed line). Recombination of electrons with atoms causes reradiation of various spectral emission lines in all directions, including earthward, causing the HII region to be visible as a glowing nebula.

The most common HII regions are clouds with masses about 10 to 1000 $M_\odot$.

In and near HII regions, electrons in various elements lead an up-and-down life. On the one hand, they are likely to be hit by an energetic photon from the nearby hot star and knocked out of the atom. But on the other hand they tend to recombine with atoms and cascade down through the energy levels to reach the ground state. Each time they cascade down from a higher to a lower energy level, they lose energy by radiating light of a certain color. For this reason, gas in the HII regions emits light of varying colors. Especially strong is the bright red Hα light emitted when electrons cascade down from the $n = 3$ to $n = 2$ levels in hydrogen atoms (see Figure 16-7). Also strong is invisible UV light emitted when electrons go from the $n = 2$ to $n = 1$ levels. *Therefore HII regions are also glowing nebulae called emission nebulae.* Many HII regions are among the most impressive features of the sky, as illustrated by the great swirls of gas in Figures 20-3 and 20-10 or the great red cloud known as the North American Nebula (Figure 20-12).

Stars with strong winds can push matter outward, making spheroidal expanding nebulae (usually HII gas); astronomers in 1975 coined the term **bubbles** for such objects, which have stronger expansion and symmetry than ordinary HII regions. Bubbles are essentially the same as planetary nebulae, and Figure 20-13 shows an example. (A good example of a small bubble coming off a Wolf–Rayet star was seen in Figure 19-6, especially in the central, bluish-colored spheroidal cloud. Other examples are shown in Figures 19-7 through 19-9.) As such bubbles expand, they may be buffeted and torn apart by interactions with interstellar gas or stellar winds from nearby stars, just as a ring of smoke distorts in the air.

Superbubbles: Effects of Supernovae on Interstellar Material

Astronomers were surprised when X-ray telescopes in space in the 1970s revealed large clouds of unusually hot gas radiating X rays at wavelengths such as 1.0 nm. Wien's law shows that X-ray emission from such hot gas must require temperatures over 1 million Kelvin! What could cause such high temperatures? These clouds are heated by violent expansion of gas blasted out of supernovae into surrounding space. Such events can create hot superbubbles. The inset in Figure 19-14 (page 422)

Figure 20-12 The North American Nebula, an HII region cataloged as NGC 7000 in the constellation Cygnus. Like Rorschach test blots, nebulae often have forms that evoke familiar shapes. This nebula is roughly 500 pc away and 14 pc in diameter. The color comes from Hα emission. The "coastlines" are defined by closer, dark dust clouds seen in silhouette. (48-in. Schmidt telescope photo, Hale Observatories.)

showed an X-ray image of such a cloud. A superbubble in Cygnus is about 500 pc across, has a temperature of about 2 million Kelvin, contains 10^{45} J of energy, would have required about 100 supernovae to be produced, and (based on its rate of expansion) is about 3 million years old, in contrast to the 1-million-year-old group of young O and B stars inside it. From these characteristics, astronomers infer that a sequence of many supernovae expanded this bubble (Cash and Charles, 1980). Direct evidence of supernova explosions include several supernova remnants and the black hole candidate Cygnus X-1, all located inside the Cygnus superbubble. A similar bubble was shown in Figure 19-16.

Figure 20-13 The Dumbbell Nebula, named for its appearance in small telescopes, is an example of a moderate-sized bubble (also classified as a planetary nebula). It is about 220 pc away and 0.3 pc across. It was blown off the central star, whose radiation excites the atoms to glow. Red Hα radiation dominates the outer part; bluish light from ionized oxygen atoms, excited in the space closer to the star, dominates the interior. (Copyright California Institute of Technology and Carnegie Institution of Washington; by permission from Hale Observatories.)

As shown in Figure 20-14, the endless cycle of star birth and death is bound to produce superbubbles that play a key role in redistributing the interstellar material. Star formation produces a few very massive stars that rapidly explode; this heats and expands local gas, which pushes against adjacent gas. The resulting compression starts new star formation on the outskirts of the expanding cloud. This leads to new explosions and expansions until the whole star-forming complex blows itself apart.

CLASSES OF NEBULAE

By tradition, nebulae are often categorized by their superficial appearance in telescopes. This in turn depends on whether hot stars are near them and whether we happen to see them against a background of dark space or distant bright nebulae. Table 20-1 listed characteristics of some specific nebulae.

Two main categories are dark nebulae (clouds silhouetted against brighter backgrounds) and bright nebulae. Dense parts of molecular dust clouds, for example, often appear as dark nebulae (see Figure 20-6). Bright nebulae include glowing HII regions, usually reddish due to the Hα emission line (Figure 20-1). Bright nebulae also include reflection nebulae (Figure 20-7), which are often bluish with dust and gas reflecting and scattering light of a nearby star. Expanding bubbles or so-called planetary nebulae (Figure 20-13) are another class of bright nebulae, each excited by its central star.

Example: The Orion Nebula, a Nearby Star-Forming Region

In the direction of the constellation Orion lies a region about 400 pc away dominated by nebulae and star-forming activity. Orion itself is the constellation of the hunter—a great figure raising a club over his head. As can be seen in the sketch map of Figure 18-9b, three bright

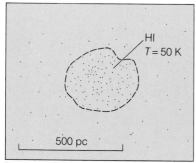

Dense, cold molecular cloud, contracting

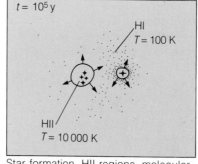

Star formation, HII regions, molecular clouds

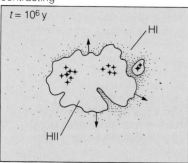

More star formation, HII expansion by heating

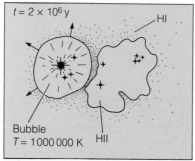

Supernova creates bubble, new star formation

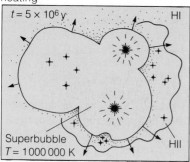

More supernovae, superbubbles, and star formation

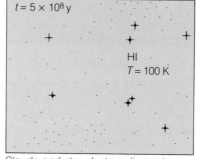

Clouds and star clusters dispersing, star formation and supernovae over

Figure 20-14 Schematic drawings showing the evolution of interstellar material in a certain region. The first box shows a random concentration of gas and dust that initiates star formation. The largest of the new stars then explode as supernovae. This expands the gas and eventually creates superbubbles of very hot expanding gas. The surrounding gas is finally blown away and the star formation ends.

stars mark his belt and three below it mark his sword. Orion's shoulder on the left side is the bright red giant star Betelgeuse, 180 pc away. His opposite knee is the blue B-type star Rigel, 270 pc away.

As shown in Figure 18-9, if you look toward Orion on a February evening, swinging your head back and forth past this part of the sky to compare it with other regions, you will see a great concentration of bright stars. Orion contains 7 of the 100 brightest stars in the sky, and except for Betelgeuse, all of them are massive, hot O- and B-type stars 140 to 500 pc away. Since massive stars are short-lived, their mere existence shows that they and other stars must have formed recently in

the nebulosity around Orion. These are well shown in Figure 20-15, an overview of Orion.

Consistent with its fame as a star-spawning region, Orion is full of vast clouds of gas and dust. Since the dust cloud temperatures are around 30 K, the dust radiates in the far infrared, where it has been imaged by satellites as shown in Figure 20-15b. Here we see the most intense dust clouds centered on the Orion Nebula in Orion's "sword" and at the left end of the "belt." By coincidence, a shell of dust has been blown outward from stars in the head region, making a ring-shaped infrared halo (invisible to the naked eye) around Orion's head.

a

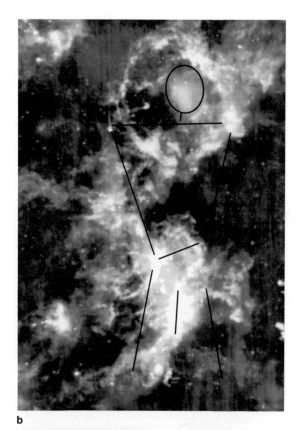

b

Figure 20-15 Three faces of the constellation Orion (see overview of the region in Figure 18-9). Stick figure of constellation gives orientation of each aspect; boxes show location of Figure 20-16a. **a** Visible-light photo shows dominance of blue-white, hot, young stars (and the prominent exception—red giant Betelgeuse in the upper left). Arrow shows pink glow of Orion Nebula in the "sword." Very faint red Hα glows can be seen left of the left star in the "belt." (Photo by author; guided 35-mm camera, 55-mm lens, f2.8, 20-min exposure on 3M ISO 1000 slide film.) **b** Same region in false color view by the IRAS satellite at infrared wavelengths of 12 to 100 μm. Redder colors are cooler; blue, hotter. Orange regions are clouds of interstellar dust. Brightest sources are Orion Nebula (bottom) and region at left end of belt. Note that most individual stars are too hot to radiate much light at these wavelengths; Betelgeuse appears as a blue spot, hotter than most dust but cooler than most stars. Blue region at top is part of background Milky Way. (NASA photo.) **c** Black and white rendition of a view at ultraviolet wavelengths (125–160 nm) emphasizing the hottest O stars, with temperatures around 20 000 K. Betelgeuse is too cool to show up at these wavelengths. (Naval Research Lab photo from rocket; courtesy George Carruthers.)

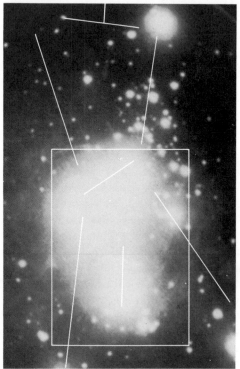

c

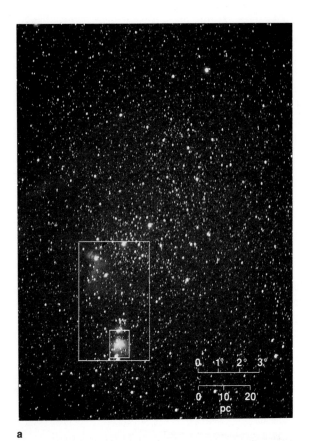

a

b

Figure 20-16 The "belt" and "sword" of Orion: heart of a star-forming region. **a** Color photo shows angular scale of region in the sky and linear scale in parsecs calculated for an average distance of 460 pc. Boxes show region of *b* and Figure 20-18a. (Photo by author, 15-min guided exposure on Ektachrome ISO 1600 film, 100-mm telephoto lens, f2.8.) **b** Color photo shows the region of the Horsehead Nebula (upper left) and overexposed Orion Nebula (bottom). (Copyright Anglo-Australian Telescope Board.)

Since O and B stars are extremely hot, they radiate prodigious amounts of ultraviolet radiation. For this reason, UV photos of Orion, such as Figure 20-15c, reveal it to be one of the most dazzling regions in the "UV sky." The massive stars have ionized much hydrogen, and the hot gas is being blown outward from the bright stars in central Orion. Some filaments of this gas show up in the outer parts of the overexposed region of the belt and sword in Figure 20-15c.

Figure 20-16b zeroes in on the star-forming area of the belt and sword. At the top we see a colorful region where the massive stars of the area, with their high temperatures and blue colors, contrast with a remarkable sheet of red-glowing gas. Silhouetted against this red curtain is the most striking dark nebula: the Horse-

head Nebula, a dust cloud resembling an ominous cosmic chess piece (Figure 20-17). At the bottom of Figure 20-16 is the still brighter Orion Nebula, which is somewhat overexposed and therefore whitish in this picture.

Figures 20-18 and 20-19 highlight the **Orion Nebula,** perhaps the most famous of nebulae. It is about 460 pc away, in the heart of the Orion star-forming area. To the naked eye it appears as the middle star in the sword, but even a small telescope or good pair of binoculars reveals it as a misty luminous haze about 5 pc across. As early as two centuries ago, English astronomer William Herschel examined this region with his pioneering large reflecting telescope and prophetically described it as "an unformed fiery mist, the chaotic material of future suns." And the details are spectacular: intermingled wisps

Figure 20-17 The Horsehead Nebula. Not readily visible to the eye in telescopes, the horsehead shape was first described from photographs around 1900 as a gap in the nebulosity. Some years later, it was recognized as an unusually dense dust cloud silhouetted in front of the Hα-glowing red cloud. Compare with Figure 20-16 to find the position of the Horsehead near the left end of Orion's belt. (Copyright Anglo-Australian Telescope Board.)

Figure 20-18 The Orion Nebula is the core of a star-forming complex about 460 pc away and 5 pc across. These images are especially processed to show faint outer details without overexposing the much brighter central region. **a** General region of the nebula. A separate star cluster and nebula lie to the north. **b** *(Opposite)* Color in the Orion Nebula includes red Hα emission and greenish tones from ionized oxygen. The brightest part of the nebula is centered on a tight cluster of four stars called the Trapezium, barely visible in the white region just to the right of the projecting dark cloud. (Both photos copyright Anglo-Australian Telescope Board, courtesy D. Malin.)

a

of red-glowing hydrogen, dark dust, and filaments dominated by pale greens of ionized oxygen and colors of other excited atoms. Here new stars are being born.

Figure 20-20 shows a cross section of the inner Orion region. On the far side from the solar system is an especially massive (about 1000 $M_\odot$) HI cloud of dust and gas, rich in molecules. According to one interpretation, expanding HII gas from the main nebula runs up against the HI cloud, limiting the size of the ionized region (Zuckerman, 1973). Especially dense clots of dust and gas were discovered inside the HI cloud by infrared observers in the 1960s. Figure 20-19 reveals one of these infrared-glowing clouds that is hidden in visible images. One of these, the Kleinmann–Low Nebula, may mark a collapsing cloud about to form a whole cluster of stars. Various clouds move at about 8 to 10 km/s relative to each other, and backward tracing of their motions suggests some may have formed within the last 100 000 y.

Radio astronomers have estimated that some 110 000 $M_\odot$ of HI and HII are involved in the expanding gas shell around the extended Orion star-forming region. Backward tracing of the motions indicates that the shell started to expand about 6 million years ago, when very massive, hot stars must have formed, heated nearby gas, and created a giant HII region that has been expanding ever since.

A time-lapse movie of Orion beginning about 6 to 10 million years ago, with frames every 10 000 years, would reveal star formation followed by a cosmic explosion of gas in the region of Orion's sword. The movie would also show certain stars racing out from central Orion like sparks from a blast. These high-speed stars are called **runaway stars.** Three bright O and B stars, for example, are racing out from the nebula at 70 to 130 km/s, and Betelgeuse itself is moving away from a region

near Orion's Belt. Their paths all point back to the region of dense gas around the nebula. Something unknown happened in that region to create high-velocity stars. They may have been accelerated during near encounters between new stars in clusters or during disruptions of coorbiting pairs of stars.

Orion has changed dramatically in the last few million years, since humanlike creatures emerged on the plains of Africa. New stars have blazed up, clouds of hydrogen have been expelled, and star formation is apparently continuing there today.

Example: Eta Carinae, a Nebula Around a Strange Star

Not many degrees from the Coal Sack Nebula (shown in Figure 17-4, page 370) lies an object that was recorded in the 1600s as an ordinary 4th-magnitude star, which

b

came to be called Eta Carinae.[2] It lies in a distant star-forming region about 2800 pc away. But in the 1800s its behavior was extraordinary. In 1827 it brightened to 1st magnitude and then dimmed. During a few weeks in 1837, English astronomer John Herschel, observing in South Africa, saw it become brighter than magnitude zero, finally reaching −0.7 in 1843! After observing with a large telescope, Herschel wrote in 1847 (quoted by Lovi, 1972):

It would . . . be impossible by verbal description to give any just idea of the capricious forms and irregular gradations of light affected by the different branches and appendages of this nebula. . . . Nor is it easy . . . to convey a full impression of the beauty and sublimity of the spectacle it offers.

---

[2]Eta Carinae lies at declination −59° and hence rises above the horizon only for observers south of 31° north latitude.

By 1857 the object had faded to +1 magnitude; by 1900 it had dropped to +8. It has remained variable, rising to +6 in 1967.

Spectra and photos from 1892 to the present show that the nebula within about 0.05 pc of the central star is blasting outward at 630 km/s, apparently from the eruptions observed from 1827 to 1843. Several parsecs from the star, ionized gas is still expanding at about 30 km/s (Walborn and others, 1984). Neutral gas expands less rapidly.

Modern photographs show the beautiful nebula described by Herschel (Figure 20-21). It exceeds 1° width and is roughly 60 pc across. Its red color comes from the strong light of the Hα emission line at 656.3 nm. The nebula is surrounded by an enormous hydrogen complex, rich in hot, blue stars, with luminosities around 5 million $L_\odot$. Only a few hundred parsecs from the nebula is a cluster of at least 130 O and B stars, known as NGC

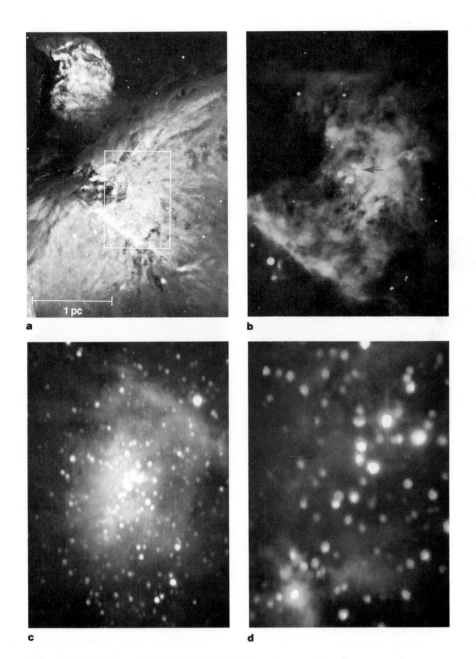

Figure 20-19 In the midst of the Orion Nebula lie many secrets. **a** The core of the nebula, with scale. Box shows size of next image. **b** The central region, showing the quartet of newly formed stars called the Trapezium (arrows). **c** Same region as *(b)* but imaged at infrared wavelengths and rendered in false color. The infrared images are made at wavelengths 1 to 5 μm and reveal a compact cloud of infrared-radiating dust (orange), undetected in visible light. Also seen are many stars that are too cool and faint to be prominent in *(b)*, but warmer (hence rendered bluer) than the dust cloud. **d** Enlarged view shows the Trapezium (upper right) and "hotspots" in the infrared nebula. The infrared nebula is probably a mass of dust and gas that is contracting and fragmenting into a group of protostars. (Photos *a, c,* and *d* copyright Anglo-Australian Telescope Board, courtesy D. Malin; photo *b,* Allegheny Observatory, courtesy W. A. Feibelman.)

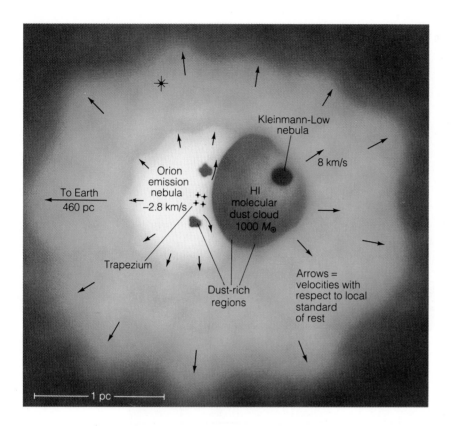

Figure 20-20 A hypothetical side view of the inner Orion Nebula showing the Trapezium, dust clouds, and expansion.

3293 and estimated to be only 8 million years old. The nebula is also bright in infrared light radiating from abundant warm (200 K) dust, probably grains condensing in material blasted out of the central star. The grains dim the star's light, but the total infrared and visual luminosity is 5 million $L_{\odot}$, like that of the nearby O stars.

Like the Orion region, the Eta Carinae region seems to involve star formation. But much of the **Eta Carinae Nebula** seems to have been thrown off a recently formed massive star that has already reached an unstable, old-age state. What kind of star? In the center of the nebula is a tiny reddish cloud about 0.02 pc across that may contain an unstable star nearing the nova stage. Its surface temperature has been estimated at 30 000 K. However, the variable-star expert D.J.K. O'Connell notes (quoted by Lovi, 1972):

Eta Carinae is certainly not an ordinary nova. Its behavior is indeed not paralleled by any other known star. Further observations are badly needed. . . . It may once again become one of the brightest stars in the sky.

Example: The Gum Nebula, an Ancient Spectacular

In the 1950s, a graduate student named Colin Gum surveyed the Southern Hemisphere sky in a search for HII nebulae. He used photographs sensitive to hydrogen alpha radiation and discovered a ring of nebulosity with the amazingly large angular diameter of 60°, which has come to be called the **Gum Nebula,** shown in Figure 20-22. It has been identified as an expanding bubblelike shell of gas from an ancient supernova explosion in the Southern Hemisphere constellation Vela.

Figure 20-23 shows the Gum Nebula in cross section. Its front side lies only about 100 pc from the solar system, and its center only about 460 pc away, accounting for its large angular size. The supernova remnant, a pulsar, has been found in the center of the expanding bubble.

Studies of the expansion rate suggest the explosion might have been witnessed around 9000 B.C. Because

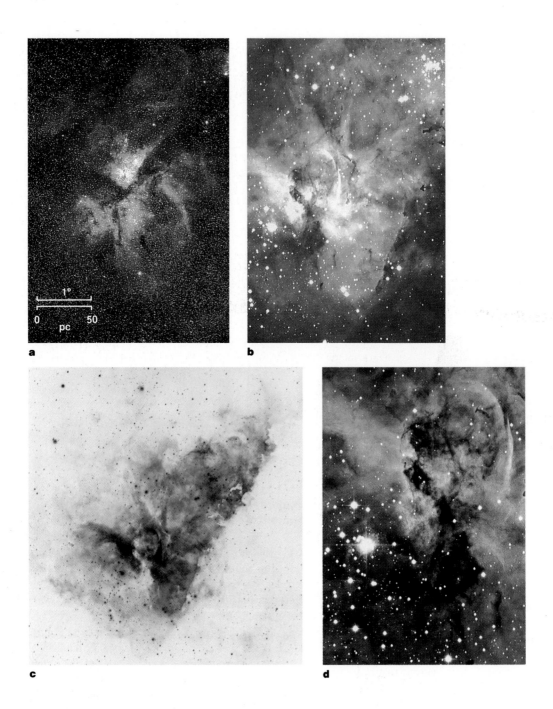

Figure 20-21 Zeroing in on the Eta Carinae Nebula. **a** Overall view shows the full nebula with about 1° width—twice the apparent size of the full Moon. **b** The bright core of the nebula, the tip of the V-shaped region above center in *(a)*. **c** False color view of *(b)*, a negative image with the sky light and the bright parts dark. Image has been processed so that regions emitting light from ionized oxygen are green, while those emitting light from ionized sulfur are pink. **d** Detailed view of the bright core region, exposed to show varied colors emitted by different elements. The star Eta Carinae is the bright object at left center. (Images *a* and *c*, copyright National Optical Astronomy Observatories/Cerro Tololo; images *b* and *d*, copyright Anglo-Australian Telescope Board.)

Figure 20-22 Remnant of a supernova: the Gum Nebula. These wispy clouds, in the southern constellation of Vela, are probably the expanded remnants of gas blown out of an ancient supernova. A pulsar—the star core left by the supernova—is located near the bright wisp and star cluster at the left center edge. The filamentary clouds are part of ragged ring around the pulsar, with apparent angular diameter about 5°. (Copyright Anglo-Australian Telescope Board.)

the explosion was so close (about one-third the distance of the Crab supernova) and had such high energy, it was probably extremely brilliant, reaching an estimated apparent magnitude of −10 (about as bright as the first quarter moon). This raises the intriguing prospect that late Stone Age humans (south of latitude +47°) may have seen a brilliant celestial beacon blaze forth for a year or two, rivaling the Moon. What effect might this have had on their concepts of the heavens and gods in the sky?[3]

[3]The effect of the Vela supernova may have been more than a dramatic skyshow. Theorists have concluded that the gamma-ray and X-ray radiation from such a close supernova would deplete the Earth's ozone layer, increase the nitrogen oxide production, and cool the atmosphere. The fallout of the NO in rain would have a fertilizing effect, increasing photosynthesis and deposition of organic sediments. G. R. Brackenridge (1981) describes eleven dramatic cases of black, organic sediments deposited around 9000 B.C., and suggests that they were caused when the Vela supernova produced a worldwide but brief climatic change. The Earth may be affected more than we know by "winds" that blow from interstellar space.

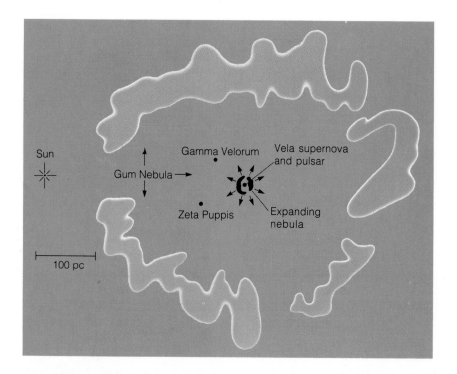

Figure 20-23 A side view of Gum Nebula geometry, showing relative positions of the Sun, Vela pulsar, and stars Zeta Puppis and Gamma Velorum (compare with Figure 20-22).

SUMMARY

Space is not empty but thinly filled with atoms and molecules of gas, grains of dust, and possibly bigger debris, which partly obscure distant stars and redden their light. Concentrations of these materials are nebulae. Some are excited to fluorescence by nearby stars, forming emission nebulae; some merely reflect the light of nearby stars; and some are dark silhouettes against distant backgrounds. Various processes of emission, Rayleigh scattering, and absorption give nebulae different colors. Various processes of expansion and turbulence give nebulae different shapes.

Nebulae are the material from which stars are born and into which the larger stars blow some of their material when their fuel runs out. Their existence shows that matter in our galaxy has not dispersed uniformly and stably, but rather is continually stirred, formed into clouds, dispersed, and disturbed by influences such as the formation of new stars, the explosions of old stars, and movements of all local material around our galaxy's center.

From nebulae and the general interstellar complex we know that all local material visible from the Earth has participated in a vast cosmic recycling. Interstellar matter forms clouds; clouds may contract to form stars; young and old stars blow out their material, creating a new interstellar medium. Nebulae also reveal clear evidence of cosmic events that have markedly changed our celestial, and perhaps even terrestrial, environment within the last million years—a period less than 0.01% of cosmic time.

CONCEPTS

Messier number
NGC number
IC number
nebula
excitation
ionization
Rayleigh scattering
interstellar reddening
interstellar atom
metastable state
forbidden line
21-cm emission line
interstellar molecule
amino acid

interstellar grains
interstellar obscuration
interstellar snowball
molecular cloud
HI region
HII region
superbubble
Planck's law
Wien's law
bubble
Orion Nebula
runaway star
Eta Carinae Nebula
Gum Nebula

PROBLEMS

1. Since sunlight is white (a mixture of all colors), why does the Sun look red at sunset? What happens to the blue light? Why does the part of the sky away from the Sun look blue?

2. Why was the sky red as photographed on Mars by Viking cameras?

3. How do massive stars help keep interstellar gas stirred up?

4. How do interstellar molecules illustrate the fact that complex organic chemistry is likely elsewhere in the universe?

5. How do masses of prominent nebulae compare with masses of single stars—are stars more likely to form singly or in groups?

6. Why are O-type supergiant stars likely to be associated with large emission nebulae, whereas solar-type stars are not?

7. Why do planetary nebulae often have simple, nearly spherical forms, whereas typical large emission nebulae, such as the Orion Nebula, are ragged, irregular masses?

ADVANCED PROBLEMS

8. If the Sun's material were redispersed into space at a density of 10 atoms/cm^3, typical of some clouds, how big a cloud would it make? (*Hint:* The Sun contains about 10^{57} atoms. 1 pc = 3×10^{16} m = 2×10^5 AU.)
   a. Compare with the size of the solar system.
   b. Compare with prominent nebulae.

9. Suppose spectral lines from gas on the near side of the Crab Nebula have a Doppler blue shift of 0.5%.
   a. At what velocity is it expanding?
   b. If the Crab Nebula is 1500 pc away (or 4.5×10^{19} m), use the small-angle equation to derive the expansion velocity of the nebula if it is observed to expand at an angular rate of 0.2 second of arc per year. (*Hint:* 1 y $\approx \pi \times 10^7$ s.)
   c. Confirm that the above two methods of estimating expansion velocity in the Crab Nebula give consistent results.

10. Suppose a tight cluster of newly formed O stars totals 100 solar masses within a region a parsec wide (like the heart of the Orion Nebula). Several of them explode as supernovae, creating a superbubble. The bubble expands to a point 10 pc from the center of the cluster and has a temperature of 1 million Kelvin. Compare the thermal velocity of a hydrogen atom at the edge of the bubble with the velocity it would need to escape from the gravitational pull of the star cluster into interstellar space. Would you expect the bubble to continue to expand and dissipate, assuming it did not run into dense clouds of adjacent material?

PROJECTS

1. Observe a cloud of cigarette or match smoke illuminated by a single light source, preferably a shaft of sunlight in a darkened room or a strong reading lamp. Compare the color of light transmitted through the smoke (by looking into the beam) with the color of light scattered out of the smoke (by looking at right angles across the light beam). Are there any differences? What can you conclude about the size of particles in the smoke cloud, assuming that the light wavelength is mostly 400 to 800 nm? Compare forms in the drifting smoke cloud with forms of nebulae illustrated in this book.

2. In a dark area away from city lights, by naked eye, observe and sketch the Milky Way in the region of Cygnus. Can you observe the dark "rift" that divides the Milky Way into two bright lanes in this region? The rift is caused by clouds of obscuring interstellar dust close to the galactic plane that are between us and the more distant parts of the Milky Way galaxy.

3. Observe the Orion Nebula with a telescope. Sketch its appearance. Locate the Trapezium (four stars near the center). The dark wedge radiating from the Trapezium is a dense mass of opaque dust. If a large telescope (50–100 cm) is available, look carefully for color characteristics. Generally, the eye is unresponsive to colors of very faint light, but large telescopes gather enough light so that colors can sometimes be perceived, especially with fairly low magnifications, giving a compact, bright image.

4. Observe the Ring Nebula in Lyra or other nebulae such as the Crab (in Taurus) or the Trifid (in Sagittarius). Comment on differences in form and origin.

CHAPTER 21

Companions to Stars: Binaries, Multiples, and Possible Planetary Systems

We have been soft-pedaling a fundamental fact: Most stars are not solitary wanderers in space. As seen in Table 17-1, four of the first six stars we encounter beyond the solar system have known companion stars. Surveys of this kind suggest that more than half of all "stars" are really coorbiting star systems.

Each pair of coorbiting stars is called a **binary star system,** or sometimes just a binary star. Each system with more than two stars is called a **multiple star system.** How do these systems affect our understanding of the universe? For one thing, we would like to know if our own multiple system, the Sun and its planets, is related to other multiple star systems, or whether it is a different kind of phenomenon. Second, we must be sure that our theories of star formation and evolution account for binaries.

A review of the preceding chapters shows that the theories are consistent with the existence of binary and multiple systems. Chapter 18 showed that stars must form in groups, and Chapter 20 gave examples of groups of young stars (such as the Orion Nebula's Trapezium) embedded in star-forming nebulae. Chapter 19 cited companion stars to explain certain facts of late evolution, such as the transfer of mass in binary systems that produces ordinary novae (not supernovae) and X radiation caused by dumping of mass from giants onto coorbiting dense stars or black holes.

OPTICAL DOUBLES VS. PHYSICAL BINARIES

To understand binaries and multiples in more detail, we face an observational problem. Among the star pairs that appear to be close together in the sky, some are at different distances and merely aligned by chance (as seen from Earth)—they are not actually coorbiting. Such star pairs are called **optical double stars.** They can be identified by the absence of any orbital motion around each other (revealed by photos or Doppler shifts) and are of little consequence in astronomy.

Pairs that are close enough to orbit around each other are called **physical binary stars.** They are the ones that clarify our knowledge of star properties and are the ones we will discuss here.

Early observers thought that *all* close star pairs were merely optical doubles, but in 1767 John Michell pointed out that so many chance alignments were unlikely. He proposed a physical association. This was confirmed in 1804 when William Herschel discovered that Castor (brightest star in the constellation Gemini) has a companion orbiting around it. In 1827 F. Savary showed that the orbit of the physical binary Xi Ursae Majoris is an ellipse fitting Kepler's laws. These results marked the *first discovery of gravitational orbital motion beyond the solar system,* an important confirmation that gravitational relations are universal.[1]

Among physical binaries, the brighter star is usually designated A and the fainter star B—for instance, Castor A and Castor B. Analysis of orbits reveals which is the more massive star, and it is usually called the *primary.* The less massive one is called the *secondary.* Normally, the primary is also the brighter, or star A.

TYPES OF PHYSICAL BINARIES

Astronomers classify physical binaries according to how they are detected. To understand what we can learn from binaries, it helps to understand these different methods of detection.

A **visual binary** is a physical binary in which both members can be resolved (seen separately) with the eye, the telescope, or the camera. Some 65 000 have been studied, and a good example is shown in Figure 21-1.

In a **spectroscopic binary,** orbital motion is revealed by periodic Doppler shifts in the spectral lines, but the individual stars cannot be resolved. There are two sub-

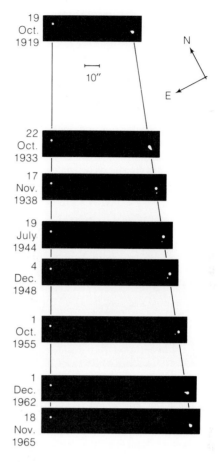

Figure 21-1 Forty-six years' photography of the visual binary Krüger 60. Binary pair Krüger 60, on the right, is moving past a background star, on the left. The photos show the orbital motion of one component of the binary around the other with a period of 45 y and a separation of 1.4 to 3.4 seconds of arc. The two stars have magnitudes about 9 and 11 and masses of 0.3 and 0.2 $M_{\odot}$. (Leander McCormick Observatory and Sproul Observatory, after Wanner, 1967.)

types: those in which only *one* spectrum can be detected (Figure 21-2), and those in which *two* sets of spectral lines are seen (Figure 21-3). The latter, displaying lines of both stars, yield more information. About 1000 pairs have been measured.

An **eclipsing binary** is a binary pair (generally unresolved) whose orbit is seen nearly edgewise and is revealed by eclipses. Because our line of sight lies in, or nearly in, the orbital plane, the stars alternately eclipse each other. (Such events are more properly called *occultations* and *transits,* but the term *eclipse* is widely used.) The eclipses are detected by plotting **light**

---

[1]Herschel realized that this discovery was much more important than his original goal of simply measuring distances, because the orbital motions reveal many properties of the stars. This illustrates how important, unexpected scientific results often derive from mundane research on another topic. Herschel reportedly likened himself to the biblical character Saul, who went out to find his father's mules and discovered a new kingdom.

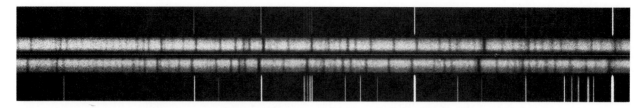

Figure 21-2 A portion of the spectrum of a single-line spectroscopic binary, Alpha Geminorum. The bright vertical lines at top and bottom are reference emission lines produced in the spectrograph; the middle two bright spectra crossed by dark, vertical absorption lines are spectra of the star on two dates. Offsets of these lines to the red and then the blue show that the star is receding and then approaching because it is orbiting around another star. (Lick Observatory.)

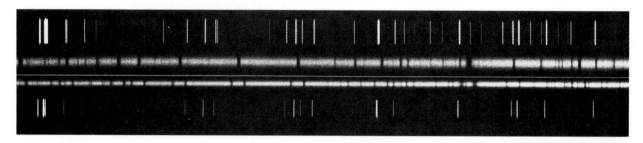

Figure 21-3 A portion of the spectrum of a double-line spectroscopic binary, Kappa Arietis. Arrangement as in Figure 21-2. Two sets of absorption lines in the top stellar spectrum reveal two stars, one receding (red-shifted) and one approaching (blue-shifted). The lower stellar spectrum, when orbital motions are perpendicular to the line of sight, shows lines merged, with no Doppler shift. (Lick Observatory.)

curves, or plots of brightness versus time, as shown in Figure 21-4. Depending on the relative brightness and size of the stars, the eclipse of the primary may produce a marked, short-term decrease in brightness. The star then returns to its normal brightness. The most famous example is Algol (β Persei), discovered in 1699. Algol dims to about one-third its normal brightness every 2.9 d.

Most eclipsing binaries are really **eclipsing–spectroscopic binaries,** in which both Doppler shifts and eclipses can be detected. This is the most informative type of binary, permitting very detailed analysis of the motion, mass, and size of the stars.

An **astrometric binary** is one detected not necessarily by sight, Doppler shifts, or eclipses, but by its orbital motions with respect to those of the background stars. According to Kepler's laws, stars in a coorbiting system revolve around their center of gravity, as described in Chapter 16. Each star therefore describes an oscillating path in three-dimensional space, moving back and forth around the system's center of gravity. Because *astrometry* is the study of stellar positions and motions, these systems are called astrometric binaries. Figure 21-5 shows a visual binary that is also a suspected astrometric binary: 61 Cygni. For some years, 61 Cygni was reported to have an unseen companion only eight times as massive as Jupiter. More recent studies (Heintz, 1978) question the mass and even reality of the companion. Until the case is settled, 61 Cygni remains a beautiful system with an aura of mystery.

Spectrum binaries make up a relatively unimportant class, where the presence of an unresolved binary is revealed by a unique spectrum consisting of lines from stars of two different temperatures, but with no detectable Doppler shifts.

Table 21-1 lists examples of the most important types of binary and multiple stars. The famous star Mizar, in the middle of the Big Dipper's handle, illustrates several types of doubles. Mizar forms an optical double with a fainter star, Alcor, less than a degree away. The

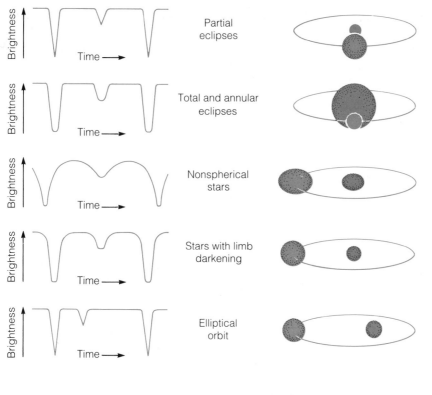

Figure 21-4 Different types of light curves (left) from eclipsing binaries reveal different geometric properties of the eclipsing stars and their orbits (right), even though the stars themselves cannot be resolved with the telescope.

a   b

Figure 21-5 Two views of the visual binary star 61 Cygni. Taken 35 years apart with the same telescope, these two photos show the motion of this relatively nearby star pair past our solar system, relative to the background stars. One of the closest star systems, 61 Cygni in 1838 became the first star to have its distance measured. An unseen astrometric companion has been reported but questioned. View **a** dates from 1916; view **b** from 1951. (Lowell Observatory photographs.)

Arabs called this pair the "horse and rider" and regarded it as a test of good eyesight. In 1650, the Italian observer Jean Riccioli discovered that Mizar itself is a physical binary with a visible 4th-magnitude companion, Mizar B, just 14 seconds of arc away. In 1889 Mizar A was found to have double spectral lines with periodic Doppler shifts, revealing that it is a spectroscopic binary.

In 1908, the same was found true of Mizar B. Thus Mizar is really a quadruple star system, with one close coorbiting pair revolving around another close coorbiting pair.

One interesting facet of binary stars is that very different types of stars can be paired: massive and not so massive, giant and dwarf, red and blue. An imaginary

TABLE 21-1

Selected Binary and Multiple Stars

System Name	Component A Distance (pc)	Component A Mass ($M_\odot$)	Component B Separation from A (AU)	Component B Mass ($M_\odot$)	Eccentricity	Component C Separation from A (AU)	Component C Mass ($M_\odot$)	Eccentricity
Visual Binaries								
α Centauri	1.3	1.1	24	0.9	0.52	10 000	0.1	
Sirius	2.6	2.2	20	0.9	0.59			
Procyon	3.5	1.8	16	0.6	0.31			
Eclipsing Binaries								
α Coronae Borealis	22	2.5	0.19	0.9?				
Algol (β Persei)	27	3.7	0.73	0.8	0.04	2?	1.7	0.13
β Aurigae	27	2.4	0.08	2.3				
ε Aurigae	1350	15?	35	15?				
Eclipsing–Spectroscopic Binaries								
β Scorpii	118	13	0.19	8.3	0.27			
η Orionis	175	11	0.6	11	0.02			
Astrometric Binaries								
Krüger 60	4.0	0.28	9.5	0.16	0.41			
Barnard's star	1.8	0.14	2.7?	0.001?				
Visual and Astrometric Binaries								
L 726-8	2.6	0.11	5.3	0.11				
61 Cygni	3.5	0.6	83	0.06			0.0008?	
BD + 66°34	10	0.4	41	0.13	0.05	1.2	0.12	0.00
Solar System								
Sun, Jupiter, Saturn	0.0	1.0	5.2	0.001	0.05	9.5	0.0003	0.06

Source: Heintz (1978); Popper (1980).

example is seen in Figure 21-6. The variety offers challenging opportunities for both measurement and imagination.

WHAT CAN WE LEARN FROM BINARY STARS?

The various types of physical binaries yield much information. *The most important application of binary star studies is in determining star masses.* As explained in

Chapter 16, the Kepler–Newton laws of orbital motion allow us to calculate the masses of astronomical bodies from measures of their revolution periods and orbit sizes (page 355). In the case of many binaries, very accurate orbits have been measured during many years of work (Figure 21-7).

As an example of a different application, consider an eclipsing–spectroscopic system. Suppose one star moves in a circular orbit at 100 km/s, and the time it takes to pass behind the other star (eclipse duration) is 3 h, or about 10 000 s. Then the diameter of the larger

Figure 21-6 Binary stars show a wide variety of pairings. Here a white dwarf orbits around a red giant that is shedding mass into a disk of gas and dust around itself. The white dwarf is a pinpoint of light relative to the swollen giant in the background. (Painting by author.)

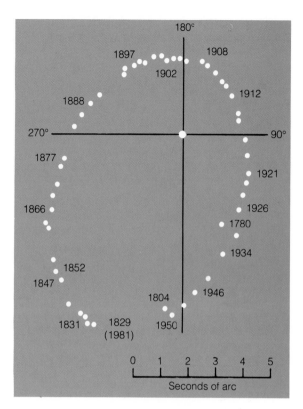

Figure 21-7 Observed positions of the fainter component of the visual binary Xi Boötes from 1780 to 1950, plotted relative to the brighter component, reveal Keplerian elliptical motion. (The apparent ellipse is the projection of the true elliptical orbit in the plane of the sky.) (After a diagram courtesy *Sky and Telescope*.)

and obscuring star is 1 million kilometers (10^2 km/s $\times$ 10^4 s = 10^6 km), or about the size of the Sun. This type of observation is the most accurate measure of stellar size—and hence one of the most basic checks on theories of stellar structure.

The same eclipse provides another example of how binaries contribute to knowledge of stellar structure. As the smaller star is eclipsed, the larger star successively obscures different parts of the smaller star's surface. It first covers part of the limb, then sweeps across the center parts, and finally covers the other limb. Later, as the eclipse ends, the process is reversed. The larger star acts as a screen, successively covering and uncovering different parts of the other star. Thus, by studying how the total brightness of the system changes during the eclipse, we can study the brightness *distribution*

across the face of the hidden star, thereby learning more about its atmospheric structure.

HOW MANY STARS ARE BINARY OR MULTIPLE?

This question is a challenge to observers. The nearest stars are easiest to observe but give too small a statistical sample to be reliable. At greater distances there are more stars, but faint companions might not be detected. Spectroscopic binary statistics are biased toward pairs with small separation distances, because according to Kepler's laws these have the fastest velocities and greatest Doppler shifts, thus being the most likely to be discovered. Visual binary statistics are biased

TABLE 21·2

Incidence of Multiplicity Among Stars (Estimated Fraction of Systems Containing *n* Members)

n	25 Systems Within 4 pc	Average of 7 Estimates by Various Authors[a]	Estimate by Batten (1973)[b]
1	0.48	0.41	0.30
2	0.36	0.41	0.53
3	0.12	0.14	0.13
4	0.04[c]	0.03	0.03
5	—	0.01	0.008
6	—	—	0.002

[a]Data from Batten (1973); Abt (1983, p. 345).

[b]Batten's estimates attempt to average over all stars, using data from various sources. Differences between the estimates are measures of our uncertainty about multiple systems.

[c]Solar system.

toward wide separation distances, which make the two stars easier to resolve. All these biases, which tend to make the data nonrepresentative of the whole population, are called **selection effects.**

Table 21-2 lists three estimates of the **incidence of multiplicity** for systems ranging from single to sextuple. By the time we reach six-member systems, definitions of multiple systems become hazy. It is unclear whether close groupings like the Trapezium in the Orion Nebula should be counted as multiple systems. The 1982 Yale Catalog of 9096 prominent stars lists multiple systems ranging as high as one system with 17 members. Such systems would present an interesting spectacle from nearby (Figure 21-8). We don't know if there is a physical relation between such large multiple systems and small clusters or if they are different phenomena. While Table 21-2 illustrates our uncertainty about the frequency of binaries, it does show that single stars are a minority.

Kitt Peak astronomers Helmut Abt and Saul Levy (1976) surveyed stars to determine the incidence not only of multiplicity but also of different masses. They found that about two-thirds of all stars have detectable companions, consistent with Table 21-2. But from statistics of companions' masses, they estimated that the

other seemingly single stars probably all have companions too small to detect! Some companions might be brown dwarfs with only a few percent of a solar mass; still smaller ones may be planets. According to this estimate, virtually all stars have at least one companion.

Thus although many people mistakenly assume that most stars in the night sky are single, *binary and multiple systems are more common than single stars.* Obviously, then, we must understand the origin of systems of two, three, four, and more stars in order to claim any understanding of stars in general. We will return to the problem of origin after reviewing some related observational data.

EVOLUTION OF BINARY SYSTEMS: MASS TRANSFER

The evolution of individual systems provides a natural way to classify them. The theory of dynamics of binary stars indicates that a system of two coorbiting stars can be pictured as containing an imaginary **Lagrangian surface** of special importance (often called the Roche surface). Its cross section is a figure 8, with one lobe around each star, as shown in Figure 21-9a. Slow-moving material inside either lobe orbits around the star in that lobe. Material that moves out of either lobe is not gravitationally bound to either star individually. It may transfer from one to the other or, with enough velocity, eventually leave the system entirely. When either star evolves toward a red giant state, it expands until it fills its lobe, as in Figure 21-9b. Its outer layers then assume the peculiar teardrop shape of the lobe, and the Lagrangian surface becomes the real surface of the star. Therefore three **classes of binaries** exist, as shown in Figure 21-9:

1. Systems in which neither star fills its Lagrangian lobe

2. Systems in which one star fills its Lagrangian lobe

3. Systems in which both stars fill their Lagrangian lobes; that is, stars in contact with each other (**contact binaries**)

Evolution can take a single binary through all three types. The pair starts as class 1. The more massive star swells into a giant as it evolves. This giant may fill its Lagrangian lobe, creating class 2. Any further tendency to expand causes matter to be shed, mostly through the point common to the two lobes, as shown in Figure

Figure 21-8 A hypothetical view within the sextuple star system Castor—brightest star in the constellation Gemini. What appears to the naked eye as a single star is in fact three pairs of close binaries, all orbiting around each other. Here we are on an imaginary planet orbiting the faintest pair, two low-mass red stars of about 0.6 $M_\odot$ each. More than 1000 AU away are the two other pairs (left). All four of the distant stars are whitish A stars about 50% more massive than the Sun and averaging 10 times as luminous. Although astronomers are searching near other stars for planets like that depicted in the foreground, none has been positively identified. (Painting by Ron Miller.)

21-9b. If the giant were a single star, it would be spherical and what little mass it did lose would stream off in all directions. But the companion's gravity forces the bloated giant to lose gas through its distorted, pointed tip. Like sand in an hourglass, this gas enters the lobe of the second star. Most of this gas spirals around the small star and crashes on it, making it gain mass as the large star loses mass. Obviously, the evolution of both stars will be altered, compared to their destinies had they been single. Later, the second star evolves to the giant state, partly through its normal evolution, but partly because of the added mass. Both stars now fill their lobes, producing a strange class 3, a contact binary (Figure 21-9c). The two stars actually touch as they orbit around each other! Such a configuration would make a strange "sun" (Figure 21-10)! Alternatively, if tidal forces drive the stars closer, the contact binary may merge into a single flattened star with a dual core. But if the two stars start very far apart, the giant stage may never fill either Lagrangian lobe, and the binary may never evolve out of class 1.

The grouping of stars in binary and multiple systems with different masses and different states of evolution in classes 1, 2, and 3 ensures that binary stars provide a rich variety of types.[2] Even backyard telescopes can reveal groupings with beautiful color combinations.

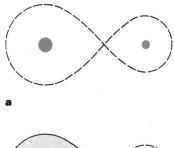

a

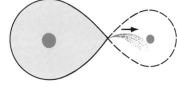

b

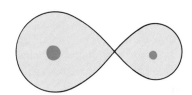

c

Figure 21-9 Evolution of a binary pair is related to the configuration of Lagrangian lobes. **a** In the first class of systems, neither star fills its lobe (dashed line). **b** After the larger star expands to become a giant, it may fill its lobe, taking on a teardrop shape and perhaps ejecting some mass through the tip, which interacts with the second star. **c** In the third class of systems, both stars fill their lobes.

[2]The colorful nomenclature of star types encourages shameless wags to concoct such descriptions as "degenerate dwarf in the company of a young starlet."

Figure 21-10 Hypothetical view of a contact binary system as seen from a distant planet. The planet is imagined to circle a third star in the system, another red giant offstage to the right. (Painting by author.)

NOVAE: EXPLODING MEMBERS OF BINARY PAIRS

Consider the case of a class 2 binary when the smaller star is a white dwarf of nearly $1.4\ M_\odot$. If the red giant's atmosphere dumps unburned hydrogen onto the surface of a white dwarf, the hydrogen may ignite in nuclear reactions that blow excess gas outward. This type of explosion is called a **nova** (plural: *novae*). An example appears in Figure 21-11.

Some novae are believed to occur in cycles. Hydrogen may accumulate on a dwarf until nova explosions occur, perhaps 10 000 y apart. Since the individual explosions blow off only a small fraction of the star's mass, leaving the rest intact, the process can start over. Among closer novae, the cloud of expanding debris can

sometimes be seen in telescopes a few years after the explosion, as shown in Figure 21-12.

Novae vs. Supernovae

Historically, the term nova was introduced long before the term supernova. *Nova* comes from the Latin root for *new* and was the term used for all "new stars" that appeared in the sky in past centuries. In ancient Chinese records, the novae were called "guest stars." In this century, astronomers discovered from their rate of light variation and other properties that there were two types of "new stars"—the novae and a much more energetic type that came to be called supernovae. Novae were found to be much more common. Of the 100 billion stars in our galaxy, 30 to 50 explode as novae each year, but supernovae occur only every century or so. Table 21-3 lists some novae and supernovae that have reached 2nd magnitude or brighter since A.D. 1000. These exploding stars were important in humanity's discovery that the heavens are not immutable, as once thought, but are actually evolving, even as we watch.

In the 1960s, observers proved that novae involve mass transfer onto white dwarfs in binary systems. Supernovae, however, came to be recognized as involving several types of explosions. As mentioned in footnote 3 in Chapter 19, one type also involves mass transfer in binary systems. Recall that above the Chandrasekhar limit, a white dwarf can't be stable. If the red giant dumps too much gas on the white dwarf too rapidly, the dwarf will exceed the limit and collapse to form a neutron star, blowing off a fraction of the excess mass in a titanic supernova explosion. A second type of supernova, emphasized in Chapter 19, is the natural outcome of evolution of very massive single stars and need not involve a binary system.

An Example: The 1975 Nova in Cygnus

The best-observed nova in history was Nova Cygni, which provided the constellation Cygnus the Swan with a prominent extra star for a few days in 1975. Most novae go unseen during the first hours of their explosion, but during the Cygnus event, amateur astronomers in California and Maryland were photographing the sky and actually recorded the nova as it brightened. As shown in Figure 21-13, the nova doubled in luminosity every 2 h during the explosion. Measurements of

Figure 21-11 Nova Herculis, showing the decline in brightness from March 10 to May 6, 1935. Later studies showed that this and other novae are binary stars. (Lick Observatory.)

Figure 21-12 Gaseous debris, blasted out of Nova Persei during an explosion observed in 1901, formed this expanding nebula around the star. This photo was made 58 y after the observed outburst. (Lick Observatory).

Doppler shifts show that the ejected gas expanded at about 1000–2000 km/s!

CONTACT BINARIES AND OTHER UNUSUAL PHENOMENA ASSOCIATED WITH BINARIES

Among the most unusual stars are the contact binaries, stars of class 3 in the evolution sequence described above (Figure 21-9c). The most famous of these are the **W Ursae Majoris stars,** named after the prototype in the constellation Ursa Major (better known as the Big Dipper). These consist of stars of rather similar mass, both filling their Lagrangian lobes. They have total masses ranging from 0.8 to 5 $M_\odot$, and

because they are so close together, their common revolution periods are very short, less than 1.5 d.

How did such a pair form? Some theorists believe they formed from protostars rotating so rapidly that they split, or *fissioned,* into two components. In contrast to the widely separated types of binaries, which have mass ratios expected from random pairings of field stars, close binaries show a tendency toward mass ratios of 1:1 (Abt, 1979), consistent with the fission theory of origin. Theorists have also proposed mechanisms that might cause widely spaced pairs to evolve into contact binaries.

Possibly related are stars with very close orbits and extremely short orbital periods. One example is the peculiar helium-rich star AM Canum Venaticorum. In this irregular variable, hydrogen-rich outer layers have apparently been transferred from one star to a dwarf companion and "blown away" in nova explosions. As mass was blown out of the system, the semimajor axis and period decreased (by Kepler's laws). As both stars lost mass, their helium-rich cores were revealed. AM Canum Venaticorum's revolution period is only 17^m31^s. The shortest known period for any binary, only 11 min, was discovered in 1986 by the European orbiting X-ray observatory, Exosat (Thomsen, 1986). The binary is apparently a neutron star orbiting a white dwarf, cataloged as 4U1820-30 and about 6000 pc away. Theoretically, mass loss could produce contact binaries with periods as short as 2 min! Somewhere in our galaxy there might be a planet in whose sky is a giant glowing figure 8 doing cartwheels like some bizarre advertising gimmick, as was illustrated in Figure 21-10.

Observations suggest that binary evolution accounts for not only novae and short-period contact binaries but

Selected "Guest Stars" (Novae, Supernovae, and Variables That Have Become Brighter Than Apparent Magnitude 2)

Star	Date Observed	Maximum Brightness (apparent magnitude)	Type
Lupus supernova	1006	−5?	Supernova
Crab Nebula explosion	1054	−2 to −6?	Supernova
Tycho's star	1572	−4?	Supernova
Kepler's star	1604	−2?	Supernova
Cassiopeia A[a]	1680?	+2 to +5?	Supernova
Eta Carinae	1843	−0.8	Nova?
T Corona Borealis	1866	+1.9	Recurrent nova[b]
GK Persei	1901	+0.2	Nova
DQ Herculis	1934	+1.3	Nova
Nova Cygni	1975	+1.9	Nova
o Ceti (Mira)	—	+1.0	Long-period variable (period 331 d)

Source: Data from Glasby (1968); Lupus data from *Sky and Telescope* (July 1976).

Note: "Guest Star" is the ancient Chinese term for temporarily visible stars. Fourteen supernovae ranging back 2000 y have been tentatively identified from ancient records. Certain expanding nebulae must also mark prehistoric supernovae.

[a]Cassiopeia A is the brightest radio source, and its expanding cloud has been imaged by X-ray telescopes in space. It may be the transient "star" observed by Flamsteed in 1680 [*Nature 285* (1980): 132]. See Figure 19-14.

[b]Second recorded flare-up in 1946 reached only +3 magnitude.

also other unusual stars. Gamma-ray bursts, X-ray bursts, and irregular brightness fluctuations are probably associated with mass transfer from an evolving member of a binary pair to a companion, for example (Ventura and others, 1983). Many irregular variables are binary systems in which matter has streamed from a giant into a disk surrounding a smaller companion. Several stars once thought to be short-period variables have been recognized as probable short-period binaries in which matter is streaming from one star to another, causing irregular variations in brightness. In some systems, matter streaming onto one side of a secondary star makes a hot, bright spot on its surface, causing unusual effects. Certain peculiar stars of spectral type A having unusually strong spectral lines of metals—and therefore known as Am stars—are mostly binaries. Tidal effects in these pairs apparently slow the rotation and cause unusual rates of diffusion among metal atoms in the stars' atmospheres. As we learn more, we may find that many peculiar types of stars are explained by strange processes of evolution in binary or multiple systems.

EXAMPLES OF BINARY AND MULTIPLE SYSTEMS

Algol and Similar Systems: Altered Masses

Algol, a variable star with a regular period of 2.9 d, is an eclipsing–spectroscopic binary. The light curve, with its dramatic dimming during eclipse, is seen in Figure 21-14. The primary is a hot, blue B8 star of about 5 $M_\odot$. The secondary is a K0 giant of about 1.0 $M_\odot$. Here lies a paradox: The more massive B or A star should be the one to expand first; yet the less massive star is the more evolved giant. Why? Is there a fundamental mistake in our ideas of stellar evolution?

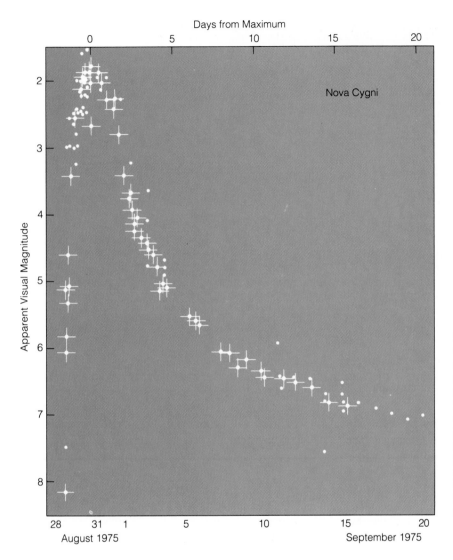

Figure 21-13 The brightness fluctuations of well-observed Nova Cygni in 1975 include observations made both before and after its discovery on August 29. Small dots are visual estimates of brightness by amateur and professional astronomers; larger symbols are more accurate photoelectric or photographic data. Nova Cygni is estimated to be 1300 pc away. (From International Astronomical Union data.)

Theoretical studies have resolved the paradox: What is now the smaller star was *originally* the more massive star. Reaching the giant stage first, it transferred so much mass to the other star that it became the less massive of the pair. In one theoretical study of a similar system, the more massive star starts out with 9 $M_\odot$ and is reduced to a mass of about 2 or 3 $M_\odot$ in only about 50 000 y! During this very rapid mass exchange, the stars reached more nearly similar masses.

SS-433: A Bizarre X-Ray Binary

As emphasized in Chapter 19, observations from space have revealed binary systems that strongly emit X rays

and gamma rays, probably from gas heated to extremely high temperatures as it transfers from one evolved star onto the surface of a dense companion, such as a neutron star. One of the most exciting is SS-433, a binary in which mass transfer not only heats the gas but also somehow accelerates the atoms into two jets that shoot out of the system in opposing directions at 26% of the speed of light—an extraordinarily high speed unprecedented in most stellar phenomena.

The story of SS-433 illustrates the sometimes laborious process of astronomical discovery. SS-433 was first cataloged in the 1950s and 1960s as one of many stars with visible emission lines as well as radio emissions; but its special nature was not recognized. Satellites in the 1970s showed X-ray radiation from this area,

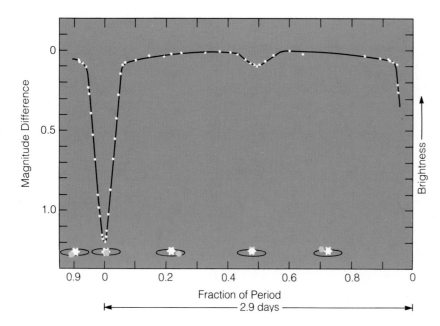

Figure 21-14 Light curve of the eclipsing–spectroscopic binary Algol. The curve plots brightness versus time and shows that Algol's brightness drops about 1.4 magnitude when the small, bright B star passes behind the fainter, cooler K giant.

and in 1978 astronomers realized that the radio waves and X radiation were all coming from a single star with a peculiar spectrum. Now the race was on to figure out its nature. Detailed spectra showed strange emission lines eventually identified as lines shifted far out of their normal positions by Doppler shifts much larger than ever found in any other star. Analysis revealed that the emission lines originate in two jets of 20 000-K hydrogen and helium gas, one approaching us (blue-shifted) and one receding (red-shifted). (SS-433 has 100 times the blue shift of any other known object in the universe!) Further study showed that the line positions oscillate in a 13-d cycle, as in a spectroscopic binary, proving that the star producing the jets is orbiting around another star every 13 d.

Figure 21-15 shows a hypothetical view of the system as interpreted by some astronomers. Gas is being transferred from one normal, evolved star into an accretion disk surrounding a dense stellar core, probably a neutron star (Margon, 1980, 1982).

Astronomers continue to study SS-433 with considerable excitement. Some models call for central objects as small as a 1-$M_\odot$ neutron star. However, Harvard astronomer J. E. Grindlay and colleagues (1984) studied X rays from SS-433 with the Einstein Orbiting Observatory and concluded the central object is a black hole of 10 $M_\odot$, surrounded by a gaseous accretion disk some 10 to 100 AU across. The Einstein Observatory's X-ray image actually shows, just barely, the two rela-

tivistic jets (Figure 21-16). Time will tell who is right!

The most important thing about SS-433 is that it somehow produces two 100-AU-long, brilliant, narrow, high-speed jets shooting out of the accretion disk. As mentioned in Chapter 18 (Figures 18-13 and 18-14), diffuse bipolar jets have been found emanating from various accretion disks, but nothing like the tightly beamed, ultra-high-speed jets of SS-433. What mechanism creates the jets? In 1984, using gamma-ray observations from a satellite telescope, Iowa astronomer Richard Lamb announced the discovery that thermonuclear reactions of the carbon cycle (see page 403) may occur in the jets. Could the jet source regions contain dense matter in which these reactions occur, as in stellar interiors? How does nature accelerate material to a sizable fraction of the speed of light without producing a diffuse explosion of million-degree gas? The answers aren't known.

Epsilon Aurigae: The Strange Case of a Dust-Cloud Companion

Epsilon Aurigae is a strange pair whose brighter component, an F-type supergiant, undergoes an eclipse every 27.1 y by an invisible companion. This eclipse takes 714 d and absorbs only about half the light of the F star. It is "so peculiar as to defy interpretation in terms of any conventional model" (Kopal, 1971). The two stars are estimated to be 35 AU apart, and the primary supergiant is about as big as Mars' orbit. The secondary is a

Figure 21-15 Hypothetical view of the strange binary SS-433. Interpretation of spectra suggests that gas is flowing out of an evolved star (left) into an accretion disk of hot gas surrounding a neutron star (right). Gas spiraling into the center of the disk is intensely heated and somehow ejected into 78 000-km/s jets moving outward in two opposing directions. Hydrogen alpha emission from the glowing jets is Doppler red-shifted to a deep red color in the receding jet (bottom) and blue-shifted to a more yellowish color in the approaching jet (top). (Painting by author.)

fuzzy disk of dust. During the eclipses, the F star's light shines dimly through the disk, which seems composed of large dust grains. But what maintains such a disk of dust? The eclipse of 1982–1984 was studied carefully and provided a possible answer. The supergiant F star was once a giant that dumped mass onto a secondary. But in this case, the secondary was not one star, but a close binary, now embedded in the dust. The motions of these two small stars around each other keep the disk of dust stirred up. As in the solar nebula, the disk cooled and condensed into many dust grains that might aggregate into planets. Meanwhile the F star is still contracting toward white dwarf state.

THE SEARCH FOR ALIEN PLANETS

Astronomers are interested not only in stars as companions to other stars, but also in brown dwarfs and planets as possible companions to other stars. One of the most provocative searches in modern astronomy is the search for such alien planets. The answer will tell us about the role of our solar system and ourselves in the cosmic scheme of things. The question of whether true extrasolar planets exist has important implications. For example, it represents a possible extension of the Copernican revolution in which we learned that the Earth is not located at the center of the universe. Perhaps we are not the center, but are we unique? One philosopher might stare at the starry night and speculate that the

Figure 21-16 An X-ray image from an orbiting telescope, called the Einstein Observatory, shows the two jets (top and bottom) emitted from SS-433 (center). X rays emitted by hot gas in the SS-433 system were used to form this image. (Harvard/Smithsonian Center for Astrophysics, courtesy Fred Seward.)

solar system is a special creation—the only place in the universe where a rare accident allowed formation of planets that could support life. Another might counter that the process of planet formation, as far as we can tell from evidence in meteorites and planetary samples, was a normal part of the Sun's birth and that other "single" stars probably had similar histories that yielded many planets—and at least a few habitable ones resembling Earth. But without data there would be no way to settle this intriguing argument. And we have no data one way or the other because we haven't been able to look at a large sample of stars with sensitive enough equipment to detect Jupiters, or Saturns, or Earths near them.

New techniques are being developed, however, that should make it possible within a decade to detect such planets, if they exist. First is improvement in *astrometry,* the technique for making precise measures of a star's position in the sky. If a star has an unseen Jupiter-like planet moving around it, the star itself "wiggles" back and forth due to gravitational pulls of the planet first on one side of its orbit and then on the other. Thus the star is revealed as a class of astrometric binary (see page 464), and in some cases the mass of the unseen companion can be estimated. A mass smaller than 2 $M_{Jupiter}$ could be considered a planet, according to our discussion on page 390. A slightly larger mass, from 2 to 85 $M_{Jupiter}$, would be classified as a brown dwarf.

A second technique is extremely precise measurement of Doppler shifts. If the same "wiggle" motions are detected in this way, the star would be a class of spectroscopic binary. Again, an estimate of the companion's mass might be possible, especially if this technique is combined with one of the other techniques. And the mass estimate might reveal the companion to be a planet.

A third technique is to seek eclipses. If a large planet crossed in front of the star, as seen from Earth, it would block a fraction of the star's light. This would cause a slight dimming of the star once during each orbit, and the star would be a class of eclipsing binary. Such repeated "micro-eclipses" would give at least a hint of a planet's presence, and the star could then be studied by the other techniques to confirm the planet.

The fourth promising technique is to detect the infrared radiation coming from the planet. You might think direct photography with giant space telescopes could reveal a planet, but generally the glare of the star is expected to be too bright for that technique to work (see Figure 14-1, page 299). In the infrared part of the spectrum, however, the hot star is putting out less radiation than at visible wavelengths, while the planet is putting out its peak radiation. (Review Wien's law, page 97, if this is not clear.) Thus astronomers can look for excessive infrared emission that might be coming from a planet at a few hundred degrees Kelvin. Already this technique has led to discovery of systems of dust particles, possibly from asteroids or comets, near other sunlike stars, as shown in Figure 18-12 (page 398)—a promising harbinger of possible planets.

Orbits and Relationships Between Binaries and Planetary Systems

One of the interesting aspects of the search for planetary systems is the relationship, if any, between multistar systems and planet systems. If we find a star system with a central star of 2 $M_\odot$, orbited by a small star of 100 $M_{Jupiter}$ and a brown dwarf of 10 $M_{Jupiter}$, is that fundamentally different from our own solar system? Did it form in a similar way or a totally unrelated way? One test is to see if the orbits are circular like those in our own solar system. Orbits in our solar system (Figure 21-17a) were forced to be circular because particles had to move in parallel tracks due to friction with the thick dust and gas medium of the solar nebula. As shown in Figure 21-17b and c, however, most orbits of binary and multiple star systems are not circular but elliptical. This is a clue that the formation processes of typical multistar systems and planetary systems probably differ.

Another test of the relation lies in the inclinations of orbits in multiple star and planetary systems. Among the nine planets of the solar system, Pluto has the highest inclination to the common plane, about 17°. None of the other eight is inclined more than 7°. In other words, as Figure 21-17a shows, planets' orbits are highly *coplanar.* Of 10 triple and quadruple star systems studied, all orbit pairs have inclinations of more than 18°, and half have more than 40° (Batten, 1973). Another system, BD + 66°34, may have three low-mass stars in coplanar orbits (Hershey, 1973). Nevertheless, while most multiple star orbits are corevolving, they are usually not coplanar (Heintz, 1978), again suggesting that typical multiple star systems are unlike planetary systems.

According to Kepler's laws, the motions of coorbiting bodies are related to the masses of the bodies around which orbital motion occurs. In many multiple systems the stars are of comparable mass, thus disturbing each other's motions. Multiple orbits of similar sizes would

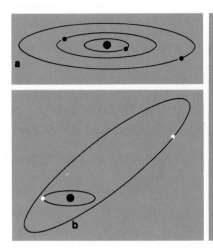

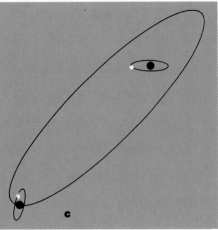

Figure 21-17 Oblique views of different types of orbital arrangements. **a** In the solar system, orbits are circular and coplanar, each about twice as far from the Sun as its inner neighbor. **b** In triple systems, a distant companion often accompanies a close pair in an elliptical, inclined orbit. **c** Quadruple systems often have two remote pairs in elliptical, inclined orbits. Differences between the planetary orbits and the multiple star orbits suggest different origins.

result in unstable systems, raising the possibility that perturbations or collisions would change the system. The most commonly observed arrangements in multiple systems are thus not concentric but hierarchical. For example, in triple systems, such as the Alpha Centauri system, two stars commonly revolve in a *close* orbit, and the third star revolves around the pair at a *great distance,* almost always more than eight times the separation distance of the close pair (Heintz, 1978), as seen in Figure 21-17b. In quadruple systems, such as Mizar, two close pairs are likely to revolve around each other, with the pairs widely separated (Figure 21-17c).

Thus in orbital shape, inclination, and arrangement, widely separated binaries appear to be fundamentally different from planetary systems. Abt (1979) found that while masses of widely separated pairs have the same statistical distribution as masses of ordinary representative stars, masses of close companions (within a few tens of astronomical units, or solar system dimensions) have a different mass distribution. Abt concluded that the wide pairs probably formed by separate collapses of two protostars; they were either near each other initially or they were captured into orbits around each other during a near approach in the crowded environment of the parent star cluster. The close pairs formed by a separate mechanism. Probably they resulted when a single protostar had too rapid a spin (too much angular momentum) to collapse into one star and split instead into two.

The transition between the wide types and the close types occurs around separations of about 10 to 50 AU. Possibly our solar system is an unusual example of the close type, where the mass ratio of primary to secondary (Sun: Jupiter) is about 1:1000. Was our Sun a pro-

tobinary that didn't make it? We don't yet know. Perhaps further studies of binary and multiple star systems will help us understand how planetary systems form.

Planet Formation vs. Star Formation

By contrasting planet formation (Chapter 14) with star formation (Chapter 18), we can see the problem we face in interpreting possible future discoveries of low-mass companions around stars. Planets apparently formed primarily by a condensation of dust and ice particles followed by accumulation during their collisions; the surrounding gas medium eventually blew away. This happened inside a cocoon nebula near a star. Most stars formed when the entire contents of an interstellar region of gas and dust collapsed gravitationally, trapping all the gas, dust, and ice particles in the region. But suppose we find an intermediate-sized object of, say, 3 M_{Jupiter} orbiting a star. Did it form by dust-particle aggregation or by gravitational collapse? Could there be some unusual planets and brown dwarfs that formed by collapse? Could there be some unusual companion stars in binary systems that formed by particle aggregation within the cocoon nebula? The answers are completely uncertain to today's astronomers, but the search for planetary systems and brown dwarf companions may clarify the situation.

Examples of Extrasolar Planetary Systems?

If we want to confirm that planetary systems can form elsewhere in the universe, and if we want to understand how they differ from multistar systems, we will have to do more than simply detect objects with mass less than about 2 $M_{Jupiter}$. We will also need to build up statistics on brown dwarfs to find out if there is a smooth transition from planets to brown dwarfs or whether most brown dwarfs as well as stars form by a different process. And we will need statistics on orbits, especially in systems with more than two bodies, to see if the orbits are coplanar (as in Figure 21-17a).

From the 1970s onward, several observers have claimed detection of objects within or approaching the planetary mass range of less than 2 $M_{Jupiter}$, but subsequent searches by other observers have not confirmed the results. Many astronomers would like to be first to deliver the revolutionary proof of planets beyond the Sun. Perhaps some have been hasty in announcing questionable detections. Barnard's Star was once said to have one or more Jupiter-scale companions, but this result seems to have been an error caused by random "noise" in the astrometric data (Gatewood and Eichhorn, 1973). In a more recent case, astrometry in 1983 indicated an unseen companion to a star known as VB8 (star number 8 on a list by a Belgian-American astronomer named Van Biesbroeck). In 1984 Arizona astronomers reported direct detection of the companion, found to be as small as 5 $M_{Jupiter}$ and having a temperature of about 1400 K (2060°F)—a brown dwarf. Other astronomers in 1986 were unable to find the brown dwarf, however, and this case remains uncertain.

The only certain detections of unseen companions by astrometry have been, generally, of relatively high-mass objects around 10 to 100 $M_{Jupiter}$. The star BD +66°34, for example, which is 10 pc from Earth, reportedly consists of two main-sequence M-type stars of mass 0.4 $M_{\odot}$ and 0.13 $M_{\odot}$, with a third unseen object of mass 0.12 $M_{\odot}$, or 120 $M_{Jupiter}$. The third object would be classed as a small star, but the three bodies may have *coplanar orbits*. Thus this system might have formed by solar-system-like processes in a cocoon disk, rather than the normal multistar-forming process that yields inclined orbits.

With each year, the size of detected objects seems to get nearer the intriguing planet/brown dwarf boundary. Some interesting results have come from the infrared technique. The famous newly forming star T Tauri has been known since the 1970s to have a surrounding cocoon nebula containing silicate dust grains from which a planet could form. In 1982, Hawaiian observers announced discovery of a faint infrared companion interpreted as an accreting brown dwarf of mass 5 to 80 $M_{Jupiter}$ and temperature 800 K (Dyck and others, 1982; Hansen

and others, 1983). It is reportedly 80 to 150 AU from T Tauri, about two to four times as far as Neptune and Pluto from our Sun.

In 1987, Canadian observers announced still more exciting results after monitoring 16 nearby stars by the Doppler shift technique. They reported possible unseen companions of only 1 to 10 $M_{Jupiter}$ around seven stars! Two cases were especially provocative. The nearby solar-type star Epsilon Eridani reportedly has a companion of only 2 to 5 $M_{Jupiter}$. A more distant solar-type star, Gamma Cephei, seems to have a tiny companion of about 1.7 $M_{Jupiter}$. If confirmed by others, this might be listed as the first true extrasolar planet. An interesting feature of these two detections is that both stars resemble the Sun, being of spectral class K2 and K1, respectively, and having around 80 or 90% of the mass of the Sun. Perhaps many "single" solar-type stars will turn out to have systems of planets!

Finally, as discussed in more detail on pages 394–395, astronomers in the mid-1980s found abundant evidence of systems of dust particles around young and middle-aged stars. This discovery gives considerable confidence that planetary matter—the dust from which planets can form—does exist around many other stars. This in turn spurs astronomers to believe they will find extrasolar planetary systems if they look hard enough.

THE ORIGIN OF BINARY AND MULTIPLE STARS

Clearly there are many questions about the origin of multistar systems. What determines whether a single star forms or one with companions? What determines the distribution into systems of two, three, four, or more members? (This may be a question of the angular momentum and its distribution in the original collapsing protostellar cloud.) And what determines whether a companion to a star is a planet, a brown dwarf, another star, or a mixture of these? Whatever the answers, most astronomers believe there may be several processes at work to form binary and multiple systems. Different processes—fission, capture, subfragmentation—may produce different types of systems.

Fission Theories that picture a fast-spinning protostar as splitting in two are called **fission theories.** Many researchers believe that very close pairs such as the W Ursae Majoris contact binaries are formed in this way.

Capture According to **capture theories,** systems of widely separated stars that share weak gravitational attraction are chance configurations arising when one star approaches another. The chance of encounters among random field stars is far too slight to explain the observed numbers of binaries. Furthermore, randomly paired stars would be of widely different ages, but this is not observed among binaries. Where could stars of similar ages interact in a closely packed group? In a newly formed star cluster. Massachusetts astronomers T. Arny and P. Weissman (1973) showed that fully half the protostars in a cluster probably undergo collisions or close encounters. Therefore astronomers believe at least some binaries and multiples formed inside newly formed open clusters as protostars approached, and, by various gravitational interactions, became bound in orbit around one another. Many wide pairs probably formed in this way.

Subfragmentation During Gravitational Collapse
Depending on rotational properties, subfragmentation of a prestellar cloud may produce multiple systems with either two high-mass stars or several low-mass stars in coplanar orbits. Some moderately close systems may have formed in this way. According to **subfragmentation theories,** the prestellar gravitationally contracting cloud shrinks because of its own gravity, but instead of forming a flat disk with a central star, its mass distribution or angular momentum distribution may make it split into two or more coorbiting clouds. These clouds then collapse independently into stars.

Each of these three theoretical processes may produce binaries of a certain type. A complication in sorting out binaries of these different types is that orbits of binary and multiple stars evolve through gravitational influences. Mass transfer in close pairs can alter orbits. Widely spaced pairs formed inside larger star clusters can evolve into closely spaced pairs as the clusters break apart. In one theoretical study of some 800 imaginary triple-star systems, about 97% were found to be gravitationally unstable, eventually kicking out one star and becoming binary systems. Sometimes the ejection velocities are quite high; 58 km/s is a typical example. This in turn helps explain the runaway stars observed speeding out of some young star-forming areas, as described in Chapter 20. Thus the observed statistics of binaries and multiples may not reflect their original characteristics.

Nonetheless, astronomers who have attempted to correct for selection effects have found true statistical differences between the orbits of closely spaced systems and widely separated systems. These two kinds of systems probably form by two different processes.

Multiple systems, in particular, present even more questions about formation processes. Many may form when one or both protostar clouds in a protobinary subdivide, leading to a three-star or four-star system. The factors that determine whether a multiple star system forms probably include masses of the protostar clouds, their temperatures, their separations, and their rotational properties. New studies of multiple stars and searches for faint companions are under way, and in a 1987 review of these systems one astronomer predicted "We are on the edge of a boom in the detection of multiple stars."

SUMMARY

At least half of all the seemingly single stars in the sky are binaries or multiple systems. Many of these may have formed by interactions of stars in crowded new clusters, but some may have formed by other means such as fission or common condensation.

Binary and multiple systems can be detected in different ways. Binaries were once classified by these different methods of detection, which yield different types of knowledge. Examples include spectroscopic binaries, eclipsing binaries, and astrometric binaries. Certain types of binaries give the best available data on certain properties of stars, such as mass and diameter (see pages 463–464).

A classification based on evolution includes binaries that have filled neither Lagrangian lobe, one Lagrangian lobe, or both Lagrangian lobes. Once one lobe has been filled by expansion of a star to the red giant stage, gas may flow from one star to the other, causing flare-ups, X-ray emission, and nova explosions.

Companions to stars include not only other stars but also brown dwarfs and probably planetary systems. Astronomers are actively searching for conclusive examples of planetary systems, which may extend the Copernican revolution by proving that the solar system is not unique. Whether companions as small as planets form in a contracting protostar may depend on the protostar's rotational properties.

CONCEPTS

binary star system

multiple star system

optical double star

physical binary star

visual binary

spectroscopic binary

eclipsing binary

light curve

eclipsing–spectroscopic binary

astrometric binary

spectrum binary

selection effect

incidence of multiplicity

Lagrangian surface

classes of binaries

contact binary

nova

W Ursae Majoris stars

fission theories

capture theories

subfragmentation theories

PROBLEMS

1. Describe verifications of Kepler's laws other than the planets' motions around the Sun. What was the first verification outside the solar system?

2. How do binary and multiple star systems generally differ from planetary systems?

3. Give evidence that at least a subclass of binary and multiple star systems might be generically related to planetary systems.

4. How are novae related to binaries? Are supernovae related to binaries?

5. How will the evolution of a 1-$M_\odot$ star in orbit close to a 3-$M_\odot$ star differ from the evolution of a 1-$M_\odot$ star by itself?

6. Why are binaries more likely to have formed in star clusters than as isolated field stars?

ADVANCED PROBLEMS

7. If a star of low mass (about 0.05 $M_\odot$, for example) were orbiting in the Earth's orbit around the Sun, what would be its period of revolution?

8. Use the circular velocity equation to derive the orbital velocity of a Jupiter-sized body around a 1-$M_\odot$ star if the separation distance is 5.2 AU (7.8×10^{11} m). (*Hint:* 1 $M_\odot = 2 \times 10^{30}$ kg.)

a. How does this compare with the actual orbital velocity of Jupiter?
b. What would be the orbital velocity if the central star had two solar masses?

9. Suppose you observe a 0.5-$M_\odot$ star in circular orbit around a 5-$M_\odot$ star and can tell from Doppler shifts that the orbital velocity is 47 km/s. What would you conclude is the separation distance between the stars in astronomical units?

10. A star of mass m is in a circular orbit around a star of mass M = 10 $M_\odot$. Star M explodes and rapidly blows away 0.6 of its mass.
a. What is the orbital velocity of m before the explosion?
b. How does the total force of attraction on m from M change during the explosion?
c. Describe the future history of m. Will it remain in orbit around the remnant of M? (*Hint:* This sort of event has been suggested as a source of runaway stars; see page 454.)

PROJECTS

1. Observe Mizar and Alcor with the naked eye. They are the middle "star" (actually a close pair) in the handle of the Big Dipper. Can you see the faint star Alcor? Sketch its position. (Inability to see Alcor may be due to insufficiently keen eyesight, a hazy sky, or a sky illuminated by city light.)

2. Observe the eclipsing binary Algol with a telescope or binoculars each evening for 10 to 20 d in a row. (This can be done as a class project with rotating observers.) Using neighboring stars as brightness reference standards, estimate the brightness of Algol. Can you detect the eclipses, which occur at 2.9-d intervals?

3. The star Epsilon Lyrae is famous as the "double double." It consists of a binary pair 208 seconds of arc apart, easily seen in a small telescope. But each of these is a binary only 2 to 3 seconds apart. These pairs are a test of good optics and good atmospheric observing conditions. Does your telescope reveal the two close pairs? Sketch them.

CHAPTER 22

Star Clusters and Associations

If you could roam through space to ever greater distances, you would eventually lose track of individual stars and see the galactic disk defined primarily by clusters of stars. Writing about clusters in 1930, Harvard astronomer Harlow Shapley pointed out that "their problems are intimately interwoven with the most significant questions of stellar organization and galactic evolution."

Yet at the same time Shapley noted that scientific study of them had hardly begun. Nobody knew how to measure their distances or plot their distribution in space until the 1920s. Although some clusters, such as the Pleiades (or Seven Sisters), are easy to recognize with the unaided eye, others are so far away that they require large telescopes to detect. Still others are so close that they cover much of our sky and were not even recognized until recent years.

In other words, our cosmic journey has brought us to clusters so far-flung that they were recognized as a class only in this century. Clusters have been of major importance because they have revealed to us the shape and age of our own galaxy, as will be clarified in this chapter.

THREE TYPES OF STAR GROUPINGS

The three basic types of clusters are *open clusters, associations,* and *globular clusters.* Examples are listed in Table 22-1.

Open Star Clusters

Open star clusters are moderately close-knit, irregularly shaped groupings of stars. They usually contain 100 to 1000 members and are usually about 4 to 20 pc in diameter. Our Sun is possibly inside or on the edge

TABLE 22·1

Selected Star Clusters and Associations

Name	Distance (pc)	Z^a (pc)	Diameter (pc)	Estimated Mass ($M_\odot$)	Estimated Age (y)
Open Clusters					
Ursa Major	21	18	7	300	2×10^8
Hyades	42	18	5	300	5×10^8
Pleiades	127	54	4	350	1×10^8
Praesepe	159	84	4	300	4×10^8
M 67	830	450	4	150	4×10^9
M 11	1 900	99	6	250	2×10^8
h Persei	2 250	156	16	1 000	1×10^7
χ Persei	2 400	167	14	900	1×10^7
O Associations					
I Orionis	470	150	?	3 000	?
I Persei	1 900	164	?	180	?
T Associations					
Ori T2	400	132	28	800	?
Tau T1	180	52	?	50?	?
Globular Clusters					
M 4	2 000	550	9	150 000	$\sim 1.4 \times 10^{10}$
M 22	3 500	460	9	530 000	$\sim 1.4 \times 10^{10}$
47 Tuc	4 400	3 100	5	1 600 000	$\sim 1.4 \times 10^{10}$
M 13	6 600	4 300	11	660 000	$\sim 1.4 \times 10^{10}$
M 5	7 600	5 500	12	850 000	$\sim 1.4 \times 10^{10}$
M 3	10 000	9 900	13	1 100 000	$\sim 1.4 \times 10^{10}$

Source: Globular cluster data from C. J. Peterson (1986, private communication).
^aZ = perpendicular distance north or south of Milky Way plane.

of a loose open cluster centered only about 22 pc away toward the constellation Ursa Major, many of whose stars belong to this cluster, as shown by Figure 22-1. The best-known clusters, the Hyades and the Pleiades (see Figure 18-9, page 394), lie 12° apart in our winter evening sky, about 42 and 127 pc away, respectively. Figure 22-2 shows the relation of the Hyades and Pleiades in the sky, while Figure 22-3 shows the Pleiades in more detail. About 900 open clusters are concentrated along the Milky Way band, indicating that they lie in the plane of our galaxy. (Open star clusters are sometimes called *galactic clusters* for this reason.)

As described in Chapter 21, stars form in open clusters, and most open clusters have prominent young stars or associated clouds of star-spawning gas. Then why aren't all stars in clusters? The reason is that most open clusters break apart into individual stars within only a few hundred million years because of dynamic forces acting on them. In comparison with most cosmic lifetimes, open clusters are short-lived.

Associations

Associations are cousins of open clusters. They often have fewer stars, but are larger in size and have a looser structure. Some large associations include an open star cluster within them. They may have 10 to a few hundred members and diameters of about 10 to 100 pc. They

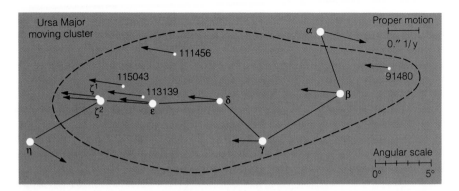

Figure 22-1 Some members of the closest open cluster, the Ursa Major cluster, form part of the familiar figure of the Big Dipper (solid lines). Arrows show proper motions of the stars, or the rates of angular motion relative to distant background stars. All but two of the stars in the figure are moving together at almost exactly the same rate, thus defining the cluster (dashed line). Stars at each end of the dipper are moving in different directions and are not true cluster members. Compare with Figure 1-4.

Figure 22-2 Two open clusters are prominent in this portion of the January evening sky. The Pleiades is the tight group in the inner box (disregard the outer box), and is about 127 pc away. Closer to us is the Hyades, a V-shaped group 12° to the upper left, about 42 pc away. The brightest star, at the upper left tip of the Hyades, is the 1st-magnitude red giant Aldebaran. The inner box shows the region of Figure 22-3. For general region, compare with lower right corner of Figure 18-9, page 394. (Ten-min exposure with 35-mm camera, f1.4, 50-mm lens, 2475 recording film.)

are rich in very young stars, such as O and B stars (which burn their fuel too fast to last long), or T Tauri stars (which evolve toward the main sequence too fast to last long). Associations are classified as **O associations** or **T associations,** depending on whether the prominent stars are O and B blue stars or T Tauri variables. Some 70 O associations were listed in a 1970 catalog. The general region of Figure 22-2 includes a T association.

The smallest associations grade into small, multi-ple-star-like groups, such as the Trapezium in the Orion Nebula. Sometimes called *trapezium systems,* these might be a link between multiple stars and small clusters. Like open clusters, associations are involved with regions of recent star formation and are short-lived.

Globular Star Clusters

Globular star clusters are quite different from the other two types. They are much more massive, more

Figure 22-3 The Pleiades, also known from mythology as the "Seven Sisters," is an open cluster of young stars approximately 80 to 150 million years old, according to estimates in 1990. It is about 127 pc away and about 2 pc in diameter. The brightest stars are massive, hot, blue stars. Dust and gas clouds around them scatter blue light, creating wispy blue reflection nebulae. (Copyright Anglo-Australian Telescope Board, courtesy D. Malin.)

tightly packed, more symmetrical, and very old. Figure 22-4 emphasizes the remarkable symmetry and compactness of globulars. They typically contain 20 000 to several million stars, although many of these stars crowd too close to be resolved by Earth-based telescopes, especially in the central regions.

Typical diameters of the central concentrations range from only 5 to 25 pc. To imagine conditions inside a globular cluster, picture 10 000 stars placed around the Sun at distances no farther than Alpha Centauri, our nearest star!

Even the nearest globular clusters are thousands of parsecs away from us. It is only because they have so many and such very bright stars that we see them at all. Yet a modest backyard telescope can reveal many prominent examples. The total of known globulars around our galaxy is about 150, and the total around our sister galaxy, M 31 in Andromeda, is at least 200. In three-dimensional space they are not confined to the galactic plane but are distributed in a spherical *halo* surrounding our galaxy. Similar distributions are found around other galaxies. They are among the oldest objects in our Milky Way galaxy, with ages estimated around 14 billion years.

DISCOVERIES AND CATALOGS OF CLUSTERS

Recognition of clusters is an enterprise spread out over many centuries, as shown in Table 22-2.

When the French astronomer Charles Messier tabulated nebulae in 1781, he included many star clusters that are still known by their "M numbers." Between 1864 and 1908 many clusters were cataloged in the New

Figure 22-4 Globular star cluster M 5 (NGC 5904). The cluster is estimated to be about 12 pc across and 9200 pc away. It contains roughly 60 000 stars. (Kitt Peak National Observatory.)

General Catalog (NGC) and the Index Catalog (IC) and became known by their NGC and IC numbers. Not until the 1920s was the difference between various types of clusters and the still more remote galaxies understood, and only then could reasonable research begin. Shapley's 1930 book was the first substantial work on star clusters.

MEASURING DISTANCES OF CLUSTERS

Because of clusters' enormous range of distances, astronomers have devised varied and sometimes ingenious methods to determine their distances. Details of all the methods are beyond the scope of this book, but a few examples can be given.

Parallax The most basic method of measuring star distances, the measurement of parallax (described in Chapter 16), is almost useless on clusters because it is not very accurate for distances beyond about 20 pc. Parallaxes give checks on distances of some stars in the Ursa Major cluster, but they mostly provide a foundation on which other distance-measuring methods are built.

Star Luminosity If observers can determine the luminosity of any star in a cluster, the distance from

TABLE 22-2

First Recognition of Selected Clusters

Cluster	Type	First Recognition
Pleiades	Open	Prehistoric[a]
Hyades	Open	Prehistoric[a]
ω Centauri	Globular	Ptolemy, c. A.D. 140[b]
Praesepe	Open	Galileo, c. 1611
M 22	Globular	Ihle, 1665
M 11	Open	Kirch, 1681
M 5	Globular	Kirch, 1702
M 13	Globular	Halley, 1714
27 globular clusters and 29 open clusters, cataloged		Messier, 1781
I Persei	Association	Ambartsumian, 1949

Source: Data from Shapley (1930).
[a]First recognition as a star cluster similar to telescopic examples was in Messier's 1781 list.
[b]Not resolved into stars, but listed as a bright, fuzzy patch.

the Earth to the cluster can be estimated if there are no complications from obscuring interstellar matter. (This method is similar to the example in Chapter 16 of estimating the distance of a certain light if we know that it is a candle and that there is no intervening fog.) One way to determine luminosity is simply to use spectra to identify spectral types and to estimate luminosities from H–R diagrams that plot luminosity versus spectral type, as derived from nearby stars.

Cluster Diameter Once the linear diameters (in parsecs) of a few nearby clusters were known and found to be roughly comparable, remote clusters could be assumed to have the same linear diameter. Then the distances of remote clusters were estimated from their angular diameters (seconds of arc) by applying the small-angle equation and assuming that all clusters of similar appearance have similar linear size. If a globular cluster 10 pc across subtends $\frac{1}{10}°$, for example, it must be about 6000 pc away.

The Complication of Interstellar Obscuration

Before 1930, these methods were carried out under the assumption that all intervening interstellar space is transparent. However, early estimates of distances, brightnesses, and sizes of clusters gave inconsistent results. In 1930 Robert Trumpler used the cluster results to show that the problem lay in the erroneous assumption that space is clear. He showed that diffuse interstellar dust dims stars and clusters that are more than a few dozen parsecs away. This throws off the estimates of stars' luminosities and distances. Fortunately, the total amount of dimming can be estimated by measuring the amount of interstellar reddening, or color change caused by the dust. The more reddening is observed, the greater the degree of dimming. Once the obscuration is measured and taken into account, distances can be accurately measured if the luminosity of any star or class of stars in the cluster is known.

Cepheid Variables as Distance Indicators

Luminosity is the key to distance measurement, and this brings us to the most useful technique for measuring distances of remote objects. As described in Chapter 19, Henrietta S. Leavitt discovered in 1912 that the periods and luminosities of **Cepheid variable stars** are correlated. The Cepheid variables have periods— the duration of their brightness variation—of 1 to 50 d. Once you measure the period of one of them, say 5 d, that tells you the luminosity of the star.

By the 1950s a small complication was discovered and overcome. The Cepheids found in open clusters form one group, called Type I Cepheids, and have a special period/luminosity relation. Those in globular clusters and in our galaxy's central bulge form another group, called Type II Cepheids, and have a somewhat different period/luminosity relation.

Thus measuring the distance of a cluster is a process requiring several steps. A Cepheid in the cluster must be located, its period and type measured, and its luminosity read from the appropriate period/luminosity diagram and combined with the apparent magnitude (corrected for interstellar obscuration) to calculate the distance. Few astronomers are all-equipped to carry out all the necessary observations, such as measuring light variations and periods, determining spectral properties, measuring absolute magnitudes, and measuring interstellar reddening and obscuration. Thus astronomers have specialized. Some study periods of variable stars; some make photometric measures of absolute brightness; some study interstellar reddening. The simple statement that cluster A is x parsecs from the Earth may represent years of work by many astronomers.

DISTRIBUTION OF CLUSTERS

Once the distances of clusters could be measured, their positions in three-dimensional space could be mapped simply by plotting their distances and directions. This process began in the late 1920s and 1930s, and striking results began rolling in at once. The distribution of clusters is not random, but defines the shape of our galaxy. The open clusters lie in a disk, and the globulars form a spherical cloud around the disk. Shapley, a pioneer in these measurements, said that the clusters reveal "the bony frame of our galaxy."

Even two-dimensional maps of cluster positions on the sky make the point. Figure 22-5 shows a map of the sky with the positions of open clusters marked. The open clusters stretch along the horizontal line that marks the Milky Way—our view along the disk of the galaxy. Figure 22-6 is a similar map of globular cluster positions. It shows that globulars lie in a ball-shaped swarm centered (in 3-dimensional space) on the galaxy's center. Most globular clusters thus lie in the part of our sky

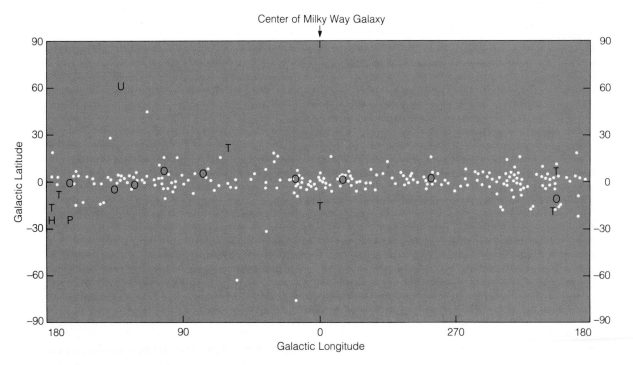

Figure 22-5 Map of the sky showing locations of selected open clusters (dots), O associations (O), and T associations (T). The Ursa Major, Hyades, and Pleiades clusters are indicated by U, H, and P, respectively. Open clusters concentrate along the Milky Way, whose equator is the zero-latitude line here.

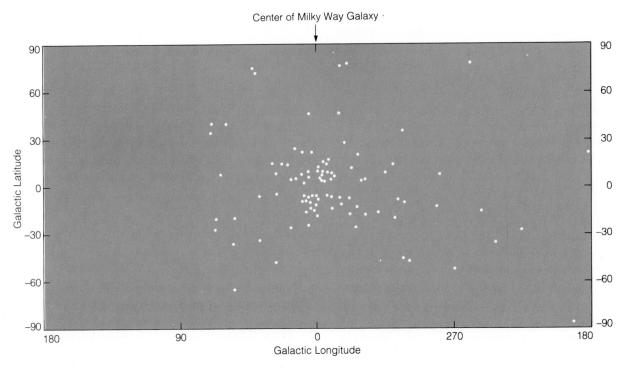

Center of Milky Way Galaxy

Figure 22-6 Map of the sky showing locations of the 93 most prominent globular star clusters. The coordinates of this map are the same as those in the preceding figure, with the center of the galaxy lying at the center of the figure. Globular clusters form a spherical swarm centering near the galaxy's center, as Shapley found using the same kind of data in 1918.

toward the galactic center. A good example is seen in Figure 22-7.

THE NATURE OF OPEN CLUSTERS AND ASSOCIATIONS

The preceding discussion mentioned that open clusters and associations are groups of stars formed from the contraction of large gas clouds, which later break apart into individual stars. We will now describe observations and theory shedding more light on the ages, evolution, and ultimate fate of these star groups.

Open Clusters: Ages and Ultimate Disruption

It is relatively straightforward but time consuming to construct the **H–R diagram of a cluster.** The apparent magnitudes and color temperatures of hundreds of stars must be measured. If the distance to the cluster is known, the apparent magnitudes can be converted to absolute magnitudes, or luminosities. Astronomers and their graduate students have plotted the H–R diagrams of many clusters.

The importance of such work, of course, lies in the evolutionary information it yields. Figure 22-8 brings together data on several open clusters. Marked along the main sequence is the age at which various stars evolve off the main sequence. (Compare with Figure 19-2, where the dashed lines show predicted positions of stars of various ages.) As discussed in Chapter 19, the point at which stars have left the main sequence in a cluster is a measure of the cluster's age. Note in Figure 22-8 that some clusters, such as NGC 2362, are so young that hardly any stars have evolved off the main sequence. **Ages of open clusters** range from about 1 or 2 million years for NGC 2362 to as much as 5 billion years for NGC 188 and 6 or 7 billion years for a cluster called Melotte 66 (Anthony-Twarog and others, 1979).

Figure 22-7 A globular cluster peeks through the clouds. Much of this picture is dominated by multicolored emission and reflection nebulae around stars in the constellation Ophiuchus. These are foreground clouds, lying between the solar system and the Milky Way's center. Through a gap in these clouds, we see one of the swarm of globular clusters (top edge, left center) that form a halo around the Milky Way's center. (Copyright Anglo-Australian Telescope Board.)

Among published ages for 27 open clusters, about half (55%) are less than 100 million years old. The solar system is nearly 50 times as old as this. It is interesting to think that we can examine a randomly chosen open cluster, such as that in Figure 22-9, and eventually state that it is only ⅟₅₀ (or some other fraction) as old as our own solar system.

These results confirm that in cosmic terms, *open clusters are mostly young,* and that (as stated in Chapter 18) star formation is continuing in open clusters to this day. Many open clusters, as in Figure 22-10, are associated with dense nebulae where star formation is still proceeding. Often their brightest stars are the hot, blue O and B stars, which formed recently, burn their fuel fast, and cannot last long. In Chapter 18 two groupings were mentioned in which the least massive stars have

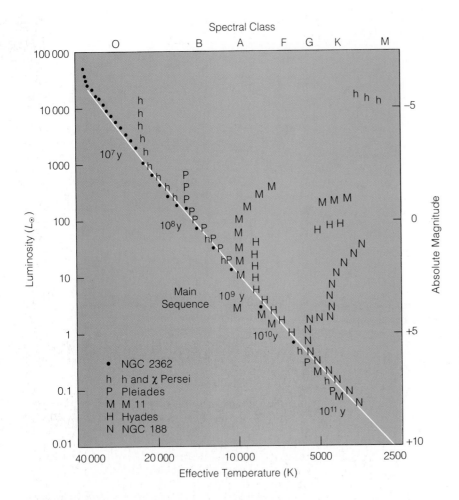

Figure 22-8 An H–R diagram for several open clusters. Numbers along the main sequence give ages in years for groups of stars turning off the main sequence. Clusters can be dated by measuring these turnoff points. (Based on the work of A. R. Sandage, O. J. Eggen, and L. Aller.)

Figure 22-9 Open cluster NGC 3293 in the constellation Carina. The cluster is about 8 minutes of arc across and contains about 50 stars brighter than 13th magnitude. At least two of the brightest are red giants, indicating that enough time has passed since the cluster's formation for a few of the more massive stars to evolve into giant status. (Copyright Anglo-Australian Telescope Board.)

not yet even evolved onto the main sequence. These examples, the Trapezium association (Figure 20-19b) and cluster NGC 2264 (Figures 18-17 and 18-18), are only about 1 and 3 million years old, respectively. In slightly older clusters some of the more massive stars may have evolved into red giants, adding interesting color effects. The cluster in Figure 22-9, for example, has two bright orangish giants among a field of young blue stars.

Studies of the Pleiades suggest ages averaging 2.5 $\times 10^8$ y for the low-mass stars and only 7×10^7 y for the high-mass stars. Some researchers believe that low-mass stars formed first in open clusters, with formation of massive stars continuing or occurring later. Such studies may be able to resolve the exact sequence of star-forming processes as a giant contracting cloud breaks up into a cluster of stars.

Velocities of cluster stars, measured by Doppler

Figure 22-10 Because stars form in clusters, many open clusters are still surrounded by the nebulosity that gave them birth. Here the impressive Lagoon Nebula (M 8 in the constellation Sagittarius) is a 25-pc-wide cloud punctuated by the bluish-white brightest stars of the open cluster to which it is giving birth. The complex is about 1200 pc away. (Copyright Association of Universities for Research in Astronomy; NOAO.)

shifts, indicate that some clusters are expanding or losing members, or both. Some high-velocity stars exceed the escape velocity of their parent clusters. Thus these clusters are breaking apart as we watch. Clusters tend to disperse after a few hundred million or a billion years. This **disruption of open clusters** occurs by several simultaneous mechanisms. Fast-moving stars, or stars that are accelerated by interaction with others, may escape. This reduces the mass and gravitational self-attraction of the cluster, so that other stars can escape. According to Kepler's laws, stars on the side of the cluster closer to the galactic center orbit around the galactic center faster than stars on the other side, thus stretching and shearing the cluster. The Hyades, for

example, are barely stable now, and the outer parts of the Pleiades are already dissipating, though the central, tighter grouping may be stable.

Associations: Ages and Ultimate Disruption

Like open clusters, associations are young. Because of their loose structure, they may break up even faster than ordinary open clusters. For example, a T association of eight T Tauri variables, about 100 pc away and 25 pc across, has been estimated to be only about 10 million years old; it may soon break apart because of the tidal gravitational forces of the galaxy acting on the

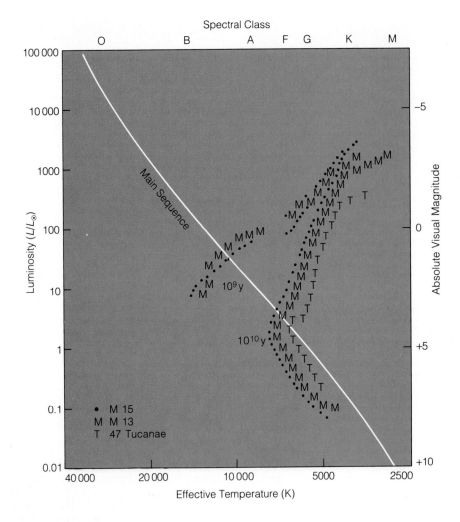

Figure 22-11 An H–R diagram for three selected globular clusters. Estimated main-sequence turnoff positions are indicated for ages of 10^9 and 10^{10} y. Because globular clusters' stars have fewer heavy elements than stars in open clusters and near the Sun (see next chapter), their main sequences are shifted leftward from the reference line, which is derived from open clusters and nearby stars. (Based on data of A. R. Sandage.)

cluster. Most associations cannot last more than a few tens of millions of years, because of the disruptive tidal forces and the tendency for their stars to follow individual orbits around the center of the galaxy.

In many associations there are large masses of neutral hydrogen gas, some of which exceed the mass of the stars. This material adds more gravitational attraction for the member stars and may help hold the associations together somewhat longer than would otherwise be the case. The hydrogen may be debris left over from the formation of incorporated stars or it may be material ejected from the fastest-evolving stars.

THE NATURE OF GLOBULAR CLUSTERS

The technique of H–R diagram analysis is especially well suited to determining the ages of globular clusters

because they have well-defined turnoff points along the main sequence. It turns out that globulars are extraordinarily old. Star formation has ceased in them. In all globulars of our galaxy, the O, B, and A stars have already evolved off the main sequence and have become red giants, as seen in Figure 22-11. For this reason, *the bright stars in almost all globulars have a reddish color and there are no bright blue stars.*

Assigning numerical ages to globulars must be done with some care, because spectra show that their stars contain fewer heavy elements than the more familiar stars of the solar neighborhood. This means that the details of their inner processes of energy generation and transport are different. Since the evolutionary time scales may thus be different, different calculations are used to determine the ages corresponding to various main-sequence turnoff points.

The best techniques set the **age of globular clusters** at about 14 ± 4 billion years (Jones and Dem-

Figure 22-12 Globular cluster M 22 (NGC 6656) in Sagittarius. It is about 9 pc in diameter and 3000 pc away—one of the closest globular clusters. It has an estimated mass of 7 million solar masses, making it one of the most massive clusters. (Observatory of New Mexico State University.)

arque, 1983; King, 1985; VandenBerg and Durrell, 1990). There has been some argument about whether they show a range of ages, from about 11 to 17 By, but the 1990 results cited above suggest that they all date from a single, brief era when our whole galaxy formed. We will return to this important relation between globulars and our galaxy in the next chapter.

Shapes of Globular Clusters

Why are globular clusters globular in shape, as seen in Figures 22-4 and 22-12? Many cosmic systems form flat disks because of their rotation—for example, the rings of Saturn, the solar system, the primeval solar nebula, and even the galactic disk itself. *A disk is the final stable state for a contracting, rotating, self-gravitating system with substantial initial rotation.*

The final shape of a system is determined by how much rotation (or more exactly, angular momentum) it

had when it started to form. The real surprise would be if it had zero rotation, causing a perfectly spherical shape. Globular clusters are not spherical, however, but are slightly flattened. They rotate slowly with rotation axes distributed approximately at random with respect to the galaxy.

These results imply that globulars formed from gas clouds with random directions of rotation and less rotational motion (less turbulence?) than in the present interstellar gas. This difference may in turn be due to differences between the present galactic gas and the conditions some 14 billion years ago when the globulars, and the Milky Way galaxy as well, were still forming.

Inside a globular cluster the orbits of individual stars must be very complex. The cluster's overall gravity field, the spatial distribution of its stars, their relative speeds, and the effects of near encounters among stars are all-important in determining how an individual star orbits in complex loops around the globular's central regions. Even in these crowded conditions, however, actual collisions between two stars are very rare or nonexistent.

Furthermore, as the cluster orbits the galaxy, it passes through the galactic disk every 100 million years or so. We can imagine that such passages might allow spectacular close-up views of globular clusters from planets in the galactic disk, as shown in Figure 22-13. During such a passage, shock waves from collisions with galactic gas and dust cause important effects, such as sweeping gas out of the globular. (The galactic disk, being much larger, is little affected, though some gas may be dragged out of the galactic plane.)

Because of their dense population of stars, globular clusters are too tightly bound by gravity to have dispersed within the history of the universe. Nevertheless, a few stars probably do escape from their outer regions. These account for some, but not all, of the scattered individual stars above and below the galactic plane.

Globular Clusters as X-Ray Sources

Satellites above the Earth's obscuring atmosphere have mapped celestial sources of high-energy X radiation. Most sources are in the galactic disk, perhaps associated with binary pairs in which one star dumps material onto another. But by 1975 at least five globular clusters had been found to be sites of strong X-ray radiation. Although the spectra of the X radiation are similar to those found in other sources, the **globular cluster X**

Figure 22-13 Imaginary view from a planet about 60 pc from a globular cluster. In such a case, the globular cluster would dominate the sky in a blaze of pinkish stars. Such a planet is unlikely to circle a star within the cluster, since globular clusters are deficient in the heavy elements that form solid planets; however, such scenes may occur as globular clusters pass through the galactic disk near stars that may (hypothetically) have planets. (Painting by Chesley Bonestell.)

radiation does not have the smooth periodic variations that are caused by orbital motions in binary pairs. Instead they have irregular variations, sometimes over weeks or months, but sometimes doubling in intensity within a few minutes. In NGC 6440, rates of energy radiation in the X rays alone reach as much as 7×10^{30} J/s (17 000 times the Sun's total luminosity).

These surprising findings led to new thinking about conditions inside globular clusters. An early idea was that crowded stars in the cores of globular clusters might collapse and merge together into giant black holes whose accretion disks emit the X rays. Generally, however, the X-ray sources are not exactly at the centers of clusters. They seem to come from binary systems. Studies in the 1980s showed that inner core regions of globulars are even more crowded with stars than originally thought and that binary pairs may form more often than usual under these conditions. One idea is that globulars may contain many neutron stars and other "burnt-out" stellar remnants and, because of the overcrowding, these may often capture passing stars into binary orbits. As the captured star evolves and blows off mass, gas is transferred onto the dense stars, and the resulting high-energy impact of gas onto the neutron star or its accretion disk may produce X rays.

There is little doubt that the dense packing of stars in the central few cubic parsecs of a globular cluster make these regions extraordinary stellar environments. New X-ray satellites and other instruments will clarify these conditions and try to establish more clearly the exotic processes occurring inside globular clusters.

ORIGIN OF CLUSTERS AND ASSOCIATIONS

Figure 18-2 and its accompanying theory of gravitational collapse give an excellent introduction to understanding how different types of clusters formed. The gas in halos around newly formed galaxies was extremely thin, with densities perhaps 10^{-22} to 10^{-26} kg/m^3. This means that large masses around 10^{35} kg became gravitationally unstable and began to contract into discrete entities. These masses—around 10^5 $M_\odot$—became globular clusters on the fringes of galaxies, probably around 14 billion years ago. In cool dust clouds in the disks of galaxies, gas densities were higher, around 10^{-17} kg/m^3. This led to the contraction of clouds containing around 10^{32} kg. These masses—a few hundred solar masses—became open clusters and associations strewn through galactic disks. Although most open clusters are young due to their rapid breakup, the oldest examples are around 8 billion years old (Jones and Demarque, 1983); thus we can be sure open clusters began forming at least that long ago. Contracting clouds of these masses ultimately broke up into individual stars rather than a single short-lived superstar of, say, 10^4 $M_\odot$.

The American theorists P. Peebles and R. Dicke (1968) and their Russian colleague T. Ruzmaikina (1972) have suggested that many globular clusters may have formed as part of the process of collapse of protogalaxies. More recent studies have showed that the fraction of heavy elements in stars of globular clusters is near

zero for the furthest globulars, out to at least 100 kpc, and rises as we consider globulars with orbits closer to the Milky Way galaxy (Pilachowski, 1984). As we will see in more detail in the next chapter, the earliest star-forming gas had the least heavy elements. Thus this evidence implies that globular clusters did begin forming during or before the earliest stages of protogalactic contraction, but the formation of globulars continued as the galaxy contracted toward its present disk shape.

SUMMARY

Although three types of star groups have been defined, they can be grouped in two main categories. The open clusters and associations are young groups of about 10 to 1000 newly formed stars located in the galactic disk. The globular clusters, which are old groups of 20 000 to 1 million stars, are located within a spherical volume above and below the galactic plane, which is centered on the galaxy's center. They formed about 14 ± 4 billion years ago.

Star clusters have three major areas of importance. First, they have played an extremely important role in mapping our galaxy and clarifying its history. Second, they clarify stellar evolution by presenting groups of stars formed at about the same time, so that their H–R diagrams clarify evolutionary tracks. Third, their stars reveal two different stellar populations in our galaxy. Stars in the disk and in open clusters and associations have solar-type compositions, with a few percent heavy elements, a certain type of Cepheid variable, and other distinctive properties. Stars in globular clusters have very few heavy elements, a different type of Cepheid, and other distinctive properties. The importance of these properties of clusters will be clarified in the next chapter as we turn our attention to our galaxy as a whole.

CONCEPTS

open star cluster	H–R diagram of a cluster
association	age of open clusters
O association	disruption of open clusters
T association	age of globular clusters
globular star cluster	globular cluster X radiation
Cepheid variable star	

PROBLEMS

1. Why are O and B stars the brightest in open clusters? Why are red giants the brightest stars in globulars?

2. If you saw the galaxy from a great distance, which would be brighter, open or globular clusters? Which redder? Which farther from the galactic disk?

3. Sketch the H–R diagrams of open and globular clusters and associations.

4. Describe a view of the sky near the center of a globular cluster.

ADVANCED PROBLEMS

5. If the Pleiades have 350 stars in a diameter of 4 pc, how many stars per cubic parsec are there? Roughly how far apart are the stars? Compare these numbers with those in the neighborhood of the Sun. (*Hint:* Volume of a sphere $= \frac{1}{3} \pi r^3$.)

6. Assuming globular cluster M 3 has 200 000 stars in a diameter of 13 pc, make the same comparison with the solar neighborhood as in Problem 5.

7. If a telescope could resolve 1 second of arc, what would be the smallest details it could reveal in:
   a. An open cluster 1000 pc away?
   b. A globular cluster 10 000 pc away?

8. A star is in a circular orbit around a globular cluster:
   a. What would be its orbital velocity? Assume a clus-

ter mass of 300 000 $M_\odot$ (6×10^{35} kg) and a radius of 5 pc (1.5×10^{17} m).
   b. How might this velocity be measured with a spectrometer?

9. Use the distance and angular scale given in the caption of Figure 22-2 to get a rough confirmation of the size of the Pleiades diameter quoted in the caption of Figure 22-3.

PROJECTS

1. Observe the Pleiades with your naked eye and make a sketch. How many stars can you count in the group? Can you see all "Seven Sisters"? (The number of stars seen depends on keenness of vision, darkness of the observing site, and the clarity of the atmosphere.)

2. Observe the Pleiades and Hyades or open clusters h and χ Persei in a telescope. Move the telescope and compare star fields in and out of the cluster. Estimate how many times more stars are in the cluster than in the background region.

3. Locate a globular cluster with the telescope. Make a sketch. Can you resolve individual stars? Compare the view in the telescope with photos, where the central region is often overexposed and "burned out."

Galaxies

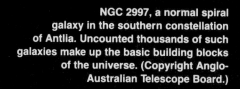

NGC 2997, a normal spiral galaxy in the southern constellation of Antlia. Uncounted thousands of such galaxies make up the basic building blocks of the universe. (Copyright Anglo-Australian Telescope Board.)

The Milky Way Galaxy

Our exploration of space has taken us out to distances of a few thousand parsecs. By looking at the distribution of stars and clusters throughout volumes of this size, we begin to perceive the **Milky Way galaxy.** To a remote observer the Milky Way would be a disk with a central bulge. The disk is about 30 000 pc across, 400 pc thick, and packed with open clusters, individual stars, dust, and gas, mostly arranged in ragged spiral arms. Globular clusters surround the disk in a spherical swarm concentrated toward the center of the disk.

These distances are difficult to comprehend. In a model of the Milky Way galaxy the size of North America, stars like the sun would be microscopic specks less than a thousandth of a centimeter across and scattered a block apart. The solar system would fit in a saucer.

The view from the Earth is not from the outside, of course, but from the inside of the galaxy. From a point partway out in the disk, we see a band of unresolved, faint, distant stars when we look out along the plane of the disk. This is the Milky Way, shown in a wide-angle photographic view in Figure 23-1.

DISCOVERING AND MAPPING THE GALACTIC DISK

Even before the invention of the telescope, people could plainly see a band of light arching across midnight skies at certain seasons.[1] Democritus (c. 400 B.C.) correctly

---

[1]The naked-eye prominence of the Milky Way is unknown to most modern urbanites. It can't be overemphasized that faint celestial displays must be viewed away from urban lights. My own astonishment at the Milky Way's clarity was greatest when I was living in a deserted region at about 3 km altitude on Mauna Kea volcano in Hawaii, prior to the establishment of the observatory at that site. One night when my eyes were already dark-adapted and I stepped outside, I thought the sky was partly cloudy, because I

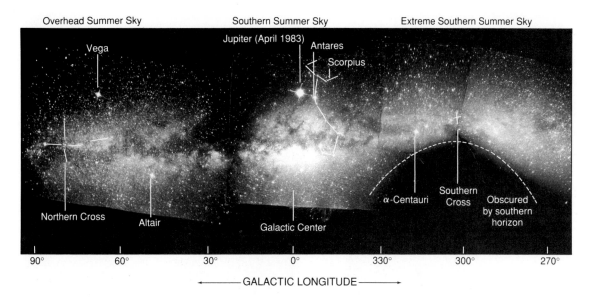

Overhead Summer Sky   Southern Summer Sky   Extreme Southern Summer Sky

Jupiter (April 1983)   Antares

Vega   Scorpius

Northern Cross   Altair   Galactic Center   α-Centauri   Southern Cross   Obscured by southern horizon

90°   60°   30°   0°   330°   300°   270°

←——— GALACTIC LONGITUDE ———→

Figure 23-1 A panorama of the inner Milky Way galaxy from our position near the outer edge. This 180° view clearly shows the bright central bulge and the ragged dust clouds that lie along the plane of the galaxy between us and the center. The galaxy's nucleus is hidden behind dark dust clouds at galactic longitude 0°. Seasons of visibility for the different regions, as seen in U.S. evening skies, are at the top. On the date of the photo, Jupiter happened to lie north of the galactic center. (Mosaic of photos from Hawaii by author, exposures 32 to 77 min with 24-mm wide-angle lens, f2.8, 35-mm camera with commercially available 2475 Recording film.)

attributed this glow to a mass of unresolved stars, which came to be called the *Via Lactea,* or Milky Way. In 1610 Galileo turned his telescope on the Milky Way and confirmed Democritus' idea. The proof is seen in Figure 23-2. Galileo wrote (quoted by Shapley and Howarth, 1929):

The galaxy is nothing else but a mass of innumerable stars planted together in clusters. Upon whatever part of it you direct the telescope, straightaway a vast crowd of stars presents itself to view.

In 1750, the English theologian Thomas Wright correctly hypothesized that the galaxy must be a slablike arrangement of stars. Other theoreticians, such as

Immanuel Kant, analyzed this idea more mathematically with Newton's laws in the 1750s and 1760s.[2]

Herschel's Star Counts

Around 1773, a German-born composer and musician in England, William Herschel, bought some astronomy books and began building his own telescopes. Fascinated by the stars, he finally shifted his career from music to astronomy, and within a few years built a telescope with a 1.2-m mirror (a size not surpassed until the 1840s). Backed by what we would call federal support—an annual stipend from King George III beginning in 1782 (after Herschel discovered Uranus)—Her-

could see dark patches blotting out parts of the softly glowing Milky Way. Then I realized the air was crystal-clear. The dark patches were clouds, all right, but instead of being clouds of water droplets 1 km away, they were clouds of dust grains 100 million billion kilometers away! Such a view makes one believe we really do live in an immense, disk-shaped system of dust, luminous nebulae, and distant stars!

_____

[2]In an excellent, nontechnical account of these early theories, C. A. Whitney (1971) points out that Wright's work is now little remembered because it contained much metaphysical speculation. Whitney calls both Newton and Wright "astrotheologians. . . . Newton wished to discover God by studying the universe; Wright wished to discover the universe by studying God."

Figure 23-2 Proof that the Milky Way is composed of individual stars. To the naked eye, the Milky Way in this region of the constellation Sagittarius (looking toward the galactic center) displays fuzzy glowing clouds. This telescopic photo reveals that the clouds comprise innumerable distant stars. The central brightest star cloud, cataloged as M 24, is part of a spiral arm 5000 pc away, between us and the galactic center. At the top is the red-glowing nebula M 17, about 1800 pc away. The vertical height of the picture is 5°. (Copyright Anglo-Australian Observatory.)

a

b

Figure 23-3 A comparison of views looking "up" out of the galactic disk toward the north galactic pole, and across the disk toward the galactic center. The disk is so thin that the view "upward" out of our region shows few stars and no star clouds. **a** Region of north galactic pole in the constellation Coma Berenices. **b** Region of the galactic center, about 7500 pc away, but obscured by gas, dust, and stars. (Both exposures have the same angular width, about 40°, and were 15-min exposures on Ektachrome ISO 1600, 55-mm lens, f2.8, by author.)

schel used the following method to figure out what he called "the construction of the heavens."

He swept the skies, counting stars in each direction. He assumed that the fainter the stars he could see in any direction, the farther away they were. He found far more faint stars in the Milky Way than in other directions (Figure 23-3). He correctly took this as evidence that stars are scattered farther from us along the Milky Way than in other directions. He correctly mapped the

Milky Way system as a disk of stars, with us inside it, but he did not know how big the disk was.

Further Studies of the Galactic Dimensions and Halo

Several historical studies of the Milky Way's dimensions are summarized in Figure 23-4. In 1918, the Dutch astronomer J. C. Kapteyn tried to refine Herschel's star

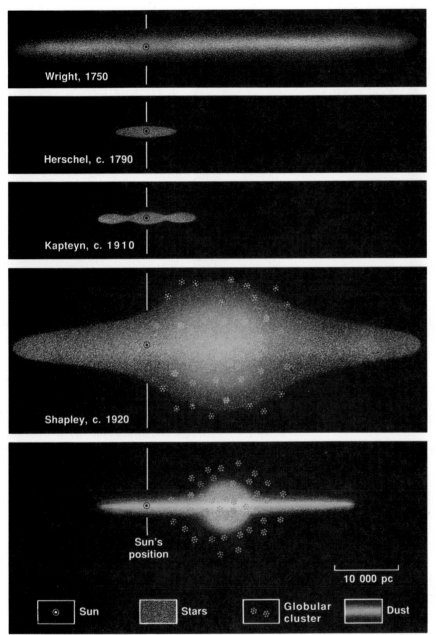

Wright, 1750

Herschel, c. 1790

Kapteyn, c. 1910

Shapley, c. 1920

Sun's
position

10 000 pc

Sun   Stars   Globular cluster   Dust

Figure 23-4 Two centuries of conceptions of our galaxy, showing approximate relative sizes of edge-on galactic models and the position attributed to the Sun.

counts, but he did not realize that interstellar dust prevented him from seeing beyond a few thousand parsecs. Thus his derived galactic disk was too small and erroneously Sun-centered.

The next major advance came as astronomers began to apply Henrietta Leavitt's 1912 discovery of the relation between the period and luminosity of Cepheid variable stars. This allowed them to estimate the distances of globular clusters, as described in the last chapter. By 1918 Harvard astronomer Harlow Shapley showed that globular clusters are distributed in a spheroidal swarm extending above and below the disk. This swarm of clusters, which also includes some sparsely scattered individual stars and gas, is called the **galactic halo.** It is shown in the lower parts of Figure 23-4. Shapley also showed that the halo is centered not on the Sun but on

a **b**

Figure 23-5 Comparison of a wide-angle view of the Milky Way extending from the horizon (*a*) and a telescopic view of a distant edge-on galaxy (*b*), showing similarity in form as an observer looks edgewise through a galactic disk. **a** About 50° of the Milky Way near the central region in Sagittarius; compare with lower left of Figure 23-1. (35-mm camera; 24-mm lens at f2.8; 16-min exposure on 2475 Recording film; photo by author.) **b** A portion of the galaxy NGC 55. (Cerro Tololo Inter-American Observatory.)

a distant point in the disk in the direction of the constellation Sagittarius. He correctly hypothesized that this point is the center of the galaxy.

Further work refined estimates of the **dimensions of our galaxy.** According to current estimates, the disk is about 30 000 pc across with the Sun about 7500 pc from the center.

Correcting the Galaxy's Size

As described on page 443, Trumpler discovered the effects of dust obscuration in the Milky Way about 1930. Astronomers quickly corrected distance estimates for this and other effects, and as early as 1935 astronomers were deriving approximately the correct dimensions. According to current estimates, the Sun is about 6500 to 8500 pc from the center (Racine and Harris, 1989), and the overall diameter of our Galaxy is about 30 000 pc.

Proof of Other Galaxies

Another important step was recognizing other galaxies like our own. Until 1924, many astronomers thought that certain disk- or spiral-shaped glowing patches were nebulae of gas, like the Orion Nebula. Many were incorrectly called "spiral nebulae." In 1924, however, the new 2.5-m telescope on Mt. Wilson in California produced photos showing that at least one of these objects, the spiral-shaped object in the constellation Andromeda, consisted of very faint, distant stars, not gas. Most impor-

tantly, some of the stars were very faint Cepheids, which could be used to estimate the Andromeda galaxy's distance. The results revealed that the Andromeda object was far beyond any stars in our own galaxy. The Andromeda object and others were soon recognized to be complete galaxies like our own. These discoveries allow us to perceive galactic shapes all at once, instead of probing through the murk from inside the system. As shown in Figure 23-5, photos of other galaxies seen edge-on reveal striking similarity to our own view of our galaxy, seen edge-on from the inside.

Galactic Latitude and Longitude

A galactic coordinate system has been defined to help describe the sky and galaxy as seen from Earth. The **galactic equator** runs along the center of the Milky Way band. See the star maps following the index. The galactic north pole lies in the hemisphere of the sky containing the North Star. **Galactic longitude,** designated l, measures the angular distance around the Milky Way, starting from a zero point defined to lie at the galactic center in the constellation Sagittarius. The direction $l = 90°$ lies toward the constellation Cygnus, near the top of the Northern Cross. Longitude $l = 180°$, away from the center, lies toward Taurus, near the Pleiades and Hyades clusters; $l = 270°$ is somewhat south of Canis Major. **Galactic latitude** is designated b. Its zero value lies in the Milky Way on the galactic equator. The value $b = +90°$ is at the north galactic pole in a direction "straight up," out of the disk.

THE ROTATION OF THE GALAXY

We have noted that cosmic systems of particles tend to become flattened if they are rotating. The galaxy's flattened shape suggests that it, too, is rotating. All stars, including the Sun, are in fact orbiting around the massive central bulge. But which way is the galaxy turning? Are stars in our area moving toward Cygnus or toward Canis Major?

To answer, we need some frame of reference outside the galaxy itself. The distant galaxies provide such a frame, and their Doppler shifts reveal a systematic motion of the Sun and nearby stars toward Cygnus ($l \approx 90°$) and away from Canis Major ($l \approx 270°$). The velocity of the Sun in this direction is about 220 to 250 km/s. This is the orbital velocity of the Sun and nearby stars around the galactic center.

If the Sun travels at about 235 km/s, then the **Sun's revolution period,** or time required to travel all the way around our circular orbit, which has a radius of about 7500 pc, is nearly 200 million years.

Nearby stars are at nearly the same distance from the center as we are, and so they are moving around the center at almost the same speed as our Sun. Orbital speeds around the center are different at different distances from the center.[3] Both the linear speeds and angular speeds around the center vary at different distances, which means that no galaxy rotates as a solid disk. No spiral galaxy, for instance, rotates like a pinwheel painted on a phonograph record; rather, the inner parts turn faster than the outer parts. This difference in speed at different distances is called **differential rotation** of the galaxy. Differential rotation, together with random motions of stars (typically about 20 km/s), means that the stars of the galaxy do not move smoothly together but are ceaselessly changing their positions relative to each other as they move around the center.

THE AGE OF THE GALAXY

In the last chapter we saw that globular clusters have an average age of around 14 ± 4 billion years. Because globulars are believed to have been among the first objects formed as the galaxy itself took shape, this age of roughly 14 billion years is believed to be the approximate **age of the galaxy.**

MAPPING THE SPIRAL ARMS

As seen in the Part G opening photo on page 497, many galaxies have beautiful spiral patterns. **Spiral arms** of galaxies are spiral-shaped patterns formed by the brightest hot stars and their associated, bright emission nebulae.

Does our galaxy have spiral arms like the Andromeda galaxy and some other nearby galaxies? Yes. The objects that need to be mapped in order to seek **evidence of spiral structure** in our galaxy lie in the disk, not in the halo. Therefore, astronomers began mapping positions of bright, young stars of spectral types O and B, young clusters, and O associations, which are bright, easiest to detect over large distances, and good markers of star-forming sites. For example, if a cluster was found to be 1000 pc away toward longitude $l = 100°$ and latitude $b = 0°$, it could be plotted on a map of the galactic plane.

By the 1950s, maps similar to Figure 23-6 showed that open clusters, O and B stars, nebulae, and other objects lay in bands interpreted as local pieces of our galaxy's spiral arms. Modern studies of this type reveal parts of four nearby spiral arms, lying at an angle to the Sun–center line.

The spiral arms are named for the constellations in the directions of prominent features in each arm. The next arm beyond us (sometimes designated $+1$) is called the Perseus arm. Our arm (sometimes designated 0) is the Orion arm, or sometimes the Cygnus arm. The next arm in toward the center (-1) is the Sagittarius arm, and a suspected arm beyond it (-2) is the Centaurus, or Norma–Centaurus arm.

The View from a Spiral Arm

The maps show us to be located on the inner edge of a spiral arm. On an evening with a clear sky, rural observers have a commanding view of the galaxy from our home in the Orion arm. If you look at the brightest parts of the Milky Way in the southern summer sky,[4] as in

---

[3]The relation of orbital speed to distance departs from Kepler's third law, because the galaxy's mass is spread throughout many stars in the large central bulge, rather than being concentrated in one central object at the center, as in the case of the solar system.

---

[4]The seasonal and directional references are for Northern Hemisphere observers.

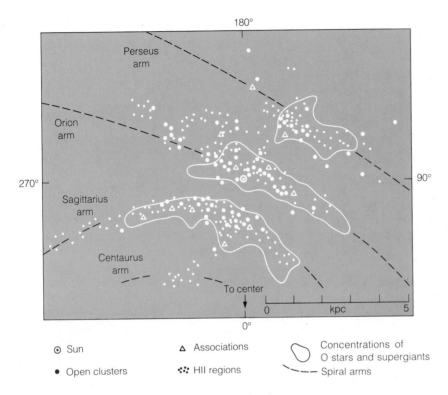

Figure 23-6 Evidence of spiral arm features. Galactic longitude (*l*) is plotted at the edges. Clusters, associations, HII gas clouds, and young stars are concentrated in spiral arms around a distant center toward *l* = 0°. (Adapted from data of Klare and Neckel, 1970; Moffat and Vogt, 1973; Walborn, 1973.)

Figure 23-7 This 60° view of the summer Milky Way shows the dark rift in the region of Cygnus (Northern Cross, outlined), caused by obscuring dust clouds between us and the distant parts of the Milky Way. Names identify bright stars comprising the "summer right triangle." Compare with the left end of Figure 23-1. (35-mm camera; 24-mm lens at f2.8; 16-min exposure on 2475 Recording film; photo by author.)

Figure 23-8 A 120° panorama of the winter Milky Way, looking away from the galactic center. Sirius and the Orion star-forming region lie in the direction looking down our local spiral arm. The Hyades and Pleiades clusters are in our arm in a direction opposite from the center. The *W*-shaped constellation Cassiopeia lies to the right. (35-mm camera; 15-mm fish-eye lens at f2.8; 25-min exposure on 2475 Recording film; photo by Floyd Herbert and author.)

the lower left of Figure 23-1, you are looking toward the center of the galaxy, or $l = 0°$, between the constellations Sagittarius and Scorpio. The true center itself is hidden behind 7500 pc of intervening dust and gas.

Higher in the sky, stretching toward Cygnus, the Milky Way is divided by the Great Rift, a band of nearby dark dust clouds that lie in the plane and obscure background stars. These can be seen in Figure 23-1 and in Figure 23-7 in a more detailed view. Overhead on a summer evening, the bright star clouds around Cygnus (sometimes called the Northern Cross) mark our view down our own spiral arm, about 70° from the center.

This is somewhat short of the 90° longitude because the arm is spiral, winding in toward the galactic center in this direction.

Because we live on the inner edge of a spiral arm, our view away from the galactic center toward $l = 180°$ directly crosses our arm. The view in this direction is spectacular on a winter evening, as seen in Figure 23-8. In this direction are open clusters of the Pleiades and Hyades—regions of recent star formation within our Orion arm. The concentration of bright, bluish stars in the region from the Pleiades through Orion to Sirius helps convince us that we are looking into a dense star

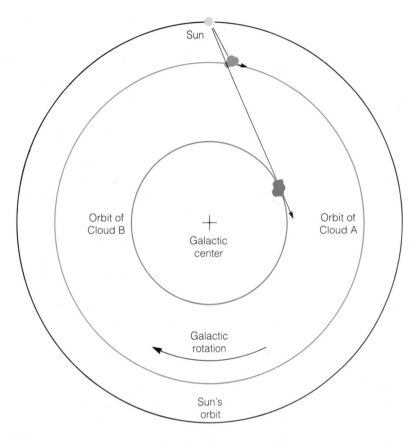

Figure 23-9 A view of the galactic plane from the north shows two hydrogen clouds, *A* and *B*, lying in nearly the same direction from the solar system. Different Doppler shifts, caused by different velocities V_A and V_B, allow radio astronomers to distinguish the positions of the clouds.

swarm—our spiral arm—in this direction. The central Orion Nebula and neighboring star associations are located about 500 pc away down the arm.

Mapping Distant Hydrogen Clouds by Radio

All the mapping just described reveals only the closest parts of nearby spiral arms of our galaxy—the regions not obscured by the interstellar dust. What about the rest of the galaxy? Is there any way to map the spiral features farther around the disk, such as on the other side of the center? Visual light is useless, because we can see only a few thousand parsecs through the dust. But radio waves pass much farther through the interstellar dust. The **21-cm radio waves** produced by neutral hydrogen, or HI, are especially useful for galactic mapping, because they allow us to detect HI clouds, which are concentrated in the spiral arms.

Suppose we start scanning along the Milky Way with a radio telescope tuned to the 21-cm wavelength. If we find a strong signal near, say, $l = 30°$, as in Figure 23-9, a concentration of HI must lie in this direction. To plot it on a map, we need its distance. We don't know whether it is nearby (cloud A) or far away (cloud B). Once again, the Doppler shift comes to the rescue, as seen in Figure 23-9. All clouds in the disk are revolving around the center, and objects nearer the center orbit somewhat faster than objects in our neighborhood. Thus the velocity of cloud B away from the Sun would be greater than that of cloud A. This effect is enhanced because V_B is directed along the line of sight, while V_A is not.

Radio astronomers have combined theoretical dynamical laws with observations of objects in our galaxy and in other galaxies to estimate the rotation rates in our galaxy as a function of distance from the center. These models give the velocity V_B of a cloud at position B, or V_A for a cloud at position A, and velocities for objects at all other positions in the galaxy. Thus, once a cloud's velocity is measured, its position can be plotted.

In this way HI clouds have been mapped over a large part of the galaxy. Their positions roughly define the spiral arms and offer our first view of our galaxy's spiral shape. Most of the galaxy's material is concentrated in a thin disk only about 200 to 400 pc thick. Some hydrogen clouds "above" or "below" the disk are falling toward the disk and some are moving away. They might be material dragged out of the disk by passing globular clusters or blown out of the chaotic central regions.

Radio astronomy roughly triples the distances to which we can probe in the galactic disk. We can map structures that lie some 15 000 pc away, even farther than the galactic center. However, there are regions directly on the far side of the center that we cannot map. Therefore, we cannot tell exactly how tightly wound the arms are or where the Milky Way falls in the range of forms of other galaxies. Nonetheless, using Figure 23-10 as a guide, we can say that if our galaxy could be seen from "above," it would probably look something like the view in Figure 23-11. The figures in the next chapters show that many other galaxies have a similar appearance.

Why Does the Galaxy Have Spiral Arms?

This question has puzzled astronomers for many years. Some plausible "common sense" explanations do not fit well with observations. For example, a common sense analogy might be a rotating garden sprinkler, where the water jets spray out in spiral arm forms. The trouble with this model is that most material in the galactic spiral arms is not moving radially outward from the central region, as the water droplets do.

A better analogy is a stirred cup of coffee with a few drops of cream. Because the coffee surface rotates faster at the center than at the rim, which slows it, the cream is sheared into long spiral streamers that look like galactic arms. The trouble with this model is that if arms are primordial features of the galaxy, twisted by rotation, they should be very old. Since the age of the galaxy is about 14 billion years and the rotation time is nearly 200 million years, there has been time for spiral arms to be twisted into about 70 complete windings. In contrast, observations show that spiral arms of various galaxies (including ours) consist of the youngest stars and clusters. They rarely show more than one complete winding.

Though the final answer is still uncertain, two modern theories help explain some features of galaxies. The **density–wave theory** emphasizes that spiral arms may not be fixed features of specific star groups, but rather waves in the galactic material. The crest of an ocean wave consists of certain molecules at one moment and certain other molecules the next, yet an outside observer sees a single wave that seems to have a history of its own. Just so, spiral arms may be persistent concentrations of material, with individual stars entering an arm, passing through, and finally emerging on the other side. The galactic gas tends to pile up in the spiral arms, reaching densities about 10% greater than that between the arms. Higher densities trigger more gravitational collapse, thus explaining why star formation occurs mainly in the arms. This in turn explains the prominence of the arms, since newly formed massive stars are a galaxy's brightest stars. According to this view, the pattern of spiral arms does not rotate at the orbital speed of its constituent stars, but more slowly, with different stars defining the arms at different times. In 1959, the Swedish astronomer Bertil Lindblad found that the spiral pattern of our galaxy rotates in about 480 million years in the spiral arm region beyond 4000 pc from the galaxy's center—a period about twice that of the Sun's orbital journey around the center. Because interstellar gas and dust overtake the arm and enter it on the inner edge of the spiral pattern, this theory also explains concentrations of star formation along the inner edges of arms.

A second theory is a modified coffee cup model,[5] which we might call the **chain reaction theory.** It is based on the fact that star formation has not happened smoothly and continuously since the galaxy formed, but rather in chain reaction bursts. As discussed in the last chapter, star formation occurs in open clusters, and (as shown in Figure 22-10, page 491) the expanding gas from massive supernovae in one cluster compresses neighboring clouds of gas and dust, initiating formation of new, adjacent clusters. Therefore, during a period of up to 100 million years, a large region of new clusters containing brilliant, hot, massive, short-lived stars may be produced in one part of a galaxy. During the galaxy's rotation in 200 million years or so, the inner edge of this region pulls ahead of the outer edge because of differential rotation (compare the rates of clouds A and B in Figure 23-9), and the region of bright, new stars is sheared into a spiral-trending segment. Following a few hundred million years, this arm segment runs out of new, young stars and fades, which explains why the spiral pattern rarely achieves more than one winding. This theory also explains why spiral arms frequently appear to be made up of bright segments, instead of a

---

[5]It was cumbersomely called the "stochastic, self-propagating, star-formation model" or "SSPSF model," by its original authors.

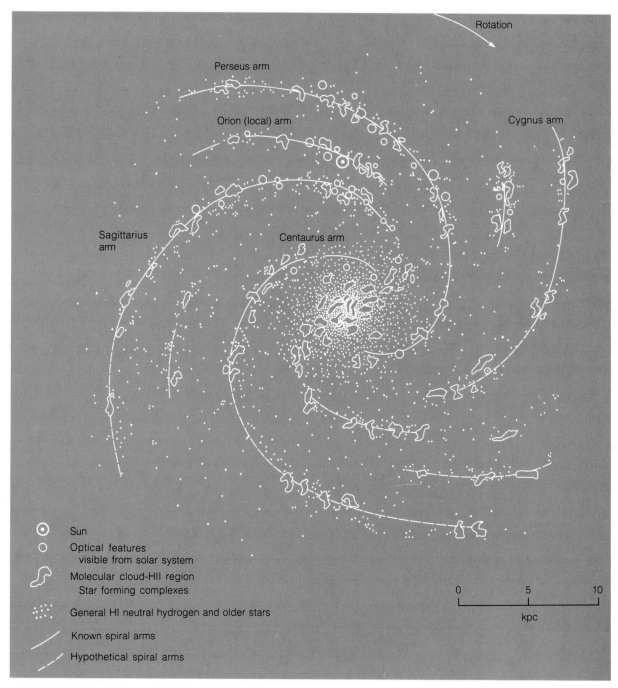

Figure 23-10 Currently known and estimated features of the Milky Way galaxy viewed from the north. Features nearest the Sun are the most certain; features on the far side are unmapped.

Figure 23-11 Our home, the Milky Way galaxy, as it might be seen from "above" the plane of the disk. The orientation is the same as in the preceding figure. The solar system would be only one microscopic dot among the stars of the Orion arm in the upper left center. The spiral arms are bluish Population I stars. The central region and a foreground globular cluster (upper right) are marked by reddish Population II stars. Brightest object is the nucleus. (Painting by author.)

continuous winding from the center outward to the edges of the galaxy.

This concept resembles the coffee cup model, but with new tiny droplets of cream being added intermittently instead of all the cream being added at the beginning. This theory also explains certain correlations of spiral shapes and rotation speeds observed in other galaxies.

In summary, both theories suggest that spiral arms are shifting features associated with star formation; their individual member stars come and go, but the pattern persists.

MEASURING THE GALAXY'S MASS

Knowing the Sun's velocity around the galactic center (about 235 km/s) and its distance from the center (about 9000 pc, or 2.7×10^{17} km), we can calculate the amount of mass in the central bulge around which we are orbiting. This is a simple application of the circular velocity equation (Chapter 4), giving a result of 4×10^{41} kg, or

200 billion solar masses in the central region, within our orbit.

Until the 1980s, this was thought to represent most of the mass of our galaxy. In the early 1980s, however, increasing evidence indicated that much more mass lies in a halo surrounding the galaxy. Recent estimates of the **mass of the galaxy** amount to 1000 billion (10^{12}) $M_{\odot}$ or even more. Discovery of a distant star falling at about 465 km/sec into the galaxy from 45 000 pc above the plane and 59 000 pc from the center gives an estimated total galactic mass of some 1400 billion $M_{\odot}$ (Hawkins, 1983). Much of this unseen halo mass may be gas and dust, with only 200–400 billion $M_{\odot}$ being luminous stars. In any case, we live in an enormous system!

COMPREHENDING GALACTIC DISTANCES

We have been casually discussing distances of thousands of parsecs, but it is good to pause and consider what these distances mean. Table 23-1 lists some distances to various objects in our own galaxy. To put them

TABLE 23·1

Distances to Selected Destinations in the Milky Way Galaxy

Destination	Distance (pc)	Travel Time of Light from Object to Earth (y)[a]
Nearest star beyond Sun	1.3	4.2
Sirius	2.7	8.8
Vega	8.1	26
Hyades cluster	41	134
Pleiades cluster	125	411
Central part of our spiral arm (Orion arm)	400	1 300
Orion Nebula	460	1 500
Edge of galactic disk in Z direction (perpendicular to plane)	1 000	3 300
Next-nearest spiral arm (Sagittarius arm)	1 200	3 900
47 Tucanae globular cluster	4 600	15 000
Center of galaxy	9 000	29 000
M 13 globular cluster[b]	11 000	36 000
Far edge of galaxy	24 000?	78 000?
Full diameter of galaxy	30 000?	98 000?

[a]These numbers, by definition, also equal the distance as expressed in light-years.

[b]Target of first beamed radio transmission from Earth directed to hypothetical extraterrestrial civilizations, November 1974. Signal will reach the cluster in A.D. 38 000.

in more meaningful terms, the right column lists the distances in light-years—the time their light takes to reach us (or, conversely, the time it would take us to reach these objects if we could travel at the speed of light). While we could reach the nearby stars in a few years even at less than the speed of light, we would have to build spacecraft to accommodate several generations of travelers to reach well-known open clusters. To reach distant spiral arms, globular clusters, or the galactic center would require more than the entire recorded history of humanity so far![6] If twentieth-century physics is correct in saying that no matter or energy can ever travel faster than the speed of light, then prospects for astronautical exploration of the rest of our galaxy seem indeed dim.[7]

When we consider objects in distant parts of our galaxy, we are dealing with distances of thousands of parsecs. For this reason, astronomers often use a unit of distance still larger than our earlier units of astronomical units, light-years, and parsecs. This unit is a **kiloparsec (kpc)**—note the use of the same prefix,

---

[6]As the spaceship approached the speed of light, the elapsed time experienced by the astronauts would decrease according to effects predicted by the theory of relativity. In any case, the sponsoring civilization back on Earth would have to wait centuries or millenia for the spaceship or its messages to return from such distant sites.

[7]But maybe twentieth-century physics is not the last word. After all, anybody in any other century who claimed to know the ultimate truth about physics and astronomy would later have been proved wrong. Historically, it has been risky to predict that some deed is impossible.

TABLE 23·2

Stellar Populations and Their Properties

Property	Extreme Population I	Intermediate Populations[a]	Halo Population II
Orbits	Circular	Elongated, perturbed by galaxy	Elliptical
Distribution	Patchy, spiral arms	Somewhat patchy	Smooth
Concentration toward galactic center	None	Slight	Strong
Typical Z range (pc)[b]	120	400	2000
Heavy elements (%)[c]	2–4	0.4–2	0.1
Total mass ($M_\odot$)	2×10^9	5×10^{10}	2×10^{10}
Typical ages (y)	10^8	10^9	10^{10}
Typical peculiar velocities (km/s)	10–20	20–100	120–200
Typical objects	Open clusters, associations, gas and dust, HII regions, O and B stars	Sun, RR Lyrae stars $(P < 0.4$ d),[d] A stars, planetary nebulae, giant stars, novae, long-period variables	Globular clusters, RR Lyrae stars $(P > 0.4$ d),[d] Population II Cepheids

Source: Data based on tabulations by D. O'Connell, A. Blaauw, J. Oort, C. Allen, and others.
[a]Includes older Population I and intermediate Population II stars.
[b]Z = distance above or below galactic plane.
[c]Elements heavier than helium, sometimes loosely referred to as metals.
[d]P = period of light variation.

kilo-, that is used throughout the metric system to indicate 1000:

$$1 \text{ kpc} = 1000 \text{ pc}$$

Galactic astronomers say, for example, that we are 9 kiloparsecs from the galactic center, and that our galaxy is roughly 30 kiloparsecs in diameter.

THE TWO POPULATIONS OF STARS

One of the most startling discoveries about our galaxy is that it contains a range of star types of different composition, age, distribution, and orbital geometry. These are commonly divided into two major groups called *Populations I and II*. **Population I stars** have composi-

tions similar to the Sun's, are relatively young, and are distributed in nearly circular orbits in the galactic disk. A few percent of the mass of Population I stars consist of "heavy elements"—the elements heavier than helium, including carbon, oxygen, silicon, and iron. **Population II stars** are nearly pure hydrogen and helium with only a fraction of a percent of heavy elements. They are old and are associated with the bulge of stars in the center of our galaxy and with globular clusters that have orbits taking them far above and below the galactic plane. You can recall the two types more easily by remembering that Population I stars were the *first* group astronomers became familiar with, since they are the type located near the Sun; Population II stars were discovered *second*. Some aspects of the two populations are shown in Table 23-2.

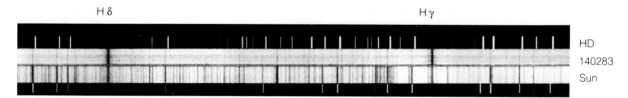

H δ   H γ   HD 140283 Sun

Figure 23-12 Comparison of spectra of a Population II star (upper, HD 140283) and a Population I star (lower, the Sun) of similar spectral type. (Bright lines at top and bottom are matching comparison spectra produced in the laboratory.) Both stars have prominent absorption lines of hydrogen (Hδ and Hγ), but the Sun has many additional absorption lines caused by various heavy elements. In the Population II star these lines are very weak or absent, indicating that the heavy elements are virtually absent from its gases.

The evidence for chemical difference between the two populations is seen in Figure 23-12, which compares spectra of a Population I G-type star (the Sun) and a Population II G-type star. They have hydrogen lines of equal strength, but the latter has very weak lines of metals and other heavy elements, showing that Population II stars have very low abundances of heavy elements.

Population II stars include a useful type of variable related to Cepheids but not found in Population I. They are called **RR Lyrae stars** and are useful because they all have a distinctive period around 0.3 to 1 d and a luminosity around 100 times that of the Sun (100 $L_\odot$). This means that they can be quickly recognized, and their apparent brightness can be used to measure the distance of the cluster or other system in which they lie.

Ages of Populations I and II

Ages of populations are especially intriguing. Population I stars, like the Sun and other stars in the galactic disk, have varied ages, from billions of years to zero age in regions where stars are still forming. But studies of H–R diagrams of Population II star clusters show that these stars are all around 14 ± 4 billion years old—an indication that the globular clusters and other objects in the halo formed during one era roughly 14 billion years ago.

Motions and Orbits of Populations I and II

Consider stars in our local region of the galactic disk. These Population I stars can be thought of as a swarm

having nearly random velocities of about 20 km/s, but all moving together around the distant galactic center at a speed of about 235 km/s. We might represent this by analogy with a swarm of gnats randomly darting among each other at 2 m/s, while the whole swarm is moving through the air at 20 m/s. These stars can be represented by the nearly concentric orbits shown in the upper part of Figure 23-13.

As early as 1785, William Herschel discovered the random motions of stars in our local swarm and measured the direction of the Sun's motion relative to the average of all other motions in the swarm. This drift of the Sun with respect to the local swarm is called the motion of the Sun toward the **solar apex,** a point in the sky about 10° from the star Vega. The Sun is moving in that direction at 19.7 km/s. In the gnat analogy, if you (representing the Sun) stood still in the middle of a hovering swarm of gnats (stars), you would see them darting about in random directions. But if you started walking toward the east (the solar apex), you could detect your motion by a net drift of gnats past you toward the west (away from the solar apex).

Once these measures were made, it was natural to measure the velocities of other stars relative to our **local standard of rest,** or **LSR**—an imaginary point moving in a circular orbit around the galactic center in such a way that it stays in the middle of our local swarm. The velocity of a star relative to the LSR is called its **peculiar velocity.**

Measurement of peculiar velocities led to the discovery of Population II stars. Since Population I stars move in circular orbits with the LSR, they have low peculiar velocities—random speeds of about 20 km/s. But as shown in the lower part of Figure 23-13, Population II stars do not share these motions; they have highly inclined, elongated orbits. As a result, astro-

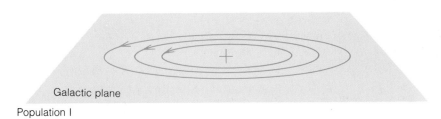

Galactic plane

Population I

Galactic plane

Population II

Figure 23-13 Comparison of orbits of disk Population I and halo Population II stars.

metric astronomers in the early 1900s began detecting a few stars speeding through the solar neighborhood with peculiar velocities sometimes exceeding 120 km/s. Prominent among these so-called **high-velocity stars** are the RR Lyrae variable stars mentioned above, now known to be associated with Population II. Why are Population II stars perceived as high-velocity stars? We (the Sun and LSR) move around the galactic center at 230 km/s (Figure 23-13, top). A Population I star moving with us seems to have low velocity, relative to us (like a neighboring car on a racetrack). But a Population II star moving at 230 km/s on an inclined or retrograde orbit (Figure 23-13, bottom) seems to zip by at high speed (like a car moving the wrong way around the track).

Mapping and Subdividing the Populations

By the 1930s, Harlow Shapley had shown that Population II consists of a diffuse swarm of individual stars and globular clusters orbiting in the regions near the galaxy, but usually outside the disk. In their journey around the galactic center, these stars pass through the disk twice on each orbit.

Figures 22-5 and 22-6, in the last chapter, revealed the two populations by plotting the distributions of different types of galactic objects on maps of the sky. The newly forming open clusters (Figure 22-5) cling tightly to the galactic equator since they belong to the disk's Population I, while the globulars (Figure 22-6), in Population II, swarm around the galactic center.

Many objects do not share the extreme properties of "pure" Population I or II. Therefore, as shown in Table 23-2, astronomers have subdivided the populations into additional intermediate subclasses.

Origin of Populations I and II

The **composition of Population II stars** *includes only a fraction of a percent of elements heavier than hydrogen and helium.* Yet such elements comprise about 2% of the Sun and other Population I stars. These elements

a

b

Figure 23-14 Views toward our galaxy's center, at three different wavelengths. Each has a scale calculated for the distance of the galactic center, though foreground stars and dust clouds also appear. **a** True color wide-angle view in visible light with an ordinary 35-mm camera shows the bright bulge of stars and nebulae 7500 pc away at the center, partly obscured by ragged, dark dust clouds between us and the center. (Angular width about 85°. Compare with narrower-angle view in Figure 23-3; 24-mm lens, f2.8, 16-min exposure on Ektachrome ISO 1600, by author.) **b** False color, even wider-angle view in near-infrared wavelengths of 1.2 to 3.4 μm. Light at these wavelengths passes through the dark clouds and reveals the thin disk and central bulge. Redder colors in this image represent longer wavelengths (redder light), generally associated with light that has passed through dust clouds. This view, from the Cosmic Background Explorer satellite, strikingly resembles edge-on views of other spiral galaxies. This is the Milky Way view we would see if our eyes were sensitive to somewhat redder light. A few foreground stars appear. (Angular width about 140°. NASA Goddard Space Flight Center, courtesy Nancy Boggess.) **c** Wavelengths 12 to 100 μm, still further into the infrared, are radiated primarily by cold dust at only 20 to 100 K. Thus the image shows no stars, but only dust tightly concentrated along the Milky Way plane. Red tones are coolest; yellow and green tones are warmer. The view, made by the Infrared Astronomical Satellite, shows a concentration of warmer dust at the center, heated by the central concentration of stars. Large and small boxes give approximate positions of Figures 23-16 and 23-17, respectively. (NASA.)

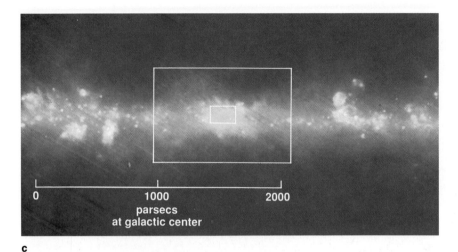

0   1000   2000
parsecs
at galactic center

c

include such important planet-forming materials as silicon, oxygen, nickel, and iron. Astronomers believe that Population II stars formed from gases with few heavy elements. Recall from the last chapter that the globular clusters in the farthest parts of the galactic halo have the least heavy elements. Thus the parent gas of their extreme Population II stars must have been a nearly pure mixture of 75% hydrogen and 25% helium. This conclusion is especially provocative since Population II objects are older than Population I objects.

What do these facts mean? Astronomers have agreed on the logical implication: When Population II formed some 14 billion years ago, the galaxy (or protogalaxy) must have been an extended spheroidal system consisting of about 75% hydrogen (by mass), 25% helium, and virtually no heavy elements. The volume of this system must have matched the volume now occupied by the most remote globular clusters in the outer halo.

These two conclusions—that the protogalactic cloud was a nearly pure hydrogen/helium mix and that later stars have more heavy elements—have a striking connection with the subject of element formation inside stars, described in Chapter 19. Nuclear reactions inside stars fuse light elements into heavier ones through successive stages of stellar evolution. Heavy elements are therefore *created* inside stars and then blasted by disruptive events (such as supernova explosions) from old stars into interstellar space. Then later generations of stars form from this heavy-element-enriched interstellar gas and create still more heavy atoms. As the protogalaxy contracted and the earliest stars evolved, more heavy elements were added to the gas. The last globular clusters to form had more heavy elements than the first

and finally, when the galaxy reached its present disk shape, nebulae and later generations of stars in the disk thus slowly grew richer and richer in heavy elements.

The two sets of data fit beautifully together. Observations show that the recently formed population contains many more heavy elements than the older population, and theories of stellar interior processes show why this is so. Observation of star populations thus confirms theories of stellar evolution.

PROBING THE GALACTIC CENTER

In many galaxies, most of the light comes from a brilliant central core, or **galactic nucleus,** buried in the heart of a large central bulge. The concentrated nucleus of the nearby Andromeda galaxy, for example, is as small as 5 pc, but it outshines the rest of the 40 000-pc-diameter disk! What mysteries are hidden in a galactic nucleus? Does our galaxy have such a nucleus? A large telescope does not help us to find out, because from our vantage point in a spiral arm, our galaxy's center is hidden behind kiloparsecs of dust, as seen in Figures 23-14 and 23-15. But the answer seems to be yes.

In the 1930s Bell Telephone Laboratories put the young researcher Karl Jansky to work on sources of radio static interfering with long-distance radio signals. Jansky built, in effect, the first radio telescope and discovered that one major source was a steady hiss from the Milky Way galaxy (Jansky, 1935). The technique was next taken up not by traditional optical astrono-

a

Figure 23-15 Color views toward the center of the galaxy emphasize vast star clouds and ragged intervening dust clouds. **a** This view shows about the same area as Figure 23-14a, revealing some red Hα nebulae and a general reddening of starlight due to the intervening dust. Brightest regions are somewhat overexposed, giving a whitish color. (Copyright Anglo-Australian Telescope Board.) **b** Shorter exposure emphasizes reddened color of the bright galactic center protruding from behind intervening dust. This is the brightest region at left side of a. (National Optical Astronomy Observatories.)

b

mers, but by an amateur astronomer and radio buff, Grote Reber, who built a radio telescope in his backyard and made the first radio maps of the Milky Way. He established that the strongest radio emission comes from the galactic center (Reber, 1944). This type of mapping succeeds in revealing the nucleus because radio waves (and some infrared wavelengths) can penetrate all the way from the galactic center, even though visible light is blocked by dust.

Subsequent radio, infrared, and gamma-ray studies identified a unique radio source at the center of our galaxy. It is called Sagittarius A—the brightest radio source in the constellation Sagittarius. A region about 150 by 300 pc on the east side of Sagittarius A emits nonthermal radiation presumed to be synchrotron radiation, which results from the interaction of ionized gas and strong magnetic fields (see Chapter 11). This region may be a supernova remnant. A magnetic field of about 10^{-5} gauss—two to five times stronger than the interstellar field in our region—has been estimated. A region about 35 by 80 pc centered on the west side of Sagittarius A is a concentration of dust clouds, HII regions, and molecular clouds rich in hydroxyl, ammonia, and other molecules, as sketched in Figure 23-16. The region is also crowded with red giants.

The western part of Sagittarius A (often called Sgr A West) itself is about 10 pc across with a still brighter core. This complex resembles the nucleus of the Andromeda galaxy. It contains some 60 million stars, together with HII regions and dust. Its gas clouds are ionized by radiation like that of stars as hot as 35 000 K (Lacy and others, 1980). Its central 1-pc core contains an amazing 2 to 3 million stars along with gas and dust (Lacy and others, 1980; Genzel and others, 1984)! This compares with 10 000 stars in the central parsec of a globular cluster and only one star within about a parsec of the Sun! In the galactic center stars may average only some 1600 AU apart; the sky would be ablaze with stars as bright as our moon!

ENERGETIC EVENTS IN THE GALACTIC NUCLEUS

The uniqueness of the nucleus is shown not only by the high density of stars and dust mentioned earlier, but also by evidence for the release of prodigious amounts

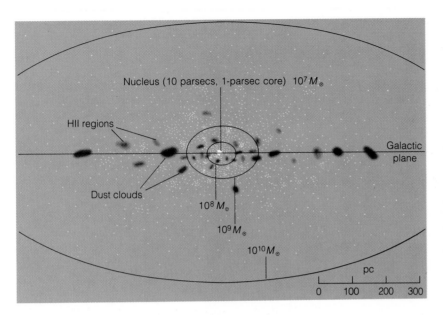

Nucleus (10 parsecs, 1-parsec core) $10^7 M_\odot$

HII regions

Dust clouds

$10^8 M_\odot$

$10^9 M_\odot$

$10^{10} M_\odot$

Galactic plane

pc

0   100   200   300

Figure 23-16 A schematic composite map of features near the center of the Milky Way. Contours contain the stated number of stars, indicating the highest densities of stars, dust, and gas clouds near the center. Nature of central 1-pc nucleus is still a mystery.

of energy. For example, a mass of hydrogen that resembles a spiral arm is expanding from the nucleus at a distance about 3000 pc from the center. This cloud, called the **3-kiloparsec arm** (or 3-kpc arm), contains about 10 million solar masses of neutral hydrogen. It revolves around the center at about 210 km/s, but also rushes outward toward the Sun at about 53 km/s. Thus to set this mass of gas in motion required about 10^{46} joules (J) of energy—over 1000 times the energy production of the Sun during its lifetime! Yet the 3-kpc arm is only about 10 to 100 million years old. Apparently a large energy release occurred "recently."

A second line of evidence for energetic events in the nucleus is simply the high luminosity of the nucleus, much of it in the form of infrared radiation from hot dust clouds. Some energy source must be available to heat these clouds.

A third line of evidence comes from other galaxies. The Andromeda nucleus is bright but relatively stable; yet certain galaxies emit strong radio radiation and jets of gas, indicating occasional explosions with energies comparable to the 10^{46} J needed to produce our own 3-kpc arm.

What **conditions in the galactic center** could produce the strange environment and energetic events just described? No one really knows, but this area of research is very active and exciting, partly because of increasingly detailed maps of the galactic nucleus being made by radio astronomers. There are two main theories.

The "Starburst" Theory

First is the "starburst" model, which proposes that episodes of very active star formation occur as dust and gas fall into the central region. Enormous, unstable "stars" of, say, 500 or even 1000 $M_\odot$ may form. As we saw in Chapter 19, such huge bodies would soon explode, creating powerful supernovae. This could explain the energies needed to create the motions and luminosity observed in and near the nucleus—for example, in the 3-kpc arm around 10 million to 100 million years ago. Analysts have proposed explosions of short-lived objects of as much as $4 \times 10^8 M_\odot$ (Lo and Claussen, 1983). The wisps in the left part of Figure 23-17, only 40 pc from the nucleus, might be gas thrown out of such explosions and forced into arc-shaped wisps by magnetic fields. Some astronomers have suggested that the nucleus is really a mighty cluster of stars, in which the supernova explosions occurred. According to Harvard analyst Paul Ho, such an explosion 10 000 to 100 000 y ago deflected wisps of gas toward a central massive object, causing them to be sheared out into long streamers (Waldrop, 1989). Luminosities as much as 20 million $L_\odot$ were needed in the central parsec to heat the presently observed gas and dust (Waldrop, 1985); such amazing energy production may be explained by supernovae. The supernovae might also have produced many neutron stars or black holes. Indeed, as early as 1973, Italian astrophysicist L. Maraschi and his colleagues

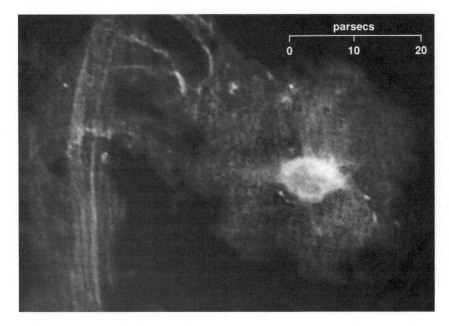

parsecs

0   10   20

Figure 23-17 Gas wisps (left) and shell surrounding the Milky Way's nucleus (blob at right). Wisps up to 50 pc long and only 1 pc wide are believed to be organized in this pattern by magnetic fields surrounding, and generated in, the nucleus. The process generating the fields, however, is uncertain. The radiation used to make this image is radio radiation at 20-cm wavelength. (National Radio Astronomy Observatory, operated by Associated Universities, Inc., under contract with the National Science Foundation. Observers: F. Yusef-Zadeh, M. Morris, D. Chance.)

proposed that 20 million neutron stars lurk within 350 pc of the nucleus—the product of intense supernova activity.

The Central Black Hole Theory

The most widely accepted theory about the nucleus of the galaxy is that it is not a star cluster, but a single huge black hole, of about a million solar masses (Lo, 1986; Genzel and Townes, 1987). If such an amazing object really exists, it would give us a good chance to learn more about black holes. The trouble is, it is hidden behind the 7500 pc of dust that lies between us and the galactic center. Nonetheless, various indirect observations support its existence.

For example, in the 1980s studies of star and gas motions near the center showed that a few million $M_\odot$ must be crammed into the central parsec (right side of Figure 23-17). Radio and infrared data indicated that the adjacent gas is not ionized enough to be explained by a hot stellar core in a central star cluster, as in the "starburst" theory. Instead, the data are consistent with a single central black hole of a few million $M_\odot$, surrounded by a huge, hot accretion disk, which is the central heat source. The accretion disk is heated by gas and dust spiraling into it at an estimated rate of 10 $M_\odot$ every million years. Turbulence and drag effects in the gas and dust around the nucleus dissipate energy, pre-

venting the gas and dust from staying in a perpetual orbit like Mercury's motion around the sun. Instead, the material spirals inward toward the nucleus. A 4-pc-wide spiral of about 60 $M_\odot$ of (infalling?) ionized gas has been discovered by radio mapping (Lo and Claussen, 1983), as shown in Figure 23-18. This may be our first view of the spiral drift of material toward the nucleus.

In support of these views are observations of gamma rays coming from the center (Levanthal and others, 1982). The gamma rays have just the wavelength created when electrons collide with positrons (positive particles of electron mass). This fits with the idea of energetic electrons and positrons boiling off the hot gas of an accretion disk surrounding a supermassive black hole.

Many astronomers believe that the black hole forming the nucleus lies in the densest core of Sagittarius A. Combination of signals from several radio telescopes shows this object is no larger than 20 AU across. It may be the accretion disk around the massive black hole (Geballe, 1979; Waldrop, 1985), but its nature is unclear and it is still being studied. Some astronomers believe that occasional clouds of material falling on it may explain the sporadic explosive outbursts in the galaxy's center.

Clearly, astronomers are still groping for explanations of what is happening in the center of the Milky Way, but new techniques of observing the center with balloon and satellite gamma-ray telescopes, ground-based radio telescopes, and infrared telescopes promise exciting results in the next few years. Meanwhile, even from

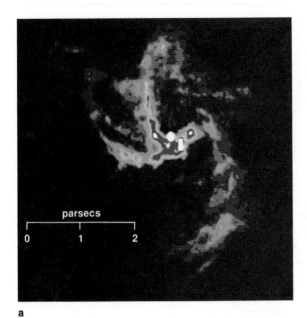

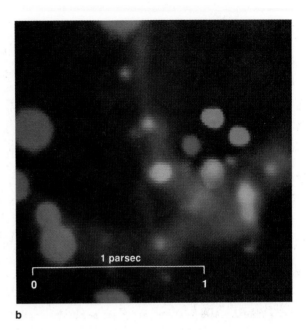

a   b

Figure 23-18 The central few parsecs of our galaxy. Radio astronomers discovered this small spiral, believed to be associated with matter spiraling into the accretion disk around the galaxy's massive nucleus (black hole?). The spiral appears tipped up toward us, out of the galactic plane, possibly indicating that the gas was chaotically disturbed (by an explosion?) before the accretion began and is now rotating in a new plane. **a** Radio image at 6-cm wavelength. Dark central features are due to image processing that contoured the image according to brightness levels; central region is actually brightest. (California Institute of Technology, courtesy K. Y. Lo.) **b** This enlarged false color image of the same region is a composite of radio data (redder tones) and shorter-wavelength infrared data (bluer tones). It reveals the radio-emitting spiral structure and several other sources that are probably infrared-emitting dust clouds. Such images allow us to map the features of the galactic nucleus at a wide range of wavelengths. (Copyright Anglo-Australian Telescope Board.)

what we know so far, we can imagine that the strange scene near the galaxy's nucleus must be extraordinary (Figure 23-19).

SUMMARY

Our discussion in the last few chapters can now be fitted into a compelling theory of the origin and evolution of the galaxy. Globular clusters give us the best estimate of the age of the galaxy—about 14 billion years. An independent method based on the decay of radioactive elements sampled in the Earth gives a similar result (Dicke, 1969).

What was happening 14 billion years ago? Evidently the galactic mass was a single cloud of hydrogen-rich gas, with probably 25% helium (by mass) and virtually no heavier elements. The mass of this protogalaxy equaled the present galactic mass, roughly 10^{12} $M_\odot$. The protogalaxy must have been rotating and contracting. When it reached

a rather spherical shape perhaps 30 to 40 kpc across, it became dense enough that separate, large, self-gravitating masses formed within it. These became globular clusters with certain characteristics we observe today:

Age: *about 14 billion years*

Composition: *few heavy elements*

Distribution: *spheroidal halo 30–40 kpc across; elongated orbits*

Mass: 10^5–10^7 $M_\odot$ *each.*

Because of rotation and collisions among the atoms and clouds of gas, the inner part of the protogalaxy, still containing some 10^{11} $M_\odot$, flattened into a disk shape. The globular clusters, too far apart to interact, were left behind and did not form a disk. Thus two populations (with some intermediate objects) arose from the earlier- and later-formed systems. The stars and gas in the disk took on the properties we see today in nearby space:

Figure 23-19 Hypothetical panorama from a position only a few hundred parsecs away from the Milky Way's nucleus. Here we are surrounded by thousands of brilliant red giants and other stars of Population II and by dense clouds of dark dust and glowing nebulosity lying along the Milky Way's plane. In the right distance is the actual nucleus—possibly a large black hole surrounded by a brilliantly glowing accretion disk and partly obscured by clouds. (Painting by author.)

Age: *youthful; a few billion years for stars; only tens of millions of years for nebulae*

Composition: *few percent heavy elements*

Distribution: *flat disk; 30 kpc across, 400 pc thick*

Mass: *stars formed in clusters of a few hundred $M_{\odot}$*

In the center, an extraordinary, massive black hole may have formed, and violent explosions may have taken place. In the disk, nebulae traveling on elliptical orbits quickly collided with other nebulae until the gas, dust, and nebulae all moved together in relatively circular orbits. The spiral arm pattern probably emerged after a number of galactic rotations, perhaps within a billion years. In the spiral arms, the densest clouds contracted and spawned associations and open star clusters. Each group broke apart into scattered stars a few hundred million years after its formation, but new star groups continued to form, so that the galaxy kept its present general appearance. Supernovae blew out gas laced with heavy elements created inside stars, so that later generations of stars had more heavy elements than the earlier stars. Perhaps 7 or 8 billion years after the galaxy's formation, in one of the spiral arms, an obscure star formed—our Sun—and in its surrounding dusty nebula, the Earth was born.

CONCEPTS

Milky Way galaxy

galactic halo

dimensions of the galaxy

galactic equator

galactic longitude

galactic latitude

Sun's revolution period

differential rotation

age of the galaxy

spiral arms

evidence of spiral structure

21-cm radio waves

density–wave theory

chain reaction theory

mass of the galaxy

kiloparsec (kpc)

Population I stars

Population II stars

RR Lyrae star

solar apex

local standard of rest (LSR)

peculiar velocity

high-velocity star

composition of Population II stars

galactic nucleus

3-kiloparsec arm

conditions in the galactic center

PROBLEMS

1. During what percent of human recorded history (define as you think appropriate) have people *not* known that we live in an isolated galaxy of stars similar to other remote galaxies?

2. From the appearance of the Milky Way, how do we know that the solar system is in the galactic disk and not far above or below it? How might the appearance of the central region differ in the latter case? (*Hint:* See Figures 23-1 and 23-14.)

3. Compare the shapes of the volumes occupied by the swarm of open clusters and by globular clusters. Relate the difference to differences in stellar populations.

4. How do we know the size of the galaxy?

5. How do we know the location and distance of the galactic center?

6. Describe evidence for spiral structure in the Milky Way. Which of the following types of objects reveal spiral structure when their positions are mapped on the Milky Way plane?
- **a.** O stars
- **b.** M stars
- **c.** HII clouds
- **d.** Open clusters
- **e.** Globular clusters
- **f.** Star-forming regions
- **g.** Supernovae

7. Will the present constellations be recognizable in the Earth's sky 100 million years from now? Why or why not?

8. Why is the term *high-velocity star* a misnomer?

9. Why are no O- or B-type stars found in the galactic halo?

10. Summarize evidence for violent, energetic activity in our Galaxy's central region.

ADVANCED PROBLEMS

11. What is the linear size, in parsecs, of a feature subtending an angle of 1 second of arc, located at the galactic center?
- **a.** Could it be seen with a telescope resolving 1 second of arc?
- **b.** Could it be seen with a radio telescope resolving 1 second of arc?

12. If the Sun moves in a circular orbit at 230 km/s and is 9000 pc from the orbit's center, calculate the time required to complete one circuit. (*Hint:* Circumference of a circle $= 2\pi r$; 1 pc $= 3 \times 10^{16}$ m; 1 y $\approx \pi \times 10^7$ s.)

13. Using the relations in Problem 12, confirm the calculation of the galaxy's mass given in the text. Why would it be incorrect (or at least not meaningful) to quote the galaxy's mass to three significant figures (such as 2.34×10^{41} kg)?

14. What percentage of the galactic diameter could be crossed in one lifetime (say, 70 y) if we could travel at the speed of light?

15. If an asteroid could be hollowed out and converted to a spaceship on which many generations of people could live (as in some science fiction stories), how many generations would live (at 20 y each) on the way to the Orion Nebula at the speed of light?

16. Suppose interstellar hydrogen atoms could fall freely onto the surface of an accretion disk 10 AU from a $10^5 \, M_\odot$ black hole at the galactic core.
- **a.** Neglecting any relativistic effect and recalling that material falling from far away falls at about escape velocity, estimate the velocity at which the gas would hit the accretion disk.
- **b.** If much gas could flow onto the accretion disk at this speed, so that particles were maintained at such velocities, estimate the approximate temperature of the gas in the impact regions.
- **c.** Show that the peak thermal radiation escaping from such regions would be extremely short-wave gamma radiation. (Other radiation from nonthermal processes would probably occur also.)
- **d.** Comment on how detection of such processes in galactic cores requires nontraditional astronomical instruments, preferably above the atmosphere (see Figure 5-16).

PROJECTS

1. Compare views of the Milky Way with the naked eye, binoculars, and a small telescope. Scan along the Milky Way with each instrument and record the number of stars per square degree in different constellations (or at different galactic longitudes). Relate these densities to the actual structure of the galaxy. Note that if observations of the summer evening Milky Way can be obtained in the regions of Scorpio and Sagittarius, the direction toward the center can be studied. Compare star counts with each instrument in the Milky Way plane with counts near a point 90° from the plane, where we are looking directly "up" out of the disk. Can you account for the differences?

2. Compare the Milky Way as seen with the naked eye or binoculars:
- **a.** In the heart of your city
- **b.** On the edge of town
- **c.** As far from lights as you can possibly get

City lights have little effect on the telescopic views of individual bright stars. But when it comes to broad areas of faint nebulosity or unresolved star clouds, even a single street light can illuminate local smog or fog and cause the iris of the eye to contract, thus destroying faint contrast and the ability to see faint glows.

CHAPTER 24

The Local Galaxies

What lies in the vastness of space beyond our own galactic disk and its surrounding swarm of globular star clusters? Our first problem is to see out of our own galaxy. The thick clouds of interstellar galactic dust keep us from seeing out along the Milky Way's plane, just as they keep us from seeing our galaxy's nucleus. At higher galactic latitudes, however, we look "up" or "down" out of the Milky Way disk, so that our line of sight passes through little dust. The view in these directions reveals galaxy upon galaxy far outside our own Milky Way galaxy, as far as we can see. Some are like our own Milky Way; some are not.

To understand the spacings of galaxies, consider a scale model where we let a 12-in. dinner plate represent the disk of the Milky Way. The closest galaxies would be irregular objects like crumpled balls of cotton 2 or 3 in. across only about 2 ft above the plate. A dozen other galaxies of various shapes, ½ to 4 in. across, would be scattered around the room. In a hallway, about 20 ft away, is the nearest galaxy resembling the Milky Way— a 16-in. platter representing the great spiral galaxy in Andromeda. A three-dimensional sketch of the situation is seen in Figure 24-1, and properties of these and other galaxies are listed in Table 24-1.

To understand the nature of galaxies, we will take a reconnaissance journey, describing these nearby galaxies in order of distance. These objects are so remote that their true nature was recognized only a few decades ago. Astronomer Allan Sandage has commented: "What are galaxies? No one knew before 1900. Very few people knew in 1920. All astronomers knew after 1924." As mentioned in the last chapter, that was when astronomers discovered Cepheid variable stars in the Andromeda galaxy, proving it was a system of stars outside our own galaxy.

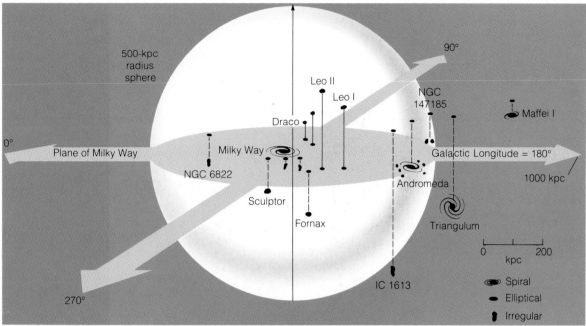

Figure 24-1 A three-dimensional plot of most of the Local Group of galaxies. Solid lines extend north of the galactic plane; dashed lines extend south. Table 24-1 lists details of these systems. Clustering of dwarf satellite galaxies around the great Milky Way and Andromeda spirals can be seen.

THE NEARBY GALAXIES

In small telescopes (up to about 30-cm aperture), most galaxies look like mere blurs of light. At the time of the Messier, New General, and Index catalogs of celestial objects, galaxies were not distinguished from nebulae, and so they received M, NGC, and IC numbers in sequence with nebulae. The Andromeda galaxy, for example, is M 31.

Later photographs made with large telescopes show some galaxies as pinwheels with beautiful spiral arms—like the Milky Way—but many others as ellipsoidal groups of stars like giant, flattened globular clusters or irregular masses of stars and nebulae. These objects were proved to be extremely remote galaxies by identifying individual stars, Cepheid variables, and supernovae in them, thus showing that they are not amorphous clouds of gas. Supernovae in other galaxies, seen in Figure 24-2, are particularly striking evidence that processes of star evolution proceed in these remote systems as they do in our own. As was shown in Figure 23-5, many of these galaxies show dark lanes of dust clouds strikingly reminiscent of those in the Milky Way.

Once these objects were identified as galaxies, their distances could be estimated. For example, Henrietta Leavitt's 1912 discovery of the Cepheid's period/luminosity relation (page 410) could be used: Periods of Cepheids in other galaxies could be observed; then their apparent brightness and known luminosity combine to give a distance, as shown in Figure 16-11. For instance, if the supernova in Figure 24-2 had an apparent brightness of the 9th magnitude, Figure 16-11 would suggest that its galaxy is about 1000 kpc away. Another method for measuring a galaxy's distance uses the galaxy's apparent angular size. For instance, a galaxy with about the size and shape of ours that subtends an angle of $2°.6$ in the sky would have to be around 670 kpc away. This is the situation with the Andromeda galaxy.[1]

---

[1] It is surprising to realize that one of the objects subtending the largest angle in our sky is not our own nearby Moon (½°), our Sun (½°), a planet (Jupiter's whole satellite system subtends $2°.1$, while the disks of Jupiter and Venus reach only about 1′), nor a cluster like the Pleiades (1½°), but the enormously remote Andromeda galaxy ($2°.7$ to $4°.5$, depending on the sensitivity of the detector).

TABLE 24·1

Known Galaxies Out to 1000 kpc

Name	Catalog Number	Distance (kpc)	Diameter (kpc)	Mass ($M_\odot$)	Absolute Magnitude (M_v)	Type[a]
Milky Way Subgroup						
Milky Way	—	9	30	2×10^{11}	−20.6	Sb
Large Magellanic Cloud	—	52	8	10^{10}	−18.4	Irr
Small Magellanic Cloud	—	63	5	2×10^9	−17.0	Irr
Draco system	DDO 208	90	1	10^5	− 8.0	Dwarf E
Ursa Minor system	DDO 199	90	2	10^5	− 8.2	Dwarf E
Sculptor system	—	71	2	3×10^6	−10.6	Dwarf E
Carina	—	90	2?	—	—	Dwarf E
Fornax system	—	150	6	2×10^7	−12.9	Dwarf E
Leo I system	DDO 74	180	1	4×10^6	− 9.6	Dwarf E4
Leo II system	DDO 93	180	2	10^6	− 9.2	Dwarf E1
Barnard's galaxy	NGC 6822	550	2	3×10^8	−15.1	Irr
WLM	DDO 221	610	?	—	−15.0	Irr
—	IC 5152	610	?	—	−14.6	Irr

Doppler shifts help us learn about galaxies' rotations. By measuring the Doppler shift on each side of a galaxy's center, we can see stars approaching on one side and receding on the other due to the stars' orbital motions around the galaxy's center. Combining the rotation speeds with the stars' distances from the galactic center, we can apply Kepler's laws and determine the mass of the galaxy, as was done for the Milky Way in the last chapter.

A second method for estimating a galaxy's mass can be used if two galaxies are adjacent and orbiting around each other. Again the velocities can be estimated from Doppler shifts and the masses estimated from Kepler's laws, as for the analysis of a binary star.

The Magellanic Clouds: Satellite Galaxies of the Milky Way

Medieval Arab astronomers recorded a glowing patch far down in the southern sky, visible from southern Saudi Arabia. Amerigo Vespucci, who sailed farther south, reported that there were really "two clouds of reasonable bigness moving about the place of the (south celestial) pole" (quoted by Shapley, 1957). Since there is no bright south polar star, the clouds helped navigators to mark the pole. Well described during Magellan's around-the-world expedition of 1518 through 1520, they came to be called the **Magellanic clouds.** They are shown in the wide-angle view of Figure 24-3.

Known Galaxies Out to 1000 kpc, *continued*

Name	Catalog Number	Distance (kpc)	Diameter (kpc)	Mass ($M_\odot$)	Absolute Magnitude (M_v)	Type[a]
Andromeda Subgroup						
Andromeda companion	NGC 205	640	3	8×10^9	-15.7	Dwarf E5
—	NGC 147	670	2	10^9	-14.4	Dwarf E
—	NGC 185	670	2	10^9	-14.6	Dwarf E
Andromeda companion	M 32 (NGC 221)	670	2	3×10^9	-15.5	Dwarf E2
Andromeda galaxy	M 31 (NGC 224)	670	40^b	3×10^{11}	-21.6	Sb
Andromeda I	—	670	0.7	—	-10.6	Dwarf E
Andromeda II	—	670	0.7	—	-10.6	Dwarf E
Andromeda III	—	670	0.9	—	-10.6	Dwarf E
Triangulum galaxy	M 33 (NGC 598)	770	18	1×10^{10}	-19.1	Sc
—	IC 1613	770	4	3×10^9	-14.5	Irr
Aquarius	DDO 210	920	?	—	-11.0	Irr
Pisces	LGS 3	920	?	—	-9.7	Irr
Maffei 1	—	1000	?	2×10^{11}	-20	S0

Source: Data from Allen (1973), van den Bergh (1972), Hirshfeld (1980), and Hodge (1987).
[a]S = spiral (subtypes 0, a, b, c—see discussion later in chapter)
Irr = irregular
E = elliptical (subtypes 0 through 7)
[b]Diameter of Andromeda based on angular diameter 3°.1.

The clouds turned out to be small galaxies, probably moving in orbits around the Milky Way. The Large Magellanic Cloud is 52 kpc away, and the Small Magellanic Cloud 63 kpc, less than three times as far as the far edge of our own galaxy.

What kind of galaxies are these companions of ours? The large cloud is only 8 kpc across, and the small cloud only 5 kpc, compared with about 30 kpc for the Milky Way. Star counts and radio measures of hydrogen indicate that these clouds are only a few percent as massive as the Milky Way.

In addition, the clouds do not show the Milky Way's beautiful spiral structure. Because they are ragged, they are classified as **irregular galaxies.** Each contains a softly glowing, barlike structure composed of stars, as shown in Figures 24-4 and 24-5. Hydrogen is concentrated in the central regions of the galaxies and also forms a gaseous bridge connecting the two.

Can we identify a nucleus in either galaxy? Somewhat off the end of the bar in the Large Magellanic Cloud is the **Tarantula Nebula,** also known as **30 Doradus** or NGC 2070. It is shown in Figure 24-6. This nebula, although 52 kpc away, can be seen with the naked eye, so it must be extremely luminous. In fact, if it were moved to the position of the Orion Nebula, it would fill the whole constellation of Orion and be bright enough to cast shadows on the Earth! In its center is a cluster about 60 pc in diameter, containing at least 100

Figure 24-2 Detection of occasional supernovae in other galaxies permits estimates of the distance of these galaxies because absolute magnitudes of supernovae correlate with their rate of brightness change. The arrow shows a supernova detected in the spiral galaxy NGC 7331 in 1959. (Lick Observatory.)

Figure 24-3 The Large Magellanic Cloud (left) and Small Magellanic Cloud (lower right). The bright star in the upper right is Alpha Eridani. The angular distance between the clouds is about 22°. (3-h exposure with 8-cm-aperture lens, made in 1934; Harvard College Observatory, Bloemfontein Station, South Africa.)

massive, bluish supergiant stars. This cluster is several hundred times brighter than ordinary globular clusters near the Milky Way.

As seen in Figure 24-6d, the cluster is surrounded by intricate wisps of red-glowing hydrogen, similar to those blown out by the Crab supernova explosion (see Figure 19-12, page 420). In 1981, Wisconsin astronomers proposed, on the basis of observations from a UV space telescope, that this cluster may contain the most massive star known—a behemoth named Radcliffe 136a, with an estimated mass of $1000-2000\,M_\odot$, temperature 56 000 K, diameter equaling that of Mercury's orbit, and a lifetime of only a million years before a supernova outburst (Humphreys and Davidson, 1984; Mathis and others, 1984). Other astronomers propose a central black hole or a cluster of massive O stars. Controversy abounds.

The perceptive reader will have noticed that we are really facing the mystery of the galactic nucleus all over again. Many observers regard the Tarantula Nebula and Radcliffe 136a as a partially formed nucleus of the large

cloud. Like our galaxy's nucleus, it contains a thermal radio source about 60 pc across, surrounded by a source of nonthermal radio radiation about 200×400 pc. The central object, in both cases, is mysterious.

We can best understand other galaxies and their evolution by considering what populations of stars they contain. As seen in Figure 24-7a, the Large Magellanic Cloud displays scattered red HII regions and masses of blue stars—the hallmarks of Population I, as found in our own spiral arm of the Milky Way. But are the clouds entirely dominated by Population I stars? The answer requires a careful look at an H–R diagram for the clouds. Figure 24-8 reveals that because the clouds are so far away, we can see only the brightest stars, those in the supergiant region of the diagram. True, we are seeing many blue stars, but as galactic expert Bart Bok (1966) commented: "The spectacular Population I components represent only the frosting on a beautiful cake." Population II is more important than it first seems. The central bars contain many Population II red giants, and Population II globular clusters have been found. There is

a

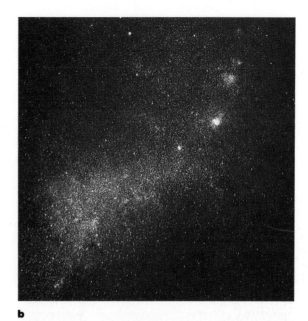

b

Figure 24-4 The Small Magellanic Cloud, about 63 kpc away and 5 kpc across, is an irregular galaxy exhibiting a barlike configuration of bluish stars. **a** Long exposure reveals extent of faint outer regions and slightly overexposes bright inner regions. (Copyright Anglo-Australian Telescope Board, courtesy D. Malin.) **b** Shorter exposure emphasizes colors of brightest blue stars and red HII regions of the inner part of the galaxy. (Copyright R. J. Dufour, Rice University.)

a

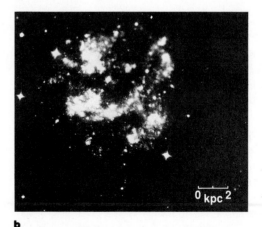

b

Figure 24-5 The Large Magellanic Cloud, about 52 kpc away and 8 kpc across, is the closest galaxy to us. **a** Color view shows a barlike distribution of bluish stars and a scattering of red Hα-glowing HII regions, including the massive Tarantula Nebula (center left). (Copyright Anglo-Australian Telescope Board, courtesy D. Malin.) **b** Image at far-ultraviolet wavelengths reveals locations of hottest, youngest stars and associated nebulae. Concentrations of Population I stars are prominent. This image was taken from the Moon's surface by an astronaut using a special UV camera. (NASA, Naval Research Laboratory.)

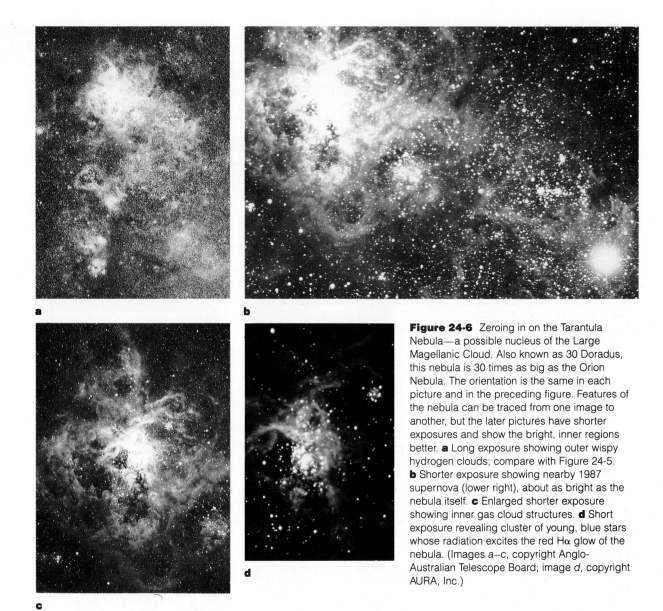

a

b

c

d

Figure 24-6 Zeroing in on the Tarantula Nebula—a possible nucleus of the Large Magellanic Cloud. Also known as 30 Doradus, this nebula is 30 times as big as the Orion Nebula. The orientation is the same in each picture and in the preceding figure. Features of the nebula can be traced from one image to another, but the later pictures have shorter exposures and show the bright, inner regions better. **a** Long exposure showing outer wispy hydrogen clouds; compare with Figure 24-5. **b** Shorter exposure showing nearby 1987 supernova (lower right), about as bright as the nebula itself. **c** Enlarged shorter exposure showing inner gas cloud structures. **d** Short exposure revealing cluster of young, blue stars whose radiation excites the red Hα glow of the nebula. (Images *a–c*, copyright Anglo-Australian Telescope Board; image *d*, copyright AURA, Inc.)

little dust in the two clouds except in the prominent young nebulae. Judging from age relations in our galaxy, we can guess that the clouds' Population II objects are many billions of years old, though the bright star-forming nebulae of Population I may be only millions of years old.

The first pulsar to be discovered outside our galaxy is in the Large Magellanic Cloud. Designated PSR 0529-66, it pulses every 1.0 s. Along with recently discovered X-ray supernova remnants in the cloud and the evidence for Populations I and II, it confirms that stellar evolution yields the same products in other parts of the universe as in our own galaxy.

Strange features of both clouds are a few blue globular clusters, unlike the red-giant-dominated clusters of our galaxy. To contain hot, blue stars (which are massive and have short lifetimes), these globulars must have formed less than a billion years ago, and others may still be forming. An example is seen in Figure 24-7b.

Astronomers in 1983 announced the unexpected result that the Small Magellanic Cloud may actually be two galaxies, one in front of the other, about 9 kpc apart. This finding was based on their stars' velocities, which seem to divide into two groups.

The Magellanic clouds are important to modern

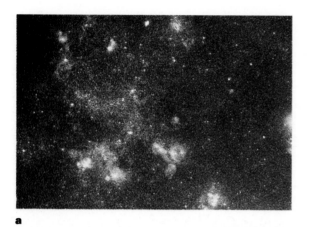

a

b

Figure 24-7 The Large Magellanic Cloud offers a ringside look at some important features of galaxies. **a** A region just above the top of Figure 24-5 displays a mixture of blue, young stars and scattered red-glowing star-forming nebulae. **b** A globular cluster exhibits bluer stars than the red-giant-dominated globular clusters of our own galaxy. The evolution of these blue globulars is not clear. (Both photos copyright Anglo-Australian Telescope Board.)

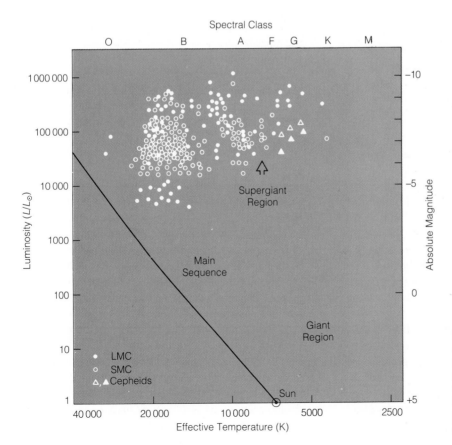

Figure 24-8 An H–R diagram of measurable stars in the Magellanic clouds. The clouds are so far away that the only stars with measurable spectra are supergiants. Main-sequence populations are assumed to exist as well.

Figure 24-9 One of the closest galaxies is this loose dwarf elliptical called Leo I, about 1 kpc in diameter. At a distance of 180 kpc it is close enough to be easily resolved as separate stars, and it looks somewhat like a very large globular cluster. (Copyright Anglo-Australian Telescope Board, courtesy D. Malin.)

Figure 24-10 The nearby galaxy NGC 6822 is a small irregular galaxy about 2 kpc across and 550 kpc away. At this relatively close distance, as galaxies go, it is easily resolved into individual stars, pinkish HII regions, and a few ring-shaped possible supernova remnants. As in other galaxy photos, uniformly scattered brighter stars are foreground stars in our own galaxy. (Copyright Anglo-Australian Telescope Board, courtesy D. Malin.)

astronomy for providing a collection of varied objects at essentially constant distances. In the words of South African astronomer A. D. Thackeray (1971):

The opportunity to compare the luminosities of supergiants, giants, main sequence stars of all types, Wolf–Rayet stars, cepheids, eclipsing variables, Mira-variables, RR Lyrae variables, globular clusters, novae, and so on, in an endless sequence all at the same distance simply never occurs in our galaxy.

A bridge of hydrogen called the **Magellanic stream** may connect the two clouds to the Milky Way galaxy. Australian radio astronomers have mapped a long filament of HI extending from the small cloud in an arc beyond the south galactic pole and in the other direction, possibly reaching into the Milky Way's plane. It resembles the bridge between the two clouds themselves. Because the two clouds are probably true satellites of our galaxy, their orbits may sometimes take them through the Milky Way disk. The Magellanic stream may thus be a tail of gas drawn out of the Milky Way during such an encounter an estimated half billion years ago (Mathewson, 1985). Some of the smaller galaxies in the Milky Way subsystem have been interpreted as local concentrations related to the Magellanic stream— concentrations of debris in the paths of the Magellanic clouds. Further studies are needed to confirm this idea.

Dwarf Elliptical Galaxies

As indicated in Table 24-1, the next seven objects beyond the Magellanic clouds are small systems of a type known as **dwarf elliptical galaxies.** These small elliptical-shaped or spheroidal galaxies resemble giant globular clusters. They are more symmetrical than the Magellanic clouds, but lack the disk shape and spiral arms of the Milky Way. Three of them are shown in Figures 24-9 through 24-11.

The dwarf elliptical galaxies are apparently the most common type of galaxy. They are dominated by old Population II stars and have little gas or dust. In most respects they are less impressive than our own giant spiral disk, with its chaotic clouds of gas and dust and regions of continuing star formation. But interest in them revived in the 1980s when studies indicated that some of their stars must be fairly young, perhaps 2 to 5 billion years old. Thus they must be somewhat more active than globular clusters, whose stars are all around 14 billion years old. Furthermore, even though they look somewhat like overgrown globular clusters, at least one (the Sculptor system) has its own swarm of globular clusters, which are about the size of our own globulars or slightly bigger. Since "dogs have fleas but fleas don't have fleas," this fact strengthens the classification of Sculptor-like dwarf ellipticals as galaxies in their own

Figure 24-11 Another "nearby" galaxy is IC 5152, roughly 610 kpc away. It is barely resolved into stars and has the appearance of a very large, flattened globular cluster. Bright image with spikes and other scattered stars are foreground stars in our galaxy. (Copyright Anglo-Australian Telescope Board, courtesy D. Malin.)

right (Hodge, 1987). Most dwarf elliptical galaxies may be satellite attendants to larger systems. Astronomer Paul Hodge calls the group near our galaxy "the Seven Dwarfs."

THE GREAT ANDROMEDA SPIRAL GALAXY

Finally, at 670 kpc, we encounter the first **spiral galaxy** truly comparable to the Milky Way, along with several of its smaller satellite galaxies. This galaxy was first recorded in a star catalog by Arab astronomer al-Sufe in A.D. 964, but it must have been known much earlier, being easy to see with the naked eye as a hazy patch on a dark, clear night. The **Andromeda galaxy** is the one that settled the argument about the nature of so-called spiral nebulae. Edwin Hubble identified Cepheid variables there and showed that it was far outside the disk of our own galaxy.[2] His announcement of this finding at the American Astronomical Society in 1924 caused a sensation and marked the first real understanding of galaxies.

[2]Even until a few years ago, a few books maintained the misnomer "Andromeda nebula," a name carried over from the 1920s when it was still thought to be only a nebula inside our own galaxy.

Similarities to the Milky Way

The Andromeda galaxy can hardly be distinguished from the Milky Way in shape or stellar content, though it may be a little bigger. The naked eye sees it as a faint patch (Figures 24-12 and 24-13), but this is really only the brightest, innermost region, a few kpc across. Telescopic photographs show that the spiral arms form a disk at least 20 kpc across (Figure 24-14), while outlying associations of B stars suggest a diameter as great as 50 kpc. As with the Milky Way, there are globular clusters and a halo of HI gas reaching perhaps 100 kpc in diameter.

Spiral Arms

The spiral arm pattern, clarified in Figures 24-14 and 24-15, also resembles that of the Milky Way, though it is much easier to observe since we see it from above the disk. *Inner arms* have conspicuous dust; *outer arms* have less dust but more HII regions and young stars. Most star formation now occurs in the intermediate and outer arms.

Stellar Populations

The Andromeda galaxy also played an important role in the discovery of the two main star populations. Hubble's photos in the 1920s with the new Mt. Wilson 2.5-m telescope revealed bright blue stars in the spiral arms (Figure 24-16) but no individual stars in the central bulge. Mt. Wilson observer Walter Baade therefore decided to look for possible bright red stars in the bulge by switching to red-sensitive photographic plates. These plates were troublesome because the exposures required as long as 9 h and tended to be fogged by the reflected skyglow of Los Angeles, below Mt. Wilson.

A world tragedy came to Baade's aid. During World War II, in 1942–1943, blackouts of Los Angeles produced exceptionally dark skies. In a few months, Baade discovered that the central bulge of Andromeda and its two nearby elliptical galaxies are dominated not by blue stars (like the young, brightest stars of the spiral arms) but by, as he put it, "thousands and tens of thousands" of Population II red giant stars (Figure 24-16). This discovery clarified the role of star populations in galaxies. It also shows that the Andromeda pattern of populations matches that of our own galaxy. *In spiral gal-*

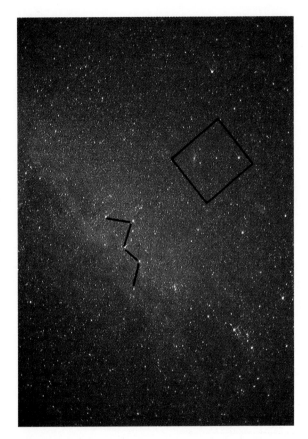

Figure 24-12 How to find the Andromeda galaxy with your naked eye. This color view shows the Milky Way extending across the region of the constellation Cassiopeia (M-shaped figure). The angular distance from the Andromeda galaxy to Cassiopeia's middle star is about 20°. Although the Andromeda galaxy is about 670 kpc away, on a clear, dark night it can be seen by the unaided eye as a faint smudge of light. This view records details somewhat fainter than the eye can see. Box shows the area of Figure 24-13. (35-mm guided camera; 24-mm lens at f2.8; 13-min exposure on 3M ASA 1000 film; photo by author.)

Figure 24-13 The region of the Andromeda galaxy, showing an angular extent approaching 3°. Vertical height of the photo is about 8°. The bright star, upper right, is Beta Andromedae. The box shows the outline of views in Figure 24-14. (35-mm camera; 135-mm lens at f2.8; 10-min exposure on 2475 Recording film.)

axies, Population I star-forming regions are concentrated in the spiral arms, which are marked by open clusters, giant nebulae, and massive blue stars. The central bulges of spirals are redder than the arms, because star forma-tion has ceased and there are no bright, young blue stars; the brightest stars are old red giants. This effect can usually be seen in color photos of other galaxies. The same applies to elliptical galaxies, which resemble cen-tral bulges of spirals, and thus are dominated by reddish Population II giants. Detailed recent studies of the Andromeda galaxy indicate it has fewer massive super-giant stars (40 to 60 $M_\odot$) than our galaxy, possibly due to different rates of recent star formation (Humphreys and others, 1990).

Dynamics and Mass

As in our galaxy, the density of stars is greatest near the center. A region of low rotational velocity and low star density about 2 kpc from the center may be a fea-ture related to our own 3-kpc expanding arm. The total mass of the main disk of the galaxy is roughly 3×10^{11} $M_\odot$, but there may be more unseen, halo mass, as in the Milky Way (Rubin, 1983).

The Nucleus

The Andromeda galaxy gives us a chance to look directly at the light of the nucleus of a galaxy like ours, rather than trying to study it through 9 kpc of dust. The results are intriguing.

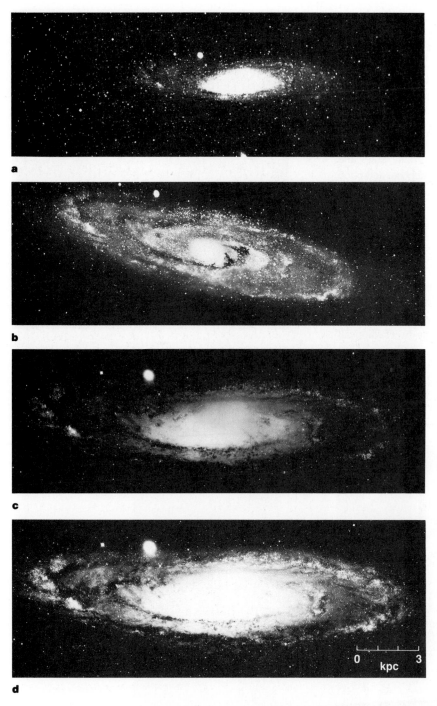

Figure 24-14 Images illustrating the varied appearance of the Andromeda galaxy with different types of photographic and telescopic equipment. **a** Short exposure with a 15-cm telescope emphasizing the nucleus and barely showing the spiral arms. (*Sky and Telescope.*) **b** Mosaic of photos with a 32-cm telescope, printed to show maximum detail in both nucleus and spiral arms. (Clarence P. Custer.) **c** Bright features of the nucleus and clusters in the arms. **d** From the same negative as c, this picture was printed to show faint details in the arms. (Steward Observatory, University of Arizona.) None of these views captures the extreme outer faint regions.

Figure 24-17 compares our moderate-angle view into the Andromeda galaxy's center to our lower-angle view of our own galaxy's dust-obscured center. Figure 24-18 shows a shorter color exposure of the central region, which reveals its yellowish and reddish tones—due partly to the Population II red giants that dominate the region. As hinted by this figure, the brightness increases toward the center of the central bulge. Most published photos of the Andromeda galaxy, such as those in Figure 24-14, overexpose this region in an effort to record the faint, outer spiral arms. But as seen in Figure 24-19, a brilliant starlike cluster exists in the center of the bulge. This is the nucleus—the brightest object in the whole galaxy.

What does Andromeda teach us about the mysterious nuclei of galaxies? Within 500 pc of the center are about a dozen X-ray sources, shown in Figure 24-20. The sharpest photos show a single bright feature about 3×5 pc in size. Interferometer techniques indicate an even smaller bright nucleus within this, no bigger than 0.4 pc and possibly much smaller, with a bluish color. As with the nucleus of our own galaxy, there are two interpretations: the nucleus is either a dense cluster of stars containing many massive young blue stars and supernova remnants, or it is an extremely hot accretion disk surrounding a black hole (Spillar and others, 1990).

SURVEYING AND CLASSIFYING GALAXIES

As seen in Table 24-1, an extension of our survey of galaxies all the way out to 1000 kpc from the Milky Way nets 26 objects, including 14 dwarf ellipticals, 8 irregulars, and 4 spirals. A good example of a dwarf elliptical, and the next spiral beyond Andromeda, is shown in Figures 24-21 and 24-22. What do the statistics of these galaxies tell us about galaxies in general?

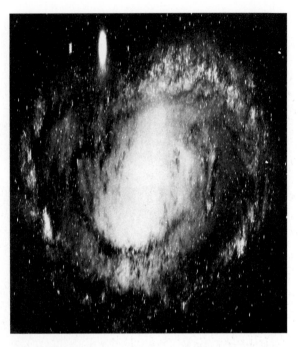

Figure 24-15 Proof of the Andromeda galaxy's spiral form. The image has been "rectified" by projection on a screen tipped 13° to the line of sight, giving an impression of the galaxy's spiral pattern as it would be seen above the plane of the disk.

Figure 24-16 Resolution of stars in two populations of the Andromeda galaxy. Left photo (blue filter) shows Population I hot, blue stars and clusters in the spiral arms. Right photo (yellow filter) shows Population II red giant stars, which dominate the elliptical galaxy NGC 205, a satellite of the Andromeda galaxy. (5-m telescope; Hale Observatories.)

a

b

Figure 24-17 A comparison between the ragged, dust-cloud-strewn heart of our galaxy and the central part of the Andromeda galaxy shows striking similarities between the two systems. **a** From our position perhaps slightly above the plane of the Milky Way, we peer over dark dust clouds in intervening spiral arms toward the glowing central bulge. **b** From a slightly higher tilt, we look at the central bulge of the Andromeda galaxy. Box shows approximate size of Figure 24-18, oval shows position of nucleus (Figure 24-19) overexposed here. High-contrast print brings out dust clouds similar to those in Milky Way. (Image *a*, 70° panorama from two wide-angle photos by author; 24-mm lens, f2.8, 40- and 46-min exposures on 2475 Recording film; image *b*, Kitt Peak National Observatory.)

First, galaxies are not randomly distributed through space. They tend to cluster in different-sized groups. As seen in Figure 24-1 and Table 24-1, for instance, most of the galaxies out to 1000 kpc are clumped in two subgroups: those around the Milky Way and those around the Andromeda spiral. The whole population of 26 galaxies out to 1000 kpc is a concentration called the **Local Group.** Just as stars form in clusters, something about the formation process of galaxies also produces clusters.

Nonluminous Mass in Galaxies

As mentioned in the case of the Milky Way and Andromeda galaxies, only a fraction of a galaxy's total material is in luminous, visible stars. Surveys confirm this result

for many other galaxies (Rubin, 1983). The rest is invisible. Where is this mass? This question is sometimes called the **problem of the missing mass.** Perhaps the missing mass is in the form of faint stars, planetary matter and dust, black holes, gas, or subatomic particles like neutrinos, but the answer is uncertain. The invisible mass has been detected by using Doppler shifts to study orbital velocities of stars around galactic centers; the velocities prove that the stars are attracted to the galaxies by more than the visible mass. As astronomer Vera Rubin (1983) concludes from her study of such velocities, "as much as 90 percent of the mass of the universe is evidently not radiating at any wavelength with enough intensity to be detected on the Earth." Solving the problem of why the missing mass is invisible

Figure 24-18 This color photo of the central bulge of the Andromeda galaxy emphasizes the great brightness of the central region relative to the faint spiral arms. The photo covers about the area of the box in Figure 24-17b; only the inner edges of the spiral dust lanes are visible. Population II stars with yellowish and reddish color dominate the central bulge. (Copyright Association of Universities for Research in Astronomy, Cerro Tololo Interamerican Observatory.)

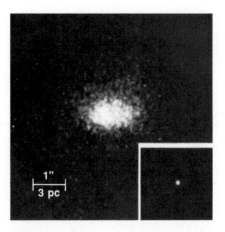

Figure 24-19 Highest-resolution photo of the bright nucleus of the Andromeda galaxy, a dense cluster of stars about 3 by 5 pc in diameter. The photo was made by a 36-in. telescope carried to 84 000 ft by a balloon. The inset at right shows the image of a star made by the same telescope. (Princeton University, Project Stratoscope, supported by NSF, ONR, NASA.)

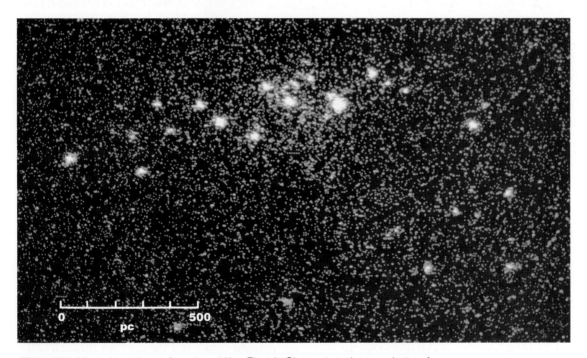

Figure 24-20 An X-ray image from the orbiting Einstein Observatory shows a cluster of X-ray sources in the central kiloparsec of the Andromeda galaxy. One of the sources (in upper center group) corresponds to the nucleus, shown in Figure 24-19. It varied in X-ray output from 10^{30} to 10^{31} watts during the 6-mo period of observation. Other sources include globular clusters and may include accretion disks around black holes. (Harvard/Smithsonian Center for Astrophysics, courtesy Leon Van Speybroeck.)

Figure 24-21 The inner core of dwarf elliptical galaxy NGC 185, about 670 kpc away. Individual stars are resolved. The prominent dark dust cloud is unusual in an elliptical galaxy, most of which are relatively dust free. (Lick Observatory.)

Figure 24-22 Central regions of the Triangulum galaxy, a spiral system about 770 kpc away. (Hale Observatories.)

is important in understanding the dynamic interactions of galaxies throughout the universe, as we will see in later chapters. Future observations from above the atmosphere may help; astronomers are especially looking forward to results from the Hubble Space Telescope in orbit.

A Classification System: Making Sense of Galaxies

There are several distinct types of galaxies. Most of the known types appear within the Local Group (Table 24-1), which takes us out to 1000 kpc. Even if we expand our horizon to, say, 15 000 kpc, as in Table 24-2, we discover only a few new types. M 87, shown in Figure 24-23, is a **giant elliptical galaxy** much larger than the ones mentioned so far; it is about 13 kpc across and more massive than the Milky Way. It contains about 4 × 10^{12} $M_\odot$, or perhaps 4000 billion stars! These *giant* ellipticals give a good reason for distinguishing the

smaller, globular-cluster-like ellipticals as "dwarf ellipticals." Another class is illustrated by M 83, a spiral galaxy whose central region is a bright, barlike object instead of an ellipsoidal bright area; such galaxies are called **barred spiral galaxies.** Good examples are seen in Figure 24-24.

How can we make sense of these different shapes of galaxies and their different types of stellar populations? The standard scientific approach is to try to classify the different types of objects and then arrange them in a system that shows smooth transitions from one type to another. This system can then be studied for possible causes of the transitions, such as evolution, mass differences, or rotational differences. For example, elliptical galaxies with different degrees of flatness seem to form an obvious sequence, perhaps related to rotation.

The best-known classification scheme was developed by Edwin P. Hubble in the 1920s and 1930s and extended by his colleagues after his death. The system

a

0 kpc 5

b

Figure 24-23 The giant elliptical galaxy M 87 (NGC 4486) is about 13 000 kpc away and 13 kpc across (including its halo of globular clusters). It is some 5 or 10 times bigger than typical dwarf elliptical galaxies. The faint, fuzzy starlike images around M 87 are thousands of globular clusters swarming in a halo like bees around a hive. M 87 is about 40 times as massive as our galaxy. Two background spiral galaxies appear in the image. **a** (*Opposite*) Color view indicates reddish colors of Population II stars. (Copyright Anglo-Australian Telescope Board.) **b** Longer black-and-white exposure indicates even more faint globular clusters in the outer extensions of the galaxy. (Courtesy Malcom Smith and W. E. Harris, Cerro Tololo Interamerican Observatory and McMaster Observatory.)

TABLE 24·2

Selected Galaxies Out to 15 000 kpc

Catalog Number	Distance (kpc)	Diameter (kpc)	Mass ($M_\odot$)	Absolute Magnitude (M_v)	Type[a]	Radial Velocity (km/s)
NGC 55	2 300	12	3×10^{10}	−20	Sc	+ 190
NGC 253	2 400	13	10^{11}	−20	Sc	− 70
M 82 (NGC 3034)	3 000	7	3×10^{10}	−20	Irr	+ 400
M 81 (NGC 3031)	3 200	16	2×10^{11}	−21	Sb	+ 80
M 83 (NGC 5236)	3 200	12	10^{11}	−21	SBc	+ 320
M 51 "Whirlpool" (NGC 5194)	3 800	9	8×10^{10}	−20	Sc	+ 550
NGC 5128 (Centaurus A)	4 400	15	2×10^{11}	−20	E0p[b]	+ 260
M 101 "Pinwheel" (NGC 5457)	7 200[c]	40	2×10^{11}	−21	Sc	+ 402
M 104 "Sombrero" (NGC 4594)	12 000	8	5×10^{11}	−22	Sa	+1050
M 87 (NGC 4486)	13 000	13	4×10^{12}	−22	E1	+1220

Source: Data from Allen (1973).
[a]S   = spiral (subtypes 0, a, b, c—see discussion in chapter)
 Irr = irregular
 SB = barred spiral (subtypes 0, a, b, c—see discussion in chapter)
 E   = elliptical (subtypes 0 through 7)
 p   = peculiar
[b]Centaurus A is a strong radio source, appearing as an elliptical galaxy with a peculiar dense dust lane across its face.
[c]M 101's distance was revised by A. Sandage and G. Tammann (1974) from 3800 to 7200 ± 1000 kpc.

a

Figure 24-24 Two barred spiral galaxies display a broad central bar instead of the normal spheroidal central bulge. Both examples show yellowish colors in the central bar due to Population II stars; bluish colors in the spiral arms are due to hot, young Population I stars forming there. **a** NGC 4650, a type SBa barred spiral, in the constellation Centaurus. (National Optical Astronomy Observatories.) **b** NGC 6744, a type SBc barred spiral, about 11 Mpc from the Milky Way galaxy. (Copyright R. J. Dufour, Rice University.)

b

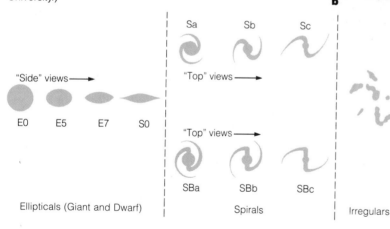

Figure 24-25 The simplified classification scheme for galaxies known as "Hubble's tuning fork diagram" after the originator of the system.

is based purely on the apparent shapes of the galaxies, not on measured properties such as spectra or mass. The scheme, shown in Figure 24-25, involves three main types of galaxies: *ellipticals,* both dwarf and giant, (designated E), *spirals* (S), and *irregulars* (Irr). Each type is subdivided according to form. For example, as shown in Figure 24-25, ellipticals are subdivided according to apparent flattening, which may be due either to our angle of view or to true flattening of the system.[3]

[3]The index number is $10(a - b)/a$, where a is the major axis and b the minor axis. E0 is circular; E7, the flattest known.

The general appearance of spirals is strongly dependent on the angle of their tilt toward us, since the disk can appear either face on or edge on. This is shown by the pair of galaxies in Figure 24-26. Edge-on views of spiral galaxies show that many of them are very thin. The subclassification of spirals in the tuning fork diagram, however, is based on the relative sizes of the disk and central bulge and on the tightness of winding of the arms. Forming a transition between ellipticals and spirals is a special class, S0, resembling a spiral's disk shape but lacking clear spiral arms. Sa spirals have the tightest arms and Sc spirals the loosest. Spirals are

a

b

Figure 24-26 Spiral galaxies (type Sc) face on and edge on. **a** NGC 2997 is seen nearly face on, emphasizing the spiral arm pattern. **b** NGC 253 is seen nearly edge on, emphasizing the thick lanes of dust in the spiral arms that cross in front of the nucleus. (Both photos copyright Anglo-Australian Telescope Board.)

subdivided into two parallel sequences—barred and normal spirals—giving Figure 24-25 its common designation as the **"tuning fork" diagram.** Thus a tight spiral with a central bar is designated SBa, instead of Sa. (Some astronomers use a still finer subdivision of spirals to indicate the presence or absence of a ringlike configuration of the innermost spiral arms.) Irregular galaxies are subdivided into Ir I (prominent O and B stars, emission nebulae, and Population I) and Ir II (amorphous appearance, no sharp nebulae, rich in Population II).

This classification system correlates well with certain physical properties of galaxies, as summarized in Table 24-3. For example, from ellipticals through irregulars there is a progression from older to younger populations of stars and from less to more dust and gas.

The implication is that star formation is still occurring in spiral arms and some irregulars, but not in ellipticals. Another systematic trend is that the amount of mass required to produce a fixed amount of stellar radiation, called the **mass/luminosity ratio,** is high in ellipticals and low in irregulars. This ratio implies that much of the mass in ellipticals is tied up in low-luminosity stars. The central bulges of many spirals bear a striking resemblance to ellipticals. A few galaxies have unusual qualities that seem to violate the pattern of the classification system, as suggested by Figure 24-27.

The statistics of the different types of galaxies surprise many people who tend to think of all galaxies as beautiful spirals. Just as giant and supergiant stars are overrepresented if we scan the prominent stars in the sky, spirals are also overrepresented because they are bright and spectacular. For example, an early survey of prominent galaxies by Hubble netted 80% spirals, and a sample of 100 photos in three recent astronomy texts included 67% spirals. But in fact the actual distribution of galaxies is more like the following:

Dwarf ellipticals	50%
Giant ellipticals	5%
Spirals	20%
Irregulars	25%

That is, while perhaps four-fifths of the *big* galaxies are spirals, only about one-fifth of *all* galaxies are spirals.

COLLIDING GALAXIES

As can be calculated from Table 24-1, the Milky Way and Andromeda galaxies are separated by a distance amounting to only 15 or 20 of their own diameters. Other galaxies, drifting through space, are separated by similar relative spacings. If you tried to keep a roomful of frisbees aloft—enough that they were only 15 or 20 diameters apart on the average—many of them would collide. Galaxies collide, too.

This is an extraordinary situation, very different from that of stars. Stars are separated typically by 20 million of their own diameters; therefore, in spite of their random motions, few if any stars in a galaxy have collided during the whole history of the universe. If you made a model of our galaxy the size of North America, stars would be microscopic dust grains a city block apart. You can imagine that dust grains a block apart could drift around a long time without hitting each other. But many galaxies have collided with other galaxies during

TABLE 24·3

General Characteristics of Galaxies

			Spirals			
Characteristic	Ellipticals	S0	Sa	Sb	Sc	Irregulars
Dust	None	Little	Some			Some
Percent neutral hydrogen (mass HI/total mass)	0	0	1	3	9	20
Most prominent populations	II	II	II (halo, center); I (arms)			I
Typical rotation periods (million years)	?	59	63	140	200	300?
Dominant color	Red	Red	Red (halo, center); blue (arms)			Blue
Spectrum of central regions	K	K	K–G	G	F	F–A
Luminosity ($L/L_\odot$)	Giants $\leqslant 10^{11}$ Dwarfs $\geqslant 10^5$		10^8–10^{10}			10^7–10^9
Diameter (kpc)	Giants $\leqslant 200$ Dwarfs $\geqslant 1$		5–50			1–10
Mass ($M_\odot$)	Giants 10^9–10^{12} Dwarfs 10^6–10^9	?	5×10^{11}?	3×10^{11}	10^{11}	10^9
Mass/luminosity ratio (solar units)	Giants 5–80 Dwarfs 1–5	50	20	10	5	3

Source: Data from Allen (1973) and de Vaucouleurs (1974).

their lifetimes. Indeed, we can actually see this process occurring among numerous sets of close galaxies, as shown in Figure 24-28.

The idea that galaxies collide was first proposed in 1940 by a Swedish astronomer, Erik Holmberg. For many years such events were considered mere quirks of galactic history. But in the 1980s astronomers began to realize that collisions between galaxies may be fundamental to the evolution and shapes of many galaxies.

Results of Galaxy Collisions

What happens when two galaxies interpenetrate? Think of a galaxy as a giant swarm of gas with stars embedded in it. When the random motions of galaxies cause them to penetrate each other, individual stars are unlikely to collide. Even if you squashed the Milky Way and Andromeda together, the number of stars per cubic parsec would increase only by a factor of 2, and the individual stars would still be many millions of diameters apart— like the dust grains in the example discussed above.

Even though the individual stars don't hit each other, there are four important consequences of collisions between galaxies. The first is that the gas components of the two galaxies *do* interact. The dusty gas in each galaxy is more like a continuous medium, so that when two galaxies collide at many tens of kilometers per second, there is a dramatic interaction of the two gas swarms. You might think of it as if the contents of two

a

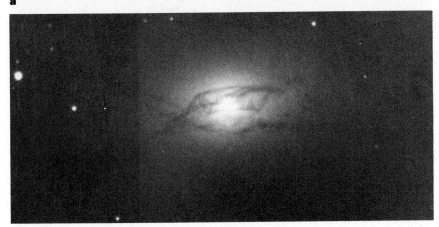

b

Figure 24-27 Some galaxies depart from the trends in the "tuning fork" classification system. **a** NGC 3718 has a bar or disk of dust across the bright central region—but usually the bar in a barred spiral is a luminous band of stars. (Kitt Peak National Observatory.) **b** NGC 4753 is an irregular galaxy with an old stellar population but laced with dust lanes typical of regions of new star formation. (McDonald Observatory, courtesy G. de Vaucouleurs.)

buckets of water were sloshed toward each other—the two masses of water would hit and make a splash. In fact, two galactic masses of gas do make a splash. And when the gas atoms collide, the gas and its entrained dust get heated. The gas atoms get excited (electrons kicked to higher energy levels) or even ionized (electrons knocked off). Hydrogen is the most abundant gas, of course. Thus colliding pairs of galaxies are often *intense sources of radio radiation from excited hydrogen* and infrared radiation from the hot dust (see Figure 24-28a). The first important consequence of colliding galaxies, therefore, is to create intense sources of radio emission. Indeed, such galaxies are among the strongest radio sources in the sky; they are called **radio galaxies.** Some astronomers have long believed that in addition

to simply exciting the galactic gas, galaxy collisions also somehow trigger formation of energetic nuclei in galaxies, perhaps extraluminous accretion disks around massive black holes (Gunn, 1979; Simkin and others, 1987). We will return to this possibility in the next chapter.

The second important consequence of colliding galaxies is distortion of shapes and creation of certain classes of peculiar galaxies that don't fit in the "tuning fork" scheme of classification. A galaxy's shape is determined by its gravitational field acting in conjunction with its spin rate and total mass. If another galaxy with comparable or larger mass approaches, that galaxy's gravitational forces will disturb the equilibrium shape of the first, just as the Moon raises a tidal bulge in the other-

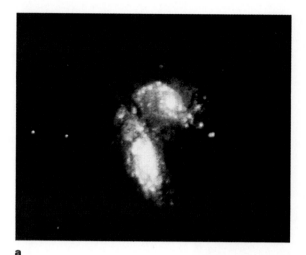

a

b

Figure 24-28 Colliding galaxies. **a** In the collision of NGC 4567 and 4568, we see spirals with their yellowish Population II centers and bluish Population I spiral arms. But the brightest HII regions of excited, red-glowing hydrogen appear to be on the outskirts where gas is being heated in the contact region between the colliding galaxies. (National Optical Astronomy Observatories.) **b** Three galaxies, including NGC 2992 and 2993, appear to be interacting. All three are connected by bridges of faint luminous material—probably drawn out by gravitational tidal forces as the various pairs passed close together. (Copyright Anglo-Australian Telescope Board.)

wise spherical shape of the Earth's oceans. The closer the two galaxies approach, the greater the distortions. If two comparable-sized galaxies draw close, the shape of each may be wildly distorted from the initial disk shape. Streams of gas may be drawn out of the pair, forming a connecting bridge, as seen in Figure 24-28b. Streamers of gas may be ejected, too, forming galaxies with long tails. Dynamicists have been able to calculate results of such encounters with amazing success. As shown in Figure 24-29, computer simulations beautifully predict some shapes actually observed among pairs of galaxies. A 1987 study showed that 14% of all galaxies show distorted shapes attributed to past collisions or close encounters; in contrast, 70% of strong infrared galaxies do (Armus and others, 1987).

A third result of galaxy collision is **starbursts**—intense bursts of star formation and supernova explosions (resulting from blowup of the most massive, most short-lived stars created during the star-forming activity). Evidence for this phenomenon comes from radio telescopes and orbiting infrared telescopes. Often 99% of the radiated energy from these galaxies comes in the infrared part of the spectrum from warm dust heated by the activity.

Galactic Cannibalism: Formation of Giant Elliptical Galaxies

The fourth and most important result of galaxy collision is that it may explain the formation of giant elliptical galaxies (Silk, 1987; Schweizer, 1986). When two galaxies collide, depending on the circumstances, one galaxy may merge with the other instead of passing on by. The larger galaxy gobbles up the smaller, so to speak, and the phenomenon is called **galactic cannibalism.**

Dynamical studies in the 1980s suggest that galactic cannibalism explains the formation of giant elliptical galaxies. As shown in Figure 24-30, after galaxies have formed in a cluster, two disk-shaped rotating protogalaxies may collide. According to recent views, if the conditions of collision are right, the disk shapes may often be destroyed, producing instead a huge ellipsoidal mass of stars. Starbursting may result in much of the gas and dust being consumed to form stars. After the burst of star-forming activity, therefore, there is little gas and dust to form later generations of stars. Star formation stops. An elliptical galaxy forms. An observer who compares this elliptical galaxy with a spiral galaxy, a few billion years later, sees a galaxy with less gas and

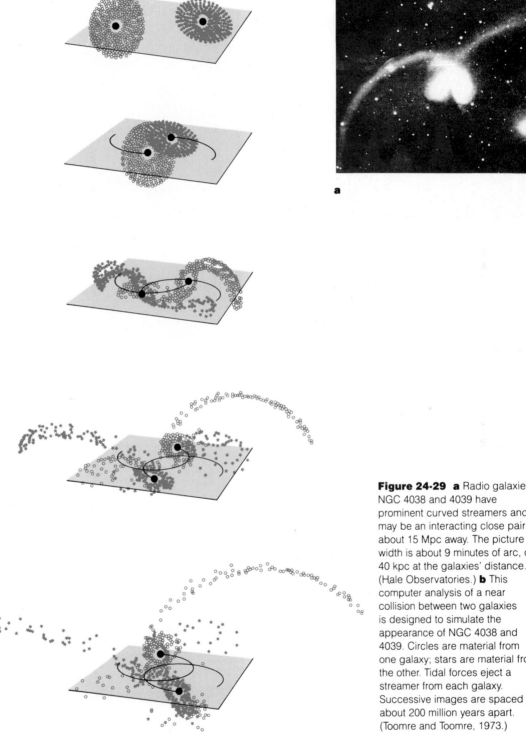

Figure 24-29 a Radio galaxies NGC 4038 and 4039 have prominent curved streamers and may be an interacting close pair about 15 Mpc away. The picture width is about 9 minutes of arc, or 40 kpc at the galaxies' distance. (Hale Observatories.) **b** This computer analysis of a near collision between two galaxies is designed to simulate the appearance of NGC 4038 and 4039. Circles are material from one galaxy; stars are material from the other. Tidal forces eject a streamer from each galaxy. Successive images are spaced about 200 million years apart. (Toomre and Toomre, 1973.)

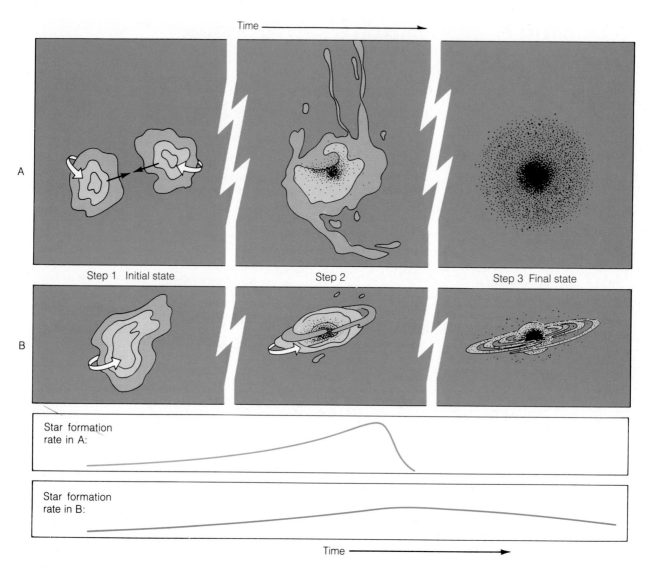

Figure 24-30 Possible formation processes leading to creation of a giant elliptical galaxy (A) and a spiral (B). In case A, two protogalaxy clouds have collided and coalesced. This excites and concentrates the gas, leading to a starburst of star formation (shown in bottom graphs). In case B, a rotating protogalaxy cloud contracts to form a disk with more leisurely star formation. In the final stages, cloud A has formed a full-fledged giant elliptical with little remaining star formations but cloud B has formed a spiral. Each type of galaxy is surrounded by a halo of globular clusters formed from initially outlying material. (Loosely based on a diagram by Silk, 1987.)

Figure 24-31 A collision of galaxies producing a giant elliptical galaxy. This photo shows NGC 7252—a peculiar galactic system with a single nucleus and two tails that are telltale signs of a collision (see Figure 24-29). The main body of the galaxy is dominated by young A stars and contains an 8-kpc-wide disk of ionized gas, apparently all formed during the collision an estimated $\frac{1}{2}$ to 2 billion years ago. The system is now believed to be reforming itself into a giant elliptical. Several distant galaxies can be seen in the background (fuzzy and elliptical images). (Cerro Tololo 4-m telescope photo, courtesy Francois Schweizer.)

dust and a population of old stars with few young stars. In other words, the observer sees Population II, the typical content of a giant elliptical. Figure 24-31 shows a galaxy pair that is believed to have recently collided and is now metamorphosing into a giant elliptical.

EVOLUTION IN GALAXIES

Colliding galaxies thus provide an important clue about the evolution of at least one class of galaxies: giant ellipticals. But what about the larger question? How do galaxies form and evolve into different types? The answers are still not clear.

Lurking in the back of astronomers' minds when they invented the "tuning fork" classification scheme was the idea that if we just arranged galaxies in the right order by shape, we would probably have an evolutionary sequence. Maybe galaxies evolved from irregulars to ellipticals or vice versa. Today this order-by-age idea seems much too simple, since all types of galaxies have at least a few old Population II stars. Furthermore, no known concentrations of intergalactic gas are big enough

to indicate current galaxy formation. Thus no single class of galaxy could have formed recently from intergalactic gas.

Current evidence suggests that most galaxies formed some 10 to 18 (probably closest to 14) billion years ago during a restricted interval. Galaxies formed by gravitational collapse (described in Chapter 18) of huge clouds of hydrogen and helium called **protogalaxies.** This explains why the oldest star groups, such as the 14-billion-year-old globular clusters, consist of hydrogen-rich Population II stars with very few heavy elements. The collapse of protogalaxy clouds was relatively fast, taking about 10^8 y for a Milky Way–size galaxy (Strom and Strom, 1982).

The various galactic types observed today are believed to have evolved due to differences in initial conditions such as mass, density, and rotation rates as well as due to the cannibalistic collisions that may have yielded the giant ellipticals. Many galaxies during collapse may have resembled an extended elliptical galaxy. According to recent theories, the smallest protogalaxies (only 10^5 to $10^9 M_\odot$) produced large globular clusters or dwarf ellipticals. Some elliptical systems may have

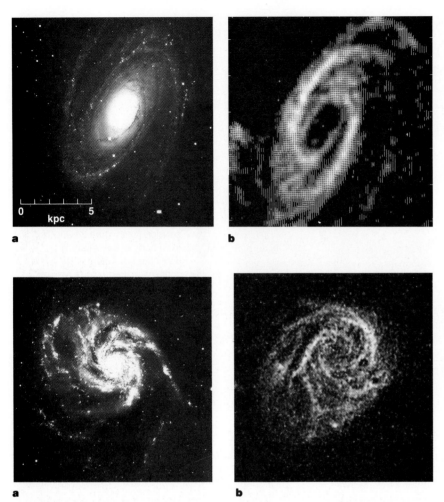

Figure 24-32 Optical and radio images of spiral galaxy M 81 (NGC 3031), about 3200 kpc away. **a** The optical image shows stars and star clusters. Scale bar applies in unforeshortened direction. **b** The radio image shows radiation from neutral hydrogen gas. Scale is about the same in both images. Important results are that (1) neutral hydrogen gas coincides with the position of Population I stars in the spiral arms, (2) there is virtually no neutral hydrogen in the center, and (3) the hydrogen spiral pattern is much larger than the visible star pattern. (Leiden Observatory.)

Figure 24-33 Optical **a** and radio **b** images of the spiral galaxy M 101 (NGC 5457), the Pinwheel galaxy, about 7200 kpc away. Comparison shows the concentration of gas in the arms and the lack of neutral gas in the center. (Image a, Kitt Peak National Observatory; image b, Westerbork Synthesis Radio Telescope, radiograph by R. J. Allen of Kapteyn Astronomical Institute, with assistance of E. B. Jenkins of Princeton University Observatory.)

spun off material that formed surrounding disks and spiral arms, hence becoming central bulges in spiral galaxies (Strom and Strom, 1982). Theoretical studies suggest that fast rotation favored formation of a bar-shaped region during collapse of the protogalaxy, possibly explaining barred spiral systems. And, as we've noted, growing evidence suggests that giant ellipticals formed by collision, though some may come from unusually large protogalaxies (around $10^{12}\ M_\odot$).

Subsequent **evolution of galaxies** and the particular **stellar populations in galaxies** depended on the initial number of massive stars formed. These stars formed as the gas in the protogalaxy subdivided into cluster-sized clouds. The initial gas was apparently all hydrogen and helium (judging from the absence of heavy elements in the oldest stars). It produced Population II stars. In the central Population II bulges of spiral gal-

axies, apparently the gas was used up in the first generation of stars. The evidence for this is seen in Figures 24-32 and 24-33, where comparisons of optical and radio images show most gas in the arms and little in the center. If the initial generation of stars in the arms contained many massive stars of $10\ M_\odot$, $20\ M_\odot$, or even more, then these would quickly explode as supernovae (see Chapter 19), spewing out gas containing heavy elements manufactured by nuclear reactions in these stars. The further evolution of the galaxy now depended on the fate of this gas (Ostriker, 1981). The heavy elements would condense into dust grains. In intermediate-mass disk-shaped galaxies, the dust would accumulate in the plane of the disk. The gas and dust would form nebular clouds and produce later generations of stars with heavy elements (Population I stars). Computer studies, as shown in Figure 24-34, have shown

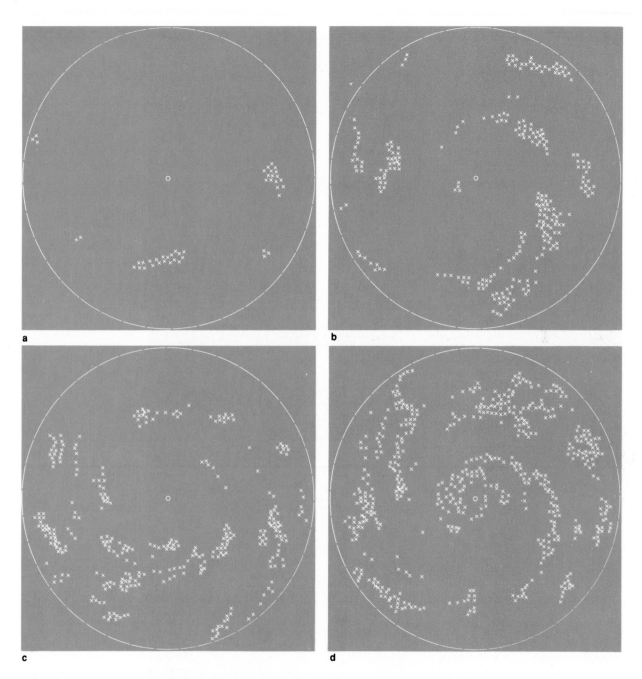

Figure 24-34 Theoretical model of spiral galaxy evolution: *x*'s represent regions of recent star formation (open clusters and concentration of bright, bluish stars). This model starts with a centrally concentrated disk of gas. Probability of star formation is related to the local gas density and the number of nearby young stars, which trigger adjacent star formation by blowing out material that compresses adjacent gas. Regions of star formation rapidly shear into spiral arms due to orbital motions of the material around the center at different rates. The model allows gas to be consumed in formation of young stars, but also returned to the interstellar medium as older stars expel gas. This example shows how a rotating, galaxy-sized gas disk develops familiar features of a spiral galaxy in only about ½ billion years. (After computer models by W. Freedman and B. Madore, 1983.)

Figure 24-35 The great barred spiral galaxy M 83 (NGC 5236) beautifully displays the consequences of stellar and galactic evolution. Here the yellowish Population II central regions are surrounded by spiral arms dominated by bluish Population I stars, massive and short-lived, forming in clusters. Scattered through the arms are red-glowing HII regions associated with local areas of star formation. M 83 is a type SBc barred spiral about 3 Mpc away from us and about 12 kpc in diameter. (Copyright Anglo-Australian Telescope Board.)

how disk-shaped dusty galaxies of this type could quickly produce spiral arms of young, Population I star clusters in only ½ billion years or so.

If the gas and dust were, instead, blown clear out of the system, no disk or spiral arms could form. Similarly, if certain galaxies produced no massive first-generation stars, supernovae would not have occurred, and there would be no dust or Population I stars. Again there would be no dusty disk. We might have a dwarf elliptical galaxy lacking Population I stars.

Study the galaxy in Figure 24-35. Nuclei of spirals have the pale yellowish-orange color of the old, Population II giants. But the spiral arms are made out of bluish clusters of young, Population I O and B stars, along with red-glowing HII regions scattered like rubies in a jewelled brooch. Thus we can actually *see* evidence of evolution among star populations in galaxies simply by looking at color pictures of galaxies!

Additional recent studies have shown that forms of some galaxies may be affected by interactions with their nearest neighboring galaxies. For instance, small galaxies in a cluster may avoid collisions, but they may be disrupted by tidal forces during close approaches to larger neighbors. Others may undergo collisions without full-scale merging, but they may be stripped of gas or altered in shape. These processes probably explain many of the distorted, peculiar galaxies.

Currently astronomers are making strides toward learning the reasons for the variety of shapes among galaxies—by means of better observations of galaxies' properties, better dynamical theories, and better computer modeling.

SUMMARY

The Local Group of galaxies, out to about 1000 kpc, contains a variety of types and sizes. Combining observations of these galaxies with data from more distant galaxies, we find that most galaxies can be classified as elliptical, spiral, or irregular. A more elaborate classification scheme, the "tuning fork" diagram, subdivides them, including barred as well as unbarred spirals. Most galaxies lie in groups called clusters.

The observations discussed here combine with the theories of stellar evolution, discussed in Chapters 17 to 19, to explain how some galaxies or parts of galaxies have evolved and used up their gas and dust, thus producing different stellar populations in galaxies.

The most massive galaxies are giant ellipticals with Population II stars and little gas or dust. Spirals appear to be intermediate-sized galaxies with "skeletal frameworks"

resembling elliptical galaxy systems—central, flattened bulges of Population II stars surrounded by a spheroidal halo of globular clusters; however, gas and dust blown out of evolved stars have also formed disks of Population I stars. Other galaxies of intermediate and small size have elliptical and irregular shapes. Some galaxies' shapes may have been affected by tidal forces or collisions.

As we look at the familiar rocks, mountains, buildings, and trees on our silicate planet, it is strange to imagine that our silicate-, carbon-, and iron-rich materials may have become concentrated only by certain processes of galactic evolution in which debris from the insides of ancient stars aggregated into star-spawning clouds in the dusty disk of a spiral galaxy. As Harlow Shapley commented, "We are brothers of the boulders, cousins of the clouds."

CONCEPTS

Magellanic clouds	giant elliptical galaxy
irregular galaxy	barred spiral galaxy
Tarantula Nebula (30 Doradus)	"tuning fork" diagram
	mass/luminosity ratio
Magellanic stream	radio galaxy
dwarf elliptical galaxy	starburst
spiral galaxy	galactic cannibalism
Andromeda galaxy	protogalaxy
Local Group	evolution of galaxies
problem of the missing mass	stellar populations in galaxies

PROBLEMS

1. Which type of galaxy tends to be biggest? Brightest? To contain fewest young stars? (See Tables 24-1 and 24-2.)

2. Of giant ellipticals, ordinary ellipticals, spirals, barred spirals, and irregulars, which type is most common? (See Tables 24-1 and 24-2.)

3. How many years would a radio signal take to reach the Andromeda galaxy?

4. Two very faint and distant galaxies were detected on photos but were too distant to allow identification of spiral arms and other typological features. If spectra showed that one was reddish with a spectrum of K-type stars, while the other was bluish with a spectrum of B and A stars, what types would you expect the two galaxies to be?

5. If you were to represent the Milky Way and Andromeda galaxies in a model by two cardboard disks, how many disk diameters apart should they be to represent the true spacing of the two galaxies?

6. In the Earth's geography, we usually designate the "up" direction as the northern rotation axis. When discussing the solar system, we use the northern ecliptic pole defined by orbital revolution. When discussing the Milky Way, we use the northern galactic pole defined by galactic rotation. Has any asymmetry or special direction appeared in this chapter that would define a preferred orientation for discussion of the distant galaxies outside the Milky Way? If so, describe it.

7. Why are small ellipticals not found in catalogs of the more distant galaxies, such as Table 24-2?

8. Irregular galaxies are dominated by stellar associations, open clusters, and gas and dust clouds, all of which indicate stellar youthfulness. Does this prove that the galaxies themselves have only recently formed?

9. Using principles of star formation and stellar evolution from Chapters 17 and 18, explain why the prominent light from star-forming regions in galaxies comes from massive, hot, blue stars.
    a. Why are these stars not seen in regions where star formation has ended?
    b. What population is indicated by hot, blue stars?

ADVANCED PROBLEMS

10. Use the small-angle equation to solve the following problems:
    a. What is the angular diameter of the main stellar part of the Andromeda galaxy if it is 40 kpc across and 670 kpc away? How does this compare with the angular size of the moon?

    b. What is the angular diameter of the 3-pc-diameter bright nucleus at the center of the Andromeda galaxy?

11. If spectral lines of stars observed on the right side of a galaxy were red-shifted 0.165 nm relative to those at the galaxy's center, while those on the left side were blue-shifted by the same amount, what would you conclude to be the rotational velocity of these stars? Assume that the spectral lines normally occur at a wavelength of 500 nm.

12. If the stars in Problem 11 were measured to be on the outer edge of the galaxy, at a distance of 4 kpc from its center, what would be the mass of the galaxy? (*Hint:* Use the circular velocity equation.)

PROJECTS

1. Locate the Andromeda galaxy by naked eye. Compare its visual appearance with its appearance in binoculars and telescopes of different sizes. Across what diameter can you detect the galaxy? ($1° = 12$ kpc at the Andromeda galaxy's distance.) Why do most photographs show a large central region of constant brightness, while visual inspection reveals a sharp concentration of light in the center? Can dust lanes or spiral arms be observed? (Check especially with low magnification on telescopes with apertures of 0.5 to 1 m [about 20 to 40 in.], if available.)

2. If photographic equipment is available, take a series of exposures with different times, such as 1 min, 10 min, and 100 min. Describe some of the differences in appearance. What physical relations are revealed among different parts of the galaxy?

3. Make similar observations of other nearby galactic neighbors of the Milky Way.

CHAPTER 25

The Expanding Universe of Distant Galaxies

The galaxies we have seen up to now vary in size, shape, and distance, but they are basically understandable. They have the same two populations of stars found in our galaxy, and we can at least partially explain their varied forms by imagining different conditions of mass and rotation in various regions of the early universe, as matter broke into protogalactic clouds.

But as we go still farther away from the Earth, we discover strange objects. As we will see, we find that all the clusters of galaxies are rushing away from each other. Remember that we are now looking at objects billions of parsecs away; thus the light from them has been traveling for billions of years. This means that we are seeing them as they were soon after their formative era. Thus, study of the most distant galaxies brings us face to face with questions about the origin and large-scale structure of the universe itself.

THE LARGEST DISTANCE UNIT IN ASTRONOMY

In dealing with stars, we used the parsec as a distance unit. In dealing with the Milky Way and nearby galaxies, we used the kiloparsec, equal to 1000 pc. The distances to remote galaxies are so great that we must use the **megaparsec** which is a million parsecs, or 1000 kpc. A megaparsec (Mpc) is about 3×10^{19} km, or 30 000 000 000 000 000 000 km!

CLUSTERS AND SUPERCLUSTERS OF GALAXIES

Among distant galaxies, as among our own Local Group, clustering is the rule. As shown in Figures 25-1 and 25-2, galaxies often appear in groups ranging from pairs

Figure 25-1 A pair of spiral galaxies, NGC 5432 and 5435. The difference in orientation shows that the planes of their disks are not parallel. Faint filaments connect them, due possibly to tidal interactions. (Lick Observatory.)

Figure 25-2 A cluster of at least five galaxies, including NGC 6027, in the constellation Serpens. (Hale Observatories.)

to collections of several hundred, with ellipticals, spirals, and irregulars mixed together in a volume about 5 Mpc across. Such groups are called **clusters of galaxies.** Mapping of galaxies, both on the plane of the sky and in terms of distance in three-dimensional space, shows that even the clusters of galaxies are clumped together. These clumps of clusters are called **superclusters of galaxies.** Although clusters and superclusters are among the most distant known entities in the universe, they may cover an angle of many degrees in our sky! For example, the Coma cluster of galaxies is about 150 Mpc away and about 8 Mpc across, and moves away from us at about 7000 km/s. Recent studies show that it is only the center of a supercluster of galaxies moving with it, which covers an area 16° or more across. The supercluster's diameter is more than 40 Mpc. Similarly, the Virgo cluster (Figure 25-3), located about 180 Mpc away, is the center of a supercluster as much as 50° (170 Mpc) across (Chincarini and Rood, 1980). As we will see in the next chapter, mapping of superclusters indicates that they themselves may be part of a larger-scale clumpy or filamentary distribution.

A cluster of galaxies is typically held together by the gravitational attraction of its member galaxies and is thus likely to be stable throughout the history of the universe. The stability of superclusters is less certain. Some of them may be expanding, and the member clusters may be insufficient in number or closeness to hold the superclusters together indefinitely.

DISCOVERING THE RED SHIFT

Routine cataloging of the Doppler shifts of galaxies led to an unexpected and amazing finding. By 1914, Doppler shifts had been published for 13 of the brightest galaxies. Strangely, most were red shifts rather than a random mixture of red and blue shifts. This phenomenon, called the **red shift of galaxies,** indicates that most galaxies are moving away from us.

Around 1920, Edwin Hubble began further Doppler shift observations with the 2.5-m reflector at Mt. Wilson, and he soon made new findings. By 1925, about 45 galactic Doppler shifts had been published, and Hubble (1936) noted that, again, red shifts completely dominated the list. He wrote:

The numerical values of the new velocities were found to be surprisingly large and of an entirely different order from those of any other known type of astronomical body.

Red shifts corresponding to recession speeds as high as 1800 km/s appeared on this early list. As Table 25-1 shows, dramatically higher speeds have since been measured.

In 1928, American physicist H. P. Robertson discovered another curious fact: The more distant the galaxy, the greater the red shift. And the relation is linear; the red shift is directly proportional to the distance.

Figure 25-3 A portion of the Virgo cluster of galaxies. This cluster is located an estimated 19 Mpc from the Milky Way. It is about 4 Mpc across and contains thousands of galaxies, any one of which might contain stars with planets similar to Earth. Several kinds of galaxies, including dusty spirals and giant ellipticals, can be seen. (Copyright Anglo-Australian Telescope Board.)

TABLE 25·1

Characteristics of Selected Clusters of Galaxies

Cluster Name	Estimated Distance (Mpc)	Diameter (kpc)	No. of Galaxies Counted	No. of Galaxies/Mpc³	Recession Velocity (km/s)
Local Group	0.6	1000	~27	50	0
Virgo	19	4000	2500	500	+1180
Pegasus I	60	1000	100	1100	+3700
Cancer	70	4000	150	500	+4800
Perseus	80	7000	500	300	+5400
Coma	100	8000	800	40	+7000
Hercules	130	300	300	20 000	+10 300
Ursa Major I	190	3000	300	200	+15 400
Leo	240	3000	300	200	+19 500
Gemini	290	3000	200	100	+23 300
Boötes	450	3000	150	100	+39 400
Ursa Major II	460	3000	200	400	+41 000
Hydra	700	?	?	?	+60 600
3C 123 and cluster	1500	?	?	?	+135 000?

(This finding applies at distances out to at least 1000 Mpc.) These statements are supported by a variety of observations. For example, Figure 25-4 shows five galaxies, their distances, and their red shifts. Additional studies show that the greater the red shift, the fainter the stars in the galaxy (indicating greater distance), the smaller the angular diameter of the galaxy (again indicating greater distance), and the smaller the angular size of the cluster of galaxies in which the observed galaxy lies.

INTERPRETING THE RED SHIFT: AN EXPANDING UNIVERSE?

According to everything we have learned so far, a red Doppler shift means that the source is moving away from us. Therefore, astronomers have concluded that all distant galaxies are moving away from us, and the farther away they are, the faster they are receding. In 1933, English astronomer Arthur Eddington titled his book on this subject *The Expanding Universe,* the name that has been given to this concept ever since.

However, the phenomenon might better be called the **mutual recession of galaxies.**

Hubble's Relation and Recession Velocities

Hubble's relation expresses how the recession velocity increases with distance. Hubble found a constant proportion between velocity and distance, known as **Hubble's constant** H. Measurements suggest H lies in the range of 50 to 100 km/s per megaparsec, and for years astronomers have been split in two camps favoring either the higher or lower value. However, an average of 11 measurements from 1980 to 1985 gives $H = 77 \pm 14$ km/s per megaparsec (Hodge, 1984; Bartel and others, 1985). This means that galaxies 1 Mpc away recede from us at 77 km/s on the average. Galaxies twice as far away recede twice as fast. Galaxies at 10 Mpc recede at about 770 km/s, and so on.

Conversely, now that this relation has been measured, we can use it to estimate distances. A galaxy with a red shift corresponding to recession at 7700 km/s is estimated to be about 100 Mpc away.

Relation Between Red Shift and Distance for Remote Galaxies

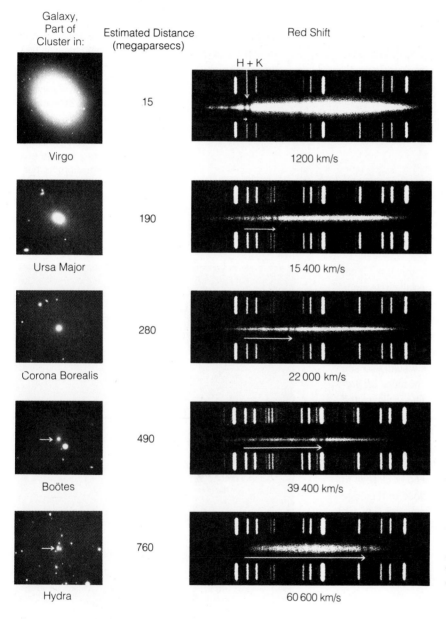

Figure 25-4 Photographic proof of red shifts of remote galaxies. Left column shows galaxies (note decreasing angular size, interpreted as caused by increasing distance). Right column shows spectra: White lines at top and bottom of each spectrum are emission lines produced in the instrument for comparison, being similar in each spectrum. A pair of dark absorption lines (the H and K lines of gaseous calcium) can be detected in each galaxy's spectrum, above the head of the white arrow. This pair is farther right (red) in each succeeding galaxy. The center column shows the inferred distance. (Hale Observatories.)

Hubble's constant H is one of the most important parameters in astronomy—not only because it helps us estimate galaxy distances, but also because it helps us measure the age of the universe (as will be clearer in Chapter 27). Modern astronomers are expending much effort with very large telescopes to observe faint galaxies and pin down a more precise value of H. Some results during the 1980s suggest a trend toward higher

H values as we go from nearer to farther galaxies—a possibility currently under vigorous study.

Are We at the Center of the Universe?

If all distant galaxies are moving away from us, does this mean that we are located at the *center* of a vast

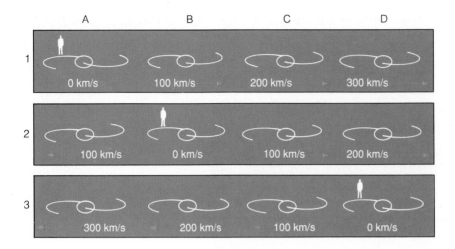

A B C D

1 0 km/s 100 km/s 200 km/s 300 km/s

2 100 km/s 0 km/s 100 km/s 200 km/s

3 300 km/s 200 km/s 100 km/s 0 km/s

Figure 25-5 The mutual recession of galaxies. Rows 1, 2, and 3 are three equivalent sketches of the same situation. In row 1, an observer in galaxy A sees the other three galaxies moving away at the speeds shown. But an observer in galaxy B would perceive the situation shown in row 2, with his own galaxy considered stationary and *other* galaxies moving away. An observer in galaxy D would draw the situation as shown in row 3. Each observer would perceive all other galaxies moving away from his/her own galaxy, and in all three cases the recession speed is proportional to distance.

explosion—that is, at the *center* of the universe? That concept would mean that we have come full circle since the Copernican revolution removed us from a special position at the center of our solar system. If we are at the center of the universe, we would occupy a special position. The answer is that *we are not necessarily at the center.* Observers in any other galaxy would also see galaxies moving away from them, too, because *all galaxies are moving away from each other.* This is illustrated in Figure 25-5.

In other words, we may very well be riding out the aftermath of a vast explosion, but the view from one flying spark is like the view from any other. There is no outside reference point to define who is closest to the center, unless we found we were close to the edge of the swarm of sparks. But no astronomer has yet found an edge to the swarm of galaxies. They seem to go on and on.

Red Shift Theory I: Cosmological Red Shifts

Everything we've said so far assumes that the observed red shifts should truly be interpreted as **Doppler shifts.** The Doppler effect has certainly been confirmed with planets and with stars. But the idea that it explains *all* red shifts of galaxies is only an assumption, called the **theory of cosmological red shifts.** This assumption, incidentally, underlies the last column of Table 25-1. But there are other possibilities.

Figure 25-6 Stephan's quintet, a cluster of galaxies (NGC 7317–20), gives evidence that some red shifts may not be directly related to distance. The galaxies are still believed to lie at the same distance, which can be estimated from various indicators, such as angular size. Red shifts of the four small galaxies, however, differ from that of the large galaxy (upper left) and are larger than expected from distance estimates. (Lick Observatory.)

TABLE 25·2

Discrepancies in Red Shifts Within Galactic Clusters

Cluster	Number of Galaxies	Lower Red Shift[a] (no.)		Higher Red Shift[a] (no.)		Discrepancy (km/s)
Stephan's quintet	5	0.0027	(1)	0.02	(4)[b]	+5 200
Seyfert's sextet	6	0.015	(5)	0.066	(1)	+15 300
VV172 system	5	0.053	(4)	0.12	(1)	+20 000
NGC 7603 pair	2	0.0266	(1)	0.053	(1)	±8 000
NGC 4319/Makarian 205	2	0.006	(1)	0.07	(1)	±19 000
NGC 2903	4	0.002	(1)	0.01–0.03	(3)	±6 000

[a]Data from Balkowski and others (1974), Hoyle (1972), and Hodge (1972). Red shift = shift in wavelength/wavelength.

[b]Radio and other studies indicate that the correct distance of Stephan's quintet is small, corresponding to a low red shift. Thus it would appear that in all of the clusters, the higher red shift is the anomalous one, even when several galaxies have higher red shifts.

Red Shift Theory II: Noncosmological Red Shifts

Physical effects besides motion away from us could also cause red shifts of some galaxies. Then the shifts would be ambiguous: Part might be caused by recession and part by something else. In fact, we know at least one other possible cause of red shifts. The theory of relativity shows that light emitted from regions of very high gravity, such as the environs of a black hole, will be redshifted. Thus at least some of the shifts we observe may be **gravitational red shifts.**

The minority of astronomers who argue that red shifts of some galaxies are nongravitational point to certain curious examples in the sky. Sometimes, for instance, astronomers find a small galaxy with a large red shift that seems to be closer to us than some distant, less red-shifted galaxies. For example, in Figure 25-6 we see a cluster of galaxies that seem to be interrelated and close together, yet one of them has a radically different red shift. Table 25-2 lists six instances in which one galaxy in a cluster differs from the other members in "velocity" (calculated from the theory of cosmological red shifts) by 5000 to 20 000 km/s. All these examples suggest to some astronomers that there may be something else affecting red shifts besides simple Doppler shifts. This theory—that some red shifts of galaxies are not Doppler shifts—is called the **theory of noncosmological red shifts.**

Most astronomers, however, are convinced that most galactic red shifts are Doppler shifts due to recession. The main reason is Hubble's and Robertson's discovery that among the nearby galaxies, red shift is correlated primarily with distance as judged by angular size and other properties, as in Figure 25-4. The red shift does not seem correlated with mass, gravity, or other properties. Some of the nearest galaxies even show blue shifts; blue shifts can't be explained by the theory of noncosmological red shifts, but they are easily explained as Doppler shifts due to approach.

INTENSELY RADIATING GALAXIES

As we move to further and further distances in intergalactic space, and to greater red shifts, we encounter seemingly endless numbers of ordinary galaxies. At very great distances, most are too faint to detect. Nonetheless, there are a few superluminous, intensely radiating galaxies that can be spotted at very great distances.

These intensely radiating galaxies give off stupendous amounts of energy. Thus they are interesting for two reasons: They can be detected at extremely large distances, and they raise important questions about how they can radiate so much energy. The extra energy, compared to normal galaxies, may appear as a combination of excessive visible light, infrared emission, radio emission, X-ray emission, and other forms of radiation. As we saw in the last chapter, the galaxies that are peculiarly luminous at radio wavelengths are called **radio galaxies.** They are just one example of intensely radiating galaxies. In the 1950s, early radio astronomers discovered that the radio emission from many radio galaxies was **synchrotron radiation,** a type of radiation produced in very hot gases when ions in magnetic fields are accelerated to nearly the speed of light. Very energetic processes must be occurring in these galaxies to get the gas so hot. While normal galaxies emit about 10^{31}–10^{33} watts at *radio* wavelengths, radio galaxies may emit 10^{33}–10^{38} watts—hundreds to millions of times more energy per second. The big question in research on intensely radiating galaxies is: What processes cause such tremendous amounts of radiation from certain galaxies?

Colliding Galaxies

As mentioned in the last chapter, one subclass of radio galaxies is colliding galaxies. As noted in that chapter, the gas throughout much of these galaxies is excited and stimulated to radiate when they crash into each other. But in this chapter the focus will be on the other class of radio galaxies—still more exotic objects where the radiation comes from mysterious galactic nuclei with extraordinary luminosities.

Galaxies with Active Nuclei

In this second subclass of intensely radiating galaxies, called **galaxies with active nuclei,** most of the excess luminosity (beyond that of normal galaxies) comes from the nucleus. The nucleus may be visibly much brighter than the nucleus of an ordinary galaxy. At least 5% of all galaxies have active nuclei, according to a 1974 study by French-American astronomer Marie-Helene Ulrich.

In these galaxies, the intense radiation must be associated with some extraordinarily strange process occurring in the galaxies' nuclei. This process is not well understood but is the subject of challenging research. It seems to work in varying degrees, producing a range of galaxy types. Some galaxies merely have an unusual, bright nucleus; others have active nuclei that produce enormous explosions and blow out clouds of gas or strange jets of material that speed out of the galaxy in opposite directions. Many are strange in appearance (for example, Figure 25-7). The most extreme cases are the weird objects known as quasars.

Before describing each type, we note that collisions between galaxies may play a role in "turning on" an active nucleus, perhaps triggering formation of giant black holes with their surrounding, high-temperature accretion disks (Gunn, 1979). This theory has been difficult to prove, but it continues to gain evidence. A 1987 study revealed an example of a galaxy with a modestly bright nucleus that has been distorted in shape by a close encounter with another galaxy (Simkin and others, 1987). Another 1987 theoretical study concluded that a crowded cluster of stars at the center of a galaxy could collapse to produce a gigantic black hole with ten thousand to many millions of solar masses—enough to power a typical active galaxy (Kochanek and others, 1987). Perhaps a collision between two galaxies could trigger such a collapse. But perhaps some galaxies evolve to this state on their own when supernovae or other events in the nucleus trigger the formation of a giant black hole or periods of extreme luminosity. In any case, the various active-nucleus galaxies that we will now explore are among the most interesting and mysterious objects in the universe.

Seyfert Galaxies: Weak But Active

In 1943, astronomer C. K. Seyfert studied some unusual galaxies with small, intensely bright nuclei, bright spectral emission lines, and brightness fluctuations. Seyfert concluded that the nuclei of these galaxies had hot gas in violent motion, often with velocities of several thousand kilometers per second. About 2% of all galaxies have these properties. They came to be called **Seyfert galaxies.** Figure 25-8 shows two examples.

In the 1960s Seyferts were found to have radio and infrared emission similar to that of radio galaxies. They seem to be a "missing link" between ordinary and radio galaxies. As shown in Figure 25-9, they share the general trend of increasing red shift with decreasing apparent size—strong evidence that the red shift does increase with increasing distance.

The unusually large and varying amounts of energy of Seyfert galaxies imply very energetic, explosive bursts in their nuclei. Of course, even ordinary galaxies have strikingly energetic events occurring in their centers—synchrotron radiation and expanding hydrogen clouds occur in our own galaxy as well as others in a core region about 100 pc across. But Seyferts are even more energetic, especially in emissions at radio and infrared wavelengths.

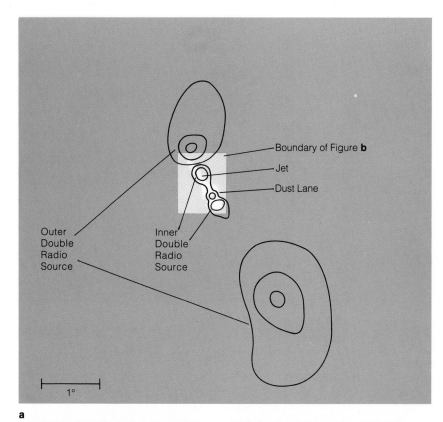

Outer Double Radio Source

Inner Double Radio Source

Boundary of Figure **b**

Jet

Dust Lane

1°

a

Figure 25-7 The unusual galaxy NGC 5128, coincident with intense radio source Centaurus A, about 4.4 Mpc away. The galaxy appears to be a giant elliptical with an unusual dust lane across its center. **a** Sketch of radio and optical features. Contoured "lobes" are clouds of radio-emitting gas on either side of the galaxy, which also emits. Inner rectangle represents area of *b*. **b** Long exposure shows the overexposed galaxy and a faintly glowing jet of gas extending along the direction of the radio lobes. (Cerro Tololo Interamerican Observatory, courtesy NOAO.) **c** Color-moderate exposure shows dust lane and features of the elliptical galaxy with its creamy Population II stars. **d** Shorter exposure of core region shows details of thick, dark dust. (Images *c* and *d*, copyright Anglo-Australian Telescope Board.)

b

c

d

Explosions and Jets in Galactic Nuclei

Certain galaxies that emit more radiation than Seyfert's have ragged filaments and long, straight jets of material speeding out of their centers. One example is radio galaxy M 82 (NGC 3034; Figure 25-10). It looks like an edge-on disk, heavily obscured by dust, but with radio-emitting, red-glowing hydrogen splattered out from explosions in the center. At a distance of "only" 3 Mpc, this nearby galaxy has been well studied, and has been called "the benchmark" in the study of disturbed, explo-

sive galaxies (Telesco, 1988). The filaments shoot out at 1000 km/s and contain an estimated 5 million $M_\odot$ of gas. Infrared studies have penetrated the dust to reveal an amazing picture of the central conditions. A rotating disk of material centered on the nucleus, with radius of only 250 pc, contains 30 billion $M_\odot$ of material, 40 young supernova remnants, and many Orion-like nebular complexes spaced only a few parsecs apart! Rapid star formation has been going on there for at least 3 My.

Somewhat similar is NGC 1275, which is an X-ray source, a Seyfert galaxy, and is also known as radio source Perseus A (Figure 25-11).

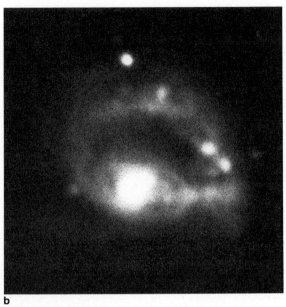

a

b

Figure 25-8 Two Seyfert galaxies—an ordinary spiral and a distorted spiral or ring. The ratio of nucleus brightness to spiral arm brightness is much greater in Seyferts than in ordinary spirals. **a** NGC 1068 (M 77) is roughly 10 Mpc away and has an angular width of about 2.5′, with some still fainter outlying regions not seen here. (Lick Observatory.) **b** VV285 shows a violently active nucleus and faint, distorted arms. (Kitt Peak National Observatory.)

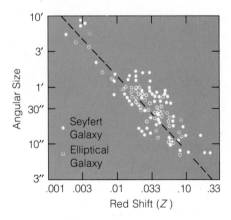

Figure 25-9 Evidence that an increasing red shift in Seyfert and elliptical galaxies corresponds to increasing distance. Z is the red shift, expressed as a fraction of the original wavelength. Galaxies with the greatest red shift have the smallest angular size, indicating they are farthest away. (After data of Khachikian and Weedman, 1974.)

In both of these galaxies, the red-glowing hydrogen filaments resemble the debris from the Crab supernova explosion. (Compare with the Crab's appearance in Figure 19-12.) This suggests that explosions in galactic nuclei may resemble superviolent supernova explosions.

Figure 25-12 shows the related phenomenon of **galactic jets**—narrow, straight beams of material shooting out of galaxies. Here we have a 1300-pc-long jet, glowing by synchrotron radiation, shooting out of the nucleus of the giant elliptical galaxy M 87. (Compare with the full view of this galaxy's outer regions in Figure 24-23.)

There are many other examples. Most common are bipolar jets extending in both directions. This feature is seen, for instance, in spiral galaxy NGC 1097, where jets extend in opposite directions from the galaxy's center (Figure 25-13). These bipolar jets are reminiscent of those we encountered in star-forming disks (page 398) and in binary star SS-433 (page 473). Assuming that the disk is about 25 kpc across, the jets extend out to 60 kpc but are only 2 kpc wide.

Recently radio astronomers have been able to use groups of radio telescopes linked in tandem to produce extremely detailed images of the radio radiation from such galaxies. Just as a photograph taken with a red filter records material emitting light at red wavelengths, these radio images record material emitting radio waves. These images show that in addition to having diffuse blobs of hydrogen blown outward from active nuclei (as was shown in Figure 25-7), many radio galaxies have remarkable radio-emitting jets streaking out in narrow beams reaching from within a few parsecs of the nucleus out to thousands of parsecs. Three examples are shown in Figure 25-14. In some of these examples, the jets are bent back—perhaps by motion of the galaxy through thin intergalactic gas, like the plume of an old-time locomotive moving forward through the air. Figure 25-15 shows an amazing example of two close galaxies, perhaps interacting, both spouting bipolar jets that shoot out, bend back, and seem to intertwine. As seen in Figure 25-14b and c, diffuse blobs of hydrogen sometimes lie a million parsecs or more beyond the ends of the jets. (See review by Blandford and others, 1982.)

What is the physical nature of these jets? How are they produced? We don't have very good answers. They are apparently narrow columns of plasma glowing by synchrotron radiation. They are very hot and have somehow been accelerated to high speeds. Some observations suggest that the material in some jets is moving outward at nearly the speed of light![1] These jet-emitting and exploding galaxies are among the most energetic known objects in the universe! Blobs of radio-emitting gas around certain radio galaxies have been estimated to contain 10^{54} joules of energy—as much as 10 billion supernova explosions—a "majestic phenomenon" indeed, in the words of astrophysicist Frank Shu (1982).

The Mystery of the Galactic Nucleus

An emerging theory that explains intense radiation, explosions, and jets in galactic nuclei involves gradual

---

[1]One such line of evidence is so-called superluminal motion: an apparent motion faster than light. This can be explained as an optical effect if the jet is aimed nearly at us and the material is actually moving a little slower than light. Then the light waves emitted from a blob in the jet at two widely separated points reach us at nearly the same time and give the impression of fast motion. Physicists still believe that no matter can travel faster than light.

a

b

Figure 25-10 Peculiar galaxy NGC 3034 (M 82) coincides with intense radio source 3C 231. This galaxy, 3 Mpc away, appears to have undergone explosive activity. **a** The galaxy displays thick, chaotic dust clouds silhouetted against a probable edge-on disk. Red-glowing excited hydrogen forms a projecting cloud near the center. (National Optical Astronomy Observatories.) **b** Semi-false color image is made by printing images made from blue light, white light, and the red Hα emission wavelength in blue, green, and red, respectively. This process reveals a bluish diffuse disk and a dramatic splatter of red-glowing, excited hydrogen gas, apparently blown out of the central regions. Hydrogen expansion speeds of 1000 km/s confirm the explosive violence and suggest major explosions as recently as 1.5 million years ago. (Original image, Hale Observatories; color image processing by Jean Lorre, Jet Propulsion Laboratory.)

Figure 25-11 Evidence for explosive activity in the nucleus of Seyfert and radio galaxy NGC 1275. In the light of Hα emission, the central region is resolved into semiradial streamers. Compare with similar patterns of eruptive hydrogen streamers in explosive objects of various sizes, such as Crab supernova cloud (Figure 19-12, p. 420), Tarantula "supernebula" (Figure 24-6, p. 528), and explosive radio galaxy M 82 (Figure 25-10b). Several apparent elliptical galaxies (fuzzy ovals) are nearby. (Sharp circular images are foreground stars in our Galaxy.) The galaxy lies about 50 Mpc away; filaments extend about 14 kpc from the galaxy. (Kitt Peak National Observatory, courtesy Roger Lynds.)

spiraling of dust and gas toward the centers of galaxies, where it forms huge clusters of stars and supermassive stars that explode into bright supernovae, with eventual collapse or merger of the star cluster to form a giant black hole. This scenario is still theoretical, but it is consistent with the evidence for a giant black hole at the center of our own galaxy. One clue about the process comes from the jets emitted from cocoon nebulae and double star SS 433. Recall that as the gas/dust mixture

spirals inward in the disk-shaped nebulae around the stars in SS 433, it gets accelerated and ejected from the disk in two jets. A similar idea is proposed for a spiral galaxy. Think of its inner region as a giant phonograph record. Dust and gas spiral inward along the grooves and accumulate in the center. Notice the difference between the slow inward spiral of the dust/gas mix and the endless elliptical orbit of a planet around the Sun. The planet repeats virtually the same orbit

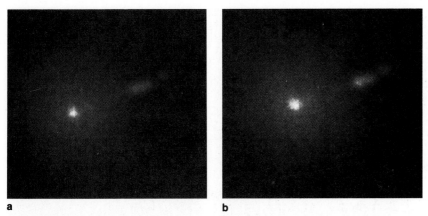

a   b

Figure 25-12 The luminous jet extending from the nucleus (left blob) of the giant elliptical galaxy M 87 (NGC 4486; compare with Figure 24-23 for full disk view). M 87 coincides with the strong radio source Virgo A. The jet and radio emission suggest violent activity in the galaxy's nucleus. The jet is bluish in color and emits 10 times as much X-ray emission as it does visible and radio emission. Comparison of these views suggests possible changes in the relative brightness of knots along the jet between 1956 and 1978. The jet is about 20 seconds of arc long, corresponding to a length of about 1300 pc, with individual knots as small as 100 pc across. (Mt. Wilson and Palomar Observatories photo, computer processed by Jean Lorre, Jet Propulsion Laboratory.)

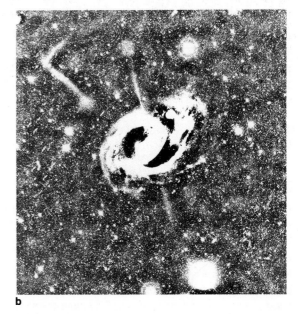

a   b

Figure 25-13 Two views of galaxy NGC 1097 and its system of jets. **a** A long-exposure view of this system, which appears to be a barred spiral galaxy roughly 10 Mpc away. Its angular size is about 9 minutes of arc. (Cerro Tololo Interamerican Observatory.) **b** A computer-enhanced version of a increases contrast and brings out faint jets and wisps extending in opposite directions out of the galaxy's disk. Origin of jets is uncertain but may involve extremely energetic events in the nucleus. (Courtesy Jean Lorre, Jet Propulsion Laboratory.)

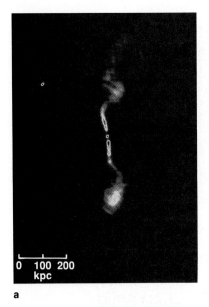

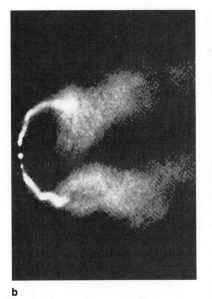

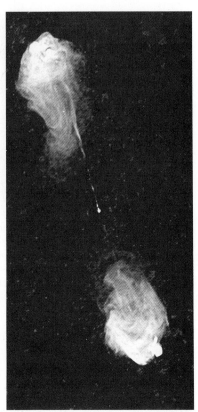

a

b

c

Figure 25-14 High-resolution radio images of jets being emitted from radio galaxies (bright central objects). The jets are hot, fast-moving gas giving off synchrotron radiation from electrons moving in magnetic fields. They expand into larger blobs of hydrogen at a distance from each galaxy. **a** Linear jets moving out from elliptical galaxy 3C 449, at an estimated distance of 66 Mpc. Approximate scale is indicated. (Computer processing yielded dark cores in images of the galaxy and jets; they are actually bright.) **b** The jets from elliptical galaxy NGC 1265 are believed to be bent to the right due to leftward motion of the galaxy at about 2000 km/s through intergalactic gas in the Perseus cluster of galaxies. **c** Linear jets and blobs streaming out of giant elliptical galaxy 3C 405, also known as Cygnus A. (National Radio Astronomy Observatory, operated by Associated Universities, Inc. under contract with the National Science Foundation; image *b*, courtesy C. P. O'Dea; image *c*, R. Perley, J. Dreher, J. Cohan.)

because there is no resisting medium to slow it or make it lose energy. But the dust/gas mix experiences turbulence and collisions among its grains and molecules, causing it to lose energy. This explains why, on each "orbit" around the center, it spirals inward. The central region thus becomes rich in star-forming material. Many stars are born, creating a unique massive cluster at the galactic center that resembles a giant globular cluster of stars. The most massive of these stars may have hundreds or thousands of solar masses; they quickly explode as supernovae, adding to turbulence and outbursts early in the galaxy's history. The most massive supernova stars, of course, leave black holes. In the central star cluster of a young galaxy, therefore, we find densely packed stars, black holes, and neutron stars, all orbiting in confused paths through masses of dust and gas. Close encounters between stars are frequent.

A star passing close to a massive black hole can be torn apart by tidal forces (Evans and Kochanek, 1989), adding to the turbulent gas. As the stars and gas lose their orbital energy because of these interactions, they continue to spiral inward, and the whole central star cluster shrinks. These effects are much stronger in the crowded central region than among the outer spiral arms, and thus only the central cluster shrinks, while the outer spiral arms are more stable during the lifetime of the galaxy.

According to various theoretical calculations (e.g., those by California Institute of Technology astrophysicist C. S. Kochanek and his colleagues, 1987), the central part of the central star cluster collapses into a supermassive black hole containing from a million to a billion solar masses, surrounded by a dense accretion disk radiating X rays, gamma rays, and ultraviolet light.

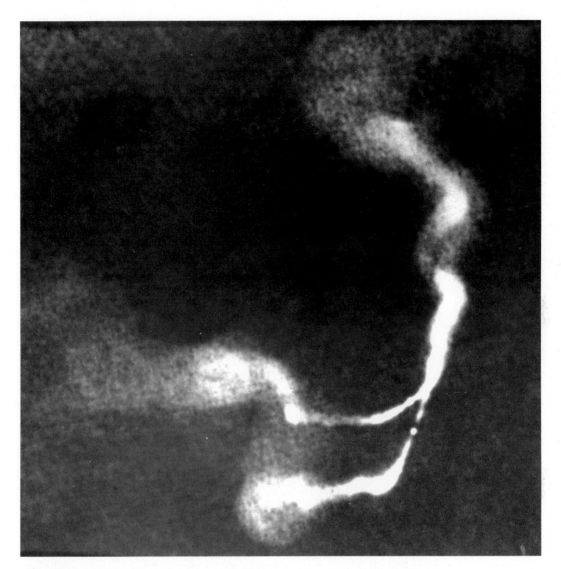

Figure 25-15 This extraordinary radio image shows two close galaxies each emitting bipolar jets of plasma giving off synchrotron radiation. The jets may interact. The two galaxies are cataloged as radio source 3C 75 in the cluster of galaxies called Abell 400. (National Radio Astronomy Observatory, AUI; image obtained by observers F. Owen, C. O'Dea, and M. Inoue.)

The "black hole feeding frenzy" occurs, as millions of stars fall into the central black hole, adding to its mass. The idea of star cluster collapse to form a huge central black hole was first developed as early as 1965 by the Soviet astrophysicist Y. B. Zel'dovich and colleagues, but only in the 1980s did it emerge as a favored theory.

Before going on to describe how this theory explains different types of active nuclei, let us mention obser-

vations that support the theory. Direct observational evidence for a supermassive object in a galactic nucleus came in 1983 from a team of European astronomers using the International Ultraviolet Explorer satellite (Waldrop, 1983). They observed spectral emission lines in the nucleus of NGC 4151, a nearby Seyfert galaxy. The lines are widened, both to the red and blue; the researchers assumed this is due to the rotation of the

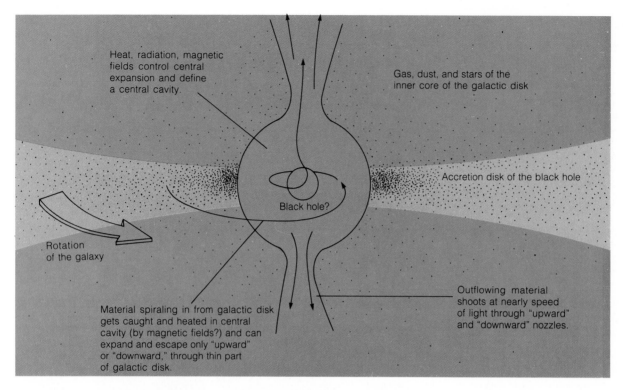

Heat, radiation, magnetic fields control central expansion and define a central cavity.

Gas, dust, and stars of the inner core of the galactic disk

Accretion disk of the black hole

Black hole?

Rotation of the galaxy

Material spiraling in from galactic disk gets caught and heated in central cavity (by magnetic fields?) and can expand and escape only "upward" or "downward," through thin part of galactic disk.

Outflowing material shoots at nearly speed of light through "upward" and "downward" nozzles.

Figure 25-16 Schematic cross section well inside the central parsec of a galaxy with jets. According to some theories, a cavity and "nozzles" may form around a central black hole. Inward-spiraling material from the accretion disk is heated and shot out through the nozzles.

brilliant nucleus. They measured the red shift from the side turning away from us and the blue shift from the side approaching us, and obtained rotation speeds such as 11 000 km/s for the gas clouds emitting a line from ionized magnesium atoms. We have seen that Seyfert nuclei are variable. By chance, this one flared up during the observations. Thirty days later, the magnesium line flared in response to the extra radiation arriving from the nucleus to the magnesium atoms. This means the distance of the cloud from the nucleus must be the distance light travels in 30 d, or about 5000 AU. Similar results were obtained for other lines, indicating clouds from about 2200 to at least 62 000 AU from the nucleus. As a result of their good fortune in observing the flare-up, the astronomers now knew the distances as well as the velocities of their clouds. Assuming the velocities were due to orbital motion around the nucleus, they could calculate the mass of the nuclear object, using Kepler's and Newton's laws (see Chapter 4). The answer came out a whopping $10^8 \ M_\odot$! Such a nucleus would

contain the equivalent of 100 million Suns in a space comparable to our solar system and its comet halo! It sounds suspiciously like the hypothesized supermassive black hole, or swarm of black holes, but it could also be a swarm of neutron stars or more conventional stars. Reporting this discovery, the journal *Science* aptly called it "The Monster in the Middle."

Additional observational techniques have been used to probe these monsters. Astronomers can use Doppler shifts to observe the orbital motions of stars within a few parsecs of the nucleus of nearby galaxies. Assuming Kepler's laws of orbital motion, they can then estimate the mass near the center, even though distances smaller than a parsec or so are too small to resolve in these galaxies. Astronomers pursuing these techniques have been suggesting that the nuclei of various galaxies are objects with a few 10^6 to $10^9 \ M_\odot$ and note that their light does not match the light expected from this many stars. Hence these astronomers interpret these nuclei as black holes with a few million to a billion solar masses

Figure 25-17 If we lived in the inner spiral arms of a galaxy with a jet, the "milky way" in our sky might present this curious appearance. Looking from a volcanic mountain at dusk, we see the "milky way" band at left, but the dim searchlight-like beam of the jet shoots across the sky at right angles to it. The picture assumes a planet at a somewhat higher distance above the galactic plane than Earth, so that we get a higher-angle view past the spiral arms into the bright region of the nucleus. (Painting by author.)

(Anderson, 1987). In particular, this work suggests that the great Andromeda galaxy has a black hole nucleus with 30 to 70 million $M_\odot$, while its little elliptical galaxy satellite, M 32, has a black hole nucleus with 8 million $M_\odot$ (Dressler and Richstone, 1988).

As gas spirals into the accretion disks around such supermassive nuclei, it is heated and ionized and may also be whipped around by magnetic fields. Theorists believe that the magnetic forces may control the gas's motions, causing it not to spiral into the black hole but to accelerate "upward" and "downward," out of the disk's plane. The region in the center of the accretion disk thus forms a double nozzle, pointing "up" and "down," with its edges defined by strong magnetic fields, as shown in Figure 25-16. Depending on physical conditions in the center, such as the mass of central black holes, the speed of sound in the central gas region, the magnetic field strength, and so on, the infalling gas may avoid crashing into the central massive black holes or stars, and instead be blown out through one of the twin exhausts of the magnetic nozzle, forming either a continuous stream of plasma blobs or a continuous narrow jet. If we lived in a galaxy with a luminous jet, the night sky might present an unfamiliar spectacle (Figure 25-17).

Calculations by Illinois astrophysicist Michael Smith and colleagues (1983) suggest three regimes of jetting activity, depending on the luminosity of the object (the

accretion disk around a monstrous, $10^9 \, M_\odot$ black hole?) in the central cavity:

1. At "low" luminosity (but already higher than in normal galaxies), small clouds of gas are ejected, but the nozzle closes after each cloud is pushed out. Seyfert galaxies are in this luminosity range.

2. At intermediate luminosity, the nozzle stays open and a narrow, steady jet forms.

3. At the highest luminosity, large clouds of gas are blown out, probably explaining galaxies with diffuse blobs of hydrogen moving out in opposite directions, as in Figures 25-7a and 25-14c.

Observations provide some support for this theory of active galactic nuclei. For example, Dutch radio astronomer J. van der Hulst and his colleagues (1983) made detailed radio maps of the center of radio galaxy NGC 5128 (Centaurus A) and found clouds of hydrogen, perhaps 10 pc across and 100 pc from the nucleus, falling into the nucleus at about 35 km/s. They believe these may produce a total accretion rate of 0.1 $M_\odot$/y falling into the nucleus and creating enough energy to fuel that galaxy's radio emission. Similarly, M. Smith and colleagues (1983) interpret the very strong radio galaxy Cygnus A as having a central mass of $10^9 \, M_\odot$, surrounded by some $10^4 \, M_\odot$ of gas, with explosive activity blowing gas *outward,* in this case, at about 10 $M_\odot$/y.

a
b

Figure 25-18 These two prints from the same negative of NGC 1398 illustrate the difficulties of analyzing galactic forms from images of the nucleus. **a** When only the bright inner structure is visible, the galaxy seems to consist of a bright bar and ring or tight spiral arms. **b** The second print with overexposed nucleus reveals faint, open spiral arms. NGC 1398 is an example of a barred intermediate spiral with a ring structure. (Original image from Hale Observatories, reprinted by author.)

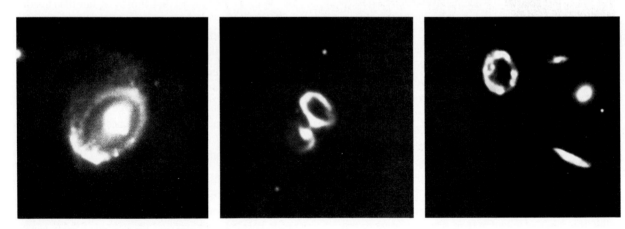

Figure 25-19 Peculiar galaxies, including examples with "missing nuclei." These forms probably result from collisions between galaxies. (Kitt Peak National Observatory.)

Before we hypothesize too much about energetic events in galactic centers, we should pause to remember that galactic nuclei are hard to study. Among remote galaxies, the nucleus may be all that is visible, or it may be partly hidden by dust. Figure 25-18 shows a difficulty in relating the nuclear structure to the outlying struc-ture. Further, as shown in Figure 25-19, there are puzzling examples of odd-shaped galaxies, including some that appear to lack nuclei altogether! These are probably more examples of how collisions can weirdly alter a galaxy's structure. The puzzling nucleus of our own galaxy—virtually on our doorstep only 0.009 Mpc away—

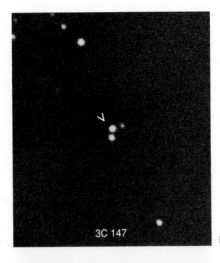

Figure 25-20 Examples of quasars. **a** 3C 48 and **b** 3C 147 are typical representatives, showing quasi-stellar images. **c** 3C 273, one of the brightest and closest quasars, is accompanied by an apparent jet of gaseous material. **d** 4C 37.43 shows some indication of nebulosity immediately adjacent to the image; the fainter image to its left is a galaxy. Distances may be around 1000 Mpc. Compare 3C 273 jet to that in Figure 25-12. (First three photos, Hale Observatories; 4C 37, Mauna Kea Observatory, University of Hawaii, courtesy Alan Stockton.)

is hidden from us by dust. Commenting on the mystery of galactic nuclei, astronomer W. C. Saslaw quotes the poet Robert Frost:

We dance around in a ring and suppose,
But the Secret sits in the middle and knows.

QUASARS: THE MOST ENERGETIC GALACTIC NUCLEI?

Quasars (an acronym for *quasi*-stell*ar* radio source)[2] are yet another strange type of object that may represent a still more energetic type of active galaxy. Quasars were discovered when optical astronomers began to photograph visible objects that could be identified with the sources of certain radio signals. By 1960,

---

[2]Many astronomers prefer QSS for *quasi-stellar source*.

although many radio sources had been photographed and identified as galaxies, a few seemed to coincide only with faint, starlike objects. Were they really stars? If so, they were unusual, sometimes with unrecognizable emission lines in their spectra and sometimes accompanied by faint wisps of nebulosity, as shown in Figure 25-20. Hence they were designated *quasi-stellar*.

A review of old photographs showed that quasi-stellar object 3C 273 (number 273 in the Third Cambridge Catalog of radio sources) varied irregularly in brightness. Palomar observer Martin Schmidt discovered in 1963 its large red shift—0.16 of the normal line wavelength. This was the first proof of *large red shifts for quasars*.

Doppler shifts of many fainter quasars are even larger. Many are larger than 1.0, which, according to Einstein's theory of relativity, corresponds to a recession velocity of 60% the speed of light. (Optional Equation IX describes the treatment of Doppler shifts for velocities approaching the speed of light, according to the theory of relativity.) A few faint quasars have enormous red shifts exceeding 4.0, which means that the *shift* in wavelength exceeds four times the original wavelength, and that the quasar is moving away from us at more than 92% the speed of light!

In 1965 other objects were found that appeared to be faint bluish stars and had spectra similar to those of quasars, but they were not strong radio sources. These were identified as radio-quiet quasars, sometimes called **QSO**s, or *quasi-stellar objects*. Over 1500 quasars and QSOs are now cataloged. We will group both types under the term *quasars*.

What are quasars? As Figure 25-21 shows, they are very difficult to observe. The interpretation is based in part on spectra. The spectra of some quasars have absorption lines as well as emission lines. Star-formed elements such as carbon and oxygen are present, suggesting that quasars contain heavy elements produced inside evolved stars. Often the absorption lines are less red-shifted than the emission lines, suggesting a surrounding cloud of absorbing gas that moves outward from the bright quasar center.

Quasars radiate at virtually all wavelengths—X ray, gamma ray, visible, and radio. Although quasars were discovered in the 1960s through their radio and optical emissions, surveys with orbiting X-ray telescopes have revealed many more; indeed, only the most powerful quasars are strong radio sources.

Observations strongly suggest that quasars are somehow related to nuclei of active galaxies. Correspondences to Seyferts, noted as early as the 1960s, include spectral features, variability, and tiny angular size of the radio-emitting region (often less than 0.001 second of arc). A breakthrough came in 1978 when South African and European observers identified a Seyfert galaxy that is also an X-ray source and has an intensely bright nucleus that is a quasar. This quasar galaxy, ESO 113-IG45, is shown in Figure 25-22. It is about 250 Mpc away and is probably a distorted spiral. Other links to active galactic nuclei are known. As seen in Figure 25-20, quasar 3C 273 has a radial jet resembling that in the brilliant radio galaxy M 87. A final breakthrough came in 1983 when Balick and Heckman showed that fuzzy regions *around* some quasars have the spectra, size, and brightness of large groups of stars; they concluded that "it seems fairly certain that quasars are the active nuclei of galaxies."

Detailed observations reveal the actual dimensions

OPTIONAL BASIC EQUATION IX

The Relativistic Doppler Shift

The text notes examples of quasars where the red shift in wavelength exceeds 4.0 times the original wavelength, λ. The perceptive reader will now have a question: How can this be true if nothing can travel faster than light? According to the Doppler equation, Optional Basic Equation VII,

$$\frac{\Delta\lambda}{\lambda} = \frac{v}{c}$$

If the shift is four times the original wavelength, then $\Delta\lambda/\lambda = 4.0$ and we might conclude the object is moving away from us at 4.0 times the speed of light!

The explanation is the important principle discovered by Einstein and his fellow physicists early in this century: Almost all the equations of classical physics (the work of Newton, Doppler, and many others) are only approximations that give accurate answers at everyday speeds; but for phenomena involving speeds near that of light, the old equations are inadequate. The latter speeds are called relativistic speeds. The equations that give correct answers (that is, answers confirmed by experiment) at both low speeds and relativistic speeds are derived from the theory of relativity. The relativistic Doppler shift equation is

$$\frac{\Delta\lambda}{\lambda} = \sqrt{\frac{1 + v/c}{1 - v/c}} - 1$$

This is more complex than our simple low-speed Doppler equation! The student should consider three cases. First, at $v = 0$, the equation reduces to $\Delta\lambda/\lambda = 0$; there is no Doppler shift at zero speed. Second, as v approaches the speed of light, $\Delta\lambda/\lambda$ becomes infinite, far exceeding the value 1 predicted by Doppler's

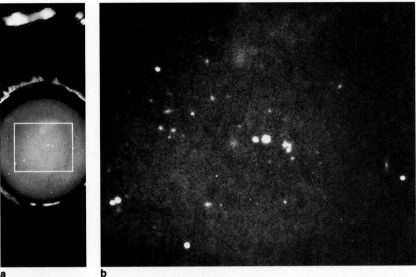

a

b

Figure 25-21 Difficulties in studying quasars are indicated by these images of double quasar 4C 11.50. The left image is at the true scale of the original 1-in.-wide photographic plate, with its circular field of view in the center: Galaxy images are less than $\frac{1}{2}$ mm across. The white box shows the portion greatly enlarged at right. Enlargement reveals that quasars (the bright pair, center) lack detail. The complex image just right of the quasars is a cluster of galaxies; this cluster and the brighter quasar have red shifts of 0.43. The left quasar has a red shift of 1.90. The discrepancy suggests that the red shifts may be noncosmological (see text). The faint honeycomb pattern is from fiber optics in the image intensifier used to make the photograph. (Mauna Kea Observatory, University of Hawaii, courtesy Alan Stockton.)

nineteenth-century understanding. Third, at very low, everyday speeds, and even speeds of nearby stars and galaxies, the relativistic equation gives nearly the same result as Doppler's nineteenth-century equation. For example, consider the Doppler shift from a star moving at 300 km/s, or $0.001c$. By substituting $v/c = 0.001$, the student can confirm that

$$\frac{\Delta\lambda}{\lambda} = 0.0010005$$

which is pretty close to the classical Doppler shift value of 0.001! At speeds up to even $1\%c$, the results are virtually the same, but at speeds over $10\%c$, the results become noticeably different, and the relativistic Doppler shift equation must be used to give a correct value of the wavelength shift.

Sample Problem. A certain quasar has a red shift of 1.0. At what fraction of the speed of light is it receding from us?

Solution. For convenience let us call the red shift R and the quantity v/c, x. Then the equation reduces to

$$R = \sqrt{\frac{1+x}{1-x}} - 1$$

Substituting $R = 1$, we get

$$2 = \sqrt{\frac{1+x}{1-x}}$$

Squaring both sides and solving for x, or v/c, we get that v/c is 0.6, or 60% the speed of light.

of some quasars. For example, quasar 3C 446 changed brightness by a factor of about 20 in a year. Others have changed by about a factor of 2 in a few months, and others by a few percent in as little as 15 min. Radio variations are also known and need not be accompanied by optical variations. Now, suppose a brightening takes 1 y. Then the region must have a front-to-back dimension no larger than 1 ly. The reasoning is as follows: Say the region were bigger. If the whole region is brightening, some physical influence must be affecting the region. This influence can propagate no faster than light and would thus take more than a year to affect a region larger than 1 ly in diameter. Thus an earthbound observer would have to wait *more* than a year after the frontside signal arrived for the backside signal to arrive. The general rule is that if a variation takes time T, the size of the varying region should be less than T times the velocity of light. By this rule, 15-min variations must originate in a region less than 1.8 AU across, or about the size of the inner solar system. The main light-emitting parts of quasars must be less than a few percent of a parsec across. These small regions are often surrounded by larger, expanding envelopes of diffuse gas and probably have intense magnetic fields and high-energy particles, which account for synchrotron radiation and spectral features. They sound suspiciously like accretion disks around supermassive black holes, based on our discussion of SS 433 in Chapter 21 and Figure 25-16's description of galactic jets.

In spite of all these observational successes, as recently as 1984, Princeton astronomer Edwin Turner lamented that in determining quasars' *physical nature* and *energy source*, "progress during the last 15 years is barely discernible!" To gain a better understanding of quasars we need to know how much energy they radiate. To deduce this, we need to know their distances—which requires us to interpret their immense red shifts. This brings us back to the problem of our two theories of red shifts.

Interpretation I: Quasars as Very Remote Galactic Nuclei

Interpretation I, the more widely accepted and conventional of the two, is based on the theory of *cosmological red shifts*. We apply the conventional conversion of red shift into distances and find that most quasars, moving away at 80 or 90% the speed of light, must be at enormous distances—thousands of megaparsecs, as shown

Figure 25-22 At least some quasars are galaxies. This spiral galaxy, ESO 113-IG45, has a red shift corresponding to a recession at 13 630 km/s and a distance of roughly 150 Mpc. It was identified as a Seyfert galaxy in 1977. Its nucleus was found to have the spectral properties of a quasar in 1978. It confirms that some quasars are bright nuclei of galaxies. (European Southern Observatory, courtesy R.M. West.)

in Figure 25-23. If they are that far away, they must have incredible luminosities to be visible to us at all. As shown in Table 25-3, radio and optical data yield luminosities of dozens to thousands of times those of normal galaxies. For example, the nearby Seyfert and quasar ESO 113-IG45 has an estimated visual brightness about 15 times that of our galaxy, without even counting its X-ray and other nonvisual radiation.

An example of an extremely luminous quasar is designated S5 0014 + 81. It has an estimated visible light output of 1.2×10^{41} watts and an additional radio output bringing the total to 1.4×10^{41} watts, or 140 trillion trillion trillion kilowatts (Kühr and others, 1983)! This is perhaps 100 000 times the luminosity of our galaxy. The red shift of 3.41 suggests recession of more than 90% the speed of light and a distance of more than 3000 Mpc! Other examples of interesting quasars and a related galaxy are shown in Table 25-4.

According to this interpretation, then, quasars have the most energetic of all known galactic nuclei. Galaxies would thus range from ordinary (nonactive) galaxies through the active Seyfert and radio galaxies to quasars.

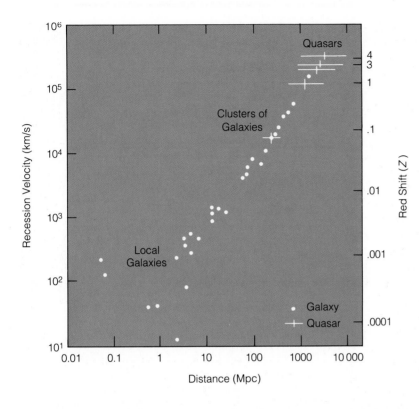

Figure 25-23 Estimated distances of galaxies and clusters of galaxies with various observed red shifts and recession velocities. Horizontal bars for quasars suggest uncertainties in their distances, based on interpretation I.

TABLE 25·3

Energy Relations Between Quasars and Galaxies

Type of Galaxy	Total Luminosity[a] (W)	Energy Involved in Gas Motions and High-Velocity Particles (J)
Normal galaxy	10^{36}–10^{37}	10^{47}–10^{48}
Strong radio galaxy or Seyfert galaxy	10^{36}–10^{38}	10^{49}–10^{53}
Quasar (if distant) (interpretation I)	10^{39}–10^{41}	10^{53}–10^{54}
Quasar (if near) interpretation II	10^{36}?	10^{49}?

[a]From radio and optical sources. Note that 1 W = 1 J of energy radiated in each second, that is, 1 J/s.

According to this view, many quasars are so far away that their light takes billions of years to reach us. Thus we would be seeing *the oldest quasars as they were billions of years ago,* perhaps at about the time our own galaxy was forming. (See review by Osmer, 1982.) Their radiation, then, may represent conditions uncommon in galaxies today, but common during the early stages of galaxy formation. For example, processes of galaxy formation may have produced many short-lived superstars with thousands or millions of solar masses. These superstars exploded quickly, as the galaxy formed, leaving massive black holes. Additional gas spiraling into

TABLE 25-4

Interesting Examples of Highly Red-Shifted Galaxies and Quasars

Object	Red Shift (Z)	Probable Recession Velocity (% speed of light)	Comments
Galaxies			
PKS 1614 + 051	3.215	89%	Most red-shifted and most distant known galaxy.
Quasars			
3C 273	0.158	15%	Quasar with greatest apparent brightness. Narrow jet 690 kpc long, <5 kpc wide.
PKS 1145 − 071	1.345	69%	Possible example of a binary pair of quasars orbiting around each other.
3C 275.1	—	—	First quasar found in a cluster of galaxies; supports identification of quasars with distant galaxies.
S5 0014 + 81	3.41	91%	Highest visible luminosity of any object known; absolute magnitude −33. (See text.)
PKS 2000 − 330	3.78	92%	Most distant quasar known by 1982.
0046-293	4.01	92%	First three quasars found with red shift exceeding 4; most distant
PC 0910 + 5625	4.04	92%	quasars known by 1987, when all three were first announced. We
Q0000-26	4.11	93%	see them as they were when the universe had only about 10% of its present age.

Source: Djorgovski and others (1987a, 1987b); Hayes and Sadun (1987); Peterson (1982); various press announcements.

accretion disks around these massive objects gets heated to extreme temperatures, which causes the quasars' extraordinary radiation. In any case, if interpretation I is correct, many quasars bring us light from the distant past. As astronomer Arthur Eddington said:

Cosmic radiation is a museum—a collection of relics of remote antiquity. These relics are stamped with an inscription indicating the dimensions of the (cosmic) world in its earliest ages. Whoever ultimately identifies the subatomic process originating the rays will be able to read the inscription.

Gravitational Lensing by Quasars: Evidence for Interpretation I

Interesting evidence in favor of interpretation I is provided by Figure 25-24, which shows a quasar with a high red shift of 1.72 whose image is distorted by an intervening galaxy. This indicates that the high red shift is indeed associated with great distance. This finding favors interpretation I.

This type of distortion is called **gravitational lensing.** If a massive galaxy (too distant and faint to be seen) lies on the line of sight between us and a distant quasar, the quasar's light rays must pass very close to the galaxy. Einstein's theory of relativity predicts that a massive object can bend passing light rays. Thus the quasar's light rays are deflected around the galaxy. As shown in Figure 25-25, this effect can create more than one image of the quasar, which may be distorted in shape and displaced slightly from its true position.

Not all examples of gravitational lensing involve quasars. In fact, in 1987 astronomers discovered strange arcs of light among certain distant galaxies, as seen in Figure 25-26. Red shifts seemed to establish that the light in the arcs is not from the visible galaxies but from

Figure 25-24 A quasar whose image is distorted by a faint galaxy (not visible here) between it and us. The geometry and identical spectra indicate that all three images seen here are formed by the gravitational bending of light rays of a single quasar as they pass around the intervening galaxy. (This "gravitational lens" effect was first predicted by Einstein in 1936.) The quasar, PG 1115 + 08, has a red shift of 1.72, suggesting that it is around 2000 Mpc away from us. The probable intervening galaxy has a red shift of about 0.4 and a distance of about 1000 Mpc. This supports the theory that most quasars are very remote. Faint grid pattern is an artifact caused by the telescope's imaging system. (Multiple Mirror Telescope Observatory, University of Arizona and Smithsonian Institution, courtesy E. K. Hege.)

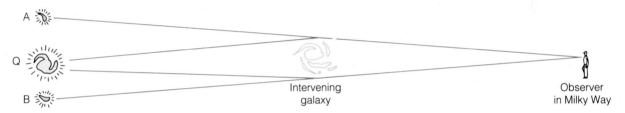

Intervening galaxy

Observer in Milky Way

Figure 25-25 The gravitational lensing effect. If a quasar (Q) is very distant, there is some chance that a galaxy will lie along the line of sight from it to an observer. The gravity of the intervening galaxy deflects Q's light rays. The observer thus sees parts of Q's light coming from directions A and B, where the observer may see distorted images of Q. The quasar itself may be visible or it may be hidden by the galaxy.

still more distant, more red-shifted, galaxies. Apparently these arcs, too, are lensed images. The spectra suggest that the very distant background galaxies are undergoing bursts of star formation, making them, perhaps, cousins of quasars.

Interpretation II:
Are a Few Quasars "Nearby"?

Supporters of the noncosmological theory of red shifts argue that some quasarlike objects are not very far away. Although interpretation II has not been proved correct, it is supported by two arguments.

First, a surprisingly large number of quasars lie within a few minutes or seconds of arc from easily recognized galaxies of much lower red shift. Some even appear to be *connected* to these galaxies by bridges of faint gas.

The second argument for interpretation II is the discrepancy in red shift among different galaxies that seem to be members of the same cluster (see Table 25-2). One plausible cause of anomalous red shift would be intense gravity. According to interpretation II, most of the light from highly red-shifted galaxies and quasars must come from regions of intense gravity—perhaps regions near extraordinarily dense and massive objects that will explode or have already exploded. Such explosions might even throw the objects out of their galaxies, producing quasars linked to galaxies by faint filamentary tails, as some observers claim.

In the 1970s, some astronomers thought all quasars might be explained by such dense objects with strong gravitational red shifts. Thus they would not be so distant and incredibly luminous as proposed by interpretation I. A problem with this theory is that their spectral

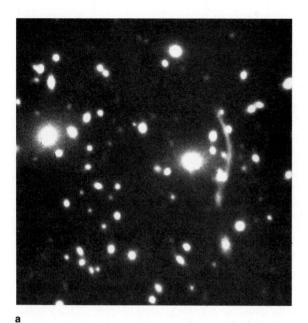 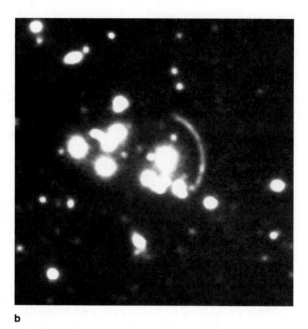

a   b

Figure 25-26 Mysterious luminous arcs in extremely distant clusters of galaxies more than a thousand megaparsecs away. Most of the images are galaxies in these clusters. The arcs are apparently ghostly mirage-like refractions of light from very distant background galaxies, caused by the lensing effect of the visible galaxies. **a** Cluster Abell 370. **b** Cluster 2242-02. (National Optical Astronomy Observatories.)

emission lines are not as broad as predicted in the case of gravitational red shifting. Also, in spite of occasional discrepancies of red shift between quasars and associated galaxies, many quasars and associated remote clusters of galaxies share common red shifts (Silk, 1980).

Thus interpretation I continues to be the favored way to interpret most quasars and highly red-shifted galaxies. It fits with the theory discussed in this chapter, of supermassive black holes in the centers of galaxies. The idea of black hole formation by collapse of a central star cluster probably explains not only quasars, but also lesser levels of activity in galactic nuclei. Rapid formation of the black hole by collapse of the cluster and its gas soon after a spiral galaxy forms would explain the galaxy's infant stage as a quasar. The black hole and its accretion disk would in turn be surrounded by the outer part of the cluster of stars, some of which would continue to fall onto the accretion disk, causing occasional flare-ups. As calculated by Soviet theorist V. I. Dokuchaev (1989), the extreme luminosity of the central quasar would heat the surrounding gas, driving it outward, and repeated close encounters of the stars with the central black hole would accelerate some of the

stars in the cluster outward. The net result would be to stop the star formation process and stop the cluster from shrinking. The cluster would expand and the black hole would end its "feeding frenzy," usually stabilizing at some millions of solar masses. This would end the quasar phase, usually early in a galaxy's history. The quasar stage was most prevalent when the universe was only about $\frac{1}{5}$ its present age, and most but not all black holes in galactic nuclei have declined in activity (Rees, 1990). The winding-down stages would explain various types of less active galaxies. Occasional fall-in of stars onto the accretion disk of the black hole, or frequent supernova blasts, would cause sporadic flare-ups and might explain Seyfert galaxies (Rafanelli and others, 1990). Similar events might explain the outburst in the center of our own Milky Way 10 to 100 My ago. Of course, at any moment, the spiral galaxy might collide with another, disrupting the central processes and converting the whole galaxy into a giant elliptical galaxy; this would explain ellipticals and some peculiar galaxies.

As shown in Table 25-4, recent discoveries have added more and more highly red-shifted quasars to the list of known objects. In 1987, the first quasar with red shift greater than 4.0 was announced, and by 1989, 10 such objects were tabulated. These objects are so far away (according to interpretation I) that the light we see today left them when the universe had only about 10% of its present age! To put it another way, we are seeing such distant quasars when their galaxies were very young. This fits with the idea that quasar activity belongs to the infant stage of galaxy formation, when the central star cluster is collapsing to form a black hole. As we will see more clearly in the next chapter, quasars probably offer our best window into the most ancient past of our universe, allowing us to study certain galaxies shortly after the primordial era of galaxy formation.

SUMMARY

Once we pass beyond the Local Group and other nearby galaxies, we encounter still more galaxies in large clusters at distances of some tens of megaparsecs. Such clusters, and perhaps clusters of clusters, extend as far as we can see, which is at least some thousands of megaparsecs. As we probe to farther distances, we encounter serious problems of interpretation, which might be summarized by the following list of statements:

Fact: *Galaxies at greater distances than a few Mpc have red shifts.*

Virtually certain assertion: *The greater the red shift, the farther the galaxy.*

Virtually certain assertion: *Among most galaxies, the red shifts indicate recession. These galaxies are moving away from us and from each other.*

Virtually certain assertion: *Some kinds of galaxies have active nuclei and emit much greater amounts of energy than normal galaxies.*

Virtually certain assertion: *Explosions or accelerations of gas, fed by enormous energy release, occur in the heart of the nuclei of some galaxies in regions less than a parsec across. They may involve super-supernovae, black hole accretion disks, or other energetic phenomena.*

Probable hypothesis: *Some faint objects with extremely high red shifts, called quasars, are intensely luminous galaxies (or the nuclei of such galaxies) thousands of megaparsecs away, sharing in the recession and moving away from us at appreciable fractions of the speed of light. They are now seen as they appeared billions of years ago, shortly after they formed. Explosive events in galactic nuclei may have been more common then than now.*

Nearly all astronomers accept the first three or four statements and agree that the universe is expanding, in

the sense that the galaxies are rushing away from each other. Astronomers are still debating whether some part of the largest red shifts might be caused by something other than recession. The most red-shifted objects are the puzzling quasars.

Most astronomers believe that quasars are remote galaxies with explosive events involving supermassive objects (possibly billion-$M_\odot$ black holes) in their nuclei. The formation of such supermassive objects may have been triggered by collisions among galaxies, or it may have been spontaneous. Formation of such objects and the resulting superluminous galactic nuclei may have been more common in primordial galaxies than in galaxies today. In any case, we would like to know more about these strange objects, which seem to offer unusual clues about the nature of the universe. In spite of all our progress in understanding quasars, Princeton astrophysicist Edwin Turner could begin the summary of his 1984 review of quasars: "Despite the expenditure of large amounts of telescope time and other resources, most of the fundamental questions concerning quasi-stellar objects (quasars) remain unanswered." Summing up this view, Joseph Silk quotes astrophysicist George Gamow:

Twinkle, twinkle, quasi-star,
Biggest puzzle from afar.
How unlike the other ones,
Brighter than a trillion suns.
Twinkle, twinkle, quasi-star,
How I wonder what you are.

As we approach the frontiers—the farthest objects that astronomers can detect—we find many strange phenomena. However we interpret them, many galaxies apparently generate extraordinary energies. Explosions in their centers, possibly occurring millions of years apart, may cause strong radio and optical radiation and may expel filaments or condensations of matter. Observing distant galaxies, we see how these objects looked billions of years ago. Interpreting these observations has created some of the most exciting controversies in astronomy today. With research on these distant frontiers, we are probing the very nature of the universe itself.

CONCEPTS

megaparsec	Hubble's relation
cluster of galaxies	Hubble's constant
supercluster of galaxies	Doppler shift
red shift of galaxies	theory of cosmological
expanding universe	red shifts
mutual recession of galaxies	gravitational red shift

theory of noncosmological red shifts

radio galaxy

synchrotron radiation

galaxies with active nuclei

Seyfert galaxy

galactic jets

quasar

QSO

gravitational lensing

PROBLEMS

1. How many miles is a megaparsec?

2. Suppose observers located in the Coma cluster of galaxies observe Doppler shifts in the spectra of our Local Group of galaxies, including the Milky Way.
 a. Would they see a red shift or a blue shift?
 b. What sizes of shift would they observe? (*Hint:* See Table 25-1.)
 c. What would they conclude about our Local Group's velocity if they believed the theory of cosmological red shift?

3. In what ways do Seyfert galaxies bridge the gap between ordinary galaxies and quasars?

4. Explain how the existence of groups of galaxies having members with widely discrepant red shifts, such as Stephan's quintet, implies that quasars might not be at the extreme distances usually assumed.

5. Since our galaxy's nucleus is a radio source, why is the Milky Way not considered to be a typical radio galaxy?

6. Why is the study of the most distant galaxies we can see related to the study of conditions around the time that our galaxy (and perhaps others) was forming?

7. Study Figure 25-5 and draw the situation that would be perceived by an observer in galaxy C.

ADVANCED PROBLEMS

8. Explain why Doppler shifts don't cause significant drifting or detuning of car radios as one drives toward or away from radio station transmitters.

9. Suppose a galaxy of stars emitted most of its light at wavelength 500 nm, where the eye is sensitive. Suppose it was as far away as the quasar OH 471, which has a red shift of 3.4 (*shift* in wavelength = 3.4 × original wavelength).
 a. Assuming that red shifts are cosmological, find the wavelength at which most of its light would appear.
 b. If it was an ordinary galaxy, would it look brighter or fainter than quasar OH 471?
 c. Comment on how the results of parts *a* and *b* would affect attempts to detect the galaxy.

10. A certain faint galaxy is found to have a recession velocity of 5700 km/s. How far away is it if its red shift is cosmological?

11. Using the small-angle equation, estimate the angular size of the galaxy in Problem 10 as seen from the Earth if it has a diameter of 30 kpc, like the Milky Way.

12.a. Compile a table showing $\Delta\lambda/\lambda$ values predicted by the classic Doppler shift equation and the correct, relativistic Doppler shift equation, for speeds of 1, 10, 50, 90, and 99% of the speed of light.
 b. At what speeds does the classic equation begin to be more than a few percent in error?

13. A red shift, $\Delta\lambda/\lambda$, reported for a quasar is 4.0. Calculate the recession velocity of this quasar. If you cannot solve the relativistic Doppler equation algebraically for *v*, then find the answer to two significant figures by successive approximations, guessing a value of *v/c*, using it to solve for $\Delta\lambda/\lambda$, assuming a new value of *v/c*, and so on.

A special place in the universe. The evolution of life as we know it apparently depends on water and a changing but relatively stable planetary climate. Modern astronomy leads us to the question of whether planets exist near other stars and whether life has evolved on them. (Photo by author.)

CHAPTER 26

Cosmology: The Universe's Structure

We are approaching the limits of our ability to probe the skies. The last chapter took us to galaxies so remote that they are not only almost too faint to see but also so red-shifted that much of their radiation is infrared or radio radiation. What can be said about still more remote regions? Are they like ours? Do galaxies exist there? These questions belong to the field of **cosmology**— the study of the structure or "geography" of the universe as a single, orderly system.

In the opening pages of this book we compared humanity to explorers on a strange island in an unknown sea. To extend the analogy, this chapter finds us at a point in our explorations where we know the island is made of grains of sand, we know how big it is, we know that the sea is large, and we know that there are many other islands out there in the distance. And now we ask: Do the islands go on forever? Is the world flat with islands dotted uniformly or in clusters? Does the world have an edge, or is it infinite? Or is the world round with no edges?

The same kinds of questions are asked in cosmology. Are galaxies dotted across space indefinitely, as photos such as Figure 26-1 seem to imply? Does space itself go on indefinitely, or does it come to some kind of end? Although these are exciting questions to debate, they are risky, because they tempt us to leapfrog beyond the limits of available observations and speak of such concepts as infinity. Cosmology tempts scientists to speculate about the **universe,** defined as all matter and energy in existence anywhere, observable or not. Yet scientists cannot be sure how concepts of infinity apply to the real universe. And, as J. D. North (1965) remarked in his history of cosmology, "It is easy to speak of the infinite . . . but it is difficult to speak of it meaningfully."

Why should normally cautious scientists attempt such speculations? There are several reasons. People have been asking cosmological questions and related religious questions for thousands of years. It is valid to do our

Figure 26-1 More galaxies. This image shows numerous elliptical and spiral galaxies in the Fornax cluster of galaxies, estimated to be 17 Mpc from the Milky Way. (Copyright Anglo-Australian Telescope Board.)

best to answer these questions. Innate in the character of curious, restless humanity is the desire to know our surroundings. Unfounded speculation should not be part of science, but scientists can legitimately take the principles we have learned about matter and energy in our part of the universe and then ask: "What would happen if these principles apply in all space and time?" or "What would happen if certain principles were different in early times or in distant space?" By such questions, cosmologists have been led to some startling ideas about the universe as a whole—ideas that lie far outside everyday experience.

THE MOST
DISTANT GALAXIES

How do we really know that a certain galaxy is 1000 Mpc away? To say we know by the red shift is not a fundamental answer, because no one has been able to

prove conclusively that the Hubble relation of red shift and distance applies to all galaxies—although most astronomers think that it does. Furthermore, the Hubble constant rests on the interpretation of Cepheid variable stars and other distance indicators. And these, in turn, rest on other techniques, primarily trigonometric parallaxes of stars close to the solar system. In other words, we have built a whole ladder of distance indicators, each rung resting entirely on the security of the preceding rung, as shown in Figure 26-2. We have been tacking the ladder together as we climb it. We use this ladder because we think it is sound, but we use it only with healthy skepticism.

As noted in the last chapter, looking at distant galaxies involves looking far back in time, because their light takes a long time to reach us. Table 26-1 shows the estimated time required for light to reach us from different places in the universe. We see the Andromeda galaxy as it was over 2 million years ago. We see the

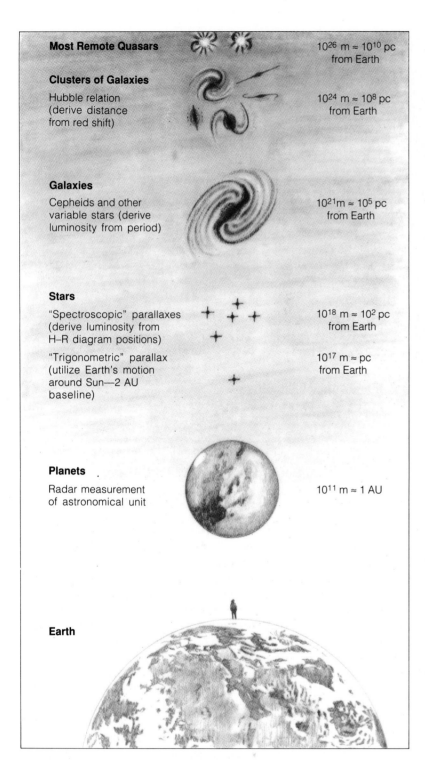

Most Remote Quasars 10^{26} m $\approx 10^{10}$ pc
from Earth

Clusters of Galaxies

Hubble relation
(derive distance 10^{24} m $\approx 10^{8}$ pc
from red shift)   from Earth

Galaxies

Cepheids and other 10^{21}m $\approx 10^{5}$ pc
variable stars (derive   from Earth
luminosity from period)

Stars

"Spectroscopic" parallaxes 10^{18} m $\approx 10^{2}$ pc
(derive luminosity from   from Earth
H–R diagram positions)

"Trigonometric" parallax 10^{17} m $\approx$ pc
(utilize Earth's motion   from Earth
around Sun—2 AU
baseline)

Planets

Radar measurement 10^{11} m ≈ 1 AU
of astronomical unit

Earth

Figure 26-2 The astronomical distance scale. The accuracy of distance measurement at each level depends on the accuracy of measurements at lower levels. Some methods of distance measurement are listed at left.

TABLE 26·1

Travel Times of Light from Distant Objects

Source of Light	Distance	Light Travel Time (y)
Typical visible stars	100 pc	326
Center of Milky Way	9 kpc	29 000
Magellanic clouds	60	196 000
Andromeda galaxy	670	2 200 000
Edge of Local Group	1000	3 300 000
M 51 "Whirlpool" spiral	3800	12 000 000
Centaurus A radio elliptical	4400	14 300 000
M 87 elliptical	13 Mpc	42 000 000
Virgo cluster	19	62 000 000
Coma cluster ($Z = 0.02$)[a]	110	368 000 000
Hydra cluster ($Z = 0.2$)[a]	700	2 300 000 000
Extremely distant galaxy 0902 + 34 ($Z = 3.4$)	3400?	12 000 000 000?
High-Z quasar OQ172 ($Z = 3.5$)	4900?	13–16 000 000 000?

[a]Z = red shift expressed as fraction of original wavelength.

Virgo cluster as it was when dinosaurs were dying. We see the Coma cluster as it was when early fishes were appearing in murky seas. We see the Hydra cluster as it was soon after the last lavas erupted in the Sea of Tranquillity on the Moon. We see some galaxies as they appeared before the solar system formed.

But now suppose we look for a galaxy (quasar) older than the last one in Table 26-1. We run up against a very fundamental limit to our ability to observe for two reasons. First, the light we are looking for would have left more than 13 billion years ago. We already saw in Chapter 23 that our own galaxy was just forming at this time, and we will shortly discuss evidence that the other galaxies formed at about this time (see Chapter 27). Therefore, if we try to look farther away in distance and further back in time, we may see no galaxies because none had yet formed. A second reason stems from the incredible recession velocities, which approach the speed of light at this distance. This means that the light from the galaxy would be extremely red-shifted. To look for objects much farther away, we would have to look for most of their light at infrared or radio wavelengths, where the photons have little energy and detection is difficult.

EARLY COSMOLOGIES

The preceding discussion shows how cosmology, the study of the universe's structure, is related to another area—**cosmogony,** the study of the universe's origin. One might think that we could study the present structure entirely separately from the ancient history of the system. In this chapter we will attempt to do this as much as we can, focusing mostly on the "geography" of space. But always in astronomy, the farther we probe in distance, the further back we look in time. Ultimately, mapping the remotest regions means mapping the early history of the system. For this reason, the remote cosmic horizon contains important clues about the origin of our universe.

To understand how cosmologists have arrived at their present conceptions of the universe, we will review a series of cosmological theories developed at different times. These demonstrate best how various religious, philosophical, physical, mathematical, chemical, and astronomical studies have forced us toward our present conceptions. In each case, we will first present the model and then discuss some of its limitations, which led to new models.

Cosmology 1:
Ancient Models (c. 3000 B.C.)

Even the earliest known philosophers 5000 years ago speculated about the nature of the universe. Knowing nothing of the physical laws that govern matter, they tried to make sense of the universe by discussing non-material attributes, or "essences," of things, as we saw in Chapter 1. They often gave these essences names, imaginary personalities, or abilities to control the affairs of the universe.

Early writers also speculated on the arrangement of the universe. According to a tradition in India, for example, the universe is a giant egg containing land, waters, animals, gods, and so forth, all created in primordial waters by Prajapati. A Tahitian tradition says that a creator, Taaroa, existed in the immensity of space before any universe existed, and that he later constructed the heavens and rocky foundations of the Earth. A well-known book compiled around 500 B.C. considers both origin and structure:

In the beginning God created the heavens and the earth. The earth was without form and void, and darkness was on the face of the deep; and the spirit of God was moving over the face of the waters. And God said, "Let there be light!"

Thus many old traditions trace both the structure and the origin of the universe back to some underlying early spirit or principle. Sometimes this spirit was given personality and described as animate gods or one god. Sometimes it was described as a fundamental principle from which all else followed, as in the Book of John (c. A.D. 100): "In the beginning was the Word." *Word*, in this passage, was the Greek term *logos*—a principle of rational logic—a concept common in Indian and Mesopotamian philosophy. So this famous cosmological thought might be translated, "In the beginning, underlying everything, was rational order." In many ancient cosmologies based on these ideas, the structure and events of the universe (including those involving living beings) were viewed as controlled by the characteristics of this initial creative essence.

Limitations These cosmologies have retained their appeal for thousands of years and provide us with an enduring link to our early ancestors. Historically, they helped set the stage for later investigations about initial conditions in the universe and about immaterial phenomena such as energy. They have the virtue of bringing us to the realization that there are forces in the universe greater than ourselves. The power of their poetry instills a beneficial respect for the sheer vastness and mystery of the universe.

But this kind of cosmology can produce the stultifying belief that all that *can* be said about the universe *has* been said. Such a belief can rob entire cultures of the motivation to question, explore, and see what they can learn for themselves. We humans have learned that we can deepen our understanding of the universe around us if we insist that assertions about nature be testable by observation, experiment, or calculation. But it is difficult to test the early cosmological ideas in this way or determine which is most accurate. Also, such cosmologies are of little value in predicting or interpreting the phenomena that we actually see in the distant universe. Ever-improving astronomical instruments reveal that much more can be said about the universe.

Cosmology 2: Newtonian–
Euclidian Universe (c. 1700)

Around 1680, when Isaac Newton described how every particle in the universe gravitationally attracts every other particle, he realized that this principle might allow a simple description of the structure of the whole universe. First, he assumed that the principles of Euclid's geometry, such as the relations between angles, lines, and planes, would work just as well over the vast distances of universal space as they do in the farmyard or among the nearest stars. Euclidean geometry and Newton's gravitational law, then, allowed description of the separations and forces between particles. Galaxies and even clusters of galaxies can be considered as particles. Newton pictured the universe as infinite in extent and filled with these randomly moving "particles," a view called the **Newtonian–Euclidean static cosmology**.

In 1755, Immanuel Kant realized that although the Newtonian–Euclidean universe remained constant in the long term, many individual stars or galaxies might come and go. Kant thought that the *present* system of stars would burn out and pass away, but others might form from the debris of former systems, just as we have described stars forming from the debris of earlier stars. The universe was thus endlessly recycling—"a Phoenix of nature, which burns itself only in order to revive again . . . through all the infinity of times and spaces." The Newtonian–Euclidean universe was thus static (not expanding or contracting), but evolving.

Limitations Newton was right in predicting many suns far beyond our own, all obeying gravitational relations. And Kant was right in imagining worlds forming and reforming. But these ideas do not adequately describe the universe as we know it today. First, the recession of distant galaxies was neither predicted nor assumed in this theory. Second, why should matter in the universe remain dispersed? Gravitational attraction between particles might cause all the mass in a given region to collapse into a single star or galaxy. Mathematicians in the 1800s studied this problem extensively and even talked of a hypothetical non-Newtonian repulsive force that might be important only over long distances, thus holding the particles apart and keeping the universe static. But the insurmountable objection is that the Newtonian–Euclidean cosmology fails to explain a problem known as Olbers' paradox.

OLBERS' PARADOX

Sometimes the simplest questions promote the most profound thoughts. The simple question, "Why is the sky dark at night?" leads to an astonishing paradox. For example, the night sky in the Newtonian–Euclidean universe ought to be ablaze with light! Astronomers are not sure who first realized this, but Edmond Halley indicated he had heard the idea as early as 1720. The idea was more carefully developed in 1823 by the German astronomer Wilhelm Olbers, whose name finally became attached to it. (See North, 1965, for a complete history.)

Olbers' paradox arises as follows. Assume that space extends indefinitely and is filled only with stars resembling the Sun. If we look at the Sun, we see that each unit of angular area of the Sun's surface (a square second of arc, for example) is intensely bright. If we now gaze in some other direction, our line of sight must ultimately intercept the surface of another star, because stars dot space to infinity. Each unit of angular area in that direction, therefore, would have about the same surface brightness as the Sun, since *surface brightness* (brightness per unit of angular area) does not depend on a star's distance. The whole day or night sky should look as bright as the surface of the Sun!

What is wrong with this argument? One early suggestion was that the dust in space simply obscures the distant stars. But interstellar dust does not explain Olbers' paradox, because the dust should absorb the stars' radiation and heat up. According to the Stefan–Boltzmann law of Chapter 16, the dust reradiates, no matter what its temperature. Its temperature keeps increasing until the amount of radiation emitted equals the amount absorbed and a constant equilibrium temperature is attained. Even if the reradiated radiation were not visible light, there would be infrared radiation that we would sense as intense heat. Because the dust is not radiating this intensely, this explanation fails.

The modern response to Olbers' paradox is more subtle, invoking the age and recession of distant galaxies. The most important factor is galaxies' ages. For several different cosmological models of the universe, astrophysicists have found that the total number of photons emitted by the galaxies in their finite lifetimes is too low to create the kind of pervasive bright glow described by Olbers (Wesson and others, 1987). To look at it another way, there is a "horizon" distance corresponding to a light-travel-time of 12 or 14 billion years, beyond which we see no galaxies because none had formed that long ago. The intense glow proposed by Olbers would apply only if the galaxies had been in existence for vastly longer times and our line of sight intercepted galaxies to virtually infinite distance. Additional calculations have suggested that the time required for the radiation to permeate the available volume, heat the interstellar dust and gas, and achieve the equilibrium assumed by Olbers would be around 10^{23}—more than a billion times the actual age of the universe! (See Harrison, 1974; Raychaudhuri, 1979.)

A secondary effect that helps explain Olbers' paradox is the recession of the galaxies. The red shifts of their light cause the apparent energies of the photons we receive from them to be reduced from high energies (short wavelengths) to low energies (long wavelengths) and cause them to be spread more thinly through space. Thus photons received from galaxies receding at almost the speed of light are strongly reduced in apparent energy.

In short, Olbers' paradox can be explained by modern discoveries concerning the age and recession of galaxies, but it can't be explained by the Newtonian–Euclidean model of the universe: extremely ancient, nonexpanding, static, infinite, and filled uniformly with stars.

MODERN MATHEMATICAL COSMOLOGIES

In most branches of astronomy, theoretical studies develop in response to, or at least hand in hand with, observations. For example, theories of stellar evolution developed to explain observations of various star types and their positions on the H–R diagram. Cosmology is different: The outskirts of the universe are so far away and so hard to measure that only a few basic observations are available to distinguish correct models of the universe from incorrect ones. Fortunately, most cosmological theories *do* make predictions that may be testable in coming decades with space telescopes and other improved equipment. But for the moment, cosmological theorizing has certain aspects of a mathematical game. The rules are (1) invent some hypothetical natural laws and (2) show that they could, in theory, govern a complete universe, but (3) also show that they do not disagree with any of the few, available observations. More than most theories, modern cosmologies involve many assumptions, because so few statistics are available about the universe out beyond about 1000 Mpc. Many modern cosmologies are essentially deductions from initial postulates, rather than experimental or observational tests of nature. As one researcher (Schatzman, 1965) has noted:

Once the basic ideas are understood, the deduction of their consequences is a simple exercise in geometry or algebra, which does not increase our knowledge of the properties of matter.

Non-Euclidean Geometries

In the late 1800s, mathematicians in Germany, Italy, and Russia became fascinated with geometries quite different from those of everyday experience. These so-called non-Euclidean geometries served as stepping stones to modern cosmologies.

Non-Euclidean geometries can be described by an analogy. Suppose we represent the three-dimensional volume of space by the two-dimensional surface of a chessboard. Instead of being able to move in any of the three dimensions of ordinary space (north–south, east–west, and up–down), light waves, sound waves, and inhabitants in the chessboard world could only travel along the two-dimensional (north–south and east–west) surface of the board. Imagine that the board is vast, covering many square kilometers. Inhabitants of this world might be visualized as tiny ants who can move and see only along the surface of the chessboard. To the ants, the chessboard represents all space.

Because the chessboard is flat, the ants would find that the principles of Euclid's plane geometry are satisfied. For example, parallel lines would never meet, and the sum of the angles in a triangle would be 180°.

But now suppose that the chessboard covered the whole surface of the Earth. It can still be called a two-dimensional surface, because the ants (and light waves) can travel only along the *surface* in two dimensions (north–south and east–west). As long as the ants operate only in a small region, the surface would seem flat to them, and Euclidean geometry would seem true within the limits of the accuracy of their observations. But if the ants probed large enough regions, they would eventually discover that their surface, or "space," is curved. For example, as shown in Figure 26-3, they would discover that the triangle defined by the equator and two latitude lines 90° apart has three 90° angles whose sum is 270°!

If the chessboard covered the whole Earth, the ants might discover another interesting thing. They would find that their universe, instead of being infinite (as the ant-Newtons and the ant-Kants might have supposed), was a finite amount of space,[1] even though it had no edge. The earth has a finite area of about 500 million km^2, but no ant crawling around the surface could identify an edge or boundary.

Mathematically, all these ideas can be applied to real three-dimensional space. Following Euclid's laws of geometry, one can write equations describing spatial relationships for either two-dimensional or three-dimensional worlds. For example, in a two-dimensional plane, the Pythagorean theorem gives the length of the hypotenuse x in a right triangle with sides a and b by the famous rule

$$x^2 = a^2 + b^2$$

And, similarly, in ordinary three-dimensional space, the length of a diagonal x in a box with sides a, b, and c is

$$x^2 = a^2 + b^2 + c^2$$

What the nineteenth-century mathematicians did was to generalize three-dimensional geometry to allow for

---

[1] The total area would be $4\pi r^2$, where r is the radius of curvature (radius of the Earth, in this example).

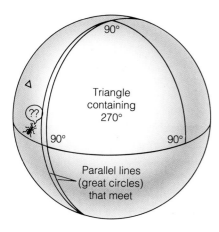

Figure 26-3 Ants living on a large spherical world would find that small figures, such as the small triangle, approximately obey Euclid's laws of geometry. But large figures, such as the large triangle, violate these laws. In the same way, studies of very large volumes of intergalactic space might reveal departures from Euclidean geometry.

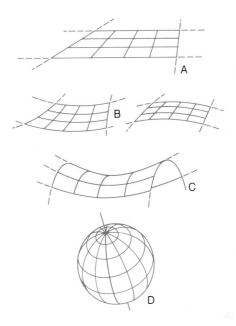

Figure 26-4 Examples of different curvatures of space, using surfaces as analogs to the volume of three-dimensional space. A is uncurved and infinite in extent; B and C are curved and infinite; D is curved and finite.

the mathematical possibility of **curved space.** If the surface is curved, the Pythagorean rule would not be true. (Draw a large right triangle on a globe, measure it, and see for yourself.) Similarly, in a curved space the second equation is no longer true. Different mathematical relations apply. Geometries of curved surfaces or curved spaces are called **non-Euclidean geometries.** After some controversy, cosmologists accepted the proposition that the real universe might actually be curved, or non-Euclidean.

Everyday experience gives us no clue as to whether or not space is curved. We normally interact with far too small a region to detect a slight curvature. The evidence for slight curvature of intergalactic space would come only from studying a very large part of space, including the remotest galaxies, just as the ants would need data from much of the Earth to test for curvature of their "space."

Furthermore, different kinds of curvature are possible, each with its own properties. Examples of different types, shown in Figure 26-4, are based on analogies with surfaces. Part A shows a surface representing uncurved space, which could extend to infinity in all directions (dotted lines). Euclidean geometry would apply everywhere. Part B shows spaces with a slight curvature. These spaces could also extend to infinity. Part C

shows an example of complex curvature, which could also extend to infinity. These types of space could have infinite volume.

Part D is an especially interesting case in which the curve has closed on itself, therefore having finite rather than infinite volume. If space is really curved in this way, the universe could have finite volume but still no boundaries, like the universe of the ants who lived on a globe. A traveler who went far enough in one direction would eventually come back to his starting point.

Space with infinite volume is called **open space.** Space that is curved and has finite volume is called **closed space.** Thus in Figure 26-4, A would be called flat and open, B and C curved and open, and D curved and closed.

Cosmology 3: Static, Curved Universes (1917)

In 1917, Albert Einstein, who had just developed the theory of general relativity, tried to see how it applied to the universe as a whole. He made some basic assumptions that seemed reasonable at the time:

1. The universe is **homogeneous** (meaning that all its particles—stars, galaxies, clusters of galaxies, or whatever—are uniformly distributed on a moderately large scale) and **isotropic** (meaning that the view is similar in all directions from all galaxies). This assumption is sometimes called the **cosmological principle.**

2. Space is curved, as in non-Euclidean geometry, and the curvature is constant. (The surface analog would be a sphere.)

3. The universe is static, which means that galaxies stay at constant distances apart; mean density of matter is constant and the radius of curvature does not change.

But when he tried to solve the equations representing such a universe, Einstein found that it was impossible! *No static solution fitted the principles of general relativity.* Gravitational attraction between galaxies would tend to make matter in such a universe collapse, instead of remaining in static dispersal.

Still assuming that the universe must be static, Einstein introduced a so-called cosmological constant—a hypothetical repulsive force between material particles, important only over long distances. The repulsion, similar to that suggested for Newtonian–Euclidean theories, could be chosen to balance the gravity of whatever mass was in the universe, so that a static state resulted, but only if there were no disturbances.

The mathematician W. de Sitter added some interesting variations, also in 1917. The amount of curvature and other properties were related to the mean density of matter in the universe. A universe with no matter in it could easily be static, he found. (A universe with no matter in it may not seem relevant to the real world, but don't laugh—see cosmology 8!) With the hypothetical repulsive force, a few particles (galaxies) put into such a universe would accelerate away from each other.

Limitations The original static model was scrapped in 1929 when Hubble proved that the galaxies were rushing away from each other. The de Sitter model with a repulsive force and a few galaxies in it actually resembles the observed recession! However, there is no independent evidence for the repulsive force that the model requires. And as we will see next, Einstein, de Sitter, and others soon dropped the idea of repulsive force.

Cosmology 4: Friedmann Evolving Universes (1922—1924)

From 1922 to 1924, the Russian mathematician Alexandre Friedmann studied a model that was evolving, not static. He simply assumed that the curvature of space varies with time. For example, space could increase

in volume if the radius of curvature increased. In such geometries, the distances between all points continually increase, and galaxies would be seen to recede. This model was not widely discussed until Hubble published his discovery of actual galactic recession in 1929. What had seemed like an abstract theoretical model now seemed to be supported by astronomical observation— galaxies are moving apart! Einstein, de Sitter, Eddington, and others soon concluded that the Friedmann model might account for an expanding universe without any repulsive force. Einstein had been on the wrong track when he introduced his hypothetical repulsive force to keep the static-model universe from collapsing. (He later called it the greatest mistake of his life.) The right track would have been to abandon the idea of the universe being static—galaxies are rushing apart and evolving.

An analog of Friedmann's expansion model would be the surface of an expanding sphere, like a balloon. Galaxies could be imagined as dots on the surface of the balloon. The analogy is not perfect, because the dots would expand with the rubber, whereas real galaxies could maintain their integrity as units of matter by their internal gravitational forces. Antlike observers on any dot would see other dots receding from them as the balloon expanded, just as we would see galaxies recede if the radius of curvature of space increased. In addition, no ant or any observer in any galaxy could say he or she lived at the center. Ants could move from dot to dot around the curved surface of the balloon, but no one dot (or galaxy) would be "centered" more than any other.

Limitations When Hubble discovered that galaxies are actually receding from each other, theorists were delighted that observations seemed to support the expanding model. Here was an interesting case where abstract mathematical theorizing seemed to be relevant to a subsequent direct observation of nature! In retrospect, however, astronomers do not say that the expansion of the universe was *predicted* in the normal scientific sense for two reasons. First, the observed expansion was only one of several possible types of Friedmann evolution dependent on the initial assumptions; second, the Friedmann model was developed almost completely outside the normal astronomical context. It was almost a completely mathematical model, with little to say about physical questions such as the distribution of galaxies or the density of matter in space. Thus the next step was to connect this exciting and provocative model more closely with astrophysical reality.

Cosmology 5:
The Big Bang (1927–)

Around 1929, a Belgian priest trained in mathematics and astrophysics, Georges Lemaître, began to put more astronomy and physics into cosmology: He pointed out that if galaxies are now flying away from each other, there probably was a time in the past when all matter was closer together. He suggested that all matter exploded from this condition of high density. Lemaître thus became known as the father of what is popularly called the **big bang theory** of cosmology and cosmogony, which is the most widely accepted theory today.

The moment of maximum density is known as *the big bang*. Many people regard this as the actual creation of the universe. Modern evidence suggests that this moment occurred around 14 ± 4 billion years ago. The universe was supposed to have begun from an extraordinary state in which the radius of all space was nearly zero and all matter was concentrated in a virtual point with virtually infinite density.

Lemaître's picture of a primordial expanding space filled with matter and energy excited physicists and astrophysicists, who recognized that they could analyze the evolution of matter in such a model. This is the hallmark of a useful theory: It allows predictions to be made and tested. The Russian–American physicist George Gamow and his associates began this work in 1948 and found that the observed abundances of atoms of the different elements display trends expected for element formation in an expanding, hot, ancient fireball! This is an important example of how a successful cosmological theory ties into observations of nature. (Later studies showed that additional element formation inside stars was also important, especially in forming the heavier elements.)

Limitations The big bang theory says nothing about how (not to mention why) the "primeval atom" or initial fireball came into existence. It has been very successful in saying that *if* there ever was such a fireball filling expanding space, then certain phenomena would result, and these phenomena have been *observed*. It does not deal with the philosophical question of conceiving a beginning to the universe—which is perhaps beyond astronomy or science in general.

Today, most astronomers accept the big bang theory as the best available cosmological picture of the universe. In the 1980s there has been a modification called the inflationary big bang model, which we will take up in the next chapter, but first we need to consider some additional cosmological ideas to bring our story up to date.

Cosmology 6:
Changing Constants (1930–)

In the 1930s physicists began to consider the possibility that certain fundamental properties of nature, normally considered constant, might slowly change. Examples could be the speed of light, the charge of the electron, or Newton's gravitational constant G. For example, the English physicist Paul Dirac postulated that G might decrease with time. This would weaken attractive forces between galaxies and thus allow galaxies to recede from each other, as observed.

Limitations Direct attempts to measure G and other constants have revealed no verified changes, though the proposed changes may be too small to observe. Measurements are continuing. Elements of these cosmologies, if verified, could perhaps be incorporated into a big bang model.

Cosmology 7:
The Steady State (1948–c. 1980)

By the 1940s the big bang theory was a favored model, supported by observations of galactic recession and the abundances of elements. However, partly to stimulate healthy debate, researchers such as Thomas Gold, H. Bondi, and Fred Hoyle proposed a new cosmology in 1948—the steady-state cosmology. In this cosmology, there is no big bang, no beginning. They proposed that *the universe—on the large scale—looks the same not only in each region of space, but also in different eras of time.* According to this idea, the universe has looked about the same forever. Accordingly, new galaxies must constantly be forming in the spaces vacated as old galaxies move apart. What material could they form from? The steady-state cosmologists made an additional assumption—the theory of continuous creation: New atoms of hydrogen spontaneously pop into existence from time to time, mostly in the space between galaxies. At first, some scientists saw this hypothesis as an outrageous violation of the concept of conservation of mass, but defenders of the theory pointed out that it is hardly more outrageous than imagining all the mass of the whole universe appearing at once, as in the big bang theory!

Limitations The steady-state theory became quite popular for some years, but it hit a major snag when astronomers discovered what seems to be radiation left over from the primeval fireball described by the big bang theory.[2] This radio radiation, which was predicted by the big bang theory, is inconsistent with the original version of the steady-state theory. Moreover, the observation that quasars are more numerous at great distances (hence earlier times) contradicts the theory. Therefore, the steady-state theory, though popular only a few years ago, has been virtually abandoned.

Cosmology 8: Hierarchical Universe (1970—)

Since the early days of Olbers' paradox, some astronomers have questioned an offshoot of the cosmological principle—the idea that we can measure the mean density of matter simply by making an inventory of the galaxies we see. As you will recall, the mean density is important in controlling properties such as the curvature of space.

Chapter 25 gave evidence that galaxies are clumped in clusters and superclusters. Most cosmologies assume that the superclusters are the largest cluster unit and that they are scattered more or less uniformly. But suppose *they* clump in super-superclusters and so on. This is called the theory of **hierarchical universe,** because the universe would be arranged in hierarchies of clusters. The density of matter in the region we can see might not be representative of the larger universe.

This raises the question of the density of the largest-scale part of the universe that we can see. This is illustrated in Figure 26-5, where the dashed lines are reference lines indicating three different densities 20 orders of magnitude apart: 10^{23} kg/m^3, much denser than even nuclear particles or pulsars; 1000 kg/m^3, the order of magnitude of density of familiar objects such as people, rocks, planets, and main-sequence stars; and 10^{-17} kg/m^3, the density of some nebulae and star clusters. Note that as we work our way to radii larger than 10^{10}m, matter tends to be less dense. Star clusters are less dense than stars, galaxies less than star clusters, and galaxy clusters less than galaxies.

Counts of galaxies in the largest volumes that we can explore suggest a mean density of about 10^{-27} or 10^{-29} kg/m^3 for the visible universe. But the hierarchical theory raises the possibility that if we could explore vaster spaces, the mean density might be as low as 10^{-30} or even 10^{-32} kg/m^3. Conceivably, as we consider larger and larger volumes, the mean density might approach zero, so that the curious "empty universe" of cosmology 3 might have some real meaning!

Limitations Hierarchical universes are more complex mathematically than homogeneous universes and have been little studied. Observations do not strongly support the model. On the positive side, the hierarchical theory has encouraged more careful observational tests of the assumption that the universe is uniform, leading to new data on the actual large-scale distribution of galaxies.

Cosmology 9: Is the Whole Universe a Hole Universe? (1970—)

Another peculiar possibility arises from considering Figure 26-5. The heavy line defines the density required at each mass and radius to produce a black hole. Matter located above the line would form a black hole, as seen by the possible position plotted for Cygnus X-1. An interesting fact is that the total observable mass in the universe, divided by the observable volume, gives a density very close to this line. That is, the universe as a whole seems very close to the definition of a black hole! For example, a 1982 paper on various sizes of black holes listed the whole universe as a hypothetical black hole containing $2.8 \times 10^{23} M_\odot$, with a radius of 2.8×10^4 Mpc and a mean density of 2.4×10^{-28} kg/m^3. (A large enough black hole does not have to be extremely dense.)

Limitations The physical meaning of this idea is not very clear, even though the density required over our several-thousand-megaparsec region is quite well defined. Could our "expanding universe" really be an expanding black hole filled with galaxies and expanding into some larger region from which we are forever detached? We don't know.

---

[2]This crucial discovery is discussed in detail in the next chapter.

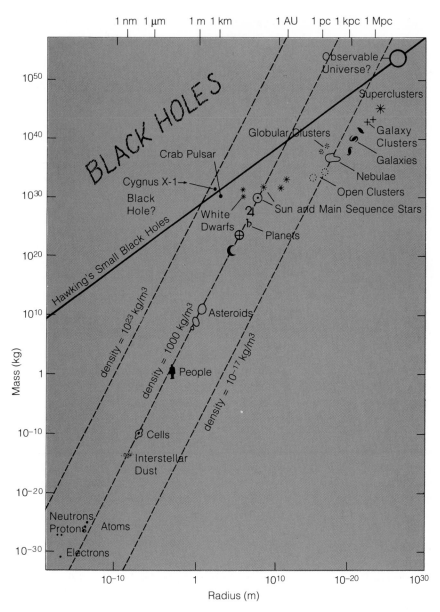

Figure 26-5 A universe of objects, ranging from subatomic particles to superclusters of galaxies, plotted on a diagram of mass versus radius. Objects above the heavy diagonal line have too much gravity for light to escape; they are black holes. Thin diagonal parallels are lines of constant density, showing that familiar objects tend to have similar densities; subatomic particles are denser and stellar groupings are less dense. Note that the observable region of the universe, treated as one object, has very low density and is close to fitting the definition of a black hole.

DIRECT COSMOLOGICAL OBSERVATIONS

Often it happens in science that out of a series of theories, such as we've just enumerated, several may contribute to our understanding. How can we pick and choose among them? We would like some observational tests to determine which theories make correct predictions. We've already mentioned that although few observational tests are currently possible, the big bang theory (cosmology 5) seems to have the most concepts matching observed nature. For instance, the galaxies do seem to be rushing apart, and the sky is dark at night instead of bright.

Our task in the rest of this chapter is to review some direct observations that clarify cosmology and to show how they fit the big bang picture. In the next chapter we will discuss the events around the time of the big bang.

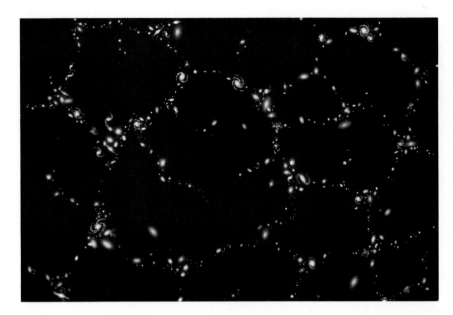

Figure 26-6 Schematic conception of the large-scale structure of the universe. Galaxies are strung through three-dimensional space in a filamentary pattern, illustrated here in a two-dimensional representation. Giant "bubbles" of relatively unpopulated space appear throughout the wispy galaxy distribution. "Nodes" at intersections of bubbles may represent clusters and superclusters of galaxies. (Painting by Dennis Davidson.)

A Bubbly Froth of Galaxies, and the "Great Wall"

We have mentioned that galaxies are distributed in clusters, and the clusters, in turn, in superclusters. But this is not the whole story. We can map the positions of galaxies in *three-dimensional* space by measuring their red shifts, calculating their distances, and then plotting their distances and directions in a 3-D model. Such mapping of about 5800 galaxies out to about 150 Mpc distance reveals that galaxies and clusters of galaxies are distributed as if on the surfaces of vast bubbles, enclosing galaxy-free voids, as suggested by Figure 26-6. This was first noticed in the 1970s but was more widely recognized only in the late 1980s (see review by Rood, 1988). This large-scale structure of the universe resembles a froth of soapsuds, where galaxies are like water molecules on the surfaces of the touching bubbles, and clusters or superclusters are like the "nodes" at the junction of several bubbles (Figure 26-6). There are interesting details in the "froth." For example, at a distance of roughly 100 Mpc in a direction between the constellations Leo and Ursa Major, mappers in 1989 found a so-called "**great wall**," an irregular sheet of galaxies stretching almost perpendicular to our direction of vision (Geller and Huchra, 1989). It is some 150 Mpc wide, but only 5 or 10 Mpc thick. Probably it is an unusually dense concentration defining the side of one

or two bubbles. Relatively empty voids up to 50 Mpc across exist on either side of the "great wall."

Astrophysicists are now trying to understand how this bubblelike pattern, this froth of galaxies, resulted from the big bang. For example, Russian cosmologist Y. B. Zel'dovich identified conditions in which the initial, expanding gas would break into huge "wisps" with about the mass of a supercluster. These wisps then contract gravitationally and break up into clusters of galaxies. Today's galaxy froth may thus be a remnant pattern of the galaxies' parent gas (Silk, Szalay, and Zel'dovich, 1983; Burns, 1986).

The "Great Attractor": A Super-Supercluster of Galaxies?

Work on the distribution of galaxies on vast scales continues to yield new and controversial surprises. In the late 1970s and 1980s, astronomers added studies of motions of galaxies to the studies of their large-scale distributions. This work seemed to show that while all galaxies share the outward motions of mutual recession, discovered by Hubble, there are localized streamings of large groups of galaxies in one direction or another. Relative to an average motion of all galaxies,[3] the group

---

[3] And relative to an average defined by the microwave 3-K radiation received from all over the sky, discussed in the next chapter.

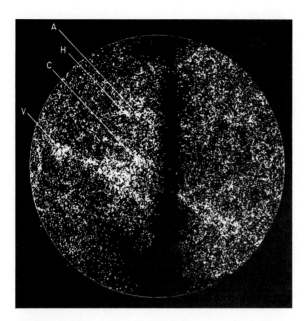

Figure 26-7 A map of hundreds of galaxies in an equal area projection of one hemisphere of the sky shows possible evidence of "the great attractor"—a hypothesized super-supercluster of galaxies. The dark vertical band is the Milky Way's disk, which obscures distant galaxies. In the rest of the sky, well-known galaxy clusters, such as those in Antlia, Hydra, Centaurus, and Virgo, are marked. A diagonal band including the Virgo and Centaurus clusters seems to mark a possible super-concentration toward which many nearer galaxies are drifting. (Courtesy Ofer Lahav, Institute for Astronomy, Cambridge, UK.)

of galaxies in our part of the universe, out to as far as 100 Mpc, appears to be drifting in the general direction of the Virgo and Centaurus clusters of galaxies at around 600 km/s. One interpretation of this drift is that there may be an unusually massive concentration of galaxies toward which we are all falling. Some astronomers have called this hypothetical super-supercluster "the great attractor." Though this idea is controversial, some astronomers have pointed to an elongated concentration of distant galaxies in this direction as possible evidence of the reality of the great attractor (Anderson, 1987). This concentration is shown in Figure 26-7. If the great attractor is confirmed as a cluster or band of distant galaxies, it may yield still a larger-scale clumping and a larger-scale gravitational control of motions than astronomers realized. This in turn would place new constraints on our theories of the large-scale structure, as well as the origin, of the universe.

WILL THE UNIVERSE KEEP EXPANDING FOREVER?

Is the expanding universe a finite closed space like an expanding balloon? Or is it infinite? Given that galaxies are rushing away from each other, we can ask another question: Can the universe recollapse? If there were enough material among the galaxies, their gravitational attraction for each other would be enough to overcome the recession; galaxies would slow down and eventually fall back together. In this case, the universe's history might be a cyclical series of big bangs! But if the density of material in the universe were low enough, gravity would be too weak to cause recollapse, and the galaxies would rush outward forever, like rockets launched from a planet at escape velocity. So the second question can be stated: Is the density of material in the universe great enough to cause recollapse?

Open Universe vs. Closed Universe

These two questions turn out to be related.[4] In general, cosmologists use the term **open universe** to refer to a universe composed of open space with infinite volume and no boundaries, in which the expansion of galaxies continues forever. Cosmologists use the term **closed universe** to refer to a universe that is composed of closed space with finite volume and that has enough density to cause a reversal of galaxies' motions and eventual recollapse. If the average density of matter in the universe is less than about 6×10^{-27} kg/m^3 (about three hydrogen atoms per cubic meter), the universe is open; if more, it is closed. This critical density is called the **closure density,** or the density required for closure of the universe.

Thus much cosmological discussion boils down to the question: Is the universe open or closed? At least three observational tests are possible. To understand the first test, which deals with whether space is curved,

---

[4]There is some confusion in the literature over the terms *open* and *closed*. Some books define an open universe as being unbounded and others define it as expanding forever. Similarly, *closed universe* is sometimes defined as finite or bounded and sometimes as destined to recollapse. In most, but not all, cosmological theories, open universes are unbounded (as in Figure 26-4B and C) *and* ever expanding, while closed universes are finite (as in Figure 26-4D) *and* will recollapse. We will thus use the terms in this latter sense.

let us return to our analogy of the ants. They ask themselves whether they live on an infinitely large flat chessboard (Figure 26-4, curve A) or on a surface more like the shape of the Earth (Figure 26-4, curve D). Suppose they know of mileposts (galaxies) scattered every mile or so at random. If they gathered data on the number of mileposts within 10 km, 100 km, and 1000 km of their colony, they could infer curvature of their space. For instance, if their chessboard were flat, the number within distance d would increase indefinitely as d got larger. But if they lived on the curved Earth, the number would start out increasing as d increased until they began tabulating mileposts from far away on the other side of the Earth. Eventually, as d approached half the Earth's circumference, they would run out of new mileposts.

Astronomers have tried to make this test by counting galaxies. As shown in Figure 26-8, they have tabulated the number of galaxies of each apparent magnitude. If galaxies were distributed uniformly with similar luminosities and with no obscuring intergalactic dust, and if space were open and flat (that is, Euclidean), then a certain smooth relation would exist between the distance and the number of galaxies within that distance. Departure from this relation would reveal curvature of space. But the test fails to give a clear-cut answer. At great distances, where the test is most sensitive, bright galaxies (quasars) are overabundant because we are seeing back to the era when many galaxies were forming and when many galaxies were brighter than they are today. Some astronomers have tried to correct for this effect, but the results are uncertain and controversial.

A second test is to plot red shifts (recession velocities) of galaxies versus their apparent brightness (distance), as seen in Figure 26-9. Different degrees of curvature give different predicted curves on this diagram. Again we run into problems with interpreting the brightness of quasars, but the data seem to indicate that the universe is quite near the critical closure condition.

The third test is to tabulate all the known matter (galaxies, clusters and gas in galactic halos, possible intergalactic gas and stars, and so on) over a very large volume and use this to estimate the average density of the universe, to see if it is less than or greater than the critical closure value. As Figure 26-1 reminds us, the test is difficult. The tabulations of detectable material give less than 10% of the amount needed for a closed universe, which would imply strong evidence that the universe is open. But this is not the whole story. Further observations connect the question of endless

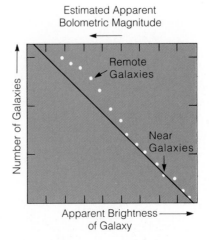

Figure 26-8 If space is Euclidean and galaxies are similar in all regions, the relation between brightness and the number of galaxies of that brightness should follow a straight line. The predicted relation is found for nearer galaxies, but not for remote ones, whose light is billions of years old. One interpretation is that space is Euclidean, but primordial galaxies billions of years ago were brighter than galaxies today. (Data after Ryle, 1968.)

expansion to one of the most important problems in modern astronomy: the problem of the missing mass.

The Problem of the Missing Mass

As early as 1933, Cal Tech astronomer Fritz Zwicky noted that in the Coma cluster of galaxies there does not appear to be enough mass to hold the galaxies gravitationally in the cluster. Many of the galaxies seem to be moving fast enough to escape from the gravity of the material we can see. Yet the cluster is still there. This suggests that an additional gravitational pull may come from missing mass that we cannot see—perhaps in the form of gas dispersed in the cluster, in the form of faint stars, or in some other form. A second line of evidence comes from studies of individual galaxies. During the 1970s, as more sophisticated studies of individual galaxies have been made, astronomers have found that even within these galaxies, orbital motions of stars around the galaxies imply the presence of more mass than we can see, especially in invisible halos surrounding the visible portions of the galaxies (Rubin, 1983; de Boer and Savage, 1982). Recall also the evidence on page 509 that our own galaxy has more mass than can be seen, probably located in an extended halo.

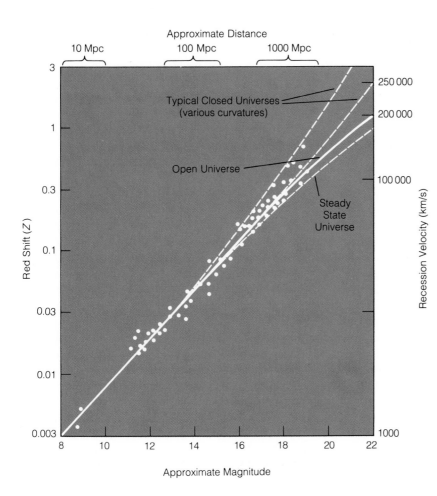

Figure 26-9 Brightness of galaxies plotted against their red shift (the so-called Hubble diagram). Only the brightest galaxies in each galaxy cluster are used, under the assumption that they are similar in intrinsic luminosity. A solid line separates the curves for open universes from those for closed universes. Data lie close to the solid line. Better data are needed.

A third line of evidence involves measuring the bending of light rays ("lensing") of remote galaxies or quasars by foreground galaxy clusters (see pages 576–577). A study of one such cluster in 1990 concluded that it caused so much lensing that it must have at least 10 times as much nonluminous dark mass as luminous mass (Tyson and others, 1990).

This so-called **problem of the missing mass**—the fact that motions of stars and galaxies imply more mass than we can see in the universe—is one of the major research problems in modern cosmology. The reason it is important is that the density of visible material in the universe is about 3×10^{-28} kg/m^3, roughly 1/25 of the critical density that divides an open from a closed universe (Shu, 1982). If we believe there is little missing mass, then the universe is open and will never recollapse. If there is a lot of missing mass, then the universe is more dense than we think and might be closed, destined to recollapse in the future.

Current observations indicate that 90 to 99% of all matter in the universe is dark and has not yet been detected by us! It is striking to realize that the majestic glow of the thousands of stars and galaxies scattered across space represents only a tiny portion of the total matter in the universe—like the bright white specks of foam floating on the surface of a vast, darker ocean.

In what form is the missing mass? The leading theory is that it is so-called cold dark matter, i.e., ordinary matter in the form of faint brown dwarf stars, interstellar snowballs, undetected interstellar clouds, or even intergalactic clouds of gas. For example, some radio astronomers have reported that unusually compact, dark interstellar clouds occasionally pass in front of quasars, distorting their light (Fiedler and others, 1987). These are yet to be confirmed.

Any form of dark matter must be able to explain the clumping of galaxies in bubbles, "great walls," and voids. Some astrophysicists have felt that there couldn't be enough cold dark matter to explain the clumping by gravitational effects. However, Princeton graduate student Changbon Park in 1990 made the largest-yet computer model of imaginary galaxy distribution (representing a 60-Mpc-wide volume with 4 million stars, galaxies, and dark clouds), and he concluded it would evolve into the kinds of bubbles and voids actually observed, supporting the theory of cold dark matter. Nonetheless, the theory that the missing mass is cold dark matter remains unproven.

Two other theories suggest that the unseen matter is hidden not in ordinary matter but in subatomic particles. The second theory is that the missing mass is in the form of neutrinos. Usually, these subatomic particles are described as massless (page 321), but some physicists have suggested that they might have a tiny mass. If they have more than 3-millionths of the mass of an electron, they would contain more mass than all the other material in the universe; if more than 49-millionths, enough to make the universe reach closure density. Recent Soviet experiments have suggested a mass in the range of 30 to 80-millionths, but American experiments indicate it is less than about 52-millionths and possibly as low as zero (Thomsen, 1987). So this hypothesis remains unresolved.

The third theory is that the missing mass consists of hitherto undiscovered subatomic particles called weakly interacting massive particles, or "WIMPs." The WIMPs, some 10 times as massive as protons, have been predicted by some theoretical models of the universe. Because they interact so weakly with other forms of matter, according to theory, they pass right through most detectors and have not yet been observed. Physicists are currently beginning experiments to determine whether they really exist and if there are enough to affect the universe's structure (Waldrop, 1986).

The final answer as to whether the universe is open or closed is still uncertain. Astronomers are devising new ways to search for nonluminous "missing mass" to see if it is really there. A good review of the situation is given by Maryland astronomer Virginia Trimble (1987).

THE FUTURE
OF THE UNIVERSE

It is possible to convert our present knowledge into some statements about the remote future. For example, the present hydrogen in stars should all be consumed within about 10^{14} y of the big bang—when the universe is some 10 000 times as old as its present age of roughly 14 billion years. (See the next chapter for more detail on the age of the universe.) Given the rate of close stellar encounters, planets' orbits should all be gravitationally disturbed, and all planets should be kicked out of their star systems in some 10^{17} y, and as much as 90% of the stars and gas of most galaxies will have been kicked out into intergalactic space in about 10^{18} y. The remaining mass in galaxies would tend to contract into black holes in galactic nuclei. Most current theories of physics predict that protons are unstable. If so, they should decay into separate subnuclear particles in about 10^{30} to 10^{32} y. (More than a dozen major experiments are now under way around the world to confirm this prediction.) Thus the fundamental particle composition of the universe will change and much intergalactic gas will eventually be composed of electrons, positrons, and other resulting particles, as well as supermassive black holes. Quantum theories of physics also predict that black holes themselves can decay into photons and other particles. The supermassive black holes would last about 10^{100} y. If the universe does not recollapse within this inconceivably long period, it should end up as a diffuse, cold cloud of electrons, positrons, neutrinos, and low-energy electromagnetic radiation of radio wavelengths, corresponding not to our current 3-K radiation, but to a radiation temperature less than 10^{-13} K. (This future is summarized in a nontechnical article by Dicus and others, 1983.)

Whether the universe lasts so long depends on whether it is open or closed. Is the difference between an open and closed universe merely an abstraction, like medieval arguments over the number of angels that can dance on the head of a pin? Yes, in the sense that it will have no immediate bearing on the Earth's evolution or human evolution within the next billion years. No, in the sense that it makes a real difference for the future of the universe we live in. Some writers have whimsically claimed to prefer an open universe because then we (or our atoms sometime in the future) would not be gobbled up in a high-temperature recollapse and a new big bang. But the laws of thermodynamics would predict that the open, ever-expanding universe would cool forever until the remaining matter consisted of black holes and burnt-out stellar debris. Our hypothetical descendants couldn't build a fire to warm up because no combustible fuel would be left. Other writers have claimed to prefer a closed universe because the recollapse might initiate a new big bang that might start a new cycle of evolution of new galaxies and new life. The subject is really too speculative to be discussed seriously, because our twentieth-century physics may not be adequate to understand it. But one might like to recall Robert Frost's lines:

Some say the world will end in fire,
Some say in ice.
From what I've tasted of desire
I hold with those who favor fire.

SUMMARY

Cosmology is a risky business, and it is unwise to put too much faith in generalizations about the universe when there is still so much more observing to be done with new and larger instruments. The French–American astronomer G. de Vaucouleurs has said:

Less than 50 years after the birth of what we are pleased to call "modern cosmology," when so few empirical facts are passably well established, when so many different over-simplified models of the universe are still competing for attention, is it, may we ask, really credible to claim, or even reasonable to hope, that we are presently close to a definitive solution of the cosmological problem?

De Vaucouleurs was healthily skeptical. What we think we know is that (1) galaxies are receding from each other through space that is either uncurved or only slightly curved and (2) some singular event, called the big bang, initiated the expansion of all matter in the universe billions of years ago. We aren't sure if the expansion will continue indefinitely. New telescopes in space may help us complete several tests that will answer these questions more definitively.

CONCEPTS

cosmology	isotropic
universe	cosmological principle
cosmogony	big bang theory
Newtonian–Euclidean static cosmology	hierarchical universe
	open universe
Olbers' paradox	closed universe
curved space	closure density
non-Euclidean geometry	great wall
open space	problem of the missing mass
closed space	
homogeneous	

PROBLEMS

1. Progress in many scientific fields, such as studies of stars, plants, or animals, has come by classification of different types of specimens, followed by comparisons of the different classes. How does a cosmologist suffer a disadvantage in this regard?

2. Telescopes much larger than present-day designs, perhaps located in space, would have much more resolving power and light-gathering ability and could reveal much fainter objects. Give examples of how this development would clarify current cosmological problems.

3. How is the real universe different from the Newtonian–Euclidean static model that dominated literary and cultural concepts in the 1700s and 1800s?

4. Why is the sky dark at night?

5. Many cosmologies, such as the big bang model, assume the cosmological principle that the universe is homogeneous at any given time, given large enough scale. Yet quasars seem to be more common per unit volume at very great distances than near our galaxy. Does this refute the big bang theory? Why or why not?

6. Give examples of observations that seem to refute:
  a. The Newtonian static model of the universe.
  b. The steady-state model.

ADVANCED PROBLEMS

7. An explosion in a very active galactic nucleus or quasar produces abundant radiating gas at a temperature of 1 million Kelvin.
  a. At what wavelength would the maximum radiation occur?
  b. Considering the types of light that are absorbed in the Earth's atmosphere, how might a large telescope in space have an advantage in studying such phenomena over a large telescope on the ground?
  c. Would a quasar-type object (or any other observable object) be likely to have enough red shift to bring this radiation to visible wavelengths?
  d. Consider 1 m^2 of the radiating surface of this gas. How much more energy does it radiate than 1 m^2 of the Sun's surface?

8. An important spectral line is produced in hydrogen when electrons make the transition from outside the atom to the lowest ($n = 1$) orbit. The wavelength of the feature in stationary hydrogen is 91 nm. Suppose this feature appears in a quasar receding from us at 90% of the speed of light. At what wavelength does the line appear to us?

CHAPTER 27

Cosmogony: A Twentieth-Century Version of Creation

How did the universe begin? When did it begin? Or do these questions have any real meaning? **Cosmogony** is the attempt to decipher the origin of the universe and its major parts, such as galaxies. *Cosmology,* as seen in the last chapter, focuses on the universe's structure but often involves cosmogony.

The big bang theory is both a cosmogonical and cosmological theory, because it accounts for both the origin and present structure of the universe. It has come to be strongly favored among all such theories because it is based on supporting observations, some of which will become clearer in this chapter. These form an interlocking pattern. Therefore, instead of comparing different theories as we did in the last chapter, we now assume that the big bang theory is essentially true and use it to describe how the universe may have formed and evolved.

All cultures have their creation myths, and the big bang idea is ours. The word *myth* does not mean a falsehood, but a scenario widely told and widely believed. We must remember, as scientists and as humans, that we don't have final, dogmatic answers about cosmogony. The big bang theory almost certainly gives us an indication of some events that actually occurred long ago, but the word *myth* should be kept in mind to remind us that this is only our best guess about a grand mystery—what one satirist has called "the Vienna Philharmonic of scientific questions."

THE DATE OF CREATION

At least three kinds of observation indicate that the universe as we know it began 10–18 billion years ago.

Age of Globular Clusters In Chapter 22 we saw that H–R diagrams allow estimates of the **ages of globular clusters**. Results indicate that the clusters formed about 14 ± 4 billion years ago. This period is believed to mark the formation of our galaxy out of hydrogen and helium gas that formed in the big bang. The big bang must have happened at least that long ago.

Hubble's Constant Because **Hubble's constant,** H, measures how fast the galaxies are now rushing away from each other, we can calculate how long it has taken them to get this far apart if they have been receding at a constant speed. We can thus compute the age of the universe from H. If the density of matter in the universe is well below the critical closure density (page 595), then the universe will have expanded since the big bang at about a constant rate, and the age since the big bang will be estimated as the distance of a galaxy divided by its speed. This turns out to be simply $1/H$. If the density of the universe is greater, the gravitational attraction of galaxies for each other will have caused a slowdown of the expansion over time. The universe would thus have been expanding faster in the past, and the age would thus be younger. If the density equals the critical closure density, it turns out that the age in most models would be estimated as two-thirds the age estimate given earlier, or $2/(3H)$. The answer given by this method depends on the value of H. Two values of H have been championed, either 50 or 100 km/s per Mpc (page 556). These could give an age of the universe as high as 20 billion years if there is no missing mass, or as low as 7 billion if there is enough missing mass to cause closure (see page 596). If we adopt $H = 77 \pm 14$ km/s per Mpc (as on page 556), the age would be 11 to 16 billion years with no missing mass, and 7 to 10 billion years with enough missing mass for closure.[1]

Ages of the Elements In Chapters 6, 7, and 13 we discussed how the radioactive elements can be analyzed to reveal the 4.6-billion-year age of the solar system. Chapters 15, 18, and 19 added evidence that the heavier elements were mostly synthesized by fusion reactions inside stars and distributed by supernova explosions during a period stretching back further than 4.6 billion years ago. The big bang theory allows physicists to

model conditions inside the primeval fireball and calculate how many elements of various types synthesized there. By comparing today's abundances of radioactive elements with their decay rates and with their calculated production rates inside stars and the primeval fireball, it is possible to calculate the total **duration of element formation** in the universe.

For example, if all uranium formed in the big bang, the universe should be about 7 billion years old to account for uranium abundances today. But we know that such a heavy element did not form efficiently in the big bang. If it all formed slowly inside successive generations of stars, the universe should be about 18 billion years old (Peebles, 1971). By this calculation, the age of the universe must be between 7 and 18 billion years.

An independent measurement using radioactive thorium-232 detected spectroscopically in stars of different ages gave more than expected. This finding implies that the stars had not existed long enough for much of the earliest thorium to have decayed, implying a younger age than expected for the galaxy. The result suggests that our galaxy is only about 10 billion years old, and the universe, 11 to 12 billion years old (Butcher, 1987). Astronomers are currently debating whether this result can be brought into accord with the somewhat greater age estimated for globulars.

From the rough agreement among these three lines of evidence, most astronomers conclude that the big bang occurred 9 to 18 billion years ago; perhaps the best estimate is 14 billion years ago, with the galaxies forming soon afterward. Some unique explosive event must have occurred at this time, apparently creating matter and sending it flying out on its expanding journey. Astronomers refer to the time interval since this event as the **age of the universe,** the age since the creation of the universe as we know it.

The term *explosive* may be misleading, because the big bang was not an ordinary explosion. No one could have floated in nearby space and watched the fireball expand, because, paradoxically, the fireball filled all space. *Space itself is viewed as contracted at the beginning.* An analog to these models would be the expanding balloon, whose *surface* represents space. An ant living on a spot on its surface could detect the expansion and other spots moving away, but the ant could not get off the surface to look at the whole balloon expanding. The ant could travel only *on* the surface and would perceive no "up and down" or any interior to the balloon—only the surface. At the beginning, the surface of the balloon would have been "small" and incandescent, but it still would

---

[1]Research around 1980 nearly doubled the best estimate of the Hubble constant from some 50 km/s to nearly 100 km/s per megaparsec, reducing the estimated age to a value around 10 billion years. This figure is difficult to reconcile with globular cluster ages of 14 billion years. The prestigious journal *Science* paraphrased Shakespeare: "Double Hubble, age in trouble."

have constituted the ant's whole universe. In this sense, the "small" balloon—the hot, high-density fireball—would fill all space perceived by the ant.

Instead of picturing an explosion, it may be more helpful to think of the big bang just as a singular moment of high-temperature compression of the universe's matter, whose earlier history, if any, we can't trace.

THE FIRST MINUTES OF THE UNIVERSE

One of the astonishing things about the big bang theory is that it lets us describe **initial conditions** during the first moments in the history of the universe. The assumptions are simple, because if we say that all the observable mass was once as concentrated as possible, then at the initial moment the density should have been nearly infinite; it has been declining ever since. Packing all mass at high density would lead to extraordinary temperatures. A simple theoretical model developed around 1950 by George Gamow assumed infinite temperature and density at the zero instant! Since we know the behavior of expanding hot material, Gamow was able to calculate subsequent conditions just as an engineer might calculate conditions in a gas expanding in a piston. After 1 s, he estimated that the temperature was 15 billion Kelvin and the density about that of the air we are breathing. This model also predicted a present temperature of about 20 K and a density of about 10^{-27} kg/m^3, which are close to the observed values.

QUANTUM MECHANICS AND THE INFLATIONARY BIG BANG THEORY

Although this early version of the big bang model had some successes, astrophysicists realized it left unanswered questions. In the first place, the universe at the instant of the big bang was viewed as a point (or as mathematicians say, a singularity). What made this point of mass explode? What was there *before* the big bang, or does this question have meaning? Moreover, observations of galaxies out to billions of parsecs indicate that the curvature of space is very small. The universe is almost flat. But the big bang models suggest the curvature could have almost any value, so that the unique, Euclidean, flat geometry seemed merely a coincidence. Astrophysicists pondering these questions in the 1970s

began to suspect that the original big bang model was too simple and too mechanistic.

At this point in the story, we must face the problem that modern cosmogony involves as much basic physics as it does astronomy. Let us pause here to consider a very brief overview of recent work. Physicists are currently working on "grand unified theories"—models that attempt to combine an understanding of matter, gravitational fields, magnetic fields, electric and other fields, on all scales. Modern quantum mechanics, which incorporates relativistic phenomena, is a key part of this effort; it is currently the best mathematical description of nature. At tiny scales, such as the world of atoms and electrons, quantum mechanics gives the best description of the behavior of nature. At larger scales, such as billiard balls and planets, the laws of Newton, which are simpler, give virtually the same answers as quantum mechanics. For that reason, engineers don't bother using complex quantum-mechanical formulas to build a bridge; they use the simpler, more familiar Newtonian principles.

But astrophysicists, who can use Newtonian concepts (sometimes modified by relativity) to trace the motions of planets and the collapse of stars, have noted that if you trace the universe back to a small enough bit of space–time—the so-called point—it behaves as a quantum-mechanical system rather than as a familiar system in Newtonian mechanics. This is equivalent to saying that while billiard balls follow Newton's laws of motion, electrons and other subatomic particles do not; they have to be described by quantum mechanics. Recent theories about the big bang have depended on quantum-mechanical descriptions of curved space or, more accurately, the space–time combination called the space–time continuum.

Starting around 1980, MIT astrophysicist Alan Guth and others described new versions of the big bang scenario in which a phase of extreme, rapid expansion occurs in the first fraction of a second; this inflates the universe to much larger scale than in the old models. This is called the **inflationary big bang theory**. In this theory, space–time is described by certain mathematical variables of quantum mechanics. They describe "fields" in which instabilities can occur, which in turn trigger explosive expansion. The "fields" are not magnetic fields or gravity fields or descriptions of matter, but more of a description of the state of space–time itself. According to the new theories, in the incredibly short instant of 10^{-43} s after the big bang was triggered, space adopted a vacuumlike quality with no particles of mass in it. This

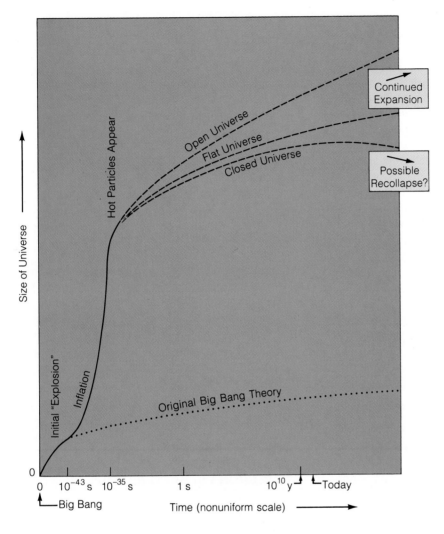

Figure 27-1 The history of the size of the universe, according to modern big bang theory. "Size" refers to any property that measures the scale of space. Using the Earth as an analogy, size could be the radius of the Earth or the distance required to traverse one degree of latitude along the curved surface. In the original big bang theory (bottom), expansion continues smoothly after the initial "explosion." In modern theories since 1980, the initial "explosion" is followed by an enormous inflation for about 10^{-35} s, after which the first energetic particles appear. The three dashed lines show possible subsequent histories. Note that it is hard in practice to distinguish between open, flat, and closed universes by present-day observations. (After diagram by Linde, 1987.)

unstable form of space did not expand at a constant rate, as space was imagined to have done in the old theories, but at an ever increasing, "exponential," rate. This process is shown in Figure 27-1. This rapid expansion, called **inflation,** continued for the first 10^{-35}s—still much less than a trillionth of a second. During this incredible inflation, the universe's volume made a sudden jump by as much as 10^{25} times! At the end of the inflation, this vacuumlike state changed and energetic subatomic particles began to appear in space. From that point on, the now-much-bigger universe continued to expand at a smoother rate, as in the older big bang theories.

These new ideas excited theorists in the 1980s. For the first time there were answers to various questions. The universe as we know it did not exist before the big bang, but one might say that the ghostly fields represented the precursor form of the universe. The universe in the new theories is much bigger than imagined in the old theories, as shown in Figure 27-1. Thus it looks fairly flat, for the same reason that the vast Earth looks flat to an ant, because the scale of its curvature is so much bigger than the ant is. The new theory also helps explain the generally uniform appearance of the universe in all directions, a property that had begun to puzzle "old-big-bang" theorists.

PREDICTING ABUNDANCES OF ELEMENTS IN POPULATION II

The big bang theory in both its original and its inflationary form succeeds in describing how the primordial particles joined together to form atoms of the various elements, and it also predicts the abundances of those elements. In the high-temperature gas of the first seconds, matter was broken down into its simplest components—not grains or molecules or even atoms, but subatomic particles such as neutrons, protons, and electrons. Calculations based on known atomic physics show how these particles would interact under conditions after 1 s, 2 s, and so on.

The first step in making atoms was to make their nuclei, or inner cores. The outer electrons could be added only later, under cooler conditions. The nucleus of a hydrogen atom is simply a proton, so it could be said that the universe was already full of hydrogen nuclei, since protons were an abundant basic particle. The next-heaviest atomic nucleus would be that of deuterium, or heavy hydrogen, which consists of a joined proton and neutron. Gamow and his associates[2] in 1948 found that protons and neutrons could collide and form deuterium:

Proton + neutron = deuterium + gamma radiation

The higher the temperature, the faster particles collided. During the first 3 min or so, collisions would have been so energetic that deuterium nuclei would have broken apart faster than they would have formed. After about 3 min, according to this model, deuterium began to accumulate. Calculations (especially by Margaret Burbidge, Geoffrey Burbidge, William Fowler, and Fred Hoyle in 1957) showed that nuclei of atomic mass 5 (heavy helium) would not form in this way and that most elements heavier than helium built up inside stars as outlined in Chapter 19. These heavy elements have slowly accumulated in the universe over billions of years—the "pollution" associated with billions of stars burning the relatively pristine hydrogen–helium mix.

Nonetheless, the big bang theory seems to account for the formation of most of the material in the universe—the hydrogen, its heavy form deuterium, and

helium—during the universe's first hour. A calculation based on these principles (Reeves and others, 1972) gives these abundances at the end of the first hour:

Element	Percent Mass
Hydrogen (^{1}H)	75
Helium-4 (^{4}He)	25
Deuterium (^{2}H)	0.1?
Helium-3 (^{3}He)	0.001?
Lithium-6, lithium-7 (^{6}Li, ^{7}Li)	0.000001?
Heavier nuclei	Negligible

Perhaps it seems presumptuous to talk of events during the first minutes some 14 billion years ago, but the results can be confirmed by observation. Remember that the oldest stars we can find, the extreme Population II stars, appear to have formed from exactly such material. This finding shows that when stars began to form, the heavy elements did not yet exist. Thus we can be pretty certain that shortly after the big bang, the universe consisted almost entirely of very hot hydrogen and helium.

THE FIRST MILLENNIA OF THE UNIVERSE: RADIATION VS. MATTER

According to the big bang model, radiation was more important than matter in the first few thousand years. According to Einstein's famous equation $E = mc^2$, energy E is equivalent to mass m (c is the velocity of light). This means that each photon of radiation, carrying a certain energy E, will have an equivalent mass $m = E/c^2$. Thus any given amount of radiation corresponds to an equivalent mass of material. At the billion-Kelvin temperatures in the first moments after inflation, the universe amounted to a **primeval fireball** full of intense radiation. Hence there were many photons per unit volume—and hence a large *equivalent* density of material if the radiation had suddenly been converted into matter. Under these conditions, the *equivalent* mass density of the radiation was greater than the density of the existing mass itself! This is indicated in the upper left part of Figure 27-2, which shows the schematic history of the universe.

---

[2]Gamow deliberately collaborated with Ralph Alpher and Hans Bethe (pronounced like *beta*) so that their paper could come under the byline Alpher Bethe Gamow, a pun on "alpha beta gamma," a phrase suitable to a milestone in cosmology. It is said that another scientist, Delter, turned down an invitation to join as fourth coauthor.

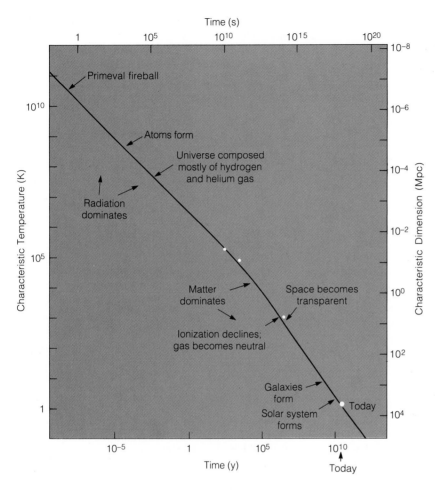

Figure 27-2 A brief history of the radiation and matter in the universe, based on current research. The logarithmic time scale (bottom) expands the early history of the universe to fill most of the graph, allowing discrimination of early events. The curve shows declining temperature (left scale) and increasing distances to which internal observers could see (right scale). Radiation dominated the early stages. Atoms formed by combinations of subatomic particles. Gravitational contraction of gas clouds led to the formation of galaxies. Characteristic temperatures corresponding to radiation have declined to 3 K at present.

Slowly the radiation converted itself into hydrogen and helium mass, as indicated in Figure 27-2. The radiation intensity declined. Today the radiation density is negligible. As Gamow pointed out, the mass of radiation in a cubic kilometer of air today is only about 10^{-17} kg. By comparison, the mass of the air in the cubic kilometer is some 10^9 kg. Even in the center of the Sun or an exploding atomic bomb, radiation amounts to only about 1 kg/m^3. But in the first tenth of a second it may have exceeded 1000 kg/m^3!

The importance of this result is twofold. Once matter appeared, the intense radiation agitated the particles of matter so violently that neither galaxies nor other clumps of matter could form. The gas was too hot. But after some thousands of years, radiation density dropped to below the density of matter. Within the first billion years, recognizable masses of material began to form.

The second consequence of the high radiation density is that in the far distant past (seen by looking toward distant, highly red-shifted regions), a different kind of radiation from that found near our own galaxy should be seen. It would be a trace of the dense radiation in the primeval fireball. The early universe was permeated by the kind of radiation that would come from high-temperature gas. As matter flew outward, the radiation was diluted and red-shifted (Figure 27-3). The red-shifting would lengthen the apparent wavelength of the radiation, and, according to Wien's law, the peak energy would thus come at a wavelength corresponding to quite a low temperature. Starting as early as 1948, this temperature has been variously predicted to be about 1 to 5 K. In other words, if the big bang theory is right, the sky should be uniformly filled with faint radio radiation resembling radiation from an object at about 1 to 5 K.

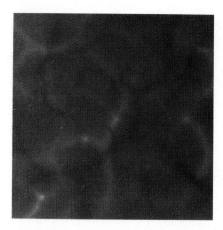

Figure 27-3 A schematic representation of a possible view in the primordial universe as matter starts to dominate over radiation. Seconds after the big bang, all of space was filled with a flash of ultraviolet light. As space expanded and the effective temperature of the radiation dropped, the light was eventually visible as a dull red glow in all directions. The first clouds of condensing matter are portrayed as luminous or nonluminous filaments silhouetted against this red glow. (Painting by Ron Miller.)

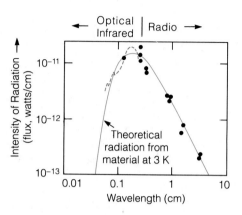

Figure 27-4 The radiation attributed to the primeval fireball of the big bang. The blue curve is the amount of radiation predicted by the big bang theory, from material at a temperature near 3 K. Observations by a satellite in 1989–1990 fit this curve, from wavelength 0.05 cm to 5 cm, to an accuracy of 1%.

This line of thought was foreseen by the father of the big bang theory, Georges Lemaître, who wrote in 1931:

The evolution of the universe can be compared to a display of fireworks that has just ended: some few red wisps, ashes, and smoke. Standing on a cooled cinder, we see the slow fading of the suns, and we try to recall the vanished brilliance of the origin of the worlds.

But why just try to recall it? Why not point detectors at the sky to see if there is long-wavelength radiation left over from the primeval fireball?

Discovering the Primeval 3-K Radiation

Indeed, direct observational evidence confirms this radiation. In the 1960s a research team at Princeton built a sensitive radio detector to search for it. The predicted radiation would be microwave radio noise. But even before the Princeton device could be applied, Bell Laboratory researchers Arno Penzias and Robert Wilson, using radio telescopes in experiments on the first Telstar communications satellite in 1965, found puzzling microwave noise. At first they thought it was a problem

in the instruments, but finally they established that it was coming from all over the sky. Analyzing these results, the Princeton physicists realized that this was the predicted remnant radiation permeating the universe from the primeval fireball. The radiation was observed exceedingly accurately in 1989–1990 from a new satellite, as shown in Figure 27-4. It corresponds exactly to radiation from material with temperature of 2.735 K. For convenience, it is labeled the 3-K radiation, and called the 3-degree radiation. Penzias and Wilson received the Nobel Prize for discovering this relic of the universe's beginning.

This discovery is the strongest confirmation of the big bang theory, especially because the **3-K radiation** seems to be uniform in all directions. Just as predicted, observations at many radio wavelengths indicate no variations in intensity within an accuracy of a few percent.

Why Do We See the 3-K Radiation?

Radio radiation from all over the sky, emanating from material that appears to be at the specific but extremely low temperature of 3 K (three degrees above absolute zero) may seem like a strange concept at first. But it is not hard to understand, given the principles we have discussed. Consider the early history of the universe, shown in Figure 27-2. For the first few hundred thousand years, the temperature was so high that the hydrogen gas was ionized and opaque, like the gas in the Sun. When the temperature dropped to around 3000 K, the electrons could join with the protons to form neutral hydrogen atoms; by a million years after the big bang, the gas was no longer ionized. We know from studying the sharply defined surface of the Sun that ionized hydrogen can be quite opaque, even though interstellar cool hydrogen between the stars is quite transparent. So, for the first million years, the universe was filled with a brilliant glowing fog of plasma, and after a million years "the fog lifted" (Silk, 1980).

Now consider astronomers looking out through space. As they look out past nearby galaxies, they look through transparent space, very thinly populated by neutral hydrogen. But as they look farther away, they look back in time. This is shown schematically in Figure 27-6 on page 608. At several thousand megaparsecs, their line of sight encounters the region where light left the brilliant fog about a million years after the big bang. This region is like a wall. It is opaque, and the astronomers can't see beyond it. But why doesn't it look like a 3000-K brilliant fog, resembling the surface of the Sun? The wavelength of the light they see from it is Doppler-shifted by about a factor of 1000. (Note that the universe has expanded in dimension by a factor of about 1000 since this era, as seen in the right-hand scale of Figure 27-2.) So the temperature they perceive (following Wien's law) is about 1000 times lower, or about 3 K. The 3-K radiation is really the light from the million-year flash that accompanied the big bang.

Evidence for Inflation?

While observing the 3-K radiation from all over the sky, astronomers identified an interesting challenge to the big bang theory. Recall from the last chapter that clusters of galaxies show a filamentary structure when mapped across the sky. This means that by the time galaxies began to form, some hundred million to a billion years after the big bang, the gaseous matter had aggregated into filaments or clouds. But the 3-K glow is virtually without fine structure anywhere in the sky, as can be seen in Figure 27-6. This tells us that at the time the glow was being emitted, only about a hundred thousand to a million years after the big bang, the glowing

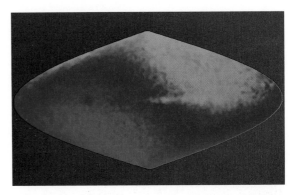

Figure 27-5 An image of the whole sky, showing the 3-K cosmic background radiation, as detected at wavelength 0.57 cm (compare wavelength scale in Figure 27-4). Slight Doppler shifting toward blue in one half of the sky and red in the other half is due to our solar system's motion through space in one direction. The Milky Way runs horizontally across the center of the picture and is faintly seen. With those exceptions, the 3-K radiation is essentially featureless, showing that the initial big bang fireball was extremely homogeneous. (NASA photo from data of Cosmic Background Explorer, or COBE, satellite.)

material was virtually uniform. As we noted earlier, this would have been at a time when the material had cooled to about 3000 K. But the original big bang theory predicted that by the time the universe reached these conditions, different regions would have had different temperatures and densities, consistent with the filamentary pattern of galaxy clusters. The theory would thus predict a spotty 3-K glow in the sky, and this is not observed! What's wrong?

During the 1980s, the inflationary big bang theory came to the rescue with new insights about the events of the first second after the big bang (Barrow and Turner, 1982). Inflation helps explain why the 3-K radiation is so uniform (Wilkinson, 1986). According to this idea, the universe we observe today arose from a volume 10^{25} times smaller, and much more uniform in density, than the universe pictured by traditional big bang theorists (cf. Figure 27-1). On the other hand, if the initial volume was *too* uniform, clumps of galaxies could not have formed. This problem was aggravated in 1989–1990 when detailed measurements of the 3-K radiation by satellite showed it to be extremely uniform, implying a very smooth early universe. The problem is still not solved. Obviously matter did begin to clump together at some stage (since we are here!), and, as we will

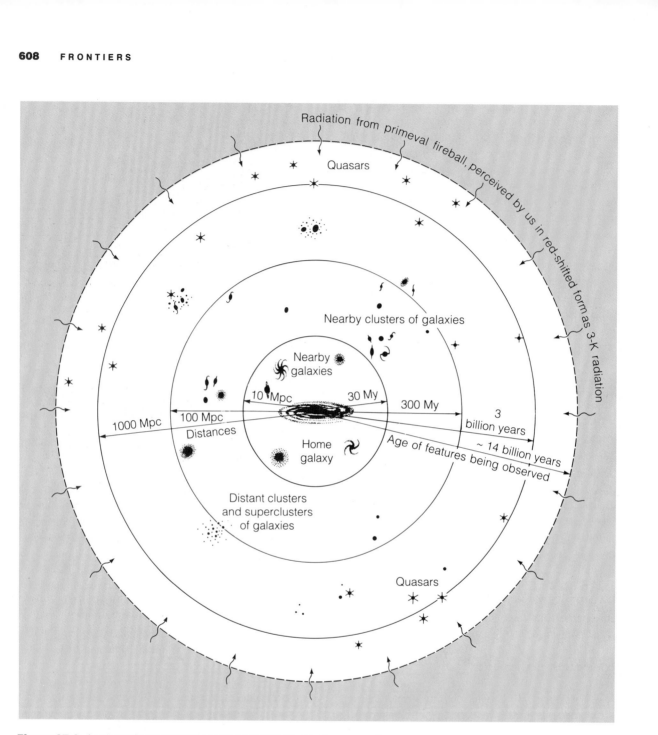

Figure 27-6 A schematic representation of the universe as seen from any galaxy, including ours. Note change of scale with increasing distance from the center. Clusters and superclusters populate space out to great distances. At very great distances, corresponding to a large red shift, we see light that was emitted billions of years ago, and the only visible objects are quasars. From an extremely red-shifted region beyond them comes radio radiation emitted during the era of the opaque fireball, about a million years after the big bang.

Figure 27-7 Galaxies without end. This very remote cluster of galaxies, the Coma cluster, is 110 Mpc away. All of the slightly elongated or fuzzy images are galaxies. The scene is dominated by two orangish giant ellipticals and one bluish foreground star (the image with diffraction spikes). Their lack of detail testifies to the difficulty of getting information about the most remote parts of the universe. (Courtesy L. Thompson, Copyright National Optical Astronomy Observatories, Tucson, Arizona)

describe next, theoreticians have been able to outline how galaxies formed after that stage.

FORMATION OF GALAXIES

According to the big bang theory, including the inflationary version, the densities of both radiation and matter decreased as material rushed apart after the big bang and the inflation. Probably within a few thousand years the density of matter became greater than that of radiation, allowing stronger gravitational forces between particles of matter. Clouds of gas formed in the bubble-like pattern detected in galaxy cluster maps. After many millions of years, temperature and density were probably such that gravitational contraction began in the denser clouds of gas, as indicated in Figure 27-2. Certain clouds or turbulent eddies had enough density and self-gravity to contract, perhaps to masses as large as entire

superclusters of galaxies. Denser subregions eventually contracted to form individual galaxies, as seen in Figure 27-7. Thus **formation of galaxies** and galaxy clusters is a predictable result of the big bang.

Galaxies in large clusters show alignment of their rotation axes—they may simply be fragments of single, rotating, primitive clouds (Ozernoy, 1974). This supports the idea that the galaxies themselves are just one more case of the fragmentation of contracting clouds (as discussed in Chapter 18).

Surprisingly, the era of galaxy formation can be dated by the statistics of quasars. A 1982 survey indicated a rapid drop-off in numbers of quasars as the red shift exceeds 3.53. This corresponds to an era about 1 to 4 billion years after the big bang, depending on the cosmological model. We see few earlier quasars. Therefore, astronomers infer that a rapid turn-on in new galaxies (some now seen as quasars) began about 1 to 4 billion years after the big bang, or perhaps 10 to 14

TABLE 27-1

Probable Origins of Selected Elements

Element	Big Bang	Supermassive Exploding "Pregalactic" Objects	Massive Unstable Stars	Ordinary Stars	Cosmic Ray Interactions
Hydrogen (^{1}H)	Yes	—	—	—	—
Deuterium (^{2}H)	Yes	?	?	—	—
Helium-3 (^{3}He)	Yes	—	?	?	—
Helium-4 (^{4}He)	Yes	?	?	?	—
Lithium-6 (^{6}Li)	—	—	—	—	Yes
Lithium-7 (^{7}Li)	?	?	?	?	Some
Beryllium-9 (^{9}Be)	—	?	—	—	Yes
Boron-10 (^{10}B)	—	—	—	—	Yes
Boron-11 (^{11}B)	—	?	?	—	Yes
Heavier elements	—	?	Yes	Yes	—

Source: Data adapted from Reeves and others (1972).

Note: "Yes" indicates strong mathematical evidence that the observed atoms formed in the given environment; "?" indicates the possibility or probability that some observed atoms formed; a blank indicates the unlikelihood that significant numbers of atoms formed.

billion years ago (Schmidt, 1982). This is shown in Figure 27-2 and is at least roughly consistent with a formation of our own galaxy roughly 14 ± 4 billion years ago.

FORMATION OF HEAVY ELEMENTS

The first galaxies and stars presumably formed from the gas produced in the big bang, about 75% hydrogen and 25% helium by mass. This statement agrees with direct observations of the ancient Population II stars. A review of the theory of stellar evolution reveals the **origin of heavy elements.** They were created later inside stars by nuclear reactions such as the s-process and the r-process, mentioned in Chapter 17.

It is the massive stars especially that synthesize heavy elements in their high-pressure interiors and then blast them into space when they explode as supernovae. When the next generation of stars forms from the resulting interstellar clouds, the new stars incorporate some of these ready-made heavy elements. Then, as they evolve, they cook up still more heavy elements in their "stellar pressure cookers."

As shown in Table 27-1, most heavy elements have probably been created in this way. The great stability of certain elements, such as iron, explains why they are especially abundant: They rarely broke apart as other atoms joined and refragmented during atomic collisions inside stars.

Thus the big bang theory, combined with our knowledge of stellar processes, also succeeds in explaining the observed abundances of elements. As Figure 27-8 shows, the heavier elements have gradually increased in relative abundance from virtually none before about 12 billion years ago to a few percent today.

A COSMIC REFERENCE FRAME

We have commented that both observations and the big bang theory imply that all galaxies are moving smoothly away from each other and that the view from any one galaxy is similar to that from any other. By measuring Doppler shifts of galaxies, we can measure our galaxy's motion relative to those of the others. For example, the "Whirlpool" galaxy, M 51, is moving away from us at 550 km/s.

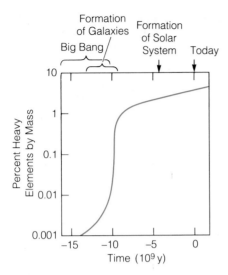

Figure 27-8 The abundance of heavy elements as a function of time. Shortly after the big bang, heavy elements constituted virtually none of the total mass. After the formation of galaxies, they increased to about 1% of total mass because they were produced inside individual stars and dispersed by supernova explosions. They have since risen to several percent due to repeated generations of star formation and disruption. (After Reeves and others, 1972.)

The discovery of the 3-K radiation gives us a larger frame of reference: We can measure Doppler shifts in the 3-K radiation in different directions. In this way, astronomers have found that our galaxy is moving at about 600 km/s relative to the average defined by the universal 3-K radiation. This is interpreted partly as a small random motion of our galaxy among the swarm of flying galaxies, just as the Sun has its own small random motion among the other stars as they all orbit around the center of the Milky Way. Similarly, observations suggest a drift of our Local Group of galaxies toward the Virgo cluster or the hypothetical "great attractor" (see page 594). Such motions might be partly a result of the motion or rotation of our supercluster of galaxies. For example, some data suggest that we belong to a supercluster system extending over about 15 Mpc, our rotational speed being around 400 km/s (Silk, 1980).

BEFORE THE BIG BANG

What existed before the big bang? No one is sure that this question has meaning or that any observations could reveal an answer.

According to classic big bang theory, the mysteri-ous explosive event at the beginning is *assumed* to have mixed matter and radiation in a primordial soup of nearly infinite density, erasing any possible evidence of earlier environments. If so, science would have no way of answering the question.

Conceivably, the big bang was not unique. This variant of big bang cosmogony is called the hypothesis of the **oscillating universe.** It can be visualized in the "closed universe" curve of Figure 27-1. In this view, after the primeval fireball explodes, gravity eventually pulls back the distant material and the universe collapses again, as seen in the turn-down at the right end of the curve. Another high-density period would follow, then another expansion, and so on. The possibilities of future recollapse or everlasting expansion were discussed in the preceding chapter.

In a variant of the oscillating universe hypothesis, the high-density era was not the infinite-density epoch of the classic big bang, but only a period of maximum, finite density, during which some structures from a preceding era might have survived. At least one rotating neutron star has been interpreted as slowing its spin at a rate that may make it older than 16 billion years, and some astronomers have speculated that it might be a survivor from an earlier era.

In the nonoscillating big bang models, the big bang is the true beginning of the universe as we know it, a creation like those in the creation myths of many cultures. The oscillating universe model—that the big bang was not a true beginning, but only the beginning of the current cycle—resembles certain Eastern mythologies.

The new inflationary big bang theories add an interesting twist to these ideas. The physicists who developed these theories note that the properties of all the particles in the universe tend to cancel each other out. For example, there seem to be as many positively charged particles as negatively charged ones, so that the net charge in the universe is zero. Similarly, the net angular momentum, and other special properties of subatomic particles, seem to add up to zero. This has led some physicists to postulate that the big bang began from a special vacuumlike energy field in which all properties added up to zero and there were no material properties. It may have been, in cosmologist Alan Guth's words, a situation that "you can't distinguish from nothing." In other words, big bang theorists can now postulate, in a sense, that the universe did start from "nothing"— the ghostly creation field that exploded in a shower of positive and negative particles leading to the state we now see.

ARE THERE "OTHER PLACES"?

We have defined the universe as all matter and energy that exist anywhere. Until recent decades, most scientists had assumed that all parts of the universe had to be continuously connected, in the sense that light or a spaceship could go from one region to any other region and back again (even if the trip took a long time). But if there are black holes, there *might* be some regions permanently detached or unobservable from anywhere else. Some scientists speculate that there might be "other regions," such as regions "inside" black holes, in which events can occur but never be seen by us.

A still more intriguing speculation is that we might be in a black hole in someone else's universe. Recent work suggests that—contrary to earlier theories that black holes could never emit anything—a black hole might be able to explode in a burst of gamma-ray radiation and subatomic particles. Thus the big bang itself might have been some kind of black hole explosion within some sort of larger universe.

Since black holes are thought to be permanently out of touch with each other and (so far) subject only to theoretical investigation, speculations on these problems are at the borders of what the scientific method can now deal with.

WHY?

Why did the universe come into being? If the preceding questions of "how?" seem difficult—if they seem at the borders of the scientific method—questions involving "why?" are clearly beyond the scientific method. Why was there a big bang? What started the bang? What was there before? What is the purpose behind it all? When thinking about these questions, we may find it helpful to recall a whimsical but deep comment by Mark Twain: "Why shouldn't truth be stranger than fiction? Fiction, after all, has to make sense. . . . " By this he meant that our cultural tradition tends to make us want to look for "sense" in a story—even the story of the universe. We like our stories to have a beginning, a middle, and an end—and perhaps even a nice moral that is consistent with what we believed before we heard the story. When we seek the story of the universe, we have no guarantee of this kind of sense. But we do have a virtual guarantee, based on experience, that if we keep making observations—for example, with improved telescopes in space—they will draw us on to new, fascinating discoveries about the universe. Perhaps they will clarify still further what the big bang was all about.

Remember that the scientific method does not answer "why?" questions, except to show how events in nature follow from laws we have already discovered. The scientific method is basically a procedure for analyzing observations and predicting phenomena slightly beyond those we know. Thus people practicing the trade of science can only ask questions that can be answered by specific observations.

To the questions of "why?", astronomers' answers may be little better than anyone else's.[3] One can always make up answers, but it is more interesting to admit that there are many things we do not understand—and things we do not even understand *how* to understand.

SUMMARY

Three lines of astronomical evidence—ages of globular clusters, the expansion age calculated from Hubble's constant, and estimated ages of elements—indicate that the formative conditions existed in the universe roughly 9 to 18 billion years ago, probably close to 14 billion years ago. Most astronomers believe that an explosionlike event called the big bang occurred then. It takes some audacity to claim that we humans know something about this mysterious event, but there are several compelling observations:

1. Because galaxies are receding from each other, they must have once been concentrated in a much smaller volume; this is consistent with the big bang model of initial high density.

2. The big bang model explains the abundances of elements that we see both in Population II stars and, with the addition of later element-forming processes inside stars, in Population I stars as well.

3. The big bang model predicted a pervasive weak radio radiation, known as the 3-K radiation, *later* found by radio astronomers.

---

[3] In *The Hitchhiker's Guide to the Galaxy,* a satire on intergalactic epics, the most giant of computers succeeded in deducing the "ultimate answer" to this ultimate philosophical question—and the answer turned out to be "42." However, the computer noted that the ultimate question itself was so vague that the computer was unable to explain the method by which it deduced this ultimate answer.

4. Theories of galaxy formation, counts of distant galaxies, and observations of the most ancient galaxies and quasars, while not uniquely predicted by the big bang theory, are consistent with it.

The big bang theory takes us close to the limits of the ability of science to explain the properties of matter and energy in the universe. Questions such as why the universe came to exist are beyond the scope of science.

CONCEPTS

cosmogony	inflation
age of globular clusters	primeval fireball
Hubble's constant	3-K radiation
duration of element formation	formation of galaxies
	origin of heavy elements
age of the universe	oscillating universe
initial conditions	
inflationary big bang theory	

PROBLEMS

1. Which phrase describes the big bang theory best?
   a. An assumption with logical consequences that turn out to be verified by observation
   b. A fact proved by repeated observation
   c. A revelation

2. Why was the discovery of 3-K radiation from all over the sky heralded as strong evidence in favor of the big bang theory? Does comparing the radiation from different parts of the sky give any evidence of asymmetry or inhomogeneity in the universe?

3. Why would planets such as Earth be unlikely to exist in globular clusters? (*Hint:* Consider the Earth's composition.)

4. How would spectroscopic observations help reveal that a cluster of very distant galaxies, which looks like a mere grouping of fuzzy stars on a photograph, is not a group of stars inside our own galaxy?

5. Why does the radiation from the primeval fireball as observed today have such long wavelength (radio waves) instead of the very short wavelengths (ultraviolet) that might be expected from the high temperatures theorized for the fireball?

6. Do you find any fundamental disagreement between the description of the universe's origin and structure, as described in this and the previous chapter, and any philosophical or religious beliefs you may hold? If so, do you believe such a disagreement might be clarified by further observations, or do you believe further observations are superfluous?

ADVANCED PROBLEM

7. Hubble's constant H can be thought of as specifying the speed at which galaxies have traveled to reach any specified distance from the (arbitrarily chosen) central site of the big bang.
   a. Show by logical deduction that $1/H$ ought to equal the age of the universe (the time since the big bang).
   b. Confirm that $1/H$ has the dimensions of 27t1time.
   c. Confirm numerically that $1/H$ equals 13 billion years, if $H = 78$ km/s per Mpc. (*Hint:* The problem really deals with conversion of units. Note that $1 \text{ y} \approx \pi \times 10^7 \text{ s}$.)

CHAPTER 28

Life in
the Universe

One of the most intriguing questions in astronomy is whether planets elsewhere in the universe harbor what we are pleased to call "intelligent life." Extraterrestrial life must either exist or not exist. Either case has striking consequences. The nineteenth-century Scottish writer Thomas Carlyle sardonically said that other worlds offer "a sad spectacle. If they be inhabited, what a scope for misery and folly. If they be not inhabited, what a waste of space."

If *intelligent* aliens exist, our society (if it survives) is someday likely to be influenced by them, for better or worse. Anthropologist D. K. Stern notes that discovering such creatures "would irreversibly destroy man's self-image as the pinnacle of creation." If alien life does not exist, we would be the only living creatures in the universe—a remarkable situation that implies a universe wide open for us to utilize and colonize.

What can astronomy say about these two possibilities? Some scientists argue that astronomy, lacking any definitive detections of extrasolar planets or life, can say nothing. However, we certainly have data related to life's origins: **Organic molecules** (complex, carbon-based molecules) have been found in interstellar clouds; meteorites containing amino acids show that organic chemistry has been active in our own solar system outside the Earth; and the paucity of organic molecules on Mars places intriguing limits on the biochemical possibilities of planets. Admittedly, discussions of the existence, intelligence, psychology, or appearance of higher alien life forms are almost entirely speculative. Nonetheless, the search for alien life seems to be a new scientific adventure in the making, and a growing number of astronomers, biologists, chemists, anthropologists, and theologians are taking an interest in this quest.

In this chapter, we consider the subject in several steps. First, we discuss what we mean by *life* and what conditions are necessary for life to exist on planets.

Second, we review the long process that led to the evolution of life on Earth, as best we understand it. Third, we ask whether there are planets that could support a similar evolution of life elsewhere in the universe. Finally, we try to estimate the probability of intelligent life actually existing on such planets elsewhere and whether we might contact it, or it, us.

THE NATURE OF LIFE

What do we mean by *life?* **Life** is not a status but a *process*—a series of chemical reactions, using carbon-based molecules, by which matter is taken into a system and used to assist the system's growth and reproduction, with waste products being expelled. The system in which the processes occur is the cell. All known living things are composed of one or more cells.[1] A **cell** is, in essence, a container filled with an intricate array of organic and inorganic molecules (protoplasm). Codes for cellular processes are contained in very complex molecules (such as the famous DNA) located in a central body called the nucleus. The elements most prominent in the organic molecules are carbon, hydrogen, oxygen, and nitrogen—all very common in regions populated by evolved (Population I) stars. Phosphorus, also important to life (in small amounts), is widely available. Carbon is especially critical because it can combine to make long chains of atoms—large molecules that encourage the complicated chemistry of genetics, reproduction, and so on. That is why the term **organic chemistry** (as well as *organic molecules*) refers not specifically to life forms but more generally to all chemistry (and molecules) involving carbon.

We often make the mistake of thinking of ourselves as static beings instead of dynamic systems. We casually assume that we are constant entities, as if our identity depended solely on the form of our bodies. But our bodies today are not the same ones we had seven years ago. Hardly a cell is still alive that was part of that body. Even our seemingly inert skeletons are living and changing, always replacing their cells. We *must* keep changing—the cells must keep processing new mate-

rials to stay alive. When the processing stops, we call it death.

The nature of living beings is illustrated in an analogy from the Russian biochemist A. I. Oparin (1962). Consider a bucket that has water pouring in at the top from a tap and flowing out at the same rate through a tap in the bottom. The water level in the bucket stays constant, and a casual observer would call it a "bucket of water." But it is not like an ordinary bucket standing full of water. The water at any instant is not the same water as at any other instant, yet the outward appearance is constant. We are like buckets with water, nutrients, and air flowing through us, but with other, much more complex attributes, such as the ability to reproduce and be affected by genetic changes that let us evolve from generation to generation.

The flow and change of material inside living organisms give us a clue to the kinds of processes involved in the origin of life. We are looking for a process in which complex, carbon-based molecules can enter cell systems and enable them to create new molecules, incorporate these molecules into new structures, eject unused material, and reproduce themselves.

Scientists therefore usually choose to define *life* by these specific carbon-based processes. Often at this point people ask, "What about some unknown form of consciousness?[2] Or what about some unknown chemistry based on other elements, such as silicon, that can form big molecules?" The answer is that we have never observed such life forms, so we can say nothing substantive about them. If we admit they are plausible, we simply increase the probability of life or consciousness in the universe. But researchers usually restrict their discussion to carbon-based life. As physicist Gary Feinberg remarked on this question, "Some of my best friends are made of carbon." We have no noncarbon friends to contemplate.

Whatever other conceptions we invent—civilization, religion, technology, art, war, love, communication—it is the chemical processes of life that define us, just as they define the spiders, sea urchins, elephants, moths, amoebas, redwoods, and all the other incredibly varied living creatures around us. To judge whether life may exist on other planets—whether other planets are

---

[1] Viruses might be an exception. They are simpler than cells, yet can reproduce themselves using materials from host cells. Biologists disagree on whether viruses should be considered a form of life.

[2] For example, in his novel *The Black Cloud* astronomer Fred Hoyle imagined an interstellar nebular cloud with matter and electromagnetic fields organized in such a way that it developed a consciousness or will of its own.

Figure 28-1 Precursors of life on Earth probably formed as a result of the interaction of environmental compounds with lightning as an energy source. Laboratory experiments simulating the early atmosphere and oceans, with sparks representing lightning, produce organic sludges such as the one pictured here floating in the Earth's primitive ocean. Any planet with such an environment would be expected to produce such organic materials. (Painting by Ron Miller.)

already "taken"—we must find out how those processes got started on the Earth and what sort of environments fostered them (Figure 28-1).

THE ORIGIN OF LIFE ON EARTH

According to the preceding discussion, understanding the origin of life requires that we understand the early history of interactions among the elements carbon, hydrogen, oxygen, and nitrogen—the most important elements in living organisms. How did these elements get together into complex molecules such as **amino acids,** the molecules that join to form **proteins,** the huge molecules in cells? And how did proteins aggregate into cells that could absorb surrounding materials, grow, and reproduce?

Production of Complex Molecules

Research in the last few decades has proved that the first steps toward life were surprisingly easy. Chapter 18 revealed that complex molecules such as ethyl alcohol (C_2H_5OH) and even the amino acid glycine ($C_2H_5O_2N$) have been discovered in molecular clouds in interstellar space. Extraterrestrial amino acids have also been found inside carbonaceous chondrites (Figure 28-2), the prim-

itive type of meteorite described in Chapter 13. That these acids originated outside the Earth is proved because the structural symmetries of their molecules differ from those of the same types of molecules on Earth.[3] Carbonaceous chondrites once contained liquid water in their carbon-rich soil; the amino acids must have formed in this moist environment in the cometlike or asteroidlike parent bodies of these meteorites.

Amino acids would have quickly formed in the Earth's early environment. This was proved by an experiment designed to simulate chemical conditions in the Earth's primitive environment. This experiment, called the **Miller experiment,** was first carried out by chemist S. L. Miller (1955), who bottled materials such as methane (CH_4) and ammonia (NH_3), representing the Earth's hydrogen-rich primitive atmosphere, together with water, representing the ocean. He then passed sparks through this material to represent lightning. After a short time, Miller found that a sludge containing amino acids, such as glycine, had formed in the flask. This experiment has been successfully repeated many times with different "primitive atmospheres" and energy sources. Even

---

[3]The atoms can link into molecules with either "left-hand" or mirror-image "right-hand" symmetry; the ratio of the two types in the meteoritic material is strikingly different from that in terrestrial material.

Figure 28-2 Fragments of a carbonaceous chondrite meteorite in which chemists have discovered extraterrestrial amino acids. Similar carbon-based compounds are important in living cells on the Earth. Although no fossils or life forms have been found in the meteorite, its dark color is due to the abundance of carbon and carbon compounds, showing that building blocks of life can form outside the Earth. This meteorite fell in France in 1964. (NASA.)

on a planet with fewer hydrogen compounds and more volcanic gases, such as carbon dioxide (CO_2), nitrogen (N_2), and carbon monoxide (CO, see Figure 6-12, page 132), sunlight tends to initiate photochemical reactions that produce water, hydrogen cyanide (HCN), and eventually cyanamide (CN_2H_2) and the amino acid glycine (Goldsmith and Owen, 1979).

In 1983 researcher C. Ponnamperuma announced that all five of the critical organic compounds—called "bases"—that are responsible for coding genetic information in the DNA and RNA of living cells were found in a single carbonaceous chondrite. At the same time, he and his colleagues were able to synthesize them in a Miller-type experiment in a primitive atmosphere. This work seems to clinch the likelihood that many of life's building blocks arose naturally in extraterrestrial bodies.

In Chapter 12 we saw that similar processes have produced a photochemical smog on the interesting satellite Titan. With its reddish clouds of organic compounds and its possible rains of liquid methane, Titan may give us an intriguing natural laboratory for further study of these processes.

This is not to say that all planets are oozing with organic slime. Sunlight encourages reactions that not only *form* organic molecules but also *break down* organic molecules. When energetic UV photons strike molecules, the molecules can break apart. On worlds like the Moon and Mars, where there is no ozone layer to block UV light, the rate of destruction is high, and gases are insufficient to provide raw materials to form new organics. As noted in Chapter 10, sunlight has apparently destroyed any and all organic molecules that may have formed on Mars in the past. Indeed, fossil and geochemical evidence suggests life did not emerge on the Earth's land surface until ocean plants, which were shielded from sunlight by seawater, emitted enough oxygen to build up an ozone (O_3) layer in the Earth's atmosphere. This ozone layer then shielded land-based life from solar UV rays.

Although the Viking landers found no organic molecules on Mars, they did reveal interesting clues about the development of such molecules on planets. Unlike the Earth, Martian soil is exposed to strong solar UV rays, which apparently produce unfamiliar chemical reactive states in minerals. The soil contains material that can synthesize organic molecules from atmospheric carbon dioxide and release gas once nutrients are added. Because the Martian soil in its natural state contains virtually no organic molecules, most scientists believe these processes are not caused by Martian life forms. In fact, laboratory experiments in 1977 duplicated most Viking results using simulated Martian soil (iron oxide minerals exposed to UV light in carbon dioxide) without any organisms. Nonetheless, the processes may indicate how chemical reactions create material from which life could form on other planets.

The Production of Cells

The next step toward life—aggregating complex organic molecules into cell-like structures—is less certain. Although some of the processes can occur in dry environments, liquid water was probably critical to extensive biochemical evolution, because it provided a fluid medium in which materials could move and aggregate. After all, we evolved from the sea, most of our body weight is water, and the earliest fossils are of sea creatures. One botanist recently commented that "all cells, of all living organisms, are strictly aquatic creatures." He described any land-based organism as "merely a protective shell" filled up with millions of aquatic cells.

Florida biologist Sidney W. Fox has shown that simple heating of dry amino acids (as might happen on a dry planet) can create protein molecules. Once water is added, these proteins assume the shape of round, cell-like objects called **proteinoids,** which take in material from the surrounding liquid, grow by attaching to each

Figure 28-3 Tidewater pools (foreground) may have facilitated the origin of life. Once organic molecules formed in seawater, evaporation of the water in stranded tidal pools would concentrate the remaining organics, promoting reactions among them and formation of more complex organic substances. Broths rich in amino acids and complex organic molecules resulted and were dumped back into the ocean at subsequent high tides, "fertilizing" the seas. (Photo by author, Baja California.)

other, and divide. Though they are not considered living, they resemble bacteria so much that experts have trouble distinguishing them by appearance.

Possibly related to proteinoids are objects discovered in the 1930s by Dutch chemist H. G. Bungenberg de Jong. When proteins are mixed in water solutions with other complex molecules, both sets of substances spontaneously accumulate into cell-sized clusters called **coacervates.** The remaining fluid is almost entirely free of complex organic molecules.

The next step toward recognizable life forms is still more uncertain. If organic molecules or coacervates are present in a pool of water, they will be left in the pool as the water evaporates. In this way, evaporation of the water in tidewater pools (Figure 28-3) probably produced high, localized concentrations of amino acids, proteins, and other molecules, allowing cell-like structures to form. The cell-like structures in the primeval pools of "organic broth" could have begun reacting with fluids in the pools and with each other, accumulating more molecules and growing more complex, as suggested by Figure 28-4. Eventually these could have evolved into biochemical systems capable of reproducing.

The Earth's Earliest Life Forms

Whatever the processes, microscopic cellular life must have arisen between 4.5 and 3.5 billion years ago, because fossils have been discovered in rocks after that period. A 1981 conference on the subject estimated 4.0 ± 0.1 billion years as the age of Earth's first life (Ponnamperuma, 1983). For example, fossils of methane-producing bacteria were found in 3.4-billion-year-old rocks from South Africa in 1977 (Gould, 1978). The precise date of the earliest life on Earth remains controversial. One review of the data (Nisbet, 1980) lists the earliest "compelling" evidence as 2.7-billion-year-old fossil **stromatolites** (colonies of blue-green algae with a cabbagelike structure) from Canada and Zimbabwe. The earliest "probable" evidence is stromatolites from Western Australia between 3.4 and 3.5 billion years old; the oldest "possible" evidence is a 3.7-billion-year-old rock containing carbon isotopes of possible biological origin from western Greenland.

Life probably couldn't have evolved much before 4.1 billion years ago because of the intense early meteoritic bombardment and the possible magma ocean covering much of the Earth's crust. Yet biochemical evolution must have begun in that period. Primitive life evidently arose "as soon as it could; perhaps it was as inevitable as quartz or feldspar" (Gould, 1978).

For the Earth's first 2 billion years, most life remained in the oceans, where liquid water provided a supporting and protective environment. Life forms were mostly soft-bodied and rarely produced fossils, so their development is hard to trace. Stromatolites became more abundant 2.5 to 2.0 billion years ago, living at the boundary of water and rock along seacoasts, as shown in Figure 28-5. Today they are found only in rare, oxygen-poor salt marshes on some seacoasts, suggesting

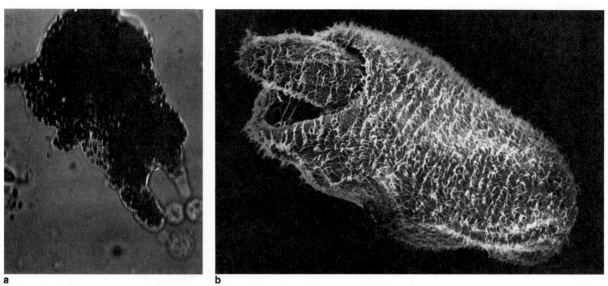

a   b

Figure 28-4 The importance of a fluid medium for primitive evolution is suggested by these photos of single-celled organisms engulfing food from surrounding fluid. **a** An amoeba flows to surround a nearby food particle. (Optical microscope photo; S. L. Wolfe.) **b** The protozoan *Woodruffia* ingests a *Paramecium*. (Electron microscope photo; T. K. Golder.)

Figure 28-5 The first signs of life encroaching on land did not occur until the Earth reached middle age. A patch of stromatolites is visible (left) at low tide, and early lichens can be seen on the rocks (right). Most of the land was still barren. (Painting by Ron Miller.)

Figure 28-6 Lichens, symbiotic combinations of algae and fungi, were probably some of the first organisms on land. As shown here, they grow on rock surfaces, weathering the rock to produce soil. They are widely adaptable, existing in both arctic tundra and hot deserts. Three lichen colonies of red, black, and gray appear in this picture, whose width is about 6 in. (Photo by author, Tsegi Canyon, Arizona.)

Figure 28-7 Fossil trilobite, a hard-bodied sea animal that appeared about 600 million years ago, providing diagnostic fossils for recognizing and dividing Cambrian time. (Gayle Hartmann and Joyce Rehm.)

that the early atmosphere in which they flourished was also oxygen poor. Stromatolites produce oxygen by photosynthesis.

It is strange to realize that during most of the Earth's history, life would have been hard to find in most *landscapes*, as Figure 28-5 suggests. Even during the Earth's middle age, the land was barren. Some areas must have looked like today's deserts or like Mars. Some areas were moist and washed by rains, but instead of luxurious forests, there were only bare acres, eroded gullies, and grand canyons. Brown vistas stretched to the sea.

The flourishing of stromatolites and early plant forms 2.5 to 2.0 billion years ago helped boost oxygen (O_2) production (Walker, 1977; Goldsmith and Owen, 1979). The atmosphere therefore evolved from more *reducing conditions* (dominated by hydrogen compounds) toward *oxidizing conditions* (dominated by oxygen compounds). Oxygen content in the initial atmosphere was probably only a few percent, but it began to rise rapidly toward the modern value of 21%. Evidence for this is that sediments that formed and were buried more than about 2.5 billion years ago are less oxidized than sediments formed today (Goldsmith and Owen, 1979). Oxidized red sediments are rare before 2 billion years ago and common afterward (Walker, 1977). Oxygen production by plants helped modify the whole environment. Solar

UV light broke some O_2 molecules, and the free oxygen atoms joined with other O_2 molecules to make *ozone* (O_3) molecules. This led to the formation of an **ozone layer** high in the Earth's atmosphere, which absorbs solar UV light and thus protects organisms on the earth's surface as mentioned before.

Living things could scarcely be unaffected by these major changes in the environment, as fossils testify. Until as recently as 1.5 billion years ago, most organisms lived best in oxygen-poor environments and were relatively resistant to genetic changes by means of UV radiation. Cells through this time period were of a type lacking a distinct nucleus (called prokaryotes); about 1.4 billion years ago, cells with nuclei (called eukaryotes) first appeared (Vidal, 1984). After 1.5 to 1.0 billion years ago, we also find new types of organisms in the fossil record, capable of oxygen metabolism, less resistant to UV damage, but able to flourish because the new ozone layer protected the surface from UV (Windley, 1977, pp. 44, 302). This was the first big step in the expansion of life from the sea onto the land (Figure 28-6)—a step as momentous as the contemplated colonization of other planets by us! The emergence of oxygen-consuming landlubbers like us was spurred by the planetary evolution of the Earth itself.

Biologists agree that the mechanism by which new

species of life forms evolved to fit the changing environment and the new environmental niches, such as land and air, was **natural selection:** Random changes in the genetic codes (as might be caused by UV irradiation or other processes) occasionally produced offspring with improved survival traits. In turn, these offspring lived longer and had more offspring of their own, promoting retention of the new trait.

A population of organisms on one arid continent, for example, might eventually evolve in a different direction from a population of the same organisms on another, wetter continent. (Witness the different animal populations on Australia and Asia.) A rapid change in global climate toward drier conditions might lead the first group to flourish while the second group died out. The first group would then spread rapidly from their home continent to other regions. The fossil record of a region thus might show the sudden emergence of the first group and extinction of the second. All these steps are probably part of natural selection, though the details of the process are still debated by biologists. Some details are uncertain because we often sample only discontinuous layers of fossil-bearing strata. Nevertheless, many fossil sequences show clear, step-by-step evolution, as with the 50-million-year sequence showing the evolution of a ½-m-tall, three-toed mammal into the modern horse.

As Darwin pointed out long ago, the overall process is like the artificial selection practiced by breeders who develop new strains by producing more offspring between individuals with traits the breeders favor.

Evidence of the adaptability of life forms is the rapid proliferation of advanced species once they evolved, as indicated in the geological time scale shown in Table 6-1. While it took about half the available time to go from complex molecules to algae and bacteria, it took only the last 12% of Earth history to go from the first hard-bodied sea creatures (such as trilobites, Figure 28-7) to humans.

Whatever the specific mechanisms, we can say from the fossil record that the Earth experienced a few-billion-year evolution from nonliving organic chemicals to small organisms, and then a much more rapid evolution to species with self-conscious intelligence.

PLANETS OUTSIDE THE SOLAR SYSTEM?

From all we have just said, we conclude that if planetary surfaces with the necessary conditions—liquid water

and the "CHON" chemicals (carbon, hydrogen, oxygen, and nitrogen)—exist long enough anywhere, life is likely to evolve. Advanced species are likely to appear eventually. But are there such planets?

Marginal Evidence for Planets Outside the Solar System

In view of the hundred billion stars in our galaxy, not to mention the innumerable other galaxies, we should not limit our search for life to the solar system alone. There are four kinds of **evidence for planets near other stars,** but the evidence is not conclusive.

First, most newly formed stars are immersed in cocoon nebulae in which mineral grains have condensed. These grains may aggregate into planetary bodies as a natural by-product of star formation.

Second, even some main-sequence stars have recently been found to have systems of dust particles orbiting around them. Vega, for example, has a Jupiter-mass or more of dust particles at least a millimeter across, revealed by infrared satellite telescope. Figure 18-12 (page 398) shows a photo of a similar dust system around the star Beta Pictoris.

Third, statistics of masses among binary and multiple stars suggest the likelihood of many stars having low-mass companions that are planets.

Fourth, unseen low-mass companions have actually been detected around a few nearby stars (Table 28-1). All are more massive than Jupiter. Many may be brown dwarfs, but the smallest may be planetlike.

At the distances of most stars, planets smaller than Jupiter are undetectable by current methods, but only marginally so. Astronomers have proposed new devices and techniques to detect even Earth-sized planets near nearby stars. Among these are space telescopes and sensitive spectrometers that would detect radial-velocity oscillations of stars, caused by planets going around them (O'Leary, 1980). The search for planets near other stars is an exciting enterprise that will be pursued in coming years; it may finally clarify whether our own planetary system is commonplace or freakish.

For the present, an informed guess might be that between 1% and 30% of all stars have at least one planet nearby.

Habitable Planets?

Even if planets do exist near some other stars, there is no guarantee that they are habitable. Astronomers have

TABLE 28·1

Components of the Nearest Systems and Other Selected Systems

Name of Largest Star in System	Distance Parsecs	Distance Light-years	10^{-4}	10^{-3} (Jupiter)	10^{-2}	10^{-1}	1 (Sun)	10
Nearest Stars								
Sun	0	0		x x			x	
Alpha Centauri	1.3	4.3				x	xx	
Barnard's star	1.8	5.9				x		
Wolf 359	2.3	7.6				x		
+36°2147	2.5	8.1			x	x		
Sirius	2.6	8.6					x	x
Luyten 726-8	2.7	8.9			xx			
Ross 154	2.9	9.4				x		
Ross 248	3.2	10.3				x		
ε Eridani	3.3	10.8		x			x	
Other Selected Systems								
61 Cygni	3.5	11.2		x			xx	
Krüger 60	3.9	12.9				xx		
Ross 614	4.0	13.0				x x		
BC +68°946	4.8	15.6			x		x	
BC +20°2465	4.9	16.0			?		x	
Stein 2051(G175-34)	5.5	17.8			x	x	x	
VB10 (Van Biesbroeck's star)	5.7	18.7			x	x		
Eta Cassiopeiae	5.9	19.2			x		xx	
VB8	6.5	21.1			?	x		
BD +66°34	10	33				x		
γ Cephei	15	48		x			x	
T Tauri	140	450		x				x

Source: Data from Allen (1973), Harrington and others, (1983), Hanson and others (1983), Strand (1977), research announced in 1987, and other sources.

proposed several conditions needed to make a planet habitable:

1. The central star should not be more than about 1.5 $M_\odot$, so that it will last long enough for substantiated life to evolve (at least 2 billion years) and so that it will not kill evolving life with too much UV radiation, which breaks down organic molecules.

2. The central star should be at least 0.3 $M_\odot$ to be warm enough to create a reasonably large orbital zone in which a planet could retain liquid water.

3. The central star should not flare violently or emit strong X rays. It should be on the main sequence in order to have been stable long enough to give its planets long-term climatic stability. Conditions 1, 2, and 3 restrict us to main-sequence stars of spectral types F, G, and K.

4. The planet must orbit at the right distance from the star so that liquid water will neither evaporate nor permanently freeze (Figure 28-8).

5. The planet's orbit must be circular—stable enough to keep it at a proper distance and prevent drastic seasonal changes. Such stable, habitable orbits are less probable, but still possible in common types of binary systems (Harrington, 1977).

6. The planet's gravity must be strong enough to hold a substantial atmosphere.

The American physicist Stephen Dole (1964) reviewed these criteria and speculated that only a few percent of all stars, mostly from 0.9 to 1.0 $M_\odot$, have habitable planets. However, the figure remains highly uncertain.

HAS LIFE EVOLVED ELSEWHERE?

If habitable planets exist, and if life evolves readily under habitable conditions, shall we immediately conclude that life must be abundant throughout the universe? There are several additional factors to consider. For one thing, planetary environments change with time, so that today's habitable planet may not be habitable tomorrow. This raises the question of the ability of life to adapt to change. The big problem is that we don't know how unusual the Earth's history has been. Has life narrowly escaped complete extinction by climatic change in the past? Or was there never any question that natural selection would allow a few species to survive each step of the Earth's evolution?

Effects of Astronomical Processes on Biological Evolution

Basic planetary or stellar processes may be involved in encouraging or hindering biological evolution. For example, the Earth's fossil records indicate an episode called the "great dying," when as many as 96% of all existing land and sea species became extinct in less than a few million years (Raup, 1979). This happened about 230 million years ago (see Table 6-1, page 137); many paleontologists believe widespread climate changes were involved.

A number of geological and astronomical processes have been identified that could cause massive climate changes that could affect the course of biological evolution. Among these are:

1. Planetary convection could not only cause plate tectonic crustal splitting, but also change sea levels, ocean currents, wind patterns, and seasonal extremes. This process has been responsible for isolating landmasses such as Australia and allowing different species to evolve there.

2. Volcanic eruptions may have spewed enough dust into the high atmosphere to reduce the sunlight reaching the surface. For example, widespread sunlight decreases of up to 25% occurred after the 1883 Krakatoa (Sumatra), 1912 Katmai (Alaska), and 1982 El Chichon (Mexico) eruptions. Rarer larger eruptions could cut sunlight for some years, lowering summer and/or winter temperatures and causing the decline of some species and ascendancy of new ones. Current discoveries of dust layers in active ice packs are beginning to allow dating and reconstruction of prehistoric volcanic cataclysms. The Mount St. Helens eruption of 1980 dropped ash in nearby states, but much of the dust also reached the stratosphere and scattered around the world.

3. Slight changes in Earth's orbit and the tip of the planetary axis to the plane of the ecliptic, caused by gravitational forces, are believed to have caused major climatic changes, such as the ice ages. (See, for example, Imbrie and Imbrie, 1980; Covey, 1984.)

4. Slight changes in the Sun's radiation may have changed climates.

5. Irradiation from a nearby supernova could have affected the climate or directly damaged organisms.

6. An asteroidal or cometary impact or atmospheric explosion could have damaged the ozone (O_3) layer, exposing organisms to enhanced radiation. Turco and

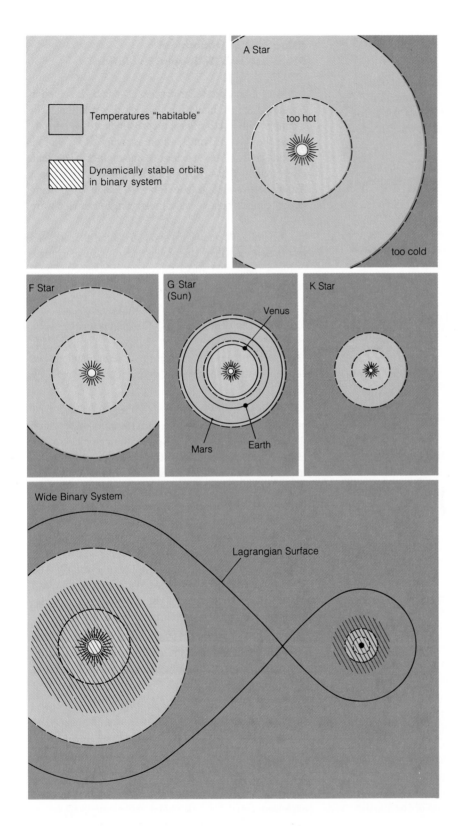

Figure 28-8 Doughnut-shaped light areas show the relative sizes where water would remain liquid in planetary systems around stars of different spectral types. In a binary system (bottom), the situation is complicated because orbits outside hatched regions experience large perturbations, causing evolution away from circular orbits, in turn causing temperature extremes. (After Huang, 1965.)

others (1981) found that the Tunguska asteroid or comet explosion of 1908 (Sekanina and Yeomans, 1984) generated as much as 30 million tons of nitrogen oxide (NO) in the stratosphere and mesosphere, reacting to deplete up to 45% of the O_3 in the Northern Hemisphere, consistent with Smithsonian measures made in 1909–1911.

7. As illustrated in Chapter 6 (page 138), a catastrophic asteroidal impact probably occurred about 65 million years ago, causing the dramatic break in the fossil record between the Cretaceous and the Tertiary periods. The idea of a large impact causing widespread extinction of species was suggested as early as 1969–1970 (McLaren, 1970) but ignored for lack of evidence. Alvarez and others presented the first evidence around 1980: They found the iridium (Ir) content of 65-million-year-old limestones equaling 20 to 160 times that of limestones of other ages. Because iridium is enriched in meteorites relative to Earth, they suggested that a 6- to 14-km-diameter asteroid struck Earth and ejected massive amounts of meteoroid-enriched dust into the atmosphere. This hypothesis has had a tremendous impact on the paleontological community and caused a burst of interesting research. Within months, other investigators confirmed the Ir excess in 63-million-year-old sediments from many parts of the world. The effects of the impact are hard to contemplate. Some 75% of species (notably dinosaurs) became extinct within a few million years, and new species (notably mammals) became dominant (Russell, 1982). Detailed studies of the fossil record show that this did not happen all at once, however, and various species were affected differently (Alvarez and others, 1984). Discovery of a worldwide soot layer in the boundary layer suggests that much of the vegetation of the world was consumed in fire (Wolbach and others, 1985)! No land animals larger than 25 kg survived, but freshwater animals and plants were scarcely affected (Russell, 1982). Ejected dust and soot may have blanketed the stratosphere and caused 3 to 6 mo of daytime darkness, consistent with a reported extinction of 44% to 49% of all genera of (light-dependent) floating marine organisms, but only 20% of (less light-dependent) land genera (Ahrens and O'Keefe, 1982). Consistent with this is a drop in calcium carbonate in the 65-million-year-old stratum, due to the disappearance of plankton that would normally drop their calcium-rich skeletons onto the seafloor. Further isotopic studies indicate a warming of surface ocean waters by as much as 10K at this time. If marine plankton were immediately decimated by the impact, they would have stopped consuming CO_2 and the consequent increase in the CO_2 of the air could have caused a greenhouse warming that led to massive land organism extinctions over the next 50 000 y or so. Because an impact into the ocean is most probable, Ahrens and O'Keefe (1982) modeled the tsunami produced; they envision a wave hundreds of meters high inundating all low-lying continental areas within 27 h of impact.

The impact that produced certain North American *tektites* may have caused the profound climatic event that ended the Eocene 34 million years ago, causing the widespread extinction of radiolarians (minute marine protozoans). An iridium excess has been found at this transition stratum (Alvarez and others, 1984).

Would such changes necessarily hinder biological evolution? Biological disasters may happen only when changes are too extreme (for instance, the freezing of all water as on Mars) or too sudden (massive infusion of dust into the stratosphere by a huge impact explosion). Many biologists are coming to believe that changes of the magnitude that have occurred on the Earth, while causing the decline of some species, have promoted the emergence of advanced species. For example, if some aspects of intelligence had first appeared in a benign, constant environment where food was plentiful, these traits would have had little value. But if ice ages destroyed the mild environment, then the more clever groups might have emerged from their previous obscurity.

Perhaps it is no coincidence, then, that the time scale of geological change is comparable to the time scale of biological change. In response to changing environments, life changes itself, on time scales as short as a million years.

Adaptability and Diversity of Life

The great variety of ancient and modern species on the Earth and the variety of environments in which they have thrived suggest that, given time, life could also have evolved to fit a wide range of conditions on other planets. Even humans have a remarkable adaptability. We thrive from equatorial wet jungles to dry deserts to arctic plains to Andean summits where air pressure is barely half that at sea level. In the past, during ice ages that glaciated New York, humans survived by migrating. We can survive a 3% variation in body temperature, from about 303 to 313 K. Simpler organisms can withstand much wider variations. Certain bacteria that appeared about 3 billion years ago can withstand 18% temperature variations, over a range of 70 K (126°F).

The environmental range covered by *all* terrestrial species is even greater. For example, some microorganisms live in Antarctic ponds that remain liquid at 228 K (−49°F) because of dissolved calcium salts, and others live in Yellowstone Hot Springs at temperatures of 363 K (194°F). Habitable pressures range over a factor of 1000. Bacteria exist at altitudes where the atmospheric pressure is only about 0.2 atm, while more advanced organisms live at ocean depths with pressures of hundreds of atmospheres.

Although sunlight is the ultimate energy source for all familiar organisms (since the ecological chain leads back to vegetable foods that flourish by photosynthesis), recent discoveries show that not even sunlight is necessary to terrestrial life. Oceanographers in 1980 announced the discovery of colonies of strange deep sea animals clustered in the darkness around seafloor volcanic vents. Their metabolism is based on volcanic heat and hydrogen sulfide (H_2S) gas, utilized by bacteria that form the bottom rung of the local food chain (Edmund and Von Damm, 1983). The local pressure is 250 atm, nearly three times that on Venus. Some bacteria recovered from these environments thrive and reproduce at 523 K (482°F), extending the temperature range for known life to a value more than two-thirds that on Venus. The research is still being debated, but the 523 K figure has been suggested as a rough upper limit for life (Brock, 1985), because known amino acids break down at higher temperatures. The seafloor vent life forms show that a pleasant, sunny planet with a clear atmosphere is not the only possible locale for life!

Could any terrestrial organisms withstand the environments of other planets? For simple organisms, probably yes. Grass seed can germinate in a variety of atmospheres composed of simple compounds of the common elements carbon, hydrogen, oxygen, and nitrogen. Eight species of insects (relatively complicated creatures) studied at different pressures behaved normally all the way down to 10 to 16% of normal atmospheric pressure on the Earth (Siegel, 1972).

For these reasons, spacecraft have been at least partially sterilized to avoid contaminating other planets. Such contamination might not only alter planetary environments but also ruin any chance for us to discover whether simple organisms or complex organic molecules ever formed independently on those planets.

It works the other way, too. Simple organisms from other planets, if they ever reached Earth, might have devastating effects on the Earth. There are historical examples of similar events. The plague caused by bacteria introduced into Europe in the 1300s killed about a quarter of all Europeans and as many as three-quarters of the inhabitants in some areas. Diseases introduced into Hawaii after the first European contact in 1778 killed about half of all Hawaiians within 50 years. Some 95% of the natives of Guam were wiped out by disease within a century of continued European contact. For these reasons, early Apollo astronauts were quarantined until it was clear they carried no lunar organisms.

With these facts as well as cultural competition in mind, anthropologist D. K. Stern (1975) has remarked: "It is likely that the meeting of two alien civilizations will lead to the subordination of one by the other." Thus change and evolution in life populations are likely not only from life's adaptability to new environments but also by the invasion and destruction of some populations by others.

Would They Look Like Us?

Opinion is divided on this point. Species with similar capabilities in similar habitats may evolve to look alike. For example, three different species of large sea creatures "designed" for fast ocean swimming look alike: an extinct reptile, the ichthyosaur; a fish, the shark; and a mammal that returned to the sea, the dolphin. The idea is that aliens living on planetary surfaces in gaseous atmospheres and using "intelligence" to manipulate their surroundings with tools might well have bilateral symmetry, appendages used as hands, a pair of eyes designed (like ours but unlike most animals') for stereo vision, and so on.

On the other hand, the famous paleontologist George Gaylord Simpson (1964, 1973) has argued that although life is likely to start, the long chain of environmental changes and evolutionary steps required to produce humans is unlikely to be approximated elsewhere, so there is likely to be a "nonprevalence of humanoids." Simpson criticized the whole attempt to estimate the probabilities of such life as nearly meaningless because of our lack of experiments or examples.

Physicist W. G. Pollard (1979) counters that we do have examples of independent evolution that show divergence toward different forms. He notes that about 180 million years ago Australia broke off Gondwanaland and can be thought of as an Earthlike planet, "A," where evolution continued independently from a primarily reptilian stock. Similarly, South America broke off Africa 130 million years ago and can be viewed as an independent planet "S." Independent evolution also continued on planet "E" (the rest of the Earth, especially Africa and the adjoining land). During the last 130 to 180 million years, independent evolution has *diverged* toward different kinds of animals on these three "planets," rather than converging. Humans appeared only on planet "E," primates on "S," and marsupials on "A." The humans on "E" appeared only about 4 million years ago and have existed for only 0.1% of the Earth's history.

In summary, natural selection seems to produce species capable of occupying any habitable environment. Thus we should not be surprised if life has evolved on another planet. But this life may look very strange to us, even if it displays recognizable intelligence. After all, if mushrooms and corals and woolly mammoths and Venus flytraps all evolved on one planet, how much greater may be the differences between life forms on two different planets? Feathers and fur and sex and seeds and symphonies may be the products of the Earth only.

Effects of Technology on Biological Evolution

If life will evolve when the right conditions exist, and if the conditions probably do exist elsewhere, then what can we predict about that life? Should we predict intelligence and civilizations? What do these terms mean? Should we assume that other civilizations will achieve spaceflight or might visit us some day? This raises the question of technology and its role in the evolution of life. Just as cosmologists have only one universe as an example, **exobiology** (the study of possible life on other worlds) has only one inhabited world as an example, so the answers to these questions are uncertain. Many exobiologists have assumed that intelligence by definition involves use of tools to modify the environment.

While limited environmental change can be helpful, environmental change that is too much or too fast can be fatal! As Pulitzer Prize-winning naturalist René Dubos points out, we are umbilically connected to the Earth, and if we alter our planet too much before acquiring an ability to leave it, we are finished. For this reason, the development of technology could actually end civilization on some worlds. This is hardly wild speculation, since we see a few nominally moral, intelligent tech-

nologists on our own world spending entire careers devising weapons solely to deal death to our own species.

Past wars did not threaten our whole species because conflicts involved only a small percentage of the world. But today, nuclear, biological, and other types of weapons could involve the whole world. For example, the radioactive strontium-90 produced by nuclear explosions has a half-life of 28 y. It was blasted freely into the atmosphere before the Nuclear Test Ban Treaty of 1963. A year after the first H-bomb tests by the United States in the Pacific, strontium-90 deposits in American soil increased soil radioactivity by ½%. A few weeks after a 1976 Chinese nuclear test, airborne radioactive debris fell onto the United States, increasing radiation levels. Obviously, a sufficiently massive nuclear exchange could devastate not only civilization but also future forms of life, whose genetic pool would be exposed to high radiation levels for decades. Humanity has thus proved that a planetary culture could wipe itself out by conscious design of weapons, as irrational as that may seem.

We have also proved that this disaster *could* happen by mistake. As our technology reaches planetary scale, our accidents can also involve the whole planet. Problems as diverse as nuclear power and aerosol spray cans illustrate the issue. The Soviet Chernobyl nuclear power plant accident, for example, affected large regions of northeastern Europe. Some currently planned nuclear power plants will produce radioactive plutonium wastes, among the most toxic of known materials. Although safe when sealed, kilograms of plutonium dust accidentally spilled into the air could devastate whole states. Yet many kilograms are already being processed, and several governments promote nuclear power plants until other energy forms become available. Accidents are not the only danger: We have seen our society spawn terrorists and madmen. With a few grams of stolen plutonium-238, such characters could threaten whole cities. Our creation of highly radioactive substances presents a technological danger that our society has allowed to develop in spite of knowing about it ahead of time.

The seemingly harmless aerosol can is a different story. In 1974, scientists realized that when Freon (the propellant gas used in such cans and the coolant in air conditioners) escapes into the air, it reaches Earth's protective ozone layer and undergoes reactions that destroy the ozone. Depletion of the ozone lets more solar ultraviolet radiation reach the surface, increasing the risks of skin cancer. Even as more Freon was being sold, a 1975 report by the National Academy of Sciences

a

b

c

Figure 28-9 Too hot, too cold, and just right. Scenes on imaginary planets emphasize that even if extrasolar planets do exist and have atmospheres and potential for life, establishing the right climate for life to evolve may be chancy. **a** The surface of this planet, too close to a pair of hot, massive stars, is a sun-blasted desert. **b** On this world, far from a red and white star pair, water is all frozen. **c** Some worlds may exist where liquid water persists and life forms have evolved. (Paintings by Don Dixon, Ron Miller, and the author, respectively.)

affirmed this effect. By the 1980s, steps were taken at international levels to substitute other gases for Freon in aerosol cans. However, good substitutes for Freon in air conditioners and refrigerators are still unavailable. The story of the lowly aerosol can is an example of a technological danger that was almost missed, and it makes one wonder about possible dangers presented by other consumer gimmicks.[4]

Still another example is the probable global warming and climate changes caused by carbon dioxide and other gases. A good review of this challenge is given by White (1990).

Our purpose here is not just to toll fashionable bells of doom and gloom. We all hope that by exercising a bit of intelligence and caution, humanity can recognize these dangers and avoid them. Nonetheless, if humanity is any example, long-term survival of a planetary culture is not assured. Although we have been around for less than 1% of the age of our planet, we are already beginning to have brushes with global disaster. Thus we can speculate that if evolution produces intelligent societies that remain tied to one planet, some may last less than a percent of the age of the universe—reducing the chance that a nearby culture will be around at the same time that we are.

---

[4]We should be haunted by the words of Harvard planetary physicist Michael McElroy, who performed some of the earliest Freon calculations: "What the hell else has slipped by?" An excellent review of these problems is given by Rowland (1989).

More optimistically, we have succeeded in identifying the **cultural hurdle** that we (and perhaps intelligent species on any planet) must surmount: the transition from scattered, competing nation-states with the capability to damage the planet to stable global or interplanetary societies of intelligence and imagination. The situation is not unlike George Lucas's conflict between the good and dark sides of "the force": We face a race between the good side of our technological abilities, which make civilization possible (plumbing, electricity, stereos, and interplanetary colonies that could bring new resources back to Earth), and the dark side of technology, which threatens civilization (carcinogenic by-products, pollution, and hydrogen bombs). If the good side can win the race, we can perhaps break out of our muddle of competitive strife over resources and ideological space on a finite planet. Perhaps some alien cultures have crossed this cultural hurdle and spread across many planets, thus ensuring their survival against ecological disaster on any one planet. Such cultures might last and be detectable for millions of years instead of thousands.

Our conclusion so far is that while some fraction of the stars have planets, most of these planets may be inclement for life, as symbolized by Figure 28-9a and b. But a fraction of the planets may have produced life, as symbolized in Figure 28-9c. Of these, a fraction may have produced either civilizations or relics of destroyed civilizations. The next question is whether any actual evidence for extraterrestrial carbon-based life as we know it exists today.

Alien Life in the Solar System?

Spacecraft exploration of the planets has virtually disproved advanced extraterrestrial **life in the solar system** and has probably ruled out any extraterrestrial life whatsoever in our system. The terrestrial planets except for the Earth seem devoid of life, according to our landers, and the outer planets seem too cold.

The most interesting environments seem to be those of Mars, Titan, the middle atmospheres of giant planets, and the sub-ice ocean of Europa. Although UV light has probably destroyed any organic material on the Martian surface, the Martian soil has the ability to synthesize organic molecules from artificial nutrients, and some Viking scientists still leave open the possibility of Martian microbes below the surface. Florida State University biologist E. I. Friedmann in 1978 reported the discovery of microorganisms thriving in pore spaces inside rocks in seemingly lifeless, Mars-like dry valleys in Antarctica, giving some support to the possibility of hidden organisms on Mars. Titan's thick atmosphere and methane chemistry remain interesting, especially if "hot springs" of any sort exist on the surface. Giant planets' atmospheres may have hydrogen-dominated layers with temperatures and pressures not too different from the atmosphere of the Earth; Carl Sagan has speculated about the evolution of organisms that could float in such atmospheres.

The discovery of the Sun-independent organisms near deep-sea vents on Earth, coupled with the discovery of tidal heating of the interiors of Jupiter satellites, has raised the possibility of organisms that might have evolved in liquid water zones under the ice of satellites such as Jupiter's moon Europa (a theme developed in Arthur C. Clarke's novel, *2010*). It might be interesting in the future if we can drill through the icy crust of this Moon and study the "buried ocean" believed to exist there.

Although we don't expect to meet creatures of any great intellect in our solar system, there are clearly places left to explore that may shed light on biochemical and biological evolution.

Alien Life Among the Stars?

There is no direct evidence on whether there is life among the stars, but American, Soviet, and other scientists in international meetings have put together a method for considering the possibilities (Sagan, 1973). The logic is to try to estimate the various fractions—of stars having planets, of those planets that are habitable, of those where conditions remain favorable long enough for life to evolve, of those where life does evolve, of those where intelligence evolves, and of the planet's life during which intelligence lasts. The product of all these fractions is the estimated fraction of all stars harboring intelligence. This statement is sometimes called the **Drake equation,** after radio astronomer Frank Drake, who first formalized this scheme for discussing alien intelligence.

Table 28-2 shows "optimistic" and "pessimistic" estimates of the various fractions, and consequent estimates of the upper and lower limits on the fraction of stars that might have civilizations today. In the optimistic case, the answer comes out a few percent; in the pessimistic case, only one star in 10^{14} (one in 100 million million). We stress that these are little more than educated guesses, because we lack the data to make better

TABLE 28·2

Estimated Fraction of Stars with Planets Having Intelligent Life

	Plausible Lower Limit	Plausible Upper Limit
Stars having planets	10^{-2}	0.3
Those stars ever having habitable conditions on at least one planet	10^{-1}	0.7
Those planets on which such conditions last long enough for life to evolve	10^{-1}	1
Those planets on which life does evolve	10^{-1}	1
Those planets on which habitable conditions last long enough for intelligence to evolve	10^{-1}	0.9
Those planets on which intelligence does evolve	10^{-1}	1
Those planets on which intelligent life endures	10^{-7}	10^{-1}
Product of fractions in column: Fraction of all stars with planets having intelligent life	10^{-14}	2×10^{-2}
Implication: Distance to nearest civilization	10^7 ly	15 ly

scientific estimates. But this need not stop us from considering the consequences of our conjectures.

For example, how far away would the civilizations be? Figure 28-10 shows the answer by plotting the radial distance required to include a given number of stars. In the first case, the nearest civilization might be only 15 light-years away—amazingly close. At the speed of light it might take only one generation to reach it. In the pessimistic case, the closest civilization would probably not be in our own galaxy, but roughly 10 million light-years away (a few megaparsecs) in a distant galaxy. Nonetheless, there are so many galaxies that it is hard to avoid the conclusion—even with the pessimistic view—that **life outside the solar system** is likely, with millions or billions of technological civilizations. If so, then at this moment intelligent creatures may be pursuing their own ends in unknown places under unknown suns.

Where Are They?

Radio astronomers have listened for radio messages from alien civilizations and heard none. Nor do our skies seem to be overrun with alien visitors trying to contact us. Nonetheless, theorists have recently pointed out that if *any one* civilization began sending out multi-generation slower-than-light starships that set up colonies, which in turn sent out new starships in a few generations to other stars, that civilization could populate all the habitable planets in the galaxy on a timescale short compared to the galaxy's age (but long compared to the timescale of evolution, so that the original civilization might produce many new species as it went). So why haven't we heard from aliens?

One answer could be that there are no other aliens—intelligent life on Earth is some sort of unique accident.

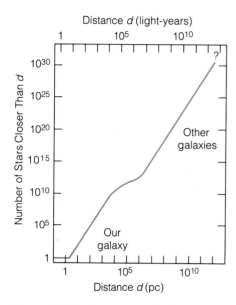

Figure 28-10 Distance required to encounter the number of stars indicated at left. If one star in 10^9 has life-bearing planets, the nearest one might be within a few thousand parsecs.

Another answer might be that we *are* being visited—witness the "flying saucer" reports. Some UFO reports are intriguing, but the verifiable evidence for actual alien spaceships is abysmally poor. (For further discussion of UFOs, see Enrichment Essay A.) If any alien visits have actually occurred, we have no proof of them.

A third possible answer is that the pessimistic figures are right and the nearest civilizations are in distant galaxies. Even their radio messages, if any, could be 10 million years old by the time we receive them, and their spaceships would be unlikely to reach the Earth if limited to speeds less than that of light, as current physics requires. However, at a 1971 Soviet–American conference on this problem, the favored estimate came out to be one civilization per 100 000 stars. This would put a million civilizations in our galaxy and the nearest civilizations only a few hundred light-years away. Why, then, are there no frequent visits to Earth?

A fourth possible answer is that biological evolution need not produce creatures who have a desire to build "civilizations" or travel through space. Not even all human societies necessarily evolve toward technology. Are humans fated to be explorers, bridgebuilders, and businesspeople rather than artists, athletes, or daydreamers? Is the stereotyped aggressive Westerner more

representative of the essence of humanity than the stereotyped contemplative Easterner? May not our aggressive technocracy be just one type of *cultural* activity rather than a universally achieved stage of *biological* evolution? Historically, patterns we once assumed to be biological have turned out to be merely cultural. (Confusion of these two has led to the racist and sexist biases we are still trying to overcome.)

If humanity is not predestined to develop a technological civilization, how much less certain is the course of social development on other worlds. It is absurdly anthropocentric to suppose that beings on other planets would resemble us physically, psychologically, or socially. Consider again the variety of highly evolved life forms on our planet alone. Ants live in ordered societies that do not appear to regard individual survival as important. Dolphins communicate and have brains that seem almost comparable to ours, but they have no manipulative organs and hence no technology. Perhaps we cannot expect aliens to be motivated by emotions that mean much to us. We certainly cannot expect, as always happens in grade C science fiction movies, that humanlike aliens will walk out of saucers and invite us to join their democratically constituted United Planets, a galactic organization structured by documents that are curiously reminiscent of the U.S. Constitution. So why assume that other civilizations would even try to visit us?

This brings us to a fifth and perhaps most significant answer: We may be farther from aliens in evolutionary time than in physical space. Even if another planet started evolving at exactly the same time as ours, and even if its biochemistry produced creatures like us, those creatures are not likely to be in a phase of evolution similar to ours. If the evolutionary "clocks" on the two planets got only 1% out of synchronization, they would be 40 million years ahead of us or behind us—as far from us evolutionarily as we are from early mammals. Thus even in the unlikely event that other planets produce civilizations we can recognize, contact might have to come in a very narrow time interval in order for us to see any recognizable common interests.

Evolution may pass through a brief **explorative interval,** during which societies on one planet try to reach other planets; beyond that stage communication or space exploration might be no more attractive than a national program on our part to communicate with chimpanzees, ants, or dolphins. To be sure, a few of our scholars do try this, but they "contact" an infinitesimal fraction of these lower creatures. By the same

Figure 28-11 Even if 10 000 alien expeditions have visited the Earth at random times during the Earth's history, visits would have averaged hundreds of thousands of years apart. Even the most recent expedition in such a scenario would therefore be unlikely to have left any readily recoverable historical or archaeological traces. (Painting by Jim Nichols.)

token, our solar system might be ignored by advanced aliens. Aliens a million years ahead of us might be no more interested in us than we are in ants.

How long might an explorative interval last? We have used tools for about 2 million years, and it appears certain that we will progress far beyond current technology in another million years, if we survive. Our explorative interval might be a few million years, then, or less than 0.1% of the history of the planet. Thus if the last factor in Table 28-2 is interpreted as "time during which intelligence is recognizable to us and interested in communicating," the optimistic, upper limit would have to be revised downward to three stars in 100 000, about the figure cited in the Soviet–American conference of 1971. If civilizations are at least 300 light years apart, then interstellar voyages and messages would take around half a millennium. There might be little incentive for the effort. Any spaceships that did arrive on the Earth could well arrive many thousands or millions of years apart.

This brings us to a sixth possible answer to the question of alien visits: They may have happened in the remote past, as symbolized in Figure 28-11. This is the "ancient astronaut" hypothesis, popularized in several pseudoscientific books (see Enrichment Essay A). There is no good evidence for this possibility. The Soviet and American collaborators I. S. Shklovskii and Carl Sagan (1966) surveyed archaeological and mythological liter-

ature even before the hypothesis was a popular fad and found no compelling evidence for ancient astronauts. Neither the Earth, the Moon, Mars, nor Venus is littered with ancient alien artifacts, and not a single mysterious artifact has been advanced as physical evidence of ancient astronauts. Recent popular books about ancient astronauts have misrepresented some ancient artifacts belonging to known terrestrial cultures; such pseudoscientific books have been resoundingly debunked.

Radio Communication

Possibly our period of isolation may be nearing an end. For half a century we have been broadcasting radio communications among ourselves. Already our unintentional but weak alert is more than 60 light years out from Earth. Aliens may one day pick up our signals and send radio messages (or an expedition?) in return. Radio astronomers in both the United States and the Soviet Union are therefore still conducting modest programs to listen for such messages with large radio telescopes. In one such search, radio astronomers listened for broadcasts near 1420 MHz (the 21-cm wavelength that marks an astronomically important radiation from neutral hydrogen atoms in space), targeting on all solar-type stars (185 in all) within 25 pc and known to be members of multiple star systems. No artificial broadcasts were identified (Horowitz, 1978). Goldsmith and Owen (1979) summarize other searches involving more than 600 stars. All were negative. But in terms of total listening time and the total number of possible wavelengths to tune in, the search has hardly begun. Larger listening instruments have been proposed in both the United States and the Soviet Union.

A message from 1000 light years away would come from a civilization 1000 y in the past, and no answers to our questions could come back for 2000 y. Such communication would be unlike dialogue but, as physicist Philip Morrison has pointed out, more like our receipt of "messages" (books, letters, plays, and art) from ancient civilizations such as Greece.

Meanwhile, we have sent a few messages of our own. Several space vehicles, such as Pioneer 10 (which flew by Jupiter and left the solar system in 1973), carry plaques designed to convey our appearance and location to possible alien discoverers of the derelict spacecraft sometime in the future. The first radio message was a test message sent from the large radio telescope at Arecibo, Puerto Rico, in 1974, beamed toward globular star cluster M 13, which is 27 000 light years away.[5]

about the reasons. Perhaps they are too far away. Perhaps most civilizations destroy themselves before successfully expanding into the universe. Perhaps evolution carries them beyond a stage where they would care to communicate with us. Perhaps they are unrecognizable.

Arthur C. Clarke has remarked that any technology much advanced beyond your own looks like magic. Perhaps we are too limited by our own concept of civilization. After all, one creature's civilization may be another's chaos, as shown by Mohandas Gandhi's remark when asked what he thought of Western civilization: He said it wouldn't be such a bad idea. It seems likely that our first contact with aliens (if they exist) might be as incomprehensible as the dramatized contact in the closing segment of Clarke's novel and the Kubrick–Clarke film, *2001.*

Clearly we have been reduced to speculation by a lack of facts. Indeed, the whole field of exobiology has been criticized as a science without any subject matter. Exobiology recalls Mark Twain's comment "There is something fascinating about science. One gets such wholesale returns of conjecture from such a trifling investment of fact." The only way to reduce the conjecture and increase the proportion of fact is to pursue research in many related fields—physics, chemistry, geology, meteorology, and biology—and listen to the skies with radio telescopes. There may be surprises waiting out there.

SUMMARY

Discovery of firm evidence for alien civilizations or even alien life forms could be a pivotal development in our view of ourselves, as suggested in Figure 28-12. Yet questions of exobiology, especially of intelligent life on other worlds, leave us with a mystery. Experimental evidence indicates rather strongly that life should start on other planets if liquid water, energy, and the right chemicals are present. Astronomical evidence suggests, but does not prove, that habitable planets ought to exist elsewhere in the universe. Biological evidence shows that life is adaptable and species can evolve to fit different environments, from ocean depths to low-pressure atmospheres of different composition. Yet advanced life has not evolved to survive on Mars.

While the limited evidence indicates that other life forms should exist, there is no evidence that they do or that they have tried to communicate with us. We can only speculate

CONCEPTS

organic molecule	natural selection
life	evidence for planets near other stars
cell	
organic chemistry	exobiology
amino acid	cultural hurdle
protein	life in the solar system
Miller experiment	Drake equation
proteinoid	life outside the solar system
coacervate	
stromatolite	explorative interval
ozone layer	

---

[5]Astronomers in these projects have been deluged with letters ranging from support to complaints about the nudity of the human figures on the Pioneer 10 plaque. One telegram read: "Message received. Help is on the way—M 13." Its authenticity might be questioned, since the round-trip message time to M 13 is 54 000 y!

PROBLEMS

1. Compare the probability of detecting radio broadcasts from intelligent creatures on Alpha Centauri, γ Cephei, and Sirius.

2. Describe several ways in which the Earth's internal evolution has affected the evolution of life.

a

b

Figure 28-12 Discovery of positive evidence for extraterrestrial civilizations could be a
pivotal and dramatic event. It could occur **a** by our going out and finding it (discovery of an
alien artifact on the Moon, as dramatized in *2001—A Space Odyssey,* © 1968 Metro-
Goldwyn-Mayer Inc.) or **b** by its arriving on the Earth (arrival of mother ship as dramatized
in *Close Encounters of the Third Kind,* ©1977 Columbia Pictures Industries, Inc.).

3. Describe several ways in which extraterrestrial events, such as solar or stellar evolution, might have affected the evolution of life on the Earth or other planets.

4. Describe ways technology could affect the survival of intelligent life on the Earth or other planets. Construct scenarios of (*a*) the possible destruction of life and (*b*) the assured survival of life. (*Hint:* In case *b*, consider the impact of space travel.)

5. In your own opinion, what would be the long-range consequences of the following:
   a. Arrival of an alien spacecraft and visitors at a prominent place, such as the UN Building
   b. Discovery of radio signals arriving from a planet about 10 light-years distant and asking for two-way communication
   c. Proof (by some unspecified means) that life existed *nowhere else* in the observable universe

6. Given the assumption that our technology has the potential for creating planetwide changes in environment, defend the proposition that *if* life on other worlds produces technologies like ours, then that life is likely either to have become extinct or to be widely dispersed among many planets by means of space travel.

7. In view of the devastation wrought on many terrestrial cultures by contact with more technologically advanced cultures, which would you say is the safest course: (1) active broadcasting of radio signals to show where we are in hopes of attracting friendly contacts; (2) careful listening with large radio receivers to see if there are any signs of intelligent life in space; or (3) neither broadcasting nor listening, but just waiting to see what happens?
   a. How would the results of our listening program be affected if other intelligent species had reached the first, second, or third conclusion?
   b. If we listen at many frequencies and pick up no artificial signals, does that prove that life has not evolved elsewhere in the universe?

8. Why would native life be unlikely on a planet associated with:
   a. An O or B star?
   b. A red giant star?
   c. A white dwarf?
   d. A pulsar?

9. For each known extraterrestrial planet in the solar system, list several environmental factors believed to be adverse to the existence of advanced life forms.

10. According to present astronomical theory, the Sun should eventually use all its hydrogen and turn into a red giant.
   a. When will this happen (see Chapter 19)?
   b. How does this compare with the time scale of biological evolution?
   c. Would you expect the human species to be recognizable by the time this happens?
   d. In what way might life that originated on Earth conceivably escape such an event?

ADVANCED PROBLEMS

11. Suppose an Earth-sized planet (diameter 12 000 km) is circling a star 1 pc away (3×10^{16} m).
   a. What would be its angular diameter when seen from the Earth?
   b. Could this be resolved by existing telescopes?
   c. If the planet orbited 1 AU (1.5×10^{11} m) from its star, what would be its maximum angular separation from the star as seen from the Earth?
   d. Could this angle be resolved?

12.a. Using the distance listed for globular cluster M 13 in Table 22-1, confirm the round-trip travel time for radio waves given in the footnote on page 633.
   b. Aside from its distance, why is M 13 a poor choice of target if we are really trying to send a message to inform carbon-based, intelligent organisms, like ourselves, of our existence?

13. If an alien spacecraft happened to pass through the solar system at 99% the speed of light, how long would it take to traverse the system (assumed to have the diameter of Pluto's orbit)?
   a. If it was broadcasting on the 21-cm radio wavelength of hydrogen, describe how its radio transmissions would be received by us, taking into account Doppler shifts.
   b. Why would detection be difficult?

The Cosmic Perspective

Just as a vacation in an interesting place can help us view our own workaday cares from a clearer perspective, our astronomical explorations to the ends of the known universe help us view our earthly cares from a clearer, cosmic perspective. In this epilogue I will give some personal thoughts about how our cosmic journey reflects on our daily lives.

The cosmic perspective has had a deep impact on humanity during the last few centuries. Astronomical experience has outmoded once popular ideas, such as:

1. The idea that there are physical heavens and hells directly above and below the Earth

2. The idea that the Earth is the universe's center, around which all other bodies revolve

3. The idea that humans are the inevitable lords of creation

4. The idea that the heavens are unchanging, or that they are constant except for occasional sudden creation of new features

5. The idea that the Earth is the sole and inexhaustible reservoir of raw materials

What has the cosmic perspective given us in exchange for these cherished ideas? Space extends without known limit around us—a new frontier. Planets circle suns; suns circle mysterious galactic cores. Humans are a special part of a great chain of life, in which many life forms have been linked. We don't know if this chain of life is strong and stretches among many planets in our galaxy, or if it is weak and exists only in very rare, scattered segments. The heavens evolve constantly. Humans have watched new starlike objects appear and old stars explode in month-long flashes that blast out clouds lasting for millennia. Even the Earth itself wobbles so that Polaris will be the North Star for only a few centuries. While we are exhausting the Earth's finite resource-rich surface layer, we may be able to utilize cosmic sources of energy and material with less damage to the Earth than is caused by our present resource-seeking and resource-consuming activities.

Some critics have said that science has only demeaned humans by putting us on a satellite of an average star on one side of an average galaxy. Certainly these ideas shocked people during the Renaissance. But as astronomer Harlow Shapley asked:

Are we debased by the greater speed of the sparrow, the larger size of the hippopotamus, the keener hearing of the dog, the finer detectors of odor possessed by insects? We can easily get adjusted to all of these. . . . We should also take the stars in stride. We should adjust ourselves to the cosmic facts.

Science has also been criticized for replacing outmoded ideas only with machines, which have brought us the evils of pollution, nuclear weapons, and a dehumanized existence. This criticism seems to me to confuse science (knowing) with technology (using knowledge and material). There is a difference between knowing and using. To me there is something appealing about knowing as much as possible and using as little as necessary. The popular criticism of "science and technology" (lumped together as if one) is really a criticism of how we have *used* knowledge and materials.

And it is justified. There is great irony in the fact that in spite of 10 000 years of effort to produce labor-saving devices, many people have trouble getting by on 40-hour-per-week jobs where they labor to make products that do not even interest them and that they themselves may label as "consumer gimmicks." This is a failure of our overall culture, rather than a failure of science or technology. The cosmic perspective is making this clearer, as more people begin to think in terms of our total planetary budget of materials, energy, and creative talent. The cosmic perspective helps us see that the political and social difficulties of maintaining a stable, workable civilization are at least as challenging as the problems of scientific exploration.

Besides giving us a new view of our place in the universe and our planetary society, the cosmic perspective gives us facts on which to base our actions. Facts, of course, are merely conclusions drawn from observations, usually verified by generations of observers. They provide a foundation for practical life in the real world. They grade indistinguishably into hypotheses about how the universe acts and how it is put together.

If a philosopher–critic argues that we can never really know facts, a scientist can agree and argue that acting on hypotheses is like betting. If you are forced to put your money on one thing or another, you try to put it on what is most likely to be successful. The world forces us to act, and we want to act on the basis of the ideas that are most likely to be true. So we *act* on our understanding of physical laws when we design bridges, and we *act* on our understanding of the solar system when we choose not to waste time or money on astrological advice. The scientific method of learning about the universe is simply a scheme for accumulating hypotheses on which to act until better information comes along. Hypotheses are useful because they suggest possibilities of new discoveries, new principles, and new places to go. They suggest goals, tests, and experiments.

This brings up the attempts, mentioned in Chapter 14, to legislate a retreat from scientific debate about our cosmic environment. According to some critics of science, the last century of scientific exploration has demeaned human and religious values, and society would be better served if the theories resulting from scientific research could be suppressed or force-fitted into one ideological mold or another. Key among the theories criticized is the theory of biological evolution. For some decades in the Soviet Union, for example, party leaders insisted on a Marxist version of the evolutionary process, and this ideological intrusion stunted Soviet biological research for years.

In the United States, lawsuits and legislation in several states have called for bans on public school teaching of evolution, or "equal time" presentations of some alternative hypothesis, usually a "creationist" hypothesis involving the sudden, one-time creation of the universe, the Earth, geological features, and species. The proposed creationist hypothesis is generally derived from fundamentalist (literal) interpretations of the Judeo–Christian Bible, raising the question of why a science teacher (if he or she *must* teach a tradition outside the scientific method) should choose this particular

creation tradition over the creation traditions of other cultures, such as Hindu, Buddhist, or native American Indian. In 1981 the legal attack on the theory of evolution shifted to the grounds that the theory of evolution itself is more like a religious faith than a scientific theory (for example, see *Science 211:* 1331; 20 March 1981). Nonetheless, "equal-time" creationist laws have been struck down by U.S. courts on the constitutional grounds that the state is not to be involved in teaching religious traditions.

These trends leave unclear whether only the theory of biological evolution of species will come under legal attack, or whether the attack will one day spread to concepts of stellar evolution, planet growth, and the evolution of the Earth's atmospheric and geological environment, since some creationist traditions propose instantaneous or seven-day formation of all present features in the physical universe. Many backers of the current "creationist" legal cases reportedly believe the world was created only 10 000 years ago.

One can sense that the intellectual thrust of this movement is to abandon our two-way dialogue with nature in favor of a one-way dialogue in which some set of ancient writings is chosen not only as a guide to ethical conduct but also the literal guide to the nature of the physical universe. To a scientist, this seems a dangerous trend. The popular press has presented it as an argument of "science versus religion," but I fail to see it in quite such an adversary way. It seems to me that real religious awe and scientific curiosity are very similar; both involve the recognition of something "bigger than we are," about which we don't have final answers. Perhaps the "something" can best be expressed as what the American naturalist Thoreau called "this vast, savage, howling mother of ours, Nature." Our philosophical ideas about life ought to mature in a way consistent with what we know about nature. When we ask the right questions, nature suggests answers that we can act on. To turn away from this dialogue with the universe and insist that science courses include someone's interpretation of old writings seems both an antiintellectual and antireligious abandonment of the talents we've been given.

Earlier chapters showed examples of civilization getting into trouble in this way before. For example, Chapter 3 described how medieval scholars accepted the views of Aristotle and Ptolemy on the Earth's place among the planets, arresting and even executing dissenters who later turned out to be correct. The Renaissance came as scholars began to debate these questions not with old manuscripts but with *new observational evidence.*

We must remember that science and science courses involve the discussion of *evidence*—that is, nature. Any hypothesis—about evolution of species, evolution of stars, the sudden appearance of humanity in the Garden of Eden, the sudden creation of everything in modern form 10 000 years ago, the big bang, relativity—can be suggested and discussed scientifically as long as the discussion centers around observation, experiment, calculation, and prediction. It is reasonable that the theories most discussed should be the ones that have emerged as consensus views from decades of scientific debate. The answer to the current attack on science is not, one hopes, to legislate that favorite theories must be taught, but rather to continue the most open, rich, vigorous discussions of hypotheses as tested against evidence. Out of such a free discussion of evidence, one is free to shape one's philosophical beliefs—a practical cosmic perspective. (See Enrichment Essay A for further thoughts on separating science and nonscience.)

Does the cosmic perspective have further bearing on economic, political, social, or philosophical problems of our civilization? Each person brings his or her own experience into answering such a question, but here are some loosely connected thoughts that might be considered:

1. A classic science fiction plot deals with a vast interstellar spaceship, kilometers long, in which generations of men and women live and die, maintaining a stable environment and culture while en route from one part of the galaxy to another. This plot turns out to be the truth. We have only recently realized that we are on board just such a vehicle—Spaceship Earth—with its finite area and finite resources.

2. There must be few places in the universe where we can walk naked away from our machines, breathe in the atmosphere, let the light of the nearest star fall on us, and find water to drink on the surface. In other words, the Earth may be a Hawaii in a universe of Siberias. Although we can take technology with us to adapt to new places (as we have adapted to the extremes of the Earth), it would be a good idea to take care of the one known place in the universe where we can do these things. Astronaut William Anders summed this up while describing his view during a trip around the Moon: The Earth "was the only color in the universe . . . very delicate indeed. . . . It reminded me of a Christmas-

tree ornament—colorful and fragile—something that we needed to learn to handle with care."

3. Twentieth-century global society is not living within its planetary means. Certain resources, such as metals, coal, and natural gas, are being consumed at rates hundreds of thousands of times faster than their replacement rates. In the past, frontiers have swept west from China, east and west from India, out from the Mediterranean, and around the world from Europe. Humanity has lived by eating its frontier. But the Earth is spherical and the frontier has closed in on itself. We are now groping our way through the required period of adjustment—which may or may not be pleasant, depending on our ability to sense the proper direction and make the necessary changes. Throughout the last decade, projections of mineral resource consumption, even assuming five times the present known supply on Earth, predict exhaustion of most metals and other resources in the next century (2000–2100), if present consumption growth patterns continue. Therefore, global social changes are in the wind.

4. It seems technologically feasible to engineer a stable society, but it also seems possible to destroy civilization and all life on the Earth either by design or by accident.

5. Pressures on humanity and the Earth's environment will increase the longer we draw on the Earth's finite resources. Even an era of conservation only stretches the time available unless new materials, energy, or ideas to solve the underlying problem of consumption in a finite world are developed. Pursuit of human capabilities in space therefore may offer a way out of these problems. A gradual shift of heavy industrial activity into space, where solar energy could be used instead of fossil fuels, could help reverse the polluting trends on the Earth today by reducing the mining of fossil fuels and some metals, reducing the emission of pollutants into the atmosphere, and fostering an ideal of Earth stewardship. Resulting pollution of the space environment would be minimal, both because of its vastness (the volume of the solar system is about 100 billion billion times that of our atmosphere) and because gases and dust are flushed out of the solar system by the solar wind. Space exploration may be a rare area of human activity that is challenging, exciting, and compatible both with environmental concerns and with human economic growth.

6. We can describe with some accuracy the arrangement and history of the planetary and galactic systems in which we find ourselves, but we do not really understand the arrangement or history of the largest-scale clusters of galaxies. The most fundamental structure of the universe is thus still a mystery.

7. We may be prohibited, by the Hubble recession and the finite speed of light, from ever obtaining information about places beyond about 10 000 Mpc.

8. To some philosophers, the only reality is the reality directly witnessed by human observers. Yet three centuries of repeated, reliable observations give us the right to talk about real events that happened over a billion years ago and real places where no person has ever walked. Time and distance can be objectively probed, giving results that remain consistent (within our limits of measurement), regardless of who does the measuring. Certain places on Mars have looked the same whether photographed in 1971 by an American spacecraft or in 1973 by a Russian spacecraft. Thus it is reasonable to suppose that there are real places where things happen, even though there are no humans to see them. In a way, it is astonishing to think that rocks roll down hills on the far side of the Moon when no one is there; that winds churn dust across empty Martian landscapes just as black winds of water churn the bottoms of our seas; that gas heaves upward in stars; that starquakes crack crusts of unseen neutron stars; that unseen cataclysms play themselves out silently in space in front of no audiences; and that waves crashed on the Earth's beaches for billions of years before there were land animals to see them.

9. No place in the universe is permanent. The seemingly changeless panorama of stars in the sky will alter in a few thousand years as the stars move with respect to the Sun. The patterns of nebulae will change as the gas clouds are torn and stirred during their 250-million-year circuits around the galactic center, just as the mountains and continents of the Earth are torn and stirred by plate tectonic processes in similar periods of time. Probably the Sun will run out of hydrogen within the next 6 to 10 billion years, expanding to the red giant state, engulfing the Earth, and finally collapsing to a dwarf. New stars will form near other stars, and perhaps other Earthlike planets will form around some of these stars.

10. As the physical universe changes, the biological universe also changes. In relatively brief periods of millions of years, new creatures may evolve, just as humanity itself appeared only in the last few million years. A philosophy that sees humanity as a permanent fixture in an unchanging environment, while it may be practical for short periods such as human lifetimes or centuries, cannot be defended on a cosmic scale.

11. In spite of our seeming sophistication and ability to modify our local surroundings, the universe remains implacable. We humans live on a small planet in the midst of forces and processes, some far greater than any we can create or modify. Known life and intelligence are adaptable and ingenious, but their present ability to affect current events in the universe is limited. Beyond our mortal processes, there are greater processes of death and birth that we have to accept as part of nature. Whole worlds are demolished and whole worlds are created.

The naturalist Thoreau commented with regret that most modern people have forgotten that the Greek word for the universe, κόσμος (cosmos), meant beauty or order. We need a sense of beauty and majesty that not only responds to a pleasant sunset but can also encompass that more permanent cosmic sunset: a supernova. Such chaotic explosions are also part of the *order of things* in the cosmos.

The question of creators is a mystery—exciting, puzzling, beyond our abilities to explain. Here astronomy establishes a point of contact with both old and new religious perceptions. Beyond this point, astronomers are no better equipped than anyone else to make definite statements.

12. Many mysteries remain, and much space is left still to explore. Such explorations would seem to be an endeavor worthy of human energy—worthier, for example, than warfare. It is good to admit that there is still mystery in the universe. There is more to life than what we see on this particular planet at this particular time.

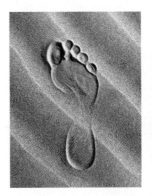 

Pseudoscience and Nonscience

When confronted with a hypothesis that sounds far-fetched, some scientists may immediately label it pseudoscience or even nonscience. This is often unfair. Any hypothesis has as much right to vie for verification as any other. The true scientific method is to test hypotheses. A good hypothesis must predict some things about nature, and if the predictions are wrong, the hypothesis must be rejected or modified. Though a hypothesis can never be proved "ultimately true," if experiments keep turning up consistent results, it is considered more and more reliable and comes to be an accepted theory or even a law.

SCIENCE AND PSEUDOSCIENCE

A **pseudoscience** is a body of hypotheses treated as if true (usually with commercial intent) but without any consistent body of supporting experimental or observational evidence. Pseudoscience may appear to be backed by the trappings of real science, such as quotations of evidence, but the evidence is often hearsay, the references are often to other poorly researched commercial books, and the work is rarely reviewed by other professional researchers before publication. Note that pseudoscience is not determined by subject matter but by its method of dealing with *evidence*.

The dangers of pseudoscience are twofold. First, its practitioners often bilk consumers of their money by falsely promising new discoveries or mystic knowledge. Second, it misrepresents real scientific discovery and often contributes to antiintellectual attitudes that exchange mysticism and magic for exploration and discovery. The exchange is a poor one because, as I hope I have shown in this book, real discoveries about the

universe are as exciting as the erroneous claims of pseudoscience.

ASTROLOGY

Astrology is a classic example of pseudoscience. We can trace its roots to ancient superstitions about the sky that developed at about the same time as ancient astronomical observing. One of a class of ancient systems for attempting to predict events from patterns in nature (such as patterns in tea leaves or flocks of migrating geese), astrology tries to predict events by using the patterns made by the moving planets against the background of starry constellations. Chapter 1 traces the roots and shortcomings of astrology in more detail.

During the tortuous evolution of history, many old superstitions have been abandoned while others have survived; it is ironic and sad to realize that while most citizens of the modern Western world reject using tea leaves to plan their day, many of them pore over an astrology column in the morning paper. Old traditions (whether sensible or not) die hard. New knowledge comes slowly, but faster in our century than ever before.

PSEUDOSCIENCE BASED ON EXTRAORDINARY REPORTS

Another form of pseudoscience has grown up around alleged eyewitness reports of strange events. Once again, it is perfectly permissible to hypothesize that strange events, such as an alien monster's visit to the Earth, will sometimes occur. We cannot dismiss the hypothesis, but we can ask if research programs have turned up verifiable evidence. To answer this question, we need to understand the processes that lead to eyewitness reports.

Perception, Conception, and Reporting

One-time events in the sky are hard to confirm. It is easier to learn from repeatable experiments in the laboratory than from sporadic views of celestial phenomena, which is one reason why less occult lore accumulates around a science like hydraulics than around astronomy.

The first step in generating an eye-witness observation is *perception*—the observer's intake of external sensory stimuli. The problem is that this perception must be converted in the observer's brain into a *conception,* a step that involves subjective factors, such as the association the person may make between the object and concepts prevailing in the culture. For example, witnesses almost invariably conceive of meteors in terms of the distances of aircraft. They say, "It landed just behind the barn," when the meteor may in fact have been hundreds of miles away.

Reporting, the third step, transmits conceptions to other people. Throughout most of history, this process has been by word of mouth, with secondhand reports and hearsay blending into long-lived oral traditions, which may incorporate extraneous incidents or myth. Even today, it is hard to transmit conceptions accurately by words. There is a UFO joke about a man seeing a large orange object on the ground, with flashing red lights, rows of windows, and small people inside. Does this account make you visualize a flying saucer? The object was a school bus. This joke shows how, in the appropriate context, simple, accurate words can lead many people to the wrong conception.

Since publications affect the beliefs of millions of people, the quality of public reporting cannot be overemphasized. Readers must retain a healthy skepticism, because many news or entertainment media thrive on sensational stories with minimal documentation. Astronomer Carl Sagan (Sagan and Agel, 1975) recounts an example of public overacceptance. He saw a woman reading a pseudoastronomy best-seller that contained many errors and flagrantly misquoted Sagan himself. Sagan asked the woman if she knew the book contained inaccuracies. "It couldn't," she said, "because they wouldn't let him publish it if it weren't true." Not so.

Many books see print simply because they *will* sell, and many bookstands carry as much fantasy as fact in their so-called nonfiction selections.

In summary, the processes of perception, conception, and reporting of rare events could in principle be a source of scientific information, but in practice these processes often produce misleading data. Reports of weird events are likely to circulate and be published whether these events really occurred or not.

UFOs

UFOs are any form of unidentified flying object reported in the sky. Undeniably, there have been thousands of reports of different kinds of UFOs, ranging from alleged metal disks to amorphous glows to unusual objects detected by radar. The previous discussion shows why they are hard to study scientifically.

Most of the literature on UFOs is pseudoscience—not because of the subject, but because of the way the evidence has been treated. Most popular UFO books abound in exaggerated claims and distortions. A 1968 University of Colorado study (in which I personally participated in hopes of discovering exciting evidence of unusual phenomena) established that many of the classic UFO photos used on covers of UFO publications are either fakes or photos of known natural phenomena. Nevertheless, these photos reappear, year after year, on new UFO publications.

A sociological, fadlike element in the UFO problem is shown by the waves of UFO reports following space events such as the first Sputniks and the first photos from Mars (see Figure A-1) and by UFO hoaxes that followed within weeks of the first "flying saucer" report in 1947. Waves of UFO reports have even occurred in earlier eras, when the UFOs were reported to look not like the saucers popularized in our movies but like images popular in those times. A UFO wave in the 1890s reported "airships" looking like Victorian dirigibles with fanlike propellers; this wave came a few years after the 1886 publication of Jules Verne's *Robur the Conquerer,* in which a mad scientist terrifies the world from . . . you guessed it . . . his newfangled dirigible. One such UFO reportedly crashed into a windmill in Texas in 1897, killing its Martian occupants. A prosaucer research organization in 1973 labeled this case a hoax.

When a satellite reentered the atmosphere and broke

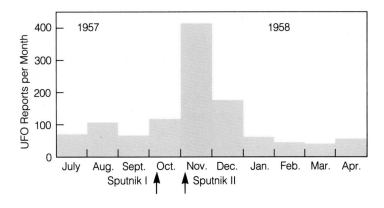

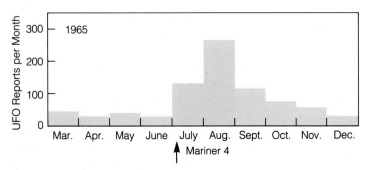

Figure A-1 Examples of sudden increases in the number of UFO reports correlated with social factors such as the first satellite launches (top) and the first close-up photos of Mars (bottom). This correlation suggests that "waves" of sightings, discussed in many popular books, have social rather than physical causes.

into a group of burning pieces in the night sky over the eastern United States in 1968, many people witnessed the phenomenon. Some correctly identified it, but many others misconceived it. Of a group of 30 extensive reports collected by the Air Force, 57% said that the objects were flying in formation, implying intelligent control. About 17% conceived that the glowing objects must be attached to an unseen black object, and they reported a "cigar-shaped" or "rocket-shaped" object that did not exist. Others conceived the glowing objects as windows and reported a dark object with glowing windows. One totally erroneous report said: "It was shaped like a fat cigar. . . . It appeared to have rather square shaped windows along the side . . . the fuselage was constructed of many pieced or flat sheets . . . with a 'riveted together look'" (Condon and others, 1969).

The abundance of such misconceptions and misreporting does not, of course, prove that "flying saucers" or alien spaceships do not exist. But it does show that proof of their existence is nearly impossible to establish from the kinds of reports available. While a few unsolved UFO cases may suggest unusual atmospheric phenomena, none give good evidence for spaceships. If there really are any spaceships around, one might predict, in

view of the increasing numbers of surveillance devices and tourist cameras, that good evidence should appear. So far, none has.

Ancient Astronauts

A similar area of pseudoscience is the literature on possible ancient astronauts. The basic technique in many of these books is to misrepresent astronomical or archaeological evidence. One best-seller, for example, claims that studies of the 1908 Siberian meteorite impact (Chapter 13) "confirmed a nuclear explosion," which is not true. The same book presents ancient lines laid out in the soil in Peru as a mystery requiring supertechnology to construct. I have visited them, and they were made by simply pushing aside an inch-thick cover of dark stones. Several scientists examined the evidence for ancient alien visitors years before these books became popular and found no convincing evidence (Shklovskii and Sagan, 1966). The ancient astronaut books have now faded from popularity after being discredited, but not until author and publisher gained considerable income from gullible readers.

Astrocatastrophes

A more interesting case was a study of ancient myths by naturalist and psychiatrist Immanuel Velikovsky. He used a pseudoscientific method of interpreting all ancient documents literally. For example, where the Book of Joshua says the Sun stood still (10:13), Velikovsky assumed that the Earth indeed suddenly stopped turning and then started again, ignoring modern evidence that this is unlikely. Compiling his results, Velikovsky (1950) concluded that two astronomical supercatastrophes occurred around 1500 B.C. and 750 B.C., during which Venus first appeared, passed near Mars, and then passed near the Earth, reaching its present orbit only after 750 B.C.

Velikovsky's work was interesting in its compilation of strange ancient myths, but when astronomers criticized his work, he was celebrated as an antiestablishment underdog and his books became best-sellers. His conclusions have been solidly refuted by scientists, however. For example, records of Venus' motions that are usually dated around 1600 B.C. and records of old eclipses show no signs of these disturbances; computer calculations that can trace planetary motions back millions of years reveal no effects suggesting the hypothesized disturbances; the lunar lava flows that Velikovsky attributed to the catastrophes date from 3 billion years ago, not 3000.

Although Velikovsky was far too literal in interpreting old records, astronomical events such as meteorite falls may indeed have influenced early cultures and myths. Scientific methods can be applied to this problem. For example, Florida geologist Cesare Emiliani and his coworkers (1975) described physical evidence of worldwide coastal flooding around 9600 B.C., which they suggest may be the basis of the worldwide flood myth.

"SCIENTIFIC CREATIONISM"

Renewed attacks on the theory of evolution raise the issue of a new kind of pseudoscience, often called *scientific creationism*. The history of this creationist movement since the 1920s is traced in some detail by Numbers (1982). It apparently grew out of a feeling of some parents that scientific evidence (usually being taught in biology classes) conflicts with the family's religious faith. This faith may be limited to the idea that humans did not evolve from other animals, but were placed on the Earth in modern form. It may also involve ideas such as the creation of the Earth only a few thousand years ago or the lack of evolution among *any* animal species.

In its most extreme form, "scientific creationism" strikes at the heart of science not by citing faulty evidence, but by proposing a "scientific" study in which evidence means nothing at all! Humanity (or the solar system or whatever) was created all at once along with fossils in the ground, radioisotopes in their present distribution, and so on. In this view, the fossils that suggest evolution have no meaning—they were put there to fool us. While this is a logically consistent hypothesis in which one might choose to put one's faith, it cannot be dealt with scientifically. Does it predict the existence of any hitherto unobserved types of fossils for which we can search? Does it encourage new exploration and theorizing? Evidently not.

This point can be understood by considering the theory of "fairies in my rose garden." Suppose I hypothesize that shy fairies live in my rose garden and help tend the roses, but are so clever that they always know if I put a camera or recorder out to document them; thus they avoid leaving any record. This explains why no evidence of fairies exists. This is a fine hypothesis, but it is not a part of rational science, because it has the built-in proviso that no meaningful evidence can be gathered. Once the assumption is made that fairies exist, many books could be published about where they might live, how they might dress, and so on. These books could be fine literature and brilliant exercises in logic, but they would be speculation, not science. (Some of the more enthusiastic books about life on other worlds have been criticized on these same grounds; but at least we have evidence of amino acids in meteorites, biochemical evidence about laboratory reactions, and astronomical evidence about the environments of other star systems!)

What is the main danger of these ideas? Why not simply integrate some of the creationist ideas into science courses, as the creationists have advocated in their lawsuits? The main problem is not with the hypotheses, whose merits should always be tested against natural evidence. The main problem is with the underlying premise that the answers are already known from ancient sources and that science merely studies nature to verify those answers. Historically, this has always been a terribly ineffective, if not dangerous, way to approach science and nature. Many scientists have fallen into the trap of becoming sure that their favorite theory is right and ended up (in the worst cases) trying to force data to fit incorrect conclusions. The business of science is an open, two-way dialogue with nature. If we ask the right questions, nature not only shows us the answers but may lead us on to the next question. The things that we discover may not only be interesting in themselves but also lead to practical applications. These results are less likely in a science that is conducted only to verify conclusions that are already "known" from ancient authorities. We commit a sin of pride if we presume to tell nature what the answer is, before we even attempt our observations!

Ideally, a science course should be a synthesis of evidence derived from nature all around us—from observation, experimentation, and calculation. How this evidence about the nature of the universe is then organized into a system of philosophical or religious thought is the subject of philosophy and religion courses—which are exciting enterprises in themselves, but not science. Many of the world's great thinkers have concluded that systems of ethical thought based not only on natural evidence but also on assumptions (for example, that human activities are all right as long as they don't damage other people or their property) are valuable in helping us define ethical conduct. But these systems of thought have always been grouped in the field of philosophy and religion, not science.

Having such systems is not inconsistent with the scientific study of nature, as realized by philosophers during the Renaissance, when application of the scientific method began to change the world. Yet in the 1980s lawsuits have attempted to force the teaching of scientific creationism under the guise of science in science courses. In the opinion of most educators, scientific creationism does not belong in a science classroom because it does not deal in a testable way with evidence. It is not part of a dialogue with nature; it is not science. It might be judged on its merits in a course on comparative religion.

HOW TO REACT TO PSEUDOSCIENCE

When considering generations of low-income people, both in America and around the world, wasting meager resources on dubious cures, astrological advice, and superstitious cults, we may grow impatient and demand that such activities be abolished. Yet, as pointed out in Chapter 1, one of the beauties of Western democracy

is that we attempt, at least, to allow freedom of speech—freedom of competing ideas—in a hope that the right ideas will emerge as the incorrect ideas are found to be flawed. We try to encourage healthy debate, and hope to find what seems closest to the truth.

Perhaps in this approach we do not pay enough attention to the fact that the human mind has an amazing ability to conceive intricate systems of thought that seem self-consistent, but in the final analysis have little to do with reality. Sometimes it is labeled fiction: Tolkien's novels of Hobbits, for example, create an intricate world of gnomes with their own language, mythology, and history—but no one claims they are reality. Sometimes we think our intricate thought systems *may* reflect reality and label it a scientific hypothesis—but then we repeatedly test it against observation and experiment to see if we can disprove it. If so, we replace it with something *better*. Sometimes we take an intricate thought system and simply claim it *is* reality, and urge people to give their hearts to it. Sometimes these ideas are untestable; uncritical acceptance may be claimed as a virtue. It is these thought systems—pseudoscience and superstition—that threaten to drain away energy, resources, and imagination.

What we can do with any thought system—fiction, scientific hypothesis, superstition, or some unclear mixture—is to use *evidence* to test it against experience and reality.

In the meantime, scientists and science teachers for their part must avoid being dogmatic about interpreting natural evidence, and people everywhere should cheerfully anticipate continuing refinement of scientific knowledge with new discoveries that deepen our understanding of the universe and our relation to it.

Astronomical Coordinates and Timekeeping Systems

Two kinds of astronomical systems are related to the rotation of the Earth. The first is the system of astronomical coordinates, whereby astronomers map positions of objects in the sky. The second is the timekeeping system employed around the world—an outgrowth of the early timekeeping and calendar systems described in Chapter 1.

RIGHT ASCENSION AND DECLINATION COORDINATES

We use the Earth's spin to define coordinate systems for locating objects both on the ground and in the sky. The ground system—the system of latitude and longitude—is based on the Earth's spin: Latitude lines are perpendicular to the axis of rotation, whereas longitude lines intersect at the poles of rotation.

A nearly identical system describes locations of stars and other celestial objects. It involves coordinates called *right ascension* and *declination* (often abbreviated "R.A." and "Dec."). *Right ascension and declination are merely longitude and latitude, respectively, projected from the center of the Earth through the Earth's surface and onto the sky and fixed with respect to the stars.* The celestial equator (see page 12) is the line of 0° declination, and the north celestial pole (see page 10) is at declination 90°N, usually written +90°; the south celestial pole is at −90°.

To visualize this system, imagine the Earth as a giant nonrotating glass sphere with only the longitude and latitude lines engraved on it, and imagine yourself at the center of the sphere. You look out and see the longitude and latitude lines, by which you can locate features on the surface of the Earth. But these lines are also the lines of right ascension and declination, and you see them projected onto the stars. The Earth's equator overlaps the celestial equator. The Earth's North Pole overlaps the north celestial pole and almost obscures the star Polaris.

Figures B-1 and B-2 help in visualizing the right ascension–declination system of sky coordinates. Figure B-1 shows two views of the northern sky an hour apart as the stars wheel around the north celestial pole. Figure B-2 shows two similar views of the eastern sky as Orion rises there.

As zero right ascension, astronomers have chosen the right ascension line running through the vernal equinox: the point where the Sun appears to cross the celestial equator going north. Instead of measuring the right ascension in degrees, astronomers measure it in hours, because they are interested in the time required for apparent motions of stars across the sky. If a star's right ascension places it 1^h below the eastern horizon, for example, astronomers know that it will rise in an hour. (The Earth turns 15° in an hour, so 1 h = 15°.)

Stars are cataloged by their right ascension and declination. Sirius, for example, lies at R.A. 6^h41^m, dec. −17°; Polaris, which is virtually on the north celestial pole, is at R.A. 1^h23^m, dec. +89°.

THE EFFECT OF PRECESSION ON RIGHT ASCENSION AND DECLINATION

Now we can more clearly describe the effects of the precession Hipparchus discovered around 130 B.C. (page 50). Because the right ascension–declination system is anchored on the Earth, and because the Earth wobbles with respect to the stars every 26 000 y, the system shifts among the stars. The coordinates of stars shift slightly every year, and astronomers must continually update the right ascension and declination positions listed in catalogs. Catalogs generally specify "epoch 1950" or

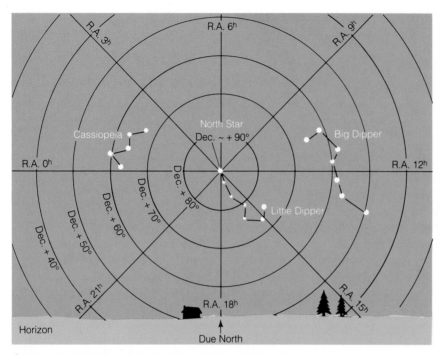

a

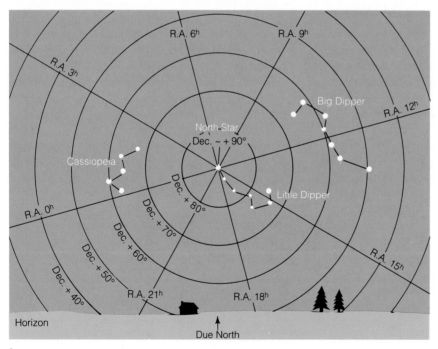

b

Figure B-1 a Schematic sky view looking north from the mid–United States on a February evening. Prominent constellations and the right ascension–declination system of imaginary coordinates are shown. Note that the leading star in Cassiopeia is nearly on right ascension 0^h and declination 60°. **b** The same view about an hour later. Note how the right ascension–declination system moves with the stars as they rise in the east and set in the west.

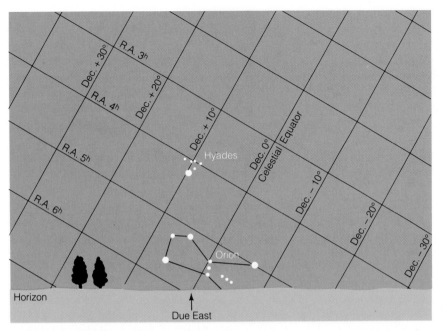

a

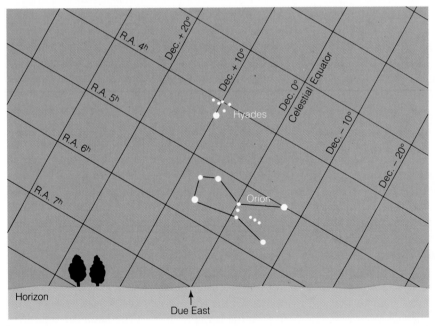

b

Figure B-2 a Schematic sky view looking east on a November evening. Prominent constellations and the right ascension–declination system of imaginary coordinates. Note that Orion lies on the celestial equator. **b** The same view about an hour later. Note how the right ascension–declination system moves with the stars as they rise in the east.

"epoch 2000" as the year for which the positions are listed. The same shift of the celestial equator and vernal equinox invalidated the original system of astrology by shifting the right ascension–anchored astrological signs away from their original constellations, as described in Chapter 1. For an example of the changing position of the north celestial pole among the stars, see Figure 2-10, page 50.

ANOTHER CELESTIAL COORDINATE SYSTEM

A system useful for describing objects in the sky during both day or night is the *altitude–azimuth system. Altitude* is the number of degrees above the horizon. *Azimuth* is the number of degrees around the horizon, starting from the north point (0°) and moving through the east (90°). Since azimuth is counted from the north to the east, an azimuth of, say, 60° is often written N60°E for maximum clarity. Thus an object on the horizon due east would have altitude 0°, azimuth N90°E. An object halfway up the sky in the west would have altitude 45°, azimuth N270°E. This system is clarified in Figure B-3, which shows two views of Orion rising in the east with altitude and azimuth lines superimposed.

An object at the *zenith* has altitude 90°. The most important azimuth line is the *meridian* (see page 11).

SYSTEMS OF TIMEKEEPING

Before clocks were invented, time was kept by watching the apparent motion of the Sun around the Earth, caused by the Earth's rotation. This is called *apparent solar time.* Noon was the moment when the Sun crossed the meridian. A.M. *(antemeridiem)* is the period before the Sun crosses the meridian; P.M. *(postmeridiem),* the period after. Once clocks were invented, it became clear that the Sun's apparent motions were not exactly uniform, because the combination of the Earth's rotation and orbital revolution produces days of varying length. Later, more careful measurements revealed slight, sudden speedups and slowdowns in the Earth's rotation due to earthquakes, tides, and winds.

For these reasons, today's clocks run on *mean solar time,* a time system based on the uniform average rate of the Sun's motion rather than its actual position in the sky. The most accurate calibration of clocks for scientific work is now based on an internationally determined definition of the second based on constant atomic processes: 1 s is 9 192 631 770 cycles in certain radiation from the cesium-133 atom. This standard is called *atomic time.*

For long-term prediction of planetary positions or eclipses, astronomers have defined an additional time system based on the motions of the planets, called *ephemeris time.*

Astronomers, who sight their telescopes on the stars every night, need a star-based time system rather than the Sun-based system of mean solar time by which most clocks run. This system, called *sidereal time,* is simply a clock reading equal to the right ascension crossing the local meridian at any moment.

TOWARD THE MODERN CALENDAR

Just as timekeeping based on the Earth's rotation has been continually refined, so has the calendar system based on the Earth's orbital revolution. Chapter 1 described how early people developed calendar systems and estimated the length of the year. The year was often officially started on an astronomically determined date, such as the summer solstice. One problem with early calendars was accurately determining the number of days in the year. We now know that the sidereal year (the Earth's motion around the Sun with respect to the stars) is 365.256 ephemeris days, and the tropical year (cycle of seasons) is 365.242 ephemeris days.

If early timekeepers adopted a calendar with, say, 365 days in a year, they discovered that the astronomically determined New Year's Day was one day in error after four years. This problem was dealt with by a process called *intercalation:* Astronomers, government officials, or priests announced extra days, often celebrated as holidays, every few years to keep the calendar in step with astronomical events.

As long as society was stable, this process could be kept up for many years. The Romans, for example, developed the ancestor of our calendar, beginning the year at the spring equinox. (The Latin-derived names *September* through *December* represent month positions 7 through 10 in that calendar.) The extra days were intercalated at the end of the last month, February, where we still insert our extra day every leap year.

By 46 B.C., however, local officials in different parts of the empire were inserting extra days as they pleased, and the calendar became confused. Julius Caesar had

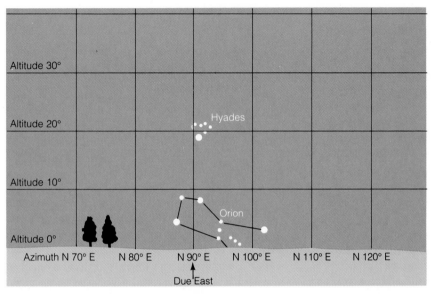

a

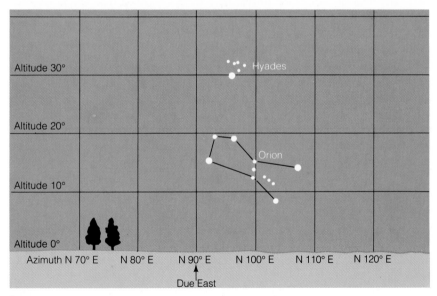

b

Figure B-3 **a** This figure is the same as Figure B-2a, but it uses the altitude–azimuth system of coordinates instead of the right ascension–declination system. In this view, the trees reach an altitude of about 5° and the Hyades are at an altitude of about 20°. **b** The same view about an hour later. Note that the altitude–azimuth system of coordinates is tied to the horizon and does not move with the stars. The trees still reach an altitude of about 5°, while the Hyades have risen to an altitude of 30°.

the calendar revised in 44 B.C. under the Alexandrian astronomer Sosigenes, and the modern leap year scheme was formalized. The fifth month of the new *Julian* calendar (Quintilis) was renamed July in honor of Julius Caesar. August was named after Augustus Caesar, who made some additional improvements in 7 B.C.

The Julian calendar worked well for more than a thousand years, but since the year is not exactly 365¼ days long, new errors crept in. In the 1200s, European scientists noted that there was a full week discrepancy in the date of the vernal equinox. After several earlier attempts, Pope Gregory XIII commissioned a new revision of the calendar during the period from 1576 to 1603.

This *Gregorian* calendar was adopted quickly in Roman Catholic countries, more slowly in Protestant countries, and still more slowly in Eastern Orthodox and other countries. England and the colonies converted in 1752, dropping 11 days. (Riots and legislation against rent abuse followed because of the resulting 19-day month.) Russia converted during the 1917 revolution and had to drop 13 days. Because of the different dates of conversion to the Gregorian calendar in different countries, scholars must beware of historical records of dates, and not compare different dates in different countries without considering the calendar in effect at the time that records were made.

To clear up historical confusion about dates, astronomers have devised a system called the *Julian day count*, which designates any date in history, anywhere in the world, by the number of days since January 1, 4713 B.C. On July 4, 1976, to pick a nonrandom date, the Julian day count was 2 443 964. Computer-calculated dates of ancient eclipses and other astronomical events are expressed in this system.

APPENDIX 1

Powers of Ten

It is no accident that the word *astronomical* has become a synonym for "enormous, almost beyond conception." Astronomy is full of extraordinary numbers designating great ages and distances. In astronomy (and other sciences), therefore, a convenient shorthand system of writing numbers is favored. In this system, an exponent, or superscript, designates the number of factors of 10 that have to be multiplied together to give the desired quantity—that is, the number of zeros in the quantity. For example:

$$1 = 10^0$$

$$10 = 10^1$$

$$100 = 10^2$$

$$1\ 000 = 10^3$$

$$10\ 000 = 10^4 \text{ and so on}$$

The most often used of these large numbers are

$$\text{one thousand} = 1\ 000 = 10^3$$

$$\text{one million} = 1\ 000\ 000 = 10^6$$

$$\text{one billion}[1] = 1\ 000\ 000\ 000 = 10^9$$

$$\text{one trillion} = 1\ 000\ 000\ 000\ 000 = 10^{12}$$

The usefulness of this system is illustrated by expressing the age of the Earth

$$4\ 600\ 000\ 000 \text{ y} = 4.6 \times 10^9 \text{ y}$$

or the distance that light travels in a year

$$6\ 000\ 000\ 000\ 000 \text{ mi} = 6 \times 10^{12} \text{ mi}$$

---

[1]Beware—the British use the term *billion* to refer not to 10^9 but to 10^{12}. This book uses *billion* to mean 10^9.

A similar system is used for very small numbers. Here the exponent is negative and refers to the number of decimal places. Thus:

$$0.0001 = 10^{-4}$$
$$0.001 = 10^{-3}$$
$$0.01 = 10^{-2}$$
$$0.1 = 10^{-1}$$
$$1.0 = 10^{0}$$

The density of gas in interstellar space, which is

$$0.000\ 000\ 000\ 000\ 000\ 000\ 000\ 002 \text{ g/cm}^3$$

can be more conveniently written as

$$2 \times 10^{-24} \text{ g/cm}^3$$

One of the beauties of the metric system (discussed further in Appendix 2) is that the units are related by powers of 10. For instance, a kilometer is 10^3 m (as opposed to our own confusing English system, where 1 mi is 5280 ft). To make the system easier to use, the International System of units (known as the SI metric system) has a standardized set of prefixes, such as *kilo-*, that represent multiples of 10^{-3}, 10^3, 10^6, and so forth. These prefixes are shown in Table A1-1.

TABLE A1·1

Prefixes Used in the SI Metric System of Units

Prefix	Symbol	Multiple
Tera- (TER-a)	T	10^{12}
Giga- (JIG-a)	G	10^{9}
Mega- (MEG-a)*	M	10^{6}
Kilo- (KILL-low)*	k	10^{3}
Hecto- (HECK-toe)	h	10^{2}
Deka- (DECK-a)	da	10^{1}
Deci- (DESS-ee)	d	10^{-1}
Centi- (SEN-tee)*	c	10^{-2}
Milli- (MILL-ee)*	m	10^{-3}
Micro- (MY-crow)*	μ	10^{-6}
Nano- (NAN-oh)	n	10^{-9}
Pico- (PEE-koh)	p	10^{-12}
Femto- (FEM-toe)	f	10^{-15}
Atto- (AT-oh)	a	10^{-18}

Note: Starred items are the most important in ordinary applications.

Units of Measurement

Most scientists express distances and other measurements in metric units, and the world is in the process of converting to this system. This conversion will save time and money, since publications, quantities of goods, and machine parts will be measured in one uniform system. The metric system is also easier to learn and use, since it expresses units in multiples of 10 (like our money system), instead of in unpredictable multiples such as 12 inches per foot, 5280 feet per mile, and 16 ounces per pound.

Table A2-1 gives some common units in both systems. This book has emphasized the International System of metric units (abbreviated "SI" for *Système Internationale*), which uses the meter, kilogram, and second as the basic measures of length, mass, and time, respectively. When using the SI system to solve problems involving complex quantities or constants, such as energy or the gravitational constant G, one must make sure that these quantities are expressed in terms of the basic units kg, m, and s. Equivalents in the centimeter–gram–second (cgs) system and in larger units are also given in Table A2-1.

While solving certain problems, such as the advanced problems in this book, we need to use certain physical constants, such as Newton's universal constant of gravitation G, described in Optional Basic Equation II on page 76. It is important to remember that in solving any such numerical problems, all measurement units must be converted into a single system of units, and any physical constants must also be expressed in that system. The SI metric system is recommended for this purpose, since it is the international standard. Table A2-2 gives some of the more useful physical constants expressed in the SI system.

TABLE A2·1

Units of Measurement

English System	SI (m-kg-s)	cgs System	Astronomical Measurements
1 inch	0.025 m	2.54 cm	—
1 foot	0.305 m	30.5 cm	—
1 yard	0.914 m	91.4 cm	—
1 mile	1609 m	1.609×10^5 cm	—
1 pound	0.454 kg	454 g	—
	1.50×10^{11} m	1.50×10^{13} cm	1 astronomical unit (AU)
	9.46×10^{15} m	9.46×10^{17} cm	1 light-year (ly)
	3.08×10^{16} m	3.08×10^{18} cm	1 parsec (pc) = 3.26 ly
	3.08×10^{19} m	3.08×10^{21} cm	1 kiloparsec (kpc)
	3.08×10^{22} m	3.08×10^{24} cm	1 megaparsec (Mpc)
	1.99×10^{30} kg	1.99×10^{33} g	1 solar mass (1 $M_\odot$)

TABLE A2·2

Some Useful Physical Constants (SI system of metric units)

Constant	Numerical Value	Optional Basic Equation)
Seconds of arc in one radian	206 265	I
G (Newton's universal gravitational constant)	6.67×10^{-11} N $\cdot$ m^2/kg^2	II, III
Wien's constant	2.90×10^{-3} m $\cdot$ K	V
k (Boltzmann constant)	1.38×10^{-23} J/K	VI
c (velocity of light)	3.00×10^8 m/s	VII, IX
σ (Stefan–Boltzmann constant)	5.67×10^{-8} W/m$^2 \cdot$ K^4	VIII
Mass of a proton	1.67×10^{-27} kg	
Mass of an electron	9.11×10^{-31} kg	
Solar constant	1390 W/m^2	
$L_\odot$ (solar luminosity)	8×10^{26} W = 8×10^{26} J/s	

Supplemental Aids in Studying Astronomy

Astronomical Calendar. Published annually by Guy Ottewell; sponsored by Department of Physics, Furman University, Greenville, SC 29613, (803) 294-2208. $15.00 each; discounts for volume purchases (1991 price including tax and postage). Star maps for each month; lists of astronomical events for each month; explanation of phenomena and various terms; astronomical glossary.

Astronomy Magazine. Published monthly; 21027 Crossroads Circle, Waukesha, WI 53186 (414) 796-8776. $24/year (12 issues; 1990 price). Popular articles on astronomical phenomena and observing techniques; astronomical paintings and photographs; monthly star charts.

Griffith Observer. Published monthly; Griffith Observatory, 2800 E. Observatory Rd., Los Angeles, CA 90027, (213) 664-1181. $12/year (12 issues; 1990 price). Popular articles about astronomy.

Mercury Magazine (Journal of the Astronomical Society of the Pacific). Published bimonthly; Astronomical Society of the Pacific, 390 Ashton Ave., San Francisco, CA 94112, (415) 337-1100. $29.50/year membership (6 issues; 1990 price). Popular articles about astronomy and the astronomical scene; news of the society, which encourages astronomy and holds occasional public functions.

Minor Planet Bulletin. Contact Derald D. Nye, 10385 East Observatory Drive, Tucson AZ 85747, for information. $7.00/year. Bulletin of the Minor Planet Section of the Association of Lunar and Planetary Observers with news of recent discoveries, predicted visibilities, and observing projects in which amateur observers can make scientific contributions. Mostly useful for advanced telescope observers.

The Planetary Report. Published bimonthly; 65 N. Catalina Ave., Pasadena, CA 91106, (818) 793-5100. Journal of a society started in 1980 by Carl Sagan and others to promote popular and congressional support for space exploration programs; news of NASA mission plans, recent discoveries about the solar system; recent spacecraft photos; paintings. Membership ($25, 1990 price) includes journal subscription.

Sky and Telescope Magazine. Published monthly; P.O. Box 9111, Belmont, MA 02178, (617) 864-7360. $24/year (12 issues; 1990 price). Popular and semi-technical articles on astronomical phenomena, history, telescope making, and observing techniques; photographs; monthly star charts.

Glossary

AU: (See *astronomical unit.*)

absolute: Intrinsic; not dependent on the position or distance of the observer.

absolute brightness: Any measure of the intrinsic brightness or luminosity of a celestial object.

absolute magnitude: The absolute brightness (luminosity) of a star expressed in the magnitude system. The Sun's absolute magnitude is +5.

absorption: The loss of photons as light passes through a medium. A photon is lost when it strikes an electron, and the photon's energy is consumed in knocking the electron to a higher energy level.

absorption band: A dark or dim region of the spectrum, caused by absorption of light over a moderate range of wavelengths, typically about 0.1 nm, usually by molecules or crystals.

absorption line: In a spectrum, a reduction in intensity in a narrow interval of wavelength, caused by absorption of the light by atoms between the source and the observer.

accretion disk: A disk of gas and dust, usually hot, surrounding a star. Usually used in the context of material that has been thrown off a nearby star and is falling into a companion star.

achondrite: A type of stony meteorite in which chondrules have been destroyed, probably by heating or melting.

age of the Earth: The period since the Earth's formation from planetesimals, estimated to be 4.5 billion years.

age of globular clusters: About 14 ± 4 billion years.

age of the (Milky Way) galaxy: Estimated to be roughly 14 ± 3 billion years.

age of open clusters: Time since formation of open clusters, judged by their H–R diagrams. Most are less than 100 million years. Ages from 1 million to a few billion years have been reported.

age of stars: Time since star formation, typically billions of years for smaller stars, but less than a million years for some massive stars in recently formed clusters. Age is difficult to measure for individual stars, but possible to measure for clusters of stars.

age of the universe: The time since the big bang, 10 to 18 billion years ago (probably close to 14 billion years ago).

airglow: Visible and infrared glow from the atmosphere produced when air molecules are excited by solar radiation.

Airy disk: In the telescopic image of a star, a small disk caused by optical effects.

Alexandrian library: The research institution and collection of ancient works preserved after the fall of Rome at Alexandria, Egypt. Alexandrian knowledge passed to the Arabs with the Arab conquest of Alexandria, and eventually back into Europe around A.D. 100–1500.

Alpha Centauri: (1) The nearest star system, composed of three members; (2) the brightest of these three.

amino acid: A complex organic molecule important in composing protein and called a "building block of life."

Andromeda galaxy: The nearest spiral galaxy comparable to our own, about 660 kpc away.

angular measure: Any measure of the size or separation of two objects as seen from a specified point, expressed in angular units (degrees, minutes of arc, or seconds of arc), but not linear units (such as kilometers, miles, or parsecs).

angular size: The angle subtended by an object at a given distance.

annular solar eclipse: An eclipse in which the light source is almost, but not quite, covered, leaving a thin ring of light at mid-eclipse.

aperture: The diameter of the light-gathering objective in a telescope.

apogee: The point in an orbit around the Earth that is farthest from the Earth.

Apollo asteroids: Asteroids that cross the Earth's orbit.

Apollo program: The U.S. program to land humans on the Moon, 1961–1972; first landing July 20, 1969.

apparent: Not intrinsic, but dependent on the position or distance of an observer.

apparent brightness: The brightness of an object as perceived by an observer at a specified location (but not measuring the object's intrinsic, or absolute, brightness).

apparent magnitude: Apparent brightness of one star relative to another as expressed in the magnitude system (see examples in Table 14-2).

apparent solar time: Time of day determined by the Sun's actual position in the sky. Apparent solar noon occurs as the Sun crosses the meridian. Apparent solar time is different at each different longitude.

apparition: The period of a few weeks during which a planet is most prominent or best placed for observation from Earth.

association: A loosely connected grouping of young stars.

asteroid: A rocky or metallic interplanetary body (usually larger than 100 m in diameter).

asteroid belt: The grouping of asteroids orbiting between Mars and Jupiter.

asthenosphere: In a planetary body, a subsurface layer that is more plastic than adjacent layers because the combination of pressure and temperature places it near (or slightly above) the melting point. Asthenospheric movements may disrupt the planet's surface.

astrology: The superstitious belief that human lives are influenced or controlled by the positions of planets and stars; this belief is rejected by modern astronomers and other scientists.

astrometric binary: A binary star system detectable from the orbital motion of a single visible component.

astrometry: The study of positions and motions of the stars.

astronomical unit (AU): The mean distance from the Earth to the Sun, about 150 million kilometers.

astronomy: The study of all matter and energy in the universe.

atom: A particle of matter composed of a nucleus surrounded by orbiting electrons.

aurora: Glowing, often moving colored light forms seen near the north and south magnetic poles of the Earth; caused by radiation from high-altitude air molecules excited by particles from the Sun and Van Allen belts.

B-type shell star: A star of spectral type B that occasionally blows off a cloud of gas, forming a gaseous shell around the star.

Balmer alpha line: A brilliant red spectral line at wavelength 656.3 nm, caused by transition of the electron in the hydrogen atom from the third-level orbit to the second-level orbit.

barred spiral galaxy: A spiral galaxy whose spiral arms attach to a barlike feature containing the nucleus.

basalt: A type of igneous rock, often formed in lava flows, common on the Moon and terrestrial planets.

basaltic rock: Igneous rock (including basalt) with a composition resembling basalt and a relatively low content of quartz (SiO_2).

basin: Large impact crater on a planet, usually several hundred kilometers across, flooded with basaltic lava and surrounded by concentric rings of faulted cliffs.

belts: Dark cloud bands on giant planets.

big bang theory: The theory that an explosionlike event that initiated the universe as we know it, probably between 10 and 18 billion years ago. Increasing evidence supports this theory.

binaries, classes of: Three categories of binary stars, depanding on whether neither, one, or both fill their Lagrangian lobes.

binary star system: A pair of coorbiting stars.

bipolar jetting: A phenomenon in which narrow streams of gas are ejected at very high speed in opposite directions from the centers of some disks of gas, perpendicular to the disk; bipolar jetting is seen both in accretion disks around individual stars and in galactic disks. The mechanism is uncertain.

black hole: An object whose surface gravity is so great that no radiation or matter can escape from it. Some black holes discussed in astronomy are collapsed stars, but much smaller ones are theoretically possible.

blue shift: A Doppler shift of spectral features toward shorter wavelengths, indicating approach of the source.

Bode's rule: A convenient memory aid for listing the planets' distances from the Sun.

body tide: A tidal bulge raised in the solid body of a planet.

Bok globule: A relatively small, dense, dark cloud of interstellar gas and dust, usually silhouetted against bright clouds.

bolometric luminosity: The total energy radiated by an object at all wavelengths, usually given in joules per second (identical to watts).

Tycho Brahe (1546–1601): Danish astronomer who recorded planetary positions, ultimately enabling Kepler to deduce laws of planetary orbits.

breccia: A rock made from angular fragments cemented together.

brecciated meteorite: A meteorite formed from cemented fragments of one or more meteorite types.

brown dwarf: A starlike object too small to achieve nuclear reactions in its center; any stellar object smaller than about $0.08\,M_\odot$.

bubble: A roughly spheroidal shell of interstellar gas blown outward from a star by a stellar explosion or strong stellar wind.

burster: A stellar object (neutron star?) that emits flashes of light lasting only seconds. Many emit gamma rays, but visible light bursters are also suspected.

Callisto: Outermost of the four large Moons of Jupiter, and most heavily cratered of the four.

Cambrian period: A period from 570 to 500 million years ago during which fossil-producing species of plants and animals first proliferated.

canals: Alleged straight-line markings on Mars found not to exist by spacecraft visits to Mars.

capture theory: A theory of origin of a planet–satellite or binary star system in which one body captures another body by gravity.

carbonaceous chondrite: A type of carbon-rich and volatile-rich meteorite, believed to be a nearly unaltered example of some of the earliest-formed matter in the solar system.

carbonaceous material: Black material rich in carbon and carbon compounds, found in carbonaceous chondrites and believed to color many black comets and asteroids of the outer solar system.

carbon cycle: A series of nuclear reactions in which hydrogen is converted to helium, releasing energy in stars more massive than about $1.5\ M_\odot$. Carbon is used as a catalyst.

Cassini's division: The most prominent gap in Saturn's rings.

catastrophic theory: A theory invoking sudden or very short (cosmically or geologically speaking) energetic events to explain observed phenomena.

catastrophism: An early scientific school which held that most features of nature formed in sudden events, or catastrophes, instead of by slow processes.

cause of eclipses: The falling of a shadow of one body onto another body.

cause of the seasons: The tilt (obliquity) of the Earth's axis to its orbit plane causes first the North Pole and later the South Pole to be tipped toward the Sun during the course of a year.

celestial equator: The projection of the Earth's equator onto the sky.

celestial poles: The projections of the two poles of the Earth's rotation onto the sky.

cell: The unit of structure in living matter.

center of mass: The imaginary point of any system or body at which all the mass could be concentrated without affecting the motion of the system as a whole; the balance point.

Cepheid variable: Any of a group of luminous variable stars with periods of 5 to 30 d (depending on their population). The periods are correlated with luminosity, allowing distance estimates out to about 3 mpc.

cgs system: Metric system of measurement using centimeters, grams, and seconds as the fundamental units.

chain reaction theory: A theory of the cause of spiral arms in spiral galaxies. Star formation in a region leads to a chain reaction of adjacent star formation, and the region of young stars gets sheared into a spiral arm.

Chandrasekhar limit: A mass of about $1.4\ M_\odot$, the maximum for white dwarfs; stars of greater mass have too great a central pressure, causing formation of a star type denser than a white dwarf.

channel: One of the river-bed-like valleys on Mars, which are possible sites of ancient Martian rivers.

chaotic rotation: A form of rotation in which dynamical forces cause the rotation period to change irregularly from one rotation to the next. Applies to Saturn's moon Hyperion.

Charon: Pluto's satellite.

chemical reaction: Reaction between elements or compounds in which electron structures are altered; atoms may be moved from one molecule to another, but nuclei are not changed and thus no element is changed to another.

chondrite: Stony meteorite containing chondrules, believed to be little altered since their formation 4.6 billion years ago.

chondrule: BB-sized spherule in certain stony meteorites, believed among the earliest-formed solid materials in the solar system.

chromosphere: A reddish-colored layer in the solar atmosphere, just above the photosphere.

circular velocity: Velocity of an object in circular orbit:

$$V = \sqrt{\frac{GM}{R}}$$

where G = Newton's gravitational constant. In SI units it is $6.67 \times 10^{-11}\ \text{N} \cdot \text{m}^2/\text{kg}^2$
M = mass of central body
R = distance of orbiter from center of central body

circumpolar zone: The zone of stars, centered on a celestial pole, that never sets, as seen from a given latitude.

circumstellar nebula: Gas and dust surrounding a star.

classes of binaries: See *binaries, classes of.*

closed space: Space that is curved in such a way that the total volume is finite.

closed universe: A theoretical model of the universe with finite volume and curved space. In most cosmological models, the universe, if closed, will eventually recollapse.

closure density: A critical density of matter in the universe, above which the universe would be closed.

cluster of galaxies: A relatively close grouping of galaxies, often with some members coorbiting or interacting with each other.

co-accretion theory: A class of theory in which two co-orbiting objects form by accreting at the same time, side by side.

coacervate: Cell-sized, nonliving globule of proteins and complex organic molecules that forms spontaneously in a water solution.

Coal Sack Nebula: A prominent dark nebula about 170 kpc away silhouetted against the Milky Way.

cocoon nebula: A dust-rich nebula enclosing and obscuring a star during its formation, but later shed.

collapse: Rapid contraction, especially of a cloud of gas and dust during star formation.

colliding galaxies: Galaxies undergoing interpenetration or close enough to cause major gravitational distortion of each other.

coma: (1) The diffuse part of the head of a comet surrounding the nucleus; (2) a type of distortion in some telescopes and optical systems.

comet: An ice-rich interplanetary body that, when heated by the Sun in the inner solar system, releases gases that form a bright head and diffuse tail. (See also *coma.*)

comet head: The coma and nucleus regions of a comet.

comet nucleus: The brightest starlike object near the center of a comet's head; the physical body (believed to be icy and a few kilometers across) within a comet.

comet tail: Diffuse streamers of gas and dust released from a comet and blown in the direction away from the Sun by the solar wind.

comparative planetology: An interdisciplinary field of astronomy and geology attempting to discover and explain differences between planets in properties such as climate and interior structure.

composition of Population II stars: About ¾ hydrogen and ¼ helium, by mass, with virtually no heavier elements.

compound telescope: A telescope that combines lenses and mirrors in the light-gathering system.

condensation sequence: The sequence in which chemical compounds condense to form solid grains in a cooling, dense nebula.

conditions in the galactic center: A dense concentration of stars (mostly Population II), dust, and gas, possibly with one or more large black holes marking the nucleus.

conduction: One of three processes that transfers heat from hot to cold regions; occurs as fast-moving molecules in the hot region agitate adjacent molecules.

conjunction: The period when a planet lies at zero or minimum angular distance from the Sun, as seen from the Earth.

conservation of angular momentum: A useful physical rule which states that the total angular momentum in an isolated system remains constant.

constellation: Imaginary pattern found among the stars, resembling animals, mythical heroes, and the like; different cultures map different constellations.

contact binary: A coorbiting pair of stars whose inner atmospheres or surfaces touch.

continental drift: The motion of continents due to (convective?) motion of underlying material in the Earth's mantle.

continental shield: A stable, ancient region, usually flat and oval-shaped, in a continent.

continuous creation theory: The hypothesis that matter is being created in interstellar or intergalactic space during the current era (and always).

continuum: In a spectrum with absorption or emission lines, the background continuous spectrum.

convection: One of three modes of transmission of heat (energy) from hot regions to cold regions; involves motions of masses of material.

Copernican revolution: The intellectual revolution associated with adopting Copernicus' model of the solar system, which displaced the Earth from the center of the universe.

core: The densest inner region of the Earth, probably of nickel–iron composition; in other planets, similar high-density central regions; in the Sun or stars, a dense central region where nuclear reactions occur; in galaxies, the densest, brightest central regions.

Coriolis drift: A departure from a straight-line trajectory, perceived by an observer in a rotating system; Coriolis effects in clouds were early evidence of the Earth's rotation.

corona: The outermost atmosphere of the Sun, having a temperature of about 1 to 2 million Kelvin.

coronograph: An instrument permitting direct observation of the solar corona without an eclipse.

cosmic fuels: Nonfossil energy sources provided by cosmic processes; for example, solar and geothermal energy.

cosmic rays: High-energy atomic particles (85% protons) that enter the Earth's atmosphere from space. Many may originate in supernovae and pulsars.

cosmogony: Any theory of the origin of the universe or one of its component systems, such as galaxies or the solar system.

cosmological principle: The assumption (unproven) that the universe is homogeneous and isotropic.

cosmological red shift: Any redward Doppler shift attributed to the mutual recession of galaxies or the expanding universe.

cosmology: The study of the structure of the universe. The term is often broadened to include the origin of the universe as well.

Cretaceous-Tertiary extinctions: Extinctions of many species 65 million years ago, probably triggered by the impact of an asteroid.

crust: The outermost, solid layer of a planet, with composition distinct from the mantle and differentiated by a seismic discontinuity.

cultural hurdle: The hypothetical survival requirement for a planetary culture between the time it achieves technology capable of quickly altering its planetary environment and the time it can establish viable bases off its planet; the uncertainty of the probability of survival affects our estimates of the probability of intelligent life elsewhere in space.

curved space: Space in which Euclidean solid geometry is not valid.

daughter isotope: An isotope resulting from radioactive disintegration of a parent isotope.

declination: Angular distance north or south of the celestial equator. (Abbreviation: Dec.)

degenerate matter: Matter in a very high density state in which electrons are freed from atoms and pressure is a function of density but not temperature.

degree: An angle equaling $\frac{1}{360}$ of a circle.

density–wave theory: The leading theory for explaining the formation of spiral arms in galaxies, which posits periodicities in star, dust, and gas motions.

deposition: The accumulation of eroded materials in one place.

desert: Any of the brighter regions of Mars.

differential rotation: The differences in speed for stars at different distances from the center of the galaxy. Orbital velocities are actually slower at 5000 pc from the center than at the Sun's distance, which is 8000 to 10 000 pc from the center.

differentiation: Any process that tends to separate different chemicals from their original mixed state and concentrate them in different regions.

diffraction: The slight bending of light rays as they pass edges, producing spurious rays and rings in telescopic images of stars.

dimensions of the galaxy: As early as 1935 astronomers agreed that the Sun is about 8000 to 10 000 pc from the center, and the overall diameter of our galaxy is about 30 000 pc.

Dione: (See Tethys, Dione, and Rhea)

dirty iceberg model: A theoretical description of a comet nucleus as a large icy body with bits of silicate "dirt" embedded in it.

disruption of open clusters: Gradual dispersion of stars from an open cluster as the cluster is sheared by differential galactic rotation, and as high-speed stars escape. Disruption time is usually a few hundred million years.

distance limit for reliable parallaxes: 20 parsecs.

Doppler shift: The shift in wavelength of light or sound as perceived by the observer of an approaching or receding body. For speeds well below that of light, the shift is given by the equation

$$\text{Original wavelength} \times \frac{\text{radial velocity}}{\text{velocity of light}}$$

Drake equation: The statement that the fraction of stars harboring intelligent life equals the number of all stars times a sequence of fractions, such as the fraction of all stars having planets, the fraction of planets that are habitable, and so on. Named after radio astronomer Frank Drake.

duration of element formation: For each element, the time interval during which it formed. Most atoms of light elements, like H, formed in the first minutes of the big bang, but heavy elements have been synthesized inside stars for billions of years.

dust trails: Toroidal lanes of dust stretching along elliptical orbits around the solar system; probably due to asteriod collisions or comet dust ejection.

dwarf elliptical galaxy: An ellipsoidal galaxy resembling a globular cluster but usually at least a few times larger.

Earth: The third planet from the Sun.

earthquake: Vibration or rolling motion of the Earth's surface accompanying the fracture of underground rock.

eclipse: An event in which the shadow of one body falls on another body.

eclipsing binary: A binary star system seen virtually edge-on so that the stars eclipse each other during each revolution.

eclipsing–spectroscopic binary: An eclipsing binary whose motions are measurable from spectral Doppler shifts. The most informative type of binary star.

ecliptic: (1) The plane of the Earth's orbit and its projection in the sky as seen from Earth; (2) approximately, the plane of the solar system.

effective temperature: Temperature of an object as calculated from the properties of the radiation it emits.

ejecta blanket: A layer of debris thrown out of a crater onto a planet's surface.

electromagnetic radiation: Light, radio waves, X rays, and other forms of radiation that propagate as disturbances in electric and magnetic fields, travel at the speed of light, and combine to make up the electromagnetic spectrum.

electron: Negatively charged particle orbiting around the atomic nucleus, with mass 9.1×10^{-31} kg.

element: A chemical material with a specified number of protons in the nucleus of each atom. Atoms with one proton are hydrogen; with two protons, helium; and so on.

ellipse: A closed, oval-shaped curve (generated by passing a plane through a cone) describing the shape of the orbit of one body around another.

emission: Release of electromagnetic radiation from matter.

emission band: Narrow wavelength interval in which molecules emit light.

emission line: Very narrow wavelength intervals in which atoms emit light.

Enceladus: One of the inner moons of Saturn, notable for its very bright, fissured surface of water ice, with several young, sparsely cratered areas.

energy: In physics, a specific quality equal to work or the ability to do work. Energy may appear in many forms, including electromagnetic radiation, heat, motion, and even mass (according to the theory of relativity).

energy level: The orbit of an electron around the nucleus of an atom.

English system: A nondecimal system of units using pounds, inches, and seconds, now being replaced by the more convenient metric system.

ephemeris: A table of predicted positions of a planet, asteroid, or other celestial body.

ephemeris time: A timekeeping system based on the motions of planets; more regular than conventional systems based on the Earth's rotation.

epicycle: A small circular motion superimposed on a larger circular motion.

epicycle theory: An early theory by Ptolemy that the planets move around the Earth in epicycles.

equatorial zone: On Jupiter, Saturn, and possibly other great planets, a bright cloud zone near the equator.

equinox: The date when the Sun passes through the Earth's equatorial plane (occurs twice annually).

erg: The unit of energy in the cgs metric system.

erosion: Removal of rock and soil by any natural process.

escape velocity: The minimum speed needed to allow a projectile to move away from a planet and never return to its point of launch. It equals $\sqrt{2}\ \times$ circular velocity. (See *circular velocity*.)

Eta Carinae Nebula: A nebula around a peculiar novalike variable star about 2 kpc away.

Europa: One of Jupiter's four large moons, notable for its smooth, bright, icy surface.

event horizon: The theoretical surface around a black hole from which matter and energy do not escape.

evidence for planets near other stars: Although planets the size of Jupiter or smaller near other stars are beyond our current detection capability, new techniques of imaging and astrometry will allow their detection within a few years, if they exist. Already some objects larger than Jupiter but smaller than stars have been found.

evidence for present-day star formation: The fact that the solar system is much younger than the galaxy; existence of young open clusters; existence of short-lived massive stars.

evidence of spiral structure (in Milky Way): Spiral distribution of hydrogen gas mapped by 21-cm radio line; spiral distribution of open clusters; spiral patterns observed in other galaxies.

evolution of galaxies: Changes in form and stellar populations of galaxies as a result of consuming gas and dust during star formation and production of heavy elements during star evolution.

evolutionary theory: A theory in which changes occur by relatively slow processes or processes commonly growing out of the initial conditions, rather than by sudden or unusual processes.

evolutionary track: The sequence of points on the H–R diagram occupied by a star as it evolves.

excitation: The process of causing an atom or molecule to go into an excited state, that is, having some electrons in elevated energy levels; the state of being excited.

excited atoms and molecules: Atoms or molecules in which electrons are not all in the lowest possible energy levels.

excited state: The state of an atom or molecule when not all electrons are in the lowest possible energy levels.

exobiology: Study of life beyond the Earth.

expanding universe: A term popularized by Eddington to describe the mutual recession of galaxies.

explorative interval: The hypothetical interval of time during which a species actively engages in the exploration of other planets.

extragalactic standard of rest: An assumed stationary frame of reference defined by using the nearby galaxies as reference objects.

fault: A fracture along which displacement has occurred on the solid surface of a planet or other celestial body.

fields: Entities dispersed in space but having a measurable value or magnitude that can be measured at any point in space. Examples are gravitational, electric, and magnetic fields.

fireball: An unusually bright meteor, which may yield meteorites.

first quarter: The phase of the Moon when it is one-fourth of the way around its orbit from new moon; the first quarter moon is seen in the evening sky with a straight terminator and half the disk illuminated.

fission theory: A theory of origin of a planet–satellite or binary star system by breakup of a single original body.

flare: (1) On the Sun, a sudden, short-lived, localized outburst of energy, often ejecting gas at speeds exceeding 1000 km/s; (2) an outburst from certain types of variable stars, sometimes called *flare stars*.

focal length: The distance from a lens or mirror to the point where it focuses the image of a very distant object, such as the Moon.

focus: One of the two interior points around which planets or stars move in an elliptical orbit.

forbidden line: Spectral line arising from a metastable state in atoms.

force: In physics, a specific phenomenon producing acceleration of mass. Forces can be generated in many ways, such as by gravity, pressure, and radiation.

formation of galaxies: Processes that led to subdivision of the universe's gas after the big bang and its collapse into individual galaxies.

free-fall: Motion under the influence of gravity only, without any other force or acceleration, such as rocket firing.

free-fall contraction: Contraction of a cloud or system of particles by gravity only, unresisted by any other force.

frequency: Number of electromagnetic oscillations per second corresponding to electromagnetic radiation of any given wavelength.

full moon: The phase of the Moon when it is closest to 180° from the Sun and therefore fully illuminated.

funneling effect: Concentration of all evolutionary tracks into the giant region of the H–R diagram.

galactic cannibalism: The absorption of one galaxy by another during a collision, forming a new, larger galaxy.

galactic equator: The plane of the Milky Way galaxy projected on the sky.

galactic halo: A spherical swarm of globular clusters above and below the galactic disk, centered on a point in the direction of the constellation Sagittarius.

galactic jets: Narrow beams of extremely hot, ionized gas shooting out of some active galaxies in two opposite directions, above and below the plane of the disk. They radiate enormous amounts of energy and their origin is uncertain.

galactic latitude: Angular distance around the galactic equator from the galaxy's center.

galactic longitude: Angular distance from the galactic equator.

galactic nucleus: The center of a galaxy.

galaxies with active nuclei: Galaxies whose centers emit more energy than other, normal galaxies. They are generally strong radio sources.

galaxy: Any of the largest groupings of stars, usually of mass 10^8 to 10^{13} $M_\odot$.

Galilean satellites: The four large satellites of Jupiter, discovered by Galileo.

Galileo Galilei (1564–1642): Italian scientist who first applied the telescope to observe other planets, discovering lunar craters, Jupiter's moons, and other celestial phenomena.

Galileo program: A NASA mission to put an unmanned probe in orbit around Jupiter in the 1990s.

Ganymede: Largest of the four Galilean satellites of Jupiter, with a fractured and cratered ice surface.

geological time scale: The sequence of events in the history of the Earth.

giant elliptical galaxy: Galaxies with diffuse elliptical form, somewhat resembling globular clusters but much larger.

giant planet: (1) Jupiter, Saturn, Uranus, or Neptune; (2) any planet much more massive than the Earth.

giant star: Highly luminous star larger than the Sun. O and B main-sequence stars are sometimes called *blue giants*; evolved stars of extremely large radius are called *red giants*.

gibbous: A phase between half-illuminated and fully illuminated, with a convex terminator (pronounced with hard g, as in "give").

globular cluster X radiation: X-ray radiation from globular clusters, discovered unexpectedly in the 1970s; it indicates energetic environments somewhere within them.

globular star cluster: A dense spheroidal cluster of stars, usually old, with mass of 10^4 to 10^6 $M_\odot$.

grain: Small (usually microscopic) solid particle in space.

granite: A rock type of modest density and high silica content, formed in association with differentiation processes and, therefore, found primarily on the Earth.

granitic rock: A silica-rich, light-colored rock type common in the Earth's continents. Granites, being low in density, tend to float to the surfaces of planets that have had extensive melting in the outer layers.

granules: Convection cells 1000 to 2000 km across, rising from the subphotospheric layers of the Sun. Each granule rises at a speed of 2 to 3 km/s and lasts for a few minutes.

gravitational contraction: Slow contraction of a cloud, star, or planet due to gravity, causing heat and radiation.

gravitational lensing: The creation of a distorted image of a distant quasar or galaxy when its light is focused by the gravity of a galaxy between it and us.

gravitational red shift: A red shift caused by light emitted from a region of very high gravity, such as a neutron star or the environs of a black hole.

gravity: The force by which all masses attract all other masses.

great circle: Any circle on the surface of a sphere (especially the Earth or sky) generated by a plane passing through the center of the sphere; the shortest distance between two points on a sphere.

Great Red Spot: A large, reddish, oval, semipermanent cloud formation on Jupiter.

greenhouse effect: Heating of an atmosphere by absorption of outgoing infrared radiation.

Gregorian calendar: Essentially the modern calendar system, introduced around A.D. 1600 under Pope Gregory XIII, and containing the modern system of reckoning leap years.

ground state: The lowest energy state of an atom, in which all electrons are in the lowest possible energy levels.

Gum Nebula: A large, relatively nearby nebula in the Southern Hemisphere sky, detected in hydrogen alpha light and formed by a supernova explosion estimated to have occurred around 9000 B.C.

HI region: Interstellar region in which hydrogen is predominantly neutral.

HII region: Interstellar region in which hydrogen is predominantly ionized.

half-life: In any phenomenon, the time during which the main variable changes by half its original value; often used loosely to indicate the characteristic time scale of a phenomenon. In radioactive decay, the time for half the atoms in a system to disintegrate.

Halley's comet: The most famous comet, which visits the inner solar system every 76 years, most recently in 1986 when close-up photos were made by space probes.

Hayashi track: A sharply descending evolutionary track in the H–R diagram covering the early period of stellar evolution from the high-luminosity phase to the main sequence.

heliacal rising: A star's first visible rising during the yearly cycle.

heliacal setting: A star's last visible setting during the yearly cycle.

helium flash: Runaway helium "burning" inside a star as it evolves off the main sequence and into the giant phase of evolution. It occurs when degenerate gas at the star's center reaches a temperature of about 10^8 K.

Helmholtz contraction: Slow contraction of a cloud or system of particles by the force of gravity, which is retarded by outward gas pressure and the limited rate at which radiation can escape.

hierarchical universe: A theoretical model of the universe organized in ever-larger clusters of galaxies, plausibly with a density that approaches zero as larger volumes are considered.

high-luminosity phase: A star's short-lived stage of maximum brightness during pre-main-sequence evolution.

high-velocity star: A star with a high velocity relative to the Sun; generally associated with the galactic halo.

homogeneous: Uniform in composition throughout the volume considered.

hour angle: The number of hours since a star (or other body) last crossed the local meridian.

H–R diagram: A technique for representing stellar data by plotting spectral type (or color or temperature) against luminosity (or absolute magnitude), named after its early proponents, Hertzsprung and Russell.

Hubble Space Telescope: A large telescope with 2.4-m (94-in.) mirror originally designed to be orbited by the Space Shuttle in 1985. Due to the Challenger disaster, its launch has been delayed.

Hubble's constant: The ratio between a galaxy's recession speed and its distance, measured to be about 50 to 100 km/s/mpc.

Hubble's relation: Expression indicating how the recession velocity of a galaxy increases with distance.

hydrogen alpha line: The designation of hydrogen's red spectral line at 656.3 nm, more properly called *hydrogen Balmer alpha.*

hydrogen Balmer series: The series of all hydrogen spectral lines from 364.6 to 656.3 nm, caused by electron transitions between the second and higher energy levels.

hyperbola: The orbital curve followed by any free-falling body moving faster than escape velocity.

Hyperion: An irregularly shaped outer moon of Saturn, noted for its chaotic rotation.

hypothesis: A proposed explanation of an observed phenomenon or a proposal that a certain observable phenomenon occurs.

Iapetus: An outer moon of Saturn noteworthy because one hemisphere has a bright, icy surface, and the other, a black carbonaceous surface.

IC number: The catalog number of a cluster, nebula, or galaxy in the *Index Catalog.*

igneous rock: Rock crystallized from molten material.

impact crater: A roughly circular depression of any size (known examples range from microscopic size to diameters greater than 1000 km) caused by a meteorite impact.

impact-trigger theory: The leading theory of the moon's origin, in which material was blasted off Earth's mantle and then reaccumulated to form the moon.

incidence of multiplicity: Among stars, the fraction of systems that contain more than one star. Probably 50 to 70% of systems have companion stars, many of these having more than one companion.

inferior planet: Mercury or Venus.

inflation: The period of extremely rapid expansion of the scale of space in the first fraction of a second after the big bang.

inflationary big-bang theory: Model of the big bang calling for a super-rapid expansion of space in the first fraction of a second after the big bang.

infrared light: Radiation of wavelength too long to see, usually about 1 to 100 μm.

infrared star: A star detected primarily by infrared light.

initial conditions: (1) In any scientific problem, the conditions that define the beginning of the process being studied; (2) in cosmogony, the conditions at the moment of, or immediately after, the big bang.

intense early bombardment: The very intensive bombardment of planetary bodies by meteorites, from 4.6 to about 4 billion years ago, following plant formation.

interferometry: A system for obtaining high-resolution astronomical observations by linking several physically separated telescopes electronically, in effect creating a single, much larger telescope.

intergalactic globular cluster: A globular star cluster or a small galaxy resembling same, found in intergalactic space.

interstellar atom: Atom of gas in interstellar space.

interstellar grain: Microscopic solid grain in interstellar space; interstellar dust.

interstellar molecule: Molecule of gas in interstellar space.

interstellar obscuration: Absorption of starlight by interstellar dust, causing distant objects to appear fainter.

interstellar reddening: Loss of blue starlight due to interstellar dust, causing distant objects to appear redder and fainter.

interstellar snowball: Hypothetical interstellar particle larger than an interstellar grain.

inverse square law: The relation describing any entity, such as radiation or gravity, that varies as $1/r^2$, where r is the distance of the entity from the source.

Io: The innermost Galilean moon of Jupiter, famous for its active volcanism, which is unique among moons.

ion: Charged atom or molecule.

ionization: The process of knocking one or more electrons off a neutral atom or molecule.

iron meteorite: Meteorite composed of a nearly pure nickel–iron alloy.

irregular galaxy: A galaxy of amorphous shape. Most have relatively low mass (10^8–10^{10} $M_\odot$).

irregular variable: A star that fluctuates in brightness irregularly.

isotope: A form of an element with a specified number of neutrons in the nucleus. Each element may have many possible isotopic forms, but only a few are stable.

isotropic: Appearing uniform no matter what the direction of view.

joule: The unit of energy in the SI metric system of units.

Julian date: The date based on a running tabulation of days, starting January 1, 4713 B.C.

Jupiter's atmospheric composition: Mostly hydrogen, with additional helium, hydrogen-based compounds, and other gases.

Jupiter's infrared thermal radiation: Infrared radiation from Jupiter due to slow contraction of its interior and exceeding the incoming solar radiation.

Jupiter's interior: Beneath Jupiter's atmosphere, a high-pressure region of liquid hydrogen, liquid metallic hydrogen, and a small central core of rocky material.

Jupiter's temperature: Temperatures around $-200°F$ are measured at and above the cloud tops in Jupiter's atmosphere, but the air temperature increases below the clouds to values around room temperature and even warmer.

Kelvin scale: The absolute temperature scale, with 0 K = absolute zero. A Kelvin degree is the same size (some temperature difference) as a centigrade degree.

Johannes Kepler (1571–1630): Astronomer who first deduced the shapes and relations of planets' elliptical orbits.

Kepler's laws: The three laws of planetary motion that describe how the planets move, show that the Sun is the central body, and allow accurate prediction of planetary positions.

kiloparsec (kpc): 1000 parsecs.

Kirchhoff's laws: Laws describing conditions that produce emission, absorption, and continuous spectra.

$L_\odot$: The luminosity of the Sun, 4×10^{26} watts.

LSR: (See *local standard of rest*.)

Lagrangian points: In an orbiting system with one large and one small body, an array of five points where a still smaller body would retain a fixed position with respect to the other two.

Lagrangian surface: An imaginary surface with a figure-8 cross section surrounding two coorbiting bodies in circular orbits and constraining motions of particles within the system.

lava: Molten rock on the surface of a planet.

life: A process in which complex, carbon-based materials organized in cells take in additional material from their environment, replicate molecules, reproduce, and do other weird things like writing books.

life in the solar system: Hypothetical biological activity on other planetary bodies, such as on Mars, under Europa's ice, or in Jupiter's atmosphere. Now regarded as unlikely.

life on Mars: Although long sought and believed possible by many scientists, biological processes on Mars were apparently ruled out in 1976 when Viking landers found no organic material there.

life outside the solar system: Hypothetical biological activity beyond the outskirts of our solar system, as on hypothetical planets around other stars. Such life is regarded as plausible, even if sparsely scattered, by most astronomers.

light curve: A plot of brightness of a star (or other object) versus time.

light-gathering power: The ability of a telescope or binoculars to gather light. It is proportional to the area of the objective, that is, the square of the aperture.

light-year: The distance light travels in one year, 9.46×10^{12} km.

limb: The apparent edge of a celestial object.

line of nodes: A line formed by the intersection of an orbit and some other reference plane, such as the plane of the solar system.

linear measure: Measurement involving linear distances, as opposed to angles or angular distances.

lithosphere: The solid rocky layer in a partially molten planet.

Local Group: The cluster of galaxies to which the Milky Way and 26 nearby galaxies belong.

local standard of rest (LSR): A frame of reference moving with the average velocity of the nearby stars (out to about 50 pc from the Sun).

luminosity: The total energy radiated by a source per second. The luminosity of the Sun ($L_\odot$) is 4×20^{26} watts.

lunar eclipse: Dimming of the Moon as it passes into the Earth's shadow.

$M_\odot$: The mass of the Sun, 2×10^{30} kg.

M giant: A giant star of spectral class M.

Magellanic clouds: The two galaxies nearest the Milky Way, irregular in form and visible to the naked eye in the Southern Hemisphere.

Magellanic stream: Gas filaments connecting the Magellanic clouds to the Milky Way.

magma: Underground molten rock.

magma ocean: Primordial layer of molten lava on the initial surface of the Moon and (by inference) planets.

magnetic braking: The slowing of rotation of a star or planet by interaction of its magnetic field with surrounding ionized material.

magnetic field: Region of space in which a compass (or other detector) would respond to magnetism of some body.

magnification: Apparent angular size of a telescope image divided by the angular size of the object seen by naked eye.

main-sequence star: One of the group of stars defined on the H–R diagram that have a relatively stable interior configuration and are consuming hydrogen in nuclear reactions; a star on the main sequence.

mantle: A region of intermediate density surrounding the core of planets.

mare (pl. *maria*): A dark-colored region on a planet or satellite; a region of basaltic lava flow on the Moon.

Mars-crossing asteroids: Asteroids whose orbits cross that of Mars.

Mars' rotation period: The day on Mars—$24^h\ 37^m$, only slightly longer than the Earth's.

Martian air pressure: The pressure exerted by the very thin Martian air. On Mars' surface, the air pressure is only about 0.7% of that at the Earth's surface.

Martian air temperature: Soil temperatures approach or exceed freezing in the day; the air is colder. Night temperatures around −123°F are recorded.

Martian atmosphere: The thin gases around Mars, composed almost entirely of carbon dioxide (CO_2).

Martian meteorites: A handful of meteorites about 1.3 billion years old, believed to have been blasted off Mars about 0.2 billion years ago.

maser (*microwave amplification by simulated emission of radiation*): (1) A device that amplifies microwave radio waves through special electronic transitions in atoms; (2) an interstellar cloud that acts in this way.

mass: (1) Material; (2) the amount of material.

mass/luminosity ratio: The mass per unit of light or total radiation emitted from an object such as a galaxy.

mass–luminosity relation: The relation between the mass of a main-sequence star and its total radiation rate; the more massive, the greater the luminosity.

mass of the galaxy: About 4×10^{41} kg, or $2 \times 10^{11}\ M_\odot$, as estimated from the circular velocity equation.

Maunder minimum: The interval from 1645 to 1715, when solar activity was minimal.

mean density: Mass of an object divided by its volume.

mean solar time: Time shown by conventional clocks, determined by the Sun's mean rate averaged over the year.

megaparsec: One million parsecs.

meridian: (1) A north–south line on a planet, moon, or star; (2) a great circle through the celestial pole and the zenith.

Messier number: The catalog number of a nebula, star cluster, or galaxy in *Messier's Catalog*.

metallic hydrogen: A high-pressure form of hydrogen with free electrons.

metastable state: In an atom, a configuration of electrons that is relatively long-lived, but is rarely found on the Earth because it is disrupted by collisions with other atoms; it may be found in interstellar atoms, creating forbidden spectral lines.

meteor: A rapidly moving luminous object visible for a few seconds in the night sky (a "shooting star").

meteorite: An interplanetary rock or metal object that strikes the ground.

meteorite impact crater: Circular depression in planetary surfaces, caused by explosions as meteorites crash into the surfaces at high speeds.

meteoroid: A particle in space, generally smaller than a few meters across.

meteor shower: A concentrated group of meteors, seen when the Earth's orbit intersects debris from a comet.

meter: 39.4 in.

Milky Way galaxy: The spiral galaxy in which we live.

Miller experiment: An experiment in which amino acids are created in laboratory conditions simulating the conditions of the early Earth.

Mimas: The innermost large moon of Saturn, icy and heavily cratered.

minerals: Chemical compounds, usually in the form of crystals, that constitute rocks.

minute of arc: An angle equaling 1/60 of a degree.

Miranda: The innermost of five large moons of Uranus, noted for puzzling fractured and grooved terrain.

missing mass: Mass suspected to exist distributed through the universe (in intergalactic space?) that, when added to the currently observed mass, would bring the universe closer to closure density.

mks system: A metric system of units expressing length in meters, mass in kilograms, and time in seconds. (See also *cgs system*.)

molecular cloud: An interstellar cloud of gas and dust, with greater than average density, dust content, and high concentration of molecules.

Moon: (1) The Earth's natural satellite; (2) any satellite.

multiple scattering: Redirection of electromagnetic radiation (such as light waves) by repeated interaction with atoms, molecules, or dust grains in space or in an atmosphere.

multiple star system: A system of three or more stars orbiting around each other.

mutual recession of galaxies: The phenomenon that all distant galaxies are moving away from us, and the farther away they are, the faster they are receding.

natural selection: The theory that states that those individuals best adapted to the ever-changing environment produce a greater number of offspring.

nebula: A cloud of dense gas and/or dust in interstellar space or surrounding a star.

negative hydrogen ions: Hydrogen atoms that have temporarily captured an extra electron, responsible for opacity in the photosphere of the Sun and many stars.

Neptune: The outermost gas giant plant in the outer solar system.

neutrino: A subatomic particle created in certain nuclear reactions inside stars. It can pass through most matter. Its mass is uncertain, being either zero or a tiny fraction of an electron's mass.

neutron: One of the two major particles constituting the atomic nucleus; it has zero charge and mass 1.6749×10^{-27} kg.

neutron star: A star with a core composed mostly of neutrons, with density 10^{16} to 10^{18} kg/m^3. Many or most neutron stars are pulsars.

new moon: The phase of the moon when it is nearest the Earth–Sun line, hence invisible from Earth because of the Sun's glare.

Isaac Newton (1642–1727): English physicist who discovered the spectrum and laws of gravitation and motion; he also developed calculus and made other discoveries. Possibly the greatest physicist of history.

Newtonian–Euclidean static cosmology: A hypothetical model of the universe with infinite volume, no expansion, and Euclidean geometry.

Newton's laws of motion: Three rules describing motion and forces. Briefly, (1) a body remains in its state of motion unless a force acts on it; (2) force equals mass times acceleration; (3) for every action there is an equal and opposite reaction.

NGC number: The catalog number of a nebula, cluster, or galaxy in the *New General Catalog*.

node: One of two points where an orbit crosses a reference plane.

noncosmological red shift: A hypothetical red shift of distant galaxies *not* caused by the Doppler effect.

non-Euclidean geometry: A hypothetical geometry in which Euclid's relations are not true; a geometry of curved space.

nonthermal radiation: Radiation *not* due to the heat of the source; for example, synchrotron radiation.

North Star: (1) Polaris; (2) any bright star that happens to be within a few degrees of the north celestial pole during a given era.

nova: A type of suddenly brightening star (from the Latin for "new") resulting from explosive brightening when gas is dumped from one member of a binary star pair onto the other.

nuclear reaction: Reaction involving the nuclei of atoms in which a nucleus changes mass.

nuclear winter: A sudden drop of temperature on Earth hypothesized to occur if extensive nuclear explosions (or an asteroid or comet impact) ejected enough dust and smoke into the atmosphere to block sunlight.

nucleus: (1) The matter at the center of an atom, composed of protons and neutrons; (2) the bright central core (or solid body) of a comet; (3) the bright central core of a galaxy.

O association: An association of O-type stars.

objective: The major light-gathering element of a telescope; the mirror in a reflector and the lens in a refractor.

oblate spheroid: The shape assumed by a sphere deformed by rotation.

obliquity: The angle by which a planet's rotation axis is tipped to its orbit.

Occam's razor: The principle that the simplest hypothesis, with the fewest assumptions, is most likely to be correct. Named after its use to cut away false hypotheses.

ocean tide: The Moon's tidal stretching of the Earth, as observed in the ocean surface (as opposed to body tide).

Olbers' paradox: The problem of why the sky is dark at night if the universe is filled with stars.

Oort cloud: The swarm of comets surrounding the solar system.

opacity: The extent to which gaseous (or other) material absorbs light.

open space: Space that is uncurved or curved in such a way as to have infinite volume and no boundaries.

open star cluster: A grouping of relatively young Population I stars (usually 10^2–10^3 $M_\odot$), sometimes called a *galactic cluster*.

open universe: A universe with infinite volume and no boundaries. In most cosmological models, the universe, if open, will continue expanding forever.

opposition: The period when a superior planet lies in an opposite direction from the Sun as seen from the Earth. At opposition, a planet appears in the midnight sky, well placed for observation.

optical double star: A pair of stars that have small angular separation but are not coorbiting.

orbital precession: A slow, cyclical change in the orientation of the plane of an orbit.

organic chemistry: Chemistry involving organic molecules.

organic molecule: Molecule based on the carbon atom, usually large and complex, but not necessarily part of living organisms.

origin of the heavy elements: The process by which light elements fused to form nuclei of heavy elements, primarily inside massive stars.

Orion Nebula: Several hundred solar masses of gas and dust composing the core of a star-forming region about 460 parsecs away. One of the most prominent nebulae in our sky, it forms the central "star" in Orion's sword.

Orion star-forming region: The larger region in which stars are forming around the Orion Nebula.

oscillating universe: A hypothetical model of the universe with cycles of contraction and expansion.

ozone layer: An atmospheric layer rich in ozone (O_3), created by the interaction between oxygen molecules (O_2) and solar radiation. On Earth, its altitude is about 20–60 km.

parabola: (1) The curved trajectory followed by a particle moving at escape velocity; (2) the curve of a Newtonian telescope's primary mirror.

parallax: An angular shift in apparent position due to an observer's motion; more specifically, a small angular shift in a star's apparent position due to the Earth's motion around the Sun. *Stellar parallax*, used to measure stellar distance, is defined as the angle subtended by the radius of the Earth's orbit as seen from the star.

parent body: A body from which a meteorite formed and later broke off as a fragment.

parent isotope: A radioactive isotope that disintegrates and forms a daughter isotope.

parsec: A distance of 206 265 AU, 3.26 ly, or 3.09×10^{13} km; defined as the distance corresponding to a parallax of 1 second of arc.

partial solar eclipse: An eclipse in which the light source is not totally obscured from an observer.

particlelike properties of light: Characteristics of light, such as concentration of energy and momentum in discrete microscopic packets (photons) that mimic the properties of particles.

Pauli exclusion principle: A principle of subatomic physics specifying that no two electrons in a very small volume have exactly the same properties of energy, motion, and so on.

peculiar velocity: A star's velocity with respect to the local standard of rest.

penumbra: (1) The outer, brighter part of a shadow, from which the light source is not totally obscured; (2) the outer, lighter part of a sunspot.

perigee: The point in an orbit around the Earth that is closest to the Earth.

permafrost: Semipermanent underground ice.

phase: The apparent shape of an illuminated body, varying with the "phase angle" from observer to body to illumination source.

Phoebe: The outermost moon of Saturn, a small dark moon believed to be captured.

photometry: The measurement of the amount of light, either total or in different specified colors, coming from an object.

photon: The quantum unit of light, having some properties of a wave. For each wavelength of radiation, the photon has a different energy.

photosphere: The light-emitting surface layer of the Sun.

physical binary stars: Two stars orbiting around a common center of mass.

Planck's law: A formula that describes the energy associated with each wavelength (color) in the spectrum. It shows that photons of blue light are more energetic than photons of red light.

planet: A solid (or partially liquid) body orbiting around a star but too small to generate energy by nuclear reactions.

"Planet X": A term sometimes used for a hypothetical tenth planet in the solar system, beyond Pluto.

planetary nebula: A type of circumstellar gas cloud that has spheroidal shape and often appears as a faint disk in telescopes; it has nothing to do with planets except for the rough resemblance to a planet's shape when seen telescopically.

planetesimal: One of the small bodies from which planets formed, usually ranging from micrometers to kilometers in diameter.

planetology: The study of the planets' origins, evolution, and conditions.

plasma: A high-temperature gas consisting entirely of ions, instead of neutral atoms or molecules. Because of the high temperature, the atoms strike each other hard enough to keep at least the outer electrons knocked off.

plate: Moving unit of the Earth's lithosphere, typically of continental scale.

plate tectonics: Motions of a planet's lithosphere, causing fracturing of the surface into plates. Primary example occurs on the Earth.

Pluto: Cataloged as the outermost and smallest planet in the solar system, but possibly one of many small worldlets on the fringe of the solar system.

Polaris: The North Star.

Population I: Stars with a few percent heavy elements (heavier than helium), found in the disks of spiral galaxies and in irregular galaxies.

Population II: Stars composed of nearly pure hydrogen and helium, found in the halo and center of spiral galaxies, in elliptical galaxies, and to a limited extent in irregulars.

powers of 10: The number of times 10s must be multiplied together to give a specific number; the exponent of 10. (Example: $10^2 = 100$; the power, or exponent, is 2; see also Appendix 1.)

precession: The wobble in the position of a planet's rotation axis caused by external forces. Also, the change in a coordinate system (tied to any planet) caused by such a wobble.

pre-main-sequence star: Evolutionary state of stars prior to arrival on the main sequence, especially just before the main sequence is reached.

primary atmosphere: The atmosphere of a planet (if any) just after planet formation. (See also *secondary atmosphere*.)

primeval fireball: The hypothetical, expanding, initial cloud of high temperature plasma during the big bang.

principle of relativity: The principle that observers can measure only relative motions, since there is no absolute frame of reference in the universe by which to specify absolute motions.

problem of the missing mass: Motions of stars around galaxies imply that these galaxies have more mass than accounted for by the visible stars. Thus there must be unseen mass in some uncertain form. Astronomers are still searching for this "missing" mass.

prograde and retrograde satellite orbits: Satellite orbits in which motion is in the same or opposite direction, respectively, as the planet's rotation.

prograde rotation: Spinning on an axis from west to east, or counterclockwise as seen from the North Pole (as in the case of the Earth).

prominence: A radiating gas cloud extending from the solar surface into the thinner corona.

prominent star: One of the brightest stars in our sky, but not necessarily one of the nearest.

proper motion: The angular rate of motion of a star or other object across the sky. (Most stars have proper motions less than a few seconds of arc per year.)

protein: Any of several types of complex organic molecules made from amino acids inside plants and animals, which are essential in living organisms.

proteinoid: Cell-like, nonliving spheroid of protein molecules created in the laboratory by heating amino acids and adding water; a possible step in the evolution of life.

protogalaxy: A gravitationally stable cloud of galactic mass contracting toward galactic dimension.

proton: One of the two basic particles in an atomic nucleus, with positive charge and mass $1.6726 = 10^{-27}$ kg.

proton–proton chain: A series of thermonuclear reactions that convert hydrogen nuclei to helium nuclei, converting a tiny amount of mass into energy.

protoplanet: A planet shortly before its final formation. Sometimes hypothesized to have a massive atmosphere and greater mass than in its present state.

protostar: A gravitationally stable cloud of stellar mass contracting in an early pre-main-sequence evolutionary state.

pseudoscience: Research that has the trappings of science but does not follow the scientific method, usually lacking review and repetition of observations by independent researchers.

Ptolemaic model: The ancient Earth-centered model of the solar system, with the Sun, Moon, and other planets moving in epicycles.

pulsar: A rapidly rotating neutron star with a strong magnetic field, observed to emit pulses of radiation.

QSO: Quasi-stellar object; faint bluish star with a spectrum similar to that of a quasar.

quantum: A small indivisible unit of some quantity such as energy or mass.

quasar: Any of a group of starlike, faint celestial objects with very large red shifts. Many astronomers believe they are extremely distant galaxies of unusually energetic form.

r-process reactions: Rapid reactions, probably occurring inside supernovae, in which heavy elements are formed as atomic nuclei capture neutrons. (See also *s-process reactions*.)

radial velocity: The velocity component along the line of sight toward or away from an observer. Recession is positive; approach is negative.

radiation: (1) Any electromagnetic waves or atomic particles that transmit energy across space; (2) one of three modes of heat (energy) transmission through stars or planets from warm regions to cool regions.

radiation pressure: An outward pressure on small particles exerted by electromagnetic radiation in a direction away from the light source.

radioactive atom: Any atom whose nucleus spontaneously disintegrates.

radio galaxy: A galaxy that emits unusually large amounts of radio radiation.

radioisotopic dating: Dating of rock or other material by measuring amounts of parent and daughter isotopes.

ray: A bright streak of material ejected from a crater on the Moon or other planet.

Rayleigh scattering: Scattering of light by particles smaller than the light's wavelength. This process favors scattering of blue light.

red giant: A post-main-sequence star whose surface layers have expanded to many solar radii and have relatively low temperatures.

red shift: A Doppler shift of spectral features toward longer wavelengths, indicating recession of the source.

red shift of galaxies: The shift toward longer wavelengths in light of distant galaxies, due to their recession from the solar system. It increases with galaxies' distances.

reflected radiation: Radiation that has arrived from outside a body and bounced off, as opposed to thermal radiation.

reflector: A type of telescope using a mirror as the light collector.

refractor: A type of telescope using a lens as the light collector.

refractory element: An element least likely to be driven out of a material by heating. These elements are usually concentrated in the last components to melt when a material such as rock is heated.

regolith: A powdery soil layer on the Moon and some other bodies caused by meteorite bombardment.

regression of nodes: A shifting of the R.A. = Dec. coordinate system, relative to the stars, as a result of the 26 000-y wobble of the Earth's rotation axis.

relativistic: Moving at speeds near that of light.

relativity: (See *principle of relativity.*)

representative stars: A sample of stars randomly drawn from the total population of stars in space.

resolution: The smallest angle that can be discerned with an optical system; for example, the eye can resolve about 2 minutes of arc.

retrograde motion: Revolution or rotation from east to west contrary to the usual motion in the solar system.

retrograde rotation: Spinning on an axis from east to west, or clockwise as seen from the North Pole (opposite to the spin direction of Earth).

Rhea: (See Tethys, Dione, and Rhea.)

rift: A major split in a planet's lithosphere due to active or incipient plate tectonic stresses.

right ascension: Longitude lines projected onto the celestial sphere. (Abbreviation: R.A.)

rille: A type of lunar valley.

Roche's limit: The distance from a large body within which tidal forces would disrupt a satellite.

rock: Solid aggregation of minerals.

rotation curve: Orbital velocity as a function of distance from the center of a galaxy.

rotational line broadening: Broadening of spectral lines due to rotation of the source.

RR Lyrae star: A type of variable star similar to the Cepheids that has been found associated with Population II and not Population I.

runaway star: A star rapidly moving away from a region of recent star formation.

Russell–Vogt theorem: The theorem stating that the equilibrium structure of a star is determined by its mass and chemical composition.

s-process reactions: Slow reactions in giant stars in which heavy elements are built up as atomic nuclei capture neutrons. (See also *r-process reactions.*)

Saha equation: An equation derived in 1920, by the Indian physicist Saha, that tells the percent of atoms in each different excited state, given the conditions in a gas. This in turn controls what spectral lines are emitted or absorbed by that gas.

saros cycle: An interval of 6585 d (about 18 y) separating cycles of similar eclipses, used by ancient people to predict eclipses.

satellite: Any small body orbiting a larger body.

Saturn: The sixth planet out from the Sun, famous for its prominent rings.

Saturn's atmosphere: The thick, cloudy gases around Saturn, composed mostly of hydrogen.

Saturn's ring system: A system of innumerable icy particles orbiting Saturn.

Saturn's satellite system: A family of at least 17 moons orbiting Saturn, ranging from 20 km diameter up to a size slightly exceeding that of the planet Mercury.

science: Study of nature using the scientific method. (See *scientific method.)*

scientific method: The method of learning about nature from making observations, formulating hypotheses, and constructing observational or experimental tests to see if the hypotheses are accurate.

seasonal changes on Mars: Changes in shape and darkness of the dusky patches on Mars from summer to winter and year to year. Once thought to indicate Martian vegetation, the changes are now known to result from blowing dust deposits.

second of arc: An angle equaling $\frac{1}{3600}$ of a degree.

secondary atmosphere: A planet's atmosphere after modification by outgassing and other processes. (See also *primary atmosphere.*)

sedimentary rock: Rock formed from sediments.

seeing: The quality of stillness or lack of shimmer in a telescopic image, associated with atmospheric conditions. If the atmosphere is very still and the image is sharp, the seeing is said to be good.

seismic waves: Waves passing through the interior or surface layers of a planet due to a seismic disturbance, such as an earthquake or large meteorite impact.

seismology: Study of vibrational waves passing through planets, revealing internal structure.

selection effect: Any effect that systematically biases observations or statistics away from a correct understanding.

Seyfert galaxy: A type of galaxy with a bright, bluish nucleus, possibly marking a transition between ordinary galaxies and quasars.

shepherd satellites: Satellites that move near planetary rings and act to confine the ring particles onto certain orbits.

short-period comet: A comet with revolution period less than about 100 y.

SI metric system: The internationally standardized scientific system of units, in which length is given as meters, mass as kilograms, and time as seconds.

sidereal: Referring to stars.

sidereal period: A period of rotation or revolution where the movement is measured relative to the stars.

sidereal time: Time measured by the apparent motion of the stars (instead of the Sun), used by astronomers to point telescopes toward celestial targets; it is the right ascension that is on the meridian at any given location.

significant figures: The number of digits known for certain in a quantity.

small-angle equation: The equation giving the relation between the distance D of an object, its diameter d, and its angular size α (expressed in seconds of arc):

$$\frac{\alpha}{206\ 265} = \frac{d}{D}$$

solar apex: The direction toward which the Sun is moving relative to nearby stars.

solar constant: The amount of energy reaching a Sun-facing square meter at a given planet's (usually the Earth's) orbit per unit time; for the Earth, it is 1390 W/m^2.

solar core: The Sun's central region of high-pressure gases where nuclear energy is produced.

solar cycle: 22-y cycle of solar activity.

solar eclipse: Partial or total blocking of the Sun's light by an astronomical body (in most usages, by the Moon).

solar nebula: The cloud of gas around the Sun during the formation of the solar system.

solar rotation: Turning of the Sun on its axis in 25.4 d.

solar system: The Sun and all bodies orbiting around it.

solar wind: An outrush of gas past the Earth and beyond the outer planets. Near the Earth, the solar wind travels at velocities near 600 km/s, sometimes reaching 1000 km/s.

solstice: The date when the Sun reaches maximum distance from the celestial equator (occurs twice annually).

solstice principle: According to this principle, the sunrise and sunset positions of the Sun on the eastern and western horizons (respectively) shift positions according to season and the observer's latitude. The principle can be used to optimize passive solar energy input into a home or other building.

space velocity: A star's velocity with respect to the Sun.

spectral class: A class to which a star belongs because of its spectrum, which in turn is determined by its temperature. The spectral classes are O, B, A, F, G, K, and M, from hottest to coolest.

spectral line strength: Measure of the total energy absorbed or emitted in a spectral line.

spectrograph: An instrument for recording a photographic image of a spectrum.

spectroheliograph: An instrument for observing the Sun in certain specified wavelengths.

spectrometer: An instrument for tracing the intensity of a spectrum at different wavelengths; the result is a graph.

spectrophotometry: The study of the amount of radiation at each wavelength in the spectrum.

spectroscopic binary: A binary star revealed by varying Doppler shifts in spectral lines.

spectroscopy: Study of spectra, especially as revealing the properties of the light source.

spectrum: Light from an object arranged in order of wavelength; specifically, the colors of visible light, arranged in this order.

spectrum binary: A binary revealed by mixture of two spectral classes in the spectrum.

spectrum–luminosity diagram: The H–R diagram.

speed of light: Designated as c, the speed of light is about 300 000 km/s and is constant as perceived by all observers.

spicule: Narrow jet of gas extending out of the solar chromosphere, with a lifetime of about 5 min.

spiral arm: In spiral galaxies, one of the arms lying at an angle to the Sun–center line. The arms contain open clusters, O and B stars, and nebulae.

spiral galaxy: A disk-shaped galaxy with a spiral pattern, typically containing 10^{10}–$10^{12} M_{\odot}$ of stars, dust, and gas.

standard time: Solar time appropriate to the given local time zone.

star: A mass of material, usually wholly gaseous, massive enough to initiate (or to have once initiated) nuclear reactions in its central region.

starburst: A relatively sudden and rapid episode of star formation in a galaxy, probably triggered in some cases by collision with another galaxy.

Stefan–Boltzmann law: A law giving the total energy E radiated from a surface of area A and temperature T per second: $E = \sigma T^4 A$. Sigma (σ), the Stefan–Boltzmann constant, equals 5.67×10^{-8} W/m$^2 \cdot$ K^4 in SI units.

stellar evolution: Evolution of every star from one form to another forced by changes in composition as nuclear reactions proceed.

stellar populations in galaxies: Groupings of stars with different composition and age. Population I consists of young stars with a few percent of heavy elements; stars near the Sun are Population I. Population II includes older stars with virtually no heavy elements.

Stonehenge: A prehistoric English ruin with built-in astronomical alignments.

stony–iron meteorite: Stony meteorite that probably comes from deep within the parent body, where melted stony and metallic material coexisted.

stromatolite: A primitive life form, colonies of blue-green algae that were among the first to appear along shorelines. They are among the most abundant fossils from 3.5 to 2 billion years ago.

subfragmentation: Breakup of a contracting cloud into smaller condensations.

subfragmentation theory: A theory of formation of binary and multiple stars by breakup of the protostellar cloud into two or more components during its collapse from nebular to stellar dimensions.

subtend: To have an angular size equal to a specified angle. For example, the Moon subtends to $\frac{1}{2}°$.

Sun: The star orbited by the Earth.

Sun's composition: 78% hydrogen, 20% helium, 2% other gases.

sunspot: A magnetic disturbance on the Sun's surface that is cooler than the surrounding area.

Sun's revolution period: The 230-million-year period taken by the Sun to complete its orbit around the Milky Way galaxy.

superbubble: A large volume of hot gas in interstellar space, formed by coalescence of bubbles blown around supernovae.

supercluster of galaxies: Cluster of clusters of galaxies.

supergiant star: An extremely luminous star in the uppermost part of the H–R diagram.

supergranulation: Large-scale (15 000–30 000 km in diameter) convective cell patterns in the solar photosphere.

superior planet: Any planet with an orbit outside the Earth's orbit.

supernova: A very energetic stellar explosion expending about 10^{42} to 10^{44} joules and blowing off most of the star's mass, leaving a dense core.

symbiotic stars: A pair of stars whose evolutions are affecting each other, especially by mass transfer.

synchronous rotation: Any rotation such that a body keeps the same face toward a coorbiting body.

synchrotron radiation: Radiation emitted when electrons move at nearly the speed of light in a magnetic field.

synodical month: One complete cycle of lunar phases, 29.53 d.

tangential velocity: The velocity component perpendicular to the line of sight.

Tarantula Nebula (30 Doradus): A huge HII emission nebula in the Large Magellanic Cloud.

T association: An association of T Tauri stars.

tectonics: Disruption of planetary or satellite surfaces by large-scale mass movements, such as faulting.

telescope: An instrument for collecting electromagnetic radiation and producing magnified images of distant objects.

temperature: A measure of the average energy of a molecule of a material.

terminator: The dawn or dusk line separating night from day on a planet or satellite.

terrestrial planet: (1) Mercury, Venus, Earth, or Mars; (2) a planet composed primarily of rocky material.

Tethys, Dione, and Rhea: Intermediate-sized icy moons of Saturn.

theory: A body of hypotheses, often with mathematical backing and having passed some observational tests; often implying more validity than the term *hypothesis*.

theory of cosmological red shifts: The theory that galaxies' red shifts are all due to recessional motion, increase with distance, and thus give an indicator of distance.

theory of noncosmological red shifts: The theory that at least some galaxies' red shifts are not Doppler shifts due to recession, but are due to some other cause.

theory of star formation: The theory that describes how stars form by the gravitational collapse of interstellar clouds of dust and gas.

thermal escape: Escape of the fastest-moving gas atoms or molecules from the top of a planet's atmosphere by means of their thermal motion.

thermal motion: Movement of atoms and molecules associated with the temperature of the material; they grow faster as the temperature increases.

thermal radiation: Electromagnetic radiation emitted by a body and associated with an object's temperature; it grows greater and bluer in color as the temperature increases.

third quarter: The phase of the Moon when it is three-fourths of the way around its orbit from new moon; the third quarter moon is seen in the dawn sky with a straight terminator and half the disk illuminated.

3-kpc arm: An inner arm of our galaxy expanding from the center at about 53 km/s.

3-K radiation: Radio radiation coming uniformly from all over the sky, believed to be a red-shifted remnant of the big bang radiation.

Titan: Saturn's largest moon, famous for its thick, smoggy-orange nitrogen atmosphere.

thrust: The force generated by a high-speed discharge, as from a rocket or airplane.

tidal recession: Recession of the Moon (or other satellite) from the Earth (or other planet) caused by tidal forces.

tide: A bulge raised in a body by the gravitational force of a nearby body.

total solar eclipse: (1) An eclipse in which the light source is totally obscured from a specified observer; (2) an eclipse in which a body is entirely immersed in another's shadow. (See also *eclipse.*)

transit: (1) Passage of a planet across the Sun's disk; (2) any passage of a body with a small angular size across the face of a body with a large angular size.

triple-alpha process: A nuclear reaction in which helium is transformed into carbon in red giant stars.

Triton: Neptune's largest moon.

Trojan asteroids: Asteroids caught near the Lagrangian points in Jupiter's orbit, 60° ahead of and 60° behind the planet.

tsunami: A large ocean wave generated by earthquake or volcanic activity (the correct name for a tidal wave).

T Tauri star: A type of variable star, often shedding mass, believed to be still forming and contracting onto the main sequence.

"tuning fork" diagram: A classification scheme for galaxies.

21-cm emission line: The important radio radiation at 21-cm wavelength from interstellar neutral atomic hydrogen.

21-cm radio waves: Produced by neutral hydrogen, these waves are especially useful for galactic mapping because they allow us to detect HI clouds, which are concentrated in the spiral arms.

ultrabasic rock: A rock of high density, low silica content, and high iron content, often derived from the upper mantle of a planet or satellite.

ultraviolet light: Radiation of wavelength too short to see, but longer than that of X rays.

umbra: (1) The dark inner part of a shadow, in which the light source is totally obscured; (2) the dark inner part of a sunspot.

universe: Everything that exists.

Uranus: The seventh planet outward from the Sun.

Uranus' rings: A system of very narrow rings of dark particles around Uranus, scarcely detectable from Earth.

Uranus' rotation axis: Notable for its almost right-angle tilt (obliquity) of 97° to Uranus' orbital plane.

Uranus' satellites: A system of five large moons discovered from Earth, and another ten discovered by Voyager 2 in 1986.

Urey reaction: Reaction by which the Earth's carbon dioxide was concentrated in carbonate rocks after dissolving in seawater.

Van Allen belts: Doughnut-shaped zones around the Earth (or another planet with a strong magnetic field) that traps energetic ions from the Sun.

variable star: A star that varies in brightness.

Venera 7: First spacecraft to land successfully on another planet; it transmitted data from the surface of Venus in 1970.

Viking 1: The first successful probe to land on Mars (July 20, 1976). It made the first surface photos and measures of soil composition.

Viking 2: The second successful probe to land on Mars (September 3, 1976).

visible light: Electromagnetic radiation at wavelengths that can be perceived by the eye.

visual apparent magnitude: An apparent magnitude estimate based only on visual radiation from an object (excluding infrared, ultraviolet, X rays, and so on.).

visual binary: A binary in which both components can be seen.

volatile element: Element easily driven out of a material by heating.

volcanic crater: A circular depression caused by volcanic processes such as explosion or collapse.

volcanism: Eruption of molten materials at the surface of a planet or satellite.

volcanoes: Sites where molten materials erupt from inside a planet or satellite.

wavelength: (1) The length of the wavelike characteristic of electromagnetic radiation; (2) in any wave, the distance from one maximum to the next.

wavelike properties of light: Characteristics of light, such as frequency and diffraction, that mimic properties of waves.

white dwarf star: A planet-sized star of roughly solar mass and very high density (10^8 to 10^{11} kg/m^3) produced as a terminal state after nuclear fuels have been consumed.

white light: A mixture of light of all colors in proportions as found in the solar spectrum.

Wien's law: A formula giving the wavelength W at which the maximum amount of radiation comes from a body of temperature T. The formula is $W = 0.00290/T$.

Wolf–Rayet star: A type of very hot star ejecting mass.

W Ursae Majoris stars: Contact binary stars.

X ray: Electromagnetic radiation of wavelength about 0.01 to 10 nm.

X-ray source: Celestial object emitting X rays; many are probably binary systems where mass is transferred.

zenith: The point directly overhead.

zero-age main sequence: The main sequence defined by a population of stars all of which have just evolved onto the main sequence. (Further evolution modifies the main sequence shape on the H–R diagram slightly.)

zodiac: A band around the sky about 18° wide, centered on the ecliptic, in which the planets move.

zodiacal light: A glow, barely visible to the eye, caused by dust particles spread along the ecliptic plane.

zone: Light cloud band on a giant planet.

zone of avoidance: A band around the sky, centered on the Milky Way, in which galaxies are obscured by the Milky Way's dust.

References

Asterisked references are less technical, more readily available, and recommended for general reading. They are good general references for preparation of term papers.

Chapter 1

*Ashbrook, J. 1973. "Astronomical Scrapbook." *Sky and Telescope 46:* 300.

Aveni, A. F. 1975. "Possible Astronomical Orientations in Ancient Mesoamerica." In *Archaeoastronomy in Pre-Columbian America,* ed. A. F. Aveni. Austin: University of Texas Press.

Aveni, A. F., S. L. Gibbs, and H. Hartung. 1975. "The Astronomical Significance of the Caracol of Chichén Itzá." *Science 188:* 977.

*Bok, B. 1975. "A Critical Look at Astrology." *The Humanist,* September/October, p. 5.

Bok, B., and M. Mayall, 1941. "Scientists Look at Astrology." *Scientific Monthly 52:* 233.

*Carlson, J. B. 1975. "Lodestone Compass: Chinese or Olmec Primacy?" *Science 189:* 753.

Castetter, E. F., and W. H. Bell. 1942. *Pima and Papago Indian Agriculture.* Albuquerque: University of New Mexico Press.

Crommelin, A. 1925. "The Ancient Constellation Figures." In *Splendour of the Heavens,* ed. T. Phillips and W. Steavenson. New York: McBride.

*Doig, P. 1950. *A Concise History of Astronomy.* London: Chapman and Hall.

Farrington, B. 1973. "Astrology." In *Encyclopaedia Britannica.* Chicago: Encyclopaedia Britannica, Inc.

*Gingerich, O. 1967. "Musings on Antique Astronomy." *American Scientist 55:* 88.

*Gingerich, O. 1984. "The Origin of the Zodiac." *Sky and Telescope 67:* 218.

Gleadow, R. 1963. *The Origin of the Zodiac.* New York: Atheneum.

*Harber, H. E. 1969. "Five Mayan Eclipses in Thirteen Years." *Sky and Telescope 37:* 72.

Hartner, W. 1965. "The Earliest History of the Constellations in the Near East, and the Motif of the Lion-Bull Contest." *Journal of Near Eastern Studies 24:* 1.

———. 1969. "Eclipse Periods and Thales' Prediction of a Solar Eclipse: Historic Truth and Modern Myth." *Centaurus 14:* 60.

*Hawkins, G., and J. B. White. 1965. *Stonehenge Decoded.* New York: Doubleday.

*Hoyle, F. 1972. *From Stonehenge to Modern Cosmology.* San Francisco: W. H. Freeman.

Humphreys, C. J., and W. G. Waddington. 1983. "Dating the Crucifixion." *Nature 306:* 743.

*Jerome, L. E. 1975. "Astrology: Magic or Science?" *The Humanist,* September/October, p. 10.

Lockyer, J. N. 1894. *The Dawn of Astronomy.* Cambridge, Mass.: M.I.T. Press. (Reprint of edition copyrighted 1893, published 1894).

Lowell, P. 1906. *Mars and Its Canals.* 2nd ed. New York: Macmillan.

Luce, G. G. 1975. "Trust Your Body Rhythms." *Psychology Today,* April, p. 52.

*Marshack, A. 1972. *The Roots of Civilization.* New York: McGraw-Hill.

Neugebauer, O. 1945. "History of Ancient Astronomy." *Journal of Near Eastern Studies 4:* 1.

*———. 1957. *The Exact Sciences in Antiquity.* Providence, R.I.: Brown University Press.

Ovenden, M. 1966. "The Origin of Constellations." *Philosophical Journal 3:* 1.

Owen, N. K. 1975. "The Use of Eclipse Data to Determine the Maya Correlation Number." In *Archaeoastronomy in Pre-Columbian America,* ed. A. F. Aveni. Austin: University of Texas Press.

*Pannekoek, A. 1961. *A History of Astronomy.* London: Allen and Unwin.

Schaefer, Bradley F. 1989. "Dating the Crucifixion." *Sky and Telescope,* April, p. 374.

Smiley, C. H. 1975. "The Solar Eclipse Warning Table in the Dresden Codex." In *Archaeoastronomy in Pre-Columbian America,* ed. A. F. Aveni. Austin: University of Texas Press.

*Stephenson, F. R. 1982. "Historical Eclipses." *Scientific American,* October, p. 170.

*Stephenson, F. R., and D. H. Clark. 1977. "Ancient Astronomical Records from the Orient." *Sky and Telescope 53:* 84.

Thom, A., and A. S. Thom. 1971. *Megalithic Lunar Observatories.* Oxford: Oxford University Press.

Chapter 2

*Gingerich, O. 1967. "Musings on Antique Astronomy." *American Scientist 55:* 88.

*———. 1986. "Islamic Astronomy." *Scientific American,* April, p. 74.

*Lewis, D. 1973. *We, the Navigators.* Honolulu: University of Hawaii Press.

Mozans, H. J. 1913. *Women in Science.* New York: Appleton.

Needham, J. 1959. *Science and Civilization in China.* Vol. 3. London: Cambridge University Press.

*North, J. D. 1974. "The Astrolabe." *Scientific American,* January, p. 96.

*Pannekoek, A. 1961. *A History of Astronomy.* London: Allen and Unwin.

Pritchard, J. B., ed. 1955. *Ancient Eastern Texts Relating to the Old Testament.* 2nd ed. Princeton, N.J.: Princeton University Press.

*Thomsen, D. E. 1984. "Calendric Reform in Yucatan." *Science News 126:* 282.

Wilson, J. A. 1951. *The Culture of Ancient Egypt.* Chicago: University of Chicago Press.

Zeilik, M. 1985. "The Enthnoastronomy of the Historic Pueblos, I: Calendrical Sun Watching." *Archaeoastronomy 8:* S1.

Chapter 3

Ball, W.W.R. 1972. "An Essay on Newton's Principia." New York: Johnson Reprint Corp.

*****de Santillana,** Georgio. 1962. *The Crime of Galileo.* New York: Time.

*****Dreyer,** J.L.E. 1953. *A History of Astronomy from Thales to Kepler.* New York: Dover. (Reprint of 1906 edition.)

*****Gingerich,** O. 1973a. "Copernicus and Tycho." *Scientific American,* December, p. 86.

*****———.** 1973b. *Crisis Versus Asthetic in the Copernican Revolution.* Cambridge, Mass.: Smithsonian Astrophysical Observatory.

*****Hartmann,** W. K. 1983. *Moons and Planets.* Belmont, Calif.: Wadsworth.

*****Lerner,** L. S., and E. A. Gosselin. 1973. "Giordano Bruno." *Scientific American,* April, p. 86.

*****Lerner,** L. S., and E. A. Gosselin. 1986. "Galileo and the Specter of Bruno." *Scientific American 255.* November, p. 126.

*****Pannekoek,** A. 1961. *A History of Astronomy.* New York: Interscience.

*****Taylor,** S. R. 1982. *Planetary Science: A Lunar Perspective.* Houston, Tex.: Lunar and Planetary Institute.

Chapter 4

*****Arnold,** J. R. 1980. "The Frontier in Space." *American Scientist 68:* 299.

*****Banks,** P., and D. Black. 1987. "The Future of Science in Space." *Science 236:* 244.

Banks, P. M., and Sally K. Ride. 1989. "Soviets in Space." *Scientific American, 260,* February, p. 32.

*****Burbidge,** E. Margaret. 1983. "Adventure into Space." *Science 221:* 421.

*****Clarke,** A. C. 1951. *The Exploration of Space.* New York: Harper & Bros.

Dessler, A. T. 1984. "The Vernov Radiation Belt (Almost)." *Science 226,* no. 4677, editorial page.

*****Dyson,** F. 1969. "Human Consequences of the Exploration of Space." *Bulletin of Atomic Scientists,* September, p. 8.

*****Hartmann,** W. K., R. Miller, and P. Lee. 1984. *Out of the Cradle.* New York: Workman Publishing.

*****Heinlein,** R. 1950. *The Man Who Sold the Moon.* New York: New American Library.

*****Lewis,** R. S. 1969. *Appointment on the Moon.* New York: Viking Press.

*****Logsdon,** J. 1970. *The Decision to Go to the Moon.* Cambridge, Mass.: M.I.T. Press.

National Commission on Space. 1986. *Pioneering the Space Frontier.* New York: Bantam Books.

Newton, I. 1962. *Principia.* A. Motte, trans. Berkeley: University of California Press. (Originally published 1687.)

*****Nicholson,** M. 1949. *Voyages to the Moon.* New York: Macmillan.

*****O'Neill,** G. K. 1977. *The High Frontier.* New York: William Morrow.

*****Verne,** J. 1949. *From the Earth to the Moon.* New York: Didear. (Orginally published 1865.)

Chapter 5

Bahcall, J. N., and L. Spitzer. 1982. "The Space Telescope." *Scientific American,* July, p. 40.

*****Hjellming,** R., and R. Bignell. 1982. "Radio Astronomy with the Very Large Array." *Science 216:* 1279.

Kristian, J., and M. Blouke. 1982. "Charge-Coupled Devices in Astronomy." *Scientific American,* October, p. 66.

Levy, G. S., and others. 1986. "Very Long Baseline Interferometric Observations Made with an Orbiting Radio Telescope." *Science 234:* 187.

Chapter 6

Alvarez, L. W., and others. 1980. "Extraterrestrial Cause for the Cretaceous-Tertiary Extinction." *Science 208:* 1095.

*****Davies,** G. L. 1969. *The Earth in Decay.* New York: American Elsevier.

Einstein, A. 1961. *Relativity.* New York: Crown.

Faggart, B. E., A. Basu, and M. Tatsumoto. 1985. "Origin of the Sudbury Complex by Meteoritic Impact: Neodymium Isotopic Evidence." *Science 230:* 436.

Ganapathy, R. 1980. "A Major Meteorite Impact on the Earth 65 Million Years Ago: Evidence from the Cretaceous-Tertiary Boundary Clay." *Science 209:* 921.

*****Hurley,** P. M. 1968. "The Confirmation of Continental Drift." *Scientific American,* April, p. 52.

*****Kerr,** R. A. 1987a. "Asteroid Impact Gets More Support." *Science 236:* 666.

———. 1987b. "Milankovitch Climate Cycles Through the Ages." *Science 235:* 973.

Kozlovsky, Y. A. 1984. "The World's Deepest Well." *Scientific American,* p. 436.

*****Levi,** B., and T. Rothman. 1985. "Nuclear Winter: A Matter of Degrees." *Physics Today,* September 58.

*****Maxwell,** J. C. 1985. "What Is the Lithosphere?" *Physics Today 38:* 32.

*****Morris,** S. C. 1987. "The Search for the Precambrian–Cambrian Boundary." *American Scientist 75:* 157.

Mutter, J. C. 1986. "Seismic Images of Plate Boundaries." *Scientific American 254:* 66.

*****Nelkin,** D. 1976. "The Science-Textbook Controversies." *Scientific American,* April, p. 33

Pepin, R. O. 1976. "The Formation Interval of the Earth." *Abstracts Seventh Lunar Science Conference,* Houston: Lunar and Plantary Science Institute.

*****Schneider,** S. H. 1987. "Climate Modeling." *Scientific American 256:* 72.

Schopf, J. W. 1975. "The Age of Microscopic Life." *Endeavour 34:* 51.

*****Silberner,** J. 1987. "Predicting Parkfield." *Science News 131:* 268.

Toon, O. B., and others. 1982. "Evolution of an Impact-Generated Dust Cloud and Its Effects on the Atmosphere." In *Geological Society of America Special Paper 190,* ed. L. Silver and P. Schultz. Boulder, Colo.: Geological Society of America.

*****Turco,** R. P., and others. 1984. "The Climatic Effects of Nuclear War." *Scientific American 251,* August, p. 33.

Wolbach, W., R. Lewis, and E. Anders. 1986. "Cretaceous Extinctions: Evidence for Wildfires and Search for Meteoritic Material." *Science 230:* 167.

Chapter 7

Darwin, G. H. 1962. *The Tides and Kindred Phenomena in the Solar System.* San Francisco: W. H. Freeman. (Orginally published 1898.)

*****Goldreich,** P. 1972. "Tides and the Earth–Moon System." *Scientific American,* April, p. 42.

Hartmann, W. K., and D. Davis. 1975. "Satellite-Sized Planetesimals and Lunar Origin." *Icarus 24:* 504.

Hartmann, W. K., R. Phillips, and J. G. Taylor. 1986. *Origin of the Moon.* Houston: Lunar and Planetary Institute.

Kaula, W. K., and A. Harris. 1975. "Dynamics of Lunar Origin and Orbital Evolution." *Review of Geophysics and Space Physics 13:* 363.

Lammlein, D., and others. 1974. "Lunar Seismicity, Structure, and Tectonics. *Review of Geophysics and Space Physics 12:* 1.

***Lewis,** R. S. 1977. "Space Prospect: Factories and Electric Power." *Smithsonian 8* (9): 94.

Lin, T. D. 1986. "Lunar Concrete." *Lunar News* (from Lunar and Planetary Institute, Houston), no. 47, p. 2.

Taylor, S. R. 1973. "Geochemistry of the Lunar Highlands." *Moon 7:* 181.

————. 1975. *Lunar Science: A Post-Apollo View.* New York: Pergamon Press.

***Thomsen,** D. E. 1986. "Man in the Moon." *Science News 129:* 154.

Toksoz, M. N., and others. 1974. "Structure of the Moon." *Review of Geophysics and Space Physics 12:* 539.

Wood, J. 1972. "Thermal History and Early Magmatism in the Moon." *Icarus 16:* 229.

Chapter 8

Chapman, C. R., F. Vilas, and M. Matthews, eds. 1987. *Mercury.* Tucson: University of Arizona Press.

***Cruikshank,** D. P., and C. R. Chapman. 1967. "Mercury's Rotation and Visual Observations." *Sky and Telescope 34:* 24.

***Moore,** P. 1954. *A Guide to the Planets.* New York: W. W. Norton.

***Murray,** B. C. 1975. "Mercury." *Scientific American,* September, p. 58.

Chapter 9

Alexandrov, Y., and others. 1986. "Venus: Detailed Mapping of Maxwell Montes Region." *Science 231:* 1271.

Bazilevskiy, A. T. 1989. "The Planet Next Door." *Sky and Telescope,* April, p. 360.

Connes, P., and others. 1967. "Traces of HCl and HF in the Atmosphere of Venus." *Astrophysical Journal 147:* 1230.

Duncombe, R. 1969. "Report on Numerical Experiment on the Possible Existence of an 'Anti-Earth.' " In *Scientific Study of Unidentified Flying Objects,* ed. E. U. Condon. New York: Bantam.

Eberhart, J. 1986. "The Night Skies of Venus: Another Kind of Aurora?" *Science News 130:* 364.

Grinspoon, D. H. 1987. "Was Venus Wet? Deuterium Reconsidered." *Science 238:* 1702.

Hunten, D. M., L. Colin, T. Donahue, and V. Moroz, editors. 1983. *Venus.* Tucson: University of Arizona Press.

Kaula, W. M. 1990. "Venus: A Contrast in Evolution to Earth." *Science 247:* 1191.

Kerzhanovich, V. V., and M. Marov. 1983. "The Atmospheric Dynamics of Venus According to Doppler Measurements by the Venera Entry Probes." In *Venus,* eds. D. M. Hunten and others. Tucson: University of Arizona Press.

***Moore,** P. 1954. *A Guide to the Planets.* New York: W. W. Norton.

Oyama, V. I., and others. 1979. "Venus Lower Atmospheric Composition: Analysis by Gas Chromatography." *Science 203:* 802.

Pieters, C. M., and others. 1986. "The Color of the Surface of Venus." *Science 234:* 1379.

Sagan, C. 1962. "Structure of the Lower Atmosphere of Venus." *Icarus 1:* 151.

Sagdeev, R., and others. 1986. "Overview of VEGA Venus Balloon in Situ Meteorological Measurements." *Science 231:* 1411.

***Schubert,** G., and C. Covey. 1981. "The Atmosphere of Venus." *Scientific American,* June, p. 66.

Sill, G. T. 1973. "Sulfuric Acid in the Venus Clouds." *Communications of the Lunar and Planetary Laboratory, University of Arizona 9:* 191.

Taylor, H., and P. Cloutier. 1986. "Venus: Dead or Alive?" *Science 234:* 1087.

Von Zahn, and others. 1983. "Composition of the Venus Atmosphere." In *Venus,* eds. D. M. Hunten and others. Tucson: University of Arizona Press.

Walker, J.C.C. 1977. *Evolution of the Atmosphere.* New York: Macmillan.

Young, A. T. 1973. "Are the Clouds of Venus Sulfuric Acid?" *Icarus 18:* 564.

Chapter 10[†]

Arvidson, R., and others. 1983. "Three Mars Years: Viking Lander 1 Imaging Observations." *Science 222:* 463.

***Bradbury,** R. 1950. *The Martian Chronicles.* New York: Doubleday.

Carr, M. H. 1981. *The Surface of Mars.* New Haven: Yale University Press.

Greeley, R. 1987. "Release of Juvenile Water on Mars: Estimated Amounts and Timing Associated with Volcanism." *Science 236:* 1653.

***Haberle,** R. M. 1986. "The Climate of Mars." *Scientific American,* May, p. 54.

***Hartmann,** W. K., and O. Raper. 1974. *The New Mars.* Washington, D.C.: National Aeronautics and Space Administration.

Kerr, R. A. 1986. "Mars Is Getting Wetter and Wetter." *Science 233:* 936.

Levin, G. V., and P. Stroat. 1981. "A Search for a Nonbiological Explanation of the Viking Labeled Release Life Detection Experiment." *Icarus 45:* 494.

***Lowell,** P. 1906. *Mars and Its Canals.* 2nd ed. New York: Macmillan.

Lucchitta, B. K. 1987. "Recent Mafic Volcanism on Mars." *Science 235:* 565.

McSween, H. Y., Jr. 1985. "SNC Meteorites: Clues to Martian Petrologic Evolution?" *Reviews of Geophysics 23:* 391.

National Commission on Space. 1986. *Pioneering the Space Frontier.* New York: Bantam Books.

Owen, T., and others. 1977. "The Composition of the Atmosphere at the Surface of Mars." *Journal of Geophysical Research 82:* 4635.

Prinn, R. G., and B. Fegley, Jr. 1987. "The Atmospheres of Venus, Earth, and Mars: A Critical Comparison." *Annual Review of Earth and Planetary Science 15:* 171.

Sagan, C., O. Toon, and P. Gierasch. 1973. "Climatic Change on Mars." *Science 181:* 1045.

Soffen, G. A., and others. 1977. Special issue on Viking results. *Journal of Geophysical Research 82:* 3959 ff.

Thomas, P., and P. Gierasch. 1985. "Dust Devils on Mars." *Science 230:* 175.

Warren, P. H. 1987. "Mars Regolith vs. SNC Meteorites: Possible Evidence for Abundant Crustal Carbonates." *Icarus 70:* 153.

***Wells,** H. G. 1898. *The War of the Worlds.* London: W. Heinemann.

[†]Note: Early results from Viking are reported in special issues of *Science,* 27 August 1976 and 1 October 1976.

Chapter 11

***Chapman,** C. R. 1968. "The Discovery of Jupiter's Red Spot." *Sky and Telescope 35:* 276.

Cruikshank, D. P., and R. E. Murphy. 1973. "The Post-Eclipse Brightening of Io." *Icarus 20:* 7.

Hanel, R., and others. 1979. "Infrared Observations of the Jovian System from Voyager 1." *Science 204:* 972.

***Hartmann,** W. K. 1983. *Moons and Planets.* Belmont, Calif.: Wadsworth.

Khare, B. N., and C. Sagan. 1973. "Red Clouds in Reducing Atmospheres." *Icarus 20:* 311.

Malin, M. C., and D. Pieri. 1986. "Europa." In *Satellites,* ed. J. Burns and M. Matthews. Tucson: University of Arizona Press.

McKinnon, W., and E. Parmentier. 1986. "Ganymede and Callisto." In *Satellites,* ed. J. Burns and M. Matthews. Tucson: University of Arizona Press.

Minton, R. B. 1973. "The Red Polar Caps of Io." *Communications of the Lunar and Planetary Laboratory, University of Arizona 10:* 35.

Morrison, David, ed. 1982. *Satellites of Jupiter.* Tucson: University of Arizona Press.

Peale, S. J., P. Cassen, and R. Reynolds. 1979. "Melting of Io by Tidal Dissipation." *Science 203:* 892.

*Stone, E., and others. 1979a. Special issue on Voyager 1 Jupiter results. *Science 204:* 945 ff.

———. 1979b. Special issue on Voyager 2 Jupiter results. *Science 206:* 925 ff.

*Wolfe, J. H. 1975. "Jupiter." *Scientific American,* September, p. 118.

Chapter 12

Alexander, A.F.O. 1962. *The Planet Saturn.* New York: Macmillan.

Burns, J., and M. Matthews, eds. 1986. *Satellites.* Tucson: University of Arizona Press.

Cruikshank, D. P., and P. Silvaggio. 1979. "Triton: A Satellite with an Atmosphere." *Astrophysical Journal 233:* 1016.

Cruikshank, D. P., R. Brown, and R. Clark. 1983. "Nitrogen on Triton." *Bulletin of the American Astronomical Society 15:* 857.

Gehrels, T., and M. Matthews. 1984. *Saturn.* Tucson: University of Arizona Press.

Goldreich, P., N. Murray, P. Longaretti, and D. Banfield. 1989. "Neptune's Story." *Science 245:* 500.

Greenberg, R., and A. Brahic. 1984. *Planetary Rings.* Tucson: University of Arizona Press.

*Grosser, M. 1962. *The Discovery of Neptune.* Cambridge, Mass.: Harvard University Press.

*Hartmann, W. K. 1983. *Moons and Planets.* Belmont, Calif.: Wadsworth.

Lebofsky, L., T. Johnson, and T. McCord. 1970. "Saturn's Rings: Spectral Reflectivity and Compositional Implications." *Icarus 13:* 226.

Lunine, J. I., D. Stevenson, and Y. Young. 1983. "Ethane Ocean on Titan. *Science 222:* 1224.

Marcialis, R., and R. Greenberg. 1987. "Warming of Miranda during Chaotic Rotation." *Nature 328:* 227.

Marcialis, R., G. Rieke, and L. Lebofsky. 1987. "The Surface Composition of Charon: Tentative Identification of Water Ice." *Science 237:* 1349.

Pollack, J. 1973. "Greenhouse Models of the Atmosphere of Titan." *Icarus 19:* 43.

*Stone, E., and others. 1981. Special issue on Voyager 1 Saturn results. *Science 212:* 159 ff.

*Stone, E., and others. 1982. Special issue on Voyager 2 Saturn results. *Science 212:* 499 ff.

Stone, E., and others. 1986. Special issue on Voyager 2 Uranus results. *Science 233:* 39 ff.

Tholen, D., and others. 1987. "Improved Orbital and Physical Parameters for the Pluto–Charon System." *Science 237:* 512.

Tholen, D., and M. Buie. 1989. "Circumstances for Pluto/Charon Mutual Events in 1990." Submitted to *Astronomical Journal* (cited in *Geophysical Research Letter 16,* p. 1205ff.).

Tombaugh, C. W. 1961. "The Trans-Neptunian Planet Search." In *Planets and Satellites,* eds. G. Kuiper and B. Middlehurst. Chicago: University of Chicago Press.

Van Flandern, T., and others. 1981. "The Renewal of the Trans-Neptunian Planet Search." Paper presented at the meeting of the American Astronomical Society, Albuquerque. Abstract in *Bulletin of the American Astronomical Society 12:* 830 (1980).

Voyager Team. 1989 "Report on Voyager 2 Encounter." Various papers in *Science 246:* 1417–1501.

Chapter 13

*Brandt, J. C., and R. Chapman. 1981. *Introduction to Comets.* Cambridge: Cambridge University Press.

Brandt, J. C., and M. Niedner. 1986. "The Structure of Comet Tails." *Scientific American,* January, p. 49.

Campins, H., M. A'Hearn, and L. McFadden. 1987. "The Bare Nucleus of Comet Neujmin 1." *Astrophysical Journal 316:* 847.

*Chapman, C. R. 1975. "The Nature of Asteroids." *Scientific American,* January, p. 24.

Chapman, C. R., and D. Morrison. 1974. "The Minor Planets: Size and Mineralogy." *Sky and Telescope 47:* 92.

Gaffey, M. J., and T. McCord. 1977. "Mining the Asteroids." *Mercury 6:* 1.

Gehrels, T. 1972. "Physical Parameters of Asteroids and Interrelations with Comets." In *From Plasma to Planet,* ed. A. Elvius, New York: John Wiley.

*Gehrels, T., ed. 1979. *Asteroids.* Tucson: University of Arizona Press.

*Hartmann, W. K. 1975. "The Smaller Bodies of the Solar System." *Scientific American,* September, p. 143.

*———. 1982. "Mines in the Sky Are Not So Wild a Dream." Smithsonian, September, p. 70.

*———. 1983. *Moons and Planets.* Belmont, Calif.: Wadsworth.

Kamoun, P. G., and others. 1982. "Comet Encke: Radar Detection of Nucleus." *Science 216:* 293.

Krinov, E. L. 1966. *Giant Meteorites.* New York: Pergamon Press.

Lebofsky, L. A., and others. 1981. "The 1.7 to 4.2-m Spectrum of Asteroid 1 Ceres: Evidence for Structural Water in Clay Minerals." *Icarus 48:* 453.

Nininger, H. H. 1938. "Meteorite Collecting Among Ancient Americans." *American Antiquity 4:* 39.

O'Leary, B. 1983. "Mining the Earth-Approaching Asteroids for Their Precious and Strategic Metals." *Advances in Astronautical Sciences 53:* 375.

*Sagan, C. 1975. "Kalliope and the Kaa'ba: The Origin of Meteorites." *Natural History 84:* 8.

Sagdeev, R. Z., and others. 1986. "Encounters with Comet Halley—The First Results." *Nature 321:* 259.

Sekanina, Z. 1983. "The Tunguska Event: No Cometary Signature in Evidence." *Astronomy Journal 88:* 1382.

Shapley, H., and H. Howarth, eds. 1929. *A Source Book in Astronomy.* New York: McGraw-Hill.

Sykes, M., L. Lebofsky, D. Hunten, and F. Low. 1986. "The Discovery of Dust Trails in the Orbits of Periodic Comets." *Science 232:* 1115.

Wasson, J. T. 1974. *Meteorites.* New York: Springer-Verlag.

*Wilkening, L., ed. 1982. *Comets.* Tucson: University of Arizona Press.

Chapter 14

Anders, E., and M. Ebihara. 1982. "Solar-system Abundances of the Elements." *Geochimica et Cosmochimica Acta 46:* 2365.

*Cameron, A.G.W. 1975. "The Origin and Evolution of the Solar System." *Scientific American,* September, p. 32.

Clark, B. C., and others. 1977. "The Viking X-ray Fluorescence Experiment: Analytical Methods and Early Results." *Journal of Geophysical Research 82:* 4577.

Fairbridge, R. W. 1972. *The Encyclopedia of Geochemistry and Environmental Sciences.* New York: Van Nostrand Reinhold.

Gehrels, T., ed. 1978. *Protostars and Planets.* Tucson: University of Arizona Press.

Greenberg, R., and others. 1978. "Planetesimals to Planets: Numerical Simulation of Collisional Evolution." *Icarus 35:* 1.

*****Grossman**, L. 1975. "The Most Primitive Objects in the Solar System: Carbonaceous Chondrites." *Scientific American,* February, p. 30.

Hartmann, W. K. 1978. "Planet Formation: Mechanism of Early Growth." *Icarus 33:* 50.

*****Herbst**, W., and G. Assousa. 1979. "Supernovas and Star Formation." *Scientific American,* August, p. 138.

Hubbard, W., and D. Stevenson. 1986. "Interior Structure of Saturn." In *Saturn,* eds. T. Gehrels and M. Matthews. Tucson: University of Arizona Press.

*****Lewis**, J. S. 1974. "The Chemistry of the Solar System." *Scientific American,* March, p. 50.

*****Mason,** B., and W. G. Melson. 1970. *The Lunar Rocks.* New York: John Wiley.

Page, T. L. 1973. "Notes on the 4th Lunar Science Conference." *Sky and Telescope 46:* 88.

Pollack, J. B., and D. C. Black, 1979. "Implications of the Gas Compositional Measurements of the Pioneer Venus for the Origin of Planetary Atmospheres." *Science 205:* 56.

Safronov, V. S. 1972. *Evolution of the Protoplanetary Cloud and Formation of the Earth and Planets.* Springfield, Va.: National Technical Information Service.

Shapley, H., and H. Howarth, eds. 1929. *A Source Book in Astronomy.* New York: McGraw-Hill.

*****Stone**, E., and others. 1979. "The Voyager 2 Encounter with the Jupiter System." *Science 206:* 925.

*—————. 1982. "The Voyager 2 Encounter with the Saturn System." *Science 212:* 499.

*—————. 1986. "The Voyager 2 Encounter with the Uranus System." *Science 233:* 39.

Taylor, S. R. 1973. "Chemical Evidence for Lunar Melting and Differentiation." *Nature 245:* 203.

*****Wetherill**, G. W. 1985. "Occurrence of Giant Impacts During the Growth of the Terrestrial Planets." *Science 228:* 877.

Chapter 15

Akasofu, Syun-Ichi. 1989. "The Dynamic Aurora." *Scientific American,* May, p. 90.

Anders, E., and M. Ebihara. 1982. "Solar-system Abundances of the Elements." *Geochimica et Cosmochimica Acta 46:* 2365.

*****Arnold**, J. R. 1980. "The Frontier in Space." *American Scientist 68:* 299.

Bahcall, John N. 1990. "The Solar-Neutrino Problem." *Scientific American,* May, p. 54.

Dicke, R. H. 1979. "Solar Luminosity and the Sunspot Cycle." *Nature 280:* 24.

Eddy, J. A. 1976. "The Maunder Minimum." *Science 192:* 1189.

Foukal, P. V. 1990. "The Variable Sun." *Scientific American,* February, p. 34.

Gibson, E. G. 1973. *The Quiet Sun.* Publication no. SP-303. Washington, D.C.: National Aeronautics and Space Administration.

Gilliland, R. 1982. "Modeling Solar Variability." *Astrophysical Journal 253:* 399.

*****Hubbert**, M. 1971. "The Energy Resources of the Earth." *Scientific American,* September, p. 61.

*****Kerr**, R. A. 1986. "The Sun Is Fading." *Science 231:* 339.

*****Kreith**, F., and R. T. Meyer. 1983. "Large-Scale Use of Solar Energy with Central Receivers." *American Scientist 71:* 598.

Lester, R. K. 1986. "Rethinking Nuclear Power." *Scientific American,* March, p. 31.

*****Lewis**, R. S. 1977. "The Space Prospect: Factories and Electric Power." *Smithsonian,* December, p. 94.

*****Meadows**, J. 1984. "The Origins of Astrophysics." *American Scientist 72:* 269.

Newkirk, G., and K. Frazier. 1982. "The Solar Cycle." *Physics Today,* April, p. 25.

*****Parker**, E. N. 1983. "Magnetic Fields in the Cosmos." *Scientific American,* August, p. 44.

*****Pasachoff**, J. M. 1973. "The Solar Corona." *Scientific American,* October, p. 68.

*****Pasachoff**, J. M. 1980. "Our Sun." In *Astronomy Selected Readings,* ed. M. A. Seeds. Menlo Park, Calif.: Benjamin/Cummings.

Pollack, J. B., and D. C. Black. 1979. "Implications of the Gas Compositional Measurements of the Pioneer Venus for the Origin of Planetary Atmospheres." *Science 205:* 56.

Schneider, S. H., and C. Mass. 1975. "Volcanic Dust, Sunspots, and Temperature Trends." *Science 190:* 741.

Shapley, H., and H. Howarth, eds. 1929. *A Source Book in Astronomy.* New York: McGraw-Hill.

*****Snell**, J. E., P. Achenbach, and S. Peterson. 1976. "Energy Conservation in New Housing Design." *Science 192:* 1305.

Sofia, S., P. Demarque, and A. Endal. 1985. "From Solar Dynamo to Terrestrial Climate." *American Scientist 73:* 326.

Thomsen, D. E. 1985. "Solar News: Convection and Magnetism." *Science News 127:* 326.

Weneser, J., and G. Friedlander. 1987. "Solar Neutrinos: Questions and Hypotheses." *Science 235:* 755.

*****Wilcox**, J. M. 1976. "Solar Structure and Terrestrial Weather." *Science 192:* 745.

*****Williams**, G. E. 1986. "The Solar Cycle in Precambrian Time." *Scientific American,* August, p. 88.

Willson, R., and others. 1986. "Long-term Downward Trend in Total Solar Irradiance." *Science 234:* 1114.

*****Wolfson**, R. 1983. "The Active Solar Corona." *Scientific American,* February, p. 104.

Wood, C. A., and R. R. Lovett. 1974. "Rainfall, Drought, and the Solar Cycle." *Nature 251:* 594.

Chapter 16

Allen, C. W. 1973. *Astrophysical Quantities.* 3rd ed. London: Athlone Press.

*****Bahcall**, J., and L. Spitzer. 1982. "The Space Telescope." *Scientific American,* July, p. 40.

Neugebauer, G., and others. 1984. "Early Results from the Infrared Astronomical Satellite." *Science 224:* 13.

*****Shapley**, H., ed. 1960. *Source Book in Astronomy.* New York: McGraw-Hill.

*****Shapley**, H., and H. Howarth, eds. 1929. *A Source Book in Astronomy.* New York: McGraw-Hill.

Weissman, P. 1984. "The Vega Particulate Shell: Comets or Asteroids?" *Science 224:* 987.

Chapter 17

*****Aller**, L. H. 1971. *Atoms, Stars, and Nebulae.* Cambridge, Mass.: Harvard University Press.

Baum, W. A. 1986. "The Role of Space Telescopes in the Detection of Brown Dwarfs." In *Astrophysics of Brown Dwarfs,* eds. M. Kafatos, R. Harrington, and S. Maran. Cambridge: Cambridge University Press.

Boesgaard, A., and W. Hagen. 1974. "The Age of Alpha Centauri." *Astrophysical Journal 189:* 85.

*****Burnham,** R. 1978. *Burnham's Celestial Handbook.* New York: Dover Publications.

Eddington, A. S. 1926. *The Internal Constitution of the Stars.* Cambridge, Mass.: Harvard University Press.

*****Humphreys,** R., and K. Davidson. 1984. "The Most Luminous Stars." *Science 223:* 243.

Iben, I. 1967. "Stellar Evolution: Within and Off the Main Sequence." *Annual Review of Astronomy and Astrophysics 4:* 171.

Kamper, K., and A. J. Wesselink. 1978. "Alpha and Proxima Centauri." *Astronomical Journal 83:* 1653.

Larson, R. B. 1969. "Numerical Calculations of the Dynamics of a Collapsing Proto-Star." *Monthly Notices of the Royal Astronomical Society 145:* 271.

Liebert, J. 1980. "White Dwarf Stars." *Annual Review of Astronomy and Astrophysics 18:* 363.

Popper, D. M. 1980. "Stellar Masses." *Annual Review of Astronomy and Astrophysics 18:* 115.

Shapley, H., ed. 1960. *Source Book in Astronomy, 1900–1950.* Cambridge, Mass.: Harvard University Press.

van de Kamp, P. 1981. *Stellar Paths.* Dordrecht, Netherlands: Reidel.

Westbrook, C., and C. B. Tarter. 1975. "On Protostellar Evolution." *Astrophysical Journal 200:* 48.

Chapter 18

Adams, M., K. Strom, and S. Strom. 1983. "The Star-forming History of the Young Cluster NGC 2264." *Astrophysical Journal,* Supplement Series *53,* no. 4.

Arny, T., and H. Weissman. 1973. "Interaction of Proto-Stars in a Collapsing Cluster." *Astronomical Journal 78:* 309.

Bailly, J., and C. Lada. 1983. "The High-Velocity Molecular Flows Near Young Stellar Objects." *Astrophysical Journal 265:* 824.

Bodenheimer, P. 1976. "Contraction Models for the Evolution of Jupiter." *Icarus 29:* 165.

*****Boss,** A. P. 1985. "Collapse and Formation of Stars." *Scientific American,* January, p. 40.

Cohen, M., and L. Kuhi. 1979. *Astrophysical Journal,* Supplement Series, December.

D'Antona, F. 1987. "Evolution of Very Low Mass Stars and Brown Dwarfs II." *Astrophysical Journal 320:* 633.

Hanson, R. B., B. Jones, and D. Ling. 1983. "The Astrometric Position of T Tauri and the Nature of Its Companion." *Astrophysical Journal 270:* L27.

Hartmann, L., and J. Raymond. 1984. "A High-Resolution Study of Herbig-Haro Objects 1 and 2." *Astrophysical Journal 276:* 560.

Hayashi, C. 1961. "Stellar Evolution in the Early Phases of Gravitational Contraction." *Publications of the Astronomical Society of Japan 13:* 450.

Henyey, L., R. LeLevier, and R. Levée. 1955. "The Early Phases of Stellar Evolution." *Publications of the Astronomical Society of the Pacific 67:* 154.

Herbig, G. 1968. "The Structure and Spectrum of R Monocerotis." *Astrophysical Journal 152:* 439.

————. 1975. "The Spectrum and Structure of 'Minkowski's Footprint': M 1-92." *Astrophysical Journal 200:* 1.

Iben, I. 1965. "Stellar Evolution I. The Approach to the Main Sequence." *Astrophysical Journal 141:* 993.

Jeans, J. H. 1902. "The Nebular Theory of the Origin of the Solar System." *Philosophical Transactions of the Royal Society of London 199:* 1. Reprinted in *A Source Book in Astronomy and Astrophysics, 1900–1975,* eds. K. Lang and O. Gingerich. Cambridge, Mass.: Harvard University Press, 1979.

*****Lada,** C. J. 1982. "Energetic Outflow from Young Stars." *Scientific American,* June, p. 82.

Lada C. J., and Frank H. Shu. 1990. "The Formation of Sunlike Stars." *Science 248:* 564.

Larson, R. B. 1969. "Numerical Calculations of the Dynamics of a Collapsing Protostar." *Monthly Notices of the Royal Astronomical Society 145:* 271.

Low, F. J., and B. Smith. 1966. "Infrared Observations of a Preplanetary System." *Nature 212:* 675.

Lunine, J., W. Hubbard, and M. Morley. 1986. "Evolution and Infrared Spectra of Brown Dwarfs." *Astrophysical Journal 310:* 238.

Mendoza, V. E. 1968. "Infrared Excesses in T Tauri Stars and Related Objects." *Astrophysical Journal 143:* 1010.

*****Neugebauer,** G., and E. Becklin. 1973. "The Brightest Infrared Sources." *Scientific American,* April, p. 28.

Poveda, A. 1965. "The H–R Diagram of Young Clusters and the Formation of Planetary Systems." *Boletin de los Observatorio Tonantzintla y Tacubaya* (Mexico City) *4:* 15.

Schmelz, J. 1984. "An Investigation of T Tauri Variability." *Astronomical Journal 89:* 108.

Schwartz, R. D. 1983. "Herbig–Haro Objects." *Annual Reviews in Astronomy and Astrophysics 21:* 209.

*****Scoville,** N., and J. Young. 1984. "Molecular Clouds, Star Formation and Galactic Structure." *Scientific American,* April, p. 42.

Stockton, A., D. Chesley, and S. Chesley. 1975. "Spectroscopy of R Monocerotis and NGC 2261." *Astrophysical Journal 199:* 406.

*****Strom,** S., and K. Strom. 1973. "The Early Evolution of Stars." *Sky and Telescope 45:* 279, 359.

Van der Linden, T., and R. Staller. 1983. "Evolution of Very Low-Mass Stars." *Astronomy and Astrophysics 118:* 285.

Walker, M. 1972. "Studies of Extremely Young Clusters. VI." *Astrophysical Journal 175:* 89.

*****Welch,** W., and others. 1985. "Gas Jets Associated with Star Formation." *Science 228:* 1389.

Wright, A. 1970. "Results of a Computer Program for Gravito-Gas Dynamic Collapse." *Proceedings of the Liège Colloquium* Mémoires de la Société Royale des Sciences de Liège 19: 75.

Chapter 19

Baade, W., and F. Zwicky. 1934. "Cosmic Rays from Super-Novae." *Proceedings of the National Academy of Science 20:* 259.

Baize, P. 1980. "Les Masses des Étoiles Variables a Longue Période." *L'Astronomie 94:* 71.

Bonneau, D., and others. 1982. "The Diameter of Mira." *Astronomy and Astrophysics 106:* 235.

*****Bova,** B. 1973. "Obituary of Stars: A Tale of Red Giants, White Dwarfs, and Black Holes." *Smithsonian 4:* 54.

*****Burnham,** Robert. 1978. *Burnham's Celestial Handbook.* New York: Dover Publications.

*****Burrows,** A. 1987. "The Birth of Neutron Stars and Black Holes." *Physics Today,* September, p. 28.

Conti, P. S., and R. McCray. 1980. "Strong Stellar Winds." *Science 208:* 9.

*****Hawking,** S. W. 1977. "The Quantum Mechanics of Black Holes." *Scientific American,* January, p. 34.

*****Helfand,** D. 1983. "Theory Points to Pulsating White Dwarfs." *Physics Today,* January, p. 21.

*————. 1987. "Bang: The Supernova of 1987." *Physics Today,* August, p. 25.

Hewish, A., and others. 1968. "Observation of a Rapidly Pulsating Radio Source." *Nature 217:* 709.

*Humphreys, R., and K. Davidson. 1984. "The Most Luminous Stars." *Science 223*: 243.

*Kaler, J. B. 1986. "Planetary Nebulae and the Death of Stars." *American Scientist 74*: 244.

Leavitt, H. S. 1912. *Periods of 25 Variable Stars in the Small Magellanic Cloud.* Harvard College Observatory Circular no. 173, p. 1.

*Liebert, J. 1980. "White Dwarf Stars." *Annual Review of Astronomy and Astrophysics 18*: 363.

Nolan, P., and J. Matteson. 1983. "A Feature in the X-ray Spectrum of Cygnus X-1: A Possible Positron Annihilation Line." *Astrophysical Journal 265*: 389.

Norman, C. A., and D. ter Haar. 1973. "On the Black Hole Model of Galactic Nuclei." *Astronomy and Astrophysics 24*: 121.

Novikov, I., and K. Thorne. 1973. "Astrophysics of Black Holes." In *Black Holes,* eds. C. DeWitt and B. S. DeWitt. London: Gordon and Breach.

Oppenheimer, J. R., and R. Serber. 1938. "On the Stability of Stellar Neutron Cores." *Physics Review 54*: 540.

*Penrose, R. 1972. "Black Holes." *Scientific American,* May, p. 38.

Pines, D. 1980. "Accreting Neutron Stars, Black Holes, and Degenerate Dwarf Stars." *Science 207*: 597.

Reddy, F. 1983. "Supernovae: Still a Challenge." *Sky and Telescope 66*: 485.

*Ruderman, M. 1971. "Solid Stars." *Scientific American,* February, p. 24.

*———. 1972. "Pulsars: Structure and Dynamics. *Annual Review of Astronomy and Astrophysics 10*: 427.

*Schorn, R. A. 1982. "The Gamma-Ray Burster Puzzle." *Sky and Telescope 63*: 560.

*Shaham, J. 1987. "The Oldest Pulsars in the Universe." *Scientific American,* February, p. 50.

Shapley, H., ed. 1960. *Source Book in Astronomy, 1900–1950.* Cambridge, Mass.: Harvard University Press.

Shu, Frank H. 1982. *The Physical Universe.* Mill Valley, Calif.: University Science Books.

*Wade, N. 1975. "Discovery of Pulsars: A Graduate Student's Story." *Science 189*: 359.

Waldrop, M. M. 1983. "The 0.001557806449023-Second Pulsar." *Science 219*: 831.

Wheeler, J., and K. Nomoto. 1985. "How Stars Explode." *American Scientist 73*: 240.

Witten, T. A. 1974. "Compounds in Neutron-Star Crusts." *Astrophysical Journal 188*: 615.

Woosley, S. E., and M. M. Phillips. 1988. "Supernova 1987A!" *Science 240*: 750.

Woosley, S. E., and Tom Weaver. 1989. "The Great Supernova of 1987." *Scientific American,* August, p. 32.

Zwicky, F. 1939. "On the Theory and Observation of Highly Collapsed Stars." *Physics Review 55*: 726.

Chapter 20

Allen, C. W. 1973. *Astrophysical Quantities.* 3rd ed. London: Athlone Press.

*Blitz, L. 1982. "Giant Molecular-Cloud Complexes in the Galaxy." *Scientific American,* April, p. 84.

Bowen, I. S. 1927. "The Origin of the Chief Nebular Lines." *Publications of the Astronomical Society of the Pacific 39*: 295.

Brackenridge, G. R. 1981. "Terrestrial Paleoenvironmental Effects of a Late Quaternary-Age Supernova." *Icarus 46*: 81.

*Cash, W., and P. Charles. 1980. "Stalking the Cygnus Superbubble." *Sky and Telescope 59*: 455.

Greenberg, J. M. 1974. "The Interstellar Depletion Mystery, or Where Have All These Atoms Gone?" *Astrophysical Journal 189*: L81.

———. 1984. "The Structure and Evolution of Interstellar Grains." *Scientific American,* June, p. 124.

Hartmann, J. F. 1904. "Investigations on the Spectrum and Orbit of σ Orionis." *Astrophysical Journal 19*: 268.

*Herbig, G. H. 1974. "Interstellar Smog." *American Scientist 62*: 200.

Johnson, H. L. 1968. "Interstellar Extinction." In *Nebulae and Interstellar Material,* ed. B. Middlehurst and L. Aller. Chicago: University of Chicago Press.

*Lewis, R., and E. Anders. 1983. "Interstellar Matter in Meteorites." *Scientific American,* August, p. 66.

Lewis, R. S., and others. 1980. "Stellar Condensates in Meteorites: Isotopic Evidence from Noble Gases." *Astrophysical Journal 234*: L165.

*Lovi, G. 1972. "Rambling Through May Skies." *Sky and Telescope 43*: 306.

Maran, S., J. Brandt, and T. Stecher, eds. 1973. *The Gum Nebula and Related Problems.* Publication no. SP-332. Washington, D.C.: National Aeronautics and Space Administration.

*Miller, J. S. 1974. "The Structure of Emission Nebulae." *Scientific American,* October, p. 34.

Oort, J., and H. C. van de Hulst. 1946. "Gas and Smoke in Interstellar Space." *Bul-*

letin of the Astronomical Society of the Netherlands 10: 187.

Poveda, A. 1965. "G. W. Orionis, a 20,000 Year Old T Tauri Star?" *Boletin de los Observatorio Tonantzintla y Tacubaya 4*: 77.

Scoville, N. Z. 1975. "Molecular Clouds in the Galaxy." *Astrophysical Journal 199*: L105.

*Scoville, N., and J. Young. 1984. "Molecular Clouds, Star Formation and Galactic Structure." *Scientific American,* April, p. 42.

Shu, F. H. 1982. *The Physical Universe: An Introduction to Astronomy.* Mill Valley, Calif.: University Science Books.

Simon, T., and H. M. Dyck. 1975. "Silicate Absorption at 18 μm in Two Peculiar Infrared Sources." *Nature 253*: 101.

Stromgren, B. 1939. "The Physical State of Interstellar Hydrogen." *Astrophysical Journal 89*: 526.

Swart, P. K., and others. 1983. "Interstellar Carbon in Meteorites." *Science 220*: 406.

Trumpler, R. J. 1930. "Preliminary Results on the Distances, Dimensions, and Space Distributions of Open Star Clusters." *Lick Observatory Bulletin 14*: 154.

*Turner, B. E. 1973. "Interstellar Molecules." *Scientific American,* March, p. 50.

Walborn, N., J. Heckathorn, and J. Hesser. 1984. "The High-ionization and Excited-state Interstellar Lines in the Carina Nebula: A Giant HII Region in Absorption." *Astrophysical Journal 276*: 524.

Whittaker, A. G., and others. 1980. "Carbynes: Carriers of Primordial Noble Gases in Meteorites." *Science 209*: 1512.

Zuckerman, B. 1973. "A Model of the Orion Nebula." *Astrophysical Journal 183*: 863.

Chapter 21

Abt, H. A. 1979. "The Frequencies of Binaries on the Main Sequence." *Astronomical Journal 84*: 1591.

———. 1983. "Normal and Abnormal Binary Frequencies." *Annual Review of Astrophysics 21*: 343.

Abt, H., and S. Levy. 1976. "Multiplicity Among Solar-Type Stars." *Astrophysical Journal,* Supplement series 30: 273.

Arny, T., and P. Weissman. 1973. "Interaction of Proto-Stars in a Collapsing Cluster." *Astronomical Journal 78*: 310.

*Batten, A. H. 1973. *Binary and Multiple System of Stars.* Oxford: Pergamon Press.

Dyck, H., T. Simon, and B. Zuckerman. 1982. "Discovery of an Infrared Companion to T Tauri." *Astrophysics Journal Letters 225*: L103.

Gatewood, G., and H. Eichhorn. 1973. "An Unsuccessful Search for a Planetary Companion of Barnard's Star." *Astronomical Journal 78:* 769.

Glasby, J. S. 1968. *Variable Stars.* London: Constable.

Grindlay, J. E., and others. 1984. "The Central X-ray Source in SS 433." *Astrophysical Journal 277:* 286.

Hack, 1984. "Epsilon Aurigae." *Scientific American,* October, p. 98.

Hansen, R., B. Jones, and D. Lin. 1983. "The Astrometric Position of T Tauri and the Nature of Its Companion." *Astrophysics Journal Letters 270:* L27.

*Heintz, W. D. 1978. *Double Stars.* Dordrecht, Netherlands: Reidel.

Hershey, J. L. 1973. "Astrometric Analysis of the Triple Star BD +66°34." *Astronomical Journal 78:* 935.

Kopal, Z. 1955. "The Classification of Close Binary Systems," *Annals of Astrophysics 18:* 379.

———. 1971. "The Eclipsing System of Epsilon Aurigae and Its Possible Relevance to the Formation of a Planetary System." *Astrophysics and Space Science 10:* 332.

MacRobert, A. 1988. "Epsilon Aurigae: Puzzle Solved?" *Sky and Telescope 75:* 15.

*Margon, B. 1980. "The Bizarre Spectrum of SS 433." *Scientific American,* October, p. 54.

*———. 1982. "Relativistic Jets in SS 433." *Science 215:* 247.

Popper, D. M. 1980. "Stellar Masses." *Annual Review of Astronomy and Astrophysics 18:* 115.

Radzievskii, V., and E. Radzievskaya. 1973. "Coplanar System of Binary Stars in Aquila." *Soviet Astronomy 17:* 239.

Smith, B. A., and Terrile, R. 1984. "A Circumstellar Disk Around Beta Pictoris." *Science 226:* 1421.

Thomsen, D. 1986. "A 'Brickbat' in the Sky." *Science News 127:* 154.

van de Kamp, P. 1981. *Stellar Paths.* Dordrecht, Netherlands: Reidel.

Ventura, J., and others. 1983. "Can X-ray Bursts Originate from Low-Mass Binaries?" *Nature 301:* 491.

Wanner, J. F. 1967. "The Visual Binary Krüger 60." *Sky and Telescope 33:* 16.

*Warner, B. 1972. "Six Ultra-Short Period Binary Stars." *Sky and Telescope 44:* 358.

Chapter 22

Ambartsumian, V. A. 1947. *Stellar Evolution and Astrophysics.* Moscow, U.S.S.R.: Erevan.

———. 1960. "Expanding Stellar Associations." In *Source Book in Astronomy, 1900–1950,* ed. H. Shapley. Cambridge, Mass.: Harvard University Press.

Anthony-Twarog, B. J., B. Twarog, and R. D. McClure. 1979. "The Old(est) Open Cluster: Melotte 66." *Astrophysical Journal 233:* 188.

*Iben, I. 1970. "Globular-Cluster Stars." *Scientific American,* July, p. 26.

Jones, K., and P. Demarque. 1983. "The Ages and Composition of Old Clusters." *Astrophysical Journal 264:* 206.

*Jones, K. G. 1969. *Messier's Nebulae and Star Clusters.* New York: American Elsevier.

King, I. R. 1985. "Globular Clusters." *Scientific American,* July, p. 79.

Peebles, P., and R. Dicke. 1968. "Origin of the Globular Star Clusters." *Astrophysical Journal 154:* 891.

Pilachowski, C. 1984. "The Chemical Composition of Globular Clusters: Global Trends." *Astrophysical Journal 281:* 614.

Ruzmaikina, T. 1972. "On the Cosmological Origin of Globular Clusters." *Astronomical zhurnal nauk SSSR 49:* 1229.

*Shapley, H. 1930. *Star Clusters.* Cambridge, Mass.: Harvard University Press.

Trumpler, R. 1930. "Absorption of Light in the Galactic System." *Publications of the Astronomical Society of the Pacific 42:* 214.

VandenBerg, D., and P. Durrell. 1990. "Is Age *Really* the Second Parameter in Globular Clusters?" *Astrophysical Journal 99:* 221.

Chapter 23

*Bok, B. J., and P. Bok. 1981. *The Milky Way.* 5th ed. Cambridge, Mass.: Harvard University Press.

Dicke, R. H. 1969. "The Age of the Galaxy from the Decay of Uranium." *Astrophysical Journal 155:* 123.

*Geballe, T. R. 1979. "The Central Parsec of the Galaxy." *Scientific American,* July, p. 60.

Genzel, R., and others. 1984. "Far-infrared Spectroscopy of the Galactic Center: Neutral and Ionized Gas in the Central 10 Parsecs of the Galaxy." *Astrophysical Journal 276:* 551.

Genzel, R., and C. Townes. 1987. "Physical Conditions, Dynamics and Mass Distribution in the Center of the Galaxy." *Annual Review of Astronomy and Astrophysics 25:* 377.

Hawkins, M. R. S. 1983. "Direct Evidence for a Massive Galactic Halo." *Nature 303:* 406.

Jansky, K. 1935. "A Note on the Source of Interstellar Interference." *Proceedings of the Institute of Radio Engineers 23:* 1158.

Klare, G., and T. Neckel. 1970. "Polarization of Southern OB Stars." In *The Spiral Structure of Our Galaxy,* eds. W. Becker and G. Contopolos. Dordrecht, Netherlands: Reidel.

*Kraft, R. P. 1959. "Pulsating Stars and Cosmic Distances." *Scientific American,* July, p. 48.

Lacy, J. H., and others. 1980. "Observations of the Motion and Distribution of the Ionized Gas in the Central Parsec of the Galaxy. II." *Astrophysical Journal 241:* 132.

Leventhal, M., and others. 1982. "Time-variable Positron Annihilation Radiation from the Galactic Center Direction. *Astrophysical Journal Letters 260:* L1.

Lo, K., and M. Claussen. 1983. "High-resolution Observations of Ionized Gas in Central 3 Parsecs of the Galaxy: Possible Evidence for Infall." *Nature 306:* 647.

Lo, K. Y. 1986. "The Galactic Center: Is It a Massive Black Hole?" *Science 233:* 1394.

Maraschi, L., A. Treves, and M. Tarenghi. 1973. "Accretion by Neutron Stars at the Galactic Center." *Astronomy and Astrophysics 25:* 153.

Moffat, F. J., and N. Vogt. 1973. "An Up-to-Date Picture of Galactic Spiral Features Based on Young Open Star Clusters." *Astronomy and Astrophysics 23:* 317.

Racine, R., and W. E. Harris. 1989. "Globular Clusters and the Distance to the Galactic Center." *Astronomical Journal 98:* 1609

Reber, G. 1944. "Cosmic Static." *Astrophysical Journal 100:* 279.

Sanders, R., and K. Prendergast. 1974. "The Possible Relation of the 3-kpc Arm to Explosions in the Galactic Nucleus." *Astrophysical Journal 188:* 489.

*Saunders, J. 1963. "The Globular Cluster Omega Centauri." *Sky and Telescope 26:* 133.

*Shapley, H. 1930. *Star Clusters.* Cambridge, Mass.: Harvard University Press.

Shapley, H., and H. Howarth, eds. 1929. *A Source Book in Astronomy.* New York: McGraw-Hill.

*Struve, O. 1960. "A Historic Debate About the Universe." *Sky and Telescope 19:* 398.

Walborn, N. R. 1973. "The Space Distribution of the O Stars in the Solar Neighborhood." *Astronomical Journal 78:* 1067.

Waldrop, M. M. 1985. "The Core of the Milky Way." *Science 230:* 158.

Waldrop, M. 1989. "Feeding the Monster in the Middle." *Science 243:* 478.

*Whitney, C. A. 1971. *The Discovery of Our Galaxy.* New York: Alfred A. Knopf.

Chapter 24

Allen, C. W. 1973. *Astrophysical Quantities.* London: Athlone Press.

Armus, L., T. Heckman, and G. Miley. 1987. "Multicolor Optical Imaging of Powerful Far-Infrared Galaxies." *Astronomical Journal 94:* 831.

*****Bok,** B. J. 1966. "Magellanic Clouds." *Annual Review of Astronomy and Astrophysics 4:* 95.

de Vaucouleurs, G. 1974. "Structure, Dynamics, and Statistical Properties of Galaxies." In *The Formation and Dynamics of Galaxies,* ed. J. Shakeshaft. Dordrecht, Netherlands: Reidel.

Freedman, W., and B. Madore. 1983. "Time Evolution of Disk Galaxies Undergoing Stochastic Self-propagating Star Formation." *Astrophysical Journal 265:* 140.

Gunn, J. E. 1979. "Feeding the Monster: Gas Discs in Elliptical Galaxies." In *Active Galactic Nuclei,* eds. G. Hazard and S. Mitton. London: Cambridge University Press.

*****Hirshfeld,** A. 1980. "Inside Dwarf Galaxies." *Sky and Telescope 59:* 287.

Hodge, P. 1987. "The Local Group." *Mercury 16:* 2.

*****Humphreys,** R., and K. Davidson. 1984. "The Most Luminous Stars." *Science 223:* 243.

Humphreys, R., P. Massey, and W. Freedman. 1990. "Spectroscopy of Luminous Blue Stars in M 31." *Astronomical Journal 99:* 84.

Light, E., R. Danielson, and M. Schwarzschild. 1974. "The Nucleus of M 31." *Astrophysical Journal 194:* 257.

Mathewson, D. 1985. "The Clouds of Magellan." *Scientific American,* April, p. 107.

Mathis, J., B. Savage, and J. Cassinelli. 1984. "A Superluminous Object in the Large Cloud of Magellan." *Scientific American,* August, p. 52.

Ostriker, J. P. 1981. "Some Thoughts on Galaxy Formation." *Bulletin of the American Astronomy Society 12:* 770.

*****Rubin,** V. C. 1973. "The Dynamics of the Andromeda Nebula." *Scientific American,* June, p. 30.

*****Rubin,** V. 1983. "Dark Matter in Spiral Galaxies." *Scientific American,* June, p. 96.

*****Sandage,** A. 1961. *The Hubble Atlas of Galaxies.* Washington, D.C.: The Carnegie Institution.

Sandage, A., and G. A. Tammann. 1974. "Steps Toward the Hubble Constant. III." *Astrophysical Journal 194:* 223.

*****Schweizer,** F. 1986. "Colliding and Merging Galaxies." *Science 231:* 227.

*****Shapley,** H. 1957. *The Inner Metagalaxy.* New Haven, Conn.: Yale University Press.

*****Silk,** J. 1987. "The Formation of Galaxies." *Physics Today 40,* no. 4, p. 28.

Simkin, S., and others. 1987. "Markarian 348: A Tidally Disturbed Seyfert Galaxy." *Science 235:* 1367.

Spillar, E., and others. 1990. "Infrared Speckle Observations of the Nucleus of M 31." *Astrophysical Journal 349:* L13.

*****Strom,** K., and S. Strom. 1982. "Galactic Evolution: A Survey of Recent Progress." *Science 216:* 571.

Thackeray, A. D. 1971. "Survey of Principal Characteristics of the Magellanic Clouds." In *The Magellanic Clouds,* ed. A. Muller. Boston: Reidel.

*****Toomre,** A., and J. Toomre. 1973. "Violent Tides Between Galaxies." *Scientific American,* December, p. 38.

van den Bergh, S. 1972. "Search for Faint Companions to M31." *Astrophysical Journal 171:* L35.

*****Whitney,** C. A. 1971. *The Discovery of Our Galaxy.* New York: Alfred A. Knopf.

Chapter 25

*****Anderson,** P. 1987. "Massive Objects in Galactic Nuclei May Be Black Holes." *Physics Today,* October, p. 22.

Arp, H. 1974. "Evidence for Non-Velocity Red-shifts—New Evidence and Review." In *The Formation and Dynamics of Galaxies,* ed. J. Shakeshaft. Dordrecht, Netherlands: Reidel.

Balick, B., and T. Heckman. 1983. "Spectroscopy of the Fuzz Associated with Four Quasars." *Astrophysical Journal 265:* L1.

Balkowski, C., and others. 1974. "Observational Evidence for Non-Velocity Red-shift in Stephan's Quintet." In *The Formation and Dynamics of Galaxies,* ed. J. Shakeshaft. Dordrecht, Netherlands: Reidel.

Bartel, N., and others. 1985. "Hubble's Constant Determined Using Very-long Baseline Interferometry." *Nature 318:* 25.

Blandford, R., and others. 1982. "Cosmic Jets." *Scientific American,* May, p. 124.

Burns, J., and R. Price. 1983. "Centaurus A: The Nearest Active Galaxy." *Scientific American,* November, p. 56.

Chincarini, G., and H. Rood. 1980. "The Cosmic Tapestry." *Sky and Telescope 59:* 364.

Djorgovski, S., and others. 1987a. "A Galaxy at a Redshift of 3.215." *Astronomical Journal 93:* 1318.

———. 1987b. "Discovery of a Probable Binary Quasar." *Astrophysical Journal 321:* L17.

Dokuchaev, V. I. 1989. "The Evolution of a Massive Black Hole in the Nucleus of a Normal Galaxy." *Soviet Astronomical Letters 15:* 167.

Dressler, A., and D. Richstone. 1988. *Astrophysical Journal 324,* in press.

Eddington, A. 1933. *The Expanding Universe.* Cambridge: Cambridge University Press.

Evans, C. R., and C. Kochanek. 1989. "The Tidal Disruption of a Star by a Massive Black Hole." *Astrophysical Journal Letters 346:* L13.

Gunn, J. E. 1979. "Feeding the Monster: Gas Discs in Elliptical Galaxies." In *Active Galactic Nuclei,* eds. C. Hazard and S. Mitton. London: Cambridge University Press.

Hayes, J., and A. Sadun. 1987. "CCD Observations of the Jet of the Quasar 3C 273." *Astronomical Journal 94:* 871.

*****Hodge,** P. W. 1972. "Some Current Studies of Galaxies." *Sky and Telescope 44:* 23.

———. 1984. "The Cosmic Distance Scale." *American Scientist 72:* 474.

*****Hoyle,** F. 1972. *From Stonehenge to Modern Cosmology.* San Francisco: W. H. Freeman.

Hubble, E. P. 1931. "The Velocity Distance Relation Among Extra-Galactic Nebulae." *Astrophysical Journal 74:* 43.

———. 1936. "A Relation Between Distance and Radial Velocity Among Extra-Galactic Nebulae." Reprinted in *Source Book in Astronomy, 1900–1950,* ed. H. Shapley. Cambridge, Mass.: Harvard University Press, 1960.

———. 1958. *The Realm of the Nebulae.* New York: Dover.

Khachikian, E. Y., and D. Weedman. 1974. "An Atlas of Seyfert Galaxies." *Astrophysical Journal 192:* 581.

Kochanek, C., S. Shapiro, and S. Teukolsky. 1987. "How Big Are Supermassive Black Holes Formed from the Collapse of Dense Star Clusters?" *Astrophysical Journal 320:* 73.

Kühr, H., and others. 1973 or 1983. "The Most Luminous Quasar: S5 0014 + 81." *Astrophysical Journal 275:* L33.

Osmer, P. 1982. "Quasars as Probes of the Distant and Early Universe." *Scientific American,* February, p. 126.

Peterson, B. 1982. "PKS 2000-300: A Quasi-Stellar Radio Source with a Red Shift of 3.78." *Astrophysical Journal 260:* L27.

Petrosian, V. 1974. "The Hubble Relation for Non-Standard Candles and the Origin of the Redshift of Galaxies." *Astrophysical Journal 188:* 443.

Rafanelli, P., D. Osterbrock, and R. Pogge. 1990. "The Optical Counterpart of the Radio Source Close to the Seyfert 2 Nucleus of NGC 5953 = Arp 91 B." *Astronomical Journal 99:* 52.

Rees, M. J. 1990. " 'Dead Quasars' in Nearby Galaxies?" *Science 247:* 817.

Sandage, A., and G. A. Tammann. 1974. "Steps Toward the Hubble Constant. IV." *Astrophysical Journal 194:* 559.

Saslaw, W. C. 1974. "Theory of Galactic Nuclei." In *The Formation and Dynamics of Galaxies,* ed. J. Shakeshaft. Dordrecht, Netherlands: Reidel.

Schneider, D., M. Schmidt, and J. Gunn. 1989. "A Study of Ten Quasars with Redshifts Greater than Four." *Astronomical Journal 98:* 1507.

Shapley, H., ed. 1960. *Source Book in Astronomy, 1900–1950.* Cambridge, Mass.: Harvard University Press.

Shu, F. H. 1982. *The Physical Universe: An Introduction to Astronomy.* Mill Valley, Calif.: University Science Books.

Silk, J. 1980. *The Big Bang.* San Francisco: W. H. Freeman.

Simkin, S., and others. 1987. "Markarian 348: A Tidally Distorted Seyfert Galaxy." *Science 235:* 1367.

Smith M., and others. 1983. "Bubbles, Jets, and Clouds in Active Galactic Nuclei." *Astrophysical Journal 264:* 432.

Telesco, C. M. 1988. "Enhanced Star Formation and Infrared Emission in the Centers of Galaxies." *Annual Review of Astronomy and Astrophysics 26:* 343.

*Toomre, A., and J. Toomre. 1973. "Violent Tides Between Galaxies." *Scientific American,* December, p. 38.

Turner, E. 1984. "Quasars and Gravitational Lenses." *Science 223:* 1255.

van der Hulst, J., W. Golisch, and A. Huschick. 1983. "The H1 Absorption on NGC 5128 (Centaurus A)." *Astrophysical Journal 264:* L37.

Waldrop, M. 1983. "NGC 4151: The Monster in the Middle." *Science 222:* 1003.

———. 1985. "Why Do Galaxies Exist?" *Science 228:* 978.

*Weymann, R. J. 1969. "Seyfert Galaxies." *Scientific American,* January, p. 28.

*Whitney, C. A. 1971. *The Discovery of Our Galaxy.* New York: Alfred A. Knopf.

Chapter 26

*Anderson, P. 1987. "Ripples in the Universal Hubble Flow." *Physics Today,* October, p. 17.

Brown, G. S., and B. Tinsley. 1974. "Galaxy Counts as a Cosmological Test." *Astrophysical Journal 194:* 555.

Burns, J. O. 1986. "Very Large Structures in the Universe." *Scientific American,* June, p. 38.

*deBoer, K., and B. Savage. 1982. "The Coronas of Galaxies." *Scientific American,* August, p. 54.

De Lapparent, V., M. Geller, and J. Huchra. 1986. "A Slice of the Universe." *Astrophysical Journal 302:* L1.

*de Vaucouleurs, G. 1970. "The Case for a Hierarchical Cosmology." *Science 167:* 1203.

*Dicus, D., and others. 1983. "The Future of the Universe." *Scientific American,* March, p. 90.

Einstein, A. 1952. "Cosmological Considerations on the General Theory of Relativity." In *The Principle of Relativity.* New York: Dover. (Originally published 1917.)

Fiedler, R., and others. 1987. *Nature,* April 16.

Geller, M., and J. Huchra. 1989. "Mapping the Universe." *Science 246:* 897.

*Gregory, S., and L. Thompson. 1982. "Superclusters and Voids in the Distribution of Galaxies." *Scientific American,* March, p. 106.

*Harrison, E. R. 1974. "Why the Sky Is Dark at Night." *Physics Today 27:* 5.

Kant, I. 1968. *Kant's Cosmology.* Translated by W. Hastie. New York: Greenwood.

Kraus, J. 1981. "Universe Beyond the Galaxies." Letter in *Science 212:* 610.

*North, J. D. 1965. *The Measure of the Universe.* Oxford: Oxford University Press.

Raychaudhuri, A. K. 1979. *Theoretical Cosmology.* Oxford: Clarendon Press.

Rood, H. J. 1988. "Voids." *Annual Review of Astronomy and Astrophysics 26:* 245.

*Rubin, V. 1983. "Dark Matter in Spiral Galaxies." *Scientific American,* June, p. 96.

Ryle, M. 1968. "The Counts of Radio Sources." *Annual Review of Astronomy and Astrophysics 6:* 249.

*Schatzman, E. S. 1965. *The Origin and Evolution of the Universe.* New York: Basic Books.

Shu, F. 1982. *The Physical Universe.* Mill Valley, Calif.: University Science Press.

Silk, J., 1980. *The Big Bang.* San Francisco: W. H. Freeman.

*Silk, J., A. Szalay, and Y. Zel'dovich. 1983. "The Large-Scale Structure of the Universe." *Scientific American,* October, p. 72.

Thomsen, D. E. 1987. "Neutrino Mass: A Tritium Disagreement." *Science News 131:* 342.

Trimble, V. 1987. "Existence and Nature of Dark Matter in the Universe." *Annual Review of Astronomy and Astrophysics 25:* 425.

Tyson, J., F. Valdes, and R. Wenk. 1990. "Detection of Systematic Gravitational Lens Galaxy Image Alignments: Mapping Dark Matter in Galaxy Clusters." *Astrophysical Journal 349:* L1

*Waldrop, M. M. 1983. "The New Inflationary Universe." *Science 219:* 375.

*———. 1986. "In Search of Dark Matter." *Science 234:* 152.

Wesson, P., K. Valle, and R. Stabell. 1987. "The Extragalactic Background Light and a Definitive Resolution of Olbers' Paradox." *Astrophysical Journal 317:* 601.

*Wilson, J. A. 1949. "Egypt." In *Before Philosophy,* eds. H. Frankfort and H. A. Frankfort. Baltimore: Penguin Books.

Chapter 27

Alpher, R. A., and R. C. Herman. 1948. "Evolution of the Universe." *Nature 162:* 774.

*Barrow, J., and M. Turner. 1982. "The Inflationary Universe—Birth, Death, and Transfiguration." *Nature 298:* 801.

*Burns, J. O. 1986. "Very Large Structures in the Universe." *Scientific American,* June, p. 38.

Butcher, H. 1987. "Thorium in G-dwarf Stars as a Chronometer for the Galaxy." *Nature 328:* 127.

Dicke, R. H., and others. 1965. "Cosmic Black Body Radiation." *Astrophysical Journal 142:* 414.

Gamow, G. 1952. *The Creation of the Universe.* New York: Viking Press.

*Hawking, S. W. 1977. "The Quantum Mechanics of Black Holes." *Scientific American,* January, p. 34.

Linde, A. 1987. "Particle Physics and Inflationary Cosmology." *Journal of Physics Today,* September, p. 61.

*Morrison, D., and N. Morrison. 1975. "An Infinitely Expanding Universe." *Mercury,* September, p. 27.

Ozernoy, L. M. 1974. "Dynamics of Superclusters as the Most Powerful Test for Theories of Galaxy Formation." In *The Formation and Dynamics of Galaxies,* ed. J. Shakeshaft. Dordrecht, Netherlands: Reidel.

Peebles, P. J. E. 1971. *Physical Cosmology.* Princeton, N.J.: Princeton University Press.

Penzias, A. A., and R. W. Wilson. 1965. "A Measurement of Excess Antenna Temperature at 4080 Mc/s." *Astrophysical Journal 142:* 419.

Reeves, H., and others. 1972. "On the Origin of Light Elements." Cal Tech Preprint OAP-296.

Schmidt, M. 1982. "Quasar Boundary." *Science 216:* 6.

*****Silk,** J. 1980. *The Big Bang.* San Francisco: W. H. Freeman.

*****Webster,** A. 1974. "The Cosmic Background Radiation." *Scientific American,* August, p. 26.

*****Wilkinson,** D. 1986. "Anisotropy of the Cosmic Blackbody Radiation." *Science 232:* 1517.

Chapter 28

Ahrens, T. J., and J. D. O'Keefe. 1982. "Impact of an Asteroid or Comet in the Ocean and Extinction of Terrestrial Life." *Lunar Planet Science Abstracts 13:* 3.

Allen, C. W. 1973. *Astrophysical Quantities.* London: Athlone Press.

*****Alvarez,** W., and others. 1984. "Impact Theory of Mass Extinctions and the Invertebrate Fossil Record." *Science 223:* 1135.

*****Berendzen,** R., ed. 1973. *Life Beyond Earth and the Mind of Man.* Publication no. SP-328. Washington, D.C.: National Aeronautics and Space Administration.

Berkner L. V., and L. C. Marshall. 1965. "On the Origin and Rise of Oxygen Concentration in the Earth's Atmosphere." *Journal of Atmospheric Science 22:* 225.

Brock, T. D. 1985. "Life at High Temperatures." *Science 230:* 132.

*****Covey,** C. 1984. "The Earth's Orbit and the Ice Ages." *Scientific American,* February, p. 58.

*****Dole,** S. H. 1964. *Habitable Planets for Man.* Waltham, Mass.: Blaisdell.

*****Edmund,** J. M., and K. Von Damm. 1983. "Hot Springs on the Ocean Floor." *Scientific American,* April, p. 78.

*****Gould,** S. J. 1974a. "The Great Dying." *Natural History,* October, p. 22.

*****———.** 1974b. "An Unsung Single-Celled Hero." Natural History, p. 33.

———. 1978. "An Early Start." *Natural History,* February, p. 10.

Hanson, R., B. Jones, and D. Lin. 1983. "The Astrometric Position of T Tauri and the Nature of Its Companion." *Astrophysical Journal 270:* L27.

Harrington, R. S. 1977. "Plantary Orbits in Binary Stars." *Astronomical Journal 82:* 753.

Harrington, R. S., V. V. Kallarkakal, and C. Dahn. 1983. "Astrometry of the Low-Luminosity Stars VB8 and VB10." *Astronomical Journal 88:* 1038.

Holland, H. D. 1962. "Model for Evolution of the Earth's Atmosphere." In *Petrologic Studies.* New York: Geological Society of America.

Horowitz, P. 1978. "A Search for Ultra-Narrowband Signals of Extraterrestrial Origin." *Science 201:* 733.

Imbrie, J., and J. Z. Imbrie. 1980. "Modeling the Climatic Response to Orbital Variations." *Science 207:* 943.

Karl, D. M., and others. 1980. "Deep-Sea Primary Production at the Galápagos Hydrothermal Vents." *Science 207:* 1345.

*****Kurten,** B. 1969. "Continental Drift and Evolution." *Scientific American,* March, p. 54.

Kvenvolden, K., and others. 1970. "Evidence for Extraterrestrial Amino Acids and Hydrocarbons in the Murchison Meteorite." *Nature 228:* 923.

McLaren, D. J. 1970. "Presidential Address." *Journal of Paleontology 44:* 801.

Miller, S. L. 1955. "Production of Some Organic Compounds Under Possible Primitive Earth Conditions." *Journal of the American Chemical Society 77:* 2351.

Nisbet, E. G. 1980. "Archaean Stromatolites and the Search for the Earliest Life." *Nature 284:* 394.

O'Leary, B. 1980. "Searching for Other Planetary Systems." *Sky and Telescope 60:* 111.

*****Oparin,** A. I. 1962. *Life: Its Nature, Origin, and Development.* New York: Academic Press.

*****Pollard,** W. B. 1979. "The Prevalence of Earth-like Planets." *American Scientist 67:* 653.

Ponnamperuma, C., ed. 1983. *Cosmochemistry and the Origin of Life.* Dordrecht, Netherlands: D. Reidel.

Raup, D. M. 1979. "Size of the Permo-Triassic Bottleneck and Its Evolutionary Implications." *Science 206:* 217.

Rowland, F. S. 1989. "Chlorofluorocarbons and the Depletion of Stratospheric Ozone." *American Scientist 77:* 36.

*****Russell,** D. A. 1982. "The Mass Extinctions of the Late Mesozoic." *Scientific American,* January, p. 82.

Sekanina, Z., and D. Yeomans. 1984. "Close Encounters and Collisions of Comets with the Earth." *Astronomical Journal 89:* 154.

*****Siegel,** S. M. 1972. "Experimental Biology of Extreme Environments and Its Significance for Space Bioscience—2." *Spaceflight 12:* 256.

Simpson, G. G. 1964. "The Nonprevalence of Humanoids." *Science 143:* 769.

———. 1973. Added comments on "The Nonprevalence of Humanoids." In *Communication with Extraterrestrial Intelligence,* ed. C. Sagan. Cambridge, Mass.: M.I.T. Press.

Stern, D. K. 1975. "First Contact with Nonhuman Cultures." *Mercury,* September/October, p. 14.

Strand, K. A. 1977. "Triple System Stein 2051 (G175-34)." *Astronomical Journal 82:* 745.

Turco, R., and others. 1981. "Tunguska Meteor Fall of 1908: Effects on Stratospheric Ozone." *Science 212:* 19.

Vidal, G. 1984. "The Oldest Eukaryotic Cells." *Scientific American,* February, p. 48.

Walker, J. C. G. 1977. *Evolution of the Atmosphere.* New York: Hafner.

Windley, B. F. 1977. *The Evolving Continents.* New York: Wiley and Sons.

*****Woese,** Carl R. 1981. "Archaebacteria." *Scientific American,* June, p. 98.

Wolbach, W., R. Lewis, and E. Anders. 1985. "Cretaceous Extinctions: Evidence for Wildfires and Search for Meteoritic Material." *Science 230:* 167.

Enrichment Essay A

*****Abetti,** G. 1951. *The History of Astronomy.* New York: Henry Schuman.

*****Condon,** E. U., and others. 1969. *Scientific Study of Unidentified Flying Objects.* New York: Dutton.

Emiliani, C., and others. 1975. "Paleoclimatological Analysis of Late Quaternary Cores from the Northeastern Gulf of Mexico." *Science 189:* 1083.

*****Gould,** S. J. 1975. "Velikovsky in Collision." *Natural History,* February, p. 20.

Mathur, K. D. 1974. "Indian Astronomy in the Era of Copernicus." *Nature 251:* 283.

*****Minnaert,** M. 1954. *Light and Colour in the Open Air.* New York: Dover.

*****Numbers,** R. L. 1982. "Creationism in 20th-Century America." *Science 218:* 538.

Pannekoek, A. 1961. *A History of Astronomy.* London: George Allen and Unwin.

Sagan, C., and J. Agel. 1975. *Other Worlds.* New York: Bantam Books.

*****Shklovskii,** I. S., and C. Sagan. 1966. *Intelligent Life in the Universe.* San Francisco: Holden-Day.

*****Stephenson,** F. R., and D. H. Clark. 1977. "Ancient Astronomical Records from the Orient." *Sky and Telescope 53:* 84.

Velikovsky, I. 1950. *Worlds in Collision.* New York: Doubleday.

von Däniken, E. 1971. *Chariots of the Gods?* New York: Bantam.

Index

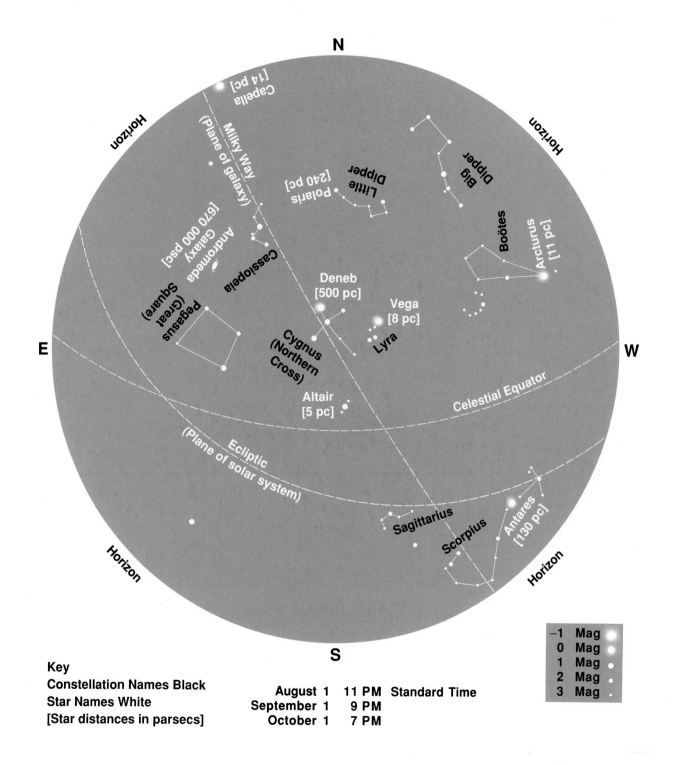

Key
Constellation Names Black
Star Names White
[Star distances in parsecs]

August 1   11 PM   Standard Time
September 1   9 PM
October 1   7 PM

-1 Mag
0 Mag
1 Mag
2 Mag
3 Mag

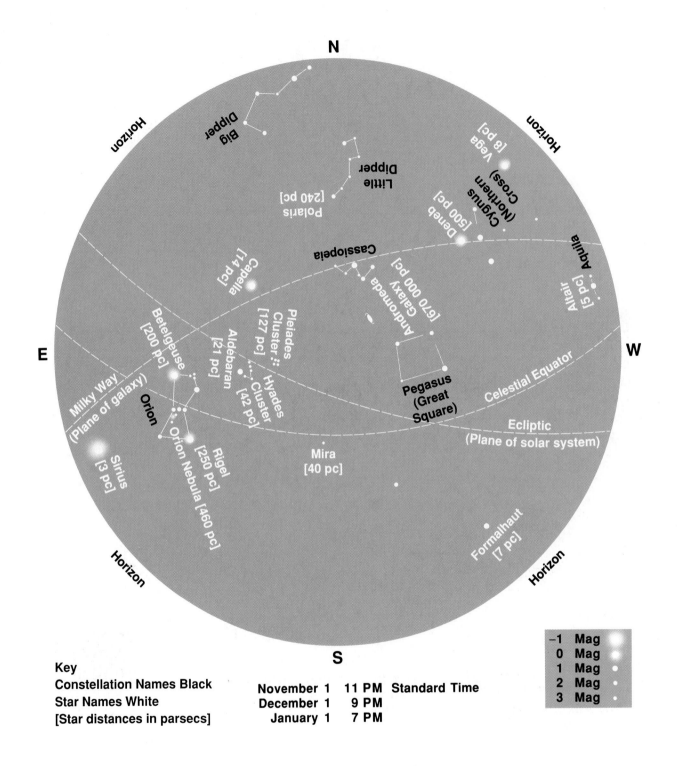

N

Big Dipper

Little Dipper

Polaris [240 pc]

Horizon

Vega [8 pc]

Horizon

Cygnus (Northern Cross)

Deneb [500 pc]

Cassiopeia

Altair [5 pc]

Aquila

Capella [14 pc]

Andromeda Galaxy [670 000 pc]

Pleiades Cluster [127 pc]

Betelgeuse [200 pc]

Aldebaran [21 pc]

Hyades Cluster [42 pc]

Pegasus (Great Square)

E

W

Milky Way (Plane of galaxy)

Orion

Celestial Equator

Rigel [250 pc]

Orion Nebula [460 pc]

Mira [40 pc]

Ecliptic (Plane of solar system)

Sirius [3 pc]

Horizon

Formalhaut [7 pc]

Horizon

S

-1 Mag
0 Mag
1 Mag
2 Mag
3 Mag

Key
Constellation Names Black
Star Names White
[Star distances in parsecs]

November 1   11 PM   Standard Time
December 1   9 PM
January 1   7 PM

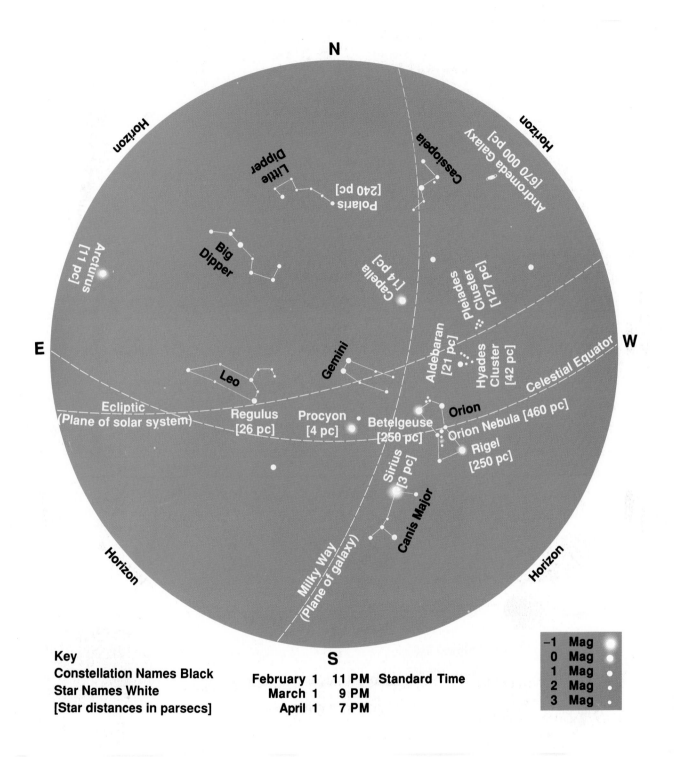

N

Horizon

Horizon

Cassiopeia

Andromeda Galaxy [670 000 pc]

Little Dipper

Polaris [240 pc]

Big Dipper

Arcturus [11 pc]

Capella [14 pc]

Pleiades Cluster [127 pc]

E

W

Aldebaran [21 pc]

Hyades Cluster [42 pc]

Celestial Equator

Gemini

Leo

Orion

Regulus [26 pc]

Procyon [4 pc]

Betelgeuse [250 pc]

Orion Nebula [460 pc]

Ecliptic (Plane of solar system)

Rigel [250 pc]

Sirius [3 pc]

Canis Major

Horizon

Milky Way (Plane of galaxy)

Horizon

S

Key
Constellation Names Black
Star Names White
[Star distances in parsecs]

February 1   11 PM   Standard Time
March 1   9 PM
April 1   7 PM

−1 Mag
0 Mag
1 Mag
2 Mag
3 Mag

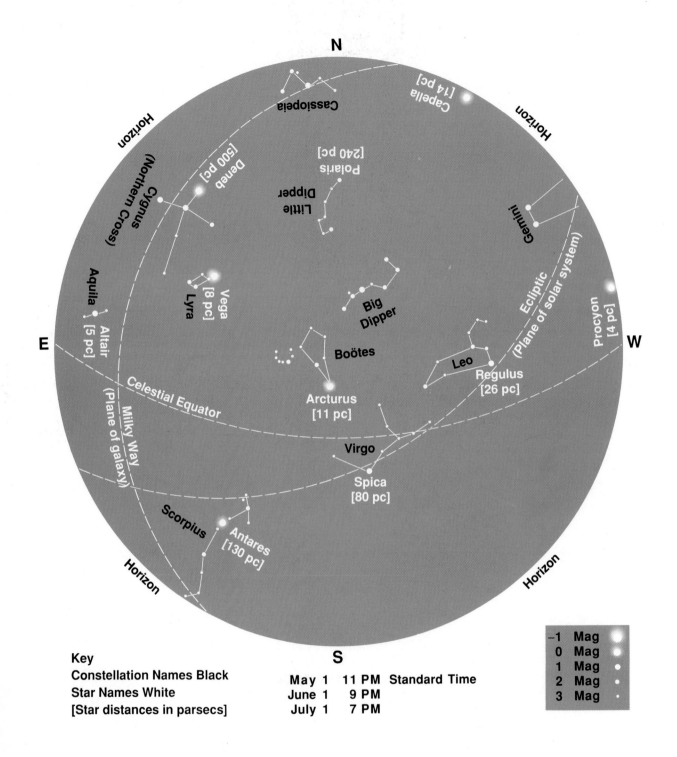

N

Horizon

Cassiopeia

Capella [14 pc]

Horizon

Cygnus (Northern Cross)

Deneb [500 pc]

Polaris [240 pc]

Little Dipper

Gemini

Aquila

Altair [5 pc]

Lyra

Vega [8 pc]

Big Dipper

Boötes

Leo

Regulus [26 pc]

Procyon [4 pc]

E

Celestial Equator

Arcturus [11 pc]

Ecliptic (Plane of solar system)

W

Milky Way (Plane of galaxy)

Virgo

Spica [80 pc]

Scorpius

Antares [130 pc]

Horizon

Horizon

S

Key
Constellation Names Black
Star Names White
[Star distances in parsecs]

May 1   11 PM   Standard Time
June 1   9 PM
July 1   7 PM

−1 Mag
0 Mag
1 Mag
2 Mag
3 Mag